the sciencebook

the sciencebook

Everything you need to know about the world and how it works

Washington, D.C.

the sciencebook

Published by the National Geographic Society

John M. Fahey, Jr.	President and Chief Executive Officer
Gilbert M. Grosvenor	Chairman of the Board
Tim T. Kelly	Executive President, Global Media Group
John Q. Griffin	President, Publishing
Nina D. Hoffman	Executive Vice President; President, Book Publishing Group

Prepared by the Book Division

Kevin Mulroy	Senior Vice President and Publisher
Leah Bendavid-Val	Director of Photography Publishing and Illustrations
Marianne R. Koszorus	Director of Design
Barbara Brownell Grogan	Executive Editor
Elizabeth Newhouse	Director of Travel Publishing
Carl Mehler	Director of Maps

Staff for this book

Judith Klein	Project Editor
Jennifer A. Thornton	Managing Editor
R. Gary Colbert	Production Director

Manufacturing and Quality Management

Christopher A. Liedel	Chief Financial Officer
Phillip L. Schlosser	Vice President
Chris Brown	Technical Director
Nicole Elliott	Manager
Monika D. Lynde	Manager
Rachel Faulise	Manager

Founded in 1888, the National Geographic Society is one of the largest nonprofit scientific and educational organizations in the world. It reaches more than 285 million people worldwide each month through its official journal, *National Geographic*, and its four other magazines; the National Geographic Channel; television documentaries; radio programs; films; books; videos and DVDs; maps; and interactive media. National Geographic has funded more than 8,000 scientific research projects and supports an education program combating geographic illiteracy.

For more information, please call 1-800-NGS LINE (647-5463) or write to the following address:

National Geographic Society
1145 17th Street N.W.
Washington, D.C. 20036-4688 U.S.A.

Visit us online at www.nationalgeographic.com/books

For information about special discounts for bulk purchases, please contact National Geographic Books Special Sales: ngspecsales@ngs.org

For rights or permissions inquiries, please contact National Geographic Books Subsidiary Rights: ngbookrights@ngs.org

Library of Congress Cataloging-in-Publication data available upon request

ISBN: 978-1-4262-0337-4 (Trade)
978-1-4262-0369-5 (Regular)
978-1-4262-0370-1 (Deluxe)

Printed in China

Staff at Peter Delius Verlag

Authors
Dr. Matthias Delbrück (Physics, Technology), Dr. Gudrun Hoffmann (Biology), Ute Kleinelümern (Earth, Biology), Martin Kliche (Chemistry), Dr. Hans W. Kothe (Biology), Dr. Martin Krause (Chemistry, Technology), Michael Müller (Universe, Technology), Uta von Debschitz (Construction), Boris Schachtschneider (Construction), Gian-Michele Tomassone (Mathematics)

Contributing Authors:
Publication Services, Inc., Champaign, U.S.A.

Translation
Julia Esrom, U. Erich Friese, Patricia Linderman, Paula Trucks-Pape

Editorial Consultants
Isobel Fleur Dumont, Dietmar Falk, Dr. Barbara Welzel

Editorial Staff
Silke Körber (Editor-in-chief)
Tanja Berkemeyer (Project Management)

Project Editors
Gigi Adair, Duncan Ballantyne-Way, John Barbrook, Tanja Berkemeyer, Uta von Debschitz, Michele Greer, Diana Leca, Natalie Lewis, Julia Niehaus, Hanna von Suchodoletz, Gian-Michele Tomassone, Marissa van Uden

Picture Research
Bettina Moll, Tilo Lothar Rölleke, Sven Schulte, Jacek Slaski, Gian-Michele Tomassone, Uta von Debschitz, Anton von Veltheim

Design
Dirk Brauns (Director of Design)
Angela Aumann, Andreas Bachmann, Markus Binner, Torsten Falke, Burga Fillery, Armin Knoll

Illustrations
Dirk Brauns, Andreas Bachmann, Burga Fillery, Uwe Gloy, Cybermedia India, Anna Krenz, Michael Römer

3D-Illustrations
Barry Croucher (The Evolution), www.the-art agency.co.uk
Mick Posen (The Human Body), www.the-art agency.co.uk

The publishers would like to express their gratitude to akg-images, corbis, Delius Producing, ESA, Flickr, fotolia, gettyimages, istockphoto, NASA, National Geographic, picture-alliance, Shutterstock, wikipedia commons, Biologie Buch/Linder Verlag, CERN, CNES/Ill. D.Ducros, David Fisher Architects, Earth Science Picture of the Day/Jens Hackmann, eMercedesBenz, GeoForschungsZentrum Potsdam, Mark Tegmark and the Sloan Digital Sky Survey (SDSS), Max-Planck-Institut für Kernphysik, National Oceanic and Atmospheric Administration, Naturkundemuseum Berlin, Naturstudiendesign, Seismological Society of America, United Nations Environment Programme/GRID-Arendal. For detailed credits and picture captions please visit the website www.TheKnowledgePage.com

FRONT COVER credits (left to right): 1, Jeffery Collingwood/Shutterstock; 2, Douglas Henderson; 3, Harold F. Pierce/NASA; 4, YKh/Shutterstock; 5, Shutterstock; 6, spe/Shutterstock; 7, NASA; 8, Jan Martin Will/Shutterstock; 9, ErickN/Shutterstock; 10, Mark Thiessen, NGP; 11, Lester Lefkowitz/Getty Images; 12, O. Louis Mazzatenta; 13, Celso Diniz/Shutterstock; 14, Karim Hesham/iStockphoto.com; 15, Sebastian Kaulitzki/iStockphoto.com; 16, iStockphoto.com; 17, Tebenkova Svetlana/Shutterstock; 18, Grune Schonheit/Fotolia; 19, Franck Steinberg/Fotolia; 20, Mark Harmel/Getty Images; 21, Shutterstock; 22, Mikael Damkier, 23, Mark Thiessen, NGP; 24, Melissa Dockstader/Shutterstock; 25, javarman/Shutterstock; 26, JoLin/Shutterstock; 27, Joel Sartore; 28, iDesign/Shutterstock; BACK COVER: 1, sinopictures/Readfoto; 2, EML/Shutterstock; 3, Sebastian Kaulitzki/Shutterstock; 4, S. Ragets/Shutterstock; 5, Popovici Ioan/Shutterstock; 6, Pchemyan Georgiy/Shutterstock; 7, NASA/JPL-Caltech/Univ. of Arizona; 8, Paul Maguire/Shutterstock; 9, Bryan Busovicki/Shutterstock; 10, David H. Seymour/Shutterstock; 11, Cornel Achirei/Shutterstock; 12, coko/Shutterstock; 13, din/Shutterstock; 14, Ilja Masik/Shutterstock, 15, Eduard Andras/Shutterstock; 16, Thomas Mounsey/Shutterstock; 17, Avesun/Shutterstock; 18, imantsu/Shutterstock; 19, Patrick Breig/Shutterstock; 20, Specta/Shutterstock; 21, Maciek Baran/Shutterstock; 22, Joel Sartore/NG Image Collection; 23, juliengrondin/Shutterstock; 24, Michele Trasi/Shutterstock; 25, Doug Stevens/Shutterstock; "I" in title, Ben Greer/iStockphoto.com.

how to use this book

The unique design and organization of this extensive volume facilitates its use and enjoyment. Opening the book to any page, the reader can quickly and easily grasp which subject and topic is being discussed.

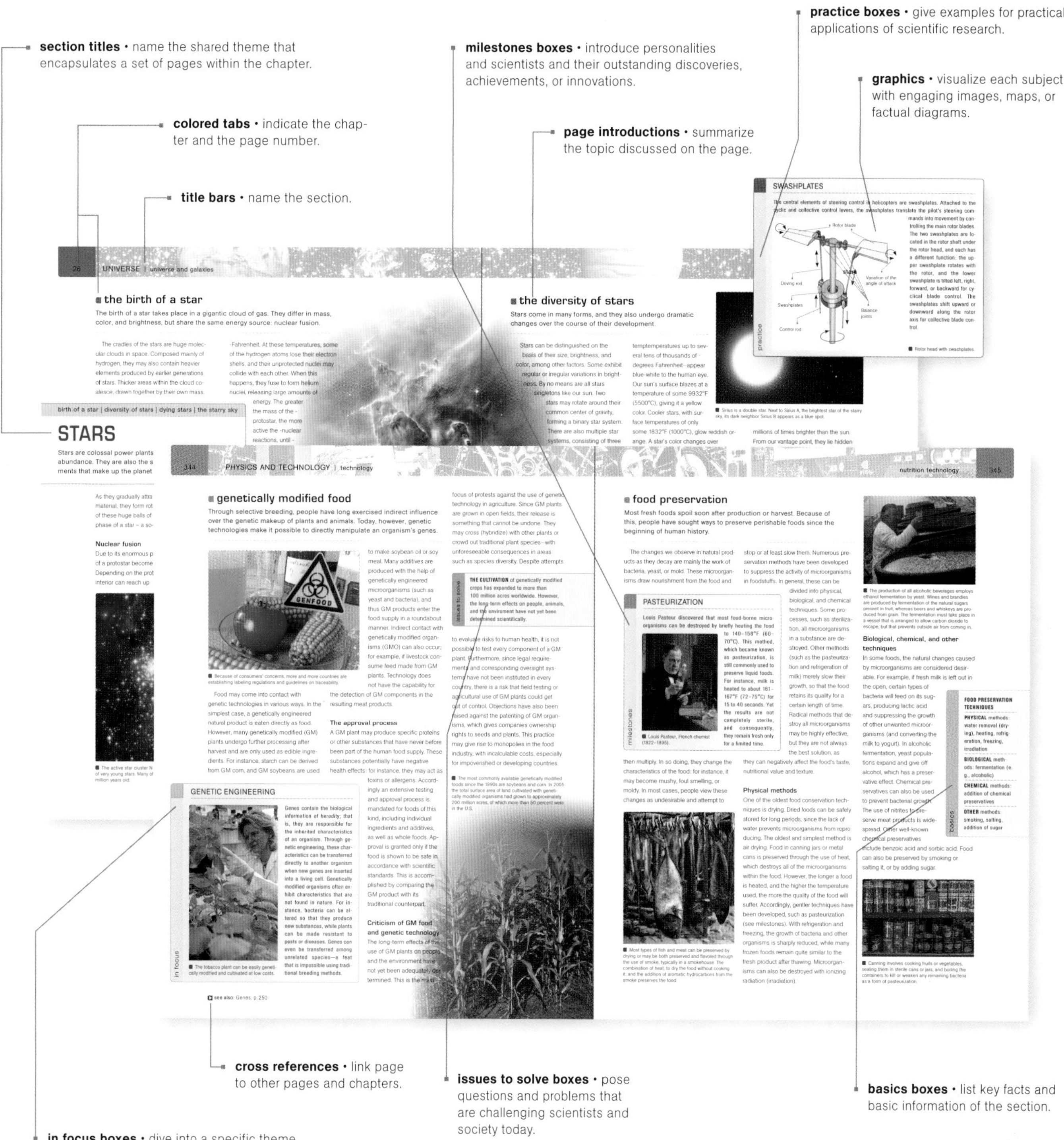

When galaxies are viewed from the side, their spiral arms are obscured, as can be seen in this image of the Sombrero galaxy. *p. 24*

Scientist Galileo Galilei improved the telescope for astronomical observations. *p. 18*

Prominences form within the ranges of sunspot groups. *p. 35*

The earliest substances in the universe were hydrogen, helium, and smaller amounts of lithium and beryllium. *p. 22*

The Giant's Causeway in Northern Ireland is made up of basalt columns. *p. 59*

The world's oceans cover about 60 percent of the Northern and more than 80 percent of the Southern Hemisphere. *p. 96*

Tropical storms can lead to significant destruction and flooding in lowland and coastal areas. *p. 117*

THE EARTH

Many factors affect the Earth's atmospheric patterns and seasonal changes. *p. 56*

The Orchid family is one of the largest families of flowering plants. *p. 189*

Among the amphibians, tree frogs are exceptionally good climbers. *p. 203*

Genes are passed on from generation to generation and cause the similarities found between relatives. *p. 250*

BIOLOGY

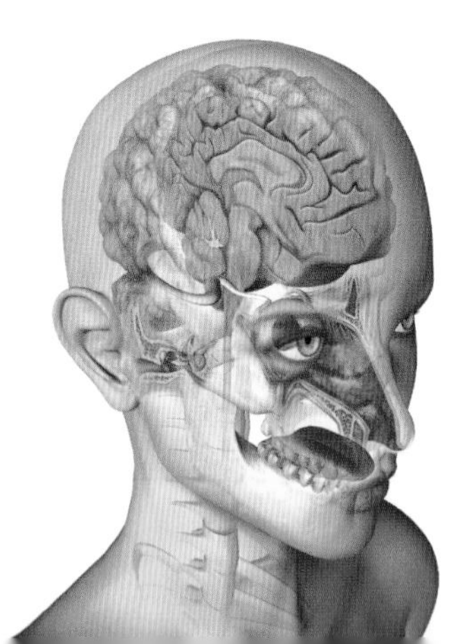
The human brain is protected by the cranial bone. *p. 220*

Hydrocarbons, such as oil, provide large amounts of energy for the world's electric generation. *p. 296*

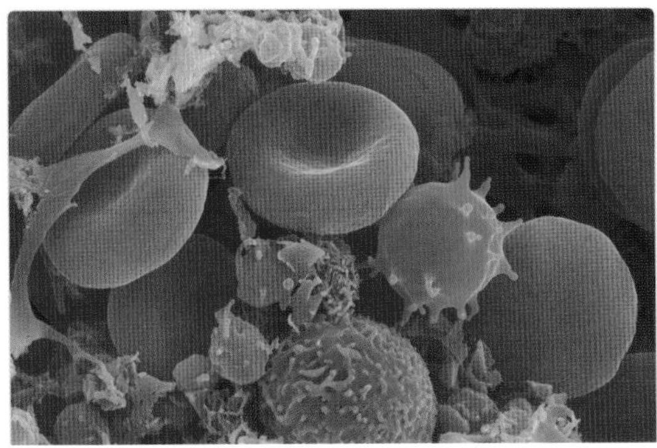
Hemoglobin is the iron-containing protein in red blood cells, responsible for transporting oxygen around the body. *p. 300*

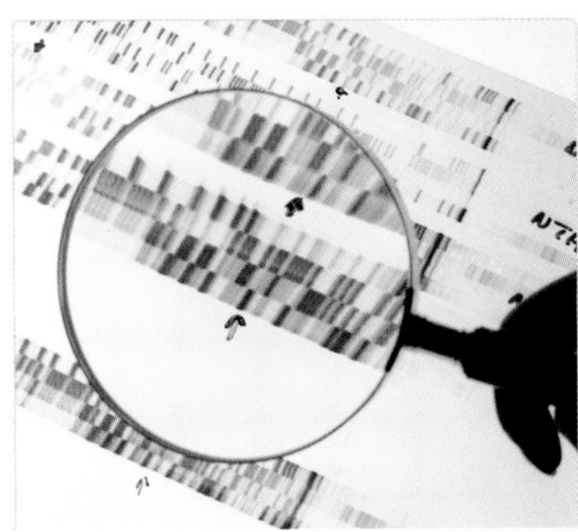
The visualization and decoding of DNA are extremely important procedures in biotechnology. *p. 303*

CHEMISTRY

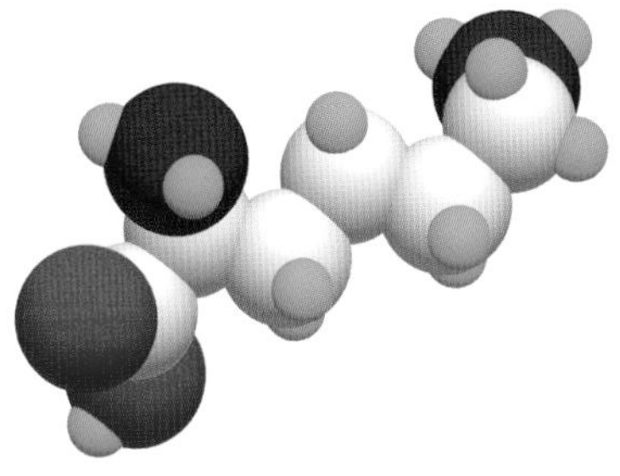
Carbon is the 13th most common element (by weight) on Earth. *p. 296*

Plastic container used for storing different gases. *p. 308*

A vibration can be observed through watching the waves that develop in a standing body of water. *p. 317*

Flight simulators were originally developed for training pilots and astronauts. *p. 380*

Robots designed to move across rough terrain are often modeled on insects. *p. 384*

PHYSICS AND TECHNOLOGY

The Loremo is one of the most environmentally friendly vehicles. *p. 357*

The basic laws of arithmetics are a fundamental part of every school curriculum. p. 411

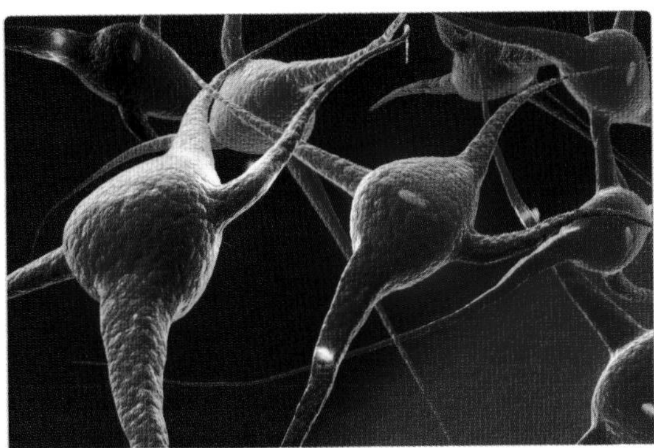
Modern mathematics can be used in various ways, for instance, to create this model of the nervous system. p. 421

Card games and roulette are forms of gambling. The outcome is simply a matter of chance. p. 417

MATHEMATICS

The abacus is an old calculating tool. p. 411

A graphic produced by a mathematical formula. p. 418

SCIENCE: THE ESSENCE OF COOL

by Marshall Brain

Have you ever spent any time thinking about science? What, exactly, is science? If someone were to ask you for a definition off the top of your head, how would you respond? I actually spend a fair amount of time thinking about it. I am the founder of a website called HowStuffWorks.com that talks about science quite a bit. I do lots of interviews where people ask about science. I work with kids in elementary school to help them get a good first impression of science. Therefore, I have a personal definition of science in my head: "Science is the coolest thing that human beings do! Science is humankind trying to figure out how absolutely everything in the universe works." It is that pure and simple. Science knows no boundaries. Science wants to take it all apart and say, "OK, we get it!" In an ideal universe, we will even figure out where our universe came from, and we will discover how to create new universes.

Scientists are curious by profession: Testing every aspect of our material world enables them to understand the order of things and then create new things.

But is that really what science is? Just to make sure, I did what any normal person would do: I consulted the dictionary (and because this is the Age of the Internet, it was an online dictionary): "Systematic knowledge of the physical or material world gained through observation and experimentation." That is so wrong! The dictionary has taken the most exciting thing human beings do and has made it as boring as possible. (This happens a lot to science, by the way, for reasons that I have never been able to fully understand.) I guess the definition is right in a technical sense, but talk about a letdown. It totally misses what science is. Science is the essence of cool, and it is also humanity's highest calling.

How cool can science be? So much so that it now envelops every one of us, and so pervasive that we take it completely for granted. For example, as I'm writing this I am sitting on an airplane. According to the pilot we are flying at 32,000 feet at 550 mph. I am looking out the window at the sun shining on the clouds far below. And right there we already have, what, like a hundred little scientific miracles that generally go unnoticed? Just think about it for a second:

First of all, I am conscious. That's a miracle because, just two inches away from my face, it is -50°F, the wind is blowing at 550 mph, and the atmospheric pressure (a measurement of air's force against a surface) is about four pounds per square inch (psi). So without the protective aluminum cocoon that surrounds me, I would pass out from lack of oxygen, freeze to death, and get dismembered by the wind. Instead, I am warm, oxygenated, unruffled, and sipping soda, with ice, from a plastic cup. How is this possible? Science! Every single bit of it.

Where did the aluminum come from? Science. The comfortable atmospheric pressure in the

During the takeoff and landing of a plane, the art of science is on display: Centuries of research and engineering have made it possible for a multiton machine to fly.

cabin? Science. The warmth? Science. The plastic in the window and the cup? Science. The ice, which was probably created on the ground even though it is -50 outside? Science. The soda, which, in a delightfully parallel way, came in its own little pressurized aluminum cocoon? Science!

So let's take the most important element—the atmospheric pressure inside the cabin. Without it, everyone on this plane would be dead in a couple of minutes. Where does the pressure come from? You can think of the airplane as a big aluminum can that acts like a pressure vessel. Even though it is four psi outside, human beings really need ten psi or higher to function normally. Yes, there may be the occasional Sherpa or two who might, perhaps, be able to hack 32,000 feet for an hour, but most folks really need ten psi or more. So what we need to do is take the outside air at four psi and boost its pressure. Here's a good mental image: We need to attach a big bicycle pump to the fuselage and pump it up to ten psi. Lacking a big bicycle pump, however, where are we going to get the pressurized air? It turns out that in the core of every turbofan engine on every airliner there is a compressor that is creating all the pressurized air we need. The engine is creating the pressurized air so that it can mix it with fuel to produce the thrust that keeps the plane moving at Mach 0.8. (You can learn all about it in the pages of this book.) But there is more than enough pressurized air coming out of this compressor to go around. We simply bleed some of the compressed air out of the engine and let it flow into the cabin.

That engine is its own scientific miracle. It is filled with alloys, lubricants, bearings, blades, shafts, shapes, and structures that have all been honed through the scientific method so that the engine runs reliably for years at a time. Science has also made sure that the engine can handle all sorts of extreme situations. What if the engine flies into a torrential thunderstorm that dumps tons of rain into the inlet? The engine can handle it. Hail? Ditto. Sandstorm? No problem. Flock of birds? Piece of cake. (It's kind of a grotesque piece of cake, but a piece of cake nonetheless.)

But what if something unexpectedly goes wrong? What if the engine dies? It's no problem, because there is a whole science of redundancy. If one engine fails, the other engine(s) can pick up the slack. Science has done the work, and we can get the pressurized air and thrust we need regardless. I know what you are thinking: What if all the engines fail? That is actually pretty unlikely, according to the science of probability, but let's say it happens. If all the engines fail, we've got a bunch of problems, because the engines provide not only the thrust and cabin pressure, but also the electricity that all the avionics in the cockpit need, plus the hydraulic pressure that lets the pilots control the plane. Science has to solve all of those problems fast. So what is going to happen? There is a rack of batteries that are going to supply backup power to the avionics. There is also a backup electric hydraulic pump that is going to create the hydraulic pressure so the pilots can fly the plane and lower the landing gear. The plane won't be arriving at the right airport, but there is a very high probability that the pilots will get the airplane safely on the ground. (Look up the Gimli Glider sometime if you want to see how interesting no-engine situations can get.)

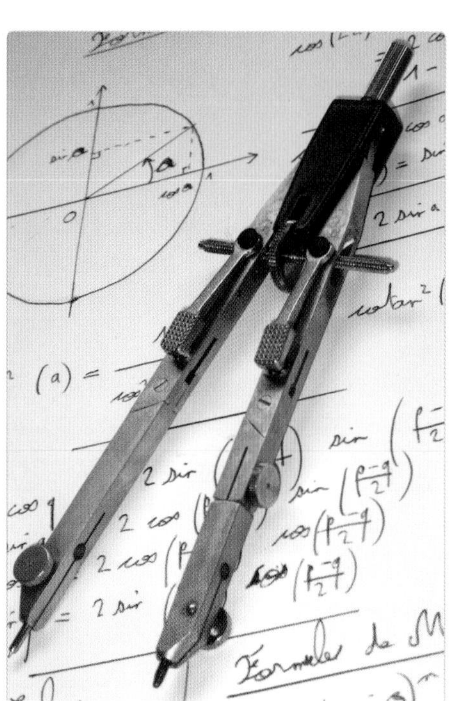

Research, creative invention, accurate planning, and control of results mark the ongoing process of science.

But what about the air pressure in the cabin? There are 200 people on the plane and a couple of really important people in the cockpit who are going to pass out very quickly without the cabin pressure that the engines were providing. It turns out the pilots have their own dedicated oxygen system up front, but the real miracle of science is going to happen in the back. How in the world can science supply 200 people with oxygen in a few seconds? Here's how: As soon as the pressure drop becomes obvious to the sensors charged with this task, a whole bunch of panels over the passenger seats are going to pop open. Plastic masks are going to drop down from the ceiling, and dozens of little oxygen canisters are going to light on fire. In the process, the heat from the chemical reactions inside these canisters is going to liberate oxygen that all the passengers will breathe. Everyone will put on their masks, and they'll be fine.

I have flown on hundreds of airplanes in my life. Only once have the masks come down. I was on a commuter jet flying to New York City, and at 32,000 feet the co-pilot's window cracked. The crack released cabin pressure in a few seconds. The masks popped down automatically, and the amazing thing is that no one really said a thing. There was no panic. We all knew what to do. We put on our masks. Then the pilot took the plane very rapidly down to about 5,000 feet and he told us what had happened. We flew to LaGuardia and landed, on time, without incident. Could science get any cooler than that? Instead of dying from asphyxiation, we all walked off the plane with a good story to tell at dinner.

Another thing that is cool about science is the way it brings people together. Think about how many different people and ideas work together to create an airplane. You've got the people who make the aluminum for the plane and the alloys for the engine (metallurgy) and the plastics for the windows and seats (organic chemistry). Then people work all those materials into their scientifically designed shapes so that the plane has lift (aerodynamics) and air flows efficiently through the engines (fluid and thermodynamics). Then there are the people who put it all together (manufacturing science), the people who design and program the electronics (computer science), the people who make the fuel and the oils (petroleum science) ... The list goes on and on.

So why do we do this? Where, in other words, does science come from? What is the science of science? And why do you care? Why, for example, have you picked up this book? This is where it gets really amazing. Here is what science tells us right now...

There was a big detonation that marked the beginning of our universe. The explosion condensed into space that was filled with immense amounts of hydrogen gas. The hydrogen atoms harbored inside of themselves an attractive force that we call gravity (a force that science does not yet completely understand). This force brought blobs of hydrogen gas together into immense spheres. The spheres were so large that the gravitational forces created gigantic pressures at their cores, igniting fusion reactions. The hydrogen atoms started fusing into heavier and heavier atoms, until eventually the fusion reactions ended and the spheres exploded, creating more and heavier atoms still. All of these atoms dispersed. Out of such dispersed debris came a cloud that, again through gravity, produced our solar system and the planet we call Earth.

On this planet of water, carbon, nitrogen, and minerals there were all the chemicals needed for uncountable chemical reactions to occur spontaneously. And there was a source of energy, in the form of the sun, at just the right distance to keep the water from freezing or boiling, as well as a handy moon to create tidal forces. The atoms and molecules chained together in such a way that they possessed the ability to reproduce copies of the chains, which is a reaction that we refer to as life. Simple living cells, through a process of mutation and selection, became more complex cells, and then multiple cells, and finally, plants and animals. Animals mutated and selected until, one day, there was a species on Earth that embodies what we call general intelligence. With that intelligence comes language, learning, logic, love, and, lo and behold, curiosity. And from that curiosity comes

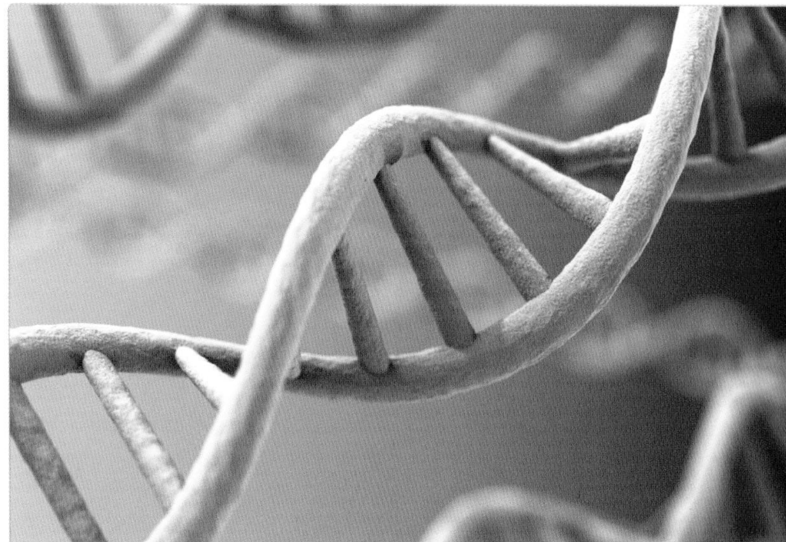

■ Deciphering the biological information encapsuled in the DNA of all living beings was a major step of scientists toward understanding life on Earth.

science—the desire to understand how stuff works. Applied carefully, over millions of people and hundreds of years, science creates the aluminum tube that I happen to be flying in today, complete with its backup oxygen supply and pressurized cans of soda.

If you think about it from that huge, broad perspective, the fact that we are sitting here today having this conversation is utterly amazing. The paper you are holding is amazing, and so is the ink, and so is the pressurized air that I am breathing, and the laptop on which I type, and so are the chemical reactions happening in the neurons of your brain, and the exploded star that produced those chemicals ...

All of it is the purview of science.

From this book you will get an amazing look at just how broad and encompassing that purview is. Science is incredibly, unbelievably, immeasurably cool, and I hope you enjoy it.

UNIVERSE

UNIVERSE AND GALAXIES

Just by looking at the stars, one can get an idea of how infinitely large the universe must be. Countless stars and galaxies are scattered all over this enormously vast space, with immense distances between one another. Today's scientists use radiation emitted from space, as well as state-of-the-art telescopes to explore the mysteries of the universe. Data is collected, analyzed, and interpreted in order to create theories and physical models to explain the universe's origins and development. The currently accepted theory assumes that the universe originated from an unprecedented, singular explosion.

beginnings | subdivisions | instruments

ASTRONOMY

The magnificent, starry sky has always inspired the human imagination. A growing understanding of geography, mathematics, and physics over the millennia has ensured that astronomy has never lost its ability to help us explore new realms of the unknown. Today, modern technology continues to drive this age-old cosmic search.

from the beginnings of astronomy

The mere observation of celestial bodies made it possible to make calculations based on calendars and navigation. With the dawn of modern physics, the first verifiable explanatory models for cosmic events were developed.

People in the ancient civilizations of Egypt, Babylonia, China, and in Middle America were already making systematic observations of celestial events in the fourth millennium B.C. Thus, they were able to create calendars and predict eclipses of the sun and moon. They also tried to use the stars to identify and understand the influence of the gods on their own fate.

16th-century scientist Galileo Galilei improved the telescope for consequent astronomical observations.

The ancient Greeks—heirs of Babylonian knowledge—continued to develop these observations in order to seek, above all, the causes for celestial events. The Greeks already knew not only that the Earth was a sphere, but also its approximate circumference, and could determine the distances and sizes of the sun and moon. Nevertheless, a heliocentric worldview, in which the sun is central, did not achieve acceptance. Instead, the Earth remained the central point of reference. The movements of planets were explained by assuming they moved in small circles, so-called epicycles. In turn, these circular movements were incorporated into their respective orbits around the Earth. In approximately A.D. 150, Ptolemy provided a comprehensive description of this epicycle theory in his manual called the *Almagest*.

The road to modern astronomy

In the 15th century, more exact measurements of the planets' orbits identified inaccuracies in the epicycle theory. For this reason, in the 16th century Nicolaus Copernicus began advocating the heliocentric worldview. Tycho Brahe made further measurements, which Johannes Kepler used to recalculate the planets' orbits. He was able to prove that the planets move in elliptical orbits around the sun. The heliocentric worldview was further supported around this time by observations made with the recently invented telescope.

In the 17th century, Isaac Newton set down the theoretical foundations for modern physics. His law of gravity also provided a scientific explanation for the elliptical orbits calculated by Kepler. A great deal of work by other researchers followed these developments, including work regarding the speed of light, the distance to the sun and the radius of the Earth.

During the 19th century, Joseph von Fraunhofer discovered the spectral lines in the spectrum of sunlight. Building on this finding, Gustav Robert Kirchhoff and Robert Wilhelm Bunsen established spectral analysis. For the very first time, scientists could investigate the chemical and physical characteristics of stars and other celestial bodies.

In the 20th century, Hans Albrecht Bethe and Carl Friedrich von Weizsäcker explained that nuclear fusion provides a source of energy for stars. Physical theories (p. 22) regarding the formation and development of the universe were articulated.

practice

THE ASTROLABE

The astrolabe was once an astronomer's most important instrument, used to calculate the positions of the stars and represent them in two dimensions. It was often made of brass, with rotating disks, degree scales, pointers, and a sight rule. It was replaced by more precise instruments and methods during the course of the 16th century.

The curved spikes on this 18th-century Persian astrolabe mark the brightest stars.

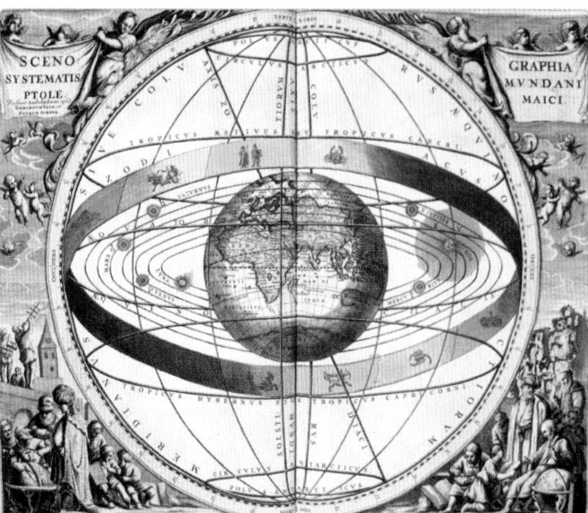

The Earth-centered system of the universe, proposed by Claudius Ptolemy in the second century A.D., was accepted as scientific wisdom until Nicolaus Copernicus proposed his sun-centered model in 1543.

see also: Exploring the universe, pp. 48–51

astronomy today

Highly technical measuring instruments, sophisticated observation procedures, physical models, and mathematical simulations have all had an impact on modern astronomy. Space travel also offers significant support.

Light radiated by celestial bodies still serves as the foundation for astronomy. In earlier times, astronomers were limited to studying visible light. Today, modern technology has substantially widened the spectrum to include particle radiation, radio signals, infrared, ultraviolet, x-ray, and gamma radiation. Other branches of natural science, such as mathematics and physics, are closely linked with modern astronomy.

The four domes of the Very Large Telescope (VLT) in Chile are capable of rendering remarkably sharp images of deep space.

Subdivisions of astronomy

The classical subdivisions of astronomy are astrometry and celestial mechanics, which are mainly concerned with measuring and calculating the positions and orbits of celestial bodies. Astrophysics examines their characteristics, such as the strength of their magnetic fields, temperatures, densities, and compositions. In particular, it concentrates on their formation and development. Cosmology looks at the formation and development of the universe as a whole.

Astronomical instruments

Astronomical instruments are necessarily varied in form. Enormous radio telescopes are used to intercept radio signals from space, and their receiving capacity can be combined to simulate an even better resolution. Radar technology is used to study meteorites or other objects in the solar system, even visual light from celestial bodies can be detected with sophisticated telescopes by using adaptive lenses that compensate for atmospheric disturbances and provide high-definition images.

The majority of radiation that comes from space is blocked by the Earth's atmosphere. However, balloons, aircraft, and rockets all offer new possibilities for high altitude observations. Developments in space travel mean specialized instruments and telescopes can even be taken into space. Beyond Earth's atmospheric disturbance, these instruments record magnetic fields, particles, and radiation from celestial bodies. They are not limited to orbiting the Earth, some orbit the sun, other planets, or even asteroids; others traverse the solar system.

The Very Large Array, an astronomical radio observatory in New Mexico, consists of 27 radio antennas. Each one measures 82 feet (25 m) in diameter.

practice

SPACE TELESCOPES

Space telescopes have their own power supply, position control, and scientific instruments. Depending on the application, these instruments may include a highly sensitive or wide-angle camera, a spectrograph, and devices to filter or measure radiation intensity. High power radiation requires specialized telescopes. For example, x-ray telescopes use a mirror system, aligned so that radiation is not absorbed, but rather glances off the surfaces.

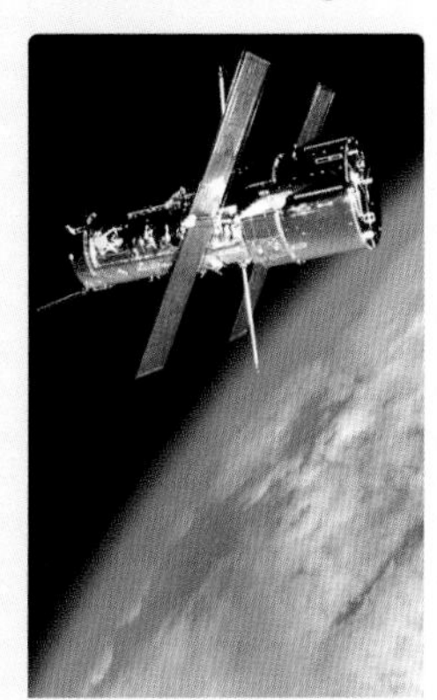

The Hubble Space Telescope has been orbiting Earth above the atmosphere.

structure | survey | the big bang | dark matter | dark energy

THE UNIVERSE

Information regarding the appearance of the universe as a whole, its origins, and how it evolved is limited, but by measuring the radiation coming from celestial bodies in combination with knowledge about the physical laws, scientists are able to create models of the universe, which are assessed by how well they match observational data.

the structure of the universe

The matter of the universe is not evenly distributed. Instead, matter is influenced by gravity and shaped in diverse ways. The space between these differently shaped structures is unimaginably vast.

Glowing agglomerations of matter are visible at night. These stars, together with our sun, make up the spiral-shaped Milky Way, which in turn is a member of a collection of galaxies called the Local Group. These, and larger accumulations known as clusters, tend to exist in even greater groups separated by huge voids of space. These so-called superclusters are vast and lumpy in shape. Their distribution throughout space resembles the structure of soapy lather, with heavier distributions along the walls or intersections of the "soap" bubbles and pockets of empty space in between. Neighboring stars are normally a few light-years apart from each other. The diameter of a large galaxy is around 100,000 light-years and the Local Group has an estimated diameter of about 10 million light-years. Superclusters can spread over several 100 million light-years. Threadlike structures consisting of these clusters encompass the empty spaces, the largest of which extend over one billion light-years.

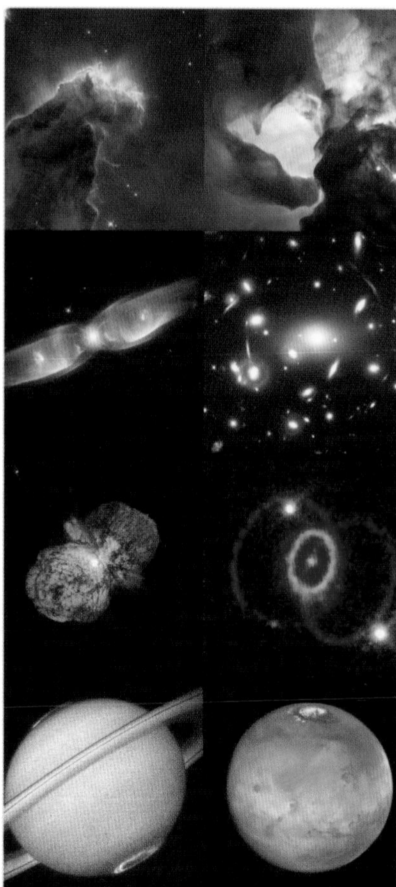

In the field of cosmology astronomers and physicists work closely together to try to explain the origin, expansion, and structure of the universe.

Modeling

Due to the huge scale of the distances involved, only indirect methods (p. 21) can be used to measure the cosmic structure of the universe. Certain assumptions must be made and adapted to interpret the astronomical observations and data. Most scientists have accepted the big bang model—that suggests the universe evolved from an extremely condensed primeval state—as a valid model (p. 22). The predominant method for measuring cosmic distances is the redshift (p. 21) of the light emitted by distant, receding objects. Both of these approaches allow scientists to develop hypotheses about the expansion of the universe and are closely related to the general theory of relativity developed by Albert Einstein (p. 330). There may be other valid models which are are also consistent with observational data.

Scientists create physical models of the universe with the aid of Einstein's general theory of relativity.

in focus

THE LIGHT-YEAR

Light travels at 186,282 miles per second (299,792 km/s) in a vacuum. Large distances can be estimated based on this information; they are simply given in terms of how far the light travels within a certain time-frame. This is a much easier way of communicating measurements on the cosmic scale. Hence, one light-year is defined as the distance which light can travel within one year. This is almost 5.9 trillion or 5,900,000,000,000 miles (9,500,000,000,000 km). The nearest neighbor to the sun is a star called Proxima Centauri, a little more than four light-years away, and the nearest galaxy, Andromeda, is two million light-years away.

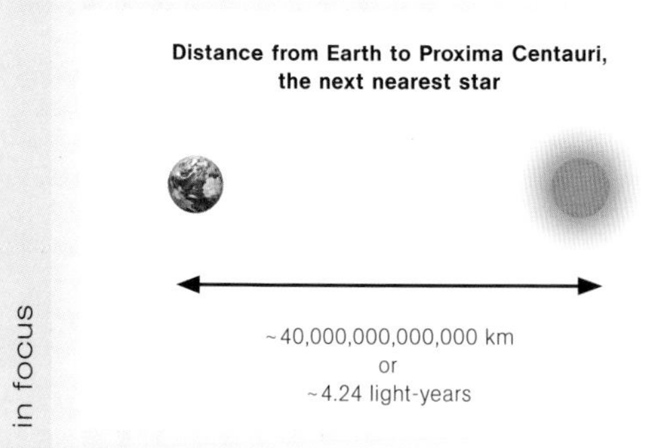

The nearest star to Earth after the sun is Proxima Centauri.

see also: Theory of relativity, pp. 330–331

survey of the universe

In order to understand the universe, it needs to be surveyed. Astronomers use various methods to measure the vast distances between stars and galaxies, and to study the behavior of these remote objects.

If two objects radiate with the same intensity (luminosity), then the closer object will appear brighter. If a star's actual luminosity is of a known magnitude, then it can be used together with the observed brightness to calculate how far away it is.

Although actual luminosity is not always obvious, astronomers are able to determine this value for certain stars (known as "standard candles") and use it to estimate their distance from Earth. Examples include Cepheids, which are giant bright stars that pulsate regularly. Their fluctuation periods are dependent on their size and luminosity; these pulsations can be measured and provide useful information about the distance of a galaxy. White dwarf stars that explode as extremely bright supernovas in binary star systems can also be used as standard candles. Their observed brightness can be used to estimate the distances of very remote galaxies.

Stars may appear brighter than others due to being closer to the observer or because the radiation intensity they emit is higher. When these two values are known, their distances can be estimated.

Redshift

Starlight can be partitioned into its prismatic colors using a spectrometer, much as sunlight is in a rainbow. However, chemical elements in the gas layers of stars absorb light of certain wavelengths, causing dark absorption lines to break up the spectrum. Almost all galaxies have absorption lines that are shifted toward the longer (red) wavelengths. The redshift effect is greater the farther away a galaxy is. This relationship can be explained by the expanding universe.

THE HUBBLE CONSTANT

milestones

Edwin Powell Hubble (1889–1953) discovered the relationship between the observed redshifts and the distances of galaxies in the 1920s. Today this is known as the Hubble constant. Modern measurements give it a value of about 44 miles per second (71 km/s) per megaparsec (about 3.26 million light-years). The escape velocity of the galaxies increases by this value per megaparsec of the distance.

Edwin Powell Hubble linked the redshift of galaxies to the expansion of the universe.

As space itself expands the wavelengths of the light are stretched. The longer light travels through the universe, the more its waves are stretched. Therefore, if the redshift of a galaxy is known, its distance can be calculated. This relationship is known as the Hubble constant. Of course, the interpretation of the redshift needs to be correct and the expansion rate of the universe must be known as precisely as possible.

There are also other effects that can cause a redshift, for example when a galaxy in space is moving away from us. However, this is not enough to explain redshift in general because it implies that our galaxy is somehow special, in that other galaxies are moving away from us but not from each other. It has also been proposed that light would lose energy on its long path and perhaps turn more red for this reason. So far, however, no generally accepted explanation for redshift has been found.

basics

THE BEHAVIOR of galaxies in the expanding universe can be imagined as raisins in rising dough, where the dough is equivalent to the space between the galaxies (raisins), which are all moving apart from each other at the same rate.

Due to the expansion of the universe, all objects are moving away from each other. This stretching of the space between objects results in wavelengths of light being stretched into longer redshifted light: pictured, a light source (3) moving to the right (2). The frequency is lower on the left (1, redshift) and higher on the right (4, blue shift).

see also: Stars, pp. 26–29

theory of the big bang

Most cosmologists accept the big bang theory as a realistic model of the origin of the universe. The model suggests that the universe evolved from an extremely condensed primeval state.

Nobody knows for sure what really happened during the first moments. The part of the universe that we can observe with our telescopes today was very condensed and most likely measured no more than a few millimeters. Radiation continuously turned into particles of matter and back.

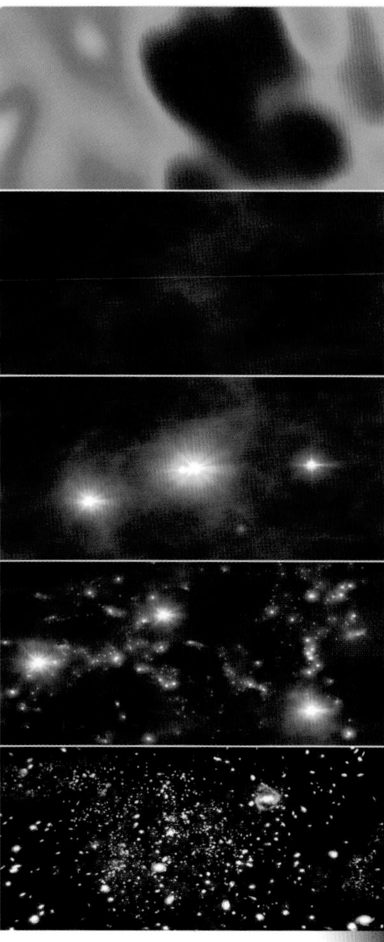

Matter distributed throughout the universe condensed to form distinct structures.

Expansion of the universe

Although the universe expanded rapidly, there was no explosion as there was no surrounding space for explosive power to go to. Space itself simply expanded.

The more the universe swelled, the cooler it became, and the less energy was emitted by radiation. The original matter slowly formed the building blocks of atoms: protons, neutrons, and electrons. About ten seconds after the big bang, protons and neutrons could combine to form the first stable and lightweight atomic nuclei. The radiant energy became too weak to separate the particles. After further cooling, these atomic nuclei were able to capture electrons, forming the first atoms a few hundreds of thousands of years after the big bang. The universe started becoming transparent. Radiation could now pass through space without barriers, since only very few electrically charged particles were still floating around.

Today it is possible to measure cosmic background radiation in any direction. This is thought to be the radiation that was originally released in the early phase of the universe. Once the radiative pressure decreased, gravity was able to take over and the first large accumulations of material could form, approximately one million years after the big bang. Later on, the first galaxies and stars began to conglomerate. The big bang theory is based on both quantum field theory and Albert Einstein's general theory of relativity, as well as the cosmological principle.

Quantum field theory deals with describing the characteristics and forces of elementary particles. Einstein's general theory of relativity attempts to explain gravity by the warping of space-time, using a mathematical model which creates a close connection between the three dimensions of space and the passage of time. Space-time is warped by material and this warp, in turn, determines the movement of material.

basics

ACCORDING TO the big bang theory, the universe rapidly expanded from a condensed state about 13.7 billion years ago.

The cosmological principle states that on the large scale, material is generally distributed evenly throughout the universe, although locally distinct structures may be obvious. Taken together, these assumptions result in the mathematical expansion of the universe.

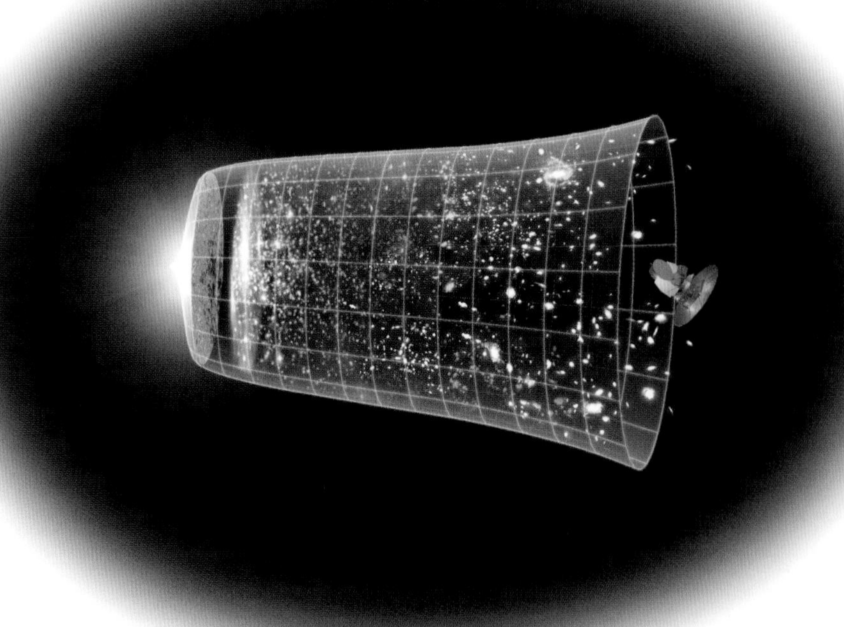

The earliest substances in the universe were hydrogen, helium, and smaller amounts of lithium and beryllium. Heavier chemical elements were produced later by nuclear reaction in the stars.

see also: Theory of relativity, pp. 330–331

dark matter and dark energy

Astronomical observations suggest the existence of an invisible material, only recognized by indirect effects of its gravity. Mysteriously, there may also be a special kind of energy accelerating the expansion of the universe.

Apparently there is a very large amount of matter between the stars. This matter does not emit or reflect light, nor does it swallow light. Its mass is by far larger than that of normal matter.

This type of matter is only obvious due to its gravity influencing normal matter, as shown by numerous astronomical measurements. For example, galaxies do not rotate the way we would expect. Their masses can be estimated from their stars and gas clouds and the results suggest that the star's velocity of circulation should decrease with increasing distance from the center. However, spiral galaxies show a constant velocity of rotation regardless of the distance from the center. This contradiction could be explained by the existence of a spherical mantle of so-called dark matter, with a mass up to ten times that of the glowing matter.

Gravitational lenses are another indicator of the existence of dark matter. There are masses, for example galaxy clusters, which bend the light of other galaxies situated far behind them, just as a magnifying glass does. The mass of a galaxy cluster can be determined by the degree of light refraction; these tend to be much greater than can be explained by their stars and gas clouds.

New structure of the universe

Dark matter also plays an important role in the big bang theory. This is due to the uniform background radiation (p. 22) which indicates a similarly uniform distribution of normal matter within the young universe. Far more condensed masses would have been necessary for galaxies to form from this. Therefore, it is thought that the dark matter might have been involved, although it remains to be seen how dark matter could have reached a sufficient level of density. The composition of dark matter is still unknown. It is possible that it consists of dark celestial bodies or unknown elementary particles.

Dark energy

Astronomers trying to determine the speed of cosmic expansion have measured the luminosity of supernovas and their redshift (p. 21). Surprisingly, these studies and other data (such as that from the space probe WMAP) suggest that the expansion of the universe is accelerating. It is not known how this may be occurring, but so-called dark energy has been suggested as a possible accelerator. This theoretical concept in fact fits mathematically with the general theory of relativity (p. 331). Some scientists even speculate that dark energy may be another natural force or fundamental characteristic of space.

in focus

THE FUTURE OF THE UNIVERSE

The expansion of the universe seems to be accelerating. If this is the case, stars that have already become extinguished and their planets will be destroyed by the decay of protons. Even black holes will vaporize because of the so-called Hawking radiation. Afterward, nothing but the infinitely diluted gas of the remaining particles will be left over. Due to the unimaginably large distances between them, forces will no longer be effective. Time will have lost its meaning.

Based on our current knowledge of the universe and its laws, various hypotheses about its future can be formulated.

basics

THE ENTIRE density of the material and energy of the universe determines its fate, whether it will continue expanding forever or, one day collapse.

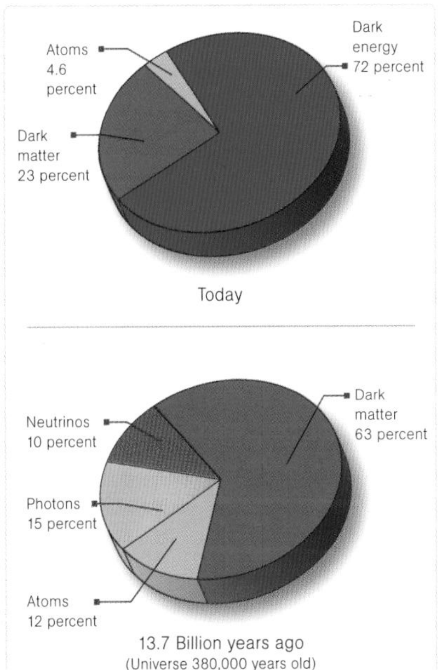

The components of the universe can be derived from data from the WMAP probe.

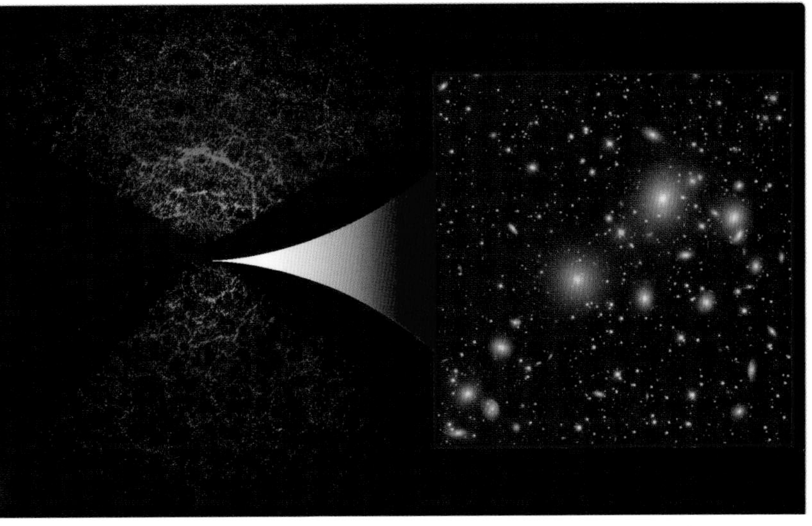

Data from surveys of the luminosities of galaxies can be combined and interpreted as three-dimensional models that show the distribution of matter throughout large sections of the universe.

spirals | ellipses | the milky way

GALAXIES

Stars are very seldom lone objects; they are almost always part of a gigantic, rotating star system. The massive dimensions of a star are virtually inconceivable for most people, and the breathtaking space between them even more so. It takes about 2.5 million years for the light from the nearest large neighboring galaxy to reach us.

spirals and ellipses

The world of galaxies is varied in form. These star systems differ markedly in structure, size, and composition.

Approximately 100 billion stars, together with gas and dust, form the enormous, rotating celestial systems called galaxies. Some are elliptical in shape, while others form spirals or irregular shapes. A galaxy is held together by the gravity associated with its component parts. These forces compel the stars and residual matter to rotate around the galactic center. The sun is a part of such a system known as the Milky Way (p. 25).

Galaxies are generally found together, either in very large numbers called clusters, or in smaller collections, which are known as groups. The spiral arms of a galaxy are formed by aggregations of stars in the galactic disk, similar to the air compressions caused by sound waves.

Matter is therefore much more concentrated in these areas, which thus results in a relatively high number of new stars being formed from the interstellar gas. For this reason, the arms appear much brighter than the surrounding areas. Although a spiral galaxy rotates around its galactic core, the arms do not actually wind inward around the center because they are not made up of specific stars. Instead, individual stars move into the spiral arm zones and back out again.

The Antennae galaxies are an example of two galaxies colliding and merging.

Interacting galaxies

When galaxies pass near each other or converge, they can be reshaped by each other's gravitational forces. When this happens, the paths of individual stars may be deflected or, in rare cases, the stars can merge.

During these interactions, a dwarf galaxy will generally dissolve into the larger galaxy, while those that are of similar size will retain their respective cores.

Spiral arms are obscured in galaxies viewed side on, such as the Sombrero galaxy, but this angle allows an insightful perspective on both the central bulge and the thin dust disk.

Interstellar matter

A very fine distribution of gas and dust exists between stars. This includes matter ejected by stars as well as that left over from their initial formation. The gas is largely made up of hydrogen and helium, while dust is mostly graphite or silicates, such as those commonly found in rocks. Clouds of this matter appear as various types of nebulas. In emission nebulas, the radiation of nearby stars heats the gas so that it emits its own light. Reflection nebulas are illuminated by starlight shining on their dust. In contrast, a nebula that absorbs background starlight simply appears as a dark cloud.

in focus

GALAXY TYPES

Only a minimum of structure can be recognized within an elliptical galaxy, and sometimes none at all. Spiral galaxies, on the other hand, have two or more spiral arms that coil around a central core. These are then divided into those with and those without a central bar through the middle. Lenticular galaxies, like spiral galaxies, have a core but no spiral arms. Irregularly shaped galaxies are rarer and have unique shapes.

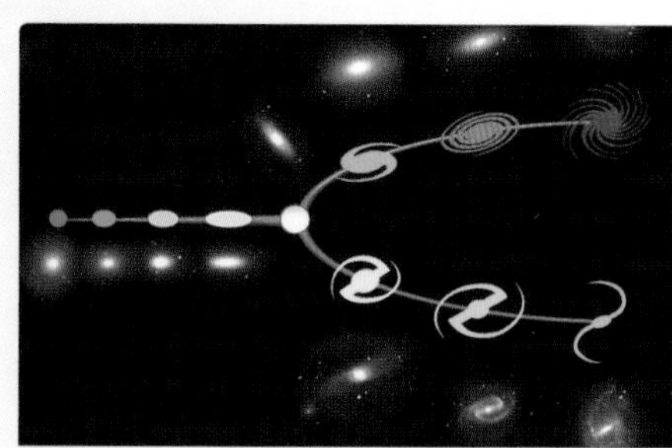

The Hubble classification scheme groups diverse galaxies by their features, such as central bars, arms, and shapes.

basics

IN EARLIER times galaxies could only be identified as nebular objects, and so their nature long remained unclear. It was not until 1923 that Edwin Hubble demonstrated that the Andromeda nebula (the Andromeda galaxy) is actually a neighboring star system.

the milky way system

The solar system is also part of a galaxy. Together with billions of stars it travels along its path around the center of the Milky Way system.

Our home star system is known as the Milky Way galaxy. On a clear night, with good visibility, we can make out a shimmering band of diffuse light crossing the sky. When this happens, we are looking directly into the disk section of our galaxy. This band of light is known as the Milky Way, but the term is also used to describe the entire galaxy, of which our sun is a part. We cannot see the Milky Way galaxy from the outside, but we can try to measure the distances between its stars with the best possible precision.

Looking out from Earth, the galactic disk appears in the sky as a band of light, known as the Milky Way.

The structure of the Milky Way galaxy

Because opaque interstellar dust clouds complicate this process, the structure of the Milky Way was not understood until the previous century when it became possible to use infrared and radio frequencies to make astronomical observations. As it turned out, the Milky Way system is a relatively large spiral galaxy.

Our solar system is located in one of the spiral arms, near the outer region. The outer layer of the galaxy is called a galactic halo, consisting of widely scattered aggregations of ancient stars, so-called globular clusters.

Star types can be distinguished based on their composition, age, and distribution in space. The youngest are located near the symmetrical sections of the spiral arms and have a relatively high concentration of heavy chemical elements. These elements were produced by older generations of stars in the fusion of their nuclear fuel. As they weakened, some of their matter was ejected into space, and was later incorporated in the formation of new generations of stars.

Middle-aged stars are found near the galactic disk. The old stars in the globular clusters contain the lowest percentage of heavy chemical elements. It is likely that they were already formed when the galactic matter aggregated.

To locate the center of the Milky Way galaxy, we have to look toward the constellation Sagittarius. The center is shrouded in thick, interstellar matter and is not visible to us. However, its structure can still be explored based on radio frequency and infrared and x-ray radiation.

Around the central area, stars are distributed more thickly than they are in the outer regions. In the direct center, it appears that the equivalent of approximately three million sun-sized masses is concentrated into a tight space.

It is suspected that an extremely large concentration of matter has accumulated here, a so-called black hole.

in focus

THE LOCAL GROUP

The Milky Way galaxy is part of a group of more than 30 galaxies held together by gravitational forces. The Milky Way galaxy and the Andromeda galaxy are the largest representatives of this so-called Local Group and are about 2.5 million light-years away from each other. They are surrounded by a cluster of smaller galaxies, including the Large and the Small Magellanic Clouds. At approximately 150,000 and 180,000 light-years away, they are relatively close to the Milky Way.

The Small Magellanic Cloud is an irregular galaxy that is part of the Local Group.

basics

THE MILKY WAY system has a diameter of about 100,000 light-years. It consists of approximately 100 to 400 billion stars. The sun is about 27,000 light-years away from the galactic center. It orbits around the center at 124 miles per second (200 km/sec). One revolution takes 240 million years.

The Milky Way is a spiral galaxy with a central bulge in the galactic center and several spiral arms made up of interstellar gas, dust, and billions of stars. One of these stars is the sun, which, with its satellites, makes up our solar system. This is located on one of the arms toward the outer regions of the galaxy.

birth | diversity | dying stars | starry sky

STARS

Stars are colossal power plants, producing energy in abundance. They are also the source of the chemical elements that make up the planets—and our own bodies.

the birth of a star

The birth of a star takes place in a gigantic cloud of gas. Stars differ in mass, color, and brightness, but share the same energy source: nuclear fusion.

The cradles of the stars are huge molecular clouds in space. Composed mainly of hydrogen, they may also contain heavier elements produced by earlier generations of stars. Thicker areas within the cloud coalesce, drawn together by their own mass. As they gradually attract more and more material, they form rotating masses. Each of these huge balls of gas is the preliminary phase of a star—a so-called protostar.

Nuclear fusion

Due to its enormous pressure, the core of a protostar becomes extremely hot. Depending on the protostar's mass, its interior can reach up to several million degrees Fahrenheit. At these temperatures, some of the hydrogen atoms lose their electron shells, and their unprotected nuclei may collide with each other. When this happens, they fuse to form helium nuclei, releasing large amounts of energy. The greater the mass of the protostar, the more active the nuclear reactions, until finally it shines as a new star.

The main sequence

After the nuclear fusion process has begun, the star settles into a stable form. Its interior pressure is high enough to counteract gravity, so that the forces both generated and released remain balanced. This relatively calm period in a star's development is known as the main sequence—the phase that our sun is currently experiencing. A star's main sequence continues until the supply of hydrogen fuel in its core has been completely exhausted.

■ The Carina Nebula is a vast, interstellar dust cloud that covers a space of 200–300 light-years and contains many areas of new star formation.

■ The active star cluster NGC 3603 includes a myriad of very young stars. Many of them are only about two million years old.

in focus

STAR MASSES AND ENERGY LEVELS

A star's mass largely determines its lifespan: the larger the mass, the shorter the star's life cycle. While a massive star starts out with a larger supply of nuclear fuel, it uses up its fuel much more rapidly, since the pressure in its core is greater. Because of this increased pressure, the core is hotter and more fusion reactions take place per second. The amount of released energy rises dramatically, and the star shines more brightly. A star ten times as massive as the sun uses its fuel 1,000 times faster than one similar to the sun. Thus, it will shine only for some 100 million years, instead of 10 billion like our sun. Stars with even lower masses use their energy reserves only sparingly and can shine thousands of times longer than the sun.

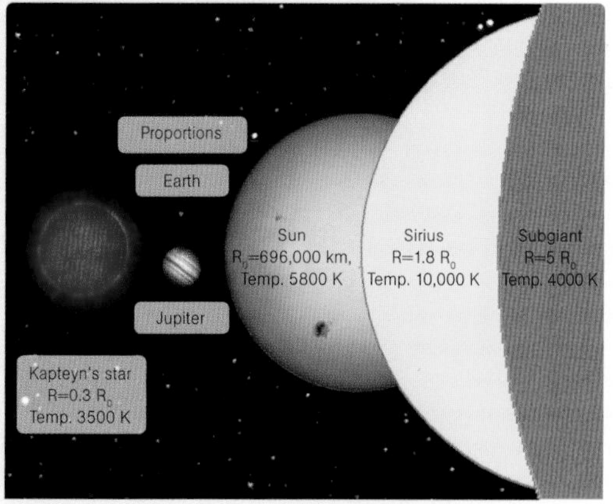

■ Stars in comparison.

the diversity of stars

Stars come in many forms, and they also undergo dramatic changes over the course of their development.

Sirius is a double star. Next to Sirius A, the brightest star of the starry sky, its dark neighbor Sirius B appears as a blue spot.

Stars can be distinguished on the basis of their size, brightness, and color, among other factors. Some exhibit regular or irregular variations in brightness. By no means are all stars singletons like our sun. Two stars may rotate around their common center of gravity, forming a binary star system. There are also multiple star systems, consisting of three or more stars. Stars can appear in clusters as well, since they often arise and develop in groups. These star clusters may contain dozens, hundreds, or perhaps even thousands of stars. If a star cluster's mass is below a certain level, its stars will gradually drift apart.

The meaning of color

Like iron in a forge glowing red or white-hot, a star's color depends on its surface temperature. According to this temperature, stars shine with varying levels of intensity at different wavelengths. Hotter stars emit more short-wavelength light than cooler ones. Particularly hot stars—with surface temperatures up to several tens of thousands of degrees Fahrenheit—appear blue-white to the human eye. Our sun's surface blazes at a temperature of some 9932°F (5500°C), giving it a yellow color. Cooler stars, with surface temperatures of only some 1832°F (1000°C), glow reddish orange. A star's color changes over the course of its lifespan. For example, a massive star first shines with a bluish light, but later appears red when it expands.

Lightweights and heavyweights

Most stars are relatively cool and dim. These so-called red dwarves, with only eight to 50 percent as much mass as the sun, use their hydrogen fuel sparingly and shine with a reddish color. Astronomical objects with even smaller masses do not become hot enough to ignite. Unable to fuse hydrogen, they become so-called brown dwarves. These dark objects primarily emit heat radiation. The upper limits of a star's mass are not reliably known. There seem to be few stars with more than 100 times the sun's mass. These extremely unstable and short-lived stars shine millions of times brighter than the sun. From our vantage point, they lie hidden behind the enormous clouds of gas they emit. It is often unclear whether they are single or double stars.

One of the most famous ring clouds is NGC 6720 in the constellation Carina. The nebula rotates around a white dwarf.

in focus

RED GIANTS AND WHITE DWARVES

A star with only a few multiples of the sun's mass ends its life cycle as a so-called white dwarf. At some point, all of the hydrogen in its core has been used up and converted to helium. No longer able to resist its own gravity, the star begins to collapse. As its mass is pressed together, it becomes hotter, and the hydrogen remaining around the edges of the core can now begin to fuse. Energy is given off once again, causing the outer layers of the star to expand. The heat is distributed over a larger volume, cooling the star's surface and changing its color: it is now a "red giant." The core temperature rises, and the helium within it fuses to form carbon and other elements. As the helium is gradually used up, the star begins to pulsate, ejecting its outer layers into space. The star's former core remains behind as a white dwarf, while the expanding shell forms a so-called planetary nebula. This will be the eventual fate of our sun, approximately five billion years from now.

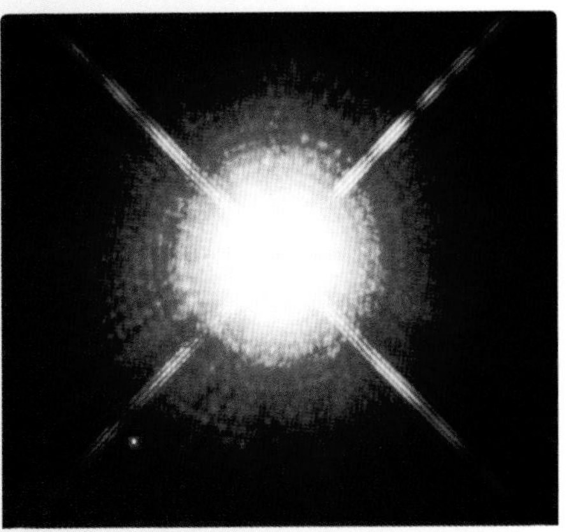

White dwarf near Sirius, in the constellation Canis Major.

dying stars

The nature of the last phase in the life of a dying star is largely determined by its mass. The star's last energy reserves are expended, which can produce stunning visual effects.

When all the hydrogen in the center of a star has been used up and fused to form helium, the core contracts and begins to heat up. The outer layers of the star also begin to undergo fusion. This consequently causes the star to inflate massively and become what is known as a red giant. Planets orbiting near this star will be swallowed up. The star glows red because its energy is distributed over a large surface area. The temperature in the center of a red giant can easily reach a remarkable 180 million°F (100 million°C).

Meanwhile, at the center of a red dwarf, helium fuses to form carbon and oxygen. The star casts off its outer gaseous envelope to form planetary nebulae and the carbon-oxygen core remains to form a white dwarf. A white dwarf, however, no longer undergoes heat-generating reactions, so it collapses to be extremely dense.

basics

DYING STARS produce heavy chemical elements.

MASSIVE STARS explode as supernovae.

AFTER A STAR DIES, white dwarfs, neutron stars, or black holes remain, depending on the mass of the original star. A white dwarf can explode as a supernova if it picks up matter from a nearby star.

RADIATION from a supernova can damage organisms on a planet, even at a distance of several light-years away.

The fate of massive stars

Stars above a certain mass (approximately ten times the mass of our sun) become so dense and hot that heavier elements undergo fusion. Carbon, neon, oxygen, and silicon undergo fusion. The formation of elements heavier than iron consumes rather than produces energy, thus hastening the star's eventual collapse.

When the energy source of the star finally dries up, it collapses and explodes in a giant supernova. For a few days, it shines brighter than its entire home galaxy. Within the remaining core of the star, the atomic components are compressed to such an extreme that electrons and protons are converted to neutrons. One teaspoonful of this matter from such a neutron star has the mass of about one billion small cars. The neutron star revolves at high speed and emits so-called pulsar radio waves; for instance, in the Crab Nebula one pulsar revolves about 30 times per second.

If the neutron star has a mass between 3 and 15 times the size of the sun, it undergoes a more radical gravitational collapse and becomes a black hole: an area of space with such enormous gravitational pull that neither matter nor light can escape from it.

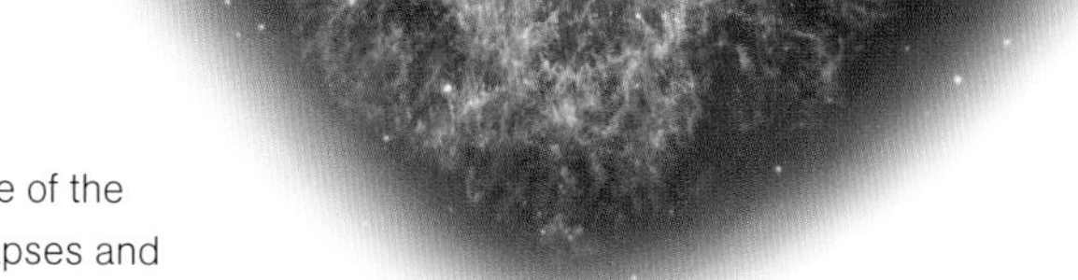

The Crab Nebula is the remainder of a supernova explosion. Chinese astronomers first documented its glow in 1054.

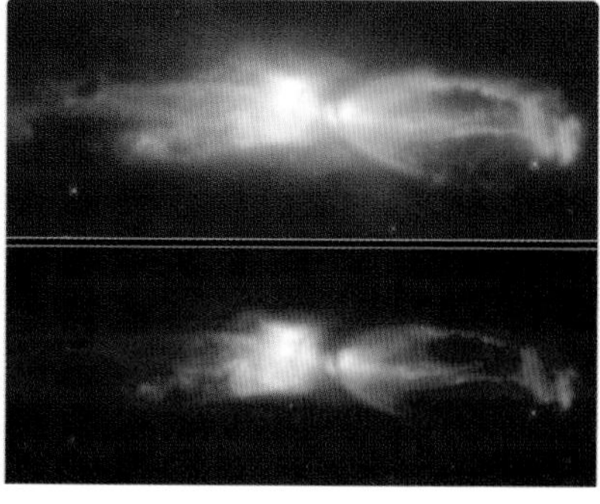

Surprisingly complex structures often develop when sunlike stars cast off their outer gaseous envelopes at the end of their lives.

in focus

ACCRETION DISKS IN DOUBLE STAR SYSTEMS

An accretion disk consists of gas or dust. In a close double star system, the gas from a red giant can form an accretion disk around a white dwarf, which slowly gathers the gas. If the white dwarf accumulates too much gas, it will collapse under its own weight and be torn apart by a supernova explosion. Before the final supernova, there can be novae outbreaks along its surface, similar to hydrogen bomb explosions.

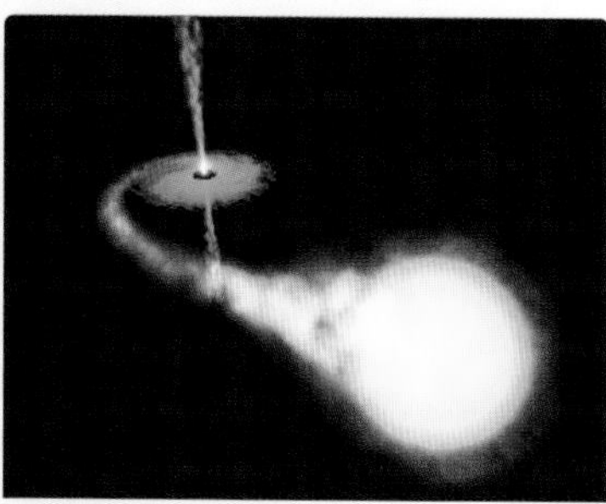

Accretion disks around massive objects are able to release a lot of energy on their own.

see also: Theory of the big bang, p. 22

the starry sky

Light given off by stars is often outshone by light sources on Earth. Yet, anyone away from the big cities and looking up to the night sky can enjoy the stellar sparkle in all its glory.

The sky is full of stars even during the day; however, only one of them—the sun—is actually visible, because it outshines all the others. The stars and planets visible in the night sky change with the annual trajectory of the Earth around the sun.

This photograph clearly indicates the apparent tracks of stars around a celestial pole.

People have long tried to recognize some order among the stars visible in the night sky. Thousands of years ago, people around the world traced imaginary lines between bright stars to create figures of their gods and mythological characters. Each culture thus developed its own constellations. Today, the International Astronomical Union (IAU) has set 88 constellations as standard. Different constellations are visible depending on the time of year and one's location.

Celestial poles

The celestial poles are points on the celestial sphere in the direction of the Earth's axis. Because the Earth revolves around this axis, the stars seem to rotate in circles around the celestial poles. At the Earth's Poles, these circles appear parallel to the horizon. Elsewhere, the stars rise and set along the horizon, just like the sun and the moon.

Stars and planets

Stars appear variously bright, because they radiate with different intensities and they are at different distances from the Earth. Often they also appear to be trembling or sparkling, with their brightness flickering rapidly. Planets are easy to distinguish from stars in the night sky, because they do not twinkle at all or only very little. The twinkling of stars is caused by schlieren—optical inhomogeneities in the Earth's atmosphere caused by air at different temperatures. Because stars are so far away, they appear to us essentially as points of light. This means that their position in the night sky can be distorted by atmospheric turbulence, which appears to us as twinkling. Since planets are much closer to us than stars are, they appear as disks in the sky. This means that variations in brightness are spread out equally over the entire disk area, negating schlieren effects. Venus, Mars, and Jupiter are particularly easy to recognize because of their brightness.

This celestial map from the 17th century shows constellations of the Northern and Southern Hemispheres. The constellations are illustrated as figures of classical Greek and Roman mythology, yet also display a high level of astronomical detail and accuracy.

practice

THE NAMES OF STARS

The brightest stars have proper names, which often come from Arabic. Later, they were assigned Greek letters—usually in alphabetical order in the sequence of their brightness—with the Latin name of the constellation added on. Roman letters and numbers have also been used. Because of the multitude of stars, most are given catalog numbers and only some are named after astronomers.

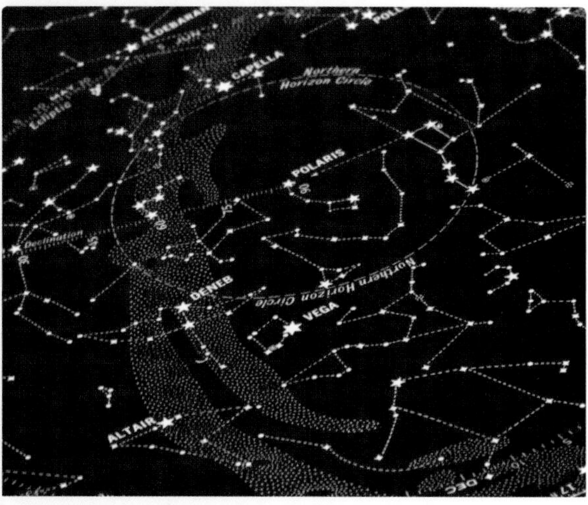

The brightest star in the Taurus constellation is Alpha Tauri or Aldebaran (Arabic: "the one who follows").

basics

STARS VISIBLE with the naked eye belong to our own galaxy, the Milky Way system.

SIRIUS is the brightest star in the night sky. It belongs to the Canis Major (Large Dog) constellation and it is 8.6 light-years away.

see also: Astronomy, pp. 18–19

UNIVERSE

THE SOLAR SYSTEM

Advances in space travel have changed the way we have come to see the Earth, its neighboring planets, and the universe as a whole. Thanks to space probes we can now more thoroughly explore our solar system, which is governed by the sun's energy. The sun forces planets and other celestial bodies into specific orbits and provides light and heat energy to the cold universe. Nevertheless, most regions of the solar system are rather inhospitable and Earth seems to be the only planet with such a vast richness of life. Other celestial bodies, however, may at least be a habitat for microorganisms. A long time ago, the sun—a stable star—formed from the same gaseous dust cloud as its surrounding planets. This process is not unique; other planets have been discovered in the vicinity of such stars. They may also show signs of life—the search has only just begun.

disks | planets | orbits

THE SOLAR SYSTEM

The Earth and many other celestial bodies orbit the sun like a merry-go-round. These celestial objects each differ in the nature of their orbits, from the close circling of the planets to the elongated paths of more distant comets.

disks of dust and planets

In certain parts of the universe, young planets have been observed to be surrounded by dust disks. Some of the events happening here are thought to be similar to the formation of the solar system.

The solar system is governed by the sun. Due to its gravity, planets and other celestial bodies are forced into more or less circular orbits around it. This system originated from an enormous cloud of gas and dust; condensed by its own weight, a star formed in its center. This was our sun (p. 34), around which the remainder of the cloud continued to circle. Due to the centrifugal force, this cloud flattened into a disk, from which the other celestial bodies of the solar system emerged. Although these protoplanetary disks have also been observed surrounding young stars, they are not present as a rule, therefore the exact process of how planets form is yet to be discovered. The dust is sometimes absorbed entirely by the star, or alternatively, is blown into space, either by the star's own radiation or by that of another nearby star.

Formation of planets

It is thought that particles in the dust disks accumulate together, gradually creating lumps that combine and attract more and more dust as they grow into planet-sized objects. Once young planets have reached several times the size of Earth, their gravity enables them to bind even gas, with the result that they will eventually become huge gas planets, much like Jupiter, Saturn, Uranus, and Neptune (pp. 44–45). Dust disks can be observed by masking out the bright glare of light from the star in the center. In this way, gaps and deformations in the disks can become visible. These paths in the dust are thought to be cleared by young planets during their formation.

Finely distributed dust is also present throughout the solar system. It extends out around the sun, mostly within the plane of the planetary orbits. The reflected light can be observed from Earth, especially from the tropics. This zodiacal light (or "gegenschein" when it is seen opposite the sun) is best viewed just before sunrise or just after sunset on very clear and dark nights.

Weak sunlight reflections on the finely spread dust of the solar system is known as zodiacal light.

WHAT IS A PLANET?

On August 24, 2006, the International Astronomical Union redefined the term "planet" and decided on the following conditions: 1. A planet orbits around a star and is neither a star nor a moon; 2. Its shape is spherical due to its gravity; 3. It has "cleared" the space of its orbit.

A "dwarf planet" such as Pluto (p. 46) only meets the first two conditions. Celestial bodies that only meet the first condition are referred to as small solar system bodies, for example, asteroids or comets.

Planet Earth, the moon, and dwarf planet Pluto compared by size: the trans-Neptunian objects (p. 46) Quaoar and Sedna are no more than dwarf planets.

in focus

Planets and other celestial bodies emerge from protoplanetary disks.

see also: Stars, pp. 26–29

orbits of celestial bodies

The objects of the solar system come in all kinds of shapes and can be classified by size, type, and orbit.

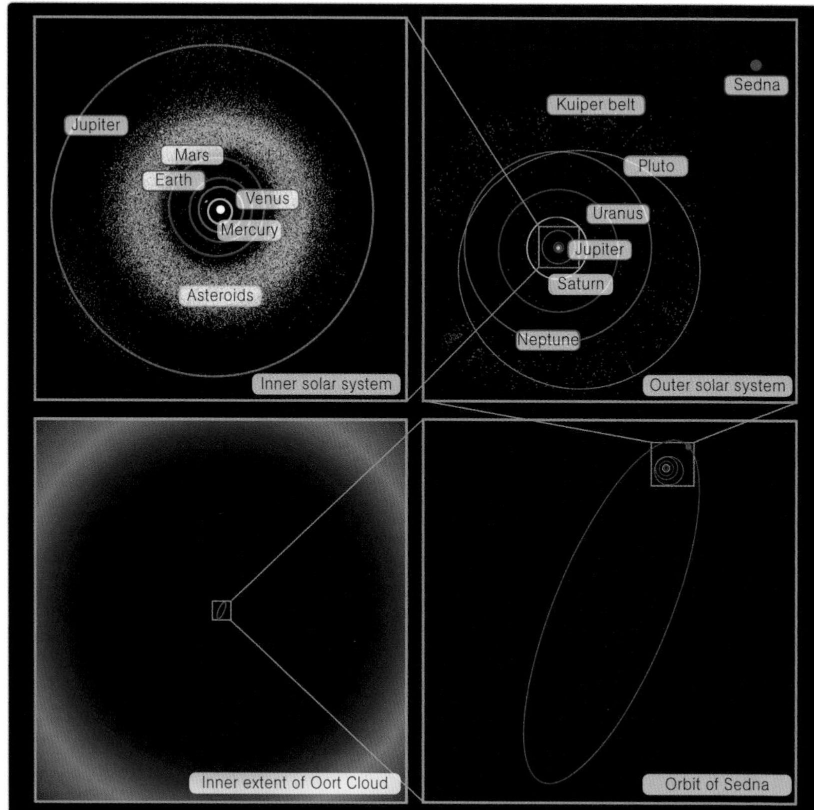

Proportions of the solar system: The inner solar system is surrounded by the outer solar system, including the Kuiper belt (p. 46). The trans-Neptunian object Sedna (p. 46) is an example of a celestial body on the outermost edge of the Kuiper belt. All of these objects are largely enclosed by the Oort cloud.

The sun (p. 34) and all the celestial objects orbiting it, including the planets, make up the solar system. Their movements are more or less in the same plane and almost circular, except Mercury, which has a more elongated orbit. A top-down view onto the solar system shows it has several well-defined regions.

basics

THE SOLAR SYSTEM emerged from a huge, circling cloud of gas and dust about 4.6 billion years ago.

THE EARTH orbits around the sun at a distance of almost 93 million miles (150 million km).

The regions of the solar system

The planets Mercury (p. 38), Venus (p. 38), Earth, and Mars (p. 41) travel along the inside orbits. These planets are mainly rocky and due to this similarity with the Earth are also referred to as terrestrial or telluric planets. None of these planets, except the Earth, offer suitable habitats for living creatures. The only planet to come close to a state of sustaining Earthlike life is Mars.

The planets traveling along the outer orbits are Jupiter (p. 44), Saturn (p. 44), Uranus (p. 45), and Neptune (p. 45). These are much larger than the terrestrial planets and consist mostly of hydrogen and helium gas. Hence, they are often referred to as the gas giants.The two planet groups are separated between Mars' and Jupiter's orbits by an asteroid belt (p. 42) of mostly irregularly shaped rocks, only a few of which are larger than about 60 miles (100 km).

The dwarf planet Pluto (p. 46) and several smaller objects circle outside of Neptune's orbit in the Kuiper belt (p. 46). Several comets (p. 43) also originate from here. Their glowing tails often become visible when they move into the inner solar system.

The Oort cloud (p. 43) is thought to be located even further outside. It encloses the planetary system like a shell, creating a boundary. The cloud consists of icy celestial bodies spread over a vast area. This is where comets accumulate, and from time to time are diverted into the solar system.

in focus

THE LIFE ZONE

The Earth orbits the sun at the "right" distance for warming: it is not too cold and not too hot. This, along with the right atmospheric pressure, allows water to occur not only as ice or vapor, but also as liquid. This is a precondition for life forms we are familiar with, although some microorganisms survive in extreme conditions. The "life zone" is the distance belt around a star that allows water to be liquid, and is dependent on solar radiation intensity.

Earth is a life-friendly planet thanks to its suitable distance from the sun.

A size comparison of the sun, the planets, and dwarf planets. The distances are much larger in reality and not shown true to scale.

anatomy of the sun

The sun's energy is generated by nuclear fusion. Its layered structure is similar to the layers of an onion. During a solar eclipse the sun's outer gas layer, the chromosphere, becomes visible.

THE SUN

The sun is the star central to both our planetary system and our lives as a huge power station providing the Earth with warmth and light. Its particle radiation generates the beautiful auroras (northern lights or southern lights) in the Earth's atmosphere, but can also cause damage to electronics, satellites, and astronauts.

The sun is a sphere full of ionized gas in the form of hot plasma. The plasma contains mostly hydrogen, and then smaller quantities of helium and other heavier chemical elements. The atoms of these gases hit each other with great force because of the extremely high temperatures. During this process, atomic nuclei and electrons are separated from each other. The sun comprises almost 99.9 percent of the total mass of the solar system. It has about 333,000 times the mass of the Earth, and 109 times the diameter.

The layered structure of the sun

The sun is held together by its own mass. This creates enormous pressure at the center, where the temperature is about 27 million°F (15 million°C). Due to the intensity of the heat, hydrogen nuclei move very quickly and can overcome their electrical repulsion to fuse together. Energy released from this fusion reaction in the core is the energy we eventually receive on Earth. The solar core is surrounded by an extensive layer referred to as the radiative zone. This is where the radiation from the core is projected back and forth between plasma particles in a process that can take up to a million years before finally passing through the radiative zone and reaching the next layer, the convection zone. Here, energy is transported to the surface of the sun by means of hot plasma movements, which cause the surface of the sun to appear to bubble.

The convection zone is surrounded by the photosphere. This layer is only a few hundred miles thick with a much cooler temperature of around 9900°F (5500°C). The photosphere emits the light that we can see on Earth. This layer is, therefore, often referred to as the sun's surface, even though a star does not have a firm surface as such. The next layer up from the photosphere is the chromosphere. Although the illumination from the photosphere normally overpowers this reddish shimmer, it can be observed for a few seconds during a solar eclipse. In the chromosphere, the temperature rises again to 18,000°F (10,000°C) while the particle density decreases significantly.

The photosphere is enclosed by the corona which is visible during a solar eclipse in the form of an aureole of varying widths. The solar matter of this layer is very thin and gradually merges into interplanetary space. In the corona, the temperature rises to about 3.6 million°F (two million°C) again. How the temperature is heated up to such a degree has not yet been explained, but a possible source of energy might be the sun's magnetic field (p. 35). Electrically charged particles are constantly released from the corona; it is these particles that are referred to as solar wind (p. 35).

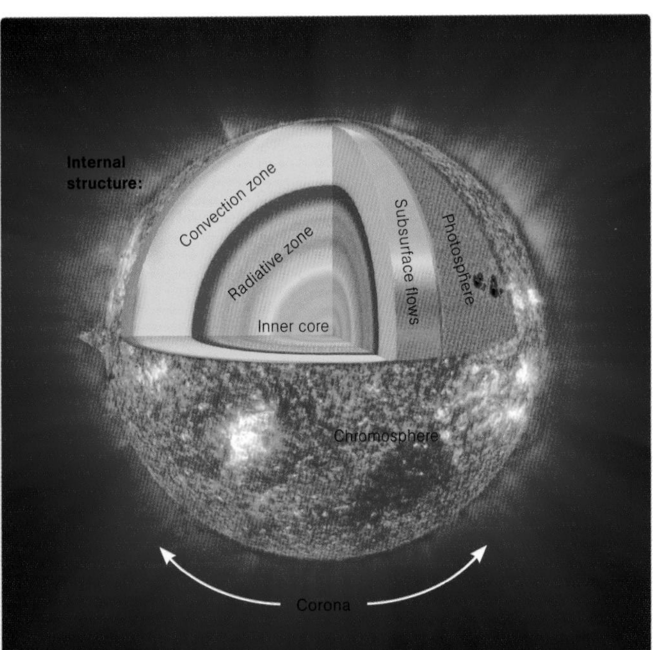

The sun's energy is produced in its center through nuclear fusion. The energy is transported between layers by radiation and convection. The visible light we see from the sun is the energy emitted by the photosphere at its surface.

POWER STATION SUN

in focus

Stars like the sun mainly produce energy during a process called proton-proton chain reaction. Hydrogen nuclei are positively charged protons. When these protons fuse, a proton is converted into a neutron, and an atomic nucleus with two particles is produced. When this merges with another proton, a nucleus of three particles is formed. Finally these nuclei can go on to form a helium nucleus with two protons and two neutrons, releasing two of the original protons. Additional particles released during this chain of reactions include positrons, neutrinos, and very high-energy gamma rays.

When the moon covers the sun during an eclipse, normally unseen aspects of the sun's gas layers can become visible.

magnetic fields and solar wind

The sun's influence extends far into space. The complex magnetic field of the sun changes continuously within cycles that last several years. These are tied into various phenomena, such as the formation of sunspots.

The hot plasma (gas) of the sun consists of electrically charged particles. The sun rotates faster near its equator than it does near its northern and southern latitudes, and hot plasma rises from the center toward the surface during this rotation, creating electrical currents. These form magnetic fields over the sun, which in turn feed back and affect the electrical currents, resulting in a complex interaction of influences. The consequence of these processes together is that the magnetic fields of the sun are continuously changing shape. Sometimes, they are similar to those of the Earth, in that the field lines run straight from pole to pole.

Sunspots appear dark because they are comparitively cooler regions caused by strong magnetic fields.

At other times, due to the sun's uneven rotation speeds between the equator and poles, the fields become more and more twisted, resulting in their effects being amplified a thousandfold in several regions of the surface. Magnetic field lines emerge in these areas, known as sunspots, while retreating in others. Sunspots appear darker in comparison to their surroundings because the emerging magnetic fields block the rising plasma, causing patches of cooler temperature where less energy is being transported to the surface from the interior.

Some plasma is guided along the magnetic field lines creating filaments; these enormous protuberances consist of the sun's material and can last anywhere from hours to months. When viewed against a dark background they appear as huge prominences that flare outward and loop back toward the surface.

Bright prominences form within the range of sunspots. The filaments made up of the sun's material flow along magnetic field lines.

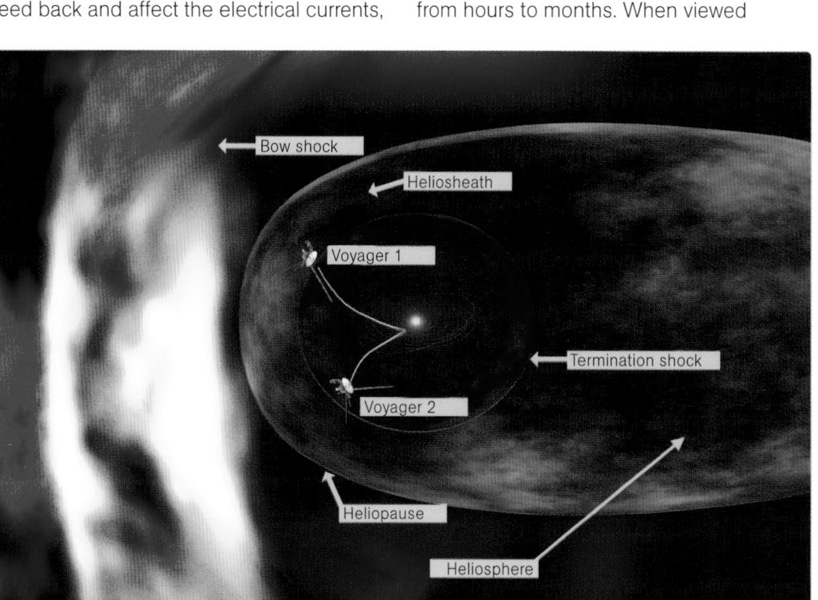

The heliosphere is a bubble-like area in space, cleared of interstellar material by the pressure of the solar wind.

The sunspot cycle

The total number of sunspots depends on the distortion cycle of the solar magnetic field. During a sunspot minimum, when field lines are straight, there are almost no visible sunspots. As the fields twist over a period of about 11 years, more develop until the sunspot maximum, the point at which the solar magnetic field reverses polarity and straightens out. The previously magnetic north pole is now the magnetic south pole and vice versa. A full cycle of magnetic repolarization therefore includes two sunspot cycles and takes about 22 years.

basics

THE SUN has a ratio of one million hydrogen nuclei to 98,000 helium nuclei, and several hundred nuclei of heavier elements. About 72 percent of the solar mass accounts for the hydrogen mass and about 26 percent for the mass of helium.

EVERY SECOND, the sun loses a mass of about one million metric tons (one billion kg) due to solar wind.

Solar wind

The thin outer layer of the sun, the corona (p. 34), reaches far into space. Due to its extreme heat, the sun's matter is continuously released as solar wind. This wind mainly consists of protons, electrons, and helium nuclei, which are found between the Earth and the sun. Solar wind often reaches speeds of about 250 miles per second (400 km/s), and sometimes double this. It creates its own magnetic fields, which deform the shape of both the sun and the Earth's magnetic fields.

see also: Magnetic field of the Earth, p. 57

cosmic weather

The sun is not merely a source of light and heat to its surrounding environment. Solar radiation is a factor that influences many cosmic and terrestrial phenomena, some of which can be harmful.

The planetary system is surrounded by the sun's magnetic field and by solar radiation. The latter includes mainly visible light, thermal radiation, ultraviolet radiation, radio waves, x-rays, and electrically charged particles in the solar wind (p. 35). The Earth's atmosphere is equipped with a relatively effective shield, which protects it from such radiation. Electrically charged particles are deflected by the magnetic field of the Earth. Similar to terrestrial weather conditions, these predominant conditions in space are called cosmic weather. They fluctuate depending on solar activity and may temporarily become harmful.

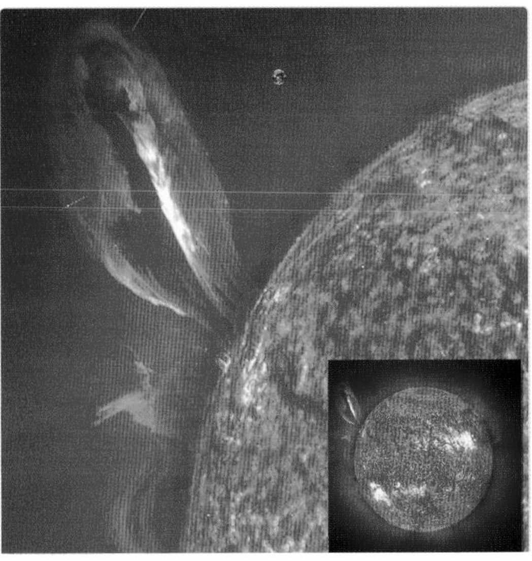

The sun can flare intensely from relatively small, localized areas in solar ejections that release extremely large amounts of energy.

The moody sun

Sometimes the sun explosively catapults solar matter into space (coronal mass ejections), and small areas of the solar surface may flare up for short periods of time. The energy released during one of these events is almost equivalent to the amount of total solar energy emitted per second. At this time, the sun also ejects high-energy electrons and protons. So far, it is only possible to partially predict these phenomena. They tend to occur at regular and observable intervals. The intensity of solar radiation and the frequency of its outbursts correspond in fluctuations with the regularly recurring sunspot cycle.

Effects of the Earth's magnetic field

Electrically charged particles from the sun and the Earth's upper atmosphere create circular radiation zones in the Earth's magnetic field. They include the Van Allen radiation belt above the Equator and the equatorial electrojet. Electric currents above the polar areas reach down to an altitude of about 330,000 feet (100 km) and result in the phenomenon known as the northern or southern lights, visible at lower latitudes. If large amounts of particles hit the Earth's magnetic field—for example, after a coronal mass ejection by the sun—then a geomagnetic storm can occur.

During times of high solar activity, electrical currents in the polar regions cause the air molecules to glow, which creates auroras.

Geomagnetic storms

Intensified particle radiation amplifies the electric currents of the Earth's magnetic field, which may overlap with the terrestrial field and cause it to fluctuate, potentially causing geomagnetically induced currents in transmission lines and transformers. For safety reasons, regional power supplies and telecommunication services may be manually shut down. Increased amounts of charge carriers in the atmosphere, as well as the intense radiation of flares, can also disrupt terrestrial radio communication, satellite communication, and navigation.

Geomagnetic storms may also damage satellite electronics. Moreover, the upper atmosphere is heated and expands, which slows down satellites that are close to Earth, potentially altering their orbit. Radiation exposure risk is slightly elevated for aircraft flying at higher altitudes, and even affects migrating birds and carrier pigeons, which can become disoriented.

in focus

THE SOLAR CYCLE

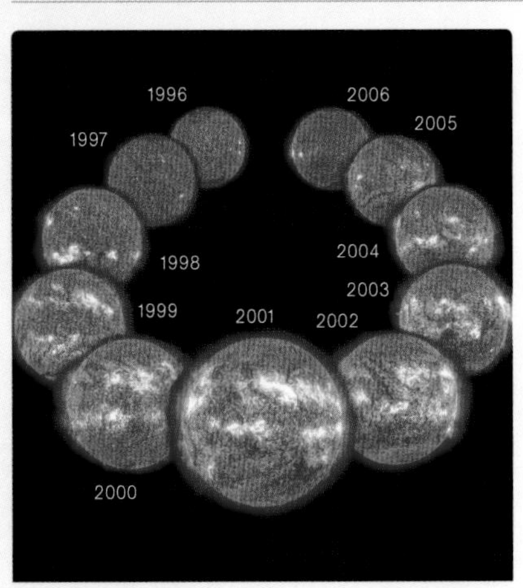

Solar activity is closely linked to the 11-year sunspot cycle. Sunspots are the result of locally amplified magnetic fields that emit extreme levels of radiation and can cause intense flares. An increase in sunspots, therefore, can be interpreted as a sign of potentially greater solar activity and consequently can act as an indicator for turbulent cosmic weather.

Cosmic weather predictions are important in the case of astronauts, who may be seriously endangered at times of sudden elevations in solar activity. In the event of increased solar activity, astronauts must seek shelter in protected areas of the space station. Efforts in finding ways of predicting solar activity similar to a terrestrial weather forecast are still ongoing.

Solar activity corresponds closely to the 11-year sunspot cycles on the sun, caused by regular magnetic field activity.

see also: Atmosphere, pp. 112–113

solar research

Many phenomena related to the sun can be observed and studied from Earth. A more detailed investigation of developing solar magnetic fields and flares is, however, made possible by the use of space probes and satellites.

Ancient cultures observed and studied the position of the sun throughout the year, recording important or unusual events such as solar eclipses. But the invention of the telescope around 1600 was the breakthrough, finally allowing systematic observation of the sun's surface. However, specialized equipment must be used for this, as observations of the sun may lead to serious eye damage or even blindness.

One of the major advances in solar research in the 19th century was the discovery of dark lines in the solar spectrum of light. In 1814 Joseph von Fraunhofer began to systematically study these lines, which were later named after him. The composition of the outer gas layers of the sun can be derived from them.

The sun was first recognized as a source of energy during the first half of the 20th century, when hydrogen fusion was discovered. Solar radio waves and x-rays were also discovered, and the first solar vibrations were measured during the second half of the century. This provided much information about the inner structure of the sun. Giant underground detectors were built to measure the level of neutrinos emitted by the sun. Neutrinos, which have no electric charge and can easily pass through solid matter, are generated during solar energy production.

Space probes such as the SOHO carry highly specialized instruments and technology in order to record observations of the sun.

Satellites and space probes

Several satellites observe the sun from orbit. They are especially useful for studying the parts of solar radiation that are blocked by the Earth's atmosphere, such as ultraviolet radiation and x-rays.

Due to the high temperatures and intensive radiation, it is technically challenging to bring space probes close to the sun. In the 1970s, Helios 1 and 2 were sent into elliptical orbits around it, and were able to reach less than one-third of the Earth's distance in proximity. The space probe Ulysses has been orbiting around the sun and across its solar poles since 1990. This allows a unique view of the solar magnetic field from the "top" and "bottom" perspectives. The Solar and Heliospheric Observatory (SOHO) has been orbiting along the Earth-sun line since 1995. The space probe is orbiting the First Lagrangian Point (L1), while the Earth simultaneously orbits the sun. A second satellite called TRACE was sent for support in 1998. Together, they track the development of magnetic fields, plasma structures, and mass ejections of the sun. They are a major contributing factor in the prediction of solar flares.

Two STEREO space probes have been in orbit around the sun since 2006. One of them is traveling ahead of the Earth, while the other one is traveling behind. This allows for three-dimensional measuring of processes taking place within the sun and solar wind.

A coronagraph blocks the brighter parts of the sun to create an image with visible coronal ejections.

basics

THE SUN'S ENERGY SOURCE was not known until the 20th century. Previously it was thought that the sun might consist of glowing coal. A further idea was that the sun might have drawn energy under compression from its own gravity.

in focus

SPECTROSCOPY

A spectrograph splits the light according to its wavelength, similarly to the way a glass prism does, but in much more detail. Many parts of sun spectra recorded in this way show dark lines. These are the result of various gases from the photosphere (p. 36) absorbing light of certain wavelengths. This allows analysis of the sun's chemical composition. The magnetic field of the sun can also be measured using the spectrograph method because it splits the lines. Even the light of other stars can be analyzed.

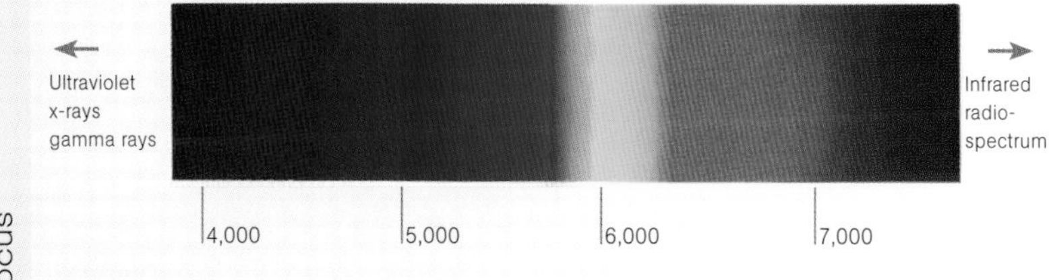

Dark absorption lines appear in the various spectra of light depending on the chemical composition of gases present in the photosphere or atmosphere of distant objects.

see also: Exploring the solar system, p. 51

INNER PLANETS

The surfaces and gaseous envelopes of Mercury, Venus, Earth, and Mars are clearly distinguishable. None except Earth can support life. The Earth is also distinguishable by the presence of its moon. Since the moon is relatively large, the Earth and moon are sometimes referred to as a double planet system.

mercury and venus

Mercury and Venus are relatively close to the sun, so both are scorching hot. Neither one has any moons. Other than that, they are very different. Venus—in contrast to Mercury—has a dense carbon dioxide atmosphere.

Mercury is the innermost and smallest planet of the solar system. Its surface is marked by giant cliffs and areas of angular terrain that extend over hundreds of miles. These features are the result of compressive stress within the crust. They may have developed after the igneous phase of the planet, when the planet cooled off and contracted. Temperatures on the surface of Mercury reach as high as 800°F (430°C) and as low as about -270°F (-170°C). This extreme difference between high and low temperatures is due to Mercury's thin gaseous atmosphere, which cannot store heat. Mercury's proximity to the sun has made investigation difficult, because space probes need to withstand both the intense radiation from the sun and its gravitational pull. In the 1970s, the Mariner 10 probe investigated Mercury as it flew past the planet. Today, NASA is preparing their space probe Messenger, which is due to orbit Mercury in 2011. The launch of the European-Japanese mission BepiColombo is planned for 2013.

The surface of Mercury has numerous meteorite craters, large mountain ridges, and plateaus—presumably of volcanic origin.

Venus is the second planet from the sun and is nearly as large as the Earth. Venus glows brightly due to a dense creamy-colored cloud cover made up largely of sulfuric acid droplets, which reflects more than 70 percent of the incoming solar radiation. By comparison, the Earth reflects only about 40 percent.

The atmosphere of Venus contains carbon dioxide, nitrogen, and traces of sulfur dioxide, water, and other substances. Venus has the densest atmosphere of all terrestrial planets: pressure on the ground is about 90 bar—the equivalent of oceanic pressure at a depth of 2,950 feet (900 m). Ground temperatures can reach 860°F (460°C). The surface of Venus resembles a rocky desert, with giant plateaus, depressions, highlands, volcanoes, and craters. Some interplanetary space probes have already brought back scientific data from Venus, and the planet has been studied by the European Venus Express since 2006.

Venus glows brightly in the night sky, which has led to it being known as the evening and morning star.

in focus

THE ROSETTE ORBIT OF MERCURY

The orbit of Mercury is neither circular nor elliptical; rather, it performs a rosette-like orbit. The deviation is small but measurable. Weaker deviations also occur among the other planets. The orbital deviations cannot be fully explained by Newton's theory of gravitation; however, Albert Einstein's theory of general relativity provides an adequate explanation.

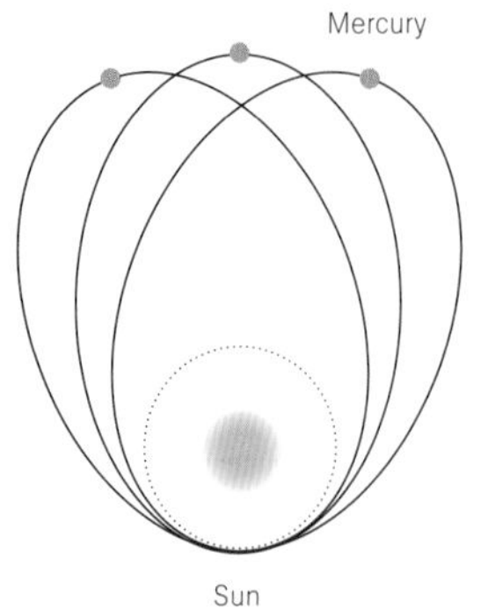

After one orbit around the sun, Mercury does not return exactly to its original starting point.

the moon

No other celestial body is as close to Earth as the moon. It has been comprehensively explored, but some of its secrets have still not been revealed.

The surface of the moon, scarred with craters, provides a glimpse of what the Earth might have looked some four billion years ago. At that time, the newly formed planet was exposed to a hail storm of asteroids (p. 42). Since then, the craters on Earth have been eroded by wind and water or filled in by terrestrial changes in the Earth's crust. In contrast, the appearance of the moon has barely changed since the formation of its lowlands.

The far side of the moon (left) is not visible from Earth. It was first seen when space probes reached the moon. The back of the moon is distinctly different from the front (right): it consists almost entirely of highlands and has many more craters.

Lunar seas and craters

Before much was known about the moon, its dark lowlands were thought to be seas. It was on the Mare Tranquilitatis (Sea of Tranquility) that the first human visitor left his footprint in 1969 (p. 48). Today we know that these lunar "seas" are solidified lava flows that have filled impact craters and depressions.

The moon does not have a protective atmosphere, meaning that large and small meteorites are not prevented from impacting it. These collisions have pulverized the rocks on the surface, creating a layer of rubble and dust called the regolith. The bright highlands of the moon were formerly thought to be continents and were called terrae. They are geologically older than the lowlands and covered with substantially more craters. Most of these craters come from asteroid impacts during the early life of the moon. They are named after astronomers, philosophers, and other scholars. Because it lacks a gaseous envelope, temperatures on the moon can vary extremely. In full solar radiation the temperature reaches approximately 265°F (130°C) and drops again to about -255°F (-160°C) during the lunar night.

Water ice

The moon is an extremely dry celestial body, especially in comparison with its nearest neighbor, the water planet Earth. Nevertheless, space probes have found evidence of possible water ice in the polar regions of the moon. For instance, ice from impacting comets could lie at the bottom of deep polar craters, where it would remain out of reach of heating by solar radiation (and thus would be protected from evaporation and escape). Such water ice could be a valuable resource for any future space stations or human settlements on the moon.

This false-color photograph shows the different surface soil composition of the moon. Red areas generally correspond to the lunar highlands, while blue to orange shades indicate the ancient volcanic lava flow of a mare, or lunar sea. Bluer mare areas contain more titanium than the orange regions.

see also: Tides, coasts, and waves, p. 99

the earth-moon system

The Earth and moon were forced into coexistence by the collision that created the moon. Since then their movements have been closely interwoven. They influence each other, creating diverse phenomena.

The diameter of the moon is four times smaller than that of the Earth. In comparison, the moons of other planets are proportionately much smaller. Because of this, the Earth and moon are sometimes referred to as a double planet system. The moon has more influence on the Earth than an average planetary moon. However, the gravitational force of the moon is too small to maintain an atmosphere.

The "dark" side of the moon is illuminated by light reflected from the Earth.

CLOSE TO THE HORIZON, the moon looks larger than when it is high in the sky. This is an optical illusion.

THE TIME BETWEEN two equal moon phases, that is from full moon to full moon, lasts 29.5 days.

THE MOON'S ORBIT around the Earth, relative to the Earth's orbit around the sun, is tilted by 5°.

basics

Origin of the moon

Based on computer simulations and examinations of moon rock, it is thought that the young Earth collided with a smaller protoplanet around 4.5 billion years ago. The rubble from this collision formed a ring around the Earth, which then agglomerated to form the moon.

Moonlight and earthlight

The side of the moon that is illuminated by the sun reflects that light back to the Earth. The amount of moon visible to the Earth depends on its position in relation to the sun. The moon changes its phases over the course of a month as it orbits the Earth. At new moon, the dark side of the moon (that part not lit by the sun) is directed at Earth. At full moon, we see the half of the moon lit by the sun and the sunlight reflected from it. Even the dark side of the moon is never totally black—it is lit by earthlight (sunlight reflected by the Earth's clouds). By measuring this ash gray light, we can recognize changes in cloud cover and the Earth's atmosphere.

Synchronous rotation

The moon always shows the same side to the Earth—this means it takes as long for one orbit around the Earth as it does for one revolution around itself. This synchronous rotation is caused by tidal periodicity. Early in the moon's history, its rotation slowed and became locked in this configuration with the Earth as a result of frictional forces caused by tidal effects from the Earth. Another consequence of the tidal interaction between the Earth and the moon is that the gravitational force of the Earth's tidal peaks pulls the moon slightly along its orbit. This causes the moon to travel faster and raise its orbit, so that the moon moves roughly 1.5 inches (3.8 cm) further away from the Earth every year.

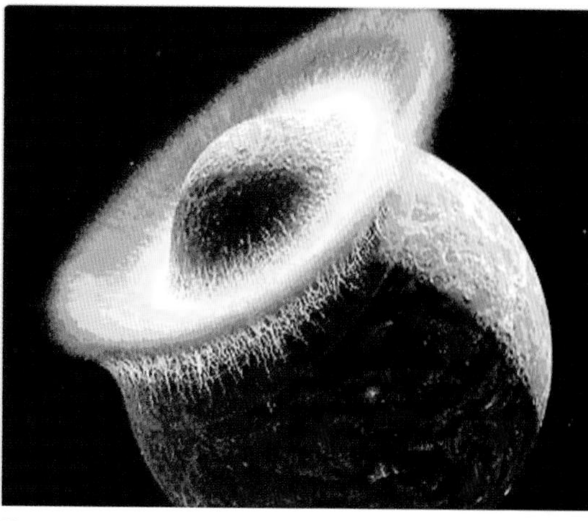

A collision between the Earth and a smaller planet created the moon. The moon would have been liquid at first, before cooling and solidifying.

in focus

LUNAR AND SOLAR ECLIPSES

The moon darkens when it passes through the shadow of the Earth. The sun darkens when the shadow of the moon falls onto the Earth. In both instances, sun, moon, and Earth are in a straight line. However, the moon orbit is inclined with respect to the Earth orbit, so that eclipses do not occur during each orbit of the moon.

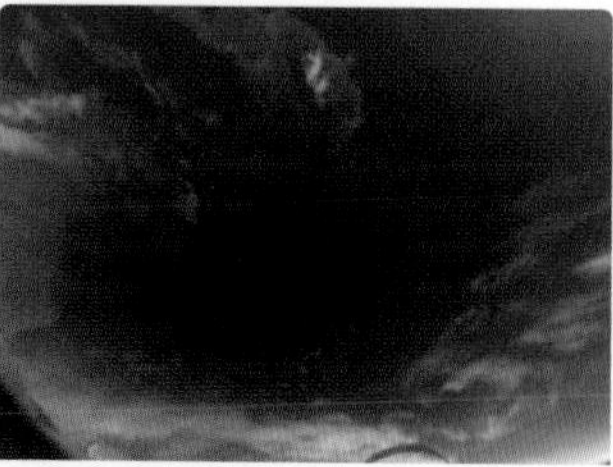

The moon's shadow is smaller than the Earth, so a complete solar eclipse can only be observed in parts of the world.

mars

Mars is only half as large as the Earth, and it is enclosed by a thin atmosphere. From all the planets of the solar system, its surface conditions are closest to those on Earth—so it has long been of great interest to science.

Mars is 1.5 times further from the sun than the Earth. It is known as the Red Planet due to its rusty coloration, as its surface is high in iron oxide. The rotation of

The largest canyon in the solar system, the Valles Marineris, is about nine times longer and four times deeper than the Earth's Grand Canyon.

Mars on its axis takes 24.6 hours. It is orbited by two moons: Phobos, with a diameter of 16 miles (27 km), and Deimos, diameter nine miles (15 km).

basics

DIAMETER of Mars: 4,221 miles (6,794 km)

DISTANCE of Mars to the sun: 142 million miles (228 million km)

ORBIT of Mars around the sun: 687 days

Atmosphere

Mars's atmosphere consists of 95 percent carbon dioxide, 3 percent nitrogen, and small amounts of argon, oxygen, carbon monoxide, and water vapor. The atmosphere is thin with a ground pressure varying between four and nine millibars. By comparison, the atmospheric pressure on Earth at sea level is 1,013 millibars; therefore without a pressurized suit and an oxygen supply a human could not survive. Only a little heat can be stored by the thin atmosphere, and temperature differences are extreme. They can vary between -112°F (-80°C) and 68°F (20°C) at the Equator.

The surface

Mars's atmosphere must once have been denser, as river valleys with streamlined islands indicate the presence of running water. Nowadays, water can only last in the form of ice or water vapor, although the presence of liquid water underneath Mars's surface is still possible. By means of radar measurements made by the European space probe, Mars Express, water ice has now been discovered at the south pole, embedded deep within the ground. The surface of Mars resembles a rocky desert. The northern plains, flattened by lava flows, contrast with the southern highlands, which have huge impact craters. The volcano Olympus Mons is the highest mountain in the solar system, reaching 16 miles (26 km) above the lowland—three times higher than Mount Everest. The Valles Marineris is part of an enormous trench system across the planet. It extends for more than 2,485 miles (4,000 km) and in places is four miles (seven km) deep.

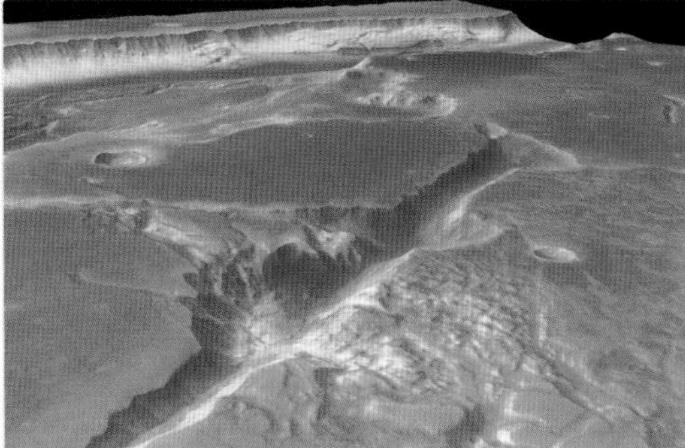

Impact craters on Mars provide scientists with a glimpse of the deeper layers of the planet, and give clues to Mars's history.

in focus

LIFE AND WATER ON MARS

Biochemical data taken by the Viking probes during the 1970s may indicate the presence of microorganisms on Mars. Another indication of life in the Mars atmosphere is the presence of methane. This could have developed geochemically or from the metabolism of microorganisms. Meteorites from Mars have also shown possible evidence of fossilized microorganisms.

Evidence suggesting the existence of life on Mars remains scientifically controversial.

Mars's red soil is caused by high levels of iron oxide. Large amounts of sulfur and silica have also been found in the surface.

see also: Exploring the solar system. p. 51

asteroids | comets

ASTEROIDS AND COMETS

Asteroids and comets are probably remnants from the time the solar system formed. Consequently, they are fascinating to science, providing clues to the early life of the solar system.

asteroids

Asteroids are irregularly shaped boulders, usually less than 60 miles (100 km) in diameter. They orbit the sun, generally in slightly elliptical orbits. A few can stray through the solar system.

Most asteroids are found in the region of the solar system called the asteroid belt, which is a ring of celestial objects of varying size, orbiting the sun between the planets Mars and Jupiter. Approximately 200 asteroids in this belt have diameters larger than 60 miles (100 km). Asteroids are remnants of material from the beginning of the solar system, not fragments of planets as previously believed.

Jupiter's gravitational force prevents the asteroids in the asteroid belt from agglomerating into planets. Sometimes the asteroids' orbits are disturbed by collisions or gravitational forces, causing them to career off through space and possibly impact with planets or their moons. Sunlight reflected by asteroids can be analyzed by spectroscopic examination, which reveals the chemical composition of their surface areas. Fragments from asteroids sometimes fall to Earth as meteorites, which can be studied. Asteroids have also been examined by interplanetary space probes (p. 43). Around 75 percent of asteroids have dark surfaces containing carbon (graphite); the rest have bright surfaces and consist of silicates or iron and nickel.

Occasionally, asteroids come close to Earth or cross its orbital path—some have even passed within the moon's orbit. As the effects of a large asteroid colliding with Earth could be catastrophic, they are systematically monitored from various positions around the globe. Although they may pose a threat now, it is a possibility that the asteroids and comet collisions with the Earth first supplied the planet with water and the chemical building blocks necessary for the later development of living organisms.

Other names for asteroids are planetoids or minor planets. The first celestial object of many to be discovered in the asteroid belt was Ceres in 1801, by the Italian astronomer Giuseppe Piazzi. Today, Ceres is the largest known asteroid with a diameter of 580 miles (933 km) and, like Pluto, is considered a dwarf planet.

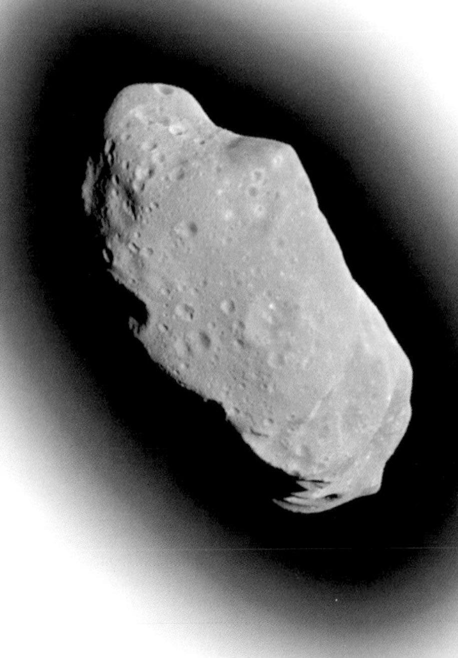

The results of many asteroid impacts can be seen in the impact craters on the moon.

ESTIMATING IMPACT RISK

practice

The risk posed by an asteroid or meteorite increases with the probability of a hit and possible damage that could be inflicted. The magnitude of damage expected depends on the speed and composition of the object (ice, porous or solid rock, metal). The impact of a 325-foot (100-m) rock could devastate an entire region, and an asteroid several miles across could trigger a global, environmental catastrophe.

THE TORINO SCALE
Assessing Asteroid/Comet Impact Predictions

The Torino Scale depicts the potential risk level of a suspect celestial object.

In its history the world has been hit frequently by asteroids. A giant impact 65 million years ago may account for the extinction of the dinosaurs.

comets

Comets are essentially chunks of dirt and ice. They have changed even less than asteroids since the beginning of the solar system. Therefore, comets can provide valuable information about their origin.

Comets are celestial bodies that are invisible until they near the sun. Some comets have short orbital periods and are thought to originate from the Kuiper belt, beyond the orbit of Neptune. Comets with orbital periods over 200 years are believed to originate from the Oort cloud that surrounds the solar system—almost out of range of the sun's gravitational force. These comets are so far out in space that one orbit can last up to 30 million years.

Comet Hale-Bopp was visible to the naked eye for 18 months, making it one of the most spectacular comets of the 20th century. It passed its closest point to the sun on April 1, 1997.

A comet awakens

The core of a comet often measures only a few miles across and is composed of ice, rock, dust, and frozen gases. When the comet penetrates the inner solar system it is warmed by the sun, the volatile components escape, and jets of gas and dust are expelled. A fog-like envelope (the coma) then forms around the core. The coma glows because it reflects sunlight and its atoms and molecules release radiation. The pressure of sunlight and solar winds loaded with electrically charged particles blow the coma away from the sun, leading to the formation of a comet "tail." This tail extends as a plasma tail of electrically charged molecules or as a curved dust tail.

The end of a comet

Each orbit around the sun causes the comet to lose matter from its coma until it breaks up. Some comets get so close to the sun that they vaporize (sun-grazers), while others hurtle directly into the sun.

The Deep Impact impactor spacecraft collided with the Tempel 1 comet in 2005, gathering data about the composition of the comet.

Meteorites

Shooting stars that flash across the sky are actually caused by dust, rock, or metal bodies burning up in the atmosphere; they can originate from planets, asteroids, or comets. Larger chunks do not burn out completely and reach the surface of the Earth as meteorites. There are also meteorite showers that can occur annually due to the Earth crossing a comet's orbit and the material lost by the comet entering the Earth's atmosphere.

in focus

MISSIONS TO ASTEROIDS AND COMETS

Numerous asteroids and comets have already been explored. In 1986, an international armada stormed toward Halley's comet. Galileo (NASA) passed two asteroids on its way to Jupiter in 1991 and 1993. Stardust (NASA) visited the comet Wild 2 in 2004 and gathered dust particles that were returned to Earth. Rosetta (ESA) is currently on its way to orbit and launch a landing probe on to the comet Churyumov-Gerasimenko, which it will reach in 2014.

Numerous space probes and missions have explored asteroids and comets.

OUTER PLANETS

The outer solar system is dominated by four gaseous planets: Jupiter, Saturn, Uranus, and Neptune. They are followed by the dwarf planet Pluto and numerous other small celestial bodies. Other stars are also orbited by planets. In the near future, intergalactic telescopes may find Earthlike planets in outer space too, and possibly other life.

jupiter and saturn

The world of the two largest planets has been brought closer to us since the 1970s, with the help of several space probes. It consists of gloriously colored gas envelopes, unique moons, and complex ring systems.

The giant planets, Jupiter and Saturn, are called gas giants because they are immersed in gas envelopes of hydrogen and helium. Unlike the rocky planets closer to the sun, these so-called Jovian planets have no solid surface as such, they simply consist of layers of atmosphere at increasingly high pressures. The deepest layers of gas may behave more like a liquid under such pressure. The cores, however, may be rocky or metallic.

Jupiter is 2.5 times bigger than all the other planets in the solar system combined. The high pressure in its atmosphere causes hydrogen to have metal-like properties and become electrically conductive. This, combined with its immense rotational speed of one cycle every ten hours, produces a powerful magnetic field and results in turbulence and storms in the atmosphere, some of which are visible even with basic telescopes. Jupiter is encircled by roughly 62 moons—the four largest were discovered in 1610 by Galileo. Ganymede is the solar system's largest moon and has its own magnetic field. Io is the most volcanically active celestial body in the solar system. It is believed that an ocean lies beneath Europa's ice crust. Jupiter's other moons are probably captured asteroids (p. 42). Of Saturn's estimated 60 moons, Titan is the largest and has a dense atmosphere. Iapetus is distinctive due to its two-tone color.

Saturn has an extensive ring system separated by gaps—the largest being the Cassini Division. This system extends over hundreds of thousands of miles into space and consists of ice crystals, dust, and rock. All gaseous planets of the solar system are surrounded by rings, but Saturn's are the most spectacular. Remarkably, Saturn is the only planet in the solar system that is less dense than water.

An enormous cyclonic storm has been seen in Jupiter's atmosphere since the 17th century. Known as the Great Red Spot, it could swallow up the Earth twice over.

The most popular theory proposes that Saturn's rings come from a former moon of Saturn that was broken up or destroyed by a collision.

THE CASSINI-HUYGENS MISSION

practice

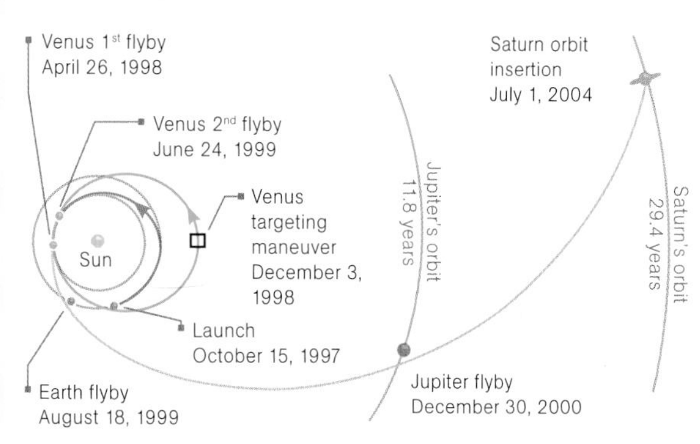

Cassini-Huygens took almost seven years to reach Saturn. It circled Earth and Venus several times to gain gravitational momentum and direction.

The double space probe Cassini-Huygens (U.S./Europe) was launched in 1997. With a weight of 5.6 tons, it is the largest space probe ever built. After four swing-by maneuvers (p. 51), it reached Saturn in the year 2004. There Huygens was released and then traveled on to the Saturn moon Titan. Three weeks later, the probe, fitted with a heat shield, penetrated the thick nitrogen atmosphere of Titan and landed by means of parachutes. Cassini continues to explore the Saturn system.

uranus and neptune

Uranus and Neptune are the two giant outer planets of the solar system. Neptune is further from the sun and is not visible with the naked eye.

Uranus is slightly larger than Neptune, and both planets are about four times larger than the Earth. Their gaseous envelopes resemble Jupiter and Saturn and consist primarily of hydrogen and helium. Their bluish green appearance is caused by methane. Both planets have rocky cores. High pressure located deep inside the atmospheres of these planets surrounds their cores with liquid mantles of water, ammonia, and methane. Due to the enormous distances to the sun, both planets receive little solar energy. However, the atmosphere of Neptune is penetrated by strong wind currents and cyclonic storms in the form of oval spots, which can be seen in Neptune's atmosphere. The rotational axis of Uranus is tilted sideways, so that it appears as if this planet is inclined toward one side. This inclination is the result of a collision with another celestial body.

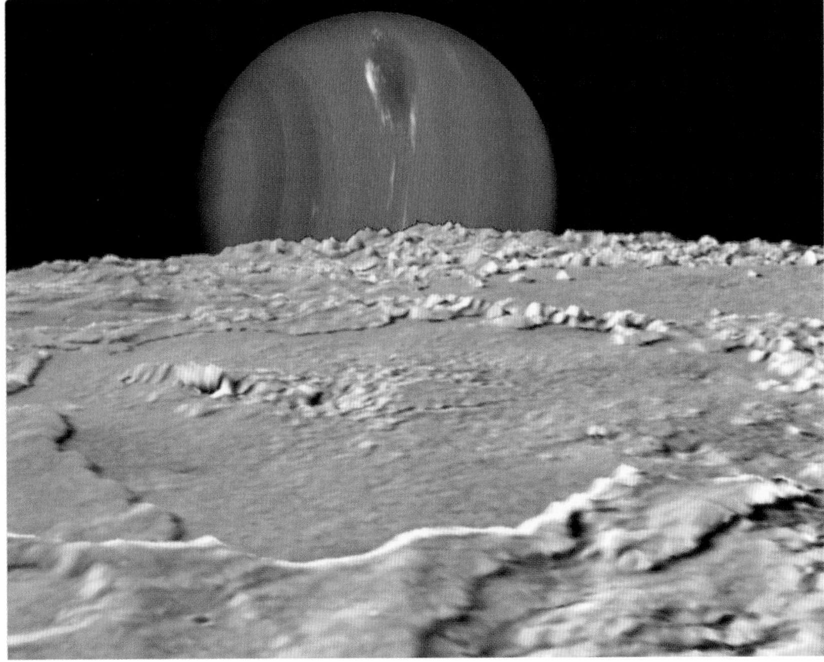

A view of Neptune from above the surface of Triton, Neptune's largest moon. The terraces visible on the surface of Triton indicate multiple episodes of "cryovolcanic" flooding.

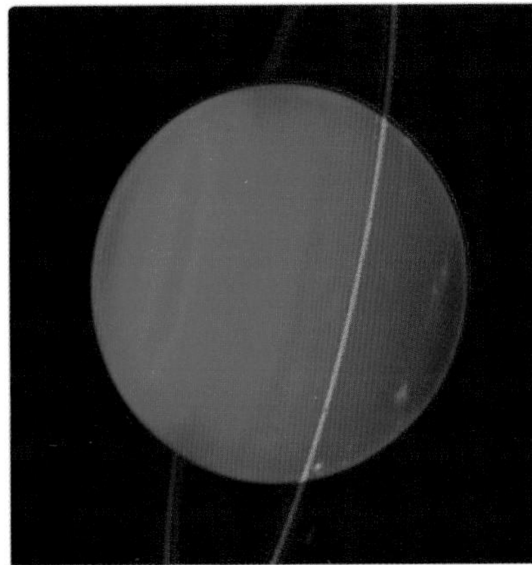

The axis of Uranus is sharply tilted—the poles are where most planets have their equator.

Rings and moons

Both planets possess dark ring systems, although they are not as well defined as that of Saturn. Moreover, the ring system of Neptune contains unusual ring arcs as well as extremely fine dust. Uranus is encircled by roughly 27 moons. The five largest are composed of ice and rock. One moon, Miranda, has a diameter of only 292 miles (470 km), but it has a unique surface—rock fragments border onto terrace-like plains and canyons up to 12 miles (20 km) deep. The small moons of Uranus seem to be captured asteroids (p. 42). Neptune has at least 13 moons. Triton, with a diameter of 1,678 miles (2,700 km), is by far the largest. Since its icy surface reflects most of what little sunlight there is, this moon, with a temperature of only -400°F (-240°C), is one of the coldest objects in the solar system. Its icy volcanoes spew liquid nitrogen, methane, and dust. This mixture freezes immediately and precipitates back to the surface as snow.

in focus

THE DISCOVERY OF NEPTUNE

Neptune was the very first planet to be found mathematically. After the discovery of the planet Uranus in 1781, measurements of the expected trajectory revealed deviations. It was presumed that this was due to gravitational interference by another planet. John Couch Adams and Urbain Jean Joseph Le Verrier each calculated the trajectory and the mass of the presumed planet, which was confirmed almost 60 years later.

Johann Gottfried Galle discovered Neptune in 1846, close to the predicted location.

basics

DISTANCE TO THE SUN: Uranus 1,785 million miles (2,872 million km), Neptune 2,793 million miles (4,495 million km)

ORBIT: Uranus 84.0 years, Neptune 164.8 years

ROTATION PERIOD: Uranus 17.2 hours, Neptune 16.1 hours

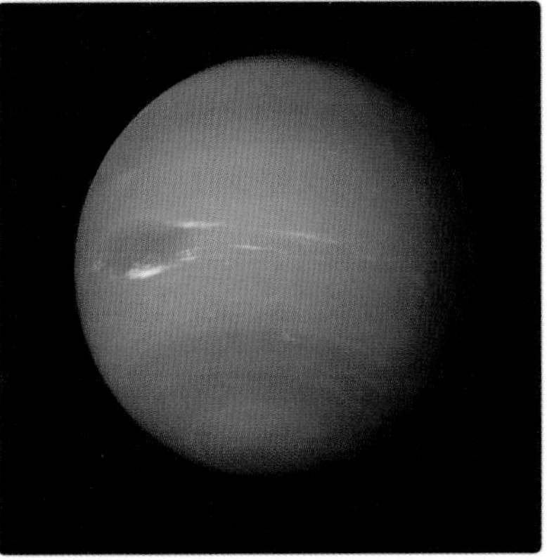

The dark spot visible on Neptune is an anticyclonic storm that circles the planet every 18.3 hours.

pluto

Pluto was discovered in 1930 in a review of astrophotographic images, but since 1992 more and more celestial bodies have been detected in the region of its orbit. Consequently, it is now considered only one of the many icy objects located on the far side of Neptune's orbit.

Pluto and Charon orbit around a common center of gravity in the space between them.

Pluto, whose diameter is only two-thirds that of the moon's, was previously known as the planet furthest from the sun—the ninth planet in our solar system—until August 24, 2006. On this date the International Astronomical Union redefined the concept of "planet" (p. 32) to include the condition that a true planet must have cleared the path of its own orbit during its development phase. Because Pluto did not do so, it is now defined as a dwarf planet, just one of many objects orbiting the sun in the so-called Kuiper belt, a disk-shaped region beyond Neptune's orbit. These celestial bodies are also called trans-Neptunian objects. In comparison with the planets in the solar system, Pluto's orbit is considerably more elongated. The largest section lies beyond the nearly circular orbit of Neptune (p. 45), while another part lies within it.

basics

DIAMETER of Pluto: 1,485 miles (2,390 km)

DISTANCE from Pluto to the sun: 2.757–4.583 million miles (4.437–7.376 million km)

TIME it takes Pluto to revolve around the sun: 247 years and eight months

TIME it takes Pluto to rotate once on its axis: 153 hours and 18 minutes

Pluto has not yet been visited by a space probe. What we know about Pluto comes largely from observations made from the Hubble telescope, the Infrared Astronomical Satellite (IRAS) and from Earth. According to previous findings, it is likely that Pluto consists of a rocky core surrounded by a mantle of ice. In addition, it appears to be covered with several layers of frozen methane, nitrogen, and carbon monoxide. When Pluto passes near the sun, these icy layers evaporate into gases, building an extremely thin atmosphere. These gas layers freeze again when Pluto's orbit takes it further away from the sun.

in focus

TRANS-NEPTUNIAN OBJECTS

Pluto and its moons belong to the group of celestial bodies in the Kuiper belt on the far side of Neptune's orbit (known as Kuiper belt objects, KBOs, or trans-Neptunian objects). Numerous objects consisting of rock and ice, possibly including many comet cores as well (p. 43), revolve within this ring-shaped area. It is likely that the KBOs were formed in parallel with the planets, and the gravitational forces associated with Neptune pulled a few of them into elongated orbits. The largest KBOs known so far have dimensions and characteristics similar to Pluto. One example is Eris, which is actually somewhat larger than Pluto. Eris has its own moon and is also classified as a dwarf planet.

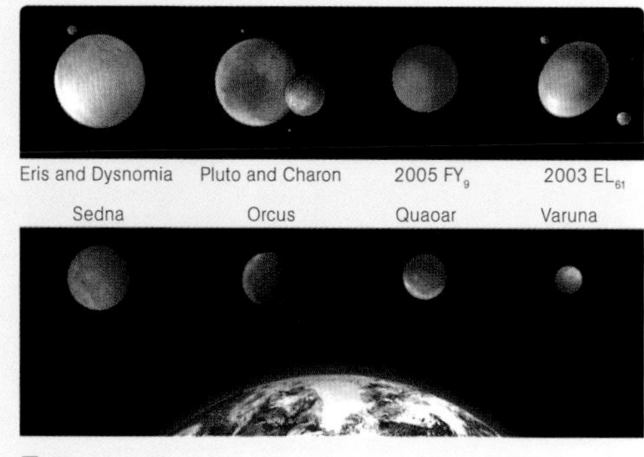

Pluto and Eris have their own moons, but are just two of many trans-Neptunian objects orbiting in the Kuiper belt.

Three moons

Pluto has at least three moons: Charon, Hydra, and Nix. Charon was discovered in 1978, but the others were not found until 2005. Charon appears to be covered with ice and is about half the size of Pluto, which is proportionally very large for a moon. These two objects revolve around a common gravitational center in the space between them, rather than a gravitational center within Pluto. Due to the brightness of the other two moons, their diameters have been estimated to be only around 105 miles (170 km) in length.

New Horizons will be the first space probe to make the journey to Pluto. The probe, which belongs to the U.S. space agency NASA, was launched in 2006 and passed very close by Jupiter the following year, using a flyby technique that utilized the planet's gravity to boost its speed. It will not reach Pluto and its moons until 2015.

As an extension of the mission, the probe will then proceed further into the Kuiper belt in order to investigate one or more of its small celestial bodies and shed light on the nature of the trans-Neptunian objects. These celestial bodies form their own group, in addition to the terrestrial (rocky) planets (pp. 38, 41) and the gas giants (pp. 44–47).

It is possible that after the results of the investigation are evaluated, we will understand more about the development of the solar system (pp. 32–37).

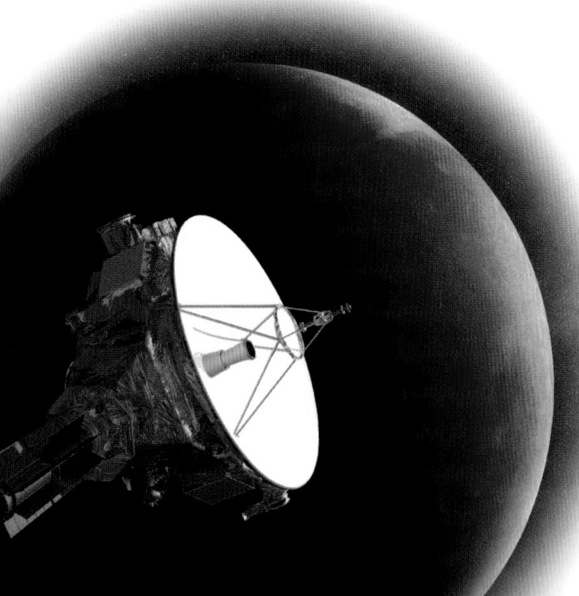

The probe New Horizons left Earth in 2006. It will take over nine years to reach Pluto and its moons before continuing on into the Kuiper belt.

see also: Astronomy today, p. 19

extrasolar planets

Our solar system is not the only one of its kind; more and more planets in the vicinity of distant stars are being found all the time. Even though finding planets similar to Earth is particularly difficult, it is an especially compelling search and remains the goal of many research projects.

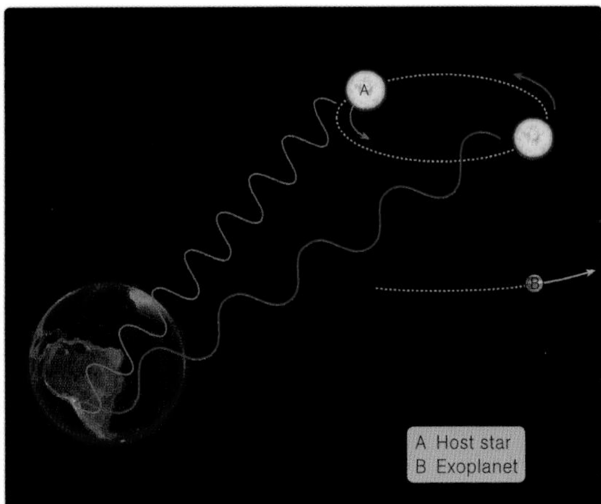

The Doppler method involves measuring the wavelength of light coming from objects moving toward or away from the Earth.

In many regions of our galaxy, debris disks can be seen revolving around stars. It is from this debris that planetary systems are formed. However, discovering individual planets in the vicinity of distant stars is significantly difficult because a star and its planet(s) appear extremely close to one another from such vast distances. Furthermore, a star like our sun outshines its planets to the extent that it is about a billion times brighter, thus the light reflected from a planet is barely detectable as a separate light source to its star. A planet only reflects, it does not generate any visible light itself. However, this difference in brightness is somewhat less in the infrared spectrum because a planet does emit some infrared light or thermal radiation, even if it is only very little in comparison with the star it is orbiting.

Naming

Extrasolar planets, also called exoplanets, are usually named after the star they revolve around. Letters are also added to their names, indicating the order in which they were discovered, beginning with "b," for instance, HD 38529 b, HD 38529 c, and so on.

Methods of proof

Most of the exoplanets discovered so far have a mass considerably larger than the Earth and are more comparable in size to the giant planets of our solar system. Furthermore, they revolve around their stars in very close orbits. Modern methods and technology aid us in demonstrating that these bodies are planets. Under certain conditions even some smaller planets with less mass have been detected. Photographs of planets near stars of normal brightness cannot yet be taken with telescopes, so proof that these bodies are planets is more indirect. Of the methods developed for this purpose, the transit method and the Doppler method are those most often applied.

The transit method is based on recording a star's small drop in brightness when a planet moves in front of it, blocking some of the light. The Doppler—or radial velocity—method is based on another effect: strictly speaking, a star and planet revolve around a common gravitational center, even if this center is located within the star itself. Within the orbit interval, the star's orbit appears to wobble somewhat, with heavier planets causing a greater deflection. This can be proven with a spectroscopic examination of the star's light. Scientists are now developing methods for masking a star's light. With these methods, they can investigate even smaller planets that may be similar in composition to the Earth, and will be able to make analysis of their atmospheres. These methods are due to be ready for use from observatories in space within the next decade. Among the related international research projects are Darwin, based in Europe, and Terrestrial Planet Finder, based in the United States.

basics

IN 1995, the first planet in the vicinity of a star similar to the sun, 51 Pegasi, was detected.

THE FIRST PLANETS orbiting other stars were discovered in 1992 and now number in the hundreds. More continue to be found with advances in technology.

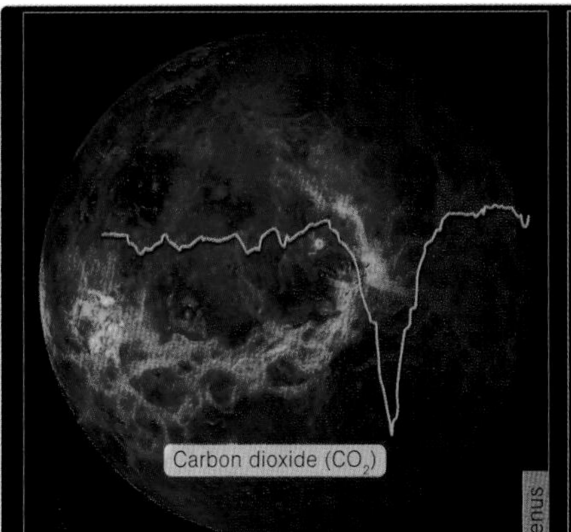

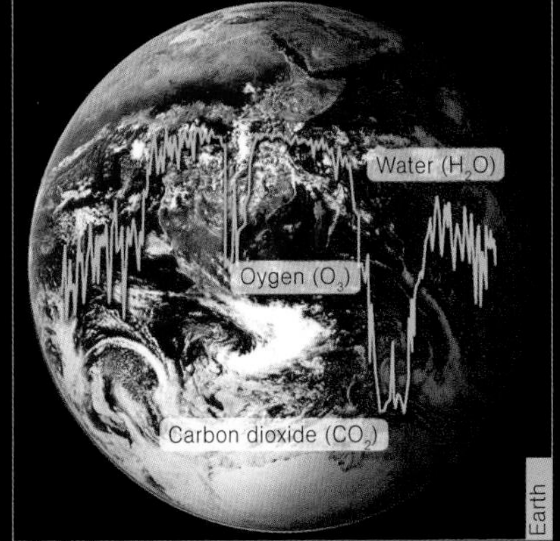

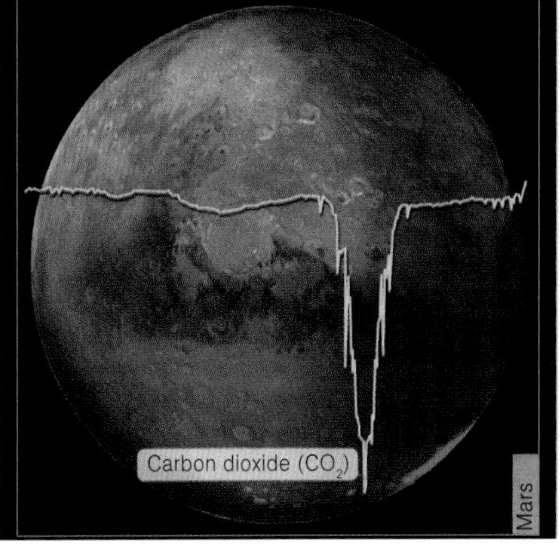

As gases and other substances absorb particular wavelengths of radiation, scientists are able to study absorbtion lines in the spectrum of light or thermal radiation from a planet's atmosphere. In this way they can identify constituents, such as water, oxygen, or carbon dioxide.

see also: Stars, pp. 26–29

the development of space travel

The dream of space travel eventually became a reality through the work of far sighted pioneers. Today, spacecraft carry out many important tasks, making them an irreplaceable—if rarely seen—component of modern life.

By 1900 the mathematical groundwork for space travel had already been laid, but it took time before the concept of the rocket was taken seriously. First, solutions to key technical problems had to be found. Military forces were also interested in rocket technology. The breakthrough of large liquid-fueled rockets occurred during the Second World War, when the German A4 (V2) rockets became the first to reach the borders of space. Although they were able to soar to heights of more than 50 miles (80 km), with a range of 185 miles (300 km), they did not significantly affect the war's outcome. When the Second World War gave way to the Cold War, rocket technology developed rapidly as the U.S. and U.S.S.R. competed for dominance.

The space age began in 1957, with the launch of the first artificial satellite into orbit. Soon after, intercontinental missiles capable of carrying atomic weapons were constructed. The first surveillance and weather satellites were launched, followed in 1961 by the first man in orbit, Yury Gagarin. The greatest triumph of the space age came soon after: walking on the moon. Later, even as the tensions of the Cold War eased, the ability to undertake missions in space remained a key element of political and military power. Although it is not immediately obvious, the modern world depends heavily on space travel. Satellites (p. 49) transmit news, telephone conversations, and computer data around the world. They assist cars with navigation, deliver weather data, and provide detailed maps of the Earth. Other satellites, such as the orbiting Hubble Space Telescope, offer glimpses into the depths of the universe. Space probes and robots explore the solar system, collecting valuable information that helps us understand the universe and our place in it. At the same time, technologies are being developed that may be used in the future to build outpost civilizations on the moon and Mars. Astronauts from many nations work together in the International Space Station. They carry out research in the weightless environment, especially in the fields of medicine, materials science, and astrophysics. Private space-travel initiatives are also underway, aiming to offer affordable technological services as well as space tourism to individuals.

space travel | satellites | astronauts | missions

EXPLORING THE UNIVERSE

Just 12 years separated the launch of the first satellites and the first manned moon landing. Today, humanity's presence in space is limited to near-Earth orbit, where astronauts carry out research in a weightless environment. Meanwhile, countless satellites circle the Earth on scientific and technical missions. The farthest reaches of the solar system are still left to unmanned probes.

In 1961 the Russian cosmonaut Yury Gagarin became the first person in space. He reached Earth's orbit and then returned safely to Earth.

in focus

THE APOLLO PROGRAM

The U.S. Apollo program carried people to the moon and brought them safely back to Earth. It also developed the largest and most powerful rocket ever built: the Saturn V. The huge rocket was 360 feet (110 m) long, including its payload, lifted a mass of 120 tons into Earth orbit, and propelled the 45-ton space module with three astronauts toward the moon. On July 20, 1969, Neil Armstrong and Edwin "Buzz" Aldrin of the Apollo 11 mission emerged from their landing module to become the first humans to walk on the moon. Five more moon landings followed by 1972. The technological advances achieved by the Apollo space program have proved even more significant than the data it collected from the moon.

Neil Armstrong and Buzz Aldrin collected moon rock samples and performed scientific measurements near the Eagle moon lander.

see also: Survey of the universe, p. 21

satellite technology

Artificial satellites are spacecraft that orbit the Earth or other bodies in space. They carry out tasks that are difficult or impossible to undertake from the Earth's surface.

Satellites are monitored and directed from ground stations. While in orbit, sensors determine their position with respect to the Earth or sun. They can then be guided or maneuvered as needed, using electrical gyroscopes and small course correction engines.

Three synchronous satellites, each rotated 120°, can reach every part of the globe except the Poles. Modern communications satellites have high-powered transmitters, whose signals can be directly received by private satellite dishes.

A satellite's electrical power is usually provided by solar cells mounted directly on the satellite or on winglike solar panels. Battery power is used when the satellite travels through Earth's shadow. Satellites are exposed to extreme temperature differences as they alternate between exposure to direct solar radiation and the shadow of the Earth. Accordingly, their instruments require protection with insulating or heat-reflecting shields.

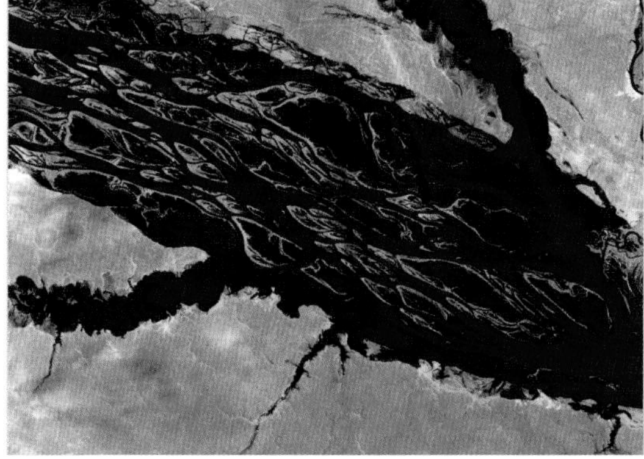
The Amazon Delta, the largest drainage basin in the world, as photographed by an Earth observation satellite.

Types of satellites

Telecommunications satellites serve as relay stations for data links and for radio, television, telephone, and fax signals. A satellite receives signals from a ground station, amplifies them, and sends them to another ground station. Earth observation and weather satellites monitor the sunlight and heat radiation reflected from the planet. Weather satellites also detect signals from the horizon, gathering data from various levels of the atmosphere. Detailed observations of the Earth's surface can be produced using Synthetic Aperture Radar (SAR), which passes through the cloud layer.

These images from a military spy satellite show the same area of Beirut: before (left) and after (right) a bombing raid by the Israeli air force.

Satellite paths

The paths followed by satellites in orbit can be circular or elliptical. An orbit of up to 620 miles (1,000 km) above ground level is considered a low Earth orbit. A satellite in low Earth orbit travels at approximately five miles per second (eight km/s). In a geostationary orbit, on the other hand, it travels at approximately two miles per second (three km/s). Many observation and weather satellites are located at this level in order to be close to Earth's surface. Orbits may be tilted with respect to the Equator and even cross the Poles. Since the Earth is always rotating under the satellites, large portions of the planet's surface can be covered. Geostationary orbit is particularly useful for telecommunications and television satellites. On this circular orbit, some 22,370 miles (36,000 km) above the Equator, the satellite's orbit takes the same amount of time as the planet's rotation. The device therefore stays over the same point on the Earth's surface. Fixed antennas can be conveniently aimed at such a satellite without needing to locate or track it.

practice

SATELLITE NAVIGATION

Navigation satellites help individuals, airplanes, cars, and other satellites determine their location. Systems include the U.S. GPS and Russian GLONASS, while the European Galileo system is under construction. Navigation satellites constantly transmit their positions in orbit over time. A navigation device receives data from several satellites. It can calculate the signal's travel time and thus its distance from the satellite. Using the positioning data from the satellites, it then triangulates its own precise location.

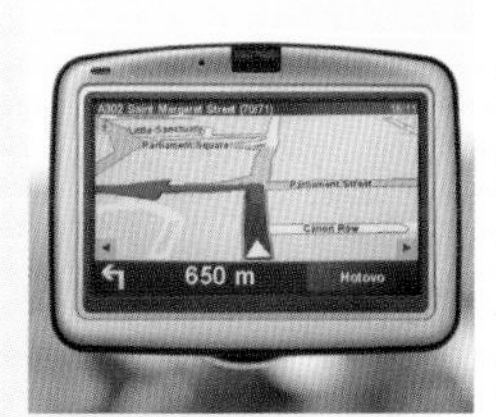

Display screen of a GPS navigation device for a car.

see also: Astronomy today, p. 19

human beings in space

Space is an unfamiliar and extremely dangerous environment for human beings. People are only able to create environments for living and working in space with extensive technical support.

An unprotected person in space would be exposed to an environment that is airless and a near vacuum. Without the pressure and warmth of the Earth's atmosphere, bodily tissues would quickly suffer serious damage; the person would soon lose consciousness and die. In addition, temperature variations are extreme because the sun's radiation is not blocked in space. Surfaces in the inner solar system can reach more than 248°F (120°C) under direct solar radiation, while temperatures in the shade fall below -148°F (-100°C). Therefore manned spacecraft, space stations, and space suits require sophisticated life-support systems.

SPACE SUITS

practice

Space suits made of numerous layers of textiles and artificial materials provide breathable air and protect astronauts from the near vacuum of space. The inner suit is lined with water tubes that help cool the body. Next come gas-impermeable layers. The outer layers are fireproof and reinforced against tearing. The air pressure inside is kept low to prevent the suit from inflating.

Astronauts move clumsily in space as the pressure of the suit causes stiff joints.

Life-support systems

In order to protect humans in space, support systems must supply air, water, food, and energy; regulate temperature and pressure; and provide bodily hygiene. Oxygen may be stored in liquid or gaseous form, or produced as required through the electrolysis of water recycled on board—for example, from washing water or urine. Nitrogen is added to the oxygen to provide Earthlike air for breathing, and the carbon dioxide breathed out by space travelers is removed from the air using special chemicals. The electricity used in space generally comes from solar cells, fuel cells, or batteries. In future bases on the moon or Mars, greenhouses and artificial ecosystems may play an important role.

Weightlessness

On Earth, gravity gives us a sense of "up and down." However, in space, astronauts experience weightlessness, which disturbs the balance organs of the inner ear. At the beginning of a space mission, this can lead to "space sickness," causing dizziness, nausea, headaches, and vomiting. After an adjustment period, astronauts usually learn to control their sense of orientation in space and these problems recede.

During longer periods of weightlessness, bone and muscle mass decline and the body's blood volume is reduced; however, these effects can be partially counteracted through exercise. By studying these symptoms in astronauts, scientists hope to gain insight into similar medical conditions on Earth. After returning from a mission, astronauts are usually able to regain their previous, normal physical state.

Humans are able to adjust their perceptions and movements surprisingly well to weightless conditions.

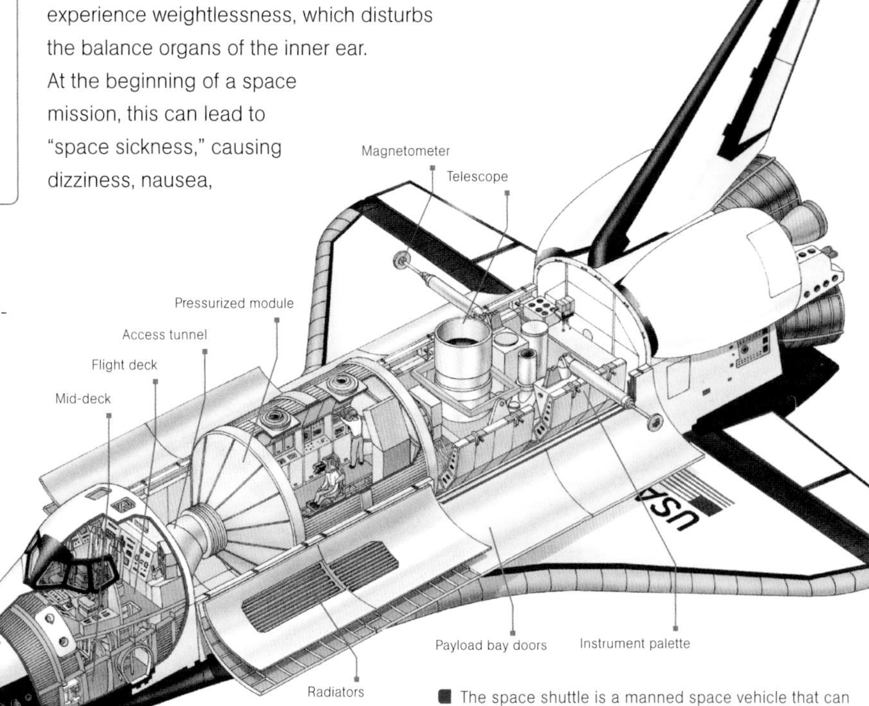

The space shuttle is a manned space vehicle that can land like an airplane. Two powerful solid rocket boosters provide lift for its vertical launch.

basics

SPACE TRAVELERS are variously known as astronauts (if American or European), cosmonauts (if Russian), or taikonauts (if Chinese).

DURING A SPACE MISSION, astronauts are responsible for carrying out a wide range of scientific and technical experiments, as well as the day-to-day running of the spacecraft.

see also: Astronomy today, p. 19

exploring the solar system

Space probes travel through space for months or years, journeying millions or even billions of miles to their destinations. Along the way, they help expand our knowledge of unexplored regions.

Space probes are unmanned spacecraft designed to gather information about the sun, planets, moons, asteroids, comets, and the space between them. Basic types include flyby, lander, and orbiter missions.

Voyager 1 was launched in 1977 to explore the outer planets. It continues to transmit data back to Earth.

Unlike a terrestrial satellite, which remains in the Earth's orbit, a space probe leaves the vicinity of the Earth to explore its target on location. Probes are generally launched from Earth on a path toward the target. On arrival, a braking maneuver must be carried out to place the probe in orbit around the faraway object.

To save fuel or allow more cargo to be carried, a probe may not take the most direct path to its destination. Instead, it takes a "swing-by," circling another planet to gain a gravity boost. If the target object has an atmosphere—such as Mars—fuel can also be saved in the deceleration process: the probe enters the atmosphere at an angle, using the resistance for aerobraking. Space probe navigation must be extremely precise; due to the vast distances within the solar system. Large antennas are used to track probes from Earth. Depending on the distance, navigational commands may take several minutes or more to reach the probe. However, modern computer technology, probes enables probes to undertake some actions independently, especially in critical situations.

Equipment

Space probes exploring the moon and inner planets usually produce their own electricity by using solar cells to convert the light energy from the sun. Yet, very little sunlight is available beyond Mars. Probes traveling further than this use energy supplied by a radioisotope generator. The emissions of a radioactive substance produce heat that is then converted into electrical energy. Depending on their mission, space probes may be equipped with cameras sensitive to various wavelengths, or with sensors that measure radiation or magnetic fields. Many orbiter probes use radars that can penetrate deep into the surface of an object in space, revealing its inner structure. Some space probes carry landing units for placement on the object's surface. Remotely controlled rovers or robots can survey a larger area than fixed landing units. Impactors or penetrators are probes that collide forcefully with an object, burrowing into it and collecting information about its constituent parts.

Mars is currently being studied by numerous orbiting probes and landing units. Since 2004, these have included the NASA Mars rovers Spirit and Opportunity.

in focus

FALSE-COLOR IMAGES

The human eye can differentiate between a significantly larger number of colors than it can distinguish degrees of light intensity. Thus, the details of a picture can be made clearer by using a visual trick: replacing the various brightness levels with different colors. This technique has proved especially useful in space science, medicine, and the evaluation of satellite images.

In this false-color view, an infrared image of the moon highlights different types of minerals.

basics

THE FIRST SPACE PROBE to leave Earth's orbit was the Soviet Union's Luna 1 mission. In 1959 it successfully traveled past the moon and began collecting data about magnetic fields, radiation, and the density of interplanetary gases.

see also: Solar research, p. 37

EARTH

ORIGINS AND GEOLOGY

Since its origin 4.56 billion years ago, the Earth has been subject to countless geological changes. Despite the apparent stability of the continents, they are in constant motion and may even break apart one day causing some mountain ranges to rise and others to disappear. This is all part of the continuous rock cycle. While most processes are so slow that humans are not even aware of them, others are abrupt with sometimes fatal consequences. Natural disasters such as earthquakes and volcanic eruptions show the power of the forces located in the Earth's interior. The Earth's outer crust is constantly affected by water, wind, and temperature fluctuations and opposing forces such as gravity. These forces are reflected in the landscape of the Earth's surface.

once upon a time, 4.56 billion years ago

In its early phase, the Earth was a place of extremes, with fire-spewing volcanoes, innumerable comet impacts, lightning, severe thunderstorms, and a toxic atmosphere. These conditions improved as the Earth cooled down.

ORIGINS

The stages of the Earth's development can be derived from hypotheses based on both good evidence and speculation about the origins of the solar system. However, one thing is certain: it was a long journey from a glowing ball to the blue planet, and at the beginning there was no hint that life would one day flourish there.

The Earth was created—along with the sun and the other planets—about 4.6 billion years ago from an interstellar gas and dust cloud. Presumably triggered by a shock wave from a supernova explosion, this cloud collapsed, started to rotate, and gradually flattened. Its matter became increasingly concentrated in the center, and eventually the increasing density triggered nuclear fusion, forming the protosun. The volatile components of the original cloud were blown toward the edges, while the firm dust particles in the interior of the cloud agglomerated into small clumps of matter. These clumps collided with each other and formed larger bodies, the planetesimals. These in turn grew into the inner planets of our solar system in a process called accretion. One of these planets was the proto-Earth. The process beginning with the fusion of the first cosmic particles and continuing on until the formation of the Earth—a body around 8,000 miles (13,000 km) in diameter—took place very quickly in astronomical terms, probably in no longer than 30 million years.

Crust formation

The cosmic dust that formed the Earth primarily contained aluminum, magnesium, iron and nickel compounds, and silica. Under the pressure of the growing mass, the interior of the Earth heated up and became a liquid, igneous mass. The decay of radioactive elements added to the heating process. This liquification caused a separation of substances: the heavier elements, especially nickel and iron, sank and formed a metallic core. Lighter substances, such as silicates, were displaced toward the surface and formed the crust.

The formation of a shell, which is known as differentiation, was probably complete about 100 million years after the birth of the Earth. The moon (p.39) was also formed during this phase, after a collision between the Earth and another protoplanet.

RADIOMETRIC DATING

The age of the Earth was unknown until the French physicist Henri Becquerel (1852–1908) discovered radioactivity in 1896. Every radioactive isotope decays at a characteristic rate (called the half-life). The remaining amount of isotopes present in rocks can therefore be used to calculate the age of the rocks and the age of the Earth.

A mass spectrometer separates the components of a substance, allowing the concentration of isotopes to be measured.

in focus

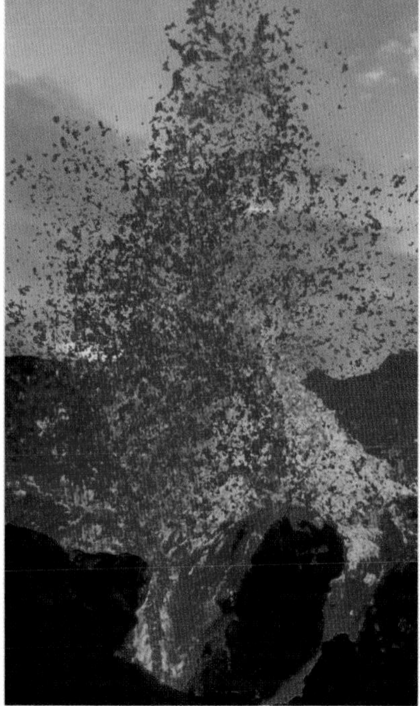

The processes of volcanism were important for the creation of the atmosphere on Earth.

Around four billion years ago, the Earth's atmosphere contained approximately 80 percent water vapor, 10 percent carbon dioxide, and small amounts of sulfur dioxide, helium, and methane.

Zircon is the oldest known mineral and is found throughout the Earth's crust.

see also: Theory of the big bang, p. 22

the origins of life

In the beginning, Earth was a boiling, fiery ball. Conditions improved gradually as it cooled down and eventually the primordial soup of inorganic compounds generated complex, organic molecules—the source of all life.

A primordial atmosphere of hydrogen and helium existed even in the earliest phase of the Earth's existence. However, due to the high temperature and low gravitational forces this atmosphere simply escaped back into space. Only once the meteorite showers reduced in intensity did

Prokaryotes lack a membrane enclosed nucleus.

Clouds discharge electricity through lightning.

temperatures drop and conditions begin to stabilize. In many places, molten material made its way from the core to the surface, through the still brittle crust. During this process large amounts of water vapor containing hydrogen (H), carbon dioxide (CO_2), ammonia (NH_3), and methane (CH_4) were released, forming a layer of gas around the Earth.

see also: Atmosphere, pp. 112–113

This secondary atmosphere was still vastly different from today's, particularly because it lacked oxygen (O_2). The production of oxygen only occurred once water molecules (H_2O) began to be broken down by the ultraviolet (UV) radiation from the sun, which easily penetrated the thin, unprotected atmosphere.

The formation of today's atmosphere

The formation of the atmosphere and oceans were closely connected. Earth's early atmosphere was about 80 percent water vapor. At saturation point, this condensed and fell as heavy rain during massive thunderstorms. As Earth's surface was very hot at this time, large amounts of acid rain simply vaporized immediately, only to condense and fall again in an ongoing cycle. This process gradually cooled the planet until water began to collect in depressions on the surface. About 3.5 billion years ago, the evolution of life began. A well accepted theory for the origin of life is that energy from lightning and the sun caused inorganic chemicals to form into simple biomolecules. From these, more complex molecules and macromolecules,

The first biomolecules may have formed in primordial oceans.

such as cyanobacteria (blue-green algae) evolved. These prokaryotes used photosynthesis to convert energy from the sun, producing oxygen as a byproduct. Thus, the oxygen content of the atmosphere

basics

PRISCOAN 4.6 to 3.8 billion years ago: Earth forms from the protoplanet and a shell-like structure develops.

ARCHEAN 3.8 to 2.5 billion years ago: The Earth's crust hardens and the temperature sinks to below 212°F (100°C). The first organic molecules form.

PROTEROZOIC 2.5 billion to 542 million years ago: The oxygen content of the atmosphere increases due to photosynthesis and a protective ozone layer develops.

slowly increased. About 570 million years ago it made up 12 to 15 percent of the atmosphere, opening the way for the evolution of oxygen-respiring organisms.

The formation of the Earth's atmosphere was a consequence of the eruption of molten material and gas from the core to the surface.

shape | magnetic field | core | mantle | crust

STRUCTURE

As insignificant as the Earth may be within the totality of the universe, it does hold a special position in our solar system, giving us our daily, yearly, and seasonal patterns. Its internal structure plays an important role in both the processes that affect its surface and its magnetic field, the invisible protective shield of the blue planet.

shape of the earth

From outer space, the Earth appears to be a uniform sphere. However, this appearance is misleading; the shape of the Earth is extraordinarily irregular. The reason for this is the unequal mass and density distribution of its interior.

Just like the other celestial planets, the Earth is never motionless. It orbits around the sun at a distance of about 93 million miles (149 million km) on a slightly elliptical trajectory. A complete revolution takes about 365 days (one year). Simultaneously, the Earth is spinning counterclockwise on its own axis (the line running through its interior between the Poles). One complete rotation takes 23 hours and 56 seconds (a sidereal day), which gives us the alternating light and dark of day and night. The annual seasons occur principally because the Earth's axis is tilted by 23.5° from the vertical, relative to its orbital plane. In relation to space, its position remains virtually unchanged during Earth's orbit around the sun, thus the angle of incoming solar radiation on any particular part of the Earth's surface changes throughout the year.

The speed that the Earth rotates depends on the respective latitude. While certain points at the Poles will always remain in the same location during Earth's rotation, a point at the Equator is moving at 1,525 feet per second (465 m/s). These high-speed revolutions cause centrifugal forces to develop, with the effect that the Earth is not strictly a sphere. Instead, it represents a rotational ellipsoid, slightly compressed at both Poles by about 13 miles (21 km) relative to the Equator.

The Earth's varying angle in relation to the sun results in the seasons.

Mathematically, the Earth cannot be described as a simple geometric figure. Once the height and weight parameters are considered, a physical model of its shape is represented as a geoid. The shape and rotation of the Earth influence its gravity, the force responsible for attracting all bodies toward its center, giving them "weight." This depends on the mass of a particular object. The rule is: force of gravity (weight) equals the mass multiplied by the acceleration of Earth. Due to the compressed shape of Earth and its centrifugal force, gravity is lowest around the Equator at 32.086 ft/s^2 (9.780 m/s^2) while at the poles it measures 32.257 ft/s^2 (9.832 m/s^2). Inside the Earth, between the crust and the core, it reaches 34.4 ft/s^2 (10.5 m/s^2).

The Earth as a geoid has a smooth but irregular surface (see picture below).

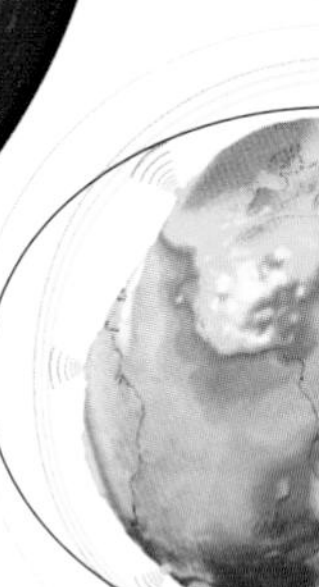

Many factors affect the Earth's atmospheric patterns and seasonal changes, in particular its position in relation to the sun.

THE EARTH IN NUMBERS

The equatorial circumference is 24,901 miles (40,075 km); the equatorial radius is 3,963 miles (6,378 km)

Its mass is 6.585 x 10^{21} tons, the volume is 260 billion cubic miles (1,083.3 billion km^3), and the mean density is 344 pounds per cubic foot (5,515 kg/m^3).

Its surface measures about 317 million cubic miles (510 million km^3); if this 224 million cubic miles (361 million km^3) are covered by water.

basics

see also: changing Earth, pp. 88–91

the magnetic field of the earth

What do a compass needle pointing north, the glowing veils of polar lights and the sense of direction of carrier pigeons have in common? All three are influenced by the invisible force of the Earth's magnetic field.

VAN ALLEN BELT Zones with a high density of electrically charged particles that surround the Earth at altitudes of 620–3,100 or 9,320–15,530 miles (1,000–5,000 or 15,000–25,000 km).

MAGNETOPAUSE The boundary of the magnetosphere, which extends about 37,300 miles (60,000 km) on the side directed toward the sun; on the other side it ends about 3.1 million miles (5 million km) from the Earth.

basics

The Earth's magnetic field is like a gigantic bar magnet tilted at 11° to the rotational axis of the Earth. Therefore the position of the magnetic poles deviates from the geographical North and South Poles. The magnetic poles are constantly moving. The magnetic north pole was on the Canadian Boothia Peninsula when it was discovered in 1831, but since then it has moved about 685 miles (1,100 km), heading with increasing speed in the direction of Siberia.

A magnetically charged compass needle aligns itself with the Earth's magnetic field.

The origin of the magnetic field lies in the interior of the Earth and is produced by enormous convection currents within the liquid outer core of the Earth. At depths between 1,800 to 3,200 miles (2,900 to 5,150 km), liquid iron churns around the solid iron inner core of the Earth. This movement creates an electric field, and thus the accompanying magnetic field—a phenomenon also referred to as a geodynamo.

Invisible protective shield

The magnetic field of the Earth extends into space, where it is called the magnetosphere. It protects the planet against cosmic radiation—especially the lethal rays of the solar wind, which are directed around the field. The magnetic field lines are compressed on the side facing the sun and extend, tail-like, on the other side.

Auroras are light phenomenons that occur near Earth's poles as a result of the magnetic field.

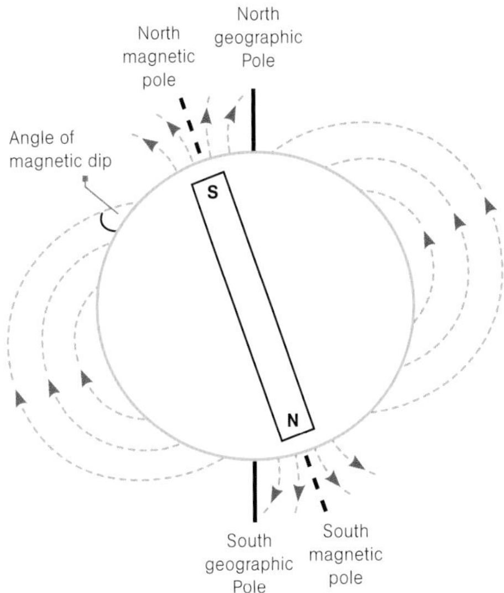

The magnetic poles are somewhat tilted against the geographic Poles of the Earth.

North becomes south

Magnetized rocks from the early history of the Earth date the magnetic field to at least 3.5 billion years. These rocks also indicate that the magnetic field has repeatedly collapsed and reversed its polarity. On average, such a repolarization occurs every 500,000 years. During the actual reversal of polarity, the protection of the magnetosphere is temporarily lost. The last time the magnetic north and south poles changed their position was about 780,000 years ago. Therefore, another reversal is already overdue.

Solidified magma, which is created on the seafloor at mid-ocean ridges in the process of plate tectonics, indicates that the magnetic field of the Earth can reverse.

see also: Forces, fields, interactions, p. 313

the core and mantle of the earth

The interior of the Earth long presented an enormous puzzle to science. Seismic measurements helped to start unraveling the mystery and assisted the exploration of the structure and contours of the Earth's interior.

The interior of the Earth is separated into three chemically distinct parts: the crust, mantle, and core, in a way similar to the composition of an egg. At the present time only the outer layer of the Earth—a relatively wafer-thin section—has been directly studied. The deepest boreholes have yet to reach one percent of the 3,959-mile (6,371 km) distance to the center of the Earth. Instead, our best source of information about the Earth's interior are the waves generated during earthquakes. The speed with which these waves travel depends on the temperature, pressure, and composition of rocks they penetrate. Wave measurements indicate that the physical characteristics of the Earth's layers change dramatically in two regions. These discontinuities mark the delineation between the crust, mantle, and core of the Earth.

The outer core of the Earth is primarily composed of molten iron.

Crust (solid): 0–62 miles (0–100 km) thick

Mantle (viscous): 1,800 miles (2,900 km) thick

Outer core (molten): 1,367 miles (2,200 km) thick

Inner core (solid): 778 miles (1,250 km) thick

Temperature, pressure, and composition interact to form the distinctive layers of the Earth and shape the forces—including plate tectonics and the Earth's magnetic field—that arise from them.

The Earth's mantle

The Mohorovičić discontinuity characterizes the geological interface between crust and mantle. On average, the mantle is 1,770 miles (2,850 km) thick and makes up about 68 percent of the Earth's mass. Its upper layer consists of rocky material, which—together with the solid crust—forms the lithosphere. Below that, with a thickness of roughly 250 miles (400 km), is the asthenosphere. When temperatures reach about 2550°F (1400°C) and pressure reaches 200–350 kbar, the rocks turn into viscous magma that has a density of about 3.3 ounces per cubic inch (3.4 g/cm^3). The increased pressure is responsible for the transition zone located at a depth of 250 to 560 miles (400 to 900 km), the mesosphere. Below the mantle, pressure rises to 1,450 kbar.

Granite forms a major part of the crust.

Despite a temperature of 4892°F (2700°C), the mantle remains solid because of the pressure and has a density of 3.3 ounces per cubic inch (5.7 g/cm^3).

The Earth's core

At a depth of about 1,800 miles (2,900 km), the Wichert-Gutenberg discontinuity forms the boundary where the solid mantle rock and the molten iron core meet. The density rises to 5.5 ounces per cubic inch (9.5 g/cm^3) and the temperature increases abruptly by about 1830°F (1000°C). In fact, heat is continuously exchanged between the core and mantle. The convection current this creates—hot material rising and cold material sinking—is thought to be the driving force of plate tectonics (p. 64). Convection currents in the outer core also create the Earth's magnetic field (p. 57). High pressure causes the liquid outer core to become solid at a depth of 3,200 miles (5,150 km). The rocks are compressed under more than 3,600 kbar, and the density is up to 7.8 ounces per cubic inch (13.5 g/cm^3). Temperatures at the Earth's center can reach 11,730°F (6500°C).

basics

P WAVES: The fast primary waves oscillate in the direction of the dispersal corridor and can disperse through solid, liquid, and gaseous material.

S WAVES: The slower secondary waves oscillate perpendicular to the P waves. They can only travel within a solid elastic mass.

see also: Plate tectonics, pp. 64–65

structure of the earth's crust

The Earth's crust is the thin and rigid outer shell of the Earth. The Earth has a continental and an oceanic crust, differing in thickness, density, and type of rock content, as well as the age and origin of the rock material.

The thin outer shell of the Earth is on average only about 22 miles (35 km) thick. It may reach 19 to 44 miles (30 to 70 km) of thickness where the continents are situated. The crust is thickest where the roots of mountain ranges reach deep into the Earth's mantle. The crust is thinnest underneath the oceans: only 3 to 5.6 miles (5 to 9 km) thick. The Mohorovičić discontinuity, which is also referred to as "Moho," separates the Earth's crust from the mantle. The temperatures in the crust and the outer mantle are not hot enough to maintain the rock material in a liquid state. These brittle rock layers combined are called the lithosphere, moving according to plate tectonics and the underlying movement of the lower layer, called the asthenosphere.

The highlands of Tibet: the result of colliding continental plates.

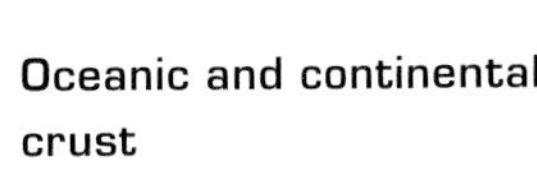

Oceanic and continental crust

One third of the Earth's crust is continental crust and two-thirds are oceanic crust. In between these two types of crust, there is a slightly broken-up seismic layer of discontinuity referred to as the Conrad discontinuity. At no point is the oceanic crust older than 200 million years as it continuously regenerates at the edges of the large continental plates in the areas of the mid-oceanic ridges. In contrast, the continental crust contains the oldest rocks ever found. They are about four billion years old. Nevertheless the continental crust changes shape all the time due to tectonics, volcanic activity, erosion, and sedimentation. The comparatively young oceanic crust mainly consists of heavy rock material such as basalt or gabbro with a high density of 1.6 to 1.7 ounces per cubic inches (2.9–3.1 g/cm³). Due to its prevalent silicon magnesium compounds, it is also called sima rock. Continental crust material is slightly lighter by 1.5 oz/in³ (2.7 g/cm³) and consists of granite with a high silicon and aluminum content. Therefore, it is often referred to as sial rock.

Basalt columns make up the Giant's Causeway.

in focus

ANDRIJA MOHOROVIČIĆ DISCONTINUITY

Andrija Mohorovičić (1857–1936), former director of the Institute for Meteorology and Geodynamics in Zagreb, started studying seismology in 1900. He produced seismic diagrams following an earthquake near the capital of Croatia in 1909 and found that seismic waves were dispersing at different speeds. From these findings, he concluded that the density must be changing within the Earth's crust and defined the limiting value between the Earth's crust and the mantle.

Andrija Mohorovičić was a pioneer in the field of modern seismology.

basics

CHEMICAL COMPOSITION
of the Earth

Crust element	Weight proportion
Oxygen	46.6%
Silicon	27.7%
Aluminum	8.1%
Iron	5.0%
Calcium	3.6%
Sodium	2.8%
Potassium	2.6%
Magnesium	2.1%

Diamonds form at great depths in the lithosphere.

see also: Earthquakes, pp. 68–71

minerals: formation

Minerals, including metals and salts, are the solid building materials of the Earth. Minerals develop through an array of geological processes, such as sedimentation, volcanic eruptions, or chemical weathering.

Minerals are solid matter with a crystalline structure. They consist of both pure elements and chemical compounds, and are classified according to the way they were formed. The different types of formation classifications include igneous, sedimentary, metamorphic, and weathered minerals. Some minerals, such as garnet, can have multiple classifications because they can be formed in multiple ways.

Igneous minerals

Igneous minerals are formed when magma from the Earth's mantle cools at a temperature of 2732°F (1500°C), causing crystallization of the molten rock to the extent that no new minerals can emerge. Examples of igneous materials include feldspar, quartz, and mica.

Sedimentary minerals

Many minerals are formed by sedimentation. Sedimentation can occur by evaporation, compression, and chemical reaction. Sedimentation by evaporation occurs when seawater evaporates and leaves minerals such as calcite, dolomite, anhydrite, gypsum, halite (rock salt), and potassium chloride. Clay is formed by the compression of loose particles by increased temperature and pressure, along with chemical reactions. Sedimentary rocks cover nearly 75 percent of the Earth's surface.

Metamorphic minerals

Metamorphic minerals are the result of changes in the original crystal lattice of other minerals. When high temperature or intense pressure is present, the original crystal lattice of minerals becomes unstable and the components will take the form of other structures to withstand the adverse conditions. The group of metamorphic minerals includes graphite, talc, and garnet.

Weathered minerals

Weathered minerals are formed by changes in the chemical composition of minerals due to outside conditions. Copper ores, when exposed to the air, oxidize and form malachite. Kaolinite is formed when clay is subjected to chemical decomposition above ground.

minerals | magmatic | metamorphic | sedimentary

ROCKS

Humans have mined rocks and minerals since the Stone Age, using them as raw materials. Gems are frequently used for jewelry because of their form, color, and luster. The study of the origin, characteristics, and composition of rocks is known as petrology, while mineralogy studies the origin, characteristics, and composition of minerals.

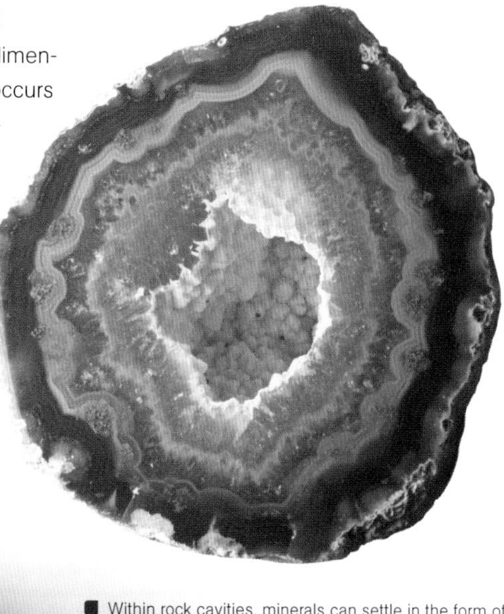

Within rock cavities, minerals can settle in the form of geodes—round, hollow rocks with inward-facing crystals.

Hot springs, such as the Mammoth Hot Springs of Yellowstone National Park, are the site of mineral formation. When hot mineral-enriched water evaporates, mineral crystals, above all carbonates, remain.

minerals: structure

Minerals—the basic ingredients of rock—provide many of our most important raw materials. They usually occur in the form of crystals, which have been treasured throughout history for their colorful variety and refractive properties.

Crystals are solid substances whose ions, atoms, or molecules are arranged in a repetitive three-dimensional lattice structure. The configuration of the lattice depends on the mineral's chemical composition. Only a few minerals, such as the precious stone opal, are not found in crystalline form; these are called amorphous minerals. Polymorphic minerals, on the other hand, share the same chemical composition but exhibit different crystal structures. For instance, diamond and graphite both result from carbon. Diamond—the hardest mineral in nature—forms in an environment of high pressure and temperature, while soft graphite arises under low pressure and temperature conditions. Thanks to this realization, diamonds can be artifically produced today. Synthetic crystals are also manufactured in the semiconductor industry, whose most important raw material is silicon. Aside from their classification into crystal systems, minerals can be distinguished based on their physical properties, such as hardness or density. Hardness is the measurement of a substance's resistance to scratching. Density, or specific weight, depends on its constituent elements and how closely the atoms within it are packed. When a mineral cracks, the breaks generally follow its atomic lattice structure. Thus mica, rock salt, and calcite tend to crack into pieces with flat, parallel surfaces. With quartz, on the other hand, the bonds between the atoms are so strong that the broken pieces are irregular. The most obvious characteristics of minerals are their color and shininess. Some are also distinguished by their magnetic, fluorescent, or radioactive properties, as well as the conditions under which they formed. Crystals may exhibit various types of symmetry. On this basis, they are divided into 32 classes, which in turn are assigned to seven different crystal systems: the cubic, hexagonal, rhombohedral (or trigonal), tetragonal, orthorhombic (or rhombic), monoclinic, and triclinic systems. The cubic system, such as table salt, rock salt, galena, and pyrite) is the easiest to recognize.

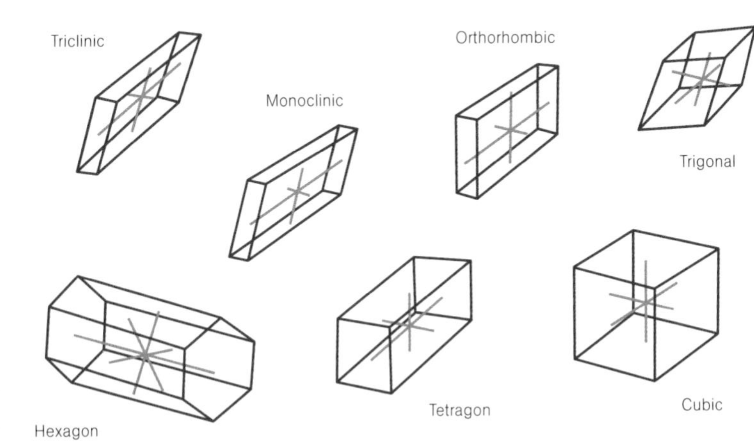

Crystals can be arranged in seven basic crystal systems. Each crystal system consists of a certain spatial arrangement of atoms upon the points of a three-dimensional lattice.

basics

HARDNESS OF A MINERAL is defined as the resistance it displays during mechanical stress. In particular, the scratch resistance is closely compared. On the ten-point Mohs' Hardness Scale (named after the German mineralogist Friedrich Mohs), a mineral can scratch any other mineral with a lower or same hardness number. Talcum, for example, has the lowest degree of hardness while a diamond has the highest; therefore, a diamond can only be scratched by another diamond.

Quartz is the most common mineral found in the Earth's crust. It has a prismatic shape with steep pyramidal terminations and exists in a variety of colors.

practice

GEMSTONES

Precious stones are minerals used as jewelry. Whether an amethyst, rose quartz, aquamarine, or garnet is used as jewelry depends on its purity, color, and translucency. Weight too plays a significant role for the value of such pieces. Diamonds are among the most precious gemstones although emeralds, rubies, and sapphires are also popular. In the past these stones were usually worn as coarse or roundly polished forms; however, nowadays many gemstones are polished in crystal-shaped forms, in order to enhance the reflection of light and the luster.

Imperial crown of the German emperor Otto I, the Great, late tenth century.

see also: Mountains, pp. 76–79

magmatic and metamorphic rocks

The rock material that makes up the Earth is a mixture of natural minerals. Some rocks are created from molten mass as it cools, others are formed by the high temperatures and extreme pressure conditions of Earth's interior.

Rocks are composed of a variety of minerals, the presence and ratios of which depend largely on their origin and the environmental conditions during their formation. In contrast to minerals, the composition of rocks can be varied. They will only ever have identical compositions to one another when they are made up entirely of one single type of mineral.

During the intense heat and pressure that sedimentary rock undergoes during metamorphism within the Earth, marble rock forms. This hard rock is quarried on the Greek island of Thassos.

Igneous rocks

Magmatites or igneous rocks are formed by the cooling of liquid magma. Minerals crystallize and amalgamate into larger rigid structures. If the process of solidification occurs deep inside the Earth, then plutonic rock is created. Plutonic rock, such as granite or diorite, has particularly large crystals. They are able to grow large due to the slow cooling process. However, only small crystals are created when magma reaches the surface of the Earth during the eruption of a volcano. Now the magma is referred to as lava. Typical vulcanites or extrusive rocks include basalt and rhyolite. Some vulcanites also have single larger crystals. This is how embedded crystals, such as diamonds, are carried up to the surface. It is also possible that no crystallization of minerals occurs at all if the lava cools down rapidly. In this case volcanic glass such as obsidian is formed instead. Loose volcanic rock that originates from bursting magma, during eruptions that involve large amounts of gas, is classified as sediment.

Metamorphic rocks

Metamorphose means transformation: metamorphic rocks form by the transition of other types of rock within the Earth's crust. This can only happen when the temperature and pressure are sufficiently high. Under these conditions the original rock material begins to melt and change structure. For example, marble forms during the transformation of sedimentary rocks or pararock. The transformation of magmatites results in so-called orthorocks such as mica schists. Metamorphic rocks show the foliated structure (schistosity) characteristic of this type of rock. These rocks can be split into thin plates along the schistosity plane.

basics

PETROLOGY is the study of the history of rocks, particularly their origin and formation processes. Rocks are classified by occurrence, mineral and chemical composition, texture, and physical characteristics.

When igneous rock is extruded, it can form into columns, such as basalt columns.

in focus

ROCKS FROM OUTER SPACE

Meteorites are fragments of rocky material that hit the Earth's surface. If they disintegrate in the atmosphere they are meteors (shooting stars). These intruders from the early days of our solar system can be extremely large and result in giant impact craters when they hit planets. However, most are no bigger than pebbles and arrive on Earth without notice. They mainly consist of silicate minerals or an alloy of iron and nickel. There are different categories of meteorites, such as stony, iron, or stony-iron.

Meteorites are mainly pieces of asteroids that have been caught by Earth's gravity.

see also: Asteroids and comets, pp. 42–43

sedimentary rocks

Sedimentary rocks are the most common type of rock on Earth. All sedimentary rocks are formed in a similar manner: by sedimentation, compaction, or cementation of many smaller particles of mineral, animal, or plant origin.

Sedimentary rocks are categorized into three groups: clastic, chemical, and biogenic. Clastic sediments, such as clay and sand, originate from fragments of rocks that have eroded or disintegrated. Among other techniques, classification of clastic rocks is based on the size of their components, found using the grain size. Rock salt and gypsum are examples of chemical sediments, which form by evaporation of aqueous solutions. Biogenic sedimentary rocks, such as coal, develop from animal or plant remnants. Petroleum originates from dead microorganisms that have been deposited, undecomposed and under the exclusion of oxygen, on the seafloor. There, they are modified through diagenesis. Limestone occurs through precipitation from solutions or from remnants of marine shellfish.

basics

COAL formation began with primeval forest remains changing into layers of peat. Oceanic incursions then covered the peat with new deposits. The ensuing rise in pressure and heat resulted in coal.

Rocks from sediments

Millions of years pass before sediments compact to form rocks. Following their deposition they compress under the increasing pressure of laminated layers above. Grains are densely packed together, becoming interlocked to form a massive unit. However, only during the cementation phase are particles "baked" into rocks. Groundwater flowing through the sediments contains calcite leached out from calcareous rocks.

Sandstone cliffs weathered by the Virgin River in Zion National Park, Utah, U.S.

The precipitation of dissolved calcite into gaps in the structure leads to grain cementation, forming sedimentary rock.

Sedimentation cycle

Weathering and erosion wear down rocks exposed to the atmosphere. Rock fragments are transported by wind, rivers, and ocean currents and deposited in layers. In time the sediment is covered over and compacts (diagenesis). Later on, through large-scale uplifts and the formation of mountains, sedimentary rocks are pushed up to the Earth's surface, and erosion starts the sedimentation cycle once again.

in focus

SOLNHOFEN LIMESTONE

Most fossils can be found in sedimentary rocks such as clay, lime, or sandstone, where organisms became entrapped during sedimentation. Some sedimentary rocks are made up almost exclusively of fossils. The Solnhofen Limestone site in Bavaria, Germany is one of the most significant fossil deposits in the world. All known specimens of *Archaeopteryx* have been found there.

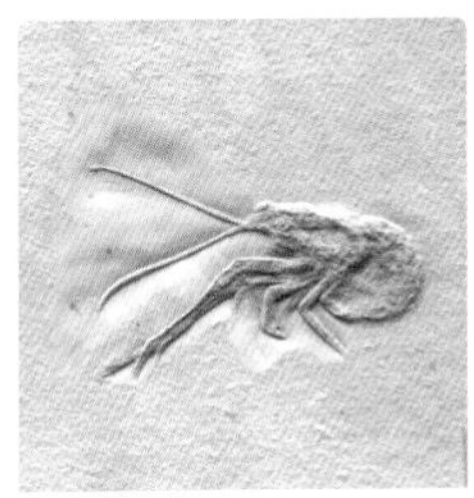

A fossil of the decapod *Mecochirus longimanatus* found at the Solnhofen site.

The cliffs of Calabria, Italy: These steep limestone cliffsides on the Tyrrhenian coast, which reach heights of around 150 feet (46 m), are home to the medieval cliff-top town Tropea.

earth in motion

How did the Atlantic Ocean spread, reaching all the way to Africa before splitting about 30 million years ago? Today the model of plate tectonics explains the movement of the Earth's crust.

Earth's uppermost mantle and crust make up the approximately 62-mile (100-km)-thick lithosphere. Earth's solid outer shell, the constitution of which varies between the oceanic and continental lithospheres, is a mosaic of plates—seven large and about ten smaller sections. Beneath the plates lies the asthenosphere. Soft and partly molten, the asthenosphere provides a base, over which these plates slowly glide. The cause of this gliding movement lies deep within the interior of the Earth.

in motion | plate boundaries | formations | early continents

PLATE TECTONICS

Earthquakes in Afghanistan, volcanoes in Java, ice-capped mountains in South America, and mid-ocean ridges and trenches are all natural phenomena caused by shifting continents or forming oceans. They are the results of hidden geological forces within the Earth.

The force behind plate tectonics

Radioactive decay heats the lower layers of Earth's mantle, causing them to rise to cooler areas. There the molten rock cools and sinks, only to rise again as part of a cycle called mantle convection. The molten mantle material, or magma, pushes up Earth's crust, ultimately breaking it open and solidifying along the resulting lines of fracture. Intense volcanic activity can cause magma to spill out onto Earth's surface, creating lava flows. Such active faults and volcanic flows have formed the Mid-Atlantic Ridge on the floor of the ocean. The solidification of magma forms new ocean crust as old cracks break open again and again, each time increasing the amount of crust.

In this way, two oceanic lithospheric plates have formed, each of which is gradually growing and drifting away from the other. In much the same way, when the crust within a continent breaks open, new oceans can eventually form. For example, in about ten million years the East African trench could divide Africa.

milestones

CONTINENTAL DRIFT

The Atlantic coasts of Africa and South America fit into each other like jigsaw pieces and are also home to many of the same sediments and reptiles. Similarly, closely related plants made their homes in India, Australia, and Antarctica 200 million years ago. These observations led German meteorologist and geoscientist Alfred Wegener to propose in 1911 that all the continents were once one big landmass that later split up. 50 years later his theory was confirmed by the model of plate tectonics.

Alfred Wegener was initially criticized, because he could not explain how the continents moved apart.

One consequence of continental drift and the movement and collision of continental plates was the creation of mountain ranges and the deepening of the ocean floor.

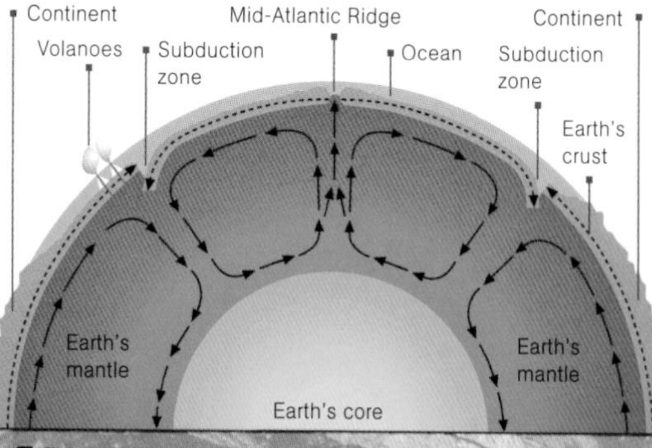

The crust is the outermost layer of the Earth: Directly underneath lies the mostly solid mantle and the partly liquid core.

plate boundaries

For new ocean crust to continue to form, older parts of the crust must first be destroyed. This happens when tectonic plates collide with each other.

When lithospheric plates meet each other, one plate can slide below the other in a process called subduction. This process, and the changes in the suface associated with it, happens at different rates.

In the Persian Gulf, the younger Arabian plate (lower left) is running up on the Eurasian plate (upper right). This area was once the site of a rift, where the two plates pulled apart from each other, and the Indian Ocean filled in the widening gap between the plates. The process then reversed, and about 20 million years ago the gulf began to close up. The collision of the two plates gives Iran its mountainous terrain.

The East Pacific Ridge, for example, grows five inches (12 cm) each year, but the Mid-Atlantic Ridge grows only two inches (five cm). Areas where the plates drift away from each other are called divergent plate boundaries, and areas where they collide are called convergent plate boundaries.

The slower the drift of an oceanic plate, the cooler and heavier it becomes. When plates collide, the heavier plate is subducted. The rocks and sediments deposited on the plate melt in the upper mantle, forming magma that rises to the surface and flows from volcanoes as lava. Volcanoes that lie above sea level are visible as chains of islands bending in the direction of the descending plate.

An oceanic trench, the point of contact of the plates, is a part of every island arc. The less extreme the angle of descent of the subducting plate, the farther away the melting zone is, forming oceanic trenches and island arcs long distances from each other, and flattening the channel between them.

The greater the angle of descent, the faster subduction takes place, creating an oceanic trench close to the island arc.

On a collision course

Ocean crusts are composed chiefly of rocks, such as basalt and gabbro, making them heavier than continental crusts, which are primarily composed of granite and gneiss. Because of this, when plate tectonic movements cause an oceanic plate to collide with a continental plate, the oceanic plate is usually subducted, causing earthquakes and volcanic eruptions in the subduction zone. When two continental plates collide, however, subduction does not take place. Instead the plates push against each other and buckle upward, forming mountains at the point where they meet.

Tarawera Mountain in New Zealand is close to the boundary of the Pacific and Australian plates.

basics

HOT SPOTS are points of volcanic activity located in the center of the Pacific Ocean. They are not situated on a tectonic boundary; rather, the Pacific plate skims over the top of the hot spots. Nonetheless, magma flows constantly rise from a depth of around 10,000 feet (3,000 m) as a result of a mantle plume—an area of unusually hot rock in the mantle. The regular lava flows continue to form islands, including the Hawaiian and Galápagos Islands.

Shifting with consequences

Sometimes neither subduction nor collision takes place. Instead plates slide against each other in a motion called shearing. In shearing zones, the plates can interlock, causing earthquakes as they tear free from each other.

The most well-known shear zone is the San Andreas Fault, but the lines of fracture in the mid-ocean ridges are also sites of frequent earthquake activity.

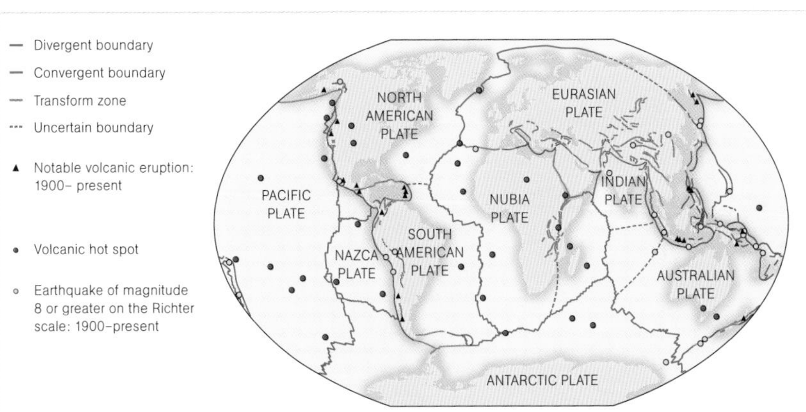

The Earth's crust consists of the following major tectonic plates: the Eurasian, African, Pacific, Antarctic, North American, South American, and Australian.

see also: Structure of the Earth's crust, p. 59

reconstruction

In order to reconstruct how and when the continents migrated into their current positions and the route they are now following, researchers have developed measuring procedures to observe the traces of rock "memory."

There are traces of ferrous silicates almost everywhere in the lithosphere. When molten rock is cooling in the crystallization phase, the magnetic iron minerals arrange themselves in the direction of the existing magnetic field, much like a compass needle. Measuring the angle of the minerals, known as the inclination, can determine where a continent formed and which movement it has since taken.

basics

GRAVIMETRY measures changes in gravitational force.

SEISMOGRAPHY analyzes the duration, form, and strength of earthquake vibrations.

GROUND-PENETRATING RADAR "illuminates" the topmost layers of the Earth with electromagnetic waves.

MAGNETOTELLURICS is a method of imaging the Earth's subsurface by using natural magnetic fields.

Geophysicists are able to retrace more than just large-scale plate movements across geological time. Thanks to more sophisticated procedures, they can also measure current changes in position, although these only amount to a few inches per year. At locations where the earthquake risk is particularly high, for example, near the San Andreas Fault in California, they have set up a network of measuring points. With laser geodesy instruments, changes in angle and distance between individual points can be precisely determined. However, the function of geodetic networks is tightly confined due to air pollution and the Earth's curvature.

GPS: global positioning system

Another procedure for calculating plate movement is the GPS measuring method. Most commonly used in navigation systems for automobiles, it consists of 24 satellites, each transmitting a distinct signal. As every point on Earth can always receive a signal from at least four satellites at once, its exact position can be determined nearly to a fraction of an inch based on the differences in signal transit time. Thus, the various gaps between individual reference points measured over a year can be used to determine the movements of the continental massifs. The same principle governs radio astronomical measuring procedures; however, in this case, the radiation from pulsars or quasars is analyzed. Specialized laser satellites are used to measure vertical changes at the edges of the plates.

Radio telescopes can "image" many astronomical objects that emit radiation at radio wavelengths.

The Global Positioning System (GPS) is a useful measuring tool for geophysical research, such as the study of earthquakes. Using GPS devices, scientists can record the Earth's movements in detail.

A panoramic space view of the Himalaya provides significant information about the geological structure of the Earth's highest and most extensive mountain range.

primeval continents

Pangaea, Laurasia, Gondwana: these names are taken from the Earth's history. Today the outlines of these three former continents can be reconstructed on the basis of models simulating continental drift.

The location of each continent today is the result of geologically recent tectonic changes. The further back an epoch is in the history of the Earth, the harder it is to imagine what the Earth would have looked like. The most recent research suggests that after the formation of the Earth's crust, continental plates must have moved much more actively than previously thought. Therefore several individual supercontinents must have formed from the landmasses. The earliest supercontinents were Kenorland about 2.5 billion years ago and Columbia about 1.5 billion years ago. The massive primeval continent Rodinia formed about 1.1 billion years ago. This supercontinent was surrounded by a gigantic ocean called Mirovia. It is thought that Rodinia disintegrated about 800 million years ago.

Evidence for continental drift: Fossils of the Lystrosaurus were found in Africa, India, and Antartica.

Canyons, such as Kings Canyon in California, were formed by erosion caused by conflicts between plates.

Gondwana and Laurentia

Toward the end of the Precambrian, South America, Africa, Australia, Antarctica, and India formed a massive coherent landmass in the Southern Hemisphere of the Earth. The supercontinent Gondwana was separated from the supercontinent of the Northern Hemisphere, Laurasia, by the Tethyan gateway, which varied in width. After the oceans retreated, the continents were flooded again. Distances between landmasses decreased once more during the Paleozoic. Toward the end of the Cambrian about 500 million years ago, fragments of the continents collided and formed giant mountain ranges. Remains of the Caledonian mountain range can still be found today in the Appalachian Mountains, the Scottish Highlands, and in Norway. Toward the end of the Paleozoic, Gondwana and Laurentia combined to form the supercontinent Pangaea. High mountains were created during the collisions, including the Variscan Belt, which has now been almost entirely eroded. The Tethys Ocean closed during this event.

Pangaea

Pangaea stretched from Pole to Pole and covered about a third of the Earth's surface. The remaining part was taken up by the primeval ocean Panthalassa that surrounded all landmasses. A few shallow inland seas still existed, but they dried out relatively quickly. About 200 million years ago, deep cracks and large rift valleys formed as the Pangaea "supercontinent" started to break up. The outlines of today's continents became recognizable. The proto-Atlantic formed as an inland sea between Africa and North America during the Jurassic. The continents began to travel toward their current locations on the Earth's surface. The denomination of the continents used today emanates from Greek antiquity; Herodotus divided the Earth into Europe, Asia, and Africa. Nowadays seven areas are regarded as continents: Asia, Africa, North America, South America, Antarctica, Europe, and Australia.

in focus

THE PRIMEVAL OCEAN

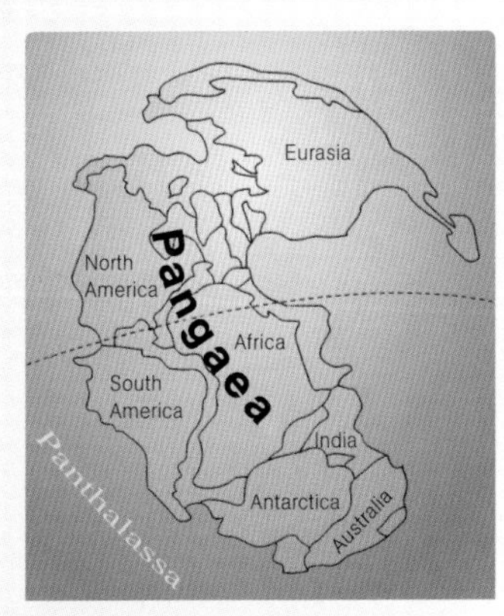

The Pacific Ocean evolved from Panthalassa after the breakup of Pangaea.

Panthalassa is a global ocean that used to surround the supercontinent Pangaea during the late Paleozoic and the early Mesozoic. Its name originates from Greek and means "all oceans." A wide gulf that was situated off the eastern coast of Pangaea called the Paleo-Tethys can still be found in today's Mediterranean. Panthalassa probably originated from Mirovia, a hypothetical primeval ocean once surrounding the supercontinent Rodinia.

see also: Earth's origins, pp. 54–55

causes | measurement | faults | prediction

EARTHQUAKES

The Earth is constantly shifting below our feet. The magnitude of this movement is usually too weak to be sensed by humans, yet various factors can result in highly destructive and intense earthquakes. In areas where this occurs frequently, safety precautions are of vital importance.

causes

The constant motion within the layers of the Earth affects the surface, sometimes in the form of sudden events such as earthquakes. In highly populated regions, this can result in a devastating natural disaster.

The Earth's lithosphere is very slowly but constantly moving; when lithospheric plates under the surface of the Earth reform, subduct, or slide, an earthquake can occur. These changes are the cause of 90 percent of all earthquakes. Different types of faults, or cracks, in the tectonic plates are under extreme tension and cause tremors when they slip. The two types of faults are dip-slip faults and strike-slip faults: the plates of dip-slip faults move up and down, while in strike-slip faults they move horizontally (left or right) across the fault plane. The point of origin of an earthquake is called the hypocenter. Typically, the hypocenter of an earthquake occurs at a depth of less than 37 miles (60 km) below the surface of the Earth. Occurrences at core depths between 190 and 430 miles (300 and 700 km) are rare. The epicenter lies on the Earth's surface, directly above the hypocenter. The highest intensity surface effects are seen at the epicenter.

In January 1995 the town of Kobe was destroyed by an earthquake that measured 6.8 on the moment-magnitude scale. Approximately 6,400 people were killed, over 400,000 hurt and more than 100,000 houses destroyed—it was the worst earthquake in Japan since 1923.

Seismic waves

Energy generated from the restructuring of the Earth's layers travels in the form of seismic waves. Seismic waves can be caused by shifts in faults and most often correlate with earthquakes. The two main types of seismic waves are "body waves" and "surface waves." Body waves form in the interior of the Earth and travel in a circular motion in all directions, including through the planet's core. Surface waves are slower and travel along just below the surface.

THE EPICENTER AND THE DISTRIBUTION OF WAVES

in focus

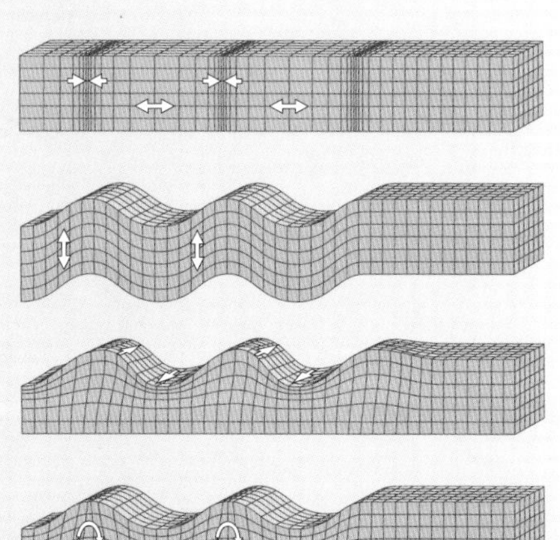

The fastest seismic waves are body waves, which include "P" or primary waves and "S" or secondary waves. P waves travel through rocks, liquids, and gases with a speed of 4–9 miles (6–14 km) per second. S waves can only be transmitted through solids and are about half as fast as P waves. Because the speed of the waves is known, the location of the exact epicenter of the earthquake can be determined using the time interval between S and P waves as they reach three separate seismological stations.

Types of surface waves include Love and Rayleigh waves, which are named after two British scientists. Love and Rayleigh waves are even slower than S waves, and can cause significant damage. Most of the shaking or tremors felt by humans during an earthquake are Rayleigh waves.

Additional causes

Not all earthquakes are caused by tectonic changes, some are caused by volcanic eruptions. The ascent of magma through the volcano and the shifting of molten rock in the magma chamber beneath the Earth's crust both cause tremors. Hundreds of smaller earthquakes will usually occur prior to an eruption.

Humans can also be the direct cause of earthquakes, for example by collapsing mines and tunnels or by conducting underground atomic bomb tests.

At 5:12 a.m. on Wednesday, April 18, 1906, an earthquake with a magnitude between 7.7 and 8.3 struck San Francisco and the coast of northern California. It is remembered as one of the worst natural disasters in the history of the United States.

measurement and consequences

It may not be possible to predict earthquakes, but by studying them scientists hope to find ways of lessening their damage.

Early in the 20th century, volcanologist Giuseppe Mercalli created a scale categorizing earthquakes based on damage incurred. This Mercalli intensity scale ranges from Level 1, which is registered only by seismographs, to Level 12, which results in severe changes on the Earth's surface and the destruction of almost all buildings. The more popular and more scientific Richter scale, developed by Charles Francis Richter, calculates the intensity of an earthquake as "magnitude" (M) on a logarithmic scale. The M-value is determined from the distance between the hypocenter of the earthquake and the seismological recording station, as well as the amplitudes recorded on seismographs. Earthquakes with magnitudes less than 2.0 are not perceivable by people. Numbers on the Richter scale represent an earthquake ten times more powerful than the number below it. Thus an earthquake of 5.0 M is ten times more powerful than one with an intensity of 4.0 M. Today, scientists use the more precise moment-magnitude scale. To calculate the M-value, this scale multiplies the area of the fault's rupture by the distance moved along the fault. The study and measuring of earthquakes is crucial in aiding scientists and engineers with planning for future occurances, especially because there could be deadly consequences. Currently, most scientists focus primarily on mitigating the hazards of earthquakes by improving structural stability of buildings.

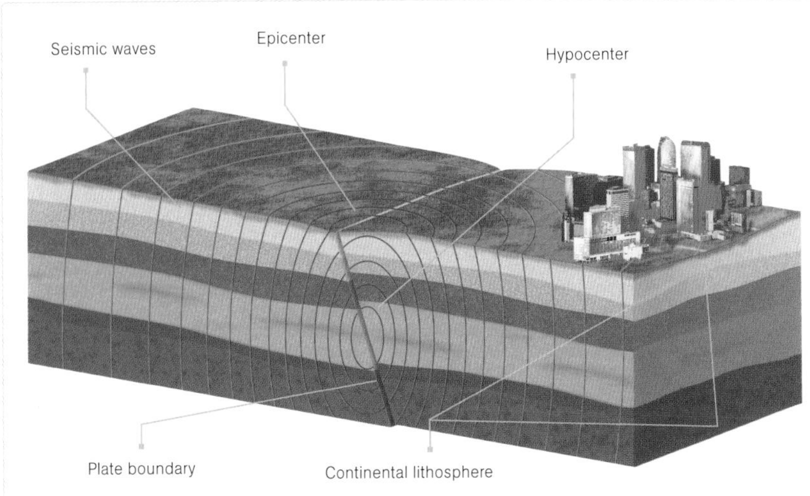

The Earth's lithosphere is a patchwork of plates in constant but slow motion. Driven by heat from within the planet's mantle and core, these tectonic movements are the source of most earthquakes. The seismic waves are the result of movement and fracturing along geological fault lines.

SEISMOGRAPHS

in focus

Seismographs, used to measure vibrations of an earthquake, come in a wide variety. Early seismographs used a weighted, hanging pendulum to etch or draw vibrations onto glass or paper. The stronger the earthquake, the greater the amplitude. Modern seismographs employ more sophisticated recording instruments with electronic sensors and amplifiers.

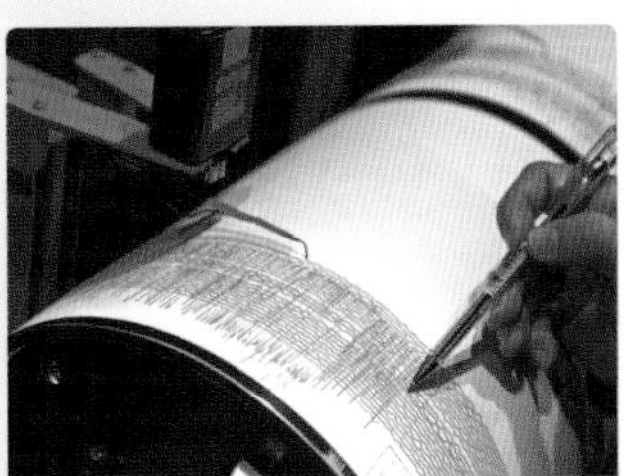

A group of seismographs can accurately locate an earthquake's epicenter.

Consequences of an earthquake

No earthquake has cost so many lives as that in the Chinese province of Shaanxi in 1556. An earthquake was recorded with a magnitude of 8.0. Approximately 830,000 people were killed—many of whom died in their beds as their houses collapsed.

Unexpected by-products of earthquakes, such as fires, floods, and lack of shelter or food, can be as destructive as the tremors themselves. After earthquakes in San Francisco in 1906 and Kobe, Japan, in 1995, survivors had to cope with extensive fires that raged for days. The capital of Portugal, Lisbon, was hit by a tsunami following a significant earthquake in 1755, submerging the city under water. After a devastating earthquake in Kashmir in 2005, difficulty in providing timely assistance to such a remote, inaccessible region as this meant that survivors faced winter without shelter or sufficient food.

In geology the visible fracturing and displacement of the Earth's surface, caused by an earthquake, is known as a fault. In the case of major earthquakes, a fault may even be as wide as several feet.

see also: Skyscrapers, p. 369

faults

Faults are ruptures or cracks in rock that often extend for miles along the upper layer of the Earth's crust where two crustal sections are displacing one another. They are usually created by earthquakes.

While we normally do not feel it, the Earth's crust is constantly moving. In most places, however, the crust only moves a few inches per year. The cracks caused by this movement (fault lines) extend for many miles and through the Earth's upper crust.

areas. However, strong shearing forces are at work at the edges of the converging plates, which can lead to the formation of faults. More specifically, these forces create strike-slip faults, which are also known as transform boundaries.

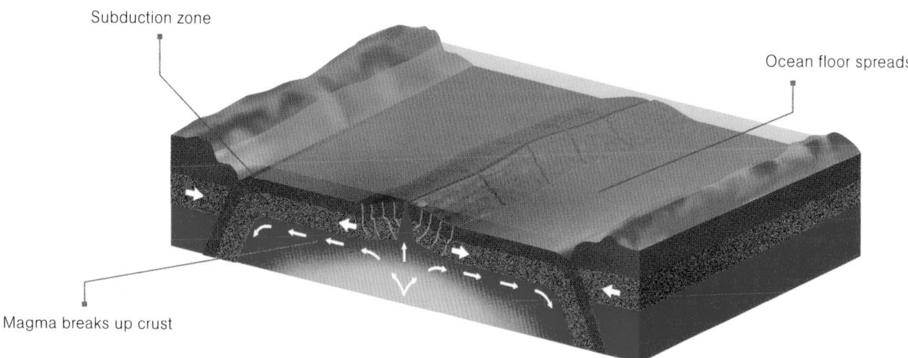

When the crust repeatedly splits open in a mid-oceanic ridge, pieces of the crust begin to break off. Over time this creates a central trench underwater.

This process is not always smooth or continuous; in fact, sometimes erratic shuddering movements occur, releasing an enormous amount of energy.

If two plates move past each other, the crustal material is neither destroyed, as with subduction, nor is it jammed together, as happens when one continent collides with another, forming high mountainous

The location of a transform boundary always indicates a high earthquake risk. A well-known example is the San Andreas Fault in California (see in focus). The North Anatolian Fault in northern Turkey is among the most active earthquake regions. Here the small Anatolian plate is moving past the huge Eurasian plate at up to seven inches at a time. About 25,000 people lost their lives in the last large earthquake that struck the city of Izmir in 1999. The Great Alpine Fault, which runs straight through the southern island of New Zealand, is one of the world's most impressive faults. Where the Australian and Pacific plates meet, the plates are not only moving past each other, but one is also moving over the other. The result of this movement is the striking elevation of the New Zealand Alps on the east side of the fault.

SAN ANDREAS FAULT

in focus

The San Andreas Fault is located in California. Here the Pacific plate is pushing past the North American plate at about one centimeter per year. Due to this movement two points that were next to each other 20 million years ago are now about 348 miles (560 km) apart. In 1906 an earthquake hit and ruptured 296 feet (477 m) of the San Andreas Fault. It happened when the two plates suddenly moved 39 feet (12 m) past each other.

The San Andreas Fault extends an approximate 8,000 miles (1,300 km) through California.

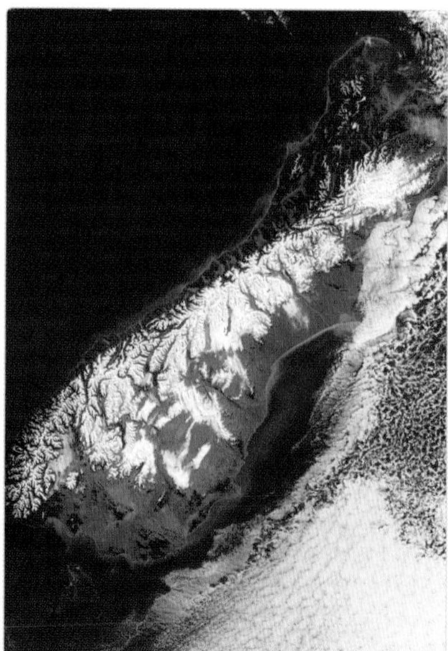

The action and movement of geological faults can create splendid rift valleys.

The mountains grow approximately 0.39 inch (0.99 cm) each year. Transform boundaries are not only present on the mainland; they can also be found on the ocean floor. At the points of numerous cracks and faults, the mid-ocean ridges are shifting at right angles to their respective longitudinal axes. The ridges are neither moving in a continuous straight line nor are they being forced apart at the same speed. At these faults, the crust is not subducted, but instead the plates are moving sideways past each other.

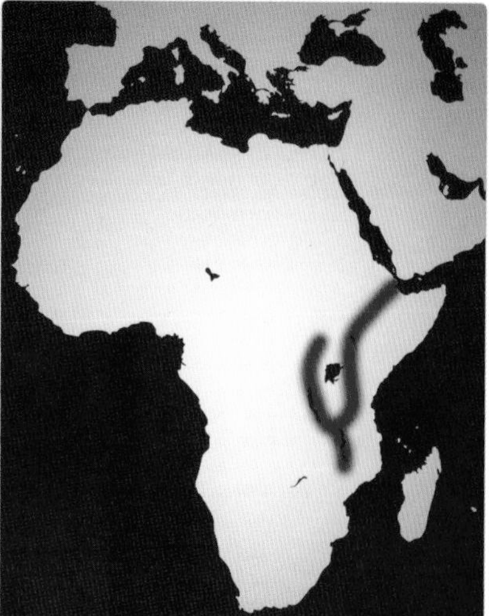

The Great Rift Valley begins in northern Syria and stretches to central Mozambique, totaling 3,700 miles (6,000 km) in length.

see also: Plate tectonics, pp. 64–65

prediction and safety

Earthquake prediction is still an inexact science; however, monitoring systems are constantly improving. With adequate preparation the worst consequences can be avoided. Without it, the scale of disaster can be enormous.

Many phenomena can precede an earthquake: Rock deformations cause changes in the active plate boundaries of Earth's crust. When water seeps from rock pores, groundwater level can change. Earth columns that open up also influence soil's electrical conductivity. Rock fissures release radioactive gases onto Earth's surface. Eventually, smaller tremors cause a large quake. All of these events can be measured by the equipment at seismological stations. Even unusual animal behavior can predict an imminent earthquake.

A high-tech buoy off the coast of Sumatra, Indonesia: monitoring systems have been installed since the disastrous 2004 tsunami.

However, such signals are not sure indicators. Earthquakes and tremors can occur without any indication at all. Prospective early warning systems are thus being investigated. Recently, activity levels of positively charged oxygen (O_2) ions have been used as predictors of seismic activity. O_2^+ ions originate in the interior of the Earth when O_2 molecules are destroyed and rise to the surface, combining with oxygen in rocks to release energy as heat. A satellite-supported system may be able to detect such emissions, providing early warning of seismic events in time to sound the alarm.

Tsunamis

Most tsunamis are triggered by seaquakes that cause vertical shifts of the ocean floor, setting the water above in motion and forming circular waves on the open sea. Unlike storm waves, in which only the uppermost layer of water is raised, all of the water in a tsunami wave is raised, making it extremely powerful. When a tsunami approaches a coastline, its waves increase in height because of the decrease in water depth and in response to the wave's decreasing speed. The rising waves that make up tsunamis cause widespread, catastrophic damage. In fact, even the wave troughs that accompany tsunamis can cause destruction, sucking out to sea anything with which they come into contact—often up to a mile from shore.

A simulated wall of a skyscraper is tested for its behavior in the event of an earthquake.

in focus

MODERN SKYSCRAPERS

Modern skyscrapers are designed to withstand earthquakes up to 8.5 on the Richter scale. Some buildings are supported by large steel and concrete columns that intertwine to form a movable "corset" that swings with the tremors instead of resisting them. Other buildings are constructed on bearings that can neutralize the effects of Earth's vibrations or use vibration compensators such as enormous hanging steel balls. The Torre Mayor in Mexico City, built in 2003, uses seismic dampers.

Construction worker on Taipei 101 mass damper.

There are 98 seismic dampers set into the reinforced concrete and dual steel structure.

anatomy of volcanoes

Volcanoes have existed ever since Earth's beginning 4.5 billion years ago. Although volcanic eruptions are destructive to the animal and plant world, lava and volcanic ash also add enriching minerals to the soil.

VOLCANOES

The active life of the inner Earth is brought to the surface and made apparent by volcanoes and thermal springs. Volcanic activity affects the global climate and can pose great dangers to the nearby population. However, volcanoes are also an important source of soil nutrients, and thermal springs have long been prized for their therapeutic qualities.

Volcanoes are fed by unseen magma chambers: pockets of molten rock that lie about half a mile beneath Earth's crust. If a chamber's pressure exceeds a certain threshold, its magma begins to rise through fissures and cracks, ultimately forming a volcanic vent. The magma may remain underground, or the vent may open at the surface, either above ground or through the ocean floor, spilling magma out of the crust as lava. This rising of molten rock to the surface is known as volcanism, and the solidification of magma under Earth is called subvolcanism. There are several different types of volcanos. Linear volcanoes (also called volcano cones or cinder cones) eject lava through linear shafts. In central volcanoes, lava flows from a central, tubelike vent. The lava flows of shield volcanoes are viscous, but those of stratovolcanoes, or composite cones, are less so.

Magma

The composition of molten rock, or magma, determines the type of each volcano's eruption. Magma with silica (SiO_2) levels greater than 66 percent is called acid magma while basic magma is about 52 percent silica. Magma releases gases as it rises, which decreases its pressure—acting in much the way carbon dioxide does when a soda bottle is opened. The higher the magma rises, the greater the amount of degasification that occurs. The enormous pressure exerted by the gas released from the magma forces the magma up the volcanic vent, blasting it out onto the Earth's surface in the form of a volcanic eruption. Because acid magma's viscosity keeps it from escaping as easily as basic magma does, eruptions of acidic volcanoes—when they do occur—are the most explosive.

Mount Bromo, also known as Gunung Bromo, is a subduction zone stratovolcano in Indonesia.

SUPER VOLCANO

in focus

Approximately 5 miles (8 km) beneath Yellowstone National Park in the United States lies a 37-mile (60-km)-long, 25-mile (40-km)-wide, and 6-mile (10-km)-deep magma chamber that contains about 5,800 cubic miles (24,000 km³) of magma. If this volcano erupted again, it could cause earthquakes and tsunamis, as well as catastrophic effects on the Earth's climate.

The last eruption of the Yellowstone Caldera about 640,000 years ago spread ash over most of the continent.

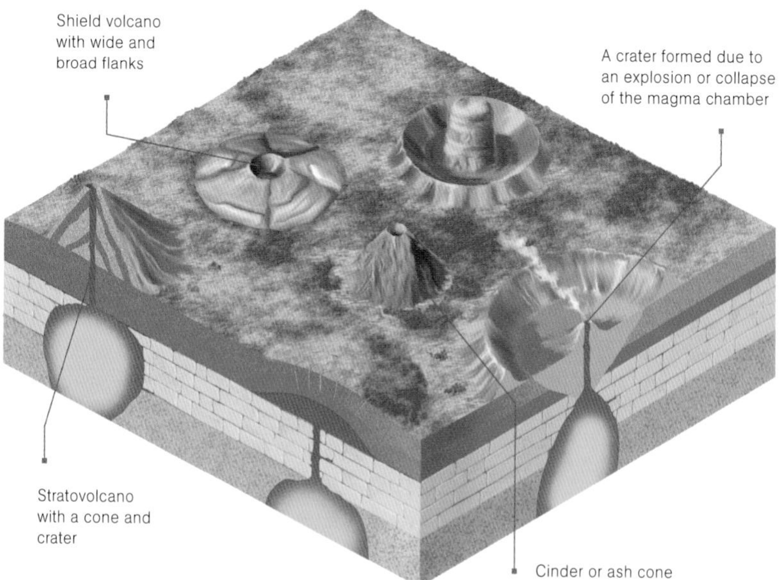

Eruptions with a Volcanic Explosivity Index of 7 or 8 are colossal events. They often form circular calderas (right) rather than cones (left) because the overlying mass collapses and fills the empty magma chamber beneath.

see also: Earth's origins, pp. 54–55

volcanic eruptions

When the huge pressure that builds beneath a volcano prior to an eruption is finally released, the results can be spectacular. Ash and pyroclastic material may tower 12.5 miles (20 km) into the sky, while molten lava pours down the mountainside.

Volcanoes eject the products of their eruptions as gases, liquids, and solids. Effusive eruptions produce basic lava flows, which are less viscous and often flood large areas before solidifying. These eruptions can last for centuries and have historically had devastating effects on Earth's climate, releasing large quantities of greenhouse gases, steam, and carbon dioxide into the atmosphere. Eruptions of viscous lava (which explodes during degasification) mingle lava with rocks. These explosive eruptions, called pyroclastic emissions, bombard the regions that surround them, rapidly burying the nearby countryside under several feet of ash. The finest pyroclastic materials can collect and form a pyroclastic cloud with a temperature of about 1800°F (1000°C) and speed up to 620 miles per hour (1,000 km/h). Explosive eruptions are usually accompanied by heavy rainfall as airborne ash causes steam to condense from the air around the eruption site. When rainwater combines with volcanic ash, large sheets of potentially destructive mud and debris, called lahars, begin to flow down the volcano's slopes to the land below.

Lava

Volcanic lava may manifest itself in many forms, according to the conditions accompanying volcanic eruptions and subsequent patterns of lava cooling. Thin smooth-flowing magma forms "pahoehoe lava," which has a smooth, billowy, undulating surface. Viscous "aa-lava," however, solidifies in sharp-edged blocks. Acidic lava that froths during degasification forms pumice, but abrupt cooling creates noncrystalline or glassy vitreous rocks, including obsidian. When lava comes into contact with water, pillow lava is formed, in rock masses that can range up to three feet (0.9 m) in diameter.

in focus

BEGINNING ANEW

Thousands of square miles of land can be destroyed by a volcano's eruption, and it can seem impossible that environments turned into such wastelands could ever recover. Yet only months after the 1980 eruption of Mount St. Helens in Washington state, the first plant growth reappeared on the nutrient-rich ash fields surrounding the mountain. Animals returned to the area soon afterward; however, the local ecosystem may need 200 years to return to its pre-eruptive state.

A fern pushes through volcanic ash and debris on Big Island, Hawaii.

Molten lava flowing from Mount Etna, Sicily: Mount Etna is one of the most active volcanoes in the world and is in an almost constant state of eruption.

Pinatubo is a stratovolcano on the island of Luzon, Philippines. Pinatubo rose about 5,725 feet (1,745 m) above sea level before the June 1991 eruption. Almost 500 feet (150 m) of the volcano was blasted away by this eruption.

life with volcanoes

Many people live with the daily threat of a volcanic eruption. It has only been in the last few years that scientists have been able to predict volcanic activity with any reliability and offer early warnings to those nearby.

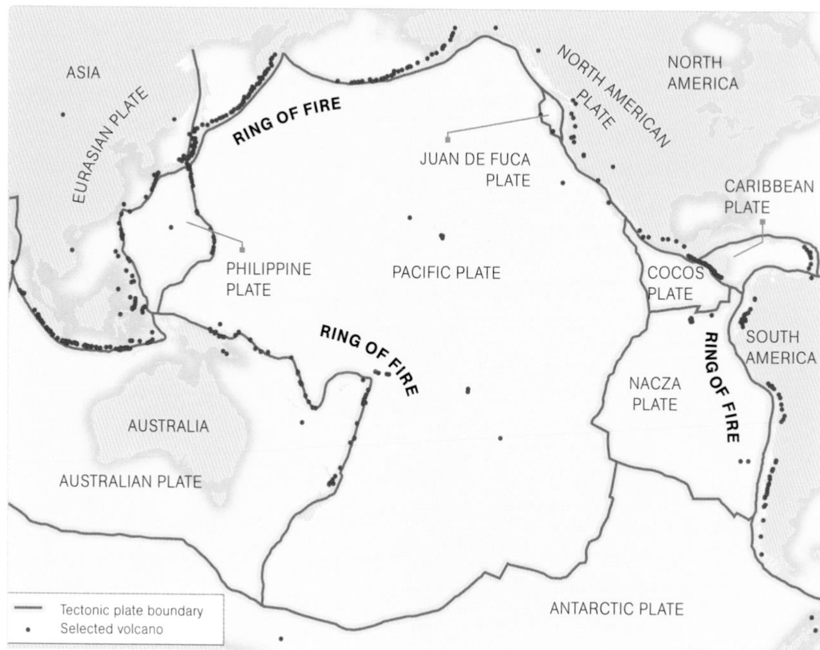

Over 80 percent of today's active volcanoes are located on the "ring of fire"—subduction zones of several tectonic plates around the Pacific Ocean.

Most volcanoes are located on the active borders of continental plates. Underwater volcanoes are found in regions where new crust is forming, along mid-ocean ridges. These volcanoes can eventually grow so large that they form islands. Land volcanoes are usually found in areas where subduction is occurring: one plate is being pushed under another at the line of collision. Many of these subduction zones are distributed around the Pacific Ocean. There are volcanoes situated in coastal regions around the Pacific, distributed much like a string of pearls on a necklace. These form the "ring of fire." Other volcanoes, such as those on Hawaii, are in "hot zones," where internal forces within tectonic plates produce magma independent of continental plate boundaries.

Many volcanic areas are very fertile, due to the mineral-rich lava and ash released by the volcano that improve nearby soil quality. Because of this, these areas are often densely populated—putting the population in great danger from future explosions. In order to evaluate the likelihood of possible eruptions, volcanologists constantly monitor active and dormant volcanoes for any volcanic activity.

Warning signals

Since all eruptions are preceded by earthquakes, scientists are able to monitor volcanic activity using seismic instruments. Another indication of an impending eruption is an increase in the concentration of sulfur dioxide in the gases that flow out of fissures and cracks. Laser-equipped measuring devices, able to measure volcanoes at very high degrees of precision, record surface deformations. Such devices measure the dilation (expansion) of volcanoes' magma chambers, a phenomenon directly related to increases of pressure within magma chambers, which can cause eruptions. If an explosion seems imminent, those living nearby will be advised to leave, thereby preventing a disaster.

in focus

VOLCANIC ERUPTIONS IN HISTORY

The eruption of Mount Vesuvius is one of the most famous in history. In 79 B.C., a pyroclastic cloud buried the Roman city of Pompeii under a thick ash cover. The eruption of Mount Tambora in Indonesia in 1815 killed about 10,000 people and caused crop failures and famine as far away as Europe. Approximately 71,000 people were affected by the eruption, in addition to those killed by the explosion itself. Conversely, the eruption of Mount Pinatubo in the Philippines in 1991 killed only 500 people due to the prediction's of volcanologists.

Pompeii was rediscovered in 1748. Its inhabitants had been preserved by the volcanic ash.

The Irazú Volcano in Costa Rica has erupted at least 23 times since 1723.

see also: Earthquakes, pp. 68–71

thermal springs and geysers

Geysers are spectacular and demonstrate the immense energy stored in the interior of the Earth. Similarly, thermal springs, fumaroles, mofettes, and solfataras hint at the dangerous potential of volcanoes.

■ The sinter terraces in Pamukkale, Turkey, were formed by deposits from calcium-rich thermal springs over hundreds of years.

Geysers are typical features of regions where volcanic activity is abating. They occur only on a few places on Earth: Iceland, the United States, Chile, New Zealand, Japan, Ethiopia, and on the Kamchatka Peninsula. Some regularly eject jets of water or steam up to 320 feet (100 m) high. The water originates from a larger reservoir deep below the surface of the Earth, where magma heats up the water gradually. The pressure created by the boiling water causes it to explosively eject through a stack from time to time. The depleted hot water is replaced by cool groundwater, which terminates the process before the cycle starts again.

■ A lava filled fumarole in a crater on the Fournaise range on the island of Réunion.

Beauty and health

Yellowstone National Park has the largest concentration of thermal springs in the world. The water that flows from the springs, which can be close to boiling point, is enriched with minerals. These precipitate as soon as the water cools down at the Earth's surface. This leads to the formation of impressive rock formations. A famous example is the sinter terraces found in Pamukkale (Western Anatolia, Turkey). These thermal springs have been famous for their therapeutic benefits for thousands of years.

Hot steam and toxic gases

Fumaroles are any cracks in the Earth's surface that emit steam and gases. They are at lower pressure than geysers, so they do not eject any water. Steam and gas from fumaroles can reach temperatures of up to 1470°F (800°C). When these hot gases meet with water upon exiting the ground, they create bubbling mud pots or mud gushers. Solfataras are fumaroles that emit sulfurous gases. They are usually somewhat cooler, and their emissions settle out next to the ejection points in the form of pure sulfur. The most dangerous of all degassing features are mofettes. Gases escape through small holes and cracks in rocks, delivering high concentrations of toxic carbon dioxide.

■ Old Faithful is a cone geyser located in Wyoming, in Yellowstone National Park in the United States. Eruptions can shoot 3,700 to 8,400 U.S. gallons (14–32 kL) of boiling water to a height of 106–184 feet (30–55 m) lasting from 1.5 to 5 minutes.

in focus

BLACK SMOKERS

Black smokers are hot springs at the bottom of the ocean. They were discovered in 1977 near the Galápagos Islands at a depth of about 8,500 feet (2,600 m). They eject water as hot as 660°F (350°C) from their chimneylike stacks. Precipitation from dissolved metallic sulfides produces plumes of black smoke. These thermal springs are fascinating for scientists as communities of living organisms survive in their dark and poisonous surroudings.

■ The water around a black smoker has a pH of approximately 2.8–similar to that of vinegar.

formation | ranges | fracture tectonics | summits

MOUNTAINS

Mountains are among the most impressive large landscapes on Earth. Young fold mountains, threatening volcanoes, and momentous fissures are all testaments to the tectonic processes that warped huge sections of the Earth's crust. The process of mountain formation is called orogeny; the physical mountains themselves are orogens.

mountain formation

On a human scale, the time period over which mountains are created seems infinitely long. Even what appears at first glance to be fixed and unalterable is subject to constant change.

The most significant mountain ranges are not simply scattered across the Earth at random. Instead, most lie along active plate boundaries and belong to one of the Earth's two large mountain systems. The Circum-Pacific system encircles the entire Pacific Ocean. It traces an arc from New Guinea, across Japan and the Aleutians, and into the American Cordilleras. The latter extend along a 9,321-mile (15,000-km) stretch from Alaska down to Tierra del Fuego in South America. The second, larger orogenic network, called the Alpine-Himalaya system, ranges east from northern Africa, over the Alps, and into the Himalaya before reaching Indonesia.

Phases of mountain formation

There have been three significant phases of mountain formation: the Caledonian, the Variscan, and the Alpine eras. The old fold mountains that were created 450 to 250 million years ago have been eroded and radically transformed by subsequent mountain formation. Of the once mighty Caledonians, only the Norwegian coastal plateau, the highlands of Scotland and Greenland, and the Appalachians remain today. Almost all young fold mountains were formed during the worldwide Alpine orogeny, which began approximately 220 million years ago and still continues today. The world's highest mountain ranges known today originated during this phase: the Alps in Europe, the Andes in South America, and the Himalaya in Asia.

Building up and breaking down

The most common high mountain formations are fold mountains, which emerge when two crustal plates come together. Under enormous pressure, sedimentary rock from the ocean floor is folded upward, pushing past younger layers of rock. This process occurs before the actual mountain range rises, and for most high mountain ranges the uplifting procedure is still occurring. However, the effects of weather and erosion gnaw away at a mountain, wearing it down even as uplifting begins. Without these effects, the Alps would be 32,808 feet (10,000 m) high today.

As erosion exposes the fold formations, the folds consisting of harder rock matter become mountain crests. In contrast, the softer rock is removed more quickly, which fosters valley and basin formation.

The Alpine-Himalaya system ranges from the Atlas Mountains to the Alps and Balkans, the Caucasus and the Himalaya, and into Indonesia.

Mountains can form through subduction, which occurs when two plates meet and one sinks beneath the other. Often an oceanic plate will sink beneath a continental or another oceanic plate. This creates orogenic or volcanic zones.

see also: Plate tectonics, pp. 64–65

mountain ranges

No mountain range is the same as any other; each has its own specific evolutionary history and unique characteristics. Nevertheless, in their genesis many follow a more or less standard formation process.

The largest mountain ranges were created through the collision of continental plates, which fold sedimentary layers upward. If the compressional movement creates enough pressure, "roofs" tear away from the lower layers. These roofs can then be pushed over one another across a distance of several miles.

The clash between the Indian and the Eurasian plates around 50 million years ago led to the lifting of the Himalaya. However, this process is not yet finished, and the "roof of the world" grows annually by about half a millimeter. The Alps are an outcome of the colliding Eurasian and African continental plates; the current form of these mountains was created by lifting that occurred about 65 million years ago. Further surface transformation happened during the Neogene ice ages.

The South American Andes demonstrate the results of a collision between an oceanic plate, called the Nazca plate, and the continental crust. Due to subduction, parts of the heavy oceanic crust were scraped away and deposited on the continental plate in a wedge formation. Simultaneously, the continental crust was shaved back. Then a belt made of volcanic and magmatic rock that had solidified in the depths was shifted inland. Many such belts, parallel to each other, were formed in this manner and are thus called cordillera, or "strings of mountains."

The Cliffs of Moher in Ireland, made of shale and sandstone, represent a slice through geological time with the oldest rocks at the bottom.

The Cuesta del Obispo in Argentina: Cuestas are ridges formed by gently titled rock layers and have steep slopes called escarpments.

Volcanic mountain ranges

When two sections of oceanic crust collide, volcanic arcs such as the Aleutians, the Kuril Islands, and the Philippines are formed in the subduction zones. When crustal plates drift away from each other, volcanic mountain ranges result; the largest of this type are found along mid-oceanic ridges. Ongoing volcanic activity also produces such mountain ranges on the continents, not only at the bottom of the sea. Movements caused by plate tectonics along the sides of the East African Rift have created huge cracks where magma wells up. The mightiest mountain in Africa ascends above this area: the Kilimanjaro massif, which is 19,340 feet (5,895 m) high.

Mount Elbrus, located in the western Caucasus Mountains of Russia, is a dormant stratovolcano that rises 18,510 feet (5,642 m) into the sky.

in focus

ISOSTASY

Mountains float on the elastic lithosphere like icebergs on the ocean. In both cases, only a part of the whole can be seen above the surface. As an iceberg melts, it floats upward, so that its height relative to the ocean's surface stays the equal. The same process occurs in mountainous regions that are worn down or where glaciers are melting and are no longer so heavy.

A popular attraction, Norway's Pulpit Rock was carved by glaciers during the ice ages.

The smallest of the giants

The low mountain ranges vary between 4,900 and 6,500 feet (1,500 and 2,000 m) high. They are often residual mountain ranges, meaning that over millions of years, the older mountains have been worn down by weather and erosion. These old, rigid remains do not yield and break under the pressure caused by expansion in the Earth's crust. Examples of these kinds of mountain ranges include the low mountains of Germany, the African Ruwenzori range (now called the Rwenzori Mountains), and the Appalachians.

fracture tectonics

Due to the movement of its plates, the Earth's crust is always subject to pressure and shear forces. When the tension in a mass of rock becomes so great that it exceeds the rock's resistance, breaks occur at the weak points.

Fractures in a rock mass that are only a few inches to a yard long are called crevices. Longer breaks are called divergent boundaries. When a fracture occurs, cold, brittle rock masses in the Earth's crust may be compressed with lateral pressure, for example, if crustal movement causes two such sections to collide. The result is that one section is pushed upward. In this way a ridge block, or massif, is lifted vertically along the fault line. Fractures can also occur if the rock expands instead of being compressed. If the angle at which the rock section is lifted is less than 45°, the occurrence is called a thrust. The term subsidence refers to the sinking of a massif.

A rift, or graben, is formed if rock breaks along two parallel boundaries and the crustal block caves in, leaving behind the steep flanks formed by the fracture. Generally, these flanks break down further into so-called staggered fractures. If rift formation occurs on a large scale (for example, if sections of the crust are drifting apart on diverging plates), elongated rifts such as the Upper Rhine Valley are formed. The opposite occurs if two crustal regions are moving closer together. Fractured massifs are then lifted to become horsts, or raised blocks. The Black Forest and the Front Range of the Rocky Mountains in Colorado were formed by such a process. Mountain ranges created by fracture tectonics are called fault-block mountains.

In addition to the forces of compression and expansion, shearing forces also impact rock. These forces are the result of sections of crusts moving past one another. Unlike the effects of thrust or subsidence, no difference in elevation occurs at these boundaries; there is, on the other hand, a lateral displacement.

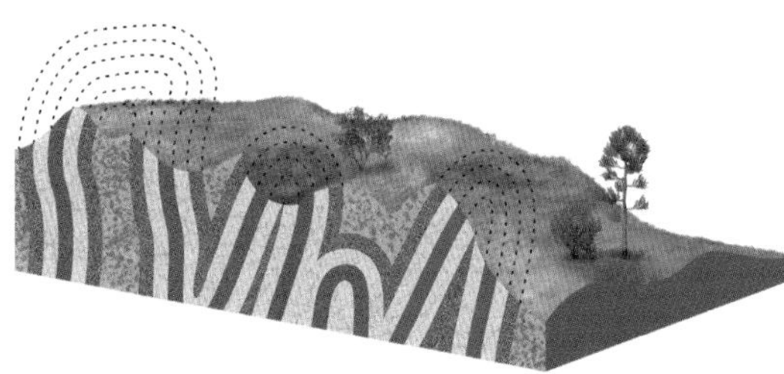

Range of mountain folds

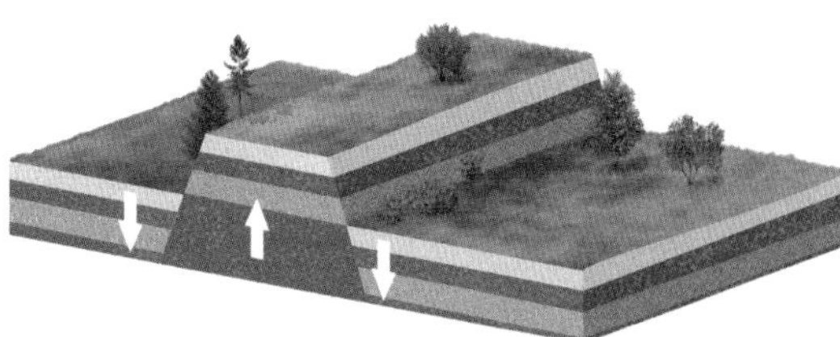

Horst

Nappe system

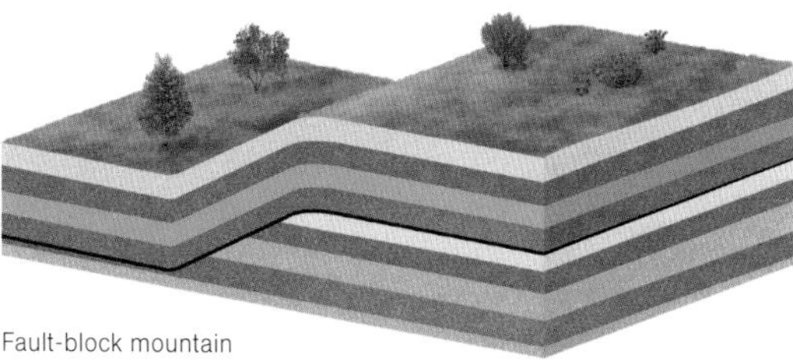

Fault-block mountain

Various geological processes can lead to the formation of mountains. From top to bottom: a range of mountain folds; a horst (raised fault block surrounded by graben, which are depressed); nappe system (body of rock moved far from its original position); and the fault-block mountain (created when normal faults fracture a section of the continental crust).

The badlands at Zabriskie Point in Death Valley: Death Valley is an example of a rift valley.

basics

FOLDS are created in the same way as fractures. They are generated by the lateral constriction of sedimentary strata, which results in a sequence of anticlines and troughs.

Magnificent mountains sprinkled with snow loom over grazing buffalo.

see also: Structure of the Earth's crust, p. 59

mountain summits

Since the beginning of time, people have had a special relationship to mountain summits. While scientists see a summit as the result of tectonic processes, climbers behold a formidable challenge.

Mount Fuji is the highest mountain in Japan and last erupted in 1707. Its snow-covered peak is fantastically symmetrical and rises 12,388 feet (3,776 m) into the sky. The mountain's name means "everlasting life."

Most mountain summits are peaks in a range of mountains, although individual summits can also ascend from a plain. Summits are normally formed when sections of the Earth's crust located at higher elevations are eroded; they can also be erected by lava that streams over the Earth's surface. Summits vary in terms of appearance: they can consist of a large, boulder-like, craggy rock or be formed of free-standing massifs with abrupt vertical drops, a volcanic cone, or domes covered in perennial ice.

Mighty mammoths

The highest summits on Earth are the "eight-thousanders," so named because they all reach heights of over 26,247 feet (8,000 m). Ten of these 14 gigantic mountains are found in the Himalaya. The summit that towers over them all at 29,029 feet (8,848 m) is Mount Everest. However, to be exact, this summit is actually exceeded by Mauna Kea in Hawaii. Although this volcano reaches only about half of Everest's height, together with its underwater base, its overall altitude is approximately 33,500 feet (10,211 m). The highest volcano on any continent, with a height of 22,615 feet (6,893 m), is Ojos del Salado, located on the border of Argentina and Chile. In Europe, the highest mountain summit is either Mont Blanc in the Alps or Elbrus in the Caucasus, depending on how the boundary between Europe and Asia is defined.

This famous monolith, named Sugarhill Mountain, is made of granite and quartz and overlooks the Brazilian city of Rio de Janeiro.

Monoliths and monadnocks

The largest monolith on Earth is Mount Augustus in western Australia. Much more famous, however, is Ayers Rock, located in the same country. Rising abruptly from their surroundings, such monadnocks, also known as inselbergs, are the remains of contiguous plateaus that have been worn down by erosion or weathering. A good example is Monument Valley in North America, with its many striking rock formations. Table Mountain in Cape Town, South Africa, was also formed from a rock layer that survived the eroding teeth of time.

Ayers Rock, or Uluru as it is known to the Aborigines, is 2.2 mi (3.6 km) long and 1,142 ft (348 m) high.

milestones

EARLY EVEREST ASCENTS

Called Sagarmatha ("king of the heavens") in Nepalese and Chomolungma ("mother goddess of the earth") in Tibetan, this Himalaya mountain got its English name from the British surveyor George Everest. The first people to conquer this summit were Edmund Hillary of New Zealand and Sherpa Tenzing Norgay on May 29, 1953. In 1975 Japan's Junko Tabei was the first woman to reach the peak. Reinhold Messner made the first ascent without an oxygen mask in 1978.

Over 200 people have lost their lives trying to climb the mightiest mountain on Earth.

types of deserts

Deserts and semideserts cover about one-third of the Earth's land surface. The best known desert, the Sahara, is also the largest. It extends over 3.36 million square miles (8.7 million km^2).

A desert is a hostile environment with little vegetation and commensurately sparse fauna. The lack of water prevents most plant growth in dry deserts. Such drought is usually caused by the absence of rain. For instance, if an area lies on the leeward side of mountain ranges—such as the South American Atacama Desert along the western side of the Andes, or the North American Mojave Desert, west of the Rocky Mountains—it receives little precipitation. The Gobi desert in Central Asia exists because of a permanently dry continental climate. Passat (trade wind) deserts, such as the Australian Simpson Desert, are created by constant subtropical high-pressure belts that inhibit the penetration of moist air masses into arid regions.

deserts | droughts | forests | habitats

ECOSYSTEMS

Animals and plants, together with the environmental conditions, soil, and climate of a place, create diverse, often sensitive ecosystems. Only small changes in the environment are required to upset the many interdependent relationships within the system, and destroy the balance that helps different species survive.

Semideserts and cold deserts

In hot deserts of the tropics and subtropics, such as the giant Sahara in North Africa, the Arabic deserts, or the Chinese Taklimakan, high evaporation rates further deplete water supplies. A lack of cloud cover leads to extreme daily temperature fluctuations. Temperatures reach up to 175°F (80°C) during the day as the sun heats the ground unimpeded. During the night, the temperature drops, sometimes below freezing. In a semidesert, like the South African Kalahari Desert, at least one wet month allows plants to grow. There are also cold deserts, such as the Wright Valley in Antarctica, where the absence of suitable temperatures above the freezing point prevent the spread of vegetation.

Sand dunes and rubble fields

Although deserts are often associated with enormous sand dunes, only three percent of them are sand deserts. The largest sand desert is the Rub Al-Khali, with dunes that cover the southern third of the Arabian Peninsula. Deserts consisting of rocky rubble fields or severely weathered mountainous regions are much more common. Gravel deserts may be formed by erosion or could be relics of glacial deposits. Salt deserts, such as the Salar de Uyuni in Bolivia or the salt flats in Utah, develop from the evaporation of salt lakes.

Death Valley has diverse life despite its hostile conditions, salinity, and high temperature.

The highest sand dunes in the world, reaching 1,250 feet (383 m) high, are found in the Namib, Namibia.

The majority of the world's deserts are scree and talus fields, where the removal of fine-grained material by the wind has exposed loose gravels.

DESERT LACQUER
Iron and manganese oxides, oozing via capillary action from rocks, covered with clay dust and spreading out like a brownish black lacquer.

DESERT ROSE
Rosette-like crystallizations, from a gypsum and sand grain mixture, formed through evaporation of water in salt lakes, but also through the weathering of gravel.

DESERT GLASS
Developed 28 million years ago through meltdown of desert sand, due to pressure and heat from a meteorite impact, into purest glass consisting of 98% silica and traces of iridium.

surviving drought conditions

Deserts may be barren, but they are nonetheless home to many plants and animals—both during the long droughts, and when there is a short-lived sea of flowers after a rare rainfall.

Numerous desert inhabitants have developed amazing survival strategies in such dry, inhospitable environments. Many trees and bushes produce long roots in order to reach the groundwater available at great depth. In order to minimize evaporation, plants breathe at night. They have a reduced leaf surface area, carry thorns instead of leaves, or exist merely as a stem, like cacti or some members of the spurge family (Euphorbiaceae).

In the Australian eucalyptus, the glaring sun light is reflected by a wax layer, or in tamarisks and some palm trees, by salt excretions. In order to utilize the precious liquid wisely, some succulents have developed large-cell storage tissue in their leaves, such as living rocks (Aizoaceae); in stalks, such as saguaros; or in stems, such as aloe trees. Some plants bridge extreme drought periods, by resting in the ground as seeds. When sufficient rainfall returns they go through their entire life cycle in an extremely short period of time.

Desert animals are primarily nocturnal. During the day, they move below ground for protection.

Desert animals

Desert fauna also displays conspicuous adaption to extreme environmental conditions. For example, the ability to move rapidly and long legs aid in protecting them against heat reflecting from the ground below.

Often they also have a reduced water metabolism. Ants and rodents get their moisture from seeds. They in turn become a source of food and water for lizards, snakes, and jackals. Other animals obtain water directly, such as the dew-drinking black beetle from the Namib, or antelope herds, which undertake long migrations to water holes.

Cactus plants store water in their stalks or stems to withstand long periods of drought.

Fertile islands

The most active life in the desert is found in the oases. These fertile islands are located where water comes to the Earth's surface. Many oases develop around artesian wells, where water rises to the surface under its own pressure. Some of these wells are fed from groundwater, others with water from artificial lakes or reservoirs or a remote river. The area around an oasis is usually densely populated and intensively cultivated. The tree most characteristic of oases is the date palm, which originally came from North Africa and India.

Welwitschia, a conifer, can live for over a thousand years. They are found in the Namib Desert.

in focus

DESERTIFICATION

In regions with a relatively dry climate, intensive human cultivation can have terrible consequences, reducing vegetation, draining water sources, and causing soil erosion and salinity. The formation of deserts threatens the existence of more than a billion people in over a hundred countries. In Africa, millions have died in the famines caused by the severe desertification occuring there.

In Africa, 46% of the land is suffering from desertification.

An oasis in the Sahara is a vital source of water for migratory and nearby animals.

see also: The evolution of new species, p. 167

types of forests

Forests cover about one-third of the Earth's land area. Acting as a sort of green lung, they produce oxygen and store large amounts of atmospheric carbon dioxide, so counteracting global warming.

Autumn colors in an old park: The fall is regarded as a time of poetic melancholy in art and literature.

Three prominent types of forest make up the green belt of our Earth. The central green zone along the Equator is formed by tropical rain forests, including the adjacent marginal tropic and monsoon forests. Trees, bushes, ferns, palms, orchids, and herbaceous plants grow in several levels over and on top of each other, as well as side-by-side. They form the most species-rich and complex ecosystem on Earth, and are home to approximately 90 percent of all animal and plant species.

basics

FOLIAGE LEAVES adjust their position relative to the sun for maximum photosynthesis. Shorter days and declining temperatures initiate leaf shedding in autumn. Nutrient material previously embedded in stems serves to initiate new leaf growth in spring.

NEEDLE LEAVES reduce the exposed surface area and protect against moisture evaporation and snow fracture. Almost all conifers retain their leaves in winter.

The rain forest in Costa Rica contains evergreen trees: those that carry leaves through the year.

Rain forests used to cover about 14 percent of the Earth's land area; now, this has halved. Toward the North Pole, the evergreen boreal coniferous forests—the Russian taiga—are the northernmost trees. Severe cold and an abundance of snow, combined with a short summer, mean that the vegetation period lasts only about 150 days of the year. Only a limited number of trees can grow in the meager and often infertile soils, and the undergrowth is also sparse. Boreal coniferous forests cover approximately 10 percent of the landmass of the Earth.

Within the wide band of temperate latitudes there are deciduous and mixed forests. Year-round rainfall facilitates the plant world, with a growth period of more than six months. In the cool temperate latitudes, deciduous forests with many tree varieties, as well as lush undergrowth, are found. The leaves they shed in autumn protect them from being killed off by frost during winter. Along the subtropical west coasts of the continents, original hard-leaf forests protect themselves against evaporation during hot summers with small, tough leaves, such as those of the holly leaf oak in the Mediterranean and the eucalyptus in Australia. Leatherlike leaves characterize laurel forests, which are scattered worldwide in perpetually damp, subtropical climates. Such moist forests are likely to be found along east coasts, such as eastern China, southern Japan, and Florida.

Sequoia trees, here in Yosemite Valley, grow 325 feet (100 m) tall.

forests as habitats and economic areas

Forests provide habitats for myriad animal and plant species and support the subsistence of millions of people. But they are in great danger. About half of the original forests worldwide have already been destroyed.

Between the subterranean roots and high-reaching crowns of the trees, a forest provides an abundance of ecological niches. In the treetop levels of the rain forest, a multitude of birds and small mammals, monkeys, and sloths make their homes. In temperate forests, tree crowns serve many birds of prey, such as hawks, as well as being a nesting area for wood warblers such as the willow warbler. Branched, leafy trees in moist tropical forests are the home of agile climbers and fliers such as butterflies, tree frogs, tree snakes, parrots, and hummingbirds.
In cooler climates typical forest birds such as chaffinch, jay, and coal tit, as well as striped and flying squirrels, are found. Numerous invertebrate animals, such as caterpillars, beetle larvae, spiders, and mites, live underneath the tree bark, attracting insect-eating birds. Owls, wild ducks, bats, or squirrels occupy natural tree hollows or those hollowed out by woodpeckers. Leaves, seeds, and fruits from trees and bushes feed many mammals, including monkeys, mice, and hedgehogs. Seeds and undergrowth provide an important nutritional basis for herbivores and omnivores such as deer, elephants, and elk. These in turn provide food for predators, such as martens, bears, jaguars, lynx, or wolves. Finally, even a dead tree provides nutrition for decay-enhancing organisms, such as insects, fungi, mosses, and microorganisms. They break down plant material, thus completing the cycle of nature.

Open forest areas have diverse undergrowth, which supports herbivores and omnivores.

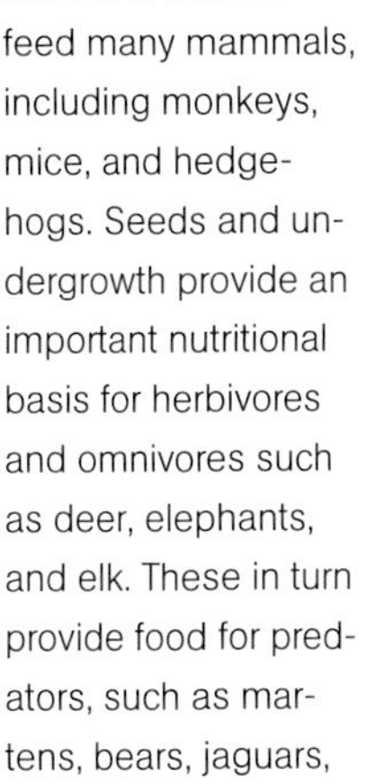

The forest floor is inhabited by many mites, insects, and earthworms. Although tiny, these creatures are vital to the forest ecosystem.

Logged trees are transported down a river in Karelien, Finland. Logging can contribute to the depletion of rich forest ecosystems.

People have long used timber as a building and heating material. Although the negative effects of overexploitation are widely known, precious rain forest woods, such as teak, are still in demand. Forests are also "farmed"—fast-growing trees, such as Douglas fir, are planted and later logged. Plantations promise quick profits, but due to their lack of a natural ecosystem, they are susceptible to massive infestations with so-called forest pests such as bark beetles.

in focus

THE RAIN FOREST LABORATORY STORE

Rain forests are gigantic reservoirs of natural medicines and other bioactive substances. For example, the bark of the cinchona tree from South America provided the effective malaria medication quinine for many years. Secretions from an Australian tree frog have yielded a new antibiotic that is effective even against resistant germs. So far only a fraction of this magnificent drugstore of nature has been explored.

Unsustainable logging could prevent new medicines from being discovered.

see also: Climate change, pp. 124–127

saltwater wetlands

The term wetland is used to describe a range of habitats that are entirely submerged or partially permeated by water. There are freshwater and saltwater wetlands. In all of these habitats, the site where the water meets the land is always shifting.

The tideland leading to Neuwerk Island in the Hamburg Wadden Sea National Park in Germany. The sea covers and exposes the sand daily.

Saltwater wetlands are located at the frontier between ocean and mainland. At these locations, saltwater from the ocean mixes with freshwater from rivers. This mixture creates a particular challenge for the plant and animal worlds, which are normally adapted to either one habitat or the other. While the number of species in general is usually low, the number of individuals of a particular species can be very high. On many flat coastal areas where rivers empty into the ocean, embankments are formed by oceanic currents and tides, as well as the sediment carried by the rivers. At the mouth of the Rhône in France, for instance, lies a motley mosaic of habitats, such as brackish water lakes, moist swamps, riparian forests, beaches, sand dunes, and salt marshes.

Coastal areas with strong tides feature a transitional zone between the mainland and the ocean: the tidal flats. The ocean washes over this area at high tide and it dries up again at low tide. Only rarely does the water reach the salt marshes at higher elevations where glasswort or cordgrass grow. True intertidal mudflats can only be found on the northwest coast of Europe and on the west coast of South Korea. However, similar conditions also exist on the east coast of North America. This unique ecosystem forms a habitat for worms, mussels, and snails. It is also a paradise for millions of migratory birds.

Camargue horses run through the salty marshes of the Camargue area in southern France. This ancient, free-ranging, and hardy breed manages to thrive in the hostile habitat at the mouth of the Rhone River.

Mangrove swamps

Mangrove swamps are located in the tidal regions on tropical coasts and at the edges of broad estuaries. Various mangrove tree and shrub species are the only woody plants that have a tolerance for salt. At high tide, often only the tree crowns can be seen above the waterline; at low tide an entanglement of roots appears, providing a habitat for mudskippers and fiddler crabs. Mangrove stocks worldwide are extremely endangered. In eastern Africa, they are used for firewood. When mangroves are removed, the protection they provide against unusually high tides is also lost.

Some mangrove trees have roots adapted to better bear water flowing in and out, while others breathe air.

in focus

THE EVERGLADES

This unique subtropical wilderness, which the Native Americans of the region called "grassy water," extends across the southern tip of Florida. What looks like a large group of lakes is actually a river fed by Lake Okeechobee, which is 50 miles wide, but only a few inches deep. Certain parts of this mosaic of swamp grass, marshlands, cypress forest, and mangrove swamps are under protection. It is home to the Mississippi alligator and the Florida manatee.

The Everglades are a haven for at least 15 endangered species.

see also: Ecosystems, pp. 276–277

freshwater wetlands

Wetlands can be found in many different climates. They make up about 6 percent of land areas and provide habitats for numerous animal and plant species. However, they are under increasing threat of destruction.

Freshwater wetlands form an interwoven network of lakes, streams, and rivers. They make a substantial contribution to groundwater replenishment and they influence the regional climate by creating humidity during drought conditions.

In the arid center of southern Africa, the Okavango River flows into a maze of lagoons, swamps, and savanna.

Riverine meadowland and freshwater deltas

The largest continuous wetlands on the mainland are the river meadows with their amphibious landscapes. A diverse ecosystem develops, with calm dead water arms and moving water surfaces, dry gravel and rubble areas, extensive reed fields, and impenetrable floodplain forests. These ecosystems are the habitat for rare animal and plant species. Many of these species are endangered because rivers have been straightened, dammed, or forced into concrete streambeds, and the fertile floodplains are being utilized for agricultural purposes or as grazing land. Along some of larger rivers, such as the Ohio, Huang, He, or Danube, only relics of the former river meadows remain. The giant floodplain at the upper Rio, Paraguay, is the home of the rare hyacinth macaw and the jaguar, and a refuge for the threatened giant river otter and the capybara. The Okavango Delta on the northern edge of the Kalahari Desert is the world's largest inland delta. This delta attracts migratory birds and animals such as elephants, rhinoceroses, giraffes, zebras, antelopes, lions, hyenas, and leopards.

Moors

Low moors usually develop through the transformation of nutrient-rich waters into land. The existence of these moors is not dependent on precipitation since there is a connection to groundwater. They can be recognized by the presence of plants such as reeds, willows, alders, and sedge. In contrast high moors are nutrient-deficient areas that rely on precipitation. The low pH and oxygen deficiency prevent the breakdown of dead plant matter and promote the formation of peat. Plants adapted to these conditions include cotton grass, heather, sphagnum moss, and the flesh-eating sundew. Animals include dragonflies, butterflies, and several rare bird species, such as snipe and black grouse.

Reeds are a characteristic plant of low moors.

basics

RAMSAR CONVENTION In 1971, delegates from 18 nations signed the Convention on Wetlands of International Importance in Ramsar, Iran. This makes it one of the oldest international agreements for environmental protection. Since the enactment of this convention in 1975, 1670 protected areas with more than 932,056 miles^2 (1.5 million km^2) have been identified in 155 countries.

Pantanal in Brazil is the world's largest area of wetlands. It houses a vast array of aquatic plants.

see also: Phosphorus cycle and water cycle, p. 279

temperate grassland: steppe and prairie

Vast areas of grassland cover much of the Earth's landmass. In tropical grassland or savanna areas, trees may be found frequently, whereas temperate or semitropical grassland areas have very little tree growth.

The term steppe, which originates from the Russian language, describes the vast plains of grassland at the center of the Eurasian continent, from eastern Europe to northern China. In North America, grasslands are called prairies. They are located in the Midwest of the United States and Canada. The life cycle of a steppe is influenced by seasons of weather extremes. Hot summers, cold winters, and limited precipitation allow for only very short growth periods of the vegetation. Today almost all steppes are cultivated and only a few are designated protected areas. Most areas are used to grow grain or sunflowers. Some parts are also used as pasture for livestock.

Plants and animals

The typical plants of a steppe are grasses with a finely interwoven root system, which creates a thick greensward. The grass buds remain below the surface, where they are protected from cold and dry conditions, frequently recurring fires, and grazing animals. Other plants only grow if there is sufficient water. The steppe becomes a colorful sea of flowers during the short blossoming time available to iris, hyacinths, crocuses, and tulips.

Temperate grasslands are an important resource for herbivores. Mammals living in this type of habitat tend to feed in herds, such as the fast Saiga antelopes in central Asia or strong American buffalo or bison. Wild horses, such as the Eurasian tarpan, used to be common but are now largely extinct. Burrowing rodents are also typical inhabitants of the temperate grasslands. Their extensive tunnel systems ensure mixing and aeration of the soil. The Eurasian counterpart to the North American prairie dog is the bobak marmot or steppe marmot. Several species of guinea pigs are native to the South American grasslands. These rodents are the preferred food of many carnivores in the grasslands, such as coyotes in North America or the long-legged maned wolf of South America.

basics

SALT STEPPES are found near salt lakes.

DESERT STEPPES are a zone of transition between grasslands and deserts.

CULTIVATED STEPPES emerge where trees have been felled for agriculture.

If sufficient water is available, some herbs, perennials, and bulbous plants may also grow on the steppe or prairie.

Western Tibet receives very little rain, due to the "rain shadow" effect, whereby precipitation is deposited on the southern side of the Himalaya.

The North American prairie dogs belong to the squirrel family, but these very social animals are named after their distinctive barking noise.

see also: The climate system, pp. 118–123

tropical grassland: savanna

Savanna grasslands are located between the humid tropical regions and dry deserts. Vast areas of this type of grassland are covered by only the occasional tree, cluster of trees, or scattered small woods.

Savannas are common in Africa, but are also found in India, Australia, and the northern parts of South America, such as the Llanos of Venezuela and Colombia. They are located in the subhumid outer tropical areas with warm air temperatures throughout the year, wet summers, and dry winters. As the distance from the Equator to the tropics grows, the number of dry months increases from less than four to about ten months and the amount of average precipitation decreases three- to sixfold.

The umbrella thorn, with its distinctive crown and feather-like leaves, is able to reach even low groundwater with its deep root system.

Savanna grasslands have different appearances according to the local climatic conditions. Savannas may be humid or semihumid, with grasses taller than humans and scattered trees, or dry savannas, with low and mostly drought-resistant grasses.

Savanna trees and shrubs show obvious signs of adaptation to the extended drought season. For example, the baobab tree is able to store large amounts of water in its trunk. Riparian forests along rivers are also part of many savannas. The African hippopotamus commonly lives in this habitat. The African savannas are home to many large terrestrial mammals, including herbivores such as elephants, giraffes, zebras, and antelope. The East African ostrich is a bird of the savanna. The equivalent bird is called the common rhea or nandu in South America and the emu in Australia.

Wildebeest migrate from the plains to the woods in May and return in November when the rains arrive.

Common carnivornes in the African savanna are lions, hyenas, and cheetahs.

Kangaroos are the largest animals of the Australian plains. Only a small number of the world's natural grassland areas remain untouched by humans. For thousands of years nomads have taken their herds of cattle, sheep, and goats into the savannas in Asia and Africa to graze and have also hunted there.

Today Africa is at risk from the desertification of major parts of dry savanna grasslands adjacent to the Sahara, due to overgrazing and severe droughts. Savannas are also threatened by increasing woody plant encroachment due to climate change.

in focus

SERENGETI SHALL NOT DIE

The "endless plain" in East Africa is at the core of the African savanna. The ecosystem depends on the herds of wildebeest or gnu passing through the grasslands every year during the dry season in search for water. The scientists Bernhard and Michael Grzimek were the first to make an effort in protecting the wild animals of the Serengeti. Their 1959 documentary *Serengeti Shall Not Die* (*Die Serengeti darf nicht sterben*) contributed greatly to increasing global awareness of the Serengeti animal park and its vulnerability to human impact.

Serengeti is home to the world's largest animal herds.

weathering

Degradation due to weathering is part of the natural cycle of rocks and a fundamental prerequisite for soil formation. Even the most massive mountains cannot withstand this unremitting force for long.

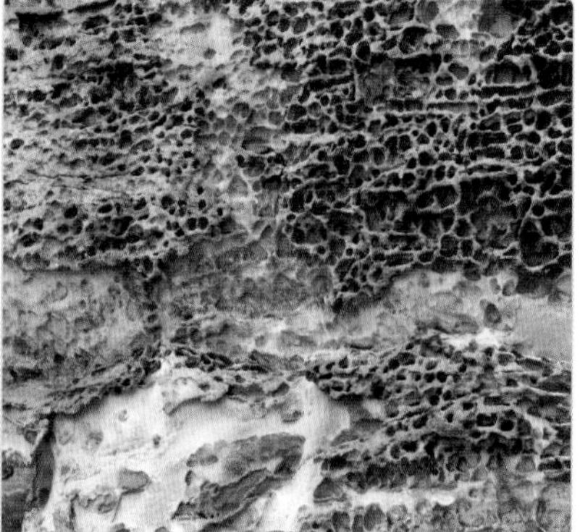

Limestone is particularly prone to chemical weathering by carbonic acid in water.

weathering | erosion | deposition | movements

CHANGING EARTH

Craggy mountains, rounded hills, wide plateaus, and endless coasts—the landscapes of the Earth are almost as varied as the outside forces that have shaped them. The current form of the Earth is only a moment in its geological history. Weathering, erosion, and precipitation are constantly changing the appearance of our world.

As soon as rocks reach the surface of the Earth they are exposed to physical and chemial processes, which change and ultimately destroy them. The effects of weathering on rocks depend on the climate and the type of rock. Pressure and temperature variations play a large role in physical weathering by causing mechanical destruction of rocks. Stones and rocks expand during the day under the influence of solar radiation, then contract again as they cool during the night. This constant fluctuation loosens the rock structure, and over time rock fragments chip off. Large boulders may split open along internal fracture lines; small stones disintegrate into grit.

Changes in temperature together with moisture lead to "frost weathering," especially in high mountainous areas and near polar regions. Water, when it freezes, increases in volume by about one-tenth. Thus, the ice that forms in the crevices of the rocks have the potential to crack them open. In arid regions, salt takes the role of ice. When saline water evaporates, the salt crystallizes outward, leading to a volume increase that cracks the rocks.

In contrast to mechanical weathering, rocks are totally disintegrated during the process of chemical weathering (decomposition). Rocks are gradually corroded by water and the dissolved salts, acids, and gases it contains.

If organisms, especially microorganisms, are involved in rock disintegration—through chemical or mechanical means—it is called biological weathering. Acids excreted by lichens, for instance, attack the crystal matrix of minerals and thus destroy the structure of rocks. The process is rather less delicate when boring clams, sponges, or worms penetrate rocks along rocky shores, or when the roots of plants force their way up into rock formations, effectively cracking open the rocks with their growth (known as root fracture).

in focus

LIMESTONE CAVES

Limestone and dolomite are rarely corroded by pure water; however, water with carbon dioxide (CO_2) dissolved in it will produce carbonic acid. This reacts with the calcium carbonate in limestone and dissolves it. Such carbonic acid weathering, especially in calcareous regions, leads to the formation of rugged rocks, sinkholes, caves, and subterranean watercourses. This is known as karst landscape, a name derived from the limestone plateau east of the Gulf of Trieste.

Crystallized limestone in Harrison's Cave, in Barbados.

Dramatic karst landscapes are created by the dissolving of solid rock over millions of years.

see also: Weather, pp. 114–117

erosion

Once rocks reach the surface of the Earth they begin to disintegrate, and can be carried away as rubble to be deposited elsewhere. All processes that contribute to this leveling of the Earth's surface are termed erosion.

Water is a significant erosive power. When precipitation falls on areas with a topographical gradient, the water will initially accumulate in small rivulets along the surface. In this process, the first soil and rock particles are pulled along and water-soluble components are leached out. This material by itself already has a substantial erosive force.

In the natural amphitheater of Bryce Canyon in Utah, wind, water, and ice erosion formed distinctive geological structures called hoodoos.

On its way downriver (downhill) it grinds away over time at massive rocky areas and digs deep furrows. The concurrent deepening of the riverbed is referred to as vertical erosion while the widening of its riverbanks is called lateral erosion. In the course of thousands of years, the erosive forces of water gradually level out landmasses. Scientists assume that high mountain ranges lose about 16 feet (five m) in height within 10,000 years. In uplands (minor mountain ranges) and lowlands, the erosion rate is distinctly less due to the reduced gradient.

GLACIAL EROSION is the effect of enormous ice flows creeping slowly over the landscape. On their way, they transport rubble in so-called moraines.

basics

Oceans also have enormous erosive powers. The movement of the surf and tides causes continuous change to our shorelines. These are barely perceptible over small time scales, but eventually cliffs are hollowed out and beaches can be washed away. Sporadically, storm waves crash like battering rams against coastlines and can alter their appearance even within a short time.

Wind erosion

Wherever a protective vegetation cover is absent, the wind picks up humus, sand, and dust particles and carries them away. In this way fertile soil is lost, especially in arid regions and during drought periods when there is insufficient moisture to hold soil particles together. Similarly, surf zones along ocean shores and dry high-water zones of rivers are areas commonly impacted by wind erosion. In dry regions without vegetation, wind can cause large sand drift (deflation), creating shallow depressions in the rock over time. Most sand grains transported by wind are dragged along close to the ground, thus causing abrasions on rocks. This tends to form distinctive wind-worn stones (ventifacts). The rocks take on bizarre, mushroom-like shapes, with narrowly rounded bases and broad upper regions.

Due to regressive erosion caused by water, the Niagara Falls are displaced upstream at a rate of about 14.6 inches (37 cm) per year.

The White Desert in Egypt is full of white chalk posts called "inselbergs"; these wind-carved eroded remnants of the Upper Cretaceous era are studied to understand similar formations found on other planets.

deposition

The surface of the Earth undergoes constant change. In this never-ending process, forces created by wind, water, or glacial ice erode the land in one place and deposit it elsewhere due to the pull of gravity.

Rivers transport loose particles and debris, and new land is created where these fragments are deposited. At river bends, material is carried away from so-called undercut slopes, while sediment creates sandbanks at slip-off slopes. In the underflow, coarse material gathers in the form of stream banks that continue onward. If the speed of the flow decreases, erosion is no longer possible and large fields of rubble build up. Finer material is often sedimented far from the mouth of the river. In some cases, sediment creates flat river deltas.

At the coast, erosion rubble is transported either by the surf or waves that contact the shore at an angle. The material deposits subsequently cause the shoreline to shift. If this debris aggregates in the ocean, it becomes a spit, which is a narrow ridge of sand. As the spit extends, it can develop into a bar. A bay that is cut off by this formation can become a backwater or lagoon; however, sand dunes are the most typical deposition landform in flat coastal areas.

The wind collects fine material from dry areas, transports it—sometimes across long stretches—and deposits it. Dust from the Sahara can reach central Europe in this way, and occasionally even the Brazilian rain forest. For most grains of sand, however, the trip ends much sooner. As soon as they encounter small obstacles, such as tufts of grass, they are deposited and become dunes. The wind also transports fine dust. Loess, a relic of the ice age found in central Europe and North America, is often deposited in vast layers. In China, loess drifts coming from dry, high-altitude areas still occur today.

Sandstorms forming in the Sahara and arid regions around the Arabian Peninsula can transport dust to the European and American continent.

in focus

LAND RECLAMATION IN THE NETHERLANDS

In the Netherlands, coastal protection has a long tradition, especially since more than one-fourth of the country lies below sea level. Dikes were built to separate the coastal shallows from the North Sea. The land was drained with a system of canals and pumps driven by wind power, creating the productive polder landscape. One of the largest projects was draining the Zuiderzee. This bay was separated from the sea in 1932 by an enclosing dam and drained to produce about 637 square miles (1,650 km^2) of dry land. The remaining expanse of water became the Ijsselmeer.

The Dutch have reclaimed fertile marshes and fenlands through the construction of polders.

A glacier tongue of Jostedalsbreen in Norway—the largest glacier in continental Europe.

Griswold Point is a sand spit located at the mouth of the Connecticut River. Its moving sand dunes are an impressive example of coastal dynamics.

Most of the material moved by glaciers is deposited gradually as the glacier melts. The characteristic rubble formations formed by the melting glacier are referred to as medial, lateral, internal, or ground moraines depending on where the material was located within or along the surface of the glacier. Deposits at the glacier tongue are known as end moraines.

see also: Sedimentary rocks, p. 63

mass movements

Avalanches, earthflows, and rockfalls: all over the Earth where the angle of the natural ground slope is exceeded, gravity makes itself felt and starts to flatten, carry away, and level the landscape.

Massive ground movements belong to the processes of large-scale erosion or denudation. This term incorporates all movement initiated by gravity, of rocks, soil, and sediments along the Earth's surface. Apart from the gradient of a slope, the type of material involved and its water content are also of significance.

Massive dry ground movements mainly occur along steep slopes, for instance in high mountain ranges. Masses of rocks can become loose and crash down into the valley as rockfalls or even as rockslides. Smaller pieces of rock that continue to break loose from rock walls are referred to as falling rocks, which lead to the formation of large taluses.

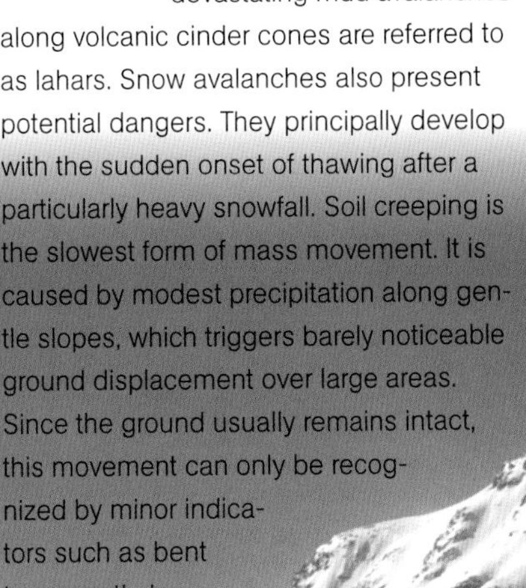

Landslides caused by earthquakes or storms can destabilize hillsides by moving large masses of mud and debris.

Mass movements are enhanced by the participation of water. Enormous mud and rubble slides occur in mountainous regions without adequate vegetation cover, especially after severe precipitation. These so-called alpine mudflows can reach speeds of several miles per hour and leave a wide strip of destruction along mountain slopes. In 1970, an earthquake in the Peruvian Andes triggered a major slide, burying 70,000 people under masses of rubble, ice, and mud. The devastating mud avalanches along volcanic cinder cones are referred to as lahars. Snow avalanches also present potential dangers. They principally develop with the sudden onset of thawing after a particularly heavy snowfall. Soil creeping is the slowest form of mass movement. It is caused by modest precipitation along gentle slopes, which triggers barely noticeable ground displacement over large areas. Since the ground usually remains intact, this movement can only be recognized by minor indicators such as bent tree growth. In permafrost regions, ground movement is referred to as solifluction. This occurs when the onset of thawing weather conditions, brought about by the warming spring sun, causes the upper layers of the ground to lose adhesion to the still frozen layers below.

Following rockslides, roads can often become impassable due to undercutting or large taluses blocking the traffic.

basics

HUMAN INTERFERENCE is partially to blame for the fact that natural events such as rockslides and mudflows increasingly turn into catastrophes. Mountain regions are being developed for tourism at ever-higher altitudes. Land clearing and road construction have destroyed stabilizing root networks along many mountain slopes.

To avoid uncontrollable avalanche catastrophes, snowslides can be artificially triggered through detonation.

LOMONOSOV RIDGE
MAKAROV BASIN
ARCTIC
MENDELEYEV PLAIN
CHUKCHI PLATEAU
CHUKCHI PLAIN
NORTHWIND ESCARPMENT
CANADA BASIN
CANADA PLAIN
CONTINENTAL SHELF
North magnetic Pole 2007
Axel Heiberg Island
QUEEN ELIZABETH ISLANDS
Ellesmere Island
Lincoln Sea
Knud Rasmussen Land
Hayes Peninsula
GREENLAND
Ice thickness 3410
Gunnbjørn 3700
Prince Patrick Island
Melville I.
Banks Island
Victoria Island
Devon I.
Baffin Bay
Baffin Island
Wrangel I.
Point Barrow
Chukchi Sea
North Slope
Beaufort Sea
Amundsen Gulf
SIBERIA
Chukchi Pen.
Brooks Range
ARCTIC CIRCLE
Bering Str.
Seward Pen.
ALASKA
Yukon
Mt. McKinley (Denali) 6194
Alaska Range
Great Bear Lake
Back
Thelon
Southampton Island
Foxe Basin
Melville Pen.
Davis Strait
Iceland
BERING SEA
Mt. Logan 5959
Gulf of Alaska
Great Slave Lake
Peace
Hudson Strait
Ungava Bay
Hudson Bay
Aleutian Islands
ALEUTIAN TRENCH
Alexander Archipelago
Queen Charlotte Is.
Churchill
Nelson
LABRADOR SEA
Gilbert Seamount
Vancouver Island
Lake Winnipeg
James Bay
Cape Farewell
IRMINGER BASIN
REYKJANES RIDGE
MAURY SEACHANNEL
ROCKALL RISE
ROCKALL TROUGH
NORTHWEST ATLANTIC MID-OCEAN CANYON
CHARLIE-GIBBS FRACTURE ZONE
FARADAY FRACTURE ZONE
Hartis Seamount 2432
TUFTS PLAIN
CASCADIA BASIN
Columbia
Missouri
NORTH
Shield
Source of the Mississippi
L. Superior
St. Lawrence
Gulf of St. Lawrence
Anticosti I.
Island of Newfoundland
Flemish Cap
Grand Banks of Newfoundland
NORTH
BISCAY
Cascade Range
Coast Mountains
ROCKY MOUNTAINS
Great Plains
L. Huron
L. Ontario
L. Erie
Lake Michigan
Nova Scotia
Bay of Fundy
Gulf of Maine
Cape Cod
Long Island
Laurentian Fan
Appalachian Mountains
Ohio
MENDOCINO FRACTURE ZONE
Cape Mendocino
Great Salt L.
Great Basin
Mt. Elbert 4399
Mt. Whitney 4418
Death Valley -86
PIONEER FRACTURE ZONE
NORTH PACIFIC
NORTHEAST PACIFIC BASIN
Ozark Plateau
Mt. Mitchell 2037
Kelvin Seamount
SOHM PLAIN
ATLANTIC
Cape Finisterre
Azores
Josephine Seamount
MURRAY FRACTURE ZONE
Moonless Mts.
Fieberling Tablemount
PATTON ESCARPMENT
Chesapeake Bay
Cape Hatteras
Bermuda Islands
Corner Seamounts
OCEANOGRAPHER FRACTURE ZONE
Madeira Is.
Midway Is.
NORTHWEST HAWAIIAN RIDGE
Musicians Seamounts
Mendelssohn Seamount
Hawaiian Islands
TROPIC OF CANCER
Guadalupe I.
Baja California
CEDROS TRENCH
Sierra Madre Occidental
Sierra Madre Oriental
Rio Grande
Mississippi
Coastal Plain
Mississippi Fan
BLAKE PLATEAU
HATTERAS PLAIN
BERMUDA RISE
NORTH AMERICAN BASIN
ATLANTIS FRACTURE ZONE
Great Meteor Tablemount
Canary Islands
MOLOKAI FRACTURE ZONE
Henderson Seamount
Gulf of California
Gulf of Mexico
MEXICO BASIN
Campeche Bank
Florida
Bahama Is.
OCEAN
CAPE VERDE PLAIN
NECKER RIDGE
Kaua'i
O'ahu
Maui
Hawai'i
Johnston Atoll
OCEAN
CLARION FRACTURE ZONE
Pico de Orizaba 5747
Yucatán Peninsula
CAYMAN TRENCH
Cuba
Jamaica
Hispaniola
Greater Antilles
Puerto Rico
PUERTO RICO TRENCH
Caribbean Sea
COLOMBIAN BASIN
AVES RIDGE
Lesser Antilles
Mathematicians Seamounts
MIDDLE AMERICA TRENCH
CENTRAL AMERICA
Krylov Seamount
Cape Verde Islands
Sénégal
CENTRAL PACIFIC BASIN
MAGELLAN RISE
Clipperton
Isthmus of Panama
Maracaibo
Orinoco
Trinidad
VEMA FRACTURE ZONE
GAMBIA PLAIN
Palmyra Atoll
CLIPPERTON FRACTURE ZONE
GUATEMALA BASIN
Cocos Island
COCOS RIDGE
Llanos
Angel Falls
Mt. Roraima 2772
Guiana Highlands
CONTINENTAL SHELF
DEMERARA PLAIN
DOLDRUMS FRACTURE ZONE
SIERRA LEONE RISE
SIERRA LEONE BASIN
Kiritimati (Christmas I.)
PANAMA BASIN
Pico da Neblina 3014
Amazon Fan
Howland I.
Baker I.
EQUATOR
Jarvis I.
COLON RIDGE
Carnegie Ridge
Galápagos Islands
Gulf of Guayaquil
Pariñas Point
Amazon
Amazon Basin
Marajó I.
CEARA PLAIN
ROMANCHE FRACTURE ZONE
Romanche Gap
CHAIN FRACTURE ZONE
Phoenix Islands
GALÁPAGOS FRACTURE ZONE
PERNAMBUCO PLAIN
Point Calcanhar
Point Coqueiros
ASCENSION FRACTURE ZONE
Ascension
Marquesas Islands
MARQUESAS FRACTURE ZONE
Nevado Huascarán 6768
La Montaña
SOUTH
Tokelau
Atafu
Swains I.
Nassau
Caroline I.
Flint I.
Stocks Seamount
Wallis
Samoa Islands
Upolu
Savai'i
Rose Atoll
Cook Islands
Tuamotu Archipelago
Pukapuka
Mataiva
Scilly
Society Islands
Tahiti
GALÁPAGOS RISE
PERU BASIN
PERU-CHILE TRENCH
Source of the Amazon
Lake Titicaca
Brazilian Highlands
BRAZIL BASIN
ST. HELENA FRACTURE ZONE
Hotspur Seamount
Martin Vaz Islands
MARTIN VAZ FRACTURE ZONE
Columbia Seamount
Fiji Islands
LAU BASIN
TONGA TRENCH
Tonga Islands
Niue
SOUTH PACIFIC
AMERICA
Salar de Uyuni
NAZCA RIDGE
Atacama Desert
Gran Chaco
Vitória
Cabo Frio
SOUTH
Rarotonga
Maria
Mangaia
Austral Islands
President Thiers Bank
Îles Gambier
Morane
Pitcairn
Isla Sala y Gómez
San Félix I.
San Ambrosio
CHILE BASIN
Iguaçu Falls
Santos Plateau
RIO GRANDE FRACTURE ZONE
RIO GRANDE RISE
TROPIC OF CAPRICORN
Rapa
Marotiri
Easter Island
Salinas Grandes
Cerro Aconcagua 6960
Paraná
Pampas
Juan Fernández Islands
ATLANTIC
Kermadec Islands
KERMADEC TRENCH
LOUISVILLE RIDGE
SOUTHWEST PACIFIC BASIN
OCEAN
CHALLENGER FRACTURE ZONE
CHILE RISE
CONTINENTAL SHELF
CONTINENTAL SLOPE
CONTINENTAL RISE
TRISTAN DA CUNHA FRACTURE ZONE
GOUGH FRACTURE ZONE
AGASSIZ FRACTURE ZONE
VALDIVIA FRACTURE ZONE
Isla Grande de Chiloé
Corcovado Gulf
Valdés Peninsula
ARGENTINE PLAIN
Gulf of San Jorge
EAST PACIFIC RISE
SOUTHEAST PACIFIC BASIN
Taitao Peninsula
Wellington I.
Patagonia
ANDES
FALKLAND ESCARPMENT
OCEAN
Chatham Is.
CHATHAM RISE
MENARD FRACTURE ZONE
Falkland Is.
FALKLAND PLATEAU
Strait of Magellan
Tierra del Fuego
Cape Horn
SOUTH GEORGIA RIDGE
South Georgia
ISLAS ORCADAS RISE
SOUTH SANDWICH TRENCH
HEEZEN FRACTURE ZONE
ELTANIN FRACTURE ZONE
THARP FRACTURE ZONE
UDINTSEV FRACTURE ZONE
HUMBOLDT PLAIN
Scotia Sea
South Sandwich Islands
Drake Passage
South Shetland Islands
South Orkney Islands
PACIFIC-ANTARCTIC RIDGE
BELLINGSHAUSEN PLAIN
De Gerlache Seamounts
AMUNDSEN PLAIN
ANTARCTIC CIRCLE
Adelaide Island
ANTARCTIC PENINSULA
WEDDELL SEA
Cape Norvegia
Scott Island
AMUNDSEN RIDGE
Marie Byrd Seamount
Peter I I.
Bellingshausen Sea
Alexander Island
Thurston I.
Siple Island
Carney I.
Amundsen Sea
Wrigley Gulf
Ellsworth Land
Sulzberger Bay
Mt. Sidley 4181
Marie Byrd Land
Vinson Massif 4897
Ronne Ice Shelf
Berkner Island
Filchner Ice Shelf
Ross Sea
Edward VII Pen.
Roosevelt I.
Ross Ice Shelf
Bentley Subglacial Trench -2550
Ice thickness 4347
Ellsworth Mts.
TRANSANTARCTIC MOUNTAINS
ANTA
Longitude West of Gre
180° 160° 140° 120° 100° 80° 60° 40°
90° 80° 70° 60° 50° 40° 30° 20° 10° 0° 10° 20° 30° 40° 50° 60° 70° 80° 90°

40° 60° 80° 100° 120° 140° 160° 180°
OCEAN
NANSEN BASIN
POLE PLAIN
MAKAROV BASIN
Franz Josef Land
Komsomolets Island
October Revolution Island
Bol'shevik Island
Cape Chelyuskin
North Land
CONTINENTAL SLOPE
CONTINENTAL SHELF
New Siberian Islands
LAPTEV SEA
EAST SIBERIAN SEA
Wrangel I.
KARA SEA
BARENTS SEA
Novaya Zemlya
Taymyr Peninsula
North Siberian Lowland
Chukchi Range
ARCTIC CIRCLE
Central Siberian Plateau
SIBERIA
Verkhoyansk Range
Cherskiy Range
Kolyma Range
Koryak Range
BERING SEA
ALEUTIAN BASIN
Aleutian Is.
ALEUTIAN TRENCH
SEA OF OKHOTSK
Kamchatka Peninsula
Sakhalin
Ural Mountains
West Siberian Plain
Lake Baikal
Stanovoy Range
Lake Onega
Lake Ladoga
Northern European Plain
Central Russian Upland
Pinsk Marshes
Carpathian Mts.
The Steppes
Kazakh Uplands
Altay Mountains
Mongolian Plateau
GOBI
Aral Sea
Caspian Depression
Black Sea
Crimea
Sea of Azov
Caucasus Mts.
El'brus 5642
Balkan Peninsula
ANATOLIA (ASIA MINOR)
Mt. Ararat 5137
Elburz Mts.
Tian Shan
Pik Pobedy 7439
Turpan Depression -154
Taklimakan Desert
Kunlun Mountains
Hindu Kush
Plateau of Tibet
HIMALAYA
Mt. Everest 8850 (29035 ft)
Qin Ling
Yellow Sea
SEA OF JAPAN (EAST SEA)
Hokkaido
Honshu
Kyushu
Shikoku
Korea
NORTH PACIFIC OCEAN
JAPAN TRENCH
Kuril Islands
KURIL TRENCH
EMPEROR TROUGH
HESS RISE
NORTHWEST PACIFIC BASIN
SHATSKIY RISE
Isakov Seamount
Makarov Seamount
Grosvenor Seamount
MAPMAKER SEAMOUNTS
TROPIC OF CANCER
EAST CHINA SEA
CONTINENTAL SHELF
Ryukyu Islands
RYUKYU TRENCH
Taiwan
PHILIPPINE SEA
PHILIPPINE BASIN
WEST MARIANA BASIN
MARIANA TROUGH
Mariana Islands
MARIANA TRENCH
Challenger Deep 10920
Guam
EAST MARIANA BASIN
MAGELLAN SEAMOUNTS
Wake I.
MID-PACIFIC MOUNTAINS
Marshall Islands
MICRONESIA
Caroline Islands
WEST CAROLINE BASIN
EAST CAROLINE BASIN
EQUATOR
Nauru
Gilbert Islands
MELANESIA
Mediterranean Sea
Crete
Cyprus
Syrian Desert
Dead Sea -416 (-1365 ft)
Mesopotamia
Zagros Mountains
Persian Gulf
Qattara Depression -133
Libyan Desert
Western Desert
Sinai
An Nafud
ARABIAN PENINSULA
Rub al Khali (Empty Quarter)
Red Sea
Nubian Desert
SAHARA
Tibesti Mts.
SAHEL
Lake Chad
Marra Mts.
Ethiopian Highlands
Gulf of Aden
Socotra
Somali Peninsula
Lake Turkana (L. Rudolf)
Lake Victoria
Kilimanjaro 5895
Sources of the Nile
Lake Tanganyika
Congo Basin
Lake Albert
Ruwenzori 5109
Cameroon Mt. 4100
Great Indian Desert
INDIA
Deccan Plateau
Western Ghats
Eastern Ghats
ARABIAN SEA
ARABIAN BASIN
BAY OF BENGAL
Ganges Fan
Andaman Islands
Nicobar Islands
Sri Lanka (Ceylon)
Indochina Peninsula
Gulf of Thailand
Hainan
SOUTH CHINA SEA
Luzon
Mt. Pinatubo 1486
Philippine Islands
PHILIPPINE TRENCH
PALAWAN TROUGH
SULU BASIN
Mindanao
Kinabalu 4101
CELEBES BASIN
Borneo
Celebes
Malay Peninsula
Strait of Malacca
Sumatra
INDONESIA
Greater Sunda Islands
Java Sea
Java
JAVA TRENCH
JAVA RIDGE
Lesser Sunda Islands
Timor
Timor Sea
Arafura Sea
New Guinea
Jaya Peak 4884
Bismarck Archipelago
New Britain
Solomon Islands
Bougainville
Guadalcanal
CEYLON PLAIN
COCOS BASIN
CARLSBERG RIDGE
OWEN FRACTURE ZONE
CHAGOS-LACCADIVE PLATEAU
Maldive Islands
Chagos Archipelago
Diego Garcia
CHAGOS TRENCH
MID-INDIAN RIDGE
MID-INDIAN BASIN
NINETYEAST RIDGE
INVESTIGATOR RIDGE
Christmas I.
Cocos Islands
SOMALI BASIN
Seychelles
Amirante Isles
MASCARENE PLATEAU
MASCARENE BASIN
Zanzibar I.
Aldabra Is.
Comoro Is.
Lake Malawi
Mozambique Channel
Madagascar
Mascarene Is.
Mauritius
Réunion
Rodrigues
Zambezi
Victoria Falls
Lake Kariba
KALAHARI DESERT
Namib Desert
Thabana Ntlenyana 3482
Drakensberg
Great Karroo
Cape Agulhas
AGULHAS PLATEAU
AGULHAS BASIN
MOZAMBIQUE PLATEAU
NATAL BASIN
MADAGASCAR BASIN
MADAGASCAR PLATEAU
INDIAN OCEAN
SOUTHWEST INDIAN RIDGE
CROZET BASIN
Crozet Islands
Prince Edward Islands
PRINCE EDWARD FRACTURE ZONE
Amsterdam
St. Paul
Kerguelen Islands
KERGUELEN PLATEAU
Heard Island
Ob Tablemount
Lena Tablemount
ENDERBY PLAIN
SOUTHEAST INDIAN RIDGE
SOUTH INDIAN BASIN
OSBORN PLATEAU
WHARTON BASIN
Wallaby Plateau
EXMOUTH PLATEAU
NORTH AUSTRALIAN BASIN
Cape Inscription
BROKEN RIDGE
EAST INDIAMAN RIDGE
PERTH BASIN
Cape Naturaliste
DIAMANTINA FRACTURE ZONE
CONTINENTAL SLOPE
Great Australian Bight
SOUTH AUSTRALIAN BASIN
AUSTRALIA
Western Plateau
Kimberley Plateau
Great Sandy Desert
Macdonnell Ranges
Great Victoria Desert
Nullarbor Plain
Lake Eyre -16
Great Artesian Basin
Central Lowlands
Great Dividing Range
Cape York Pen.
Gulf of Carpentaria
Great Barrier Reef
Darling
Murray
Mt. Kosciuszko 2228
Bass Strait
Tasmania
CORAL SEA BASIN
CORAL SEA
New Caledonia
NEW HEBRIDES TRENCH
Vanuatu
NORTH FIJI BASIN
Fiji Islands
TROPIC OF CAPRICORN
SOUTH PACIFIC OCEAN
SOUTH FIJI BASIN
Lord Howe
LORD HOWE RISE
NEW CALEDONIA BASIN
TASMAN PLAIN
TASMAN SEA
NEW ZEALAND
North Island
South Island
Aoraki/Mt. Cook 3754
EAST TASMAN PLATEAU
SOUTH TASMAN RISE
Stewart Island
CAMPBELL PLATEAU
MACQUARIE RIDGE
Macquarie I.
Cosmonaut Sea
Riiser-Larsen Peninsula
Cape Ann
Enderby Land
Prydz Bay
Cape Poinsett
South Magnetic Pole 2007
ANTARCTIC CIRCLE
Balleny Is.
Ice thickness 4776
TRANSANTARCTIC MTS.
Victoria Land
Ross Sea
Ross Ice Shelf
ANTARCTICA
Winkel Tripel Projection, Central Meridian 0°
SCALE 1:101,862,000
1 CENTIMETER = 1,019 KILOMETERS; 1 INCH = 1,610 MILES AT THE EQUATOR
0 500 1000 1500 2000 2500
KILOMETERS
0 500 1000 1500 2000 2500
STATUTE MILES
East of Greenwich

EARTH

WATER

The name of our planet may not be the most appropriate since only 29 percent of the Earth's surface consists of land, while 71 percent is covered by water. Composed of two hydrogen atoms and one oxygen atom, water is one of the most important resources on Earth and is essential for living organisms. Ocean saltwater accounts for about 97 percent of the estimated total water quantity of 34 million cubic miles (1.45 billion km^3). The oceans are the origin of all life as well as the largest biosphere on Earth. Most freshwater—the only kind of water that is drinkable—is trapped inside polar ice caps and high mountain glaciers, while some freshwater collects in standing or running water bodies. Water is continuously in motion and it circulates the planet in the form of enormous ocean currents.

the world's oceans

Globally there are 332 million cubic miles (1.384 billion km³) of water reserves, of which 97 percent makes up the world's oceans. Water is distributed more heavily toward the Earth's Southern Hemisphere.

world's oceans | physical characteristics

OCEANS

More than two-thirds of the Earth's surface is covered in water. The ocean is the largest and oldest habitat on our planet and plays a central role in shaping the climate. It provides energy, natural resources, and one of the major food sources for humans. Nevertheless, large parts of the oceans still remain undiscovered.

The interconnected world's oceans are the Atlantic Ocean, the Indian Ocean, and the Pacific Ocean. The oceans of the Northern Hemisphere are divided by continents and offshore islands. The Southern Hemisphere lacks such natural barriers. In order to geographically structure the ocean of the Southern Hemisphere, meridians running through the southern tips of the three large landmasses were defined as boundaries. Hence, the meridian running through Cape Agulhas at 20° east longitude separates the Atlantic Ocean from the Indian Ocean. The meridian running through the South East Cape of Tasmania at 147° eastern longitude is the boundary between the Indian Ocean and the Pacific Ocean. The southern boundary between the Pacific and the Atlantic Oceans passes through Cape Horn at 68° west longitude to the Bering Strait, which defines the northern boundary. Bering Strait is the channel between the Russian Cape Dezhnev, the easternmost point of the Asian continent, and Cape Prince of Wales, Alaska, which is the westernmost point of North America.

Depth affects the amount of light that reaches down into the ocean. Different organisms live at these various depths.

The Pacific Ocean is not only the oldest, but also the largest of the three main oceans and the one with the most islands. Its average depth is 12,926 feet (3,940 m). The greatest depth recorded to date is in the Mariana Trench at 36,200 feet (11,034 m) deep, whereas the deepest point recorded in the Atlantic Ocean is the Milwaukee Deep in the Puerto Rico Trench at 30,246 feet (9,219 m).

With 29 million square miles (75 million km²), the Indian Ocean is the smallest of the three major oceans. The Sunda Trench is the deepest point of the Indian Ocean, and reaches 24,458 feet (7,455 m).

Oceans cover about 60 percent of the Northern Hemisphere, while in the Southern Hemisphere they make up more than 80 percent of the Earth's surface.

Adjacent and semienclosed seas

Several adjacent seas are separate from the oceans and are located adjacent to land, hence their name. Examples include the North Sea, the Bering Sea, the Gulf of Saint Lawrence, the Irish Sea, the Gulf of California, the Sea of Japan, and the East China Sea. The so-called semienclosed seas are in fact almost entirely enclosed by landmasses, leaving only a narrow connection with the oceans. The European Mediterranean, the American Mediterranean Sea, the Baltic Sea, Hudson Bay, the Red Sea, and the Persian Gulf are all semienclosed seas. The Arctic Ocean is an adjacent sea of the Atlantic Ocean. The ocean surrounding the continent Antarctica is called the Southern Ocean, or the Antarctic Ocean. From the Antarctic, it expands to 55° south latitude.

basics

SHELF SEAS The continents are surrounded by a zone of shallow waters. These vast areas rarely reach a depth greater than 656 feet (200 m) and are known as shelf seas. The name refers to the continental shelves which are those parts of the continental plates covered by ocean water. Continental shelves vary in size from a few miles to more than 620 miles (1,000 km)—for example, in the Bering Sea.

see also: Primeval continents, p. 67

physical characteristics

Seawater may appear to the naked eye as if it were a uniform substance. However, oceans are complex environments with varying salt and nutrient contents, as well as different pressures, temperatures, and light conditions.

Different substances are continuously washed into the oceans from rivers, melting snow, precipitation, and wind. Almost all chemical elements can be found in seawater. Most of the dissolved substances are salts, mainly sodium chloride, which is more commonly known as table salt. Due to freshwater from rivers flowing into the ocean, estuaries contain fewer salts than offshore areas.

The giant Pacific octopus lives in the deep ocean, which has low temperatures, high pressure, and no light.

Ocean salt

The average salt content, or salinity, of the world's oceans is 3.5 percent, which is 0.6 ounce of salt per pint of water. When seawater evaporates, the salts dissolved into the water will remain in the ocean. The hotter and dryer the climate is, the greater the rate of evaporation and the greater the salinity of the water. This is especially the case in adjacent seas, which are less mixed due to their limited contact with the ocean. The Persian Gulf has a salinity of four percent. The polar oceans have a slightly lower salinity of 3.1 to 3.5 percent, which is due to precipitation and melting ice. The salt content of the deep sea, at a depth of more than 3,280 feet (1,000 m), is overall consistently between 3.45 and 3.5 percent.

Temperature, light, and pressure

Surface temperatures range between 28.4°F (-2°C) in the polar oceans and 86°F (30°C) in the tropics. Adjacent seas and coastal areas may reach temperatures of up to 104°F (40°C). Regional and seasonal variations only affect the temperature in the water's upper layers. The temperature in the tropics and temperate regions begins to rapidly decrease at a depth of around 656 to 3,280 feet (200 to 1,000 m). This layer of water is called a thermocline. From 3,280 feet (1,000 m) down into the deep sea, temperatures are constant, between 32°F (0°C) and 41°F (5°C). The lower limit of the light zone is at a depth of between 328 and 656 feet (100 and 200 m) depending on water conditions; in cloudy, coastal areas, light may only reach a depth of 33 feet (10 m). In the open sea, sunlight may reach a depth of up to 3,280 feet (1,000 m). Due to the increase in pressure by 1.45 pounds per square inch (10 kPa) per foot depth, there is a pressure of about 1.450 psi (10,000 kPa) in the open sea. The speed of sound in water is 4,921 feet per second (1,500 m/sec) depending on pressure, temperature, and salt content. Therefore, sound travels at four times the speed it travels through air.

issues to solve

MARINE ANIMALS AT RISK Several marine animals—including sharks, whales, and dolphins—rely on sound for orientation, as visibility is limited. However, they are severely affected by noise from engines, sonar, and military and industrial activities. These animals lose their orientation and risk getting too close to shore where they can get stranded and die.

practice

DESALINATION

The process of desalination to purify water has been copied from nature: drinking water is produced in desalination factories where salt is extracted from the seawater. Various processes are used to simulate natural separation methods. However, seawater desalination is much more expensive than accessing conventional sources of freshwater. This is mainly due to the high levels of energy used during salt extraction. Therefore desalination factories are only used where freshwater supplies are insufficient.

Salt piled into distinctive pyramids to encourage the evaporation of moisture.

ocean currents

Ocean currents, dependent upon the winds and varying salt concentrations, drive huge masses of water over long distances.

Numerous surface and deepwater currents act as giant conveyer belts, circulating the water in the world's oceans. The Earth's ocean waters can complete a turnover in a few hundred years, though it can take up to 2,000 years.

Influence of the winds

Prevailing winds drive the ocean's surface currents, circulating warm water away from the Equator and circulating cold water toward the Equator by the currents created by the circulating trade winds. Coastal desert areas are the result of cold ocean currents near the shore. Because cold ocean currents result in cold air masses directly above them, moist air does not rise, so clouds and rain do not form, creating a desert environment. Warm ocean currents also warm the air over the nearby land. Without the warm ocean currents, many areas of the Earth would have significantly lower average temperatures. One example of this system of currents is the Gulf Stream (see in focus), which forms in the Caribbean and directs warm water across the North Atlantic toward Europe, creating a relatively mild climate.

basics

OCEAN CURRENTS are measured in Sverdrup with the symbol Sv, where 1 Sv is equivalent to a volume flow rate of 10^6 cubic meters (over 264 million gallons) per second.

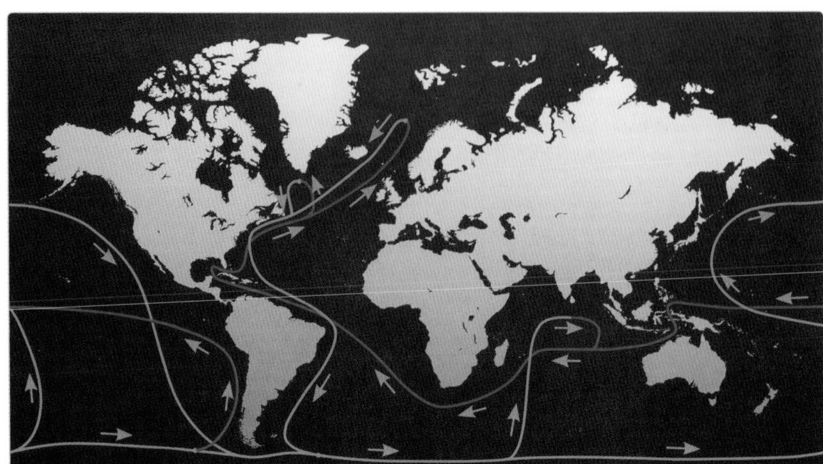

Deep ocean currents around the world: A knowledge of currents is vital to the shipping industry, as it can help to save on fuel costs.

Deep water currents

The ocean is a complex system of warm and cold currents, both on the surface and at great depths. While surface currents are driven by winds, deep currents are driven by density and temperature gradients. In the Atlantic, the Gulf Stream flows from the Caribbean across the North Atlantic toward Europe. Between Greenland and Norway, the Gulf Stream waters are cooled by the frigid winds from the North Pole, become denser, and sink deeper into the ocean until they reach the bottom. They continue to flow along the ocean floor toward the southern end of the Atlantic. The current is then channeled through the Indian Ocean to the South Pacific, gradually warming along the way and rising to the surface off the coast of South Africa, where it is picked up by a circulating current and routed back to the Caribbean to start its journey again.

in focus

THE GULF STREAM

The Gulf Stream affects more than just water temperature. When the Gulf Stream flows into the open Atlantic off the coast of North America it splits into smaller circulating streams of warm water called eddies. The warm water eddies mix with the colder surrounding water to produce the lukewarm water found off the coasts of western and northern Europe. This results in the region's unusually mild climate, as compared to other countries at similar latitudes, such as Canada—allowing palm trees to grow in Ireland, and causing fjords of the Norwegian coast to remain free of ice all year.

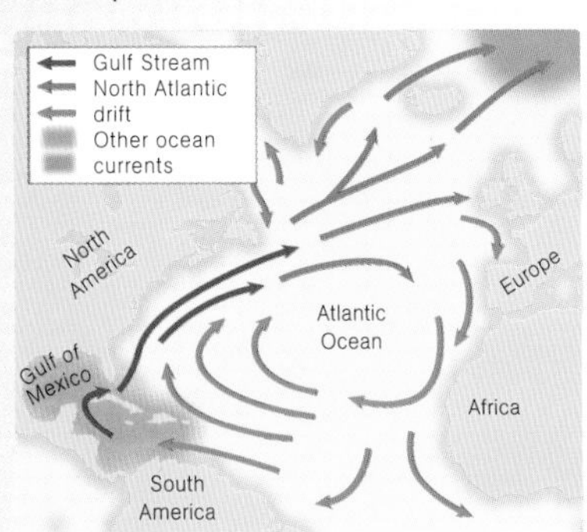

Some suggest that climate change may interrupt the Gulf Stream. The consequences for Europe would be disastrous.

The Saltstraumen sound in Norway has the strongest tidal current on Earth. Water speeds in the narrow strait reach up to 20 knots, and maelstroms and whirlpools often form.

tides, coasts, and waves

The continental coastline of the Earth is about 250,000 miles (440,000 km) long. But the area of transition between the ocean and land is constantly changing due to the tides and surge.

Coastal landscapes are shaped over time by the destructive forces of the seas that slowly but steadily eat away at rocky shores. These forces, in combination with weathering and surface-level erosion, cause shorelines to retreat further inland.

Waves and currents move large amounts of sand along shallow and sandy shorelines, thereby creating sandbanks that constantly change position. In other areas, rivers carry sediment from the mainland into the sea, where it is deposited in sediment plumes by the mouth of the river.

Coral reefs can also expand from coastal areas into the ocean and thus transform the shoreline. The fluctuations of the sea level, or rather the rise and lowering of the mainland, are of particular importance. The so-called isostatic adjustment of areas that were once covered by large ice sheets is still occurring.

Due to the persistent rise in the sea level, beaches and cliffs continue to be shifted inland while in other regions landmasses are sinking over time. The ocean penetrates valleys creating irregular coasts, or floods glacial troughs in the instance of fjord coastlines. Most coastlines are still very young due to the sea-level rise of approximately 51 inches (130 cm) over the past 18,000 years.

in focus

WAVES

Most waves are created by wind. Waves often travel across the open sea in the form of small swells until they break on the shore. Most waves reach a height of about ten feet (3 m); in exceptional cases, they can be over 98 feet (30 m) high. Waves are measured from the top, or crest, of the wave to the base of the trough. The length is the distance between wave crests.

Waves like these are surface waves that occur in the upper layer of the ocean and are caused by the wind.

The power of the moon

The gravitational pull of the moon together with the centrifugal force of the Earth create a tidal bulge on the ocean surface facing the moon. The same occurrence happens on the opposite side of the Earth, except that there, the Earth's centrifugal force is the only cause of bulging water.

A depression forms between the two tidal bulges and due to the Earth's rotation, the sea levels rise and fall twice a day as described. The ocean's retreat to its lowest point is called the ebb, and the period of rise in sea level is called a flood. The tides' amplitudes vary from region to region. For example, the narrow bays on the east coast of Canada act together as a funnel, so that the tidal range between high and low tides in the Bay of Fundy is 49 feet (15 m). By contrast, the tidal range along the North Sea coast only reaches a maximum of 11.5 feet (3.5 m).

Tidal pools are formed as a high tide comes in over a rocky shore. Water fills depressions, which turn into pools as the tide retreats.

Humans have always preferred living in coastal areas. About half of the world population is currently living no further than 60 miles (95 km) away from the ocean.

see also: The Earth-moon system, p. 40

oceanic crust

Oceanic crust covers two-thirds of the Earth's surface. Crustal rocks are relatively recent, with none older than 200 million years, and they are constantly in motion. This movement expands the oceans and shifts entire continents.

Section through the Mariana trench showing how it was formed by the movement of the Earth's crust.

The oceanic crust begins beyond the flooded part of the continental shelf. It originates at the edges of the large continental plates in the area of the mid-oceanic ridges, large mountain systems that are at a maximum 930 miles (1,500 km) wide and more than 37,000 miles (60,000 km) long traversing all the oceans. The mountain ridges rise up to almost 10,000 feet (3,000 m) above the ocean floor and occasionally reach the water surface as volcanic islands. Basaltic lava from the Earth's mantle constantly rises through central ridges 12.5 to 30 miles (20 to 50 km) wide. New crust material forms from the lava, and the seafloor expands several centimeters a year due to lateral pressure and convection currents of the Earth's mantle. This process is referred to as seafloor spreading.

Oceanic trenches located close to the continental edges are the counterpart to the mid-oceanic ridges. In these areas, old oceanic crust moves under the continental crust where it is literally swallowed (subduction). The drag of the sinking material causes trenches to develop on the seafloor. These trenches are the deepest places on the Earth's surface. The Mariana trench in the Pacific Ocean has a depth of 36,200 feet (11,034 m) and is the deepest point on Earth.

Smoking crater on Surtsey: This young island off the Icelandic coast was formed from 1963 until 1967 by a large volcanic eruption.

FORMATION OF NEW OCEANS

in focus

The oceans are continuously changing due to the formation of new oceanic crust. The Mediterranean is predicted to shrink, while the Atlantic is growing. The Red Sea will become a new ocean, dividing the African continent. The Afar Depression, also known as the Afar Triangle, is where the East African Rift meets the ridges of the Red Sea and Gulf of Aden. It is also where the African and Arabic tectonic plates drift apart at about 0.4 inch (one cm) per year.

The Afar Triangle: The meeting place of three tectonic plates

Seafloor

Most parts of the ocean floor are on average 12,234 feet (3,729 m) below sea level. They are covered by a continuously growing layer of sediment, usually a few hundred yards thick and in some places even several miles. Part of the sediment comes from the mainland, distributed by rivers, wind, or glaciers. Suspended sediments and volcanic ash are the main components of the deep-sea red clay, which covers more than a quarter of the seafloor. The majority of ocean deposits, however, are decomposed sea organisms. For example, calcareous globigerina ooze is composed of shells from unicellular planktic foraminifers, while siliceous diatom ooze consists of cell walls from dead diatoms. Radiolarian ooze is made up of the skeletons of dead radiolarians.

see also: Plate boundaries, p. 65

islands and atolls

Earth's largest island is Greenland, with an area of over 830,000 square miles. The smallest islands are just dots in the ocean. Many islands are part of a continent; others owe their existence to volcanic eruptions.

Islands are categorized into continental and oceanic islands. Continental or shelf islands were once connected to the mainland. Rising sea levels or subsiding continents have caused former mainland areas to become flooded. This explains, for instance, the island location of Great Britain or Madagascar. The long and low-humped islands off the Swedish coastline are peaks of flooded mainland areas. Dune islands are formed by wind and oceanic currents.

Oceanic islands have never been part of other landmasses. They are usually of volcanic origin. Where tectonic plates are pulling apart and then drifting away from each other, mid-oceanic ridges develop that sometimes protrude above sea level as islands, such as Iceland or the Azores. Volcanic island arches, such as the Aleutians, are formed when one oceanic plate is pushed underneath another and then continues to subside (subduction).

Islands are also formed as a result of hot spots, where magma rises from the Earth's lower mantle. When the oceanic crust breaks apart, the volcanic cone that has been created is moved and along with it the island protruding above sea level. A new volcanic cone is formed above the stationary hot spot, leading to the birth of a new island. This way, long island chains can develop, with volcanoes still active at the geologically youngest end of the chain. The Hawaiian Islands have been formed this way. In tropical waters, reef-building corals often build their colonies along rocky island coastlines in the surf zone. When the seafloor subsides due to subduction of the respective crust plate, the continuously growing reef moves further away from the coast. When the subsiding seafloor eventually submerges the entire island below sea level, the former reef continues to grow as a ring-shaped wall and then appears as an atoll. If the atoll sinks faster than the corals grow, the coral polyps end up at a depth where they can no longer survive due to the lack of sufficient light. The reef dies and the atoll disappears.

basics

RTIFICIAL ISLANDS

lore and more overrowded areas are elocating their infratructure onto water. Whether in the form f an airport in saka, a city section n Amsterdam or an xclusive tourist sland in Dubai, the end toward artificial islands seems nstoppable.

in focus

DROWNING ISLANDS

The rise of the sea level due to climate change endangers the existence of numerous Pacific islands. Based on estimates, atolls protruding only a few feet above sea level, such as Kiribati, Tuvalu, or the Marshall islands are threatened with becoming totally submerged within the next few decades. Continuous melting of ice masses would also have fatal consequences for the Maldives in the Indian Ocean. Moreover, numerous Caribbean islands would be affected by the concurrent occurrence of cyclones.

A flooded hallig (islet) among the North Frisian Islands in Germany.

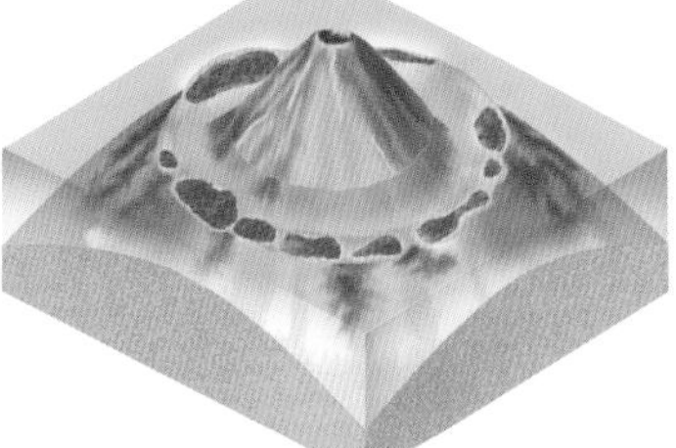

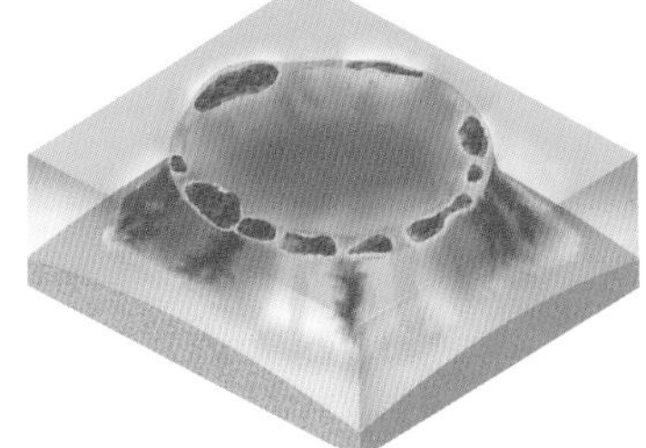

Atolls form in three stages: Corals grow around a volcanic island as a fringe reef, the volcano then subsides through erosion or subduction while the coral grows upwards. The atoll is then left with its distinctive shape, fully or partially encircling a lagoon.

The Rock Islands, a chain of over 70 small islets in the island nation of Palau in Micronesia.

see also: The climate system, pp. 118–123

coral reefs: underwater forests

Rich in life and incredibly diverse, coral reefs are the rain forests of the oceans. Coral colonies provide shelter for a myriad of fish and invertebrate species, whose diverse colors speckle the clear tropical water.

A fringe reef connected to the mountainous island of Rarotonga: Popular with tourists, the island attracts many visitors for snorkeling and sailing holidays.

Tropical coral reefs are mainly formed by corals of the genus *Madrepora*. Coral polyps take calcium and carbon dioxide from seawater and create a calcium carbonate outer skeleton. The skeletons of dead colonies are reoccupied by new generations of coral polyps. Together with the excretions of crust-forming calcareous algae, these new polyps form the skeleton of the reef body, which can result in vast coral reefs. The Great Barrier Reef off the northeastern coast of Australia stretches over 8,500 feet (2,600 m) of coastline.

Coral polyps share a symbiotic existence with the zooxanthellae algae embedded in their skin. Through photosynthesis, the algae turn carbon dioxide and water into oxygen and sugar that is used by the polyps, which in return provide nutrition and protection for the algae. The carbon dioxide that is also taken in by the algae triggers the corals to produce more calcite. Certain madrepores occur outside the tropics up to a depth of about 20,000 feet (over 6,000 m). Instead of taking advantage of a symbiosis with zooxanthellae, these corals get their nutrients directly from the water around them.

Climate change

Coral reefs are found between the Tropic of Cancer and the Tropic of Capricorn. They occur in areas where water temperatures are on average 74.3°F (23.5°C) and never below 68°F (20°C). Reef-building madrepores cannot cope with large temperature fluctuations, nor with water that is too cold or too warm as it causes them to die or degenerate. Therefore, the warming of oceans due to climate change is a serious threat to tropical coral reefs, resulting in death or "coral bleaching." Healthy reefs act as breakwaters for coastal protection; if the reefs die, many of the flat islands in tropical regions will be exposed to flooding. The sea level rise due to melting ice caps will result in the loss of several islands, whose inhabitants are already noticing the consequences of frequent flooding.

issues to solve

RESEARCHERS be lieve that aroun three-quarters of a reefs have been de stroyed or are at ris of being destroye by changing enviro mental condition tourism, motorboat over-fishing, an sewage dumping.

in focus

GREAT BARRIER REEF

The Great Barrier Reef off the northeastern coast of Australia is the largest coral reef on Earth, as well as the largest construction made by living organisms: a labyrinth of an estimated 2,500 individual reefs, lagoons, and 600 islands. It is a unique habitat for over 4,000 species of sea snails, mollusks, crustaceans, sponges, sea stars, and sea urchins, as well as about 1,400 species of fish. As a paradise popular with sailors and divers, a major part of the reef is now a national park and UNESCO World Heritage Site, protecting it for the future.

Diving in the colorful waters of the Great Barrier Reef.

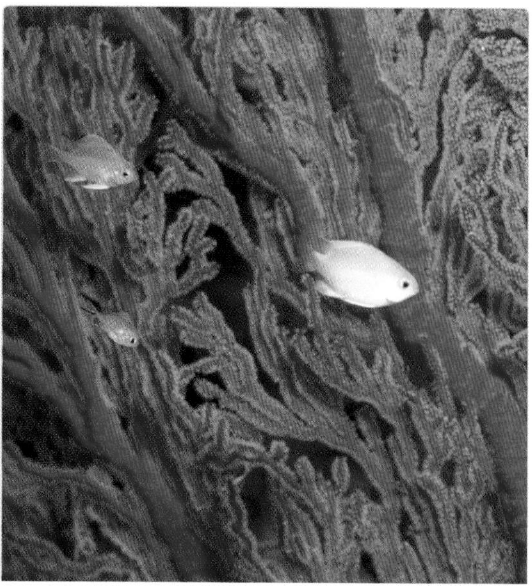

Coral reefs are the habitats of a great variety of tropical species of sea animals like clown fish, groupers, and blacktip reef sharks.

deep sea

Today, the deep sea is one of the least researched parts of our planet. Only during the last few decades have researchers had access to the technical means to reach the darkness of the deep oceans.

Biologically, the deep sea begins at a depth of about 650 feet (200 m) where the water is no longer subject to surface wind and heat variations and where photosynthesis is impossible due to the lack of sufficient light. The completely dark zone, however, only begins at a depth of about 3,280 feet (1,000 m). Humans cannot survive at such depths as the pressure is around 100 times that at sea level. The water temperature at this depth is constantly below 39.2°F (4°C).

NATURAL RESOURCES OF THE DEEP SEA

While the extraction of mineral ore such as gold, platinum, tin, and titanium in the deep seas may still be an exceptional endeavor, recent efforts are being put into exploring the potential of manganese nodules. In the Pacific Ocean, a large enough source of manganese, iron, copper, nickel, and cobalt exists to satisfy the demands of industry for this century. Another very promising source of energy is frozen gas occurring in the form of methane hydrate; however, extraction of these energy sources is not only extremely challenging, but may also pose substantial threats to the global climate.

A sea cucumber wanders over manganese nodules at a depth of 16,404 feet (5,000 m).

in focus

Life in the dark

Organisms living in the deep sea have adapted to the high pressure by storing water in their bodies, some consisting almost entirely of water. Most deep-sea organisms lack gas bladders. They are cold-blooded organisms that adjust their body temperature to their environment, allowing them to survive in the cold water while maintaining a low metabolism. Many species lower their metabolism so much that they are able to survive without food for long periods of time, as finding the sparse food that is available expends a lot of energy. Many predatory fish of the deep sea are equipped with enormous mouths and sharp teeth, allowing them to hold on to prey and overpower it. Some predators hunting in the residual light zone of the ocean have excellent visual capabilities, while others are able to create their own light (bioluminescence) to attract prey or a mating partner. Habitats near underwater hot springs require very special adaptations for organisms to survive. Many invertebrate species have mastered these unique conditions and survive on sulfur bacteria. Black smokers, a type of hydrothermal vent, were discovered in 1977 near the Galapagos Islands by the manned submersible *Alvin*.

The anglerfish, so called because of its distinctive method of catching prey.

Trips to the unknown

The first trip into the deep sea was made by the Americans William Beebe and Otis Bartin in 1934. They used a diving chamber called a bathysphere that was lowered to a depth of about 2,950 feet (900 m). The American Don Walsh and the Swiss Jacques Piccard set a world record in 1960 with the *Trieste*; they managed to reach the bottom of the Mariana Trench at a depth of 35,813 feet (10,916 m). Nowadays, there are many unmanned submersibles that explore the deep sea, forwarding data to surface research ships.

Unique marine communities survive near hydrothermal vents even in the deep sea. Because there are no plants so far from the sunlight, these tiny crabs and other organisms use chemosynthetic bacteria as their primary energy source.

catchment areas and stream courses

All types of water courses on Earth serve as natural drainage systems for surrounding land areas. They absorb excess surface water and transport it to deeper-lying areas by gravitational force.

The area drained by a river is its catchment area (basin). It is separated from the catchment area of another river system by watersheds, typically mountain ranges. River density and stream flow are dependent upon topography, climate, and vegetation cover. Main rivers that discharge into a lake or an ocean differ from tributaries that flow into main rivers. In the direction of its flow, a river is divided into upper reaches, middle reaches, and lower reaches. At a length of nearly 4,163 miles (6,700 km), the Nile River in Africa is the longest river on Earth. The South American Amazon River, although slightly shorter at 3,978 miles (6,400 km), and with its approximately 15,000 tributaries, carries the largest amount of freshwater. The Yangtze River is Asia's longer river, spreading across 3,915 miles (6,300 km).

A river can spring from a standing body of water or form from a glacial melt. Most rivers have their source in mountains, where cool, oxygenated water emerges from a spring. Through water influx and precipitation an initial rivulet gradually grows into a rapid-flowing mountain creek running straight through a mountain range with a steep gradient. Along sharply inclined stretches rapids develop; on rocky banks, which do not erode readily, water gushes down as a waterfall. At the edge of a mountain range, where gradient and current velocity abruptly decrease, a mountain creek deposits most of its carried sediments as a scree slope. Further downstream a river accumulates more water from tributaries and then becomes a calm yet large flowing stream. Under normal conditions when a narrow mountain valley broadens a river becomes wider and meanders in sweeping bends. The sediment load of the river is reduced and dissolved as sediment is carried to the ocean in larger amounts during periods of high water flow.

A swiftly flowing mountain river: Most rivers have their sources in high mountain areas.

catchment areas | streams | high and low courses

RIVERS

Rivers are the lifelines of landscapes. Prerequisite for river formation is an excess of precipitation in relation to evaporation and seepage, as well as a certain topographic gradient through the area. Although only a small part of the Earth's entire water volume circulates via rivers, all water courses—from rapid mountain stream to leisurely flowing river—contribute significantly to the structure of the Earth's surface. Additionally, they provide the source for drinking water, transportation routes, and energy.

The Amazon River is the largest river in the world.

PERMANENT (PERENNIAL) RIVERS continuously carry water at all times.

PERIODIC RIVERS are dependent upon seasonal changes between rainy and dry seasons.

EPISODIC RIVERS carry water only occasionally, such as after heavy precipitation.

ALLOCHTHONOUS RIVERS (i.e., not formed locally) originate in humid areas and flow through arid regions, thereby losing significant amounts of water.

basics

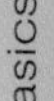

Asia's longest river is the Yangtze.

high and low courses

Nearly a third of the approximately 24,000 cubic miles (100,000 km³) of yearly precipitation flows into the oceans via creeks and rivers. During this process, rivers shape the environment through erosion and sedimentation.

A typical meander in the Saar River in Germany, known locally as the *Saarschleife* ("bend in the Saar").

The highest flow rate and strongest force of erosion occurs in the early or youthful rivers and depends on the water level, the width of the river, and the structure of the stream bed. Due to vertical erosion, a notch-shaped river valley or vale can be formed. A valley that is carved deeply into the surrounding area like a gorge but with steep slopes is called a canyon. The Grand Canyon in Colorado is the largest canyon worldwide where sudden drops in the terrain of the river bed create narrow fast-flowing streams and roaring waterfalls.

Fast moving water or rapids on the Congo River.

Meanders and the mouth of a river

As soon as the terrain flattens, the flow rate and force of erosion decreases significantly and a river is able to bypass larger obstacles. This is called a middle-aged river. As the current is stronger toward the outside of a bend, a middle-aged river begins to undercut the stream bank and widen its bed, creating an almost vertical bank. Due to the low flow rate on the inside of a bend, gravel and sand is continually deposited here. This creates a point bar with a shallow slope. Over time, meanders are formed. Named after a river with many bends in Turkey, meanders move downstream in great curves. When there is a large flow of water, a river may occasionally break through the neck of a loop, leaving behind an oxbow lake.

Where slow-flowing or older rivers drain into the ocean, large sediment plumes will often push into the ocean and form a delta. Large deltas such as the Nile Delta or the Mississippi Delta can only be generated near calm and shallow seas. Rivers draining into the sea at a coast with a large tidal range form estuaries, allowing tidal waves to reach far into the interior of the land.

High water

When heavy rain falls or rapidly melting snow surpass the capacity of a river, the river overflows and lower parts of the river valley get flooded. Human intervention has increased the risk of flooding through the destruction of vegetation, clear-cutting of forests, and the over-regulation and straightening of river courses.

The Okavango Delta has no access to the sea; instead it irrigates the desert, providing a vital water source to this arid area.

in focus

RIVERS MADE BY HUMANS

One of the most impressive inland waterways is the Saint Lawrence Seaway, completed during the 1950s in North America. Several locks allow for continuous shipping along a system 2,500 miles (4,023 km) long, from the Great Lakes in the interior all the way to the Atlantic Ocean. The Kiel Canal (in Germany Nord-Ostsee-Kanal) was completed in 1895 and is one of the world's most frequented artificial waterways.

A busy lock on the Panama Canal in Central America.

see also: Global warming, p. 126

origins | circulation

LAKES

Lakes are not simply inland water-filled depressions. They are complex ecosystems with different origins, as varied as the forces of nature. Although the water they contain covers only a small portion of the landmass, lakes can be of great importance to the local environment and climate.

origins

Lakes cover barely 2 percent of the Earth's land surface. They vary greatly in size and distribution. Together, they store a quarter of all the freshwater on Earth.

A lake can be formed nearly anywhere on Earth, provided there is an adequate surface hollow and sufficient water. Some lakes are also referred to as a sea, such as the Caspian Sea and the Dead Sea; however, neither of these is as old or as deep as Lake Baikal. The largest continuous freshwater area on Earth is made up by the Great Lakes in North America.

Young and old lakes

Most of the recently formed lakes are remnants from the last ice age, when mighty ice shields and glaciers covered vast parts of the Northern Hemisphere. As the ice masses withdrew, they left valleys, hollows, and depressions behind, where meltwater accumulated. In lowlands, debris blocked off the water and formed giant lake areas.

Lakes of older origin were formed by tectonic changes along the Earth's crust, when deep depressions, cracks, and rift valleys filled up with water. Examples are the Tanganyika and Malawi Lakes in the East African Rift Valley, the Dead Sea, and Lake Baikal. Other lakes are formed by water accumulating in craters of extinct volcanoes. Cave collapses in karst regions can give rise to doline (sinkhole) lakes.

Habitats

Lakes can be divided up into various habitats. The littoral zone is the sunlight flooded shore region, the profundal zone receives limited light, the pelagial zone is the open water area, and the benthos is the bottom of the lake. Due to the relationship between water temperature and density, lakes form layers called thermoclines. There is a constant temperature of 39°F (4°C) in the lower region (hypolimnion) of a lake, as this is the temperature at which water is densest. The surface water (epilimnion) has a larger temperature spectrum, and it can even freeze over in winter. Life continues under the ice until the spring weather warms up the surface once again.

in focus

ARTIFICIAL LAKES

Some artificial lakes are the result of the removal of raw materials such as gravel or sand. Reservoirs exist primarily along the upper reaches of rivers. They are used for drinking and commercial water storage, flood mitigation, and hydropower. The largest reservoir on Earth is Volta Lake in Ghana. One of the largest storage dams is the Three Gorges Dam in China.

Lake Kariba in Africa holds more than 6.3 billion cubic feet of water.

Crater lakes form in the calderas of extinct volcanoes. Incoming precipitation must exceed evaporation.

Lake Baikal in Siberia is the deepest on Earth. Known as the "blue eye of Siberia," it has the greatest freshwater volume of any lake in the world.

see also: Water at risk, p. 131

circulation

Geologically speaking, most lakes are both young and short-lived. Human activity also contributes to the fact that the sensitive ecological balance of many lakes is severely threatened.

Algae buildup can suffocate and kill a lake.

A lake is a water-filled hollow, which is enclosed by land and does not have a direct connection to the sea. The water may originate from precipitation, or aboveground or subterranean inflow; it is lost again through evaporation. Spring lakes are those with no surface inflow; flow-through lakes have both in- and outflow; in blind lakes, above-surface in- and outflow is absent; and terminal lakes lack an outflow. A distinction can also be made between permanent or perennial lakes, which have sufficient inflow throughout the entire year; periodic lakes, which carry water only during the rainy season; and episodic lakes, which retain water only after periods of heavy precipitation.

Large lakes can influence the local climate, as their ability to store heat tends to buffer climatic extremes. They store water during periods of precipitation, and humidify their surroundings in the dry season.

The Great Lakes between the U.S. and Canada contain roughly 22 percent of the world's fresh surface water.

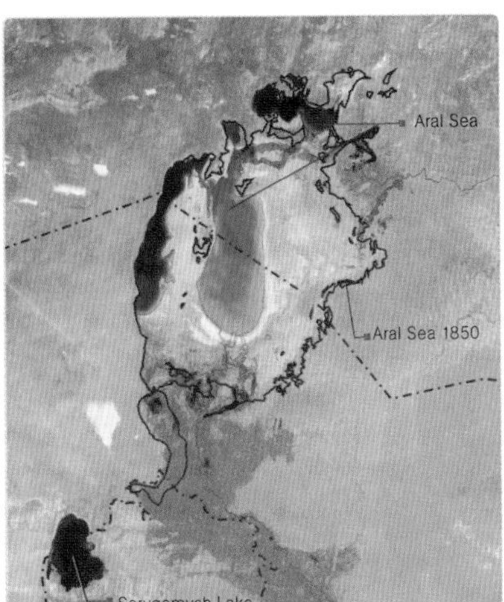

The disappearance of the Aral Sea has caused environmental and economic disaster in the region.

Transformation characteristics

In geological terms, lakes have only a short life expectancy, because they become terrestrialized due to a buildup of river sediments and plant matter. The biological aging process is influenced decisively by eutrophication. The lake is gradually enriched with nutrients from the atmosphere and inflowing water. This results in increased growth of phytoplankton and in an elevated oxygen production in the upper layers (epilimnion). Because of the breakdown of dead organic material in the lower region (hypolimnion) of the lake, oxygen is quickly used up again. Rapid exchange of water slows down eutrophication. In some Canadian lakes the rates of in- and outflow are so high that the entire water volume is exchanged within a few weeks. At the other end of the scale, it takes 700 years for the water in Lake Tahoe (in California) to be exchanged.

in focus

SALT LAKES

Most salt lakes were originally freshwater lakes. If inflow does not equal evaporation over an extended period, the concentration of salts and minerals in a lake increases steadily. In an extreme case, the salt content can exceed that of natural seawater and then salt precipitates in a crystalline crust. These types of lakes are found principally in dry regions. Some of the best known are the Salar de Uyuni in Bolivia, the Great Salt Lake in Utah, Lake Eyre in Australia, and the Dead Sea on the Israeli-Jordan border.

Great Salt Lake, Utah.

see also: Phosphorus cycle and water cycle, p. 279

formation | poles | high mountains

GLACIERS

Large areas of the Northern and Southern Hemispheres were once covered in mile-thick ice. Mighty moraines, carved-out valleys, and smooth-polished stones remain as reminders of the ice ages. Today the ice-cover of the polar areas and mountains is rapidly receding, and the full effects of global climate change are not yet certain.

glacier formation

Glaciers are masses of ice that flow slowly down a slope due to gravity. If excessive snow falls, the glacial advance is accelerated. If warm weather melts more ice than can newly form, the glacier retreats.

Mountain glaciers originate when snow and ice accumulate inside a cirque—a relatively small semicircular depression high in the mountains. In the so-called accumulation zone, above the snow line, fine snow crystals become densely packed, due to repeated melting and freezing, into coarsely grained firn ice. The constant pressure eventually results in air- and watertight glacial ice. In the Alps this process only takes a few years, but it can take up to 200 years in the colder and drier Antarctic regions. The ice becomes a glacier only once the bottom layers begin to move downhill due to the weight of the ice. If the gradient is sufficiently high, the glacier crosses the firn line and the glacier's snout pushes forward toward the valley.

Crevasses are formed when glaciers move faster than the flow rate of the ice can sustain. Transverse crevasses develop in areas of scarps and longitudinal crevasses develop in areas where glacial valleys turn wider. Radial crevasses form at the toe of a glacier, where they fan around the tip of the glacial snout.

Ice is lost through melting or evaporation in the so-called ablation zone below the snow line. It is continuously replenished by the glacial flow from the accumulation zone. The snout of the glacier moves forward when the ice flows at a higher rate than it can melt. The glacier retreats when the ice flows at a lower rate. The melted water of the ablation zone collects at the base of the glacier and exits in the form of milky white glacial creeks at the mouth of the glacial snout.

Rocks and debris that are transported by the glacier are deposited as a moraine. Solid material abraded from beneath the ice during glacial movements is carried along as a ground moraine. The material accumulates at the glacial snout and is pushed together to form an end moraine.

in focus

ICE AGES

Ice ages are geological time periods during which average temperatures were at least 8–9°F (4–5°C) lower worldwide than they are today, causing large-scale glaciations in mountain areas and toward the Poles. The last major ice age period began 2.5 million years ago and ended about 11,000 years ago. During this period glaciation periods and warmer periods called interglaciations alternated in a cycle.

Many fjords formed in the last ice age.

The Perito Moreno glacier in Argentina is fed by the Southern Patagonian Ice Field, which is the world's third largest reserve of freshwater.

ice of the poles and high mountains

The flowing motion and pressure caused by the ice caps of the Poles and high mountain glaciers not only significantly shape the landscape but are also the Earth's largest freshwater reservoir.

U.S. research base at the South Pole.

The appearance of the Earth today has been shaped by the major glaciation periods of the past two million years. More than one-third of the mainland, or about 21 million square miles (55 million km^2), were once covered by an ice sheet. Today, only 5.8 million square miles (15 million km^2) are still covered by the "eternal" ice. This accounts for about a tenth of the entire land mass.

All continents except Australia have mountain glaciers. Mountain glaciers are most common in Europe; however, even Mount Kilimanjaro, located near the Equator in Africa, is glaciated. Cold and dry regions in northern Alaska or the vast regions of Siberia do not have enough snow to form inland ice sheets. Overall, mountain glaciers cover 4 percent of the entire glaciated area; inland ice of the arctic polar regions, including Greenland, covers 11 percent; and the inland ice of Antarctica covers 85 percent.

Glaciers are collapsing and retreating at a faster rate than scientists previously predicted.

High mountain glaciers and ice sheets

Valley glaciers are the best known type of glacier. The Aletsch Glacier, for example, is the longest glacier in the Alps. The Malaspina Glacier in southeast Alaska is a typical example of a piedmont glacier, where ice masses from the mountain valleys spread onto the adjacent lowland. Ice sheets covering large areas also exist in the mountains without following the course of a valley. One such plateau glacier is the Vatnajökull Glacier on Iceland. At about 720 cubic miles (3,000 km^3), it is the largest glacier in Europe.

Continental glaciers, sometimes called inland ice or ice sheets, are even larger. At times during the last 2.5 million years, they covered vast areas of Europe and North America. Today they are only found on Greenland and Antarctica. The ice sheet of Greenland covers an area of almost 700,000 square miles (1.8 million km^2) and is more than 9,850 feet (3,000 m) thick. However, the ice sheet is melting, possibly losing up to 80 cubic miles (330 km^3) of ice per year. The ice shield of the Antarctic is even more immense, with an area of about 4.9 million square miles (12.8 million km^2) and a thickness of more than 13,120 feet (4,000 m). Entire mountain ranges are buried underneath the ice. The few bare peaks that reach through the ice are referred to as "nunataks." Large outlet glaciers move away from the inland ice toward the coast, where giant icebergs calve into the ocean. Glaciers flow at rates from several feet to a few miles per year. Temperate glaciers flow relatively quickly, due to a film of water created by the pressure on the bed of the glacier. Colder glaciers, located above the firn line, move by deformation and at a much slower rate. Alpine glaciers typically cover a distance of about 100 to 500 feet (30 to 150 m) per year. Glaciers of the Himalaya move about six to thirteen feet (two to four m) a day. In the Antarctic, the cold and stiff ice only moves just over 16 feet (five m) per year.

The Elephant Foot Glacier on the east coast of Greenland.

issues to solve

ICE-FREE FUTURE? According to the Intergovernmental Panel on Climate Change (IPCC), the area covered by snow was reduced by about 5 percent between 1980 and 2005 as a consequence of climate change. The areas of Antarctica that are covered by pack ice are diminishing at a much quicker rate than expected. If this trend continues, the North Pole may become entirely ice-free by the year 2050. It is predicted that most Alpine glaciers will disappear by 2037.

The Greenland ice sheet has been shrinking massively. The Arctic could be ice-free in summer as early as 2015, and almost certainly by 2050.

EARTH

ATMOSPHERE

The Earth is surrounded by a gaseous layer that makes life as we know it possible. The atmosphere protects us from dangerous radiation from space and provides us with breathable air. The sun is the engine of atmospheric processes. Its energy affects daily and long-term weather patterns, as well as determining the boundaries of different climate zones.

During the last few million years, the Earth's climate has undergone several significant natural changes. However, human beings have increasingly begun to play a major role in changing the climate. Few scientists today deny the reality of global warming. Nonetheless, in spite of international efforts and agreements, it is questionable whether these efforts can do more than slow down this global trend.

structure of the atmosphere

The current atmosphere is the fourth in Earth's history. It consists of 78 percent nitrogen, 21 percent oxygen, and various amounts of noble gases, carbon dioxide, water vapor, and nitrogenous and sulfurous compounds.

structure | processes

ATMOSPHERE

The atmosphere is the gaseous envelope that protects the Earth against dangerous radiation from space and enables the development and existence of life on our planet. Its formation took several billion years, but information about its structure and internal processes has only been accumulated over recent decades.

The approximately 63-mile (100-km)-thick atmosphere of the Earth is merely a thin layer, relative to the Earth's diameter of 7,926 miles (12,756 km) at the Equator. The percentage of gaseous parts varies depending on altitude, and the air pressure decreases with increasing altitude, enabling the atmosphere to be categorized as distinct vertical layers. The layer closest to the Earth is the troposphere, where practically all weather occurs. It contains more than 90 percent air and most of the water vapor in the atmosphere. At the Poles, this layer is up to 4.5 miles (7 km) high, while along the Equator it can reach up to 11 miles (18 km). The temperature of the troposphere decreases toward the upper boundary (tropopause) at a rate of about 50.8°F per mile (6.5°C/km). In the next layer, the stratosphere, the temperature increases from -140°F (-96°C) to about 32°F (0°C) at the 31-mile (50-km)-high upper boundary. This change is brought about by the warming of the ozone layer as it absorbs ultraviolet radiation.

The temperature decreases again in the mesosphere. At its upper boundary (mesopause) the temperature is about -130°F (-90°C) and its lower boundary of the polar lights (aurora borealis and aurora australis). In the 124-mile (200-km)-high thermosphere the few gaseous particles located in the upper thermosphere can reach over 1832°F (1000°C).

The last layer is the exosphere, at an altitude of about 621 miles (1,000 km). Outside the atmosphere, the Earth is surrounded by energy-rich particulate radiation (Van Allen belt). Structuring the atmosphere by gaseous electrical charge, three extra categories emerge. The neutrosphere extends up to 50 miles (80 km), becoming the ionosphere and then blending into the protonosphere toward outer space. Structuring the atmosphere based on composition, we also get the homosphere, where all gaseous components are mixed equally, and the heterosphere, where the gas mixture separates because of the diminishing attraction from the Earth.

The habitable atmosphere of the Earth is merely a thin veil, but it is where all human life exists.

The atmosphere is divided into layers that vary in chemical composition and temperature.

basics

POLAR LIGHTS are caused by charged particles from the sun, which are diverted by the magnetic field of the Earth to the polar regions. There, the particles collide with oxygen and nitrogen atoms in the upper atmosphere, creating the visible effect of colored light.

atmospheric processes

The sun is the driving force for processes in the atmosphere, fending off most of the Earth's damaging ultraviolet radiation, and permitting only the shortwave radiation of visible light to penetrate to the Earth's surface.

The radiation given off by the sun provides the energy for all meteorological processes and for a multitude of atmospheric phenomena. One-third of the radiation that penetrates to the Earth is reflected by clouds, the Earth's surface, and the air, redirecting it back into space (albedo). The atmosphere also absorbs some of the shortwave radiation. About half of the radiation reaches the Earth's surface directly or indirectly through scattering by gases and aerosols. The Earth not only absorbs radiation, it also gives off radiation in the form of longwave heat. Without the presence of the atmosphere, such terrestrial warming radiation would escape unhindered into outer space and the average global temperature on the Earth's surface would be about -0.4°F (-18°C). However, because of clouds and atmospheric gases, such as carbon dioxide (CO_2) and water vapor (H_2O), a large part of this radiation is recaptured—only a small part can escape through the so-called atmospheric window. The predominant part is radiated back to earth as atmospheric back radiation. This process is referred to as a natural greenhouse effect. This back radiation has an energizing effect on the Earth; the air close to the ground is heated on average to 58°F (14.5°C), leading to water evaporation and the motion of the air and sea.

Crepuscular rays are so called because they normally occur at crepuscular (or twilight) hours. Most people know this optical effect, caused by the scattering of sunlight by clouds, as sunbeams.

in focus: ALBEDO

Some of the received shortwave solar radiation is reflected back into space by clouds and the Earth's surface, without utilizing any of the heat. This capability to reflect is strongly dependent on the type and condition of the area that is being irradiated. While a black body absorbs all radiation and does not reflect anything, a white body produces the opposite effect. The highest albedo values are reached by areas covered with dry snow. The larger the ice cover of the Earth, the less heat is absorbed, causing a temperature drop.

Snow-covered areas cool the ground by nature of their higher reflectivity.

Radiation balance

There is an overall balance found on Earth, between radiation that is received and given off. However, depending on the latitude there are significant variations. For instance, in the polar regions there is an avarage deficit created due to the polar night, while the tropics have a radiation excess due to solar radiation throughout the entire year. In order to balance the uneven energy distribution on Earth, global wind systems circulate in the atmosphere while oceanic currents transport energy from the lower latitudes to the higher latitudes.

basics

SUNLIGHT is composed of the colors of the visible spectrum, with different wavelengths. When sunlight penetrates the atmosphere, air molecules scatter the rays. The shortwave blue light is scattered about five times as much as the red light, which is why the sky appears blue.

At sunset the low position of the sun causes increased scattering of light, with only longwave red and orange light reaching the ground.

high- and low-pressure weather fronts

The sun acts as an engine of the weather system. Its energy moves enormous air masses which travel as currents throughout the lower atmosphere. The daily change in weather usually results from events at the boundaries.

Cloud formations resulting from rising warm air. Water vapor within the clouds results in precipitation.

The air may seem weightless, but surprisingly it has a considerable weight. This weight can be measured as air pressure, using the unit hectopascal (hPa). Air pressure is created due to the uneven warming of the Earth's surface. Warm air masses rise up and leave behind a low-pressure area near the ground, while high-pressure areas develop where air masses descend after cooling. Air masses from high-pressure areas move into low-pressure areas in order to equalize differences in pressure.

In temperate zones, the weather is determined by the interaction of cold and warm air masses. When air masses of different temperatures collide, a front forms. If lighter, warmer air slides above the cold air, a warm front is created. A cold front is created by cold air pushing underneath the warm air like a wedge. Warm air rising results in cloud formations and precipitation. In a warm front, these processes happen evenly and relatively slowly while a cold front shows rapid formations of cumulus clouds and precipitation in the form of showers. Thunderstorms may also occur.

The process where a cold and warm front merges is called occlusion. At this location, a cold front moving along catches up with a warm front. Cold masses of air move underneath the warm air and push it up from the Earth's surface. The warm air that is pushed up cools, resulting in reduced wind and precipitation. The air circulation in temperate zones keeps moving due to the air pressure gradient between the subtropical high-pressure belts and the subpolar low-pressure areas. Differences in air pressure are balanced by strong westerly winds. Hot tropical air and cold air from the Poles create a boundary called the polar front, where major air turbulences result. These are called cyclones and can reach a diameter of 620 miles (1,000 km).

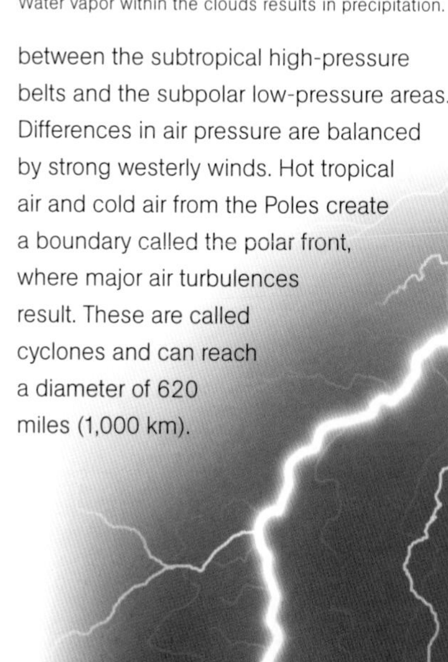

air pressure | meteorology | clouds | storms

WEATHER

All living organisms on this planet are influenced by the weather. Humans rely heavily on accurate forecasts of temperature, air pressure, humidity, and the presence or absence of wind or clouds. Due to the very complex processes happening in the lower part of the atmosphere, it is almost impossible to predict the weather more than a few days into the future as there are too many variable and uncertain factors affecting the likelihood of rain, storm, or sunshine.

in focus

JET STREAMS

Jet streams are created where very cold air collides with very warm air masses resulting in an extreme pressure drop and an exceptionally fast wind. Jet streams may reach speeds of up to 370 mph (600 km/h) at a height of 5.6 to 7.5 miles (9 to 12 km). They can be up to several miles wide and over a thousand miles long. Often jet streams can be recognized by cirrus clouds that consist of ice crystals.

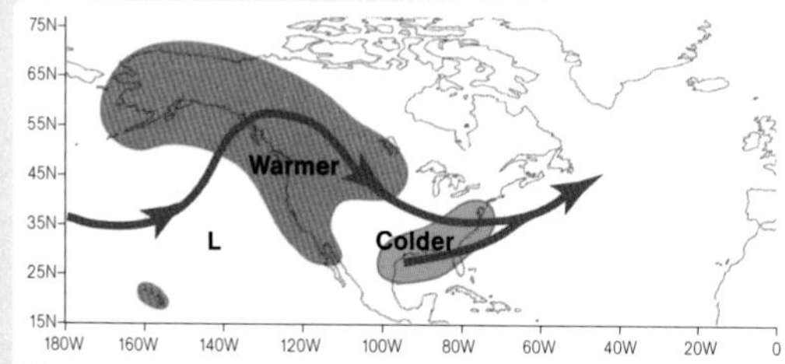

The typical meandering shape of the fast-moving jet stream.

see also: Cosmic weather, p. 36

meteorology and weather observations

People have a great interest in tomorrow's weather. The oldest written reports about weather phenomena are over 5,000 years old. The weather has been systematically recorded and analyzed for the last 150 years.

Nowadays, weather conditions are observed, measured, and recorded around the clock through a global network of data recording devices. Weather stations, measuring buoys, and weather balloons provide data on air pressure, temperature, precipitation, humidity, sun hours, wind direction, and wind speed by means of mercury barometers, hygrometers, and weather vanes. This data can be accessed at any time, even from remote areas. Planes are used for observations of the atmosphere at high altitudes, usually when dangerous tropical storms are about to form. Since the 1960s, weather satellites have also been in use, continuously sending data to receiving stations on the Earth.

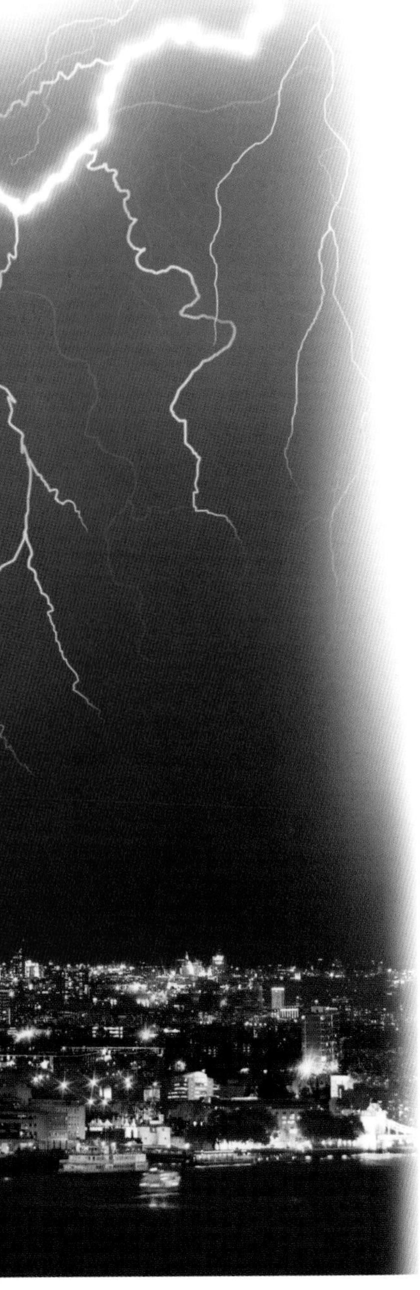

Lightning is a spectacular but sometimes dangerous weather effect. It often strikes the tallest buildings in an area. Each flash lasts a quarter of a second.

Computers are a vital tool in weather prediction, enabling ever more accurate forecasts.

Weather forecasts

Collected data is sent to various centers distributed across the globe and from there they are redirected to regional stations and national weather services. They are all members of the World Meteorological Organization (WMO), which was founded in 1951 as an agency of the United Nations. Based on the data provided, weather services produce weather maps using internationally standardized symbols. The expected weather patterns for the following few days (using numerical weather prediction) are calculated by computers using mathematical-physical formulas. Given the advances in technology, people are always hoping for more and more reliable weather forecasts. However, long-term predictions are difficult to make due to the complex processes taking place in the atmosphere. With modern computer-assisted weather models, meteorologists can reliably predict the following five days' weather, whereas ten years ago we could only predict three days ahead. Such computer-based models are also used for less accurate long-term forecasts.

Weather may seem unpredictable and random, but it follows rules and can be quite accurately predicted.

milestones

VILHELM BJERKNES

The Norwegian geophysicist and meteorologist Vilhelm Bjerknes (1862–1951) was a major pioneer of meteorology. He developed the theory of warm and cold air fronts, as well as the basic principles of numerical weather prediction. He carried out some of his most significant studies at the University of Leipzig's Institute of Geophysics in Germany. Due to the outbreak of the First World War, he was forced to return to Norway in 1917, where he founded the Geophysical Institute in Bergen.

Vilhelm Bjerknes, the founder of the practice of modern weather forecasting.

precipitation and clouds

Cloud formations and patterns have always greatly inspired humans. They are important for the global distribution of water and required for precipitation, which in turn supports life on Earth.

The Earth's water is in constant motion. Water vapor rises into the atmosphere and can travel over large distances. The colder the air temperature, the less water vapor it can retain. The point of saturation or "dew point" is when the relative humidity is 100 percent, leading to the condensation of water from vapor into liquid. This process is facilitated by tiny particles in the air, the condensation nuclei, which act as an attachment point for suspended water molecules. Below freezing point, water vapor sublimates into ice crystals. In both cases clouds become visible. Precipitation occurs when cloud particles become so heavy that they lose their buoyancy.

Clouds are categorized according to their shape and the height at which they occur. Today's international cloud classification system is based on studies published by the British pharmacist Luke Howard in 1803. There are ten main types of clouds, each with several subcategories and special types. The system for describing these cloud types is based on three cloud height levels. The two basic cloud forms are stratus and cumulus, which are found low in the sky. The horizontal dimensions of stratus clouds are larger than their vertical reach, while cumulus clouds show a large vertical reach and an almost flat lower limit. Cirrus clouds are icy clouds that, in temperate regions, occur only at very high altitudes between 23,000 and 43,000 feet (7,000 and 13,000 m). The names of the clouds situated at a medium altitude—above about 6,500 feet (2,000 m)—begin with the prefix "alto." Another type of cloud, nimbus clouds, are dense and dark. This is the type that carries a lot of precipitation.

CUMULONIMBUS, THE THUNDERCLOUD When warm, humid air close to the ground rises and forms cumulonimbus clouds, the updraft pulls positively charged droplets upwards inside the cloud, leaving negatively charged droplets near the base. As the upper regions of the cloud are altered due to ice formation, electricity is discharged in a lightning flash. The surrounding air is heated and expands, causing the distinctive sound of thunder.

basics

Raindrops, formed by small droplets of water merging, are an instantly recognizable precipitation effect.

A snowflake is a single particle of snow. It is possible to find two that are visually alike, but it is very rare.

Cumulonimbus clouds forming over the ocean: Cumulonimbus clouds are responsible for high precipitation and storms.

Precipitation

Precipitation is produced as humid air cools to below the dew point. This can occur when the air rises, such as when two fronts meet (cyclonic precipitation) or there is a ground elevation increase (relief rain or orographic rain). Humid air can also rise freely (convective precipitation). A liquid droplet is specified according to its diameter: a droplet is rain when its diameter is greater than 0.019 inch (0.5 mm), drizzle when it is less than 0.007 inch (0.2 mm), and fog when it is between 0.0003 and 0.001 inch (0.01 and 0.04 mm). Solid precipitation, which occurs at colder temperatures, includes snow, sleet, and hail.

Snow in an alpine area: Some mountains are so high that they keep a snowcap year-round.

see also: The effects of climate change, p. 127

wind, storms, and anomalies

Wind is air in motion originating from air pressure gradients in the atmosphere. It always blows from an area with high air pressure into an area with low air pressure. The greater the difference in air pressure, the stronger the wind.

There are several parallel zones on the Earth with prevailing surface winds. Warm air expands near the Equator and rises, causing a low-pressure belt called the intertropical convergence zone. This zone is also known as the doldrums. While the rising air cools slowly, it expands toward the Earth's Poles, creating convection currents. The air descends again in the subtropical horse latitudes, which are located at 30° north and south of the Equator. Here, the air merges into the trade and westerly winds.

The horse latitudes experience only light winds or calm air, similar to the doldrums. Trade winds blow near the Earth's surface and toward the Equator, replacing the rising air. They merge with the low-pressure system of the intertropical convergence zone and the cycle repeats.

Tornadoes are a swiftly rotating column of air, noted for their highly destructive energy and high wind speeds.

In contrast, westerly winds collide in the temperate regions with cold air masses called the polar easterly winds. The Poles are regions with high air pressure caused by the sinking cold and dense air.

Tropical storms can lead to significant destruction and flooding of lowland and coastal areas.

Local winds

The formation of winds is affected by local geography, such as mountains, deserts, terrestrial areas, and oceans. Therefore, many areas experience localized winds in addition to the large-scale wind systems. Coastal areas experience offshore and onshore winds depending on the daily warming of the atmosphere. Drainage or katabatic winds may occur on the lee side of mountains. In contrast, the mistral is a very cold and dry wind blowing from the high ranges of the French Massif Central. The sirocco is known as a hot and dry wind in the northern parts of the Sahara.

Hurricane Katrina, one of the most severe hurricanes in recent U.S. history, caused 1,836 deaths.

Heavy storms

Heavy storms with spiraling winds are frequently generated above warm tropical oceans. They are called hurricanes in the Caribbean, typhoons in the China Sea, and cyclones in the Indian Ocean. They move west reaching velocities of up to 186 miles per hour (300 km/h). Tornadoes are smaller and short-lived, but those generated from thunderstorm clouds are just as dangerous. The Beaufort scale, named after the British admiral Sir Francis Beaufort (1774–1857), is used to indicate the force of wind.

in focus

CORIOLIS FORCE

Winds do not blow across the globe in a straight line; they are diverted from the west to the east by the rotation of the Earth, causing the distinctive spiral effect. This rotation causes the rising warm air masses from the Equator to the Poles to shift toward the right and the descending cold air masses from the Poles to the Equator to shift toward the left. This effect was discovered by the French physicist Gaspard Gustave Coriolis (1792–1843).

Gaspard G. Coriolis: the Coriolis force is absent at the Equator and strongest at the Poles.

the global climate

Different climates exist on various parts of Earth, from the dry heats of deserts to the freezing winds of polar ice caps. All of these climates can be taken together to calculate the average global climate.

global climate | natural phenomena | tropics | subtropics

THE CLIMATE SYSTEM

Earth is home to a number of different climate zones, influenced by various factors. A commonly used system of categorization identifies zones based on their average temperatures, dividing the world into polar, temperate, subtropical, and tropical zones.

Climate is the full range of weather conditions experienced in a particular place over several decades or longer, including daily and seasonal changes. Weather, however, refers to short-term conditions during time periods ranging from as little as a few hours to a length of a couple of weeks—or, at the most, a particular season.

What determines climate?

Many conditions and processes work together to determine the climate of a particular area. Primary climatic factors include a region's latitude, its elevation, and the features of its terrain, as well as the amount of solar radiation it receives. From these primary factors arise secondary climatic factors, such as ocean currents, wind systems, and other natural cycles.

Climate observations

Depending on the size of the area under consideration, climate can be divided into microclimate (sometimes areas only a few yards or meters in size), mesoclimate (extending over several hundred miles), and macroclimate (encompassing entire continents or even Earth as a whole). Climate arises from the interaction of Earth's five "spheres": the atmosphere (air), the biosphere (living things), the pedosphere (soil), the lithosphere (rocks and minerals), and the hydrosphere or cryosphere (water, including ice). Together, these form the geosphere. A multitude of climatic factors influence each of the spheres, producing the geosphere's overall climate. In order to describe weather and climate in the atmosphere, scientists collect data about the various climatic elements, such as temperature, air pressure, humidity, wind speed, and direction. After data has been collected over a long period of time, average values can be calculated. When scientists compare these with current data, they can make predictions about long-term patterns in the climate.

in focus

NORTH ATLANTIC OSCILLATION (NAO)

The NAO index describes the difference between pressure systems over the Atlantic; a large pressure difference is a positive index while a small is a negative index. The index fluctuates, influencing Northern Hemisphere weather. It is not yet known whether the positive index over recent decades is related to unusual climate change.

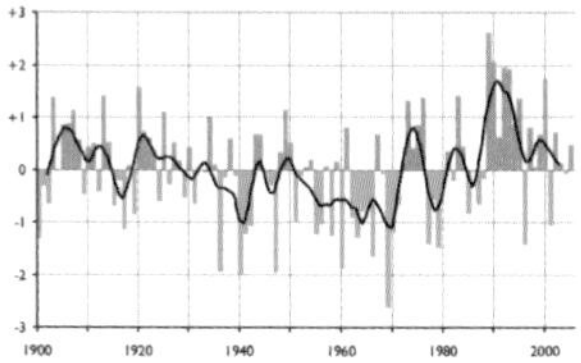

The high-pressure system that tends to form over the North Atlantic Ocean is known as the Azores High.

When cooled near the ground, water vapour in the air can condense into tiny suspended droplets, forming fog.

Various interacting factors affect the Earth's overall climate. These influences can arise from the atmosphere, biosphere, pedosphere, lithosphere, or the hydrosphere.

see also: The origins of life, p. 55

natural climate phenomena

Historically, the Earth's climate has gone through natural cycles. These can be detected in climate data of past millennia, recorded in mineral deposits and core samples of polar ice.

Fueled by the sun's energy, wind and ocean currents continually churn the Earth's troposphere and hydrosphere, affecting the Earth's climate. The so-called trade winds significantly influence global climate by carrying large amounts of ocean moisture.

El Niño and La Niña

The southeast trade winds drive the circulation of water in the South Pacific. Cool surface water off the coast of South America moves westward, warming as it goes. As this current nears Southeast Asia, it encounters cooler water, which sinks down into the Pacific and travels along the seafloor toward South America, rising again to renew the cycle. This cold current from the deep Pacific creates an extended high-pressure zone and a dry climate in western South America. The nutrient-rich water makes fishing ideal near the shores of countries such as Peru. Conversely, in Southeast Asia, warm waters produce sustained low-pressure zones regularly resulting in heavy monsoon rains in Australia and Indonesia. Every three to eight years, an unusual warming event occurs near the South American coast. One potential reason is that trade winds weaken, interrupting the water circulation pattern in the Pacific. As a result of this, heavy rains and hurricanes often ensue. Without the nutrient-laden, cold current, fish stocks drastically decline. In Southeast Asia, severe droughts cause failed harvests and forest fires. Since this event often occurs around Christmas, Peruvian fishermen dubbed it "*El Niño*," (the Christ Child). When rising pressure differences between South America and Southeast Asia allow the trade winds to gain strength, the circulation pattern resumes and the original conditions return. This is known as *La Niña* (the little sister). Scientists still do not fully understand either phenomenon.

practice

CLIMATE ANALYSIS

The Earth's climate has been influenced throughout history by natural events such as the formation of the atmosphere and new land masses, tectonic plate movement, meteorite collisions, and massive volcanic eruptions. Most climate changes were gradual, but they can still be detected from the subtle traces they have left behind. Scientists draw important information from fossils, the chemical composition of minerals, and ice deposits to gain an understanding of past climate conditions.

Because ice preserves air particularly well, polar analyses are some of the most accurate and valuable sources of past climate data.

basics

DENDROCHRONOLOGY Tree rings vary in thickness depending on how favorable the growing season was. Thus, growth rings tell us about past climate conditions as well as the tree's age.

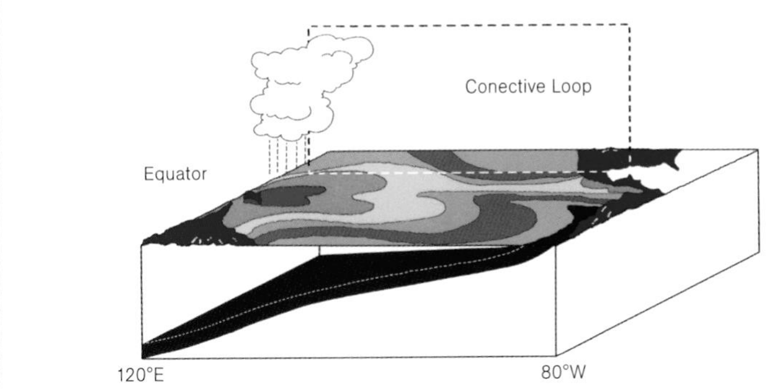

Under normal conditions, the coast recieves sufficient rainfall and cold water laden with nutrients rises from the ocean depths, replenishing nutrients to the upper layers of waters and increasing fish stocks.

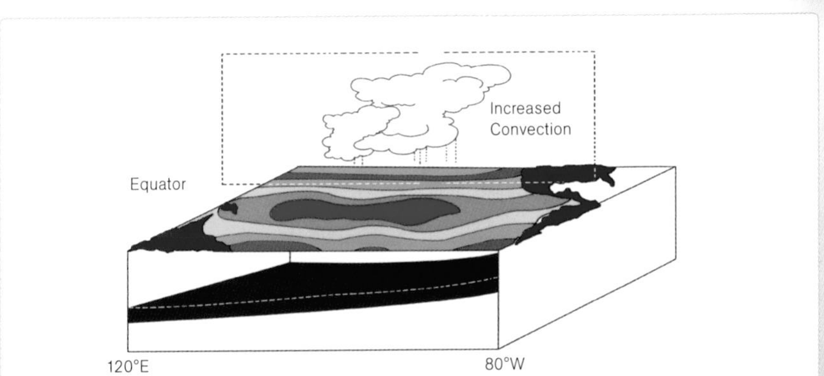

During El Niño conditions, rainfall on land is reduced and cold water circulation to the surface is interrupted by the warming pattern, causing major declines in fish stocks and other marine life.

When sunlight falls on moisture in the air, the spectrum of light appears as a rainbow.

tropical climates

The tropics are known as humid, muggy areas with evergreen rain forests rich in animal and plant species. But tropical habitats can also include the grasslands of the savannas and even dry deserts.

The term "tropics" comes from the Greek *tropos*, which means "to turn." It describes the climate zone on either side of the Equator between the Tropic of Cancer and the Tropic of Capricorn—between 23.5° north and south latitude. This zone forms a belt around the circumference of the Earth, incorporating about two-thirds of the total landmass. The key characteristics of the tropical belt are continuously high solar radiation and consistently high temperatures. At sea level the average monthly temperatures never fall below 68°F (20°C). Temperature fluctuations between day and night are greater than those over the course of the year.

Temperatures in the tropics decline with rising altitudes. The lowland tropics have a warm tropical climate and the tropical mountains have a cold tropical climate. Lowland tropical areas may be humid, with a climate typical of tropical rain forests; these are the inner tropics. On the other hand, the outer tropics have a subhumid climate typical of savannas. The transition point occurs where the annual temperature fluctuations are greater than 18°F (10°C).

From the jungle to the desert

Only the rain forests experience constantly humid conditions. They are located within the inner tropical convergence (ITC) zone. This zone covers the area between 10° north and south latitude. The annual rainfall decreases the farther away from the Equator an area is located, while the number of dry months increases. Rainfall occurs during one or several rainy seasons. The rainy seasons of south Asia are often accompanied by heavy winds called monsoons blowing in from the ocean. A long dry season follows, as dry and cold air moves in the opposite direction from the mainland toward the ocean.

The subhumid tropics are divided into vegetation zones according to the number of rainy months: the semihumid savanna has 7 to 9.5 rainy months, the dry savanna has 4.5 to 7, and the arid shrublands have 2 to 4.5. The desert or semidesert areas of the dry tropics receive less than two months of rain per year.

in focus

COASTAL CLIMATES

On the coast, the water moderates the temperature by acting as an insulator: the oceans may take longer to warm up but they also take longer to cool down. The marine air also carries plenty of water vapor. The very dry coastal deserts of Atacama and Namib are exceptions. Cold ocean currents cause short-term fog, but they simultaneously prevent the formation of rain clouds.

Coastal climates are generally characterized by balanced annual temperatures and high rainfall.

In the ITC zone daily temperatures are on average 77–80°F (25–27°C) and rainfall reaches 80–120 inches (2,000–3,000 mm) per year.

African buffalo herds roam the sun-soaked savanna.

see also: Ecosystems, pp. 80–87

subtropical climates

Subtropical areas can be found in a belt between the tropics and 45° north or south latitude. These areas are the climatic transition between tropical and temperate latitudes. The Mediterranean climate is a specific climate type located within the subtropical zone.

Shrubland communities, such as this macchia vegetation on the south coast of Crete, are widely found in Mediterranean climate zones.

The subtropics are defined as a climatic zone with high temperatures in the summer and mild temperatures in the winter caused by a seasonal shift of the subtropical high pressure areas. Temperatures and the amount of precipitation depend less on geographical latitude than on where the area is located on the corresponding landmass. Dry subtropical regions with large deserts such as the Sahara in Africa or the deserts of Australia are located at the center of these continents. Here, precipitation of less than four inches (100 mm) per year is not unusual. Even shrubs and thorn scrub will only grow in certain areas with favorable conditions. The eastern parts of these continents experience monsoons during the summer carrying heavy rainfall. This is where the subtropical moist forests are found. The western parts of these continents, however, have a subtropical climate with dry seasons in the summer and rainy seasons in the winter.

Mediterranean climate

The Mediterranean climate zone is located between 32° to 45° north and 28° to 38° south latitudes. As the name indicates, the Mediterranean climate is prevalent in the Mediterranean regions, but it also exists on the coast of California, along a coastal strip of Chile, in the Cape Town area, and on the southwest coast of Australia.

All these regions experience rainy winter seasons, sometimes with frost, yet the average temperatures of the coldest months are above 41°F (5°C). Average summer temperatures vary between 73°F (23°C) in Europe and 82°F (28°C) in Australia. The dry summer season with little or no rain lasts about four to six months. The main growth period for plants is therefore during the spring. Sclerophyll is the typical vegetation of this climate zone. Its plants have small hard leaves with a coriaceous surface that protect the plant from water loss caused by evaporation. The natural vegetation in the Mediterranean mostly consists of evergreen oak forests, with the prevalent holm oak. Many Mediterranean forests have deteriorated, however, into shrubland communities with Mediterranean heaths. This type of plant community occurs in different parts of the world and has different local names: Italian macchia, French garrigues, the fynbos of Cape Town, or the California chaparral. In Mediterranean Australia, the predominant plant of the sclerophyll forests is the eucalyptus and in Chile it is the litre tree.

In the Mediterranean, sclerophyllous holm oak have been cultivated as an efficient source for wood, truffles, and animal food since antiquity.

The koala has a specialized diet of eucalyptus native to the Mediterranean climate zone of Australia.

basics

CLIMATE CLASSIFICATIONS There are different methods for the classification of the climates on Earth. One of the most well-known methods was developed by the geographer Wladimir Peter Köppen (1846–1940). The Köppen climate classification method is based on the relationship between temperature, precipitation, and vegetation. Here, individual climate zones and types are separated by the mathematical mean of distinct threshold values. The so-called effective classification was published in 1923 and has been reworked several times since.

The subtropical, hyper-arid central portion of the Sahara mostly consists of migrating sand seas with little to no vegetation cover.

see also: Islands and atolls, p. 101

climate of the temperate zone

The temperate zone covers the Northern and Southern Hemispheres between 45° latitude and the polar circle at 66.5° latitude. It can be divided into a cool and a cold temperate zone.

The temperate zone mainly depends on the westerly winds outside the tropics which bring a maritime climate to coastal areas while inland areas experience a continental climate with decreasing annual precipitation. The average annual precipitation is around 31 inches. Overall, there are distinct seasons, slightly less so toward the Equator. Another typical characteristic is the variation of night and day hours according to the season. This variation increases with decreasing distance to the Poles. The Gulf Stream has a significant effect on the climate in central and northern Europe which would be much colder without the influence of this ocean current.

Red deer inhabit the temperate zone of Europe, the Caucasus, parts of Asia, and northwestern Africa.

CONTINENTAL CLIMATE differs from maritime climate where daily and annual temperature fluctuations are less extreme. This is due to the heat storage capacity of the oceans: Although the ocean heats up more slowly than the ground during summer, it also cools down more slowly and releases the stored heat during winter. With increasing distance from the coasts there is also a significant reduction in annual precipitation.

basics

The cool temperate zone

In the Northern Hemisphere, the warm temperate zones are located on the west coast and in the northeast of North America as well as in large parts of Europe and East Asia. In the Southern Hemisphere, they are located in Chile, southeast Australia, Tasmania, and the southern island of New Zealand. The average annual temperature is between 46.4 and 53.6°F (8 and 12°C). The actual temperatures in the coastal areas are milder with a maritime climate where monthly average temperatures hardly fall below freezing while the continental inland temperatures are very high in summer and very low in winter. The dryness of the continental climate of the temperate grasslands makes tree growth impossible. In contrast, deciduous forests with oaks, beech, hornbeam, birch, and ash trees are common in areas influenced by the maritime climate.

The cold temperate zone

If the growth period decreases to less than 120 days due to cold temperatures, deciduous woods have no chance of survival. Coniferous forests are therefore widespread in the cold temperate zones closer to the Poles. About 10 percent of this continental area is taken up by the taiga with vast swamp and raised moss areas. In the Northern Hemisphere, this type of vegetation is the most widespread. In the Southern Hemisphere, such vegetation is almost entirely absent due to the distribution of land and ocean. The average temperature of the coldest month is below 26.6°F (−3°C) and below -13°F (−25°C) in the coldest areas. The warmest month reaches on average less than 50°F (10°C).

Across the world, alpine climate exists in temperate regions that are at altitudes situated above the tree line.

The boundary between continental and hot summer continental climate is where temperatures fluctuate by more than 104°F (40°C). Precipitation falls as rain in the summer and as snow in the winter.

The taiga is the largest terrestrial ecosystem stretching across the cold temperate zone of the Northern Hemisphere. This biome does not exist in the Southern Hemisphere.

see also: Ecosystems, p. 86

subpolar and polar climate

Distinct seasons occur within the polar circles. However, during the winter, the sun never fully rises and in the extreme cases of the North and South Poles, the polar night and polar day last half a year each.

Elk, one of the Earth's largest deer species, mainly range in forest habitats of North America and eastern Asia.

The subpolar climate zone is the transition between the temperate and the polar climate zone. It is a relatively narrow belt along the coast of the Arctic Ocean from northern Scandinavia through Siberia to Alaska, northern Canada, and Greenland. The mean temperature of the warmest months during the short summer is below 50°F (10°C) and the winters are moderate to extremely cold. There is also a marked contrast between the mild oceanic climate and extremely continental climate. Precipitation usually falls throughout the year, but only little. Winds are often strong, especially in the Southern Hemisphere and close to the boundary of the West Wind Drift. Durable and low-growing plants such as lichen, moss, and grasses, herbaceous and scrub species have adapted very well to the rough climate, short growth period, and permafrost. They form the tundra, which covers large parts of the subpolar zone. The treeless tundra may even cover areas south of the northern polar circle that are elevated or exposed to the wind. Toward the north, the tundra is adjacent to the polar desert.

Polar areas

The polar areas of the Earth are cold deserts. Temperatures are constantly below freezing and solar radiation is reduced. Major parts of the Arctic are occupied by the Arctic Ocean that is covered over by pack ice. Due to the heat capacity of the water, the most extreme temperatures are not at the North Pole but in eastern Siberia where temperatures reach -94°F (-70°C). The sixth continent, Antarctica, is covered by an ice sheet of several miles thickness. With temperatures as low as -128.6°F (-89.2°C), Antarctica is the coldest place on Earth; even during the summer, the thermometer never rises above 5°F (-15°C). Temperatures are so low due to the elevation and the circumpolar current. Only little water can evaporate in such a cold climate, therefore, very little precipitation falls in this area. Annual precipitation is between 1.2 inches (30 mm) in the interior of the Antarctic and 5.9 inches (150 mm) near the coast; in the Arctic it is between 2 and 19.7 inches (50 and 500 mm). Cold katabatic winds are typical for the climate of the Antarctic. They can lead to severe storms. In this almost completely unvegetated zone, life is concentrated in the coastal areas, which are slightly warmer, similarly to the Arctic. This is where animals that have adapted to the polar climate can find sufficient food. A vast habitat for plankton and fish, the Arctic Ocean and the Southern Ocean are among the most productive regions on Earth.

in focus

MOUNTAIN CLIMATE

Climate does not only depend on latitude. High mountains are a special case in all climate zones. Temperatures drop with altitude, just like they drop from the Equator toward the Poles. They drop by about 33°F per 328 feet (0.6°C per 100 m). This results in various climate and vegetation zones. Elevation climate zones are especially distinct in the tropical mountains, where they range from tropical rain forest to forest, tree, and snow boundaries and finally to the permanent ice zone.

In elevated terrain, the tree line forms the boundary between climate zones.

Ice flows from the glacier tongues of the Greenland Ice Sheet, which covers about 80 percent of the surface of Greenland, into the Atlantic and the Arctic Oceans.

see also: Mountains, pp. 76–79

climate change

Scientists are tackling questions surrounding the causes of recent record high temperatures and a global increase in natural disasters. One possible cause of unusual climate change may be modern-day human activity.

changes | climate factors | global warming | consequences

CLIMATE CHANGE

Myriad natural factors help explain the current warming trend in the Earth's climate, in particular a natural climate cycle that occurs over thousands of years. However, the speed at which the recent change is occurring is curious and disconcerting. Scientists are exploring the possibility of human activity as one of the influences.

The alternation between cold and warm phases on Earth constitutes a natural cycle. On average, a phase lasts 100,000 years. Within a natural cycle, less significant climate changes occur in cycles of approximately 20,000 and 40,000 years. In 1920, Milutin Milankovitch, a Serbian astronomer and mathematician, realized that astronomical forces have a cyclical influence on the intensity of solar radiation reaching the Earth. The Earth rotates like a spinning top as it revolves in an elliptical path around the sun. At intervals of 20,000, 40,000, and 100,000 years, this planetary motion determines the distance between the Earth and the sun, and thus the angle at which sunlight strikes the Earth. This may serve as a climate change trigger. Milankovitch cycles are confirmed by data collected from sediments on the ocean floor and from core samples obtained by drilling into polar ice.

At this time Earth is in the midst of a warm phase within an ice age. However, this does not by itself account entirely for recent climate changes. Although it can be assumed that humans are not the sole cause of global warming, the worldwide scientific consensus asserts that human activity since the industrial revolution has certainly influenced the warming process. Statistically speaking, clusters of storms appear regularly and influence Earth's natural climate cycles. Researchers predict that such clusters and other extreme weather phenomena, such as the destructive Hurricane Katrina, which hit the U.S. Gulf Coast in 2005, may become more common, partly because of increasing global temperatures.

Storms over the Atlantic and its coastlines seem to have doubled within the last hundred years.

in focus

INFLUENCE OF NATURAL CATASTROPHES

Volcanic eruptions affect global climate by releasing sulfur dioxide (which can form sulfuric acid) and ash into the air. As aerosols (floating particles) these reflect some sunlight back into space; cooling Earth. In 1991, the Pinatubo eruption lowered average temperatures by 32.9°F (0.5°C) for two years. An asteroid strike works similarly: the impact raises clouds of dust that act as atmospheric aerosols. About 65 million years ago, a huge asteroid or comet struck Mexico, causing climatic cooling, which led to mass extinction. Of course, supervolcano eruptions and asteroid collisions are extremely rare and are not behind the recent warming trend.

Washington state's Mount St. Helens exploded on May 18, 1980, leaving its surroundings destroyed.

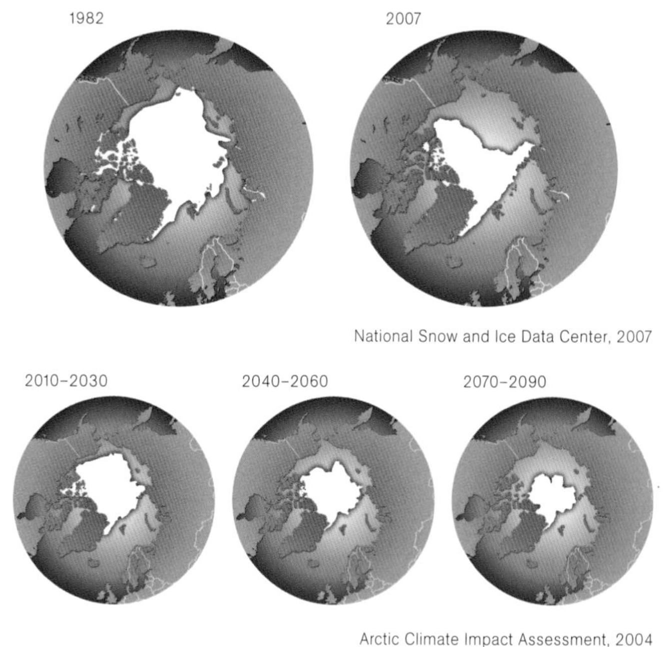

Arctic sea ice is an important factor in the global circulation systems. Satellite observations of summer ice coverage show rapid reductions and project a continuing annual loss.

see also: Environmental protection, pp. 128–137

climate factor mankind

The Earth's climatic system changes slowly. The climate trends we see today were triggered partially by events that happened decades ago.

Natural processes lead to constant climatic changes on Earth. However, since the industrial revolution human activity has been a new factor to consider. Based on past trends, it seems that historical climate data for recent centuries makes sense only if human activity is factored in—though it may not outweigh natural causes.

The major factor affecting climate change is the high-carbon lifestyle of the industrial nations. Due to its rapid industrial development, China has overtaken even the U.S. in terms of its annual carbon emission rate.

Do humans affect climate?

In the last 150 years, pollution of the atmosphere by humans has been increasing. Because the climate reacts slowly, it is uncertain whether these emissions are already affecting the Earth. However, the demand for food, housing, and energy is likely continue to rise and negatively affect the environment. Depending on the extent of the impact, the protection of the Earth's climate may demand far-reaching political, economic, and social changes.

Industrial nations: leaders or perpetrators?

Industrial nations consume nearly 75–80 percent of the fossil fuels burned each year, and they are responsible for most greenhouse gas emissions. Meanwhile, less-developed countries suffer because they lack the resources to handle natural disasters. For these reasons, industrialized countries share the responsibility of reducing emissions. The 1997 Kyoto Protocol marked the first step toward international climate protection. Some of the solutions may include using renewable energy sources and changing public policy toward motor vehicles.

Industry-driven slash and burn deforestation abolishes large woodlands and releases massive amounts of greenhouse gases into the atmosphere.

Five minutes to midnight

Even if current emissions stopped completely, it would take decades for carbon dioxide and other gases to return to the levels they were at before the industrial revolution. Reducing emissions may be the only way to prevent a possible climatic catastrophe, but the outlook is poor: The United States, one of the largest emitters of carbon dioxide, has rejected the Kyoto Protocol. Meanwhile, industrial expansion in China and India continues to generate even more emissions.

in focus

CLIMATE POLICY

The UN Kyoto Protocol aims to reduce emissions of six greenhouse gases by 5.2 percent compared to their levels recorded in 1990. This is viewed by many as a milestone in climate protection; however, many critics call it a "drop in the bucket" as long as high-emission industrial heavyweights, such as China and the U.S., resist the mandate. Also, some scientists question the effectiveness of requesting the reduction of carbon dioxide only.

Public awareness for climate issues increases internationally and pressures politicians to act.

see also: Environmentally sustainable construction, p. 370

global warming

Regardless of the degree to which human activity affects the world's climate, the fact remains: it is getting warmer. This rise in temperature can no longer be explained on the basis of natural climatic cycles alone.

Over the course of geological history, the Earth's climate has undergone numerous changes. The past million years, in particular, have been marked by continuous cycles of rising and falling temperatures.

Average global temperatures have fallen below 50°F (10°C) during cold phases, followed by increases of up to 62°F (17°C) during warm periods. Even since the most recent ice age, climatic cycles have continued, with temperatures some 1.8–3.6°F (1–2°C) above and below today's average value of 58°F (14.5°C).

Heavy logging and agricultural clearance during the 21st century have led to a sharp decrease in the acreage of the world's great rain forests.

GREENHOUSE GASES Carbon dioxide (CO_2) is one of the six types of greenhouse gas regulated within the framework of the Kyoto Protocol, along with methane (CH_4), nitrous oxide (N_2O), hydrofluorocarbons (HFCs), perfluorocarbons (PFCs), and sulfur hexafluoride (SF_6). CO_2 is emitted through the respiration processes of plants and animals, as well as the burning of wood and fossil fuels (coal, oil, and natural gas), while oceans and forests remove CO_2 from the atmosphere.

basics

Nevertheless, the unusually rapid warming of the climate in recent years is cause for concern. During the past 120 years, worldwide average temperatures have risen by some 1.26°F (0.7°C), with the most significant increases registered since the early 1970s. The direct consequences of this warming include the melting of mountain glaciers, the retreat of polar ice, increases in extreme weather phenomena, and rising sea levels—with an average increase of some 6.7 inches (17 cm) observed in the 20th century alone. According to reports from the Intergovernmental Panel on Climate Change, the main cause of the excess warming is the burgeoning concentration of greenhouse gases in the atmosphere as a result of human activity. During the 10,000 years before 1750, atmospheric carbon dioxide (CO_2) levels never exceeded 280 parts per million (1 ppm = 0.001 percent). Since 1750 this concentration has shown an accelerating rise, reaching over 380 ppm today. The upward curve is chiefly explained by the burning of fossil fuels and clearing of old-growth forests. Over the same time period, the concentration of methane, another greenhouse gas, also rose by 148 percent, mainly due to the expansion of mass livestock operations. If the proportion of these gases in the atmosphere continues to grow, a climatic catastrophe can be expected.

The production of greenhouse gases like carbon dioxide is mainly due to industrial activities in both developed and developing countries. In Russia, China, and other newly industrializing countries, the levels of air pollution and the output of greenhouse gases is increasing.

IPCC: THE UN CLIMATE COUNCIL

In 1988, the Intergovernmental Panel on Climate Change (IPCC) was founded by the World Meteorological Organization (WMO) and the United Nations Environment Programme (UNEP). The panel, staffed by experts from more than 130 nations, collects, analyzes, and evaluates the latest scientific research and international studies related to the causes and possible effects of global warming. Since 1990, its findings have been published regularly in the form of status reports, which serve as a basis for negotiations at international climate conferences.

Dr. Rajendra Pachauri, chairman of the IPCC and Nobel Peace Prize co-recipient.

in focus

One of the main problems caused by global warming is the increase of droughts, and with it the increasing threat of famines and wildfires.

see also: Enviromental exploitation, p. 130

the effects of climate change

Our blue planet is gradually growing warmer, with far-reaching consequences for human life and the environment. The full extent of these consequences cannot yet be calculated.

Climate change has begun, and it will continue throughout the 21st century. Under the most favorable circumstances, according to calculations based on the latest models, temperatures will increase by 3.24°F (1.8°C) by 2100, while in the worst-case scenario they will rise by some 11.52°F (6.4°C). Much of the divergence in the calculated scenarios depends on the future emission levels of greenhouse gases. At the same time, the unpredictability of feedback effects among the climate system's various components makes accurate forecasting extremely difficult.

Many islands in the Pacific Ocean may sink into the ocean with the rising sea level.

Experts currently point to a likely rise in temperature of about 3.6–5.4°F (2–3°C), with substantial regional differences. The most pronounced warming is expected to occur in the Arctic region. There, sea ice will disappear during summer, and Greenland's ice cap will melt away completely. Sea levels are predicted to rise between 7–23 inches (18–59 cm), flooding many coastal areas and threatening the existence of some low-lying island nations. Worldwide, inland glaciers will melt, permafrost will thaw, deserts will expand, vegetation zones will shift, and tropical windstorms will increase in strength. In North America, heat waves and forest fires will be more common, while farmlands in South America will become drier and saltier. In Western Europe, precipitation will increase in winter and decrease in summer. Around one billion people will suffer from drinking water shortages, and more than one-fourth of all plant and animal species will be threatened with extinction.

For the first time in hundreds of thousands of years, scientists expect that by 2015 the summer ice cover may disappear because of the melting ice caps.

Due to global warming, glaciers are shrinking, causing sea levels to rise.

basics

GREENHOUSE EFFECT In 1896, Swedish chemist Svante Arrhenius first calculated the warming effect on the Earth due to rising atmospheric CO_2 levels.

CLIMATE MODELS In 1967, Japanese meteorologist Syukuro Manabe performed the first climate model calculations reflecting increases in CO_2 concentrations.

WORLD CLIMATE CONFERENCE The first world conference on climate at which scientists discussed the greenhouse effect took place in Geneva in 1979.

milestones

NOBEL PRIZE FOR CLIMATE PROTECTION

Al Gore is lecturing on the topic of global warming worldwide.

Involvement in the worldwide mobilization against climate change is a contribution to world peace. With this justification, the Nobel Prize Committee honored the activities of former U.S. Vice President Al Gore and the UN's climate council (IPCC) in 2007 with the world's most prominent political award, the Nobel Peace Prize. In the same year the documentary film *An Inconvenient Truth*, which starred Gore, won an Oscar. The film's message: Humanity has only a few more years to prevent climatic catastrophe. Yet it can be done, if each individual takes personal responsibility for the effort.

EARTH

ENVIRONMENTAL PROTECTION

Environmental protection includes all measures intended to protect the natural environment from damaging influences and to improve the quality of polluted ecosystems. Such measures range from individual environmentally conscious behavior to international agreements to keep the air, water, and soil clean. One of the most important purposes of environmental protection is the prevention of negative impacts on public health. Furthermore, the basic essentials required by individuals as well as society as a whole need to be considered as much as the interests of future generations. Maintaining the basis of human life is increasingly viewed in light of sustainability principles, both on a national and an international level.

ENVIRONMENTAL EXPLOITATION

Since the beginning of the industrial age, the exploitation of the natural environment by humankind has taken a turn for the worse. Natural resources are depleted and waste is produced at a higher rate than the soil, air, rivers, and oceans can sustain.

polluted air

For a long time, pollutants such as carbon dioxide, carbon monoxide, and nitrogen oxides were blown into the air without restriction. The results are global warming, acid rain, smog, and ozone depletion in the stratosphere.

The composition of the air has changed significantly since the beginning of industrialization—unfortunately for the worse. The burning fossil fuels (such as coal and crude oil to run power stations and motor vehicles) emits particulate matter and numerous chemical compounds including sulfur, nitrogen, and carbon oxides. Once these compounds are released into the atmosphere, they react with water vapor to create sulfuric acid and nitric acid, which later return to the Earth's surface in the form of acid rain. This results in large-scale forest decline. Forests in North America and Europe have been especially affected. The reason for the trees dying is that acid rain directly attacks the leaves and needles of the trees, and this prevents photosynthesis from working effectively. In addition, dissolved heavy metals and aluminum leak into the ground. Here, they poison bacteria living in the soil and damage the fragile root tips. Trees can no longer effectively take in water and nutrients. The consequences of the increasing acidification of rivers and lakes include fish dying on a large scale as well as a reduction in biodiversity. The release of carbon dioxide has also seriously impacted our environment. By now, there is more carbon dioxide being emitted than plants can take in and convert into oxygen. The situation is worsened by ongoing massive deforestation and burning of tropical rain forests: the "green lungs" of the Earth. The concentration of carbon dioxide in the atmosphere is increasing; this is a fact that will not be changed by the odd prevention measure. It traps the heat emitted by the Earth and prevents the release of heat radiation into space. Global climate change is the consequence of the greenhouse effect enhanced by human activity. The process is further accelerated by the emission of large quantities of chlorine compounds. As a result, the ozone layer has developed holes that allow dangerous UV radiation to reach the Earth directly and unfiltered.

The increase of ground-level ozone concentrations leads to summer smog in congested urban areas.

Wildfires caused by droughts or human carelessness can have severe effects on the climate because they release large amounts of carbon dioxide into the atmosphere.

SMOG

in focus

Smog is a combined word derived from *smoke* and *fog*. Winter smog is a mixture of fine particulate matter, fog, soot, and sulfuric emissions. This condition occurs above cities and heavy industrial areas during atmospheric inversions in the winter. Summer smog is mainly formed by nitrogen oxides and hydrocarbons emitted by motor vehicles. These compounds react with the sunlight, which creates aggressively harmful ozone.

In Chinese cities, pedestrians often wear protective masks against the smog.

see also: Energy technology, pp. 346–353

water at risk

Water is one of the most valuable resources on Earth. There should be sufficient water available to cover the demands of the world population. However, there is a shortage of clean water due to the increasing pollution.

Colliding tankers, damaged platforms and pipelines, as well as the release of residual oil from oil tanks, are potential causes for catastrophes. Pictures of dying seabirds show the extent of such environmental damage. Not all environmental hazards are this obvious. Water pollution due to pesticides and heavy metals, such as mercury, lead, cadmium, and zinc, largely remains unnoticed and can be a slow and often invisible process. The dangerous prior assumption that the oceans could naturally purify themselves without limit, doubling as dump sites for industrial and nuclear waste, and as final disposal sites for chemical weapons and discarded ships, has caused possibly irreparable damage.

Not only the oceans, but many creeks, rivers, and lakes have also suffered from human interference. Many lakes have become polluted due to the chemicals added by household wastewater and the overfertilization with phosphates in agriculture. Moreover, unknown effluents from industry and acid rain add to the acidification of water bodies, which contributes to the decrease of biodiversity. Pollutants accumulate in the tissue of aquatic organisms that are part of the same food chain as humans.

Groundwater deterioration

Surface water and groundwater constantly interact with each other. Normally, groundwater is of a higher quality than surface water. This is due to the natural purification process during the passage of water through various rock layers in the ground. However, this purification effect is limited. In many areas, the groundwater is relatively close to the surface and is contaminated by nitrates from fertilizers and pesticides leaking into the soil due to their excessive usage in large areas. Pollutants may also leak into the groundwater from contaminated sites or local landfills. Furthermore, fossil water reservoirs, which have been stored underground for thousands of years—for example, beneath the Sahara—stand the risk of depletion if used extensively for irrigation or industrial purposes.

The blowout of an oil well can cause a large oil spill that leads to severe water pollution.

basics

WATER USAGE in industrialized nations is at a daily average of 38 gallons (145 l) per person, while 2.2 billion people worldwide have no access to clean water.

Excessive irrigation of farmland leads to the waste of water and chemicals and eventually pollutes the groundwater.

Inhabitants of rural African regions are frequently dependent on handpumps for their daily water supply.

soil pollution and erosion

The availability of productive land suitable for cultivation is extremely important to ensure sufficient food supply. Such land is becoming scarce due to soil contamination and deforestation which cause soil erosion.

Today's food production is increased by intensive agriculture and animal husbandry, as well as the targeted use of fertilizers and pesticides. This may not only pollute both the groundwater and surface water, especially when used in monocultures, but may also reduce the content of organic compounds, as well as the number of organisms living in the soil. As a result the natural fertility of the soil is reduced. Areas that have already been cultivated are often lost forever due to overuse, waterlogging, salinization, and desertification.

Tropical rain forests are often sites that are cleared for agricultural purposes. More often than not, cultivating such land turns out to be problematic because many of these areas are hardly suitable for intense agricultural activities. Long-term use is usually only possible by selecting adapted seeds, fertilizers, and artificial irrigation. The latter may result in entire lakes drying out. An example of this happening is the Aral Sea, which has completely dried out. The clear-cutting of rain forests for wood or cultivation leads to large-scale soil erosion.

Valuable land is increasingly lost to housing developments, road construction, and the development of recreational areas. These developments seal the soil, lowering the groundwater level, change the microclimate adversely, and deprive plants and animals of their habitat. The flood risk increases due to the lack of rainwater drainage pathways into the soil. Large quantities of waste are produced in the industrialized countries every year and often disposed of in unsafe locations. This waste includes toxic and sometimes even radioactive waste. If landfill sites are insufficiently reinforced, toxic agents may seep into the lower layers of the ground and, in the worst case, contaminate the groundwater. Appropriate and safe disposal of toxic industrial waste products is very costly. Therefore this type of waste is often "exported" to developing countries where it may not be disposed of in an ecologically safe manner.

Modern methods of fertilization have led to an increase in chemical nutrients in ecosystems.

The disposal and recycling of hazardous waste underlies national and international regulations.

The cleaning up of oil spills along corroded pipelines is important to avoid soil contamination and the destruction of animal and plant habitats.

CATASTROPHIC FLOODING IN BANGLADESH

in focus

Due to the coinciding humid monsoons of the summer season and the melting snow from the Himalaya, Bangladesh frequently suffers from flooding. Clear-cutting in large forested areas results in soil erosion. The Ganges and the Brahmaputra have to carry an increasing amount of sludge, which causes these rivers to overflow. Tropical cyclones followed by extremely high storm surges intensify the flooding that often results in a catastrophe.

Floods in Bangladesh have resulted in housing damages, land devastation, and fatal casualties.

see also: Erosion, p. 89

loss of species diversity

Ninety-nine percent of all species that once lived on Earth have become extinct. Climate change and natural catastrophes used to be the principal reasons for the decimation of biological diversity; it is now usually humankind.

Species diversity provides the foundation for individual ecosystems and thus is the prerequisite for the functioning of the biosphere. It is an immeasurable source of food and medicinal products, and an irreplaceable resource as a gene pool. Due to the increasing destruction and pollution of natural habitats, over-fishing, hunting, and lucrative trade, humans are destroying the biological multitude of life—and with that the basis of their own livelihood. The public is dismayed when when a well-known species like tiger, whale, or mountain gorilla becomes endangered; however, the majority of other cases are hardly ever noticed by the public.

Habitat destruction has led to the decimation of the orangutan population of Sumatra and Borneo.

Apart from the decline of species diversity within habitats, loss of genetic diversity within individual species has also been observed. The decimation of individual populations leads to a reduction in genetic regeneration capacity within a species. For instance selective deforestation reduces the quality of the genetic material of affected tree species, since only strong, healthy trees are logged and the weaker ones are left behind.

Human's intentional or unintentional introduction of exotic animal and plant species to new habitats plays an important role. These so-called neozoic species can become a threat to local species and lead to their complete extinction. For instance, in New Zealand the national emblem, the flightless kiwi, is threatened with extinction by introduced rats and feral cats.

Species protection

Throughout the course of evolution the extinction of species has been a natural process. The best known example is the mass die-off of dinosaurs 65 million years ago. However, the majority of species have disappeared only over the last 150 years, and at no time has the extinction rate been as high as it currently is. More than 41,000 species are on the Red List of endangered animal and plant species, which is published regularly by the International Union for the Conservation of Nature (IUCN). Trade in endangered species has been restricted and prohibited since 1973. The objectives of the Convention on Biological Diversity (CBD), signed in 1992, are both the protection of biological diversity and its sustainable utilization.

The yellow pheasant's eye (*Adonis vernalis*) is protected in many European countries.

basics

NOAH'S ARK IN PERMANENT ICE An international seed bank was established on the Norwegian Archipelago Spitsbergen in 2008. There, in the artic region of permafrost, up to 4.5 million seeds from the most important food plants on Earth are being safely stored in a bunker, where they are protected against climate change, wars, and epidemic plant diseases, and therefore secure a biodiversity for coming generations.

The baiji, a freshwater dolphin only found in the Yangtze River in China, was declared "functionally extinct" in 2006.

The polar bear is listed as a threatened species under the Endangered Species Act. Its sea-ice habitat is severely threatened by global warming.

environmental consciousness

If human beings lived in harmony with nature, the concept of environmental protection would likely be unknown. However, since we are both perpetrators and victims of environmental damage, humanity's survival ultimately depends on our ability to preserve the natural world.

For millennia, human beings have formed and changed their surroundings to suit their needs—often with irreversible consequences for the natural environment.

consciousness | actions | everyday life | saving energy

ENVIRONMENT

For the most part, mankind's awareness of the world they inhabit and the resources they are using has substantially changed over the last decades. This includes many people's behavior concerning environmental protection as well as an awareness of the consequences resulting from interventions into Earth's natural ecosystems.

The realization that we must protect nature from the effects of human activity is not a modern innovation; however, a definite environmental consciousness has arisen in Western industrialized nations only since the 1960s. In recent decades, coastlines contaminated by spilled oil, declining fish populations, smog alerts in cities, the destruction of the ozone layer, increasing desertification, and dying forests have all pointed to the tolerance limits of ecological systems—and of the planet itself. The protagonists of most environmental protection efforts include individuals, civic alliances, and political associations such as environmental and nature groups, as well as governments, businesses, and the scientific community.

All of them face the challenge of preserving the natural foundation for human life, maintaining the balance of nature, and countering environmental damage that has already occurred.

However, since the environment as a whole cannot be fully protected or restored to its original condition, environmental protection efforts always require compromises between economic, political, and social interests. Not every measure made possible by technology is truly feasible, given the various costs and trade-offs involved. Many proposals that are strongly advocated in scientific circles cannot be implemented due to economic concerns or a lack of political will. Similarly, many initiatives set in motion by governments meet with only hesitant acceptance in society.

The Yosemite National Park is located in the U.S. state of California. Designated a World Heritage Site in 1984, Yosemite is internationally recognized for its spectacular granite cliffs, waterfalls, clear streams, and biological diversity.

Sea birds, other animals, and plants struggle for life after an oil spill. The damage caused to the whole ecosystem is enormous.

in focus

GREENPEACE

The history of this international environmentalist organization began in 1971, when a small number of American and Canadian peace activists gathered to protest U.S. nuclear weapons testing in the Aleutian Islands. The group—which adopted the name Greenpeace in 1972—quickly expanded both its geographical reach and its areas of concern. Its high-impact initiatives, which often gain considerable media attention, have been aimed against the extermination of whale species, the slaughter of baby seals for their fur, global warming, the destruction of rain forests, and the use of genetic technology, and so on.

Today, the organization has representatives in some 40 countries and has become a widely recognized nongovernmental organization (NGO), enabling it to serve as an official observer and/or advisor at numerous international conferences on the environment.

Greenpeace activists are demonstrating against the transportation of radioactive waste.

see also: Human environmental impact, pp. 280–281

think globally, act locally

Environmental problems do not end at national borders; in fact, they are often worldwide in scope. Accordingly, protecting the environment requires not only local, but regional and global thought and action.

in focus

SUSTAINABLE DEVELOPMENT

This term refers to environmental, developmental, and economic policies that aim at raising the population's quality of life without compromising the future prospects of coming generations. Environmental protection and the preservation of natural resources are key requirements for the international community today.

Meeting in Berlin during the G8 conference in 2007.

The first Conference on the Human Environment was held in Stockholm in 1972, under the sponsorship of the United Nations. This international gathering led to the creation of the UN Environment Programme (UNEP). In 1992, an environmental conference in Rio de Janeiro became a turning point in international negotiations on questions of development and the environment. In addition to establishing key principles and an action plan for a worldwide sustainable development, the participating countries signed important conventions on biodiversity, deserts, and climate change. At the Third Conference of the Parties to the Framework Convention on Climate Change, held in Kyoto in 1997, representatives of 167 nations agreed to reduce greenhouse gas emissions by 2012. In 2002, ten years after the Rio gathering, the World Summit on Sustainable Development was convened in Johannesburg. Instead of binding resolutions, however, the participants reached agreement only on a plan of action to reduce species loss and allow overfished marine life to regenerate.

Since 1973, protecting plants and animals from extinction has been the aim of the Convention on International Trade in Endangered Species of Wild Fauna and Flora (CITES), also known as the Washington Endangered Species Agreement. Protecting entire habitats is the focus of the Ramsar Convention on Wetlands, which entered into force in 1975, as well as the UN's Man and the Biosphere (MAB) program and the UNESCO World Heritage List. Even broader in scope is the United Nations Convention on the Law of the Sea (UNCLOS), in effect internationally since 1994. This agreement grants coastal nations an economic zone of 200 nautical miles, which they may use but must also protect. The open sea is named the common heritage of mankind; its use is allowed only after the evaluation of potential environmental effects.

As part of the UNESCO World Heritage List the underwater landscape of the Great Barrier Reef receives special protection.

Climate change was the focus of the UN environmental conference in Bali in 2007. Once again, the representatives did not agree on binding resolutions.

basics

UNEP The United Nations Environment Programme oversees and coordinates the actions of all UN organizations dealing with the environment, while also working with politically independent nongovernmental organizations.

AGENDA 21 This is the economic and environmental action plan for sustainable global development in the 21st century, agreed upon in Rio de Janeiro in 1992. Both national and local programs of action have been developed to implement this agenda.

The panda is one of the many species threatened with extinction. It has become the symbol of environmental protection.

environmental concerns in everyday life

Climate change and environmental degradation call for global strategies to achieve sustainable development and balance economic and ecological goals. However, environmental protection begins with each individual.

Public awareness is rising that every individual can—and must—contribute to protecting the environment in order to preserve it. Surveys show that more and more people are willing to alter their daily habits in response to climate change and worldwide environmental degradation, for example, by changing their consumption patterns. In the industrialized nations, demand is expanding for organically produced and fair-trade goods.

Organic food products are grown without the use of conventional pesticides, artificial fertilizers, or food additives. They are also not genetically modified.

LOHAS "Lifestyles of Health and Sustainability," refers to choices made by a new generation of mainly well-off consumers. Unlike earlier environmentalists, their aim is not a simpler life without consumer goods. Instead, they invest in sustainable projects, use solar power, drive hybrid cars, eat organic foods, and shop in eco-boutiques. Market researchers estimate that some one-third of U.S. consumers are followers of this new "green" movement.

basics

Organic products

The most significant growth rates have been seen in organic foods, which have earned a permanent place on the shelves of most supermarkets and can now even be found at discount stores. Buyers of organic foods are opting for eco-friendly production, without chemical or synthetic fertilizers and pesticides, and with more natural living conditions for farm animals. Fair-trade products guarantee a living wage for growers in developing countries, along with basic environmental standards and labor rights. For many consumers, the search for nontoxic and "green" alternatives extends beyond food products, a trend that companies have been quick to recognize. Manufacturers are reducing unnecessary packaging, offering energy-saving appliances, or building furniture from wood grown on sustainable plantations. Clothing firms sell organic cotton T-shirts or athletic shoes made from recycled tires; paint and building supply companies highlight their products' nontoxic ingredients and environmentally friendly disposability; the IT industry offers "green" computers; and the automobile industry boasts of low CO_2 emissions rather than horsepower. "Green" labels and seals of approval are available for many products and services, such as free-range eggs, renewable electricity sources, and even hotels. Sustainable products come at a price, of course. For instance, a compact fluorescent lightbulb costs significantly more than a traditional bulb. However, its longer life and low electricity usage compensate several times over for the difference in price. Even when the cost-benefit calculation is not positive for the individual, it is still considered worthwhile for the environmental positives.

A poster about reducing the transportation carbon footprint at the Sexy Green Car Show in Cornwall, UK.

Ecotourism, or ecological tourism, is becoming more and more popular among environmentally conscious individuals. The concept seeks to minimize the negative impact of conventional tourism on the environment.

see also: Food technology, pp. 340–345

ecologically friendly consumption

The climate is warming, in part due to the accumulation of greenhouse gases. Because even small changes in our daily routines can help reduce CO_2 emissions, everyone can get involved and make a difference.

> **issues to solve**
>
> **GOOD INTENTIONS, UNINTENDED CONSEQUENCES**
> The rising use of renewable raw materials as biofuels—viewed until now as an effective weapon against climate change—has led to unexpected side effects. Food is becoming scarcer and more expensive—by around 57%, say figures from the UN Food and Agriculture Organization (FAO) for the 12-month period up to March 2008.

Fossil fuel use is a main cause of climate change and the greenhouse effect. Thus, we must significantly reduce CO_2 emissions from activities such as heating, power generation, and transportation.

Any conventional diesel engine can be converted into one that runs on vegetable oil. If previously used, the oil must be refined before use.

Reducing our ecological "footprints"

One option is for each household to switch to renewable energy sources wherever possible, such as solar, hydropower, wind, or geothermal systems. Another important step is to reduce our individual energy "footprints." For example, driving a car places a particularly high burden on the environment, due to the carbon dioxide emissions into the air. However, there are several ways to lessen the impact and still continue using a car. For instance, careful driving can help reduce harmful emissions. Car buyers can also now choose from many new energy-saving and alternative-energy models. Better still, drivers could sometimes leave their cars at home and use public transportation, ride their bicycles, or walk to their destinations. Carpooling also reduces the number of vehicles on the road.

Climate-friendly consumption

Within homes and buildings, numerous opportunities are available to save energy, for instance, by using energy-saving lightbulbs, avoiding the standby setting for electronic devices, adjusting thermostats a few degrees, and adding more effective insulation. In addition to the energy use clearly reflected on utility bills, "hidden" consumption arises from the production, transportation, and disposal of goods, which has an invisible but significant effect on the climate. By researching purchases—for example, giving preference to local and locally-grown products—we can distinguish relatively "climate-friendly" products from relatively "climate-unfriendly" ones, thus reducing carbon dioxide emissions.

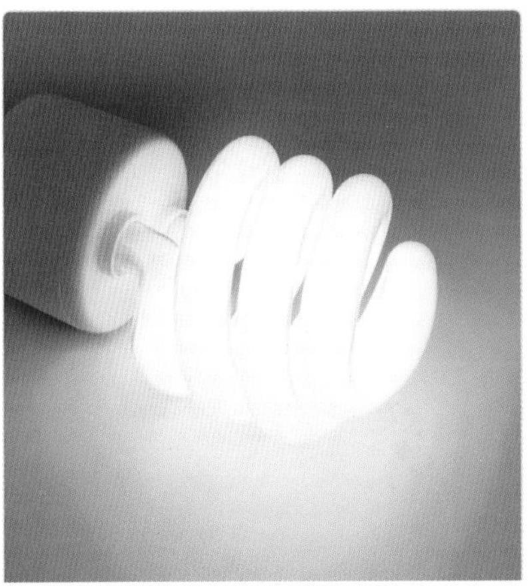

The compact fluorescent lamp is an energy-saving lightbulb that uses less power than a regular bulb.

Climate-neutral consumption

"Climate-neutral" activities are those activities for which measures have been taken to offset the carbon dioxide they generate through climate-friendly initiatives in another place. There are now "climate-neutral" household appliances, flowers, television sets, entertainment events, and even universities. Even those who need or want to travel long distances by plane can help protect the climate by voluntarily offsetting the emissions arising from their trip by purchasing carbon offsetting certificates. The funds are then invested, for example, in solar, hydropower, biomass, or energy conservation projects, to compensate for the corresponding amount of greenhouse gases produced by the flight.

Airlines such as EasyJet are offering CO_2 offsetting options that help make up for the emissions of a flight.

see also: Fuel cells and biomass, p.353

BIOLOGY

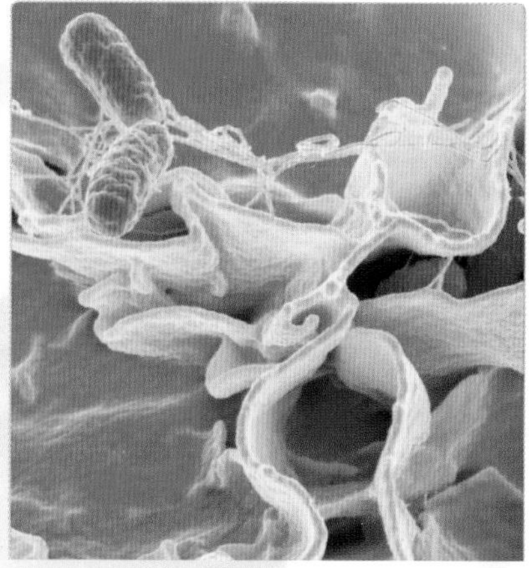

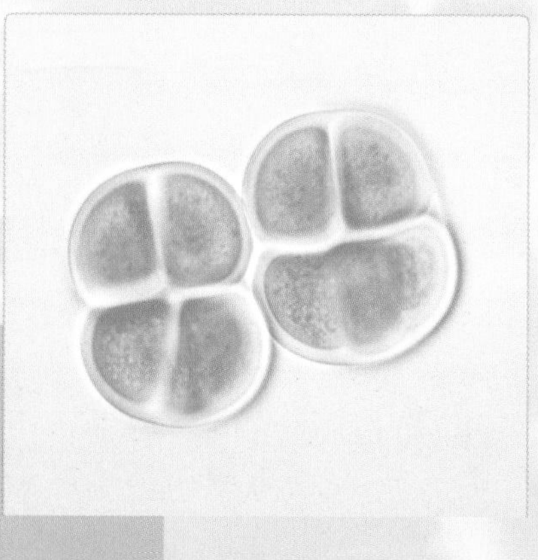

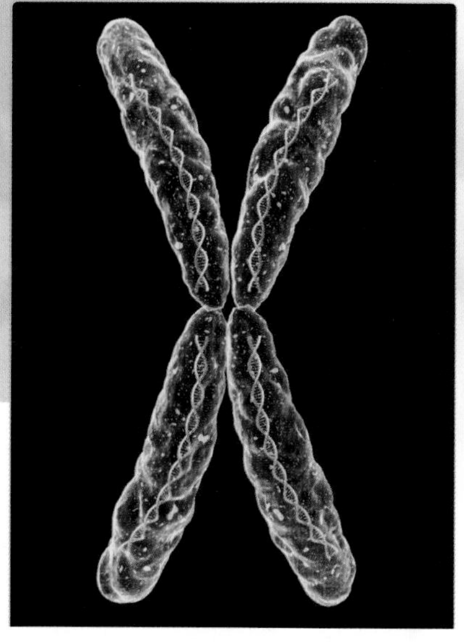

EVOLUTION

The appearance of the Earth and the organisms living here has been changing for many billions of years and will keep changing in the future. Impressive discoveries of fossils show this change in appearance and are often subject to controversial discussion.

Both the evolutionary theory proposed by Charles Darwin and the genetic principles discovered by Gregor Mendel were significant steps in tracing the development of life on Earth and the factors involved. They concluded their theories during the mid-19th century. Evolutionary theory has since gained acceptance by the scientific community and the public. Despite the increasing degree of knowledge about the development of life, many mysteries remain to be solved.

Devonian | Carboniferous | Permian

Occupation of the air by invertebrates, some giant due to the high amount of oxygen
Animals became further independent from aquatic environment
Development of seed plants

Tiktaalik (transition fish and amphibians)

Meganeura (giant dragonfly)

Ichthyostega (amphibian)

Glossopteris (gymnosperm)

Ophiacodon (reptile)

Precambrian | Cambrian | Ordovician | Silurian
Development of life: prokaryotes and increasingly complex eukaryotes
Occupation of terrestrial habitats by plants and animals
Ediacara biota (invertebrates)
Cooksonia (early land plant)
Jawless fish (vertebrate)
Asteroxylon (cryptogam)
Cladoselache (shark)
In the primordial ocean the first organic molecules formed, which gradually combined to become simple life-forms.
500
418
MILLION YEARS AGO
First periods (Precambrian, Cambrian, Ordovician, Silurian)

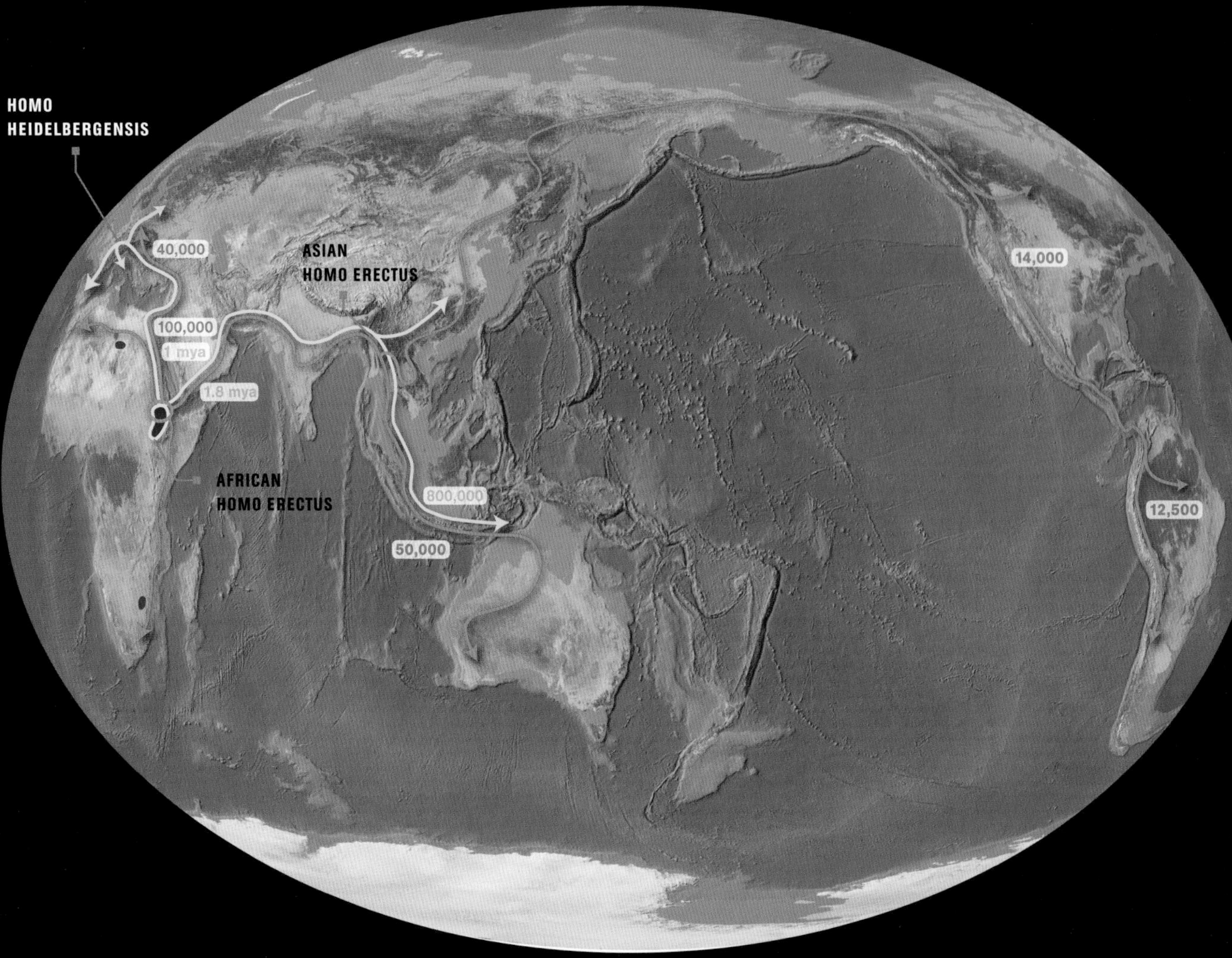

HOMO HEIDELBERGENSIS

40,000

ASIAN HOMO ERECTUS

100,000

1 mya

1.8 mya

AFRICAN HOMO ERECTUS

800,000

50,000

14,000

12,500

Early hominids (4–2.5 mya)
Prehistoric man (2.5 mya–30,000 years ago)
Homo sapiens (modern man) (200,000–0)

According to the "Out of Africa" theory, human beings originated from East Africa. From there the species spread over the rest of the world.

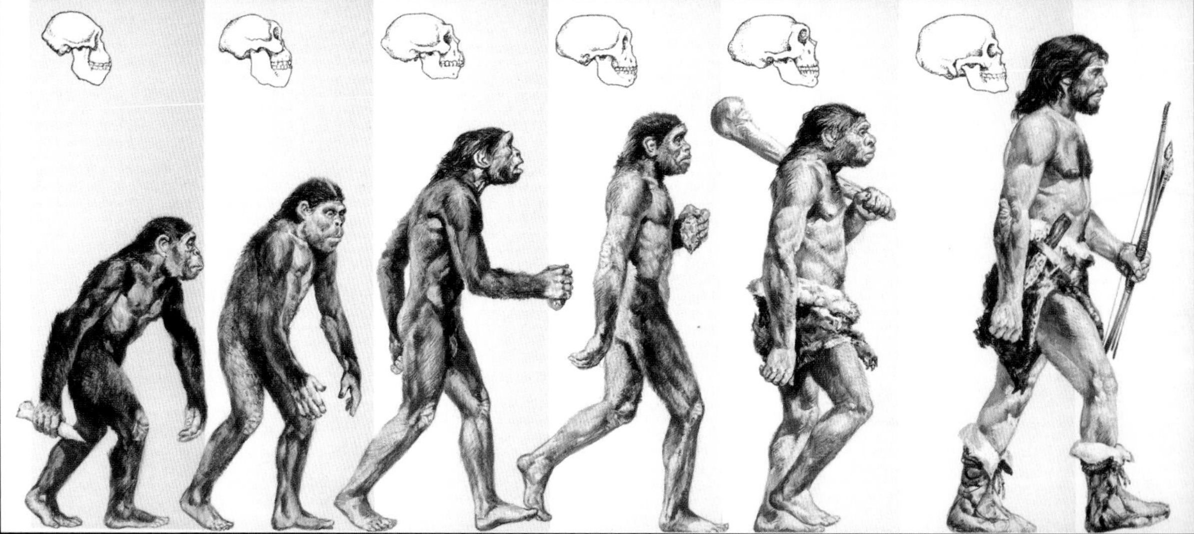

The evolution of modern humans occurred over many development stages, which partially existed in parallel with one another. From left to right: *Proconsul, Paranthropus, Australopithecus, Homo erectus, Homo neanderthalensis, Homo sapiens.*

The fungus *memnoniella echinata* uses its yellow spores for reproduction and distribution.

In plants the green pigment chlorophyll absorbs sunlight to perform photosynthesis.

Fossil findings provide important information about the origin and evolution of long extinct life forms.

EVOLUTION

About 3.5 billion years ago, the first signs of life on Earth were single-celled bacterial organisms that lived in the primeval oceans. Single-celled animals and algae, as well as all multicellular organisms, evolved from these bacteria. This evolutionary process was limited to the oceans for many millions of years. Some species developed the ability to photosynthesize, a process by which oxygen is released as a waste product. Over time, photosynthetically active organisms enriched the atmosphere with oxygen and gradually an ozone layer formed, protecting the Earth from the harmful ultraviolet radiation emitted by the sun. The conditions were now favorable for life to evolve on land. The first organisms to make the transition to terrestrial habitats were moss-like plants. Small invertebrates of the phylum Arthropoda followed. Arthropods were the ancestors of spiders and insects.

The first vertebrates were jawless fish with primitive fins. Amphibians evolved from air-breathing fish with lobe-shaped fins. These animals slowly occupied terrestrial habitats through the extensive swamps in dense forests. However, they remained in close contact with the aquatic environment, as they were forced to return to the protection of water to lay their eggs. Reptiles were finally able to live independently from bodies of water with the evolution of hard-shelled eggs that prevented the dehydration of enclosed liquids. Plants too became less reliant on aquatic environments as they evolved specific vessels that transported water and nutrients throughout the tissues of the whole plant. Eventually, they also developed seeds, which allowed for a reproduction process fully independent of bodies of water.

During the following phase, reptiles expanded into every available habitat and took advantage of every ecological niche. While dinosaurs were spreading all over the Earth, the first mammals also appeared, but for many millions of years they were a rather small and insignificant group of animals. This changed about 65 million years ago, probably due to a large meteorite hitting the Earth. The meteorite filled the atmosphere with suspended dust, which blocked the sunlight and cooled down the Earth's temperatures dramatically. The dinosaurs, as well as many other animals, became extinct. However, as devastating as this catastrophe was, some terrestrial life continued and survived the subsequent major mass extinction events. The elimination of dominant species during such mass extinction allowed other animals, like mammals and birds, to finally occupy newly vacant ecological niches, and for life to continue evolving. As warm-blooded animals that could maintain a constant, warm body temperature, mammals and birds were able to adapt to a very wide range of environmental conditions. The era of reptiles was replaced by the era of mammals whose dominant species became *Homo sapiens*. Humans evolved about 2.5 million years ago from their hominid ancestors, the australopithecines. Walking upright, this group of early humans left their original homeland in East Africa, and began their conquest of all the continents.

Paleogene | Neogene
Animals such as whales and dolphins return to aquatic habitats
Development of primates
Human evolution, humans began to prepare tools and to use fire
Purgatorius (primate)
Durodon (whale)
Australopithecus (pre-human)
Homo erectus (first human species)
Homo sapiens (modern human)
ARS AGO
Neogene

Triassic | Jurassic | Cretaceous
The age of saurians, their extinction may have been caused by the impact of meteorites
Development of birds and mammals, mammals were still small during this time
Development of angiosperms
Shastasaurus (aquatic saurian)
Megazostrodon (mammal)
Diplodocus (dinosaur)
Archaeopteryx (transition reptiles and birds)
Short-tailed, winged saurian
Credneria (angiosperm)
252
199.5
142
65
24 MILLION Y
Triassic
Jurassic
Cretaceous
Paleogene

the first biological molecules

The primordial soup theory and the biofilm theory are the most popular explanations of how molecules became organized life forms. The panspermia theory, which says that life arrived on comets, is less discussed.

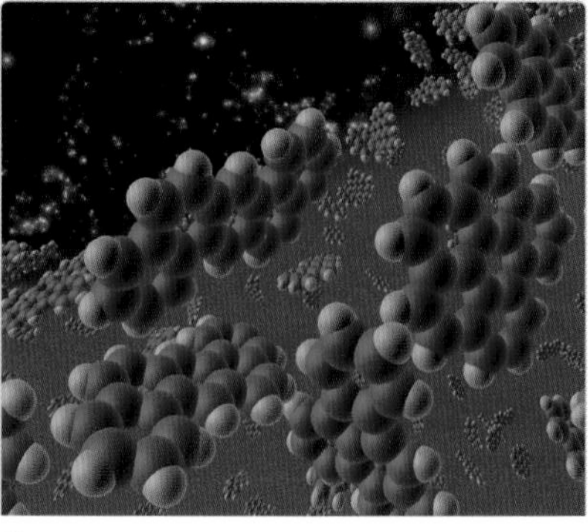
Complex sugar, amino acid, and fat molecules were probably the first steps to living things.

first molecules | eukaryotes

ORIGINS OF LIFE

The origins of life on Earth may never be proved with complete certainty. It seems likely that simple organic molecules combined under the right environmental conditions, eventually producing cell-like organisms. From these, prokaryotic and later eukaryotic cells developed. These cells were the starting point for the evolution of multicelled organisms such as today's plants, animals, and human beings.

Scientists in many fields have offered theories in their attempts to solve the mystery of the appearance of life on Earth. Since Louis Pasteur first observed bacterial cells dividing under a microscope, biology has recognized the basic concept that life arises only from already existing life. An exception to this rule would seem to be found in the first appearance of life on Earth. Physical and chemical conditions then were very different from today. The first organic molecules, from which life first developed, were seemingly produced when energy was added to nonliving material about 3.4 billion to 4 billion years ago. The primordial soup theory described the mechanisms and individual steps in this process, which was recreated in laboratory experiments in 1953. According to this hypothesis, simple organic molecules formed from chemical reactions among inorganic molecules in the primordial sea.

After the primordial soup theory, the most popular theory of life's origins is the biofilm theory. According to this theory, the first components of life, and life itself, originated around heat sources in the sea. The panspermia or exogenesis theory suggests that life first arrived on Earth via a meteorite collision. The mechanisms proposed for interstellar panspermia are hypothetical and unproven, and the theory is given little credence among scientists.

Hydrothermal vents: breeding grounds of life

The hydrothermal vents found on the ocean floor are particularly interesting to the study of the origins of life. Black smokers are those vents which emit very hot water, darkened by its high sulfide content. The mineral pyrite forms when the hot water contacts cold seawater. The surface of pyrite crystals has the special ability to form molecules that accumulate as thin skins, or biofilms. This characteristic of pyrite is now the focus of research. According to one scientific opinion, it was only a matter of time before the components connected to form organic molecules and, ultimately, a primitive form of life.

milestones

PRIMORDIAL SOUP

In 1953, American biologist and chemist Stanley Lloyd Miller astonished the experts. In a system of glass tubes, he imitated the first atmosphere, ocean, and thunderstorm. He boiled water, then added ammonium, methane, and hydrogen to the steam. The gas was exposed to an electric current. Within a few days, organic compounds like amino acids formed in the water. The mystery of the origins of life seemed to be solved. The term "primordial soup" was coined to describe this process.

Stanley Lloyd Miller was a pioneer in the search for the origins of life.

Lightning and intensive ultraviolet radiation may have supplied some of the energy needed to produce life from inorganic molecules like sugars, amino aids, and fatty acids.

see also: Evolution, pp. 140–145

first living cells

Over the past 3.5 billion years, living things have developed from simple cell structures to complex multicellular and higher organisms.

The first multicellular organism probably developed in the ocean some 600 to 700 million years ago.

The first cell-like structures were the protobionts, from which it was a clear path to prokaryotes—the first true cells.

Prokaryotes: the first cells

The first prokaryotes were very similar to today's bacteria and cyanobacteria. While the prokaryotes lacked a true nucleus, they had a cell wall that separated them from the environment and provided an internal space in which metabolic processes could take place. Some early prokaryotes used sunlight for their energy needs. They used an oxygen-free (anaerobic) form of photosynthesis, oxidizing hydrogen sulfide and giving off sulfur as a waste product. Today, purple bacteria still use this type of energy production. The next step was the adoption of oxygen-producing photosynthesis (p. 181). Using energy from the sun, carbon dioxide, and the surrounding water supply, cyanobacteria created their own nutrients. Oxygen—which was poisonous to other living things at the time—was released as a waste product. With the success of these organisms, the prokaryotes that used anaerobic photosynthesis retreated to oxygen-free habitats (such as sulfurous hot springs), leaving the way open for oxygen-breathing organisms. The oldest known prokaryote fossils are stromatolites, which date back approximately 3.4 billion years. Stromatolites are calcium carbonate, or chalk, deposits that form colonies of cyanobacteria.

The first eukaryotes

About two billion year ago, the first eukaryotes appeared. They were probably the result of symbiotic relationships among prokaryotic cells. They had a true cell nucleus and formed the foundation for the development of higher organisms. All single-celled organisms and all multicellular living things—plants, animals, and humans—are eukaryotes.

Multicellular organisms

It is probable that multicellular life may have started in colonies of single-celled organisms in which daughter cells did not separate from others following cell division. Or they could have emerged from single-celled organisms with multiple nuclei. The Precambrian period (roughly 600 to 700 million years ago) probably saw the emergence of the first multicellular organisms. There is little fossil record of these early organisms, since they did not have hard physical components (such as shells). However, the imprints of soft-bodied organisms of the Ediacaran period can be found in many locations. Later, in the early Cambrian period, many new organisms developed, during a 50 million year period known as the Cambrian Explosion. The first plants came ashore about 400 million years ago, pioneering the colonization of dry land by living organisms.

basics

ENDOSYMBIONT THEORY proposes that eukaryotic cells emerged through the fusion of prokaryotic cells into symbiotic communities that later became single organisms: eukaryotes.

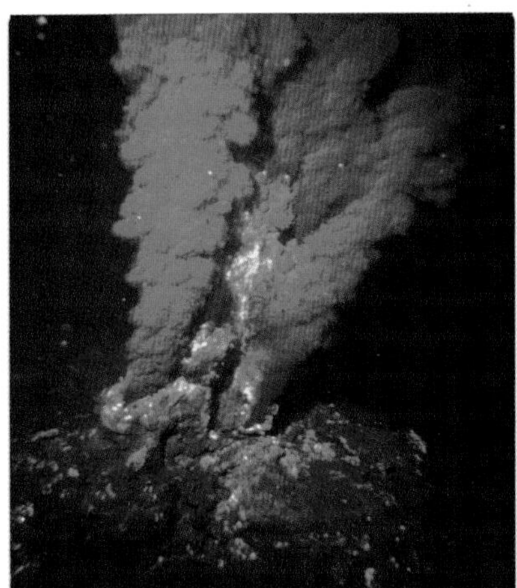

Black smokers are deep-sea hydrothermal vents. They may have played a role in the origin of life.

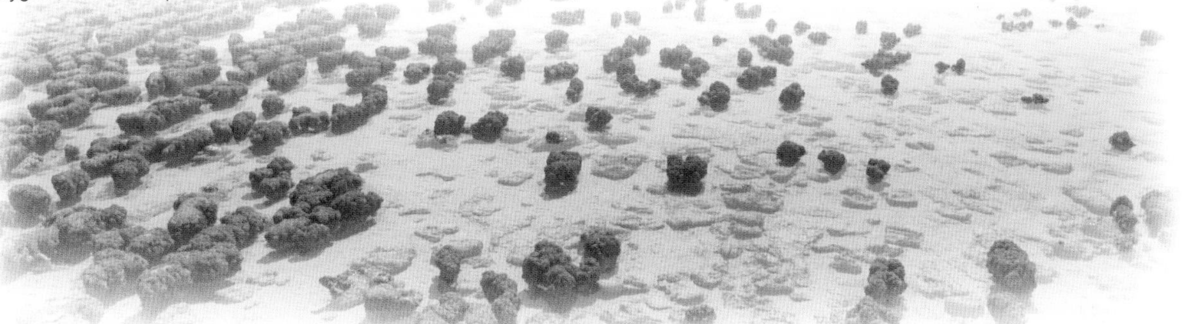

Stromatolites are chalk deposits created by cyanobacteria. Stromatolites dating back 3.5 billion years have been found. They initiated the creation of the atmosphere.

see also: Single-celled organisms, pp. 176–177

organelles | DNA | chromosomes | division | metabolism

CELLS

"Every cell comes from another cell," physician Rudolf Virchow observed as early as 1855. Biologists refer to cells as the "building blocks of life," since all living things, including animals, plants, and fungi, are made up of cells. A single cell can even form a complete unicellular organism.

cell organelles

Within eukaryotic cells, special structures called organelles provide an effective labor division. Suspended within a thick fluid called the cytosol, they form the cytoplasm, which surrounds the cell's nucleus.

The cell's complex construction becomes visible under a high-powered microscope. Every cell is protected by a plasma membrane. It consists of a double layer of lipid molecules, in which proteins are embedded. Some of these proteins help transport substances in and out of the cell. Receptors on the outside of the membrane recognize transport or signal substances approaching the cell. Within the cell's cytoplasm, biological membranes create separate compartments for various metabolic reactions. The cell's large nucleus serves as its command center. It contains the chromosomes, whose DNA (deoxyribonucleic acid) carries the necessary genetic information for the cell's maintenance and reproduction. The exchange of substances between the nucleus and cytoplasm occurs through pores in the nuclear membrane.

The capsule-shaped ribosomes translate instructions provided by the DNA for the construction of proteins. Various membrane systems, such as the rough and smooth endoplasmic reticulum, produce fats or process membrane proteins. The Golgi apparatus, in contrast, marks and activates proteins. Furthermore, it organizes membrane components such as lipids and transports these to the cell membrane. Lysosomes that contain digestive enzymes, are also created by the Golgi apparatus.

Buildup and breakdown

Animal cells derive their energy from the breakdown of food. This takes place in the mitochondria (pp. 154–155). Cells with very high levels of metabolic activity, have a correspondingly larger number of mitochondria.

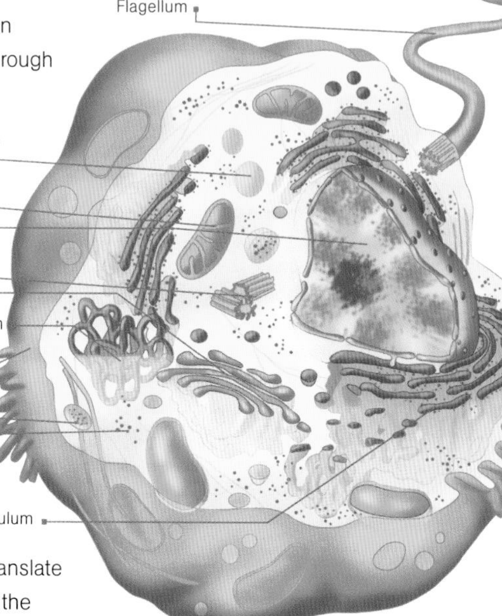

An animal cell showing the various organelles, which organize its metabolic activities.

Plant cells can use sunlight to produce nutrients with the aid of chloroplasts (p. 181). Within the cell, mitochondria and plastids, such as chloroplasts, have their own genetic material and reproduce through cell division. The cellular metabolism accumulates waste products, which can become toxic. Saclike peroxisomes and lysosomes absorb these substances, thus purifying the cytoplasm.

in focus

PROKARYOTES AND EUKARYOTES

The presence of specialized compartments within cells is a characteristic of eukaryotes. Prokaryotes, on the other hand, are microorganisms with a less differentiated cell structure. Their genetic material is not contained in a true nucleus surrounded by a nuclear membrane; instead it floats freely within the cytoplasm, usually as ring-shaped strands of DNA. Prokaryotes lack other typical organelles such as plastids, mitochondria, and Golgi bodies. However, an outer cell membrane is common to both cell types.

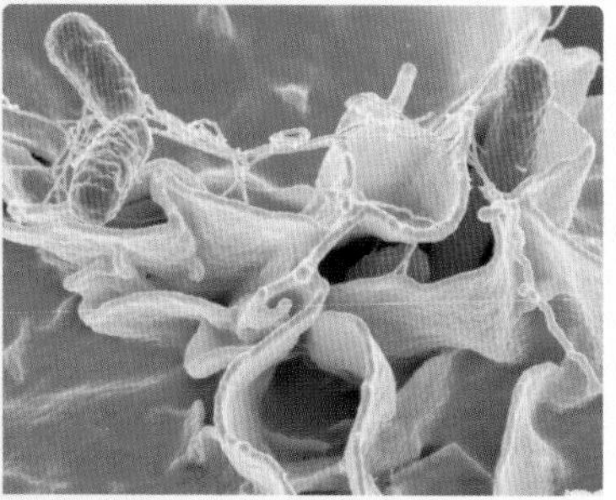

Many bacteria and disease-causing agents that attack human cells are prokaryotes.

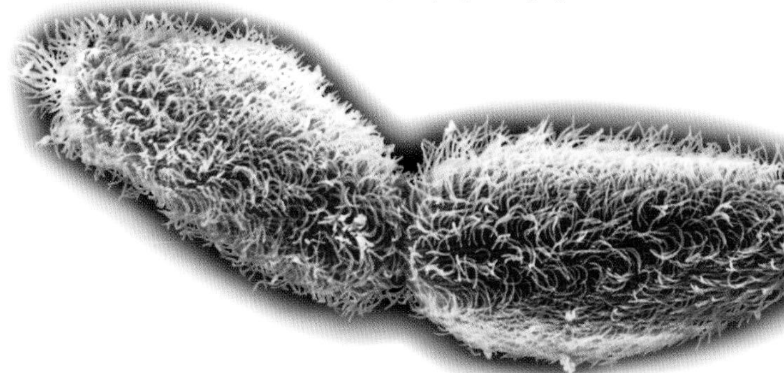

Although the paramecium is a single-celled organism, it has a highly complex structure. It reproduces through cell division, forming two daughter cells.

see also: Nerve cells and signal transfer, p. 240

differences: plant and animal cells

The eukaryotic cells of both plants and animals have a complex internal structure. Plant cells, however, tend to be significantly larger, and they have typical characteristics that make them easy to identify.

The disk-shaped structures called plastids are unique to plant cells. Even fungi, which were once grouped among the plants, lack these special organelles. Plastids include the green chloroplasts that carry out photosynthesis in a plant's stems and leaves, as well as chromoplasts, which color flowers and fruits such as tomatoes and peppers. Colorless leucoplasts store nutrients, such as the starch in a potato.

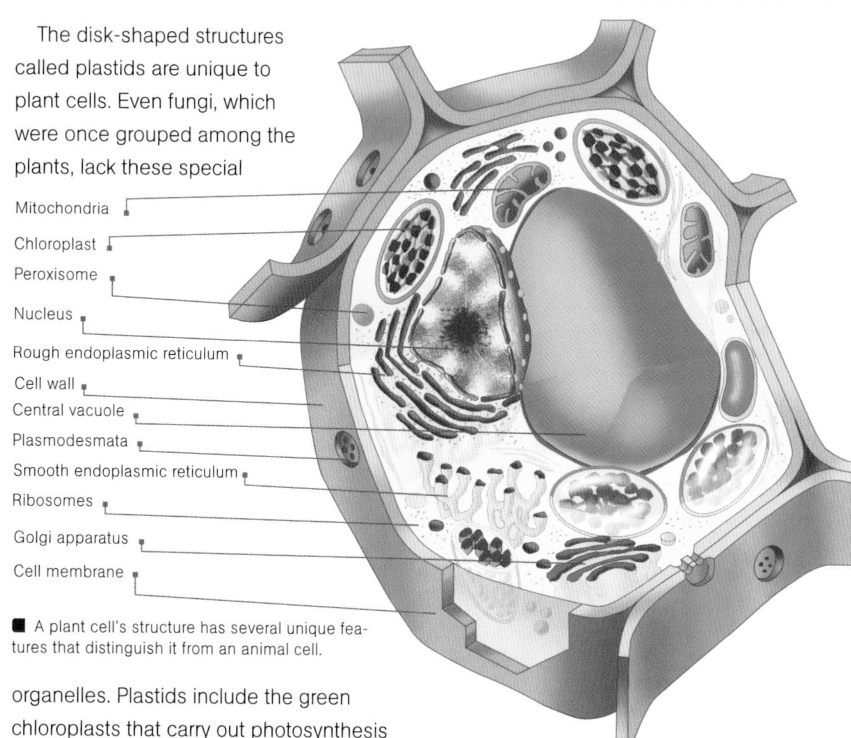

A plant cell's structure has several unique features that distinguish it from an animal cell.

Vacuoles and cell walls

Even under a simple light microscope, a plant cell's central vacuole is especially striking to the eye, since it makes up most of the cell's volume. Its liquid interior contains substances such as flower pigments, nutrients, ions, and defensive chemicals, to ward off insects, for instance. The vacuole also holds numerous enzymes, and thus it handles digestion within the cell; in animal cells, this is carried out by lysosomes. Since it is filled with liquid the central vacuole creates internal pressure that would cause the cell to burst without equivalent counter-pressure. Thus plant cells need a strong, stable protective shell: the cell wall. The cell wall lies directly outside the cell membrane. It contains countless intertwined strands of cellulose, a molecule composed of chains of up to 10,000 carbohydrate units. Along with other materials, the cellulose gives the cell wall enormous strength to resist inner or outer pressure. Since the plant's cell walls are connected to each other, they provide a kind of "skeleton" that gives shape to the plant. The cell wall space also contains signal molecules and defensive substances, which help protect the plant tissue, for instance, from attacks by insects or fungi.

Chloroplasts, found in the cells of all green parts of a plant, absorb sunlight for photosynthesis.

The cell wall is also penetrated by very fine plasmatic tubes, which connect the living cytoplasm of different cells. These tubes, so-called plasmodesmata, transport nutrients and allow communication among the various cells and tissues.

basics

AVERAGE CELL SIZES in animals, 8–20 micrometers, in plants, 100–300 micrometers (1 micrometer = 1/1000 millimeter)

A LIVER CELL contains from 500 to 2,000 mitochondria, averaging 0.5–1 micrometer each.

in focus

DETOXIFICATION WITHOUT A LIVER OR KIDNEYS

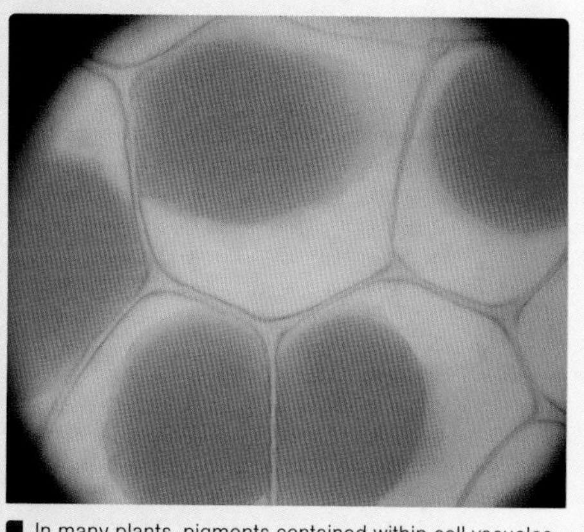

In many plants, pigments contained within cell vacuoles provide the color for flowers or leafy vegetables.

Many plants absorb harmful substances from the environment through their roots and leaves, such as salts, heavy metals, and pesticides. These substances are not only toxic to humans and animals, but can also interfere with plant metabolism. Plants, however, have no liver or kidneys to help them isolate and remove poisons. Therefore they must carry out intracellular detoxification: special transport molecules carry the poisons into the large central vacuole, where they are stored or chemically deactivated.

see also: Enzymes: active catalysts, p. 301

genetic information: dna

Genetic information is stored and passed on as deoxyribonucleic acid (DNA). The cells of all living organisms contain DNA, as do many viruses.

In eukaryotic cells, the chromosomes of the nucleus are made up of DNA. It is also found outside the nucleus in chloroplasts and mitochondria, and inside the cytoplasm in the form of plasmids. In prokaryotic cells, the DNA is always unbound within the cytoplasm. DNA molecules consist of nucleotides (polynucleotides) that are linked to each other in a chain. Each nucleotide has three components: one sugar (deoxyribose), one phosphate, and one of the four bases adenine (A), cytosine (C), guanine (G), and thymine (T). All nucleotides contain the same sugar and phosphate backbone. The genetic information is, therefore, stored in the order of the bases. The famous model of the DNA double helix constructed by Watson and Crick in 1953 illustrates this structure. Two twisted polynucleotide chains create a double helix held together by nucleotide base pairs. Periodical polyester chains between phosphates and the sugar backbone create the outer structure of the double helix. Between the strands, either adenine and thymine is paired by two hydrogen bonds, or cytosine and guanine are paired by three hydrogen bonds. Adenine can only pair up with thymine and cytosine can only pair up with guanine (the principle of base-pair complementarity). As a result, the base sequence along one polynucleotide strand determines the base sequence of the other strand. The strands of the double helix are therefore complementary, not identical, and oriented in opposite directions (antiparallel).

DNA needs not only to store genetic information, but also to identically reproduce this information in order to pass it on to the next generation. Replication of the information happens during interphase (the time period between two cell divisions).

The complementary DNA strands break away from each other. An enzyme called DNA polymerase synthesizes a new strand by binding new complementary nucleotides to the now single bases. Each of the original parent strands act as a template for the creation of a new complementary, antiparallel strand.

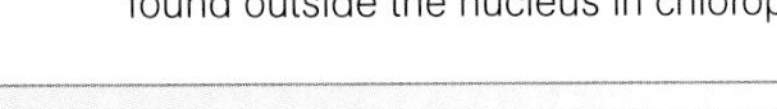

THE HUMAN GENOME PROJECT

milestones

Craig Venter was criticized for registering patents on decoded human genes.

More than a thousand scientists have taken part in this international project, initiated in 1990, to determine the entire sequence of the human genome. In 1998, the American biologist Craig Venter founded Celera Genomics to map the genome through automated sequencing with the aid of private funds, thereby competing with the international project. In June 2000, both published their versions. Data analysis is ongoing, with the goal of early disease recognition and treatment.

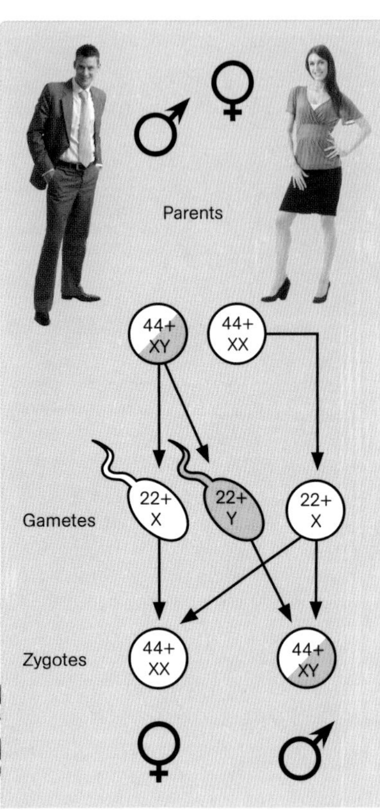

Women and men have different sex chromosomes: Women have two X chromosomes; men have one X and one Y chromosome.

The orientation of a DNA strand is defined by a 5' and a 3' end, which refer to the bonds between a free 5'-phosphate group and a free 3'-hydroxy group and two sugar molecules.

GENETIC FINGERPRINT

practice

In recent years, the so-called genetic fingerprint has received much attention. It offers a method of identifying criminals who can then be convicted on the basis of this evidence. This is made possible due to the fact that every person has a unique set of chromosomes. Even the smallest trace left behind at a crime scene (for example, hair or body fluids) allows the reconstruction of an individual DNA profile.

For DNA profiling, DNA is extracted from cells in blood, saliva, semen, or other tissues.

see also: Nucleic acids: molecular building blocks, p. 298

chromosomes

Chromosomes are threadlike structures that carry genetic information. In eukaryotes, chromosomes are located inside the nucleus. They consist of DNA and stabilizing proteins (histones).

Chromosomes continuously change shape from an uncoiled form to a condensed coiled form, depending on current functional requirements. While a cell is not in the process of division, they remain uncoiled. As soon as a cell starts dividing, they contract to form a dense and mobile shape that is visible even under a light microscope. Chromosomes consist of two identical DNA double strands (sister chromatids), which are linked together at the centromere. After cell division, a chromosome is initially composed of only one chromatid. It will eventually double so that there are two identical chromatids again.

The number of chromosomes is always the same for a certain animal or plant species. For example, there are 46 chromosomes in a human cell and 44 of these can be aligned by shape and size to make 22 pairs (homologue chromosomes). They are called autosomes, while the two remaining ones are called gonosomes (sex chromosomes). Men have one larger X chromosome and one smaller Y chromosome, and women have two X chromosomes. A diploid set of chromosomes has two copies of every chromosome. Reproductive cells, eggs, and sperm, have only a single, or haploid set of chromosomes.

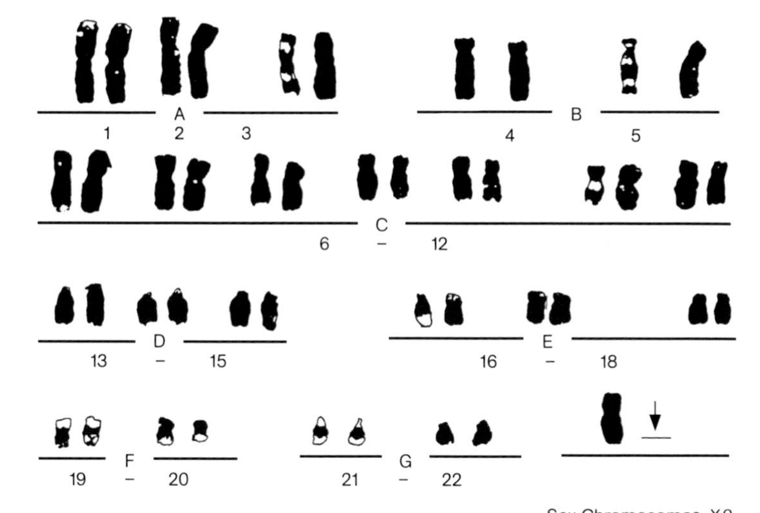

In a karyogram, the chromosomes are photographed through a microscope and ordered according to certain criteria, such as size. Any mutations, such as a missing chromosome (Turner's syndrome), can be identified.

HE NUMBER OF HROMOSOMES aries in organisms, ut this is no indication of complexity. he cells of lampreys have 174 hromosomes and umans have 46.

basics

Genetic mutations and abnormalities

Occasionally, individual chromosomes are not replicated during cell division or individual chromosome pairs are not separated for distribution into daughter cells (nondisjunction). Such a change in chromosome number is known as genome mutation. The phenomenon of one too few chromosomes is called monosomy and one too many chromosomes is called trisomy (see in focus). A change in entire sets of chromosomes is a type of euploidy. A genome mutation on any chromosome will result in a specific syndrome or condition. Some cause significant abnormalities, others may go undetected for life.

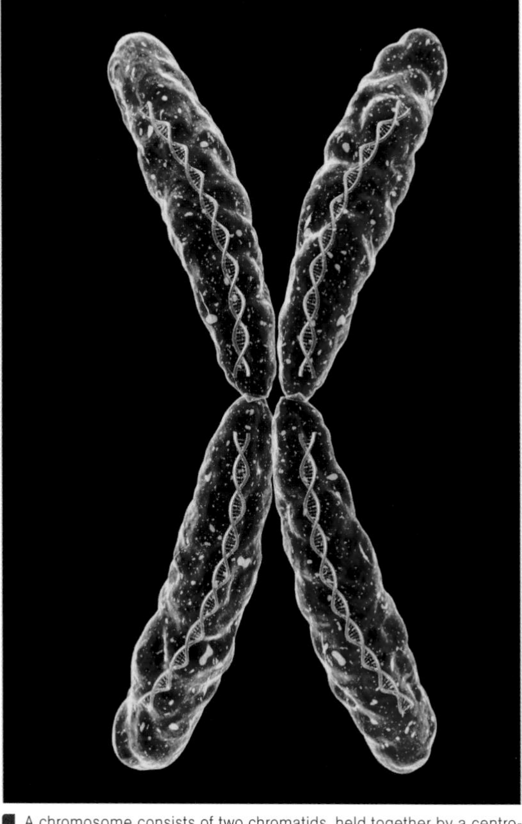

A chromosome consists of two chromatids, held together by a centromere. Every chromosome accommodates a long DNA molecule.

TRISOMY 21

in focus

Trisomy 21, or Down syndrome, is a genome mutation caused by the nondisjunction of chromosome pair number 21 during meiosis or due to a failed segregation during meiosis II that results in a gamete having two copies of chromosome 21. During fertilization, the spermatozoan adds its own chromosome number 21, so that the zygote now has three chromosomes. Such a trisomic condition results in abnormal physical and mental developments.

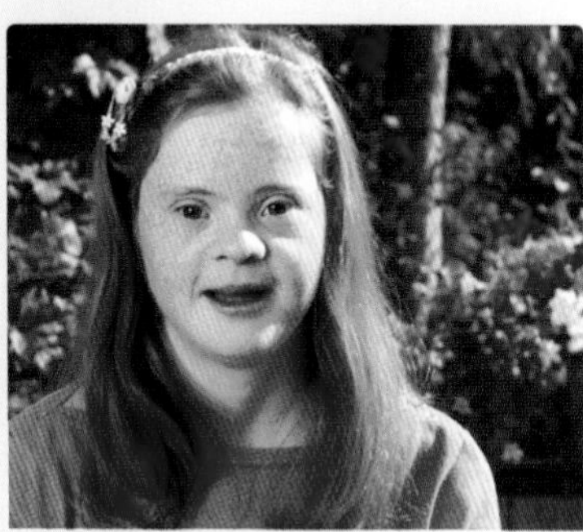

The chance of a child being born with Trisomy 21 increases with the age of the mother.

mitosis: replication of cells

Multicellular organisms consist of a very large number of cells, but they always grow from one single cell, normally a fertilized egg.

Before eukaryotic cells can replicate, their nuclei must be divided. This is what mitosis does. The genetic information of the cell is distributed evenly between the daughter cells so that, once mitosis is completed, each new cell has the same set of chromosomes as the original cell. This process may be divided into multiple phases.

During prophase, chromosomes coil up to prepare for transportation and division. The nuclear membrane begins to disintegrate. In metaphase, the chromosomes are already so condensed that they are visible in a light microscope. Tubular protein filaments (microtubules) form a spindle and the chromosomes align along the metaphase, or equatorial plate. During anaphase, the centromeres (the points where chromosomes attach to the spindle fibers) segregate and the spindles pull the chromatids of each chromosome toward opposite poles of the cell. Each pole gets one full set of chromatids. During the telophase, the spindle disintegrates, a new nuclear membrane forms around each set of chromatids, and the cytoplasm divides (cytokinesis). New cell membranes form (as do cell walls in plants) and the chromosomes reassume their functional shape. The time between nuclear or cellular divisions (interphase) allows cells to grow and reach the size of their parent cells, as well as to form organelles. Furthermore, the chromatids double (identical replication) and the nucleus begins to regulate the metabolic activities of the cell, such as protein biosynthesis, before the process of replication begins all over again.

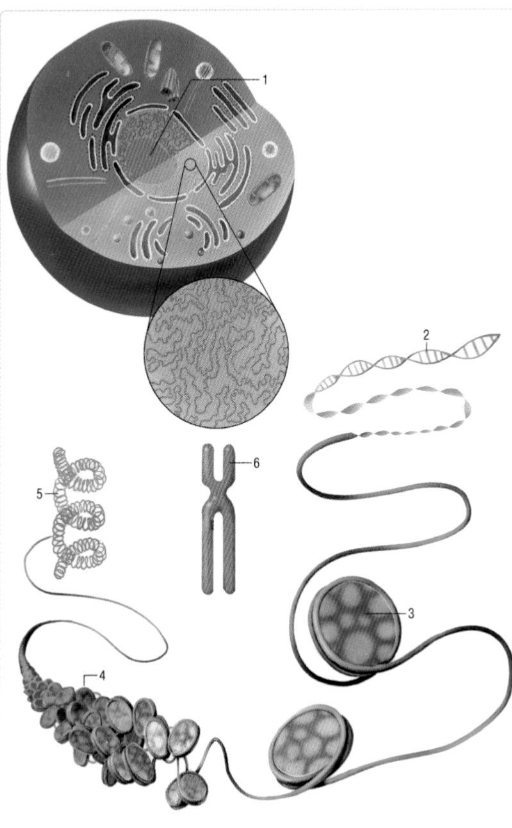

Stages of DNA compaction during eukaryotic cell (1) reproduction: DNA double helix (2), nucleosome (3), 30-nm chromatin structure (4), active chromosome (5), and metaphase chromosomes (6).

CYTOSTATIC AGENTS

practice

Cytostatic agents are used in cancer treatment to slow down the growth and division of tumor cells. Antimitotics—drugs that inhibit mitosis—are cytostatic agents that bind to the microtubular proteins (tubulin), thus temporarily blocking cell division. The use of antimitotic agents requires careful consideration: healthy cells will also be affected to some extent.

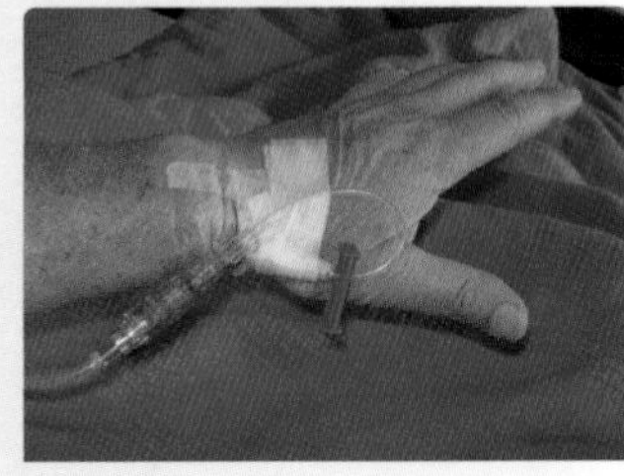
As with most cancer treatments, cytostatic agents can cause substantial side effects and also damage healthy tissue.

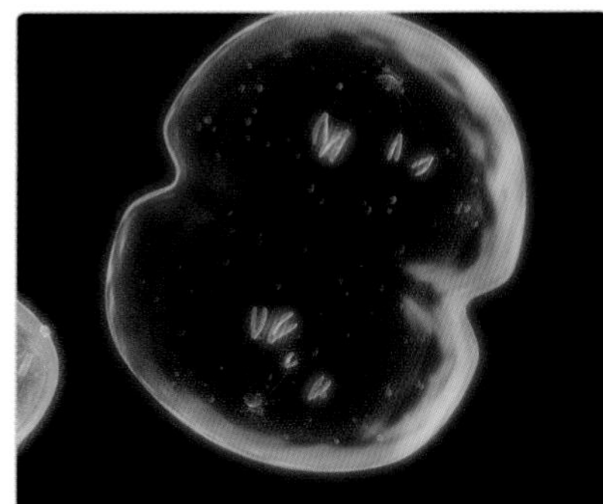
At the beginning of the replication process, a cell duplicates the chromosomes in its cell nucleus.

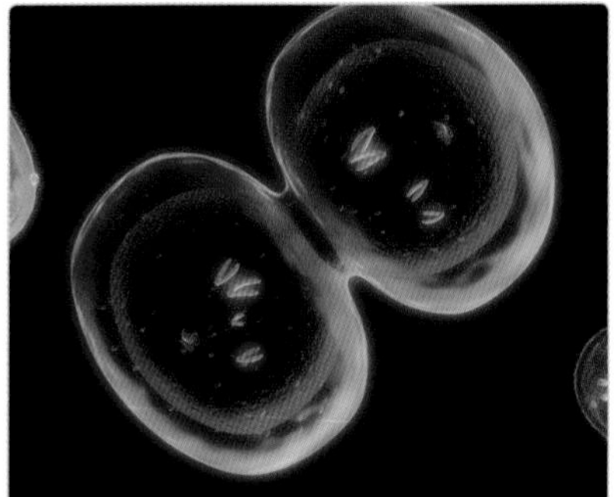
The result is two separate daughter cells, each with the same genetic material as the parent cell.

NUMBERS OF CELLS

in focus

Human bodies consist of about ten trillion (10^{13}) cells, all of which originate from a single maternal ovum that has developed into an organism by continuous cell division. In addition, the gastrointestinal tract contains about 100 trillion (10^{14}) cells of microorganisms, and there are another trillion (10^{12}) bacteria on the skin. Despite their numbers, these extra cells add only 3.5 ounces (100 g) in weight, due to the large difference in size between the eukaryotic cells of the human body and prokaryotic bacterial cells.

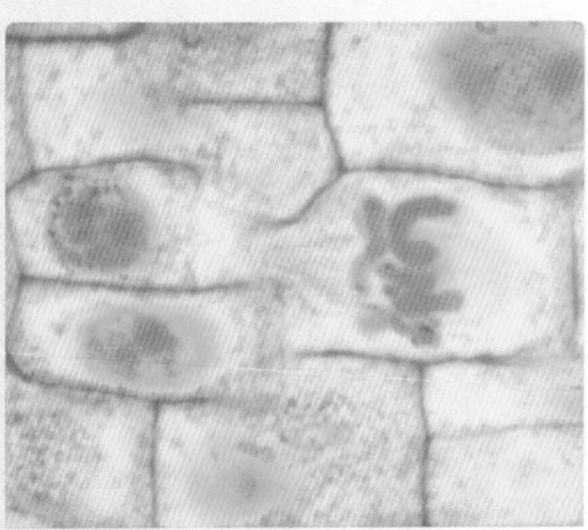
Cell division takes place continuously in the entire body of every organism.

meiosis

Meiosis is a process in eukaryotic organisms that divides the nucleus and cell and reduces a diploid chromosome set with two sets of chromosomes to a haploid set with only one set of chromosomes.

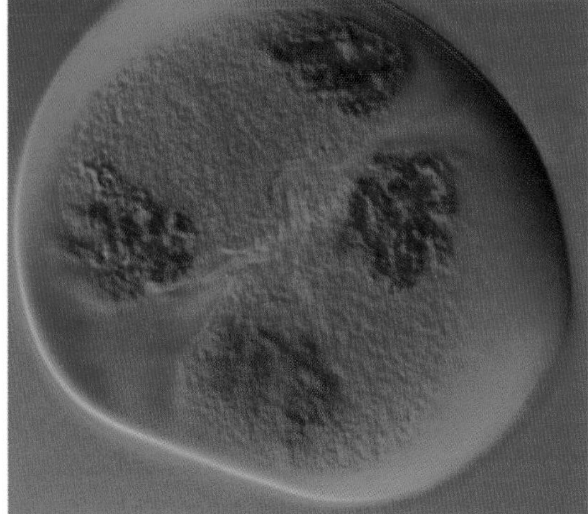

A bluebell cell replicates itself through meiosis into four different daughter cells.

In the more complex plants and animals, this process occurs when sperm and egg cells mature. In contrast to mitosis, meiosis involves two steps: meiosis I (reduction division) and meiosis II (meiotic division just like mitosis).

Meiosis I

During meiosis I, homologous chromosomes form pairs before the chromosome number is reduced. The process of meiosis can be divided into the following phases. The first step is prophase. Chromosomes coil up tightly. An undivided connection, the centromere, links the two sister chromatids of each chromosome. The homologous chromosomes align in parallel, forming a group of four chromatids (a tetrad). Exchanges of genetic material between corresponding chromatid sections may occur at this time (see in focus). In metaphase I, tetrads align at the center of the longitudinal spindle axis and the nuclear membrane disappears. The homologous chromosomes segregate in anaphase I. As the centromere remains intact, entire chromosomes travel to either pole. In telophase I, the chromosomes begin to uncoil. The nuclear membrane and nucleoli form. Homologous chromosomes are divided during this reduction division, and chromosomes from the mother and the father are distributed randomly; thus a diploid set of chromosomes has turned into a haploid set.

basics

EVOLUTION depends on meiosis. It is only during this crossing-over process that coupled genes from each parent's chromosome can randomly recombine, creating diversity in the new organism.

Meiosis II

When chromosomes have been reduced after meiosis I, meiosis II initiates the second division, which is similar to mitosis.

Here, sister chromatids of the now haploid chromosome set segregate into two separate gametes with only one chromatid each, making four gametes in total from the original single cell. After fertilization, the chromatids double through DNA synthesis, resulting in a new diploid chromosome set.

in focus

RECOMBINATION THROUGH SEXUAL REPRODUCTION

The chromosomes inherited from the parents are distributed randomly during the first reduction division of meiosis I. This leads to a new combination or recombination of the genetic material. There are two different types of combination known as interchromosomal recombination and intrachromosomal recombination. The first one involves entire chromosomes that are recombined during anaphase I. The latter involves an exchange of corresponding sections between chromatids (crossing-over).

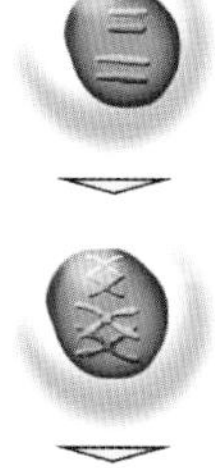

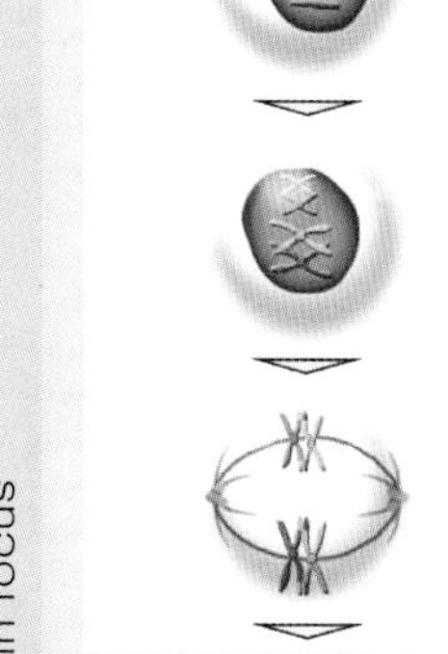

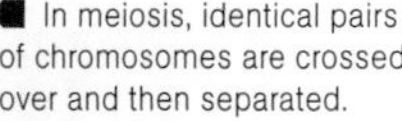

In meiosis, identical pairs of chromosomes are crossed over and then separated.

Genetic information is continuously recombined through meiosis; thus meiosis is the basic precondition for biodiversity and evolution.

general cell metabolism

The cells of all living things need enough energy to carry out their life processes. Heterotrophs (organisms that feed on other living things) gain this energy from breaking down energy-rich carbohydrates, fats, and proteins.

Aerobic energy production (cellular respiration) begins with glycolysis, which takes place in the cytoplasm of the cell. Glycolysis is a chain of reactions in which carbohydrates such as glucose are broken down into pyruvate without the need for oxygen.

ATP, THE MULTIFUNCTIONAL nucleotide adenosine triphosphate, composed of adenine, ribose, and three phosphate groups, is the most important energy storage substance in cellular respiration. The stored energy (30 kJ/mol) can later be released through the removal of a phosphate group (ATP → ADP + Pi).

NAD+ (nicotinamide adenine dinucleotide) is a hydrogen-bearing coenzyme; it acts as an electron acceptor.

basics

Glucose
Fructose-1,6-bisphosphate
2x Glycerinaldehyd-3-phosphate
2x Pyruvate
Glucose
Oxalacetate
NAD+
Citrate
NADH/H+

The citric acid, or Krebs, cycle involves a series of enzymatic reactions in aerobic organisms.

Glycolysis

At the start of glycolysis, two molecules of ATP (see basics) each contribute a phosphate group to a glucose molecule, which is a 6-carbon sugar. This produces fructose 1,6-biphosphate. This activation makes it easier to split the sugar into two nonidentical 3-carbon molecules. They are later transformed into two identical compounds (glyceraldehyde 3-phosphate). Both molecules are then oxidized to pyruvate, and some of the energy released through this process is transferred to ADP, forming the energy storage molecule ATP. In addition, some energy is transferred to NAD^+ in the form of electrons (see basics). Later, highly energy-rich ATP is formed from the hydrogen-containing compound $NADH/H^+$.

The citric acid cycle

The citric acid cycle, which is also known as the Krebs cycle, derives energy with maximum efficiency from pyruvate, the product of glycolysis, which is still quite rich in energy. It does so by breaking down the molecule completely into three molecules of CO_2 (oxidative decarboxylation), which we breathe out as a waste product. The hydrogen produced in large quantities during these oxidative steps is captured in the form of $NADH/H^+$ or $FADH_2$, which will later be used to produce energy (ATP synthesis). In addition, a molecule of an ATP analogue, GTP, is formed directly. There is an intermediate activation step between glycolysis and the citric acid cycle, in which the first CO_2 molecule is removed, and coenzyme A is added to the resulting acetic acid. The product, acetyl-CoA, now enters the multistep cycle, binding with a 4-carbon (C_4) compound (oxaloacetate), to form citrate (a C_6 compound). Some oxaloacetate also emerges from this process, reacting again with a C_2 compound. Among eukaryotes, the citric acid cycle takes place in the mitochondria (membrane-enclosed organelles); in prokaryotes, it occurs in the cytoplasm (gelatinous fluid that fills cells). Like glycolysis, it still does not require oxygen.

The electron transport chain

The resulting H_2-rich compounds $NADH/H^+$ and $FADH_2$ are now used in the last phase of cellular respiration—oxidative phosphorylation, or the electron transport chain—to store energy in the form of ATP. During this process, electrons are transferred in a cascade of reactions onto the oxygen that is inhaled by the organism. This multistage reaction process using various redox systems is important, since otherwise the oxidation of H_2 to water would occur in an explosive reaction. The electron transport chain takes place on membranes: among eukaryotes, on those of the mitochondria, and in prokaryotes, on the cell membrane.

EXPLAINING CELLULAR RESPIRATION

Among the scientists whose research significantly contributed to the understanding of cellular respiration is German biochemist Otto Heinrich Warburg (1883–1970). One of the 20th century's greatest cell biologists, he was particularly interested in cellular respiration in cancer cells. He received the Nobel Prize in physiology or medicine in 1931, for his discovery of the nature and mode of action of the respiratory enzyme (cytochrome oxidase).

Otto Heinrich Warburg: Nobel laureate and one of the 20th century's leading cell biologists.

milestones

Fitness refers to exercise that involves the body's oxygen consumption.

further metabolic processes

Cell metabolism involves the transport of electrons in a multistep process utilizing oxygen (aerobic) or in a less efficient process of fermentation without the use of oxygen (anaerobic).

The cell's metabolic processes, such as the breakdown of glucose (glycolysis) and the citric acid cycle, lead to the formation of the hydrogen-rich compounds $NADH/H^+$ and $FADH_2$. In the last stage of cellular respiration—the electron transport chain—comparatively large amounts of energy are gained in the form of ATP (an energy storage molecule as well as a building block of the nucleic acids DNA and RNA). Additional amounts of NAD^+ or FAD are also formed for reuse. During this process, electrons are transferred in a multistep process to oxygen molecules breathed in by the organism. The multistage reaction process over different redox systems (using various enzymes and cofactors) is important; otherwise the oxidation of H_2 to water would take place as an explosive reaction.

The lactic acid produced during lactic acid fermentation gives plain yogurt its sour taste.

The electron transport chain occurs on membranes: within the mitochondria among eukaryotes, and on the cell's inner membrane for prokaryotes. During some redox stages, protons are transported from the interior of the mitochondrion or cell into the space between the inner and outer membrane. This proton gradient is then used for ATP synthesis (oxidative phosphorylation).

Through the aerobic (oxygen-utilizing) breakdown of a single glucose molecule, 36 to 38 molecules of ATP are formed from ADP and phosphate.

Fermentation

Many bacteria, some fungi (such as yeast), and some animal and human cells are able to break down nutrients not only aerobically, but also anaerobically (without oxygen), although this breakdown is not as complete. This process, called fermentation, begins with glycolysis, just as aerobic metabolic processes do, with pyruvate as an end product. Since no oxygen is available, however, the hydrogen of the $NADH^+ + H^+$ cannot be oxidized to water, as it is in the electron transport chain. Instead, it is transferred to intermediate reaction products, where it serves to reduce them. In the more complete aerobic breakdown process, the energy-poor molecules CO_2 and H_2O emerge as end products. With fermentation, however, the end products still contain substantial amounts of energy, such as lactic acid in lactic acid fermentation. Much less energy is thus gained from fermentation than from aerobic processes, so that, for instance, only two ATP molecules are formed from one glucose molecule through lactic acid fermentation.

Electron micrograph of *Saccharomyces cerevisiae*, a species of budding yeast, which is the common yeast used in baking and brewing.

> basics
>
> **BIOMEMBRANES** Only about eight nanometers thick, cell membranes consist of a semiliquid double layer of phospholipids. These molecules have a polar (water-soluble) "head" and two nonpolar (water-repelling) fatty acid "tails." Other substances are also associated with the cell membrane, such as enzyme complexes of the electron transport chain, as well as special proteins.

Fatty acid breakdown

In addition to carbohydrates, cells can also break down other substances, such as fats, and use them to produce energy. After the splitting (hydrolysis) of a fat into its components of fatty acids and monoglycerides, the glycerides are used in glycolysis as C_3-compounds, while the fatty acids are attached to coenzyme A in the mitochondrial matrix and broken down in stages (ß-oxidation). Acetyl-CoA (C_2-compound) is removed during each stage of the process, until the fatty acid is completely broken down. The resulting Acetyl-CoA molecules then enter the citric acid cycle.

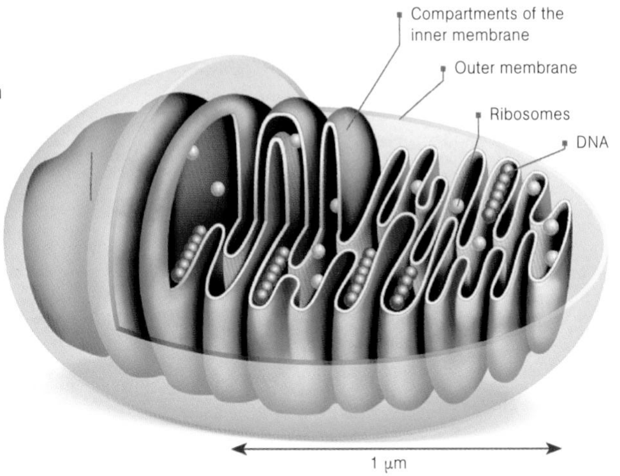

Mitochondria are sometimes referred to as "cellular power plants" since they produce most of a cell's supply of ATP.

see also: Everyday matter, pp. 304–305

fossils and fossilization

Only a tiny fraction of all plants and animals that used to inhabit Earth were preserved as fossils. The oldest fossils are microorganisms that used to live on Earth more than three billion years ago.

Fossils, such as ammonites, are important tools in the reconstruction of the evolution of life.

fossilization | fossils | living fossils

PETRIFACTIONS

On Earth, organic material decomposes and is returned to its basic inorganic elements in a continuous cycle. Occasionally, however, the remains of plants and animals from past geological eras survive through fossilization. These fossils can provide important clues to scientists about evolutionary processes and the history of the Earth.

Fossilization is an extraordinarily lengthy process. First of all, an organism's body needs to be quickly covered by sediment. This prevents decay, decomposition, or disintegration by scavengers and carrion feeders. Aquatic environments allow sediments to deposit more rapidly and continuously. Therefore, most fossils are former marine organisms. The remains of terrestrial organisms are much less likely to be covered by sediment in the fashion that would allow preservation. During fossilization, many substances that previously made up the organism are converted into stable materials. Relatively fragile materials such as bones or calcareous shells become hard and durable minerals. When wood petrifies, silicic acid enters its structure, often preserving even details such as tree rings. Fossils embedded in shale or clay often get broken during petrifaction unless they are enclosed in a mineral shell. These kinds of fossils, however, are especially valuable because the fine grain sediment is able to preserve even very fragile and soft-bodied organisms. A classic location for finding fossils of this type is the Burgess Shale located in the Canadian Rocky Mountains.

Under certain conditions, whole bodies or body parts, tracks, and also excrement can be preserved in a long multilevel process called fossilization.

Body fossils, steinkerns, and trace fossils

Entire bodies or parts of organisms—whether hard parts or rarer soft tissues–are referred to as body fossils. Internal casts (steinkerns) are created when organisms (or parts of them) leave hollow spaces in the sediment, to be filled later by other materials that harden. Occasionally, entire organisms or tissue parts are preserved, such as insects trapped in resin (amber) or mammoths found in the permafrost soils of Siberia. The traces that an organism left behind while still alive can also be preserved and are known as trace fossils. These are often footprints left in the mud or traces of motion such as digging. This type of fossil requires that the sediment containing the traces retains its structure and does not mix with the sediment deposited on top.

milestones

NICOLAUS STENO AND THE DISCOVERY OF FOSSILS

The Danish naturalist and later cleric Nicolaus Steno was the first to recognize that fossils were the remains of once living organisms. He noticed that a stone he had found in the mountains very closely resembled a shark's tooth. From then on, he dedicated all his efforts to the study of fossils. During his studies, he developed the first theory about the formation of sediment rocks and concluded that the Earth must be older than 6,000 years, which was the accepted age at the time.

Nicolaus Steno (1638–1686) significantly contributed to the development of modern crystallography and geology.

see also: Geological eras, pp. 158–161

other fossils and living fossils

Fossils are the legacies of former life. They vary in size from giant skeletons of dinosaurs to tender prints of dragonfly wings or minute traces of microscopically small bacteria.

Fossils document the Earth's history, each being associated with a geological epoch prior to the Holocene. They can be remains such as bones and teeth or traces such as footprints and feces.

Dinosaurs, the most famous animals of the Mesozoic, laid gigantic eggs in comparison with chickens.

Even fragile plant and animal parts may leave imprints. Black carbon film can preserve the shape of leaves or mollusks, and detailed organic structure can be fossilized in petrified forests through silification. Among the more spectacular fossils are dinosaur remains: footprints, bones, and even the occasional egg have been found on most continents. Fully fossilized organisms allow scientists to reconstruct extinct plants and animals as accurately as possible.

Index fossils

Fossils are also excellent tools for determining the age of rocks in a method known as biostratigraphy. The plant and animal groups best used for this—called index fossils—are those that evolved quickly to reach a high degree of diversity. Ideally, they would have existed for geologically short time periods (several hundreds or millions of years) with a wide geographical distribution. They should be abundant and easily identifiable. Most index fossils were marine organisms such as trilobites (the arthropods of the Cambrian and Ordovician periods) or ammonites and cephalopods (of the Triassic and Jurassic periods).

Living fossils

Evolutionary scientists are also very interested in living fossils. These species are not extinct but belong to genera whose basic structures have remained similar for millions of years. They live predominantly in isolated habitats that have had little to no environmental changes over over time, slowing evolutionary change. Living fossils are often the only species left in an order or belong to their own class. Examples include nautilus, coelacanth, horseshoe crabs, duck-billed platypus, and marsupials. Primeval plants that are living fossils include dawn redwood and ginkgo.

Besides fossil animals, there are also fossilized plants like this tree in Yellowstone National Park.

basics

THE TERM FOSSIL first occurred in the classic reference book *De natura fossilium* by Georgius Agricola, founder of mineralogy. The study of fossils is called paleontology. The study of decay and fossilization processes of organisms is called taphonomy.

The work of paleontologists often resembles the searching for puzzle pieces. It is comparatively rare that imprints are as completely preserved as in this case.

The coelacanth, which developed ca 400 million years ago, was believed to have been extinct until a surviving species was found.

stratigraphy | paleozoic era | mesozoic era | cenozoic era

GEOLOGICAL ERAS

In the 18th century people still thought that the Earth was relatively young. This had become common belief after an Irish bishop interpreted the Bible and concluded that the world was created on October 26, 4004 B.C. However, scientists' claim that the creation and decay of rock material was an extremely lengthy process were increasingly convincing. Within this geological time frame, the chronology of the Earth could be determined more accurately, all the way back to its formation 4.56 billion years ago.

stratigraphy

Stratigraphy is one of the oldest and most important geological dating methods. A time scale of past geological events can be produced from the sequence of rock deposits.

A Danish naturalist, Nicolaus Steno (1638–1686), was one of the first scientists to define the fundamental laws of stratigraphy. He discovered that lower rock layers are older than the layers above, as long as they remain undisturbed. Movements of the Earth, however, do happen and the stratigraphic sequence may be ambiguous and sometimes incomplete due to erosion, folding mountain ranges or other geological movements of the rock. Such ambiguities and gaps in the rock layers may not affect Steno's law of superposition but can frequently alter the layering sequence. So-called marker beds formed by events such as the ash rain from an active volcano can also help in dating rock strata or layers.

Describing the thickness and sequence of rock layers (lithostratigraphy) is not sufficient to determine the age or duration of its creation. This information is derived from fossils and other objects embedded in the rock strata (biostratigraphy). Only certain fossils are suitable for age determination. These index fossils are the remains or traces of plant and animal species that existed for only a short time (geologically speaking, which can be up to a few million years) and that were geographically wide spread and abundant These factors are important so that even rock layers at a distance from each other can be compared by identifiable, easily found fossils.

Relative age determination allows geologists to divide the Earth's history into different time intervals, called eras, periods, and epochs, to be used as a frame of reference for what order the formation of the Earth's surface and the origin and evolution of plant and animal species occurred in. The absolute age determination (a numerical measurement of time passed) of rocks and fossils was only possible with the discovery of radioactive decay by French physicist Antoine Henri Becquerel (1852–1908) for which he recieved the Nobel Prize in physics. The subsequent development of radio-dating methods also allowed a more precise age determination of geological periods.

ORGINATOR OF BIOLOGICAL STRATIGRAPHY

milestones

William Smith (1759–1839) managed to create a precise geological map of England by using different colors for different types of rocks.

The British surveyor William Smith realized during canal works that rock layers of matching ages also contained the same types of fossils. He created a geological map of Great Britain based on the fossils he found; it was published in 1815. Initially, his work was scoffed at by renowned geographers but in 1831, the scientific community finally recognized his achievements and Smith was the first to be awarded the Wollaston Medal by the Geological Society for outstanding accomplishments in geology.

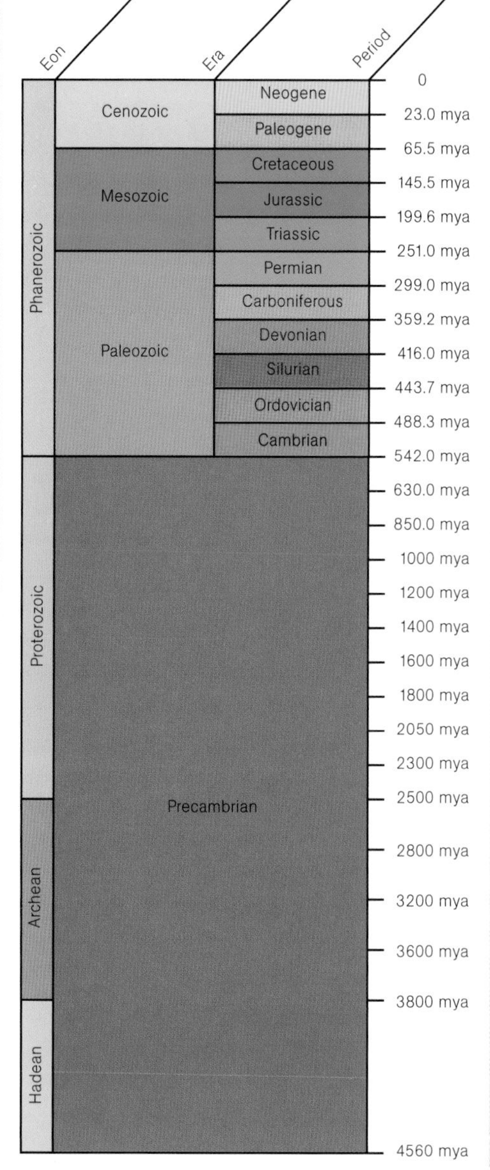

The aim of stratigraphy is to create a time scale for the dating of past geological processes, thereby dividing geology into chronological intervals.

see also: Petrifactions, pp.156–157

paleozoic era

Our detailed fossil records only begin at around 542 million years ago, in the Paleozoic era. All major animal groups, excluding mammals and birds, evolved in this era, ending in a great extinction about 250 million years ago.

The first proto-organisms with a cell-like structure appeared about 3.5 billion years ago, after which the first organisms formed relatively quickly. These single-celled, aquatic organisms, some of which had developed a nucleus, dominated life on Earth for more than two billion years. About one billion years ago the first multicellular organisms developed, along with sexual reproduction. The resulting marine organisms, such as annelids, cnidarians, and articulates, were later to become the foundation of evolution in the Paleozoic era.

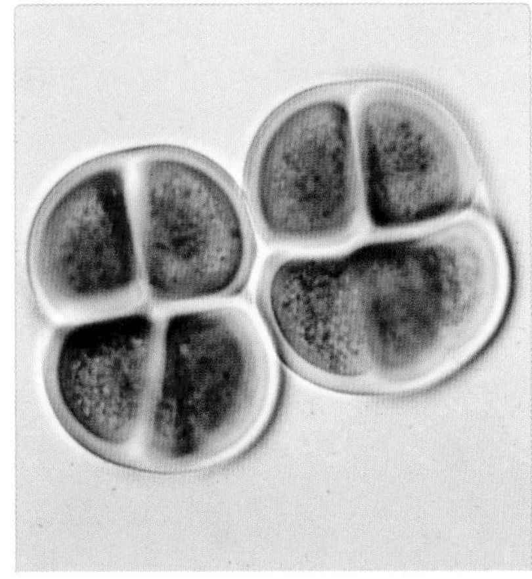

Cyanobacteria are among the oldest life-forms. They have inhabited the Earth for over 3.5 billion years and are also found in fossilized form referred to as stromatolites. This image shows the cells in a 500-fold enlargement.

From the Cambrian explosion to mass extinction

At the beginning of the Cambrian period about 570 million years ago, the first invertebrates, including trilobites and graptolites, evolved within a period of only 50 million years. The first terrestrial organisms were algae, lichen, and bacteria, which spread along the edges of shallow ponds, adding oxygen to the atmosphere. Small air-breathing animals followed, including arthropods (resembling millipedes) with hard outer skeletons that prevented them from drying out.

The first vertebrate's were oceanic fish, existing during the Ordovician period 470 million years ago. Later, during the Devonian period, placoderm fish co-existed with ray-fins, ancestors of most modern-day fish. It is thought that fish developed lungs and "legs" as a climate change reduced water levels and evolving into the first four-legged vertebrates, amphibians, from which all other vertebrates later originated.

The warm, moist climate during the Carboniferous period allowed for an abundance of vegetation—tall horsetails, club mosses, liverworts, and ferns—to florish in the swamps. Several amphibians became terrestrial and began producing eggs (protected by hard shell on land). This group later evolved into reptiles, which occupied every possible ecological niche during the Permian period. Reptiles even managed to fly, such as the *pterodactylus*. However, the masters of the air continued to be insects, some of which evolved into giants, for example, the *Meganeura*, a dragonfly with a wingspan of up to 27.5 inches (70 cm). Meanwhile the oceans were inhabited by lamp shells (brachiopods), snails, mussels and clams, bony fish, sharks, foraminifers, and numerous ammonites.

The class of trilobita has left a well-preserved fossil record. Over 15,000 species have been documented.

basics

THE MOST dramatic mass extinction in geological history occurred 250 million years ago at the end of the Permian period. Almost all marine organisms and more than three-quarters of terrestrial organisms were destroyed.

Fossil findings on all continents document the worldwide distribution of the plant genus *Archaeoptris* in the Devonian period.

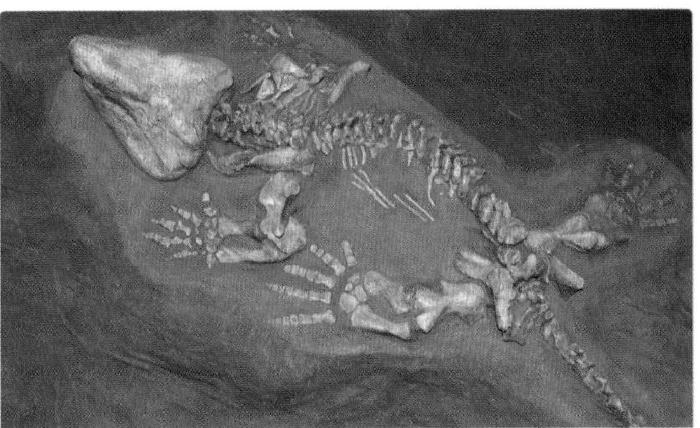

Fortune plays a significant role in searching for fossils, particularly when one finds an entire well-preserved skeleton.

see also: Bacteria, pp. 172–173

mesozoic era

A period of major mass extinctions 250 million years ago ended the Paleozoic era and began the Mesozoic, the era of the dinosaurs. Their eventual extinction, 65 million years ago, marked the beginning of the Cenozoic era.

Mass extinctions left only a few groups of animals to survive. New species evolved, adapting in response to the changing environmental conditions. Large aquatic reptiles such as *Ichtyosaurus, Plesiosaurus,* and *Pliosaurus* as well as sea turtles repopulated the oceans. Calcareous algae and madrepores built reefs in the warm, shallow oceans, forming habitats for crinoids or "sea lilies." Some fish species remained, especially cartilaginous fish such as sharks. Invertebrates were predominantly mussels, clams, snails, and, in particular, ammonites (index fossils of the Jurassic) and belemnites (index fossils of the Cretaceous period). Vegetation adapting to the warming climate became dominated by palm ferns and ginkgo plants. Angiosperms (early ancestors of today's deciduous plants) appeared during this period.

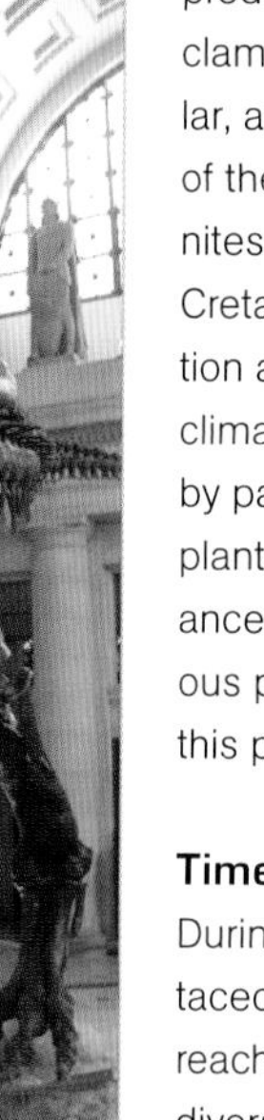

The dinosaur *Tyrannosaurus rex* is considered a carnivore, yet possibly without actively having pursued prey (scavenger).

Time of the dinosaurs

During the Jurassic and Cretaceous periods, dinosaurs reached their high-point in diversity. Herbivorores, *Apatosaurus, Brontosaurus, Barosaurus,* and *Supersaurus* are the largest terrestrial animals to ever have lived. Carnivorous dinosaurs included the giant *Tyrannosaurus rex* and the tiny *Compsognathus,* whose body structure resembled *Archaeopteryx,* the flying dinosaurs who shared air space with the *Pterosaurs.*

Other groups of animals evolved at this time whose descendents are still found today. These were mainly mammals—which played a minor role as rodents or insectivores—as well as frogs, turtles, and crocodiles. The great era of dinosaurs ended abruptly with another major mass extinction 65 million years ago, marking the end of the Cretaceous period. The currently accepted theory giving the cause of this sudden change is a meteorite hitting the Gulf of Mexico, which raised large amounts of dust, blocking out sunlight and cooling the Earth's temperatures dramatically. Many reptiles, including the dinosaurs, became extinct while ammonites and belemnites disappeared from the oceans.

in focus

ARCHAEOPTERYX

The discovery of archaeopteryx fossils at the Solnhofen limestone deposits in Germany was a great event for 19th century paleontologists, who had long predicted a link between reptiles and birds. This primeval bird was about magpie-sized, with a long reptile tail and toothed jaws. Its ability to fly would have been inferior to modern birds' due to very primitive flight muscles. The fine silt soil of Solnhofen preserved even the details of its feathers.

The archaeopteryx could only perform gliding flights from banks or trees.

Dragonflys are known since the Palaeozoic era and reached a wingspan of up to 27.5 inches (70 cm).

basics

ARCHOSAURS SPLIT into pterosaurs, crocodiles, and dinosaurs during the Triassic. Dinosaurs could be ornithischia (bird-hipped) or saurischia (lizard-hipped).

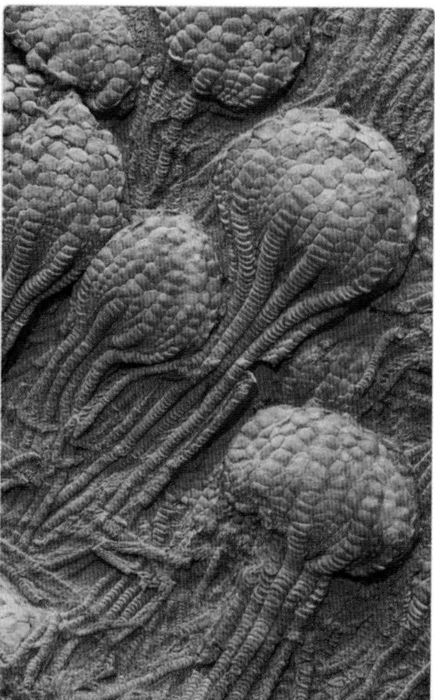

Crinoids, known as sea lillies, are an index fossil of the Paleozoic era (542–251 million years ago).

see also: Reptiles, p. 204

cenozoic era

For many millions of years, the Earth was dominated by dinosaurs. After the sudden extinction of the dinosaurs 65 million years ago, warm-blooded organisms, such as birds and mammals, began to evolve in the Cenozoic.

During the Cretaceous, plants already made the transition from the Mesozoic to the Cenozoic by evolving into angiosperms. Due to the tropical and subtropical climate of the Tertiary, flowering plants and luscious forests were growing far into the northern and southern latitudes. Bird species had reached a peak in diversity and insects were already as diverse as they are today. During this period, larger herbivores, weasel-like carnivores, pangolins or scaly anteaters, armadillos, and the first primate, *Purgatorius*, evolved. By the Eocene, all orders of mammals that exist today were already present. The evolution of the mammal species varied due to the physical isolation on separate continents. In Australia, for example, there are no placental animals whose embryos receive nutrition from the placenta. Large and quick, hoofed animals, such as the *Mesohippus*, roamed the grasslands, which expanded due to the cooling climate. Carnivorous mammals were just as quick hunting the hoofed animals and making it to the top of the food chain. Bats and flying foxes evolved from earlier night fliers. The ancestors of dolphins and whales returned to the oceans where they joined bony, scaled fish, and sharks, which were dominating the marine environment. Mussels, clams, and snails showed a great diversity among the invertebrates of the time. Many vertebrates, but also microorganisms such as large foraminifera, radiolarians, and dinoflagellates, became suitable index fossils for modern geologists.

The rat-sized *Purgatorius* is believed to be the earlist example of a primate.

The cold Quaternary

The youngest geological period is the Quaternary. It is divided into three epochs: Pleistocene, the oldest; Holocene in the middle; and the current and recently introduced epoch of human activity, Anthropocene. The Quaternary is characterized by alternating cold and warm periods. The Northern Hemisphere in particular experienced major glaciation events during the ice ages. The sea level sank considerably due to the water trapped in ice form. Land bridges emerged from the oceans providing a connection between the mainland and the islands as well as between continents. Not only plants and animals, but also humans were now able to populate new territories. Animal species dispersed according to the dramatic climate fluctuations. As temperatures were getting cooler, most mammals moved towards the lower latitudes. The tundra was still home to the woolly rhinoceros, caribou, woolly mammoth, and the musk ox, which were all adapted to the harsh life of the tundra. During warm periods, forest elephants, rhinoceros, and brown bears would migrate toward higher latitudes. The giant ground sloth and giant armadillo dispersed from South America into North America. Certain plants were unable to follow due to natural barriers and they became extinct.

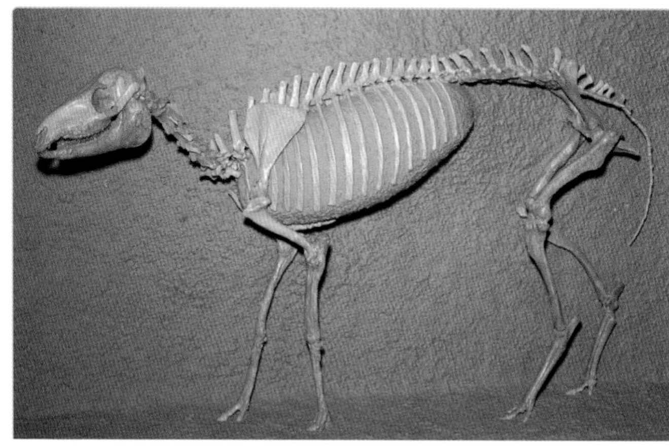

The *Mesohippus*, the predecessor of today's horse, was the size of a greyhound and native to the grass plains of North America.

Scientists have been able to reconstruct the physical shape of the mammoth due to excellently preserved fossils discovered in layers of permafrost.

see also: Evolution, pp. 140–145

the evolution of plants

The ancestors of today's land plants lived in the oceans and seas. They had to develop features to meet challenges such as dehydration in order to adapt for survival on dry land.

Fruit have qualities—such as color or flavor—to attract animals and achieve wider dispersal of seeds.

Plants are able to create their own sustenance without feeding on other organisms. They use the process of photosynthesis to capture and use the sun's light as an energy source. Plants are highly diverse in appearance, but most share characteristic structures such as leaves and roots. Over the course of evolution, they have continually adapted themselves to prevailing environmental conditions. Fossil findings make it possible to distinguish four significant developmental periods in their evolution. Each of these resulted in a new diversification of plant life.

Around 460 million years ago, the first land plants evolved from aquatic green algae in the initial period of plant development. It is thought that the periodic drying up of bodies of water encouraged adaptations that prevented dehydration. Modern plants such as mosses show examples of this transitional phase. They have a waxy layer that protects them from dehydrating in the open air.

During the next developmental stage, the first plants equipped with internal water conducting tissues emerged on coasts and other moist environments. Unlike mosses, these first vascular plants developed true roots and supportive stems, enabling them to obtain sufficient moisture outside an aquatic environment and to transport nutrients to all parts of the plant.

The first seed-producing plants (p. 188) appeared during the third developmental period. These differ from spore-producing vascular plants in that the embryo—together with a supply of nutrients—is encased in a shell: the seed. This led to the development of various kinds of plants without protective seed coverings—such as today's evergreens—called gymnosperms. Lacking the encasement of fruit, their seeds fall freely to the ground to germinate. Seed-producing plants enjoyed significant advantages in the conquest of new environments. They no longer depended on moist environments for reproduction and their embryos were more protected from adverse environmental conditions.

Ferns, horsetails, and clubmosses (p. 186) were among the first vascular plants to develop.

basics

PLANTS ARE autotrophic (self-feeding) life forms. They provide the nutritional foundation for animals and humans who cannot manufacture their own food.

Fruit-forming flowering plants called angiosperms, or plants with covered seeds, appeared during the fourth stage of development about 130 million years ago. In contrast to gymnosperms, angiosperm seeds are sealed inside chambers, or ovaries. The animals that eat the fruit end up transporting the seeds to different locations, which contributes to the enormous success of angiosperms.

Horsetails evolved during the second major developmental stage. Their supportive stems and roots allowed them to transport water and nutrients throughout their entire structure.

see also: Reptiles, p. 204

the evolution of animals

Animals—including human beings—trace their origins back to the oceans of the Cambrian period more than half a billion years ago when the first multicellular organisms that fed exclusively on other living things appeared.

An animal is a multicellular organism that cannot produce its own food molecules through photosynthesis (p. 181) as plants do, but instead it gains energy by feeding on other living things. This development in the Cambrian period opened up new resources and opportunities, leading to today's great diversity of animal species, which share typical characteristics such as sexual reproduction, a nervous system, and muscle tissue. The basic physical structures of modern animals were evident 500 million years ago. Physical, embryonic, and genetic traits are all considered in order to trace their evolutionary tree.

> **NIMALS INHABIT** early every environment on Earth, though many continue to live in water. Mammals, irds, amphibians nd reptiles, along ith insects and rachnids, have permanently conquered e land.
>
> basics

The history of animal species contains four major milestones. The first crucial step was the appearance of "true" body tissue—previously animals had only simple, saclike bodies, such as the porous sponges. The development of muscles and connective tissue was a prerequisite for specialized functions that would evolve over time, such as directional movement and breathing.

The second milestone was the development of bilateral symmetry (two halves of the body mirror each other) with an identifiable head, in contrast to radial symmetry (symmetrical in a ray pattern from the center), for example, jellyfish. This is the quality that enabled early animals to move purposefully in certain directions.

The third stage was the development of a fluid-filled cavity between the internal organs and the body wall, which is found in all "higher animals" except flatworms. This allowed the internal organs to function irrespective of the motion of the body as a whole.

The fourth milestone was in the embryonic stage of development. Vertebrates and echinoderms began to develop differentiated mouth structures from secondary openings off the blastopore (an initial opening in the embryonic cavity). This distinguishes them from mollusks and arthropods, in which this first opening becomes the animal's mouth.

The transition from water to land

The ancestor of all tetrapods (four-limbed animals) was probably a lobe-finned fish that was similar to coelacanths living today off the Comoro Island's coast. Its muscular pectoral and pelvic fins, supported by a bony skeleton, enabled it to crawl to shore. Yet, without developed lungs and the ability to keep its skin moist, it had to return frequently to the water.

Moving onto land may have been a survival strategy resulting from the need to abandon one shrinking body of water for another. Thus, individuals able to travel greater distances over land had a better chance of survival.

■ With legs, lungs, and skin adapted for aquatic life, the early vertebrates venturing onto land shared characteristics with and faced similar challenges to modern day salamanders.

■ Evolutionary trees illustrate evolution and species origin. Shared structures formed in times of similar environmental pressures, and then diversified into many unique branches.

see also: Evolution, pp. 140–145

evolution of humankind: the beginning

The origin of human evolution lies in Africa, where paleoanthropologists discovered the oldest skull bones ever to be found. These are about four million years old and point toward the early ancestors of modern humans.

In 1974, scientists in Ethiopia found the almost complete, now world-famous skeleton of a female pithecanthropid, the early ancestor of humans. Lucy, as she is named, belongs to the genus *Australopithecus*, meaning "southern ape." Their skull had a sloping forehead, a protruding brow, and a flat nose. However, in contrast to recent apes, the *Australopithecus* did not possess large canine teeth and therefore the so-called "monkey gap" was absent. Lucy's bones are about three million years old and provide clear evidence of an upright gait. This gave the savanna inhabitants of the time obvious survival advantages, such as when searching for food or detecting approaching enemies. Now that hands were no longer needed for locomotion, they could be increasingly used for other tasks.

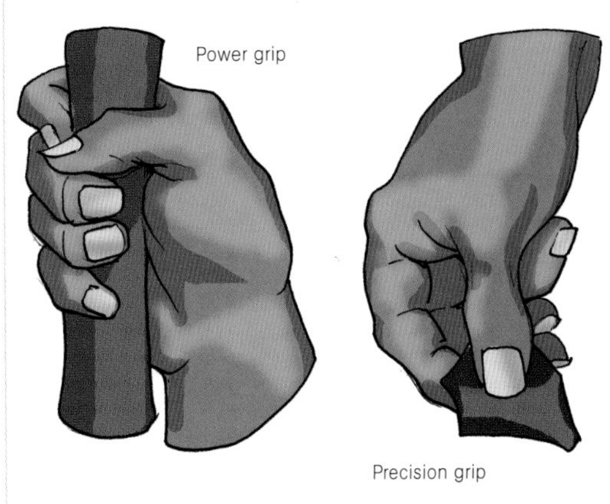

The human thumb is twisted by 130° relative to the other fingers; this enables the hand to perform a precision grip.

Presumably, one developmental line extended out from *Australopithecus* and led to the human beings of today. One of the first species of the genus *Homo* was *Homo habilis*, meaning "handy man" or "skillful person." *Homo habilis* lived more than two million years ago and was already using simple stone tools, such as chisels and scrapers. *Homo erectus* succeeded *Homo habilis* relatively quickly. With their body proportions and a height of more than 59 inches (150 cm), *Homo erectus* already resembled humans as we know them today. As the first hominid, *Homo erectus* migrated from Africa to Europe and Asia. In Europe, the species *Homo neanderthalensis* (Neanderthal man) existed around 160,000 years ago. The first *Homo sapiens* appeared about 200,000 years ago in Africa, and later encountered the Neanderthals in Europe. *Homo sapiens sapiens* spread over the Middle East and Balkan regions around 35,000 years ago at the latest. Archaeological evidence shows that the early *Homo* species lived in caves, hunted big game, and wore animal skin and fur.

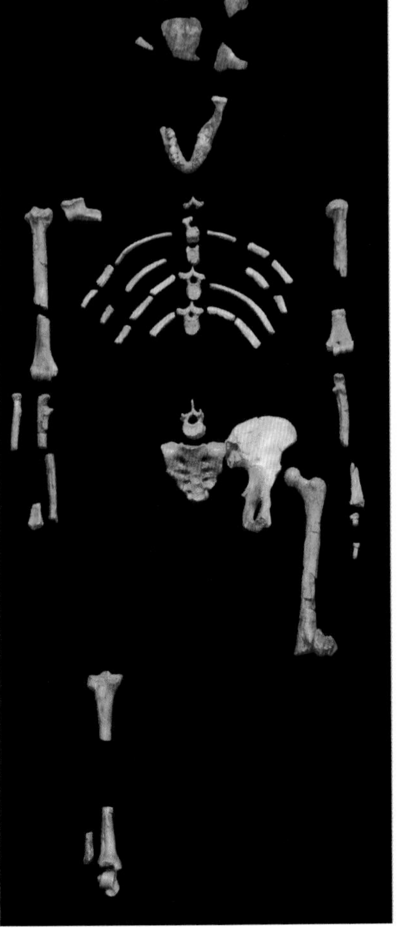

The bones of *Australopithecus* "Lucy" are 3.2 million years old and her skeleton is 40 percent intact.

BIPEDALISM

in focus

Footprints prove it: Bipedalism places *Australopithecus* into a direct, monophyletic line with modern humans. Over the course of evolution, the anatomy increasingly adapted to an upright gait. The pelvis became broader and lifted forward, and the spinal column formed into a double-S shape, acting as shock absorber. The hind extremities extended and became legs capable of running. The large toe became shorter and was no longer abductable as it is in apes, which can use their toes as gripping tools.

Footprints in hardened volcanic ash, found at an excavation site in Laetoli, Tanzania, document the upright gait of *Australopithecus*.

basics

THE BRAIN volume of *Australopithecus*, 24.4 and 42.7 cubic inches, (400 and 700 cm³) is comparable to that of today's chimpanzees or bonobos, 24.4 cubic inches (400 cm³). Humans have a brain volume of 85.4 cubic inches.

see also: Brain and spinal cord, p. 241

homo sapiens

Modern humans first appeared in East Africa around 200,000 years ago. From there they spread out to Europe, Asia, Australia, and America, and initially existed side-by-side with other hominina—until only they remained.

Homo sapiens (Latin for "wise man" or "knowing man") are today the only surviving species of the genus *Homo* and have the most highly developed brains of all living beings. With a volumetric mass of about 85.4 cubic inches (1,400 cm^3), *Homo sapiens'* brains are twice as large as those of *Homo habilis*. This volume increase was achieved in an evolutionary period of only two million years. It has given modern humans mental capabilities that have enabled them to recognize and deliberately change the world around them. The increase in brain volume is principally due to folding and surface enlargement of the cerebral cortex.

The *Homo sapiens* skull protects the highly developed brain.

Scientists consider this a side effect of the extended development period in human children. At birth, human brain capacity is only about 25 percent of its full potential, which requires substantial post-birth development of the brain. Perception of the external world using the eyes and ears, as well as the olfactory and tactile senses, strongly stimulates the formation of new brain cell linkages. During this extended period, the child is dependent on the care of adults. Therefore this phase is also a prerequisite for structuring relationships and social behavior.

basics

OWARD SAPIR, a erman-American hnologist and linuist, noted that nguage is an xclusively human, on-instinctive ethod for conveying thoughts, feelgs, and desires y means of a ystem of freely eated symbols.

The complete unfolding of the mental capacities of humans only became possible with the acquisition of language. Scientists assume that the linguistic capabilities of humans developed alongside their ability to use tools. This facilitated a cultural evolution, with progress that is unique in the animal kingdom, encompassing the processes characteristic of humans: perception, recognition, thinking, remembering, learning, and communicating about these experiences. At around 5000 to 4000 B.C., oral communication was supplemented by writing, the conveyance of information via a system of signs that had a clearly defined meaning. At that time, and independently of one another, different types of writing developed in various advanced civilizations. Probably the oldest written documents come from Mesopotamia.

These Egyptian hieroglyphs are a combination of logographic and alphabetic elements.

in focus

VOICE AND LANGUAGE

Voice and language serve human communication. The production of complex sounds requires an anatomically complex voice apparatus. In order to produce a spectrum of sounds, the cartilaginous larynx housing the vocal cords is located low in the throat. Other prerequisites are a sufficiently mobile tongue, a closed row of teeth and a highly arched roof of the mouth. Appropriate changes of the brain supported a refined control of lips and tongue; however, why and when these changes occurred is still unknown.

Human language translates thoughts and feelings into words. It forms the basis for social relationships and the development of culture.

see also: Evolution, pp. 140–145

selection

Organisms that are particularly well adapted to their environment generally have better chances of survival and reproduction, therefore prevailing over other individuals. This development is referred to as natural selection.

Among the large number of members of a population, those with the greatest likelihood of reproducing are the ones that are—based on their characteristics—best adapted to the so-called selection factors prevailing in a particular habitat. These factors can be either of an abiotic nature—relating to humidity or temperature—or of biotic nature—pertaining to other living beings, such as predators and parasites. The ability of an individual's genes to contribute to the gene pool of the next generation is referred to as fitness, while the influence of selection factors onto a population is known as selection pressure.

Stabilizing selection occurs when a population is so well adapted to its environment, that deviant mutants are constantly being eliminated due to selective pressures acting on the phenotype. Due to changing environmental conditions, one-sided selection pressure may favor individuals with certain alleles, so that their proportion increases within the population. This process is called transforming selection. In the event of splitting selection, those individuals that occur most frequently are being pushed back the strongest. Therefore individuals with specific, marginal characteristics have an advantage and are better able to assert themselves, which can ultimately even lead to a population split into two species. For instance, infectious diseases can be such a selection factor.

EVOLUTIONARY FACTORS

Following the synthetic theory of evolution, the development of life proceeds nondirectionally through the random effect of evolutionary factors. Selection and genetic drift are the most important evolutionary factors.

Genetic drift

The term "genetic drift" refers to a random change in the genetic diversity of a population. This can occur, for instance, when only a few individuals from a large population migrate into a new area. Such "founder individuals" represent a random selection of genotypes. Therefore, a random change of the gene pool becomes more likely the smaller a population is. A genetic drift can also happen when part of a population is suddenly destroyed. For instance, if both the environmentally well-adapted and less well-adapted members of a population die in a natural catastrophe, a coincidental section of individuals survives and the gene pool of the population subsequently shifts at random.

The coelacanth (*Latimeria chalumnae*) were believed to have been extinct until 1938, when a living specimen was discovered off the coast of South Africa. The species has survived in the deep sea up to the present due to the lack of selection pressure.

Ginkgo is a "living fossil." Its structure has not changed since the Permian.

in focus

DOMESTIC ANIMALS

The many races of domestic animals can serve as an example of transforming selection. This race diversity is the result of unnatural breeding conditions which tend to create genetic changes particularly fast, because only specific variants are selected that suit a particular breeding objective. Due to these selective mechanisms, a common base form, such as the wolf, has given rise to many dog races in a comparatively short period of time.

The dog, a domestic subspecies of the wolf, has developed into hundreds of different breeds.

see also: Evolution, pp. 140–145

the evolution of new species

New species evolve when groups of individual organisms of a species are isolated from each other due to barriers that prevent sexual reproduction. There are two types of speciation due to isolation: allopatric and sympatric speciation.

Allopatric speciation happens when a new species evolves due to geographic isolation over a long period of time. Geographic isolation may be the result of populations migrating onto an island, structural construction work by humans, or climate change and associated consequences (such as the glaciation of an area). Isolated populations continue the evolutionary process separately whereby new races or subspecies may evolve due to mutations. Initially, individuals of different subspecies are still able to reproduce. However, over time the gene pool (the total gene variation within a population) may not be able to mix anymore due to so-called isolation mechanisms, even if the previously separate populations repopulate a common area. If the individual organisms from different gene pools cannot reproduce anymore, they are, by definition, considered separate species.

A spotted-tailed quoll in the Cradle Mountain-Lake St. Clair National Park located in Tasmania, Australia.

There are various isolation mechanisms: isolation due to behavioral differences (such as changes in mating behavior), seasonal isolation (such as plants flowering at different times), mechanical isolation (such as anatomically altered sex organs), ecological isolation (such as population of different ecological niches as observed in Darwin's finches) and genetic isolation (such as polyploidic plants whose genetic material is made up of more than one chromosome set).

"Darwin's finches," as they are known, helped Darwin formulate his famous theory of evolution based on the different shapes of their beaks.

see also: Ethology, pp. 260–261

Sympatric speciation

This type of speciation does not involve a previous geographic isolation but is due to the genetic isolation of individual organisms occupying a common area. The genome changes in such a way that a gene exchange with other individuals of a population becomes impossible. A mechanism that can lead to sympatric speciation is a doubling of the chromosome set (polyploidy). This happens especially in plants that may, for example, self-fertilize or inbreed. A tetraploid plant (with four chromosome sets) and a diploid plant (with two chromosome sets) are unable to produce any fertile offspring.

milestones

JOURNEY TO THE GALAPAGOS ISLANDS

The English naturalist Charles Robert Darwin (1809–1881) is considered the founder of modern evolutionary theory. He stated that a visit to the Galapagos Islands inspired his hypotheses about the origin of the species which he compiled and illustrated in his chief work *The Origin of Species by Means of Natural Selection.* Conditions for speciation are unique on the Galapagos Islands as they are not only relatively far away from the mainland but the islands themselves are also more isolated from each other than other groups of islands. For this reason, Darwin observed the same plants and animals on the individual islands, but they often showed substantial variations. The most well-known example is the 14 finch species he collected on the Galapagos Islands.

Charles Darwin's theory of evolution shook the world.

basics

BIOLOGICAL SPECIES TERMINOLOGY The term species is one of the fundamental categories of biological classification. There are, however, very different definitions of what a species is. As a result, various definitions also lead to a range of different classification systems. One of the most commonly used definitions is: a species includes all individual organisms that are capable of naturally producing fertile offspring.

systematics

The aim of systematics is to describe the diversity of organisms, find distinctive characteristics, and classify them into groups of manageable size. There are two distinct methods of classification: artificial and natural classification.

systematics | evolutionary trees

CLASSIFICATION OF LIVING THINGS

Since ancient times, people have sought to systematically organize living organisms. Former classifications of the past mainly took outer features into consideration which was not necessarily very meaningful. Therefore, they often do not reflect the actual genetic relationships between organisms. Today, scientists can use modern techniques to determine genetic relationships and produce pedigrees.

In an artificial system, various organisms are assigned to a group according to similar characteristics. Here, the focus is on easily recognizable characteristics. One of the most famous artificial systems is the *Systema naturae* which was developed by the Swedish physician and naturalist Carl Linnaeus. In this classification system, he grouped all plant and animal species he knew according to morphological features. Notably, Linnaeus introduced the binomial nomenclature, which is still used up to this day. With this nomenclature each organism is given two names: a genus and a species name. This has the advantage of creating internationally standardized names which can be understood anywhere around the globe as opposed to the regional common names given to organisms. The selected name normally derives from a Latin or Greek description of characteristics. Species with common features form a genus, similar genera form a family. Today, the main purpose of such a system is the quick and reliable identification of an individual organism.

Natural classification

Modern phylogenetic systems are based on evolutionary theory, which supports the idea that all organisms have evolved over time from ancestors with a less complex phylogeny. Organisms are classified according to either common or distinctive characteristics. It is assumed that groups are more closely related to each other the more features they have in common.

J. W. Weinmann's botanical work *Phytanthoza Iconographia* is a valuable record of the plant kingdom.

Phylogenetic relationships can only be derived from homologies. These are similarities that are based on common evolutionary roots. In contrast, convergences are similarities that have evolved through adaptation to similar functions and similar environmental conditions. Such convergences are disregarded in natural classification systems. Today, actual phylogenetic relationships are sometimes determined by analyzing changes in sequence of certain nucleic acids that are determined to have a highly conserved structure.

milestones

THE BEGINNING OF CLASSIFICATION

The *Systema naturae*, first published in 1735, is considered a milestone of biological systematics. It was written by the Swedish physician and naturalist Carl Linnaeus, also known as Carl von Linné. Initially, he concentrated on the systematics of plants and used their reproductive organs as a basis of his classifications. Later, he broadened his studies and included animals and even minerals. His binomial nomenclature first described in this work remains unchanged and in use even today. After being ennobled by the Swedish king for his outstanding work in 1762, he changed his original name to Carl von Linné.

Carl Linnaeus is known as the father of modern taxonomy.

Entomology is the study of insects. Since the 16th century, scientists have classified many insect species but a large number still remains undiscovered.

see also: Petrifactions, pp. 156–157

evolutionary trees

An evolutionary tree is a schematic representation of natural relationships between organisms. Evolutionary trees are always binary, that is each node results in only two daughter branches.

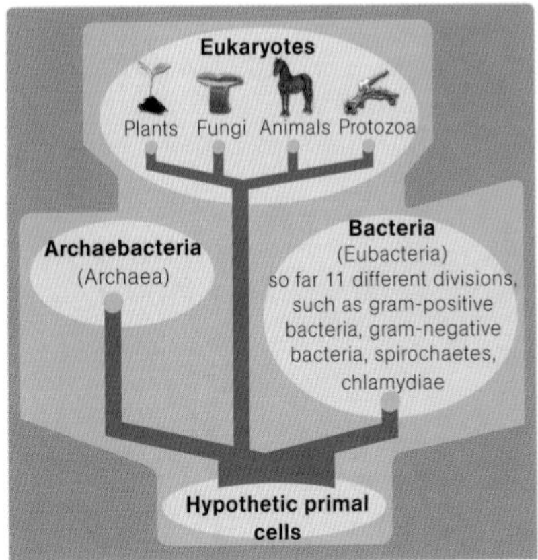

The three-domain system by Carl Woese divides cellular life into archaea, eukaryotes, and bacteria.

Evolutionary trees are comparable to relationship diagrams or cladograms whose purpose it is to show the results of phylogenetic systematics in a graphic way. They are changing all the time and often contain knowledge gaps. Therefore, they are not to be seen as a final product of classification but as an approximation of the natural system of organisms.

Nucleotide sequences of the 16S ribosomal RNA (16S rRNA) are often analyzed in order to construct modern evolutionary trees. It is thought to be especially valuable for this purpose, because the rRNA is assumed to be structurally highly conserved, as proteins are synthesized in the ribosomes, which are present in all organisms. Even minor mutations usually result in some functional loss of the ribosomes. Assuming that the number of mutations is proportional to time, differences in sequence would be a measure of the evolutionary distance between species. Evolutionary trees produced in this way are thought to be very reliable, especially because their validity has been confirmed by studies using other methods.

In order to create a phylogenetic tree, the 16S rRNA sequence data of an organism are compared to the data of other species by using computer programs. In the mid-1970s, several prokaryotes were examined using this method. The result was that the prokaryotic organisms actually did not belong to one homogeneous group but they split into two separate groups very early in evolutionary time. Therefore, archaea, single-celled microorganisms, are now considered as a phylogenetic group by itself, distinct from bacteria and the eukaryote group consisting of plants, animals, fungi, and protists. With the aid of this classification method, scientists were also able to prove that the kingdom of fungi are not part of the plant kingdom. Instead, they constitute a eukaryotic sister group to animals and plants.

basics

PHYLOGENETIC METHODS to compare species include gene sequence analysis, the comparison of nucleotide sequences of 16S rRNA, and DNA-DNA hybridization.

in focus

PRIMAL PROKARYOTES

The archaea are an evolutionarily very old group that has survived even the harshest environmental conditions. Some species, for example, live in temperatures above 212°F (100°C) or in extremely saline, acid, or alkaline environments. In certain cases they can sustain pressures of up to 1,000 bar.

Archaea are organisms that often live at extreme locations such as thermal springs, salt lakes, and geysers.

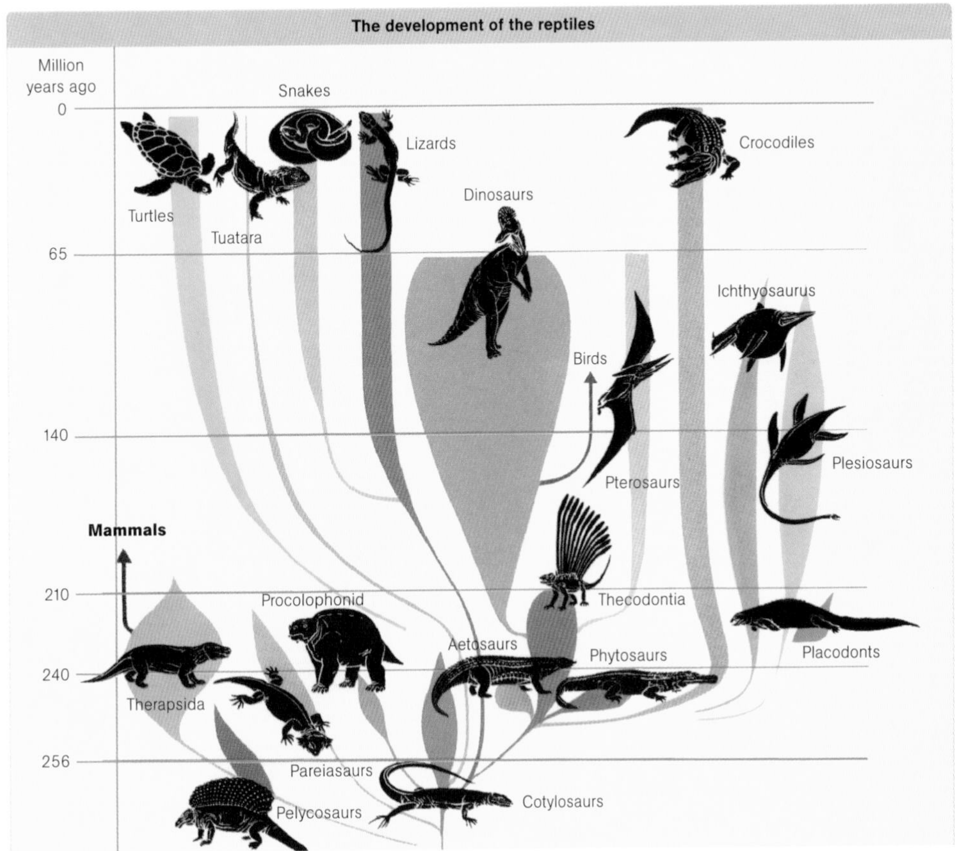

The family tree of reptiles includes five living orders—turtles, crocodiles, snakes, lizards, amphisbaenia, and the extinct order of dinosaurs, all of which share the same evolutionary source.

see also: Evolution, pp. 140–145

BIOLOGY

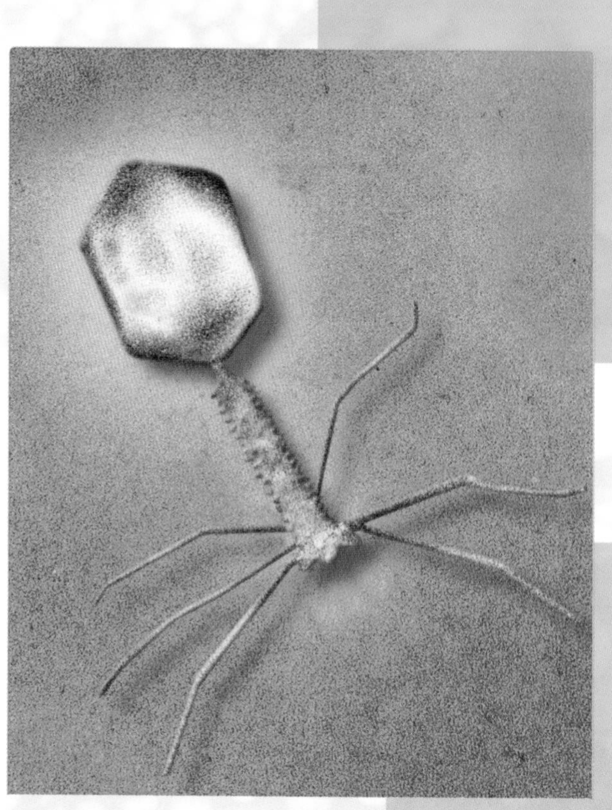

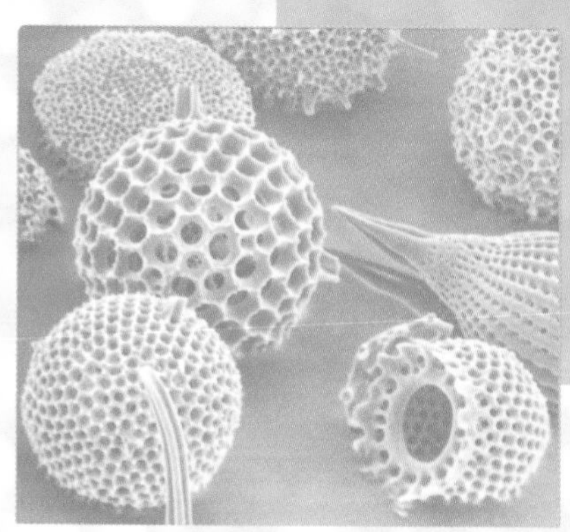

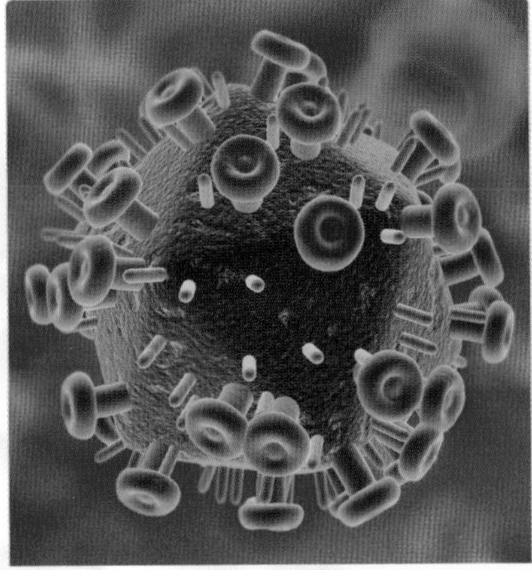

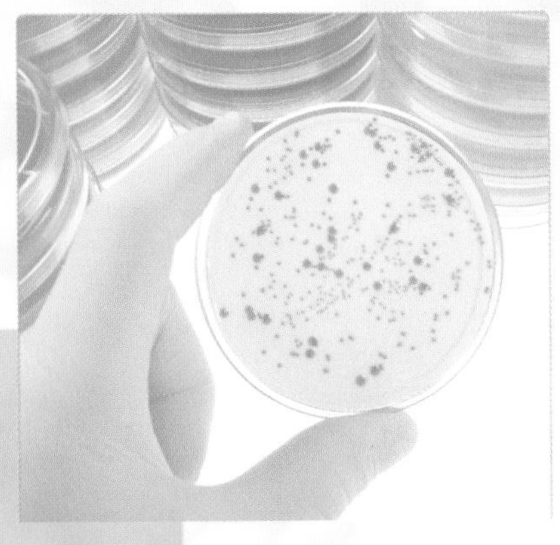

MICRO-ORGANISMS

A new fascinating world of microorganisms, invisible to the naked eye, was first made accessible to humans during the 17th century. Antoni van Leeuwenhoek was the first to observe bacteria with a microscope that he built himself.

Bacteria, viruses, as well as single-celled organisms of all kinds are part of life on Earth. They were the first living organisms and even today make up the largest part of living matter. Microorganisms can be found almost everywhere and can exist under extremely harsh environmental conditions such as in hot vents.

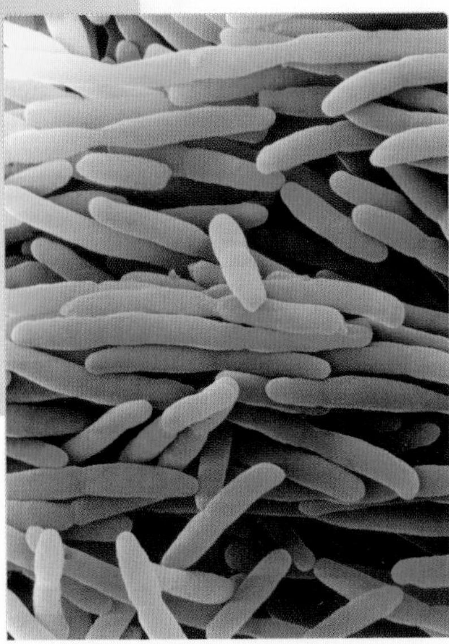

structure and metabolism of bacteria

Most bacteria are just a few micrometers in size, observable only using a microscope. So naturally, it was the microscope builder Antoni van Leeuwenhoek who discovered bacteria during the 17th century.

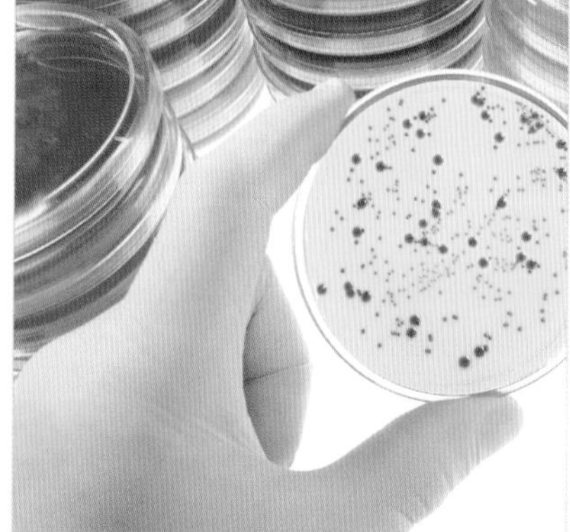

In laboratories, bacteria are often grown in a petri dish on a culture medium containing agar as a solidifier.

structure | metabolism | harmful and useful bacteria

BACTERIA

Bacteria are cellular organisms without a nucleus from the group prokaryotes. Prokaryotic cells are much smaller compared to eukaryotic cells, which contain a nucleus and several chromosomes. Bacteria are so tiny that they are invisible to the human eye, yet they can significantly impact the environment and lifestyle of other organisms.

Bacteria are typically spherical or rod-shaped. Some species consist of helical cells and others form thread-like outgrowths similar to mushroom filaments. Their size is normally between two and five micrometers (μm); in exceptional cases, bacteria may reach a length of 50 μm or even more. Bacterial cells lack nuclei as well as organelles that are typical for eukaryotic cells. Their cell walls usually consist of a rigid structure made of polysaccharide and amino acid molecules. If the cell wall is thin with a single layer the bacteria are considered gram-negative and if the cell wall is a thick multilayered mesh, then the bacteria are considered gram-positive.

IF CONDITIONS are favorable, bacteria can multiply rapidly. In extreme cases, they double every 11 minutes. After only one hour, they could weigh about 5,000 tons.

basics

Reproduction normally occurs by division. Some species have flagellae and are therefore mobile. Other bacteria produce endospores which is a dormant structure within the organism that allows survival for many years, even during unfavorable conditions.

Metabolic diversity

Different bacteria take various approaches to the production of energy. Most species use organic matter, especially carbohydrates. Certain species, however, can also generate energy from inorganic matter, such as sulfate compounds. Some bacteria use sunlight for energy production while others use the energy from chemical reactions. Bacteria can also grow without oxygen or they may live parasitically or symbiotically without access to air circulation. Furthermore, some species thrive in comparatively high temperatures while others can tolerate high concentrations of salt. Due to their rich diversity, bacteria can be found in almost all environments.

Laboratory cultures

Due to the tiny size of bacteria, close examination in the environments where they are commonly found is very difficult. Therefore, it is easier to study them in a laboratory environment. Here, pure bacterial cultures are grown on an artificially made culture medium. These can be liquid cultures or solid agars. In order to identify pathogens, it is also common practice to first set up a growth environment for a pure culture.

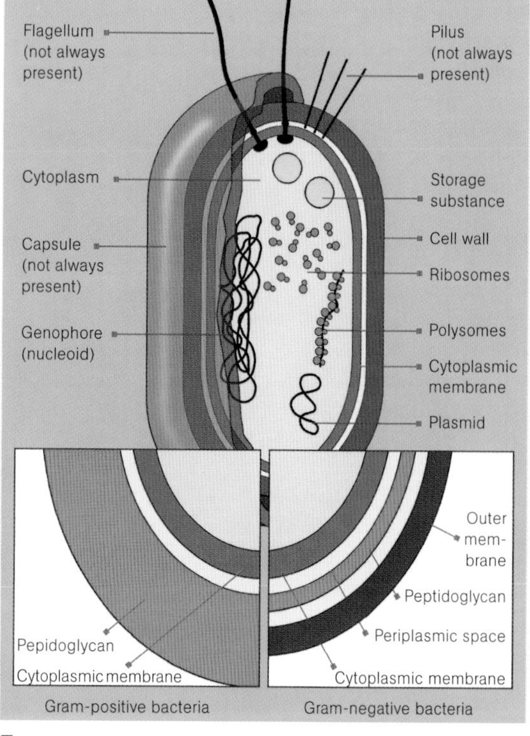

According to the categorization of Danish bacteriologist Hans Christian Gram, most bacteria are either gram-positive or gram-negative.

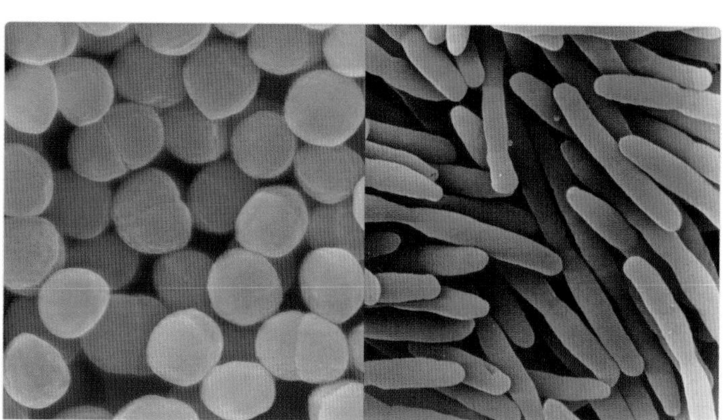

Staphylococcus and *Escherichia coli*: In contrast to their enormous physiological diversity, the variety of bacteria shapes is rather small.

see also: Cells, pp. 148–155

harmful and useful bacteria

Most people think of bacteria as pathogens carrying diseases such as tuberculosis or the plague. Contrary to this common belief, many of these tiny organisms are very useful to humans.

■ Wastewater treatment with bacteria: Sewage plants cleanse and recycle water in several stages.

Bacteria play an important role in a biological ecosystem. In many ways, they are responsible for creating a livable environment for other organisms. For example, the contribution of bacteria and fungi in the mineralization process is estimated to be 90 percent. During this process, organic carbon compounds are reduced to carbon dioxide. Without these bacteria, life on Earth would come to a stop. They are also important components of the nitrogen and sulfate cycles as well as during the self-purification of water bodies.

Bacteria may indeed carry a series of dangerous diseases, for example, the plague, leprosy, cholera, tuberculosis, diphtheria, meningitis, typhus, tetanus, or syphilis.

Many of these diseases have become less of a threat after the discovery of penicillin and other antibiotics. However, the entire world population does not have equal access to these agents. In addition, certain pathogens have become resistant to antibiotics, which has led to the revival of several diseases.

For many centuries, humans have used bacteria for processing and conserving food items. Bacterial fermentation is used for preservation, for example when pickling olives, cucumbers, or saurkraut. Products such as yogurt, acid curd cheese, or soy sauce are all made with the help of metabolically active lactic acid bacteria.

These single-celled organisms are also used on a large scale for the production of citric acid, vitamins, and antibiotics. Genetically modified strains are also used, for example, in the production of human insulin.

Bacteria also play an important role during wastewater purification, the decontamination of soils, and waste disposal. Several of these processes even produce valuable by-products, such as methane or manure.

MODERN BACTERIOLOGY

milestones

In 1905, Koch received the Nobel Prize in medicine for his discovery of the pathogenic agent causing tuberculosis.

The German physician Robert Koch was one of the founders of modern bacteriology. In 1876, he was able to multiply the pathogenic agent of anthrax (*Bacillus anthracis*) in a culture and prove its role in the development of the disease. Six years later, he discovered the pathogenic agent of tuberculosis (*Myobacterium tuberculosis*) and only a year later he found the pathogen that causes cholera (*Vibrio comma*). Koch was able to prove the existence of these pathogens by cultivating bacteria on artificial agar and using new dyeing techniques.

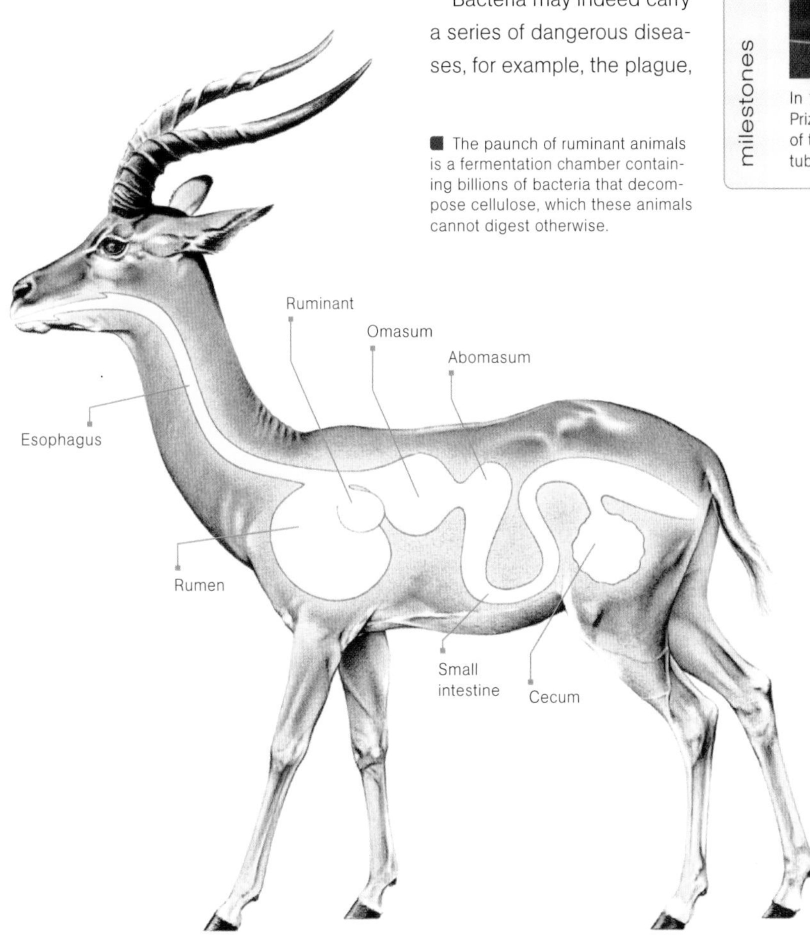

■ The paunch of ruminant animals is a fermentation chamber containing billions of bacteria that decompose cellulose, which these animals cannot digest otherwise.

basics

***CLOSTRIDIUM** botulinum* can multiply in insufficiently cooled food items and cause a life-threatening illness (botulism). This type of bacteria can survive without oxygen and produce heat-resistant spores, making even canned food items toxic.

the virus

Viruses are much smaller than bacteria, so they only become visible with an electron microscope. The smallest are only about 20 nanometers long, the largest are about 500 nanometers (one nanometer = 10^{-9} meters).

Viruses occurring outside of cells are called virions and may differ considerably in appearance. Many virions are ball or rod shaped, while others look like small lunar modules (such as the bacteriophage T4).

viruses | pathogens

VIRUSES

Viruses are tiny infectious parasites that lack their own metabolism. Therefore, viruses depend on host cells to reproduce and proliferate. Normally, viruses are strictly host-specific and affect either eukaryotes (organism with nucleus) or prokaryotes. Several viruses cause dangerous diseases, many remaining without a cure.

basics

VIROIDS ARE RNA viruses without an envelope. The smallest self-reproducing molecules, they have only been found in plants such as potatoes and tomatoes.

Genetic information is stored in the form of deoxyribonucleic acid (DNA) or ribonucleic acid (RNA). This trait is used for classifying the viruses into two groups: DNA viruses and RNA viruses. The genome of a virus is enclosed by a coat of protein molecules (capsid), and sometimes additional membranes. Some phages not only have a head filled with nucleic acid, but also complex tails for attaching themselves to bacterial cells and for injecting their genome. Phages are extremely small and require an electron microscope for study. However, their relative sizes to each other varies considerably.

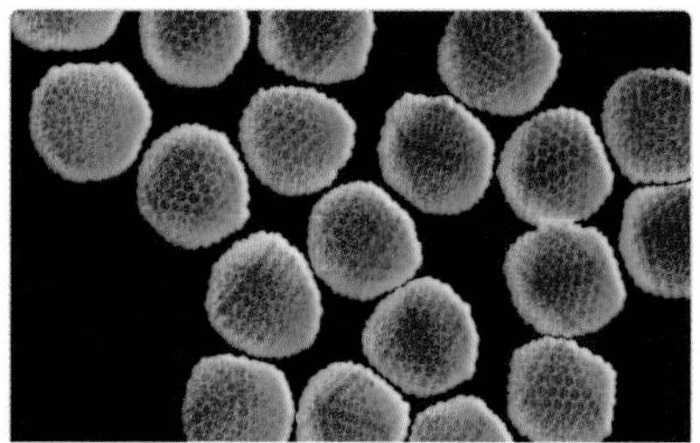

Viruses can cause various diseases in humans and animals, some of which may be fatal.

Reproduction

Reproduction of viruses can be divided into two cycles: the lytic cycle and the lysogenic cycle. The former can be found, for example, in T phages. Here, the genome of the virus is injected into the host cell. The protein coat stays behind on the surface of the cell. The viral nucleic acid then synthesizes its own proteins, which often block the activity of the bacterial DNA. At the same time, the host cell is instructed to replicate viral nucleic acid and produce proteins for a protein coat to form a nucleocapsid. New viruses develop, causing the bacterial cell to burst (lysis) and release the virions, which can then attack new host cells.

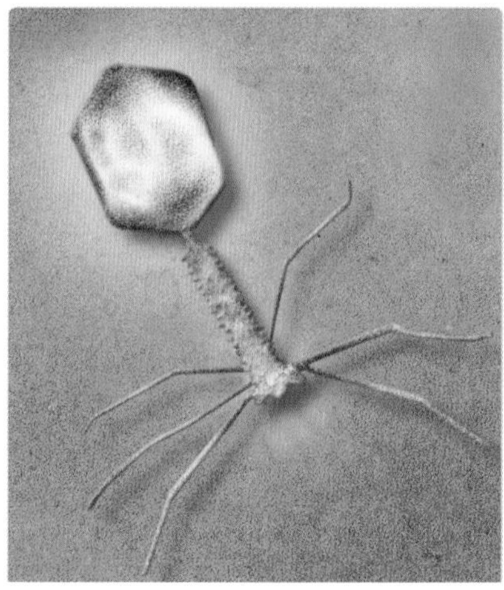

The bacteriophage T4: Bacteriophages are often classified by letters and numbers.

The lysogenic cycle differs because the viral genome is temporarily built into the DNA of the host cell. The germs remain there in the form of so-called proviruses, or prophages, and replicate passively together with the cellular genome. The creation of active viruses and cell lyses does not happen until an activation event occurs. For example, a change in temperature could trigger the lytic behavior of viruses.

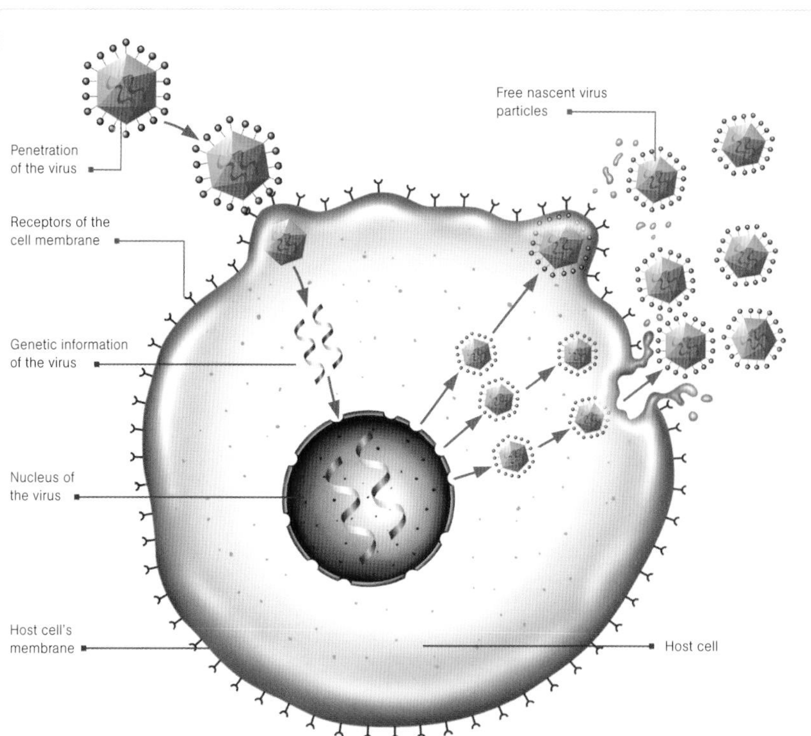

For survival and reproduction, viruses are dependent on other living organisms. They splice their genetic information into the DNA of the host cell.

see also: Genetic information: dna, p. 150

viruses as pathogens

The term virus comes from Latin and means poison. As the name implies, viruses may cause serious disease, often with no possible treatment due to the lack of effective medication.

The influenza virus exists in varying strains, but the main strains that cause the flu result in symptoms that are very similar to those of the common cold. Fatigue and pain in the limbs, headaches, and occasionally high fevers are common. The human body, which is weakened due to the influenza virus, is therefore susceptible to sometimes fatal secondary infections, for example, due to bacteria.

Very infectious strains of this pathogen may cause flu epidemics. An example of such an epidemic is the Spanish flu of 1918 to 1920. It is estimated that 25 million people died from the flu during this time. Avian flu is another extremely contagious and aggressive strain that is caused by influenza viruses. Normally these pathogens are only passed on from one bird to another, but in recent years there have been cases of viruses infecting mammals too, including humans.

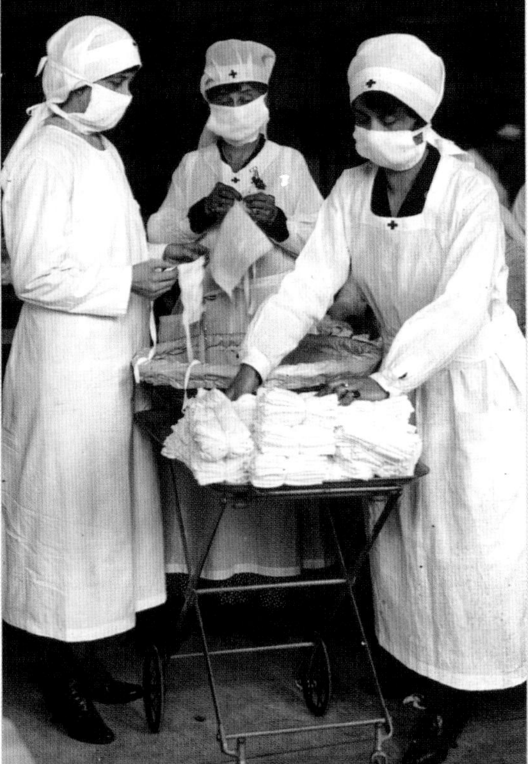

During the Spanish flu epidemic of 1918–1920, nurses manufactured protective mouth-masks for U.S. soldiers.

The human immunodeficiency virus (HIV)
Today one of the most feared viral diseases among humans is AIDS (acquired immune deficiency syndrome). It is an infection caused by the human immunodeficiency virus (HIV). The disease is almost always fatal. It is estimated that about 40 million people worldwide have been infected with the HIV virus or have fallen ill with AIDS.

The HIV pathogen is a retrovirus; it carries its genetic information in a single strand of RNA rather than DNA, and has a particular enzyme called reverse transcriptase. The enzyme is responsible for transcribing RNA into DNA during infection of a host cell. The transcribed DNA is then integrated into the genome of the host cell. This is followed by a sometimes lengthy latent period of inactivity. Once the virus becomes active it destroys important immune cells (T helper cells), which leads to a permanent deficiency of the immune system as a whole. Patients often die from diseases that their immune system would have normally been able to defend against. Despite increased research efforts into the development of medication or effective vaccination to prevent the immune deficiency, a cure still has not been found.

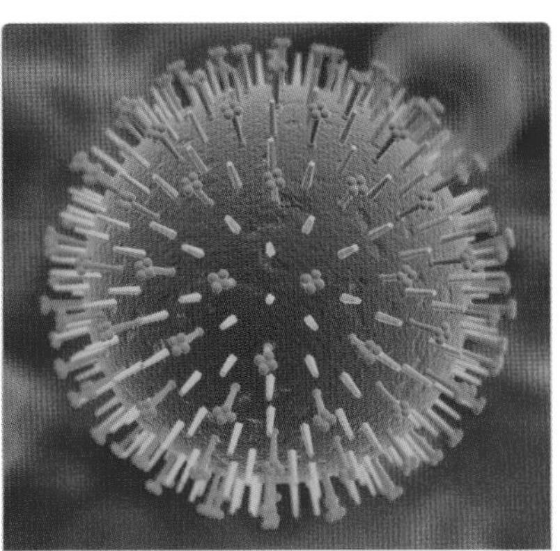

Today, computers can be used to render three-dimensional images of the influenza virus.

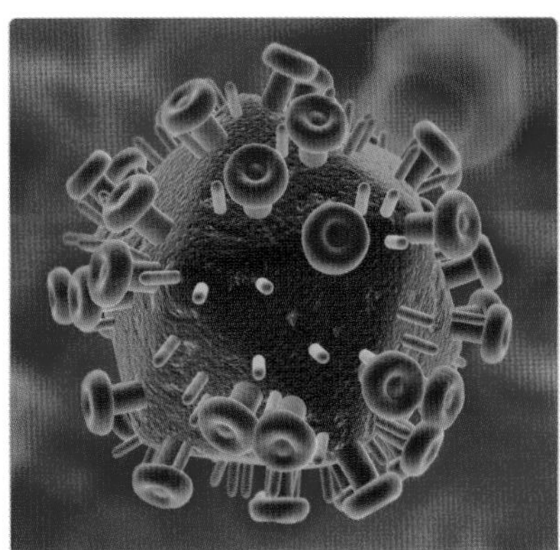

Viruses like the HIV virus can have a relatively complex structural shape.

in focus

REMBRANDT TULIPS

These tulips developed multicolored flowers that were extremely popular and valuable during the 17th century. At the beginning of the 20th century the cause for the coloring was finally discovered: a virus prevented an even distribution of pigments. Most Rembrandt tulips sold today are no longer infected with the virus, but they are produced by selective breeding.

Plants can also fall victim to virus infections, with varying effects. Rembrandt tulips are a relatively benign example.

basics

THE FIRST vaccination against the smallpox virus may have been developed about 3,000 years ago. There are accounts of Chinese healers who immunized patients with an inhalation of powder made from smallpox scabs of epidemic survivors.

see also: Immune response to disease, p. 247

heterotrophic | autotrophic

SINGLE-CELLED ORGANISMS

Protists (Protista) are a group of eukaryotes, mostly single-celled organisms, which are difficult to classify as animals, plants, or fungi. All protists have a nucleus and cell organelles as typically seen in other organisms. Species of this group are often classified according to their feeding mechanisms. However, this classification may not always correspond to their actual genetic relationships.

heterotrophic single-celled organisms

Heterotrophic single-celled organisms feed on organic material such as bacteria and other tiny organisms. Typical examples of this group are rhizopods (Rhizopoda) and ciliates (Ciliophora).

The single-celled rhizopods have no defined body shape, instead, they have false feet called pseudopods. These are lobe- or thread-shaped plasma extensions used for taking in food particles. Some species also use pseudopods for movement. Amoebas are well-known rhizopods. Most of the amoeba species live in freshwater habitats. In contrast, foraminifers are exclusively found in the sea. While amoebas are mostly bare and exposed to the outside, foraminifers have a shell with several pores, which their pseudopods can expand through. Another group of marine rhizopods are the radiolarians with a skeleton made of silicic acid and radiating pseudopods. The spherical heliozoans or sun animalcules are mainly freshwater species. The skeletons of dead radiolarians and foraminifers often build up on the ocean floor, creating thick layers of sediment. Ciliates are slightly more complex in comparison to the rhizopods. They have a mouthlike opening to take in food inside the cell, and an anal opening for excreting indigestible substances,

Reproduction may be asexual by cell division or sexual by conjugation, where two individuals temporarily connect to exchange genetic material. Another typical characteristic is the large quantity of cilia, which serve as propellers, keeping the organism in motion and swirling water and food particles into the mouth opening. The paramecium or slipper animalcule (*Paramecium* spp.) is a common example of this group. These single-celled organisms can reach a size of up to 0.01 inch (0.3 mm); they live in freshwater and are shaped like a slipper. The rhythmical beating of numerous cilia is used to move them forward. Like all ciliates they have two nuclei: a macronucleus and a micronucleus. The macronucleus regulates cellular metabolic functions, while the micronucleus plays a role in sexual reproduction.

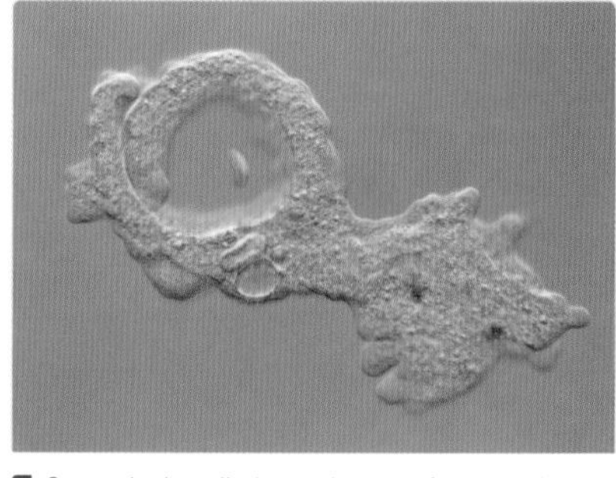

Some single-celled organisms, as for example amoebas, flow over food particles, such as ciliates and algae, and digest them inside a food vacuole.

basics

PREDATORY SINGLE-CELLED organisms may be equipped with toxicysts. These capsules have a tubular structure and are used to puncture and paralyze prey.

Radiolarians have beautiful, intricate skeletons. Their remains contribute to ocean floor sediment.

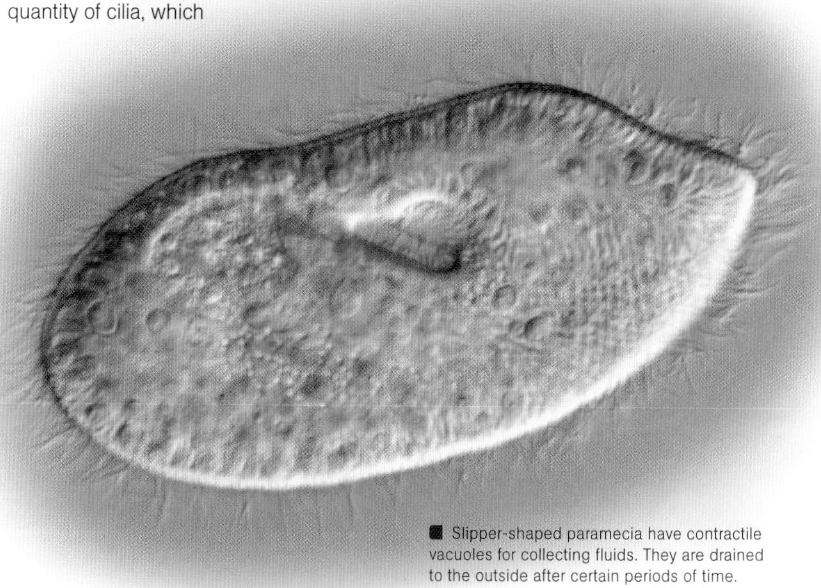

Slipper-shaped paramecia have contractile vacuoles for collecting fluids. They are drained to the outside after certain periods of time.

see also: General cell metabolism, p. 154

autotrophic single-celled organisms

In contrast to heterotrophic single-celled organisms, autotrophic single-celled organisms use the process of photosynthesis to produce their own organic molecules of nutrition from inorganic molecules.

Diatoms are a typical example of this group. They account for a major part of the world's plankton and are therefore an important part of the marine food chain. Their cell wall consists of two halves (shells), often with complex patterns. Another example of an autotrophic single-celled organism is the genus *Chlorella*, a single-celled spherical-shaped genus that lives in fresh water. They divide at very high rates and can create large amounts of biomass within short periods of time. Occasionally, *Chlorella* species are even cultivated for use in food production or the cosmetics industry. Some autotrophic single-celled organisms can switch to a heterotrophic lifestyle when light is scarce, for example, in the case of *Euglena gracilis*. This flagellated species is abundant in freshwater and is normally photosynthetically active. But when these organisms are living in low light, they are able to adopt a heterotrophic lifestyle by eliminating chlorophyll from their bodies. As a result, their appearance changes from green to transparent. As soon as light becomes available again, even if several years have passed, the chloroplasts regain functionality and photosynthesis can resume.

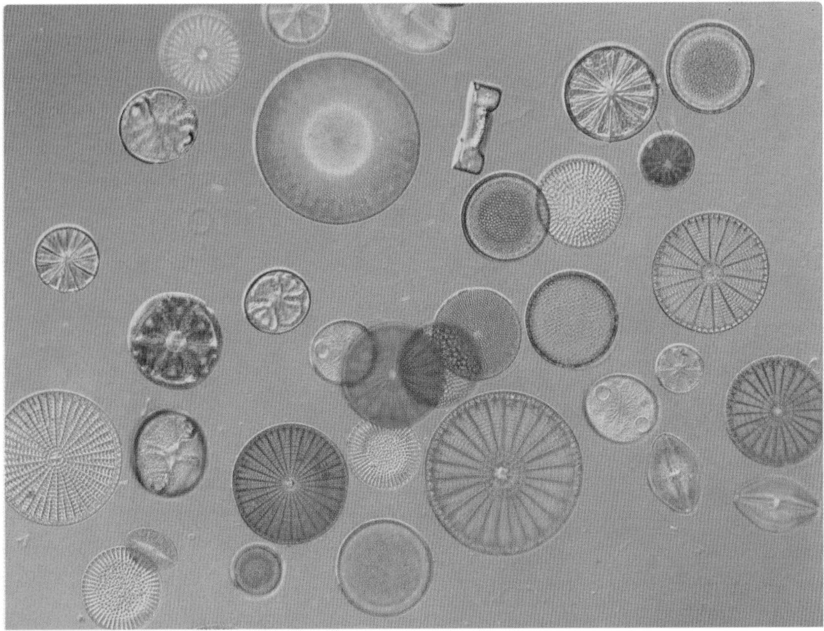

Diatoms are single-celled organisms of many varieties that make up the vast majority of the ocean's plankton biomass. They use photosynthesis to manufacture their own nutrients.

[I]MMORTAL SINGLE-[C]ELLED organisms [r]eproduce by divid[i]ng into two equiva[l]ent daughter cells. [T]herefore these [o]rganisms are con[s]idered potentially [i]mmortal. Cells that [d]ivide in this way [w]ill never die due to [o]ld age.

basics

Parasitic organisms

Certain single-celled organisms have adopted a parasitic lifestyle. Some species even carry diseases which may infect humans, for example *Entameba histolytica*, which causes tropical amoebic dysentery. When ingested through infested drinking water they can cause tissue damage with symptoms such as ulcers or other health issues that are potentially fatal if left untreated. Pathogens causing malaria *(Plasmodium* spp.) are even more dangerous. This disease is responsible for the death of one to two million people annually. Sleeping sickness is another dreaded disease common in tropical regions where an estimated 500,000 people are affected. The disease is caused by species of the genus *Trypanosoma* and carried by tsetse flies *(Glossina* spp.).

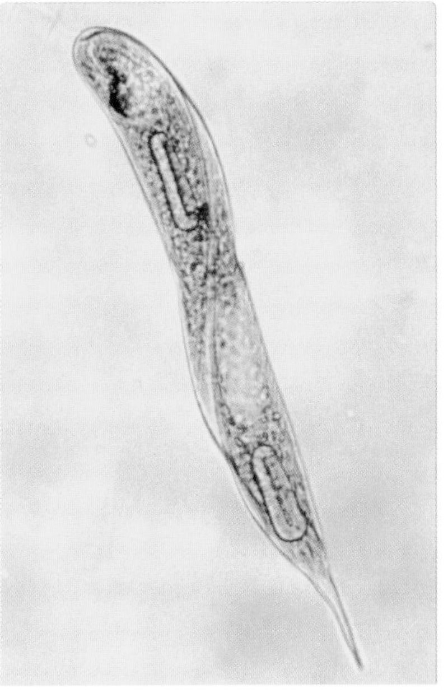

The elongated single-celled organisms *Euglena gracilis* are autotrophic.

in focus

THE FORMATION OF COLONIES

Some autotrophic single-celled organisms form colonies made up of independent cells connected by a gelatinous substance. In *Pandorina* and *Eudorina* species, 16 or 32 individual cells form a hollow sphere with their flagella to the outside. The cells are connected by channels, causing the flagella to whip at the same time. In *Volvox* species, several thousand may combine, a small fraction of which reproduce while the remaining cells take over other functions such as photosynthesis and locomotion.

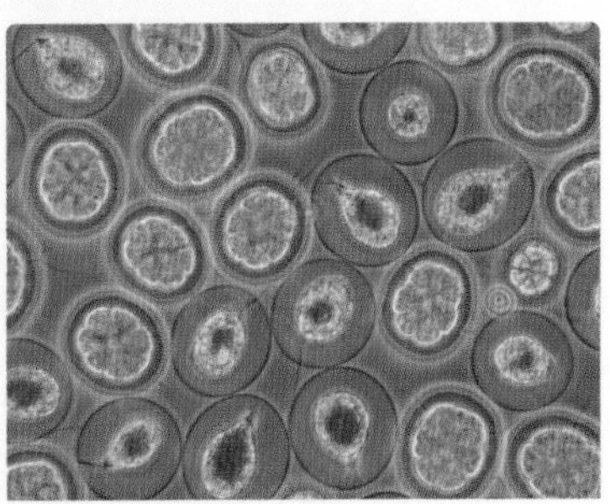

Some single-celled autotrophic species form colonies made up of individual organisms.

see also: Cells, pp. 148–155

BIOLOGY

PLANTS AND FUNGI

Just like the animal kingdom, plants and fungi also have their own independent kingdoms within the group of eukaryotes—organisms that have a nucleus. They were the first organisms that evolution brought onto land. These first terrestrial organisms evolved into an extremely diverse group with an abundance of different adaptations to local environmental conditions.

Plants include mosses as well as vascular plants such as huge sequoia trees, which may reach heights of up to about 360 feet (110 m). The most important characteristic of almost all plants is their ability to generate organic material from light and inorganic substances through photosynthesis.

With about 100,000 known species, fungi diversity is also impressive. Yeasts are considered the smallest fungi. Large fungi are much easier to see; they may reach a diameter of several hundred yards or meters.

higher plants | photosynthesis | transport | plant compounds

MORPHOLOGY AND PHYSIOLOGY

Almost all higher plants have a similar functional build, with organs for photosynthesis, to absorb nutrients and water, and for reproduction. Plants' vital processes include photosynthesis, growth, environmental adaptations, and vascular transport.

anatomy of higher plants

Nearly all plants with pathways (vessels) to carry liquids have the same basic structure: stems, leaves, and roots.

The basic components of a seed plant are the roots, leaves, stems or shoots, and—at certain times of the year—flowers and fruit. The roots function to anchor the plant in the ground. Fine root hairs absorb water and dissolve minerals from the soil. Some plants, such as carrots, have thickened roots that are used to store nutrients. The plant later draws upon these reserves to produce flowers and fruit.

The stem is an integral part of the plant and supports the leaves, flowers, and fruit, as well as transport water and nutrients upward to the leaves. These vital materials are transported through narrow tubes known as vascular bundles. Shoots generally grow toward light and, in plants such as bushes and trees, develop a woody structure.

Site of photosynthesis

The green leaves of a plant produce nutrients using photosynthesis (p. 179), a process that occurs in chloroplasts located within the leaf cells. Water and the sugar produced by photosynthesis are distributed through the vascular bundles to nourish cells. The vascular bundles can often be clearly seen as the veins on the underside of a leaf. The leaves draw carbon dioxide from the air through slit-shaped openings called stomata. These are usually found on the undersides of leaves. Typically, stomata open in the daytime to release excess water and the oxygen produced during photosynthesis. The carbon dioxide is stored in the plant's cells until it is needed. The cuticle, or upper surface of leaves, and sometimes the underside have a waxy coating to protect the plant from dehydration and the sun's harsh rays.

FLOWERS AND FRUIT

in focus

Flowers are specialized forms of leaves. The sepals—which are usually green and resemble leaves—protect the flower before it opens. Colorful petals attract insects in search of nectar and animals looking for pollen. Within the petal ring, the plant's reproductive organs (the male stamens and the female pistil) are divided into the style, ovary, and stigma. The style connects the stigma and ovary, which contains the ovules that develop into seeds. Each male stamen consists of a filament and an anther that holds pollen. After pollination, the ovule ripens into a seed; the ovary develops into a fruit. The fruit protects and disperses the seed. Its appearance is adapted to the plant's seed distribution mechanism.

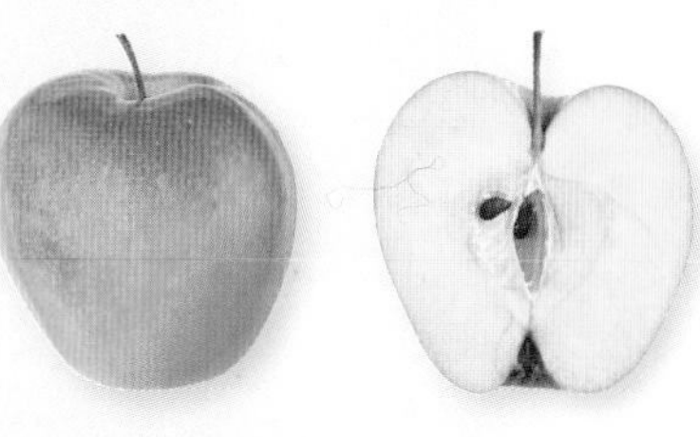

A fruit contains one or more seeds and a multilayered fruit wall called the pericarp.

The flower is the reproductive organ of a seed plant. It contains the stamen and carpels, usually surrounded by petals, which protect the reproductive structures and serve to attract pollinators.

Flower: the reproductive system of the plant.

Stem: provides structural support; filled with vascular bundles that transport water and nutrients

Leaves: carry out photosynthesis to provide the plant with energy; release wastes such as oxygen and water vapor.

Roots: absorb water and inorganic nutrients from the soil and anchor the plant in the ground.

Plants can be identified and classified as belonging to a certain plant group according to their morphology—their structural form.

see also: The evolution of plants, p. 162

photosynthesis

Plants create their own sustenance by converting sunlight into energy. Because of this ability, they are the nutritional foundation for most other living organisms.

in focus

STOMATA

Stomata are tiny pores in the epidermis of leaves, stems, or flowers that consist of two bean-shaped guard cells. Numerous stomata connect the plant interior with the outside air and allow for the exchange of carbon dioxide and oxygen. When open, the stomata also release water vapor, driving the transportation of water throughout the tissues (xylem) of the plant.

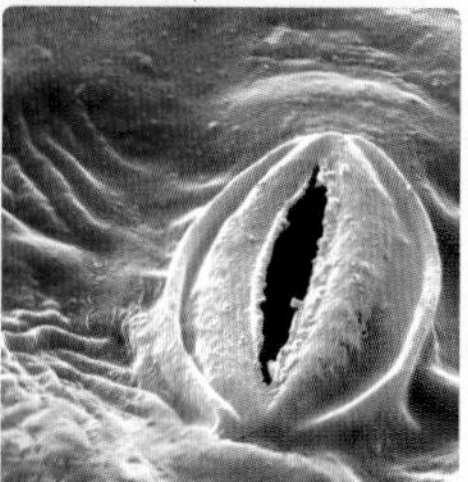

The guard cells surround an opening, or stomata, and open or close depending on the environmental conditions.

Animals feed on other organisms while plants produce their own food. Most plants do this using photosynthesis, a process in which the sun's energy is converted into chemical energy and stored.

Photosynthesis

Green plants annually remove about 200 billion tons of carbon dioxide from the atmosphere, while producing sugar and other nutrients and giving off vast amounts of oxygen as a waste product. There are two stages in this process: a light dependent stage, which requires visible light with wavelengths of between 400 and 700 nanometers and a stage that is independent of light.

Photosynthesis occurs in the chloroplasts, which are found in a plant's leaves, giving them their green coloration. In fact, in an individual plant cell there are hundreds of these lens-shaped subcellular organelles. Many flat, disk-like structures containing the green pigment chlorophyll are stacked on top of each other within each chloroplast. Chlorophyll—the engine of photosynthesis—is a molecule that can absorb sunlight and use that energy to produce other molecules. It produces the energy-carrying molecules adenosine triphosphate (ATP) and nicotinamide adenine dinucleotide phosphate (NADPH).

To capture enough light to power photosynthesis, leaves usually grow toward the sun, a habit known as phototropism. The carbon dioxide needed for photosynthesis is absorbed through microscopic openings in the leaves called stomata. The water needed for photosynthesis is taken up by the roots and transported through the plant tissues. Excess water and the waste oxygen are released via the stomata.

Light-absorbing pigments, which absorb and transfer sunlight, are primarily found in the leaves.

The transfer molecules ATP and NADPH

Photosynthesis was not fully explained until the 1960s. Melvin Calvin—who won the Nobel Prize for chemistry in 1961 for his work—saw that it was a two-stage process. In the first stage, light energy from the sun is captured within the chloroplasts and used to split water molecules and create the carrier molecules ATP and NADPH. In the second stage, the carrier molecules construct sugar molecules using carbon dioxide. The sugar is stored as a starch and transported to individual cells when energy is required. The waste product is oxygen.

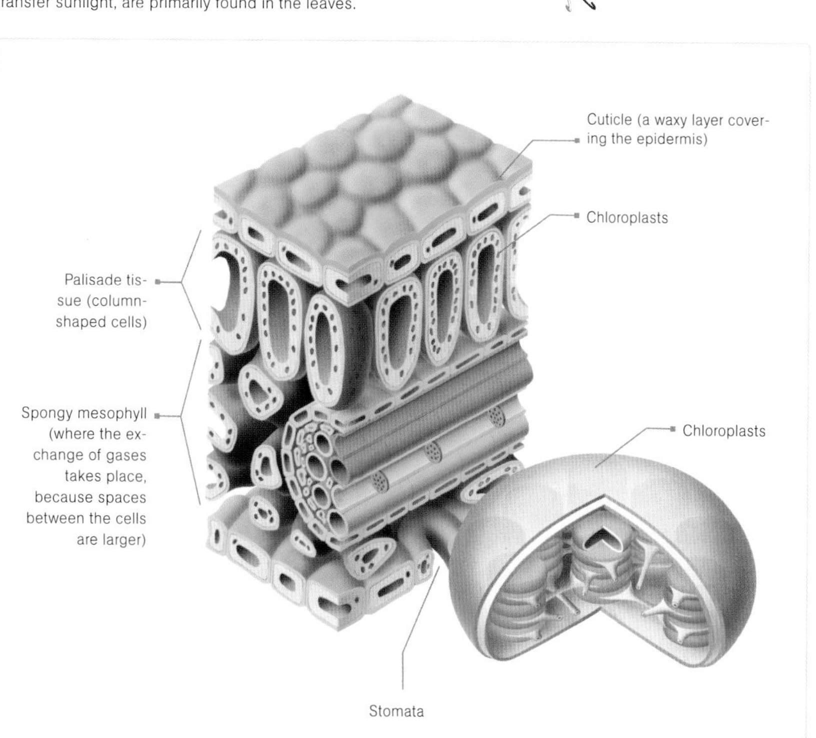

A microscopic cross section of a leaf shows that the chloroplasts required for photosynthesis are found in all leaf cells except the epidermis.

see also: Differences: plant and animal cells, p. 149

water and nutrient transport

Unlike animals, plants occupy a fixed location. In order to successfully grow and reproduce, they must adapt to local light, temperature, water, and soil conditions.

In their green leaves, plants produce organic nutrients through photosynthesis (p. 181), while the roots take up water and minerals from the soil. All of these substances are used by the whole plant. To distribute them effectively, plant tissues are permeated by transport channels: the vascular bundles.

basics

MANY PLANTS have symbiotic relationships with soil fungi that enhance the plant's ability to take up water and minerals, while receiving nutrients from the plants.

IN TREES and other woody plants, the actual wood is formed from xylem tissue, which has undergone a secondary growth process.

Xylem

In addition to water, plants need many inorganic substances for metabolism and growth, such as potassium, calcium, phosphate, magnesium, and nitrogen compounds. In higher plants, these substances are absorbed through the root hairs and transported through an internal distribution system known as the xylem. Xylem tissue includes the actual transport channels, in the form of long hollow tubes (xylem vessels) whose living protoplasm has died off. The xylem system also contains fibrous cells and storage tissues that help reinforce and strengthen the plant. This distribution system uses a passive, energy-efficient mechanism.

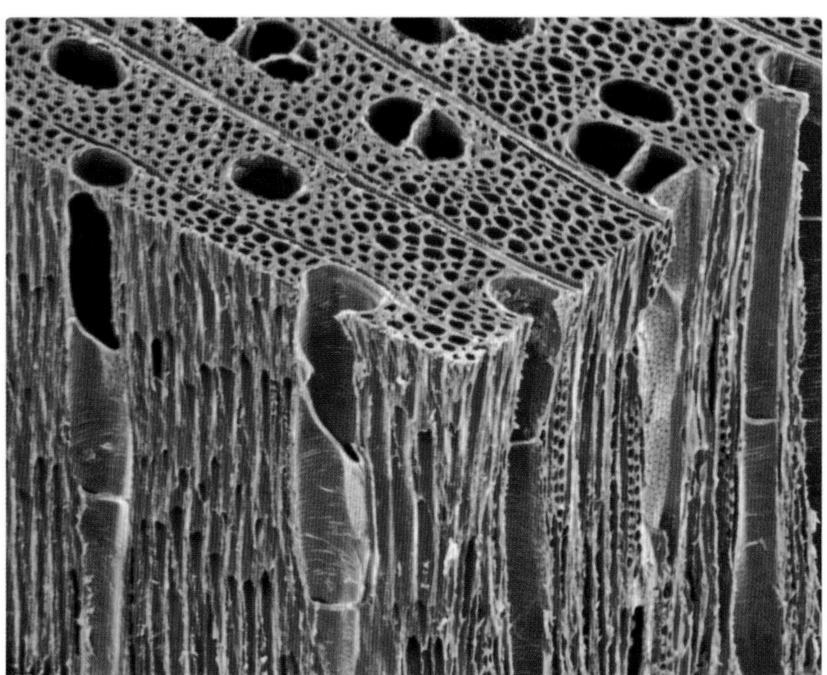

In deciduous trees, the xylem vessels can often be recognized with the naked eye as small pores, as seen here in a cross section of maple wood. The wood also contains fibrous reinforcement and storage tissues.

Leaves and phloem

In the leaves, transpiration takes place through the pores (p. 181), as water vapor evaporates into the less-humid surrounding air. This produces a suction force that draws liquid from the plant's cell walls (p. 149). Together with the xylem vessels they form a closed capillary system, which constantly draws water and minerals up from the roots.

In addition to the xylem, a plant has another transport system. It distributes organic compounds, such as carbohydrates and amino acids. These substances, produced in the plant's green tissues, must be carried to other parts of the plant. This long-distance transport takes place in tissue known as phloem. Especially high concentrations of sugars can be found in the phloem vessels, which are called sieve tubes. In the leaves, sugars are actively transported into the phloem, later being removed in the areas where they are needed.

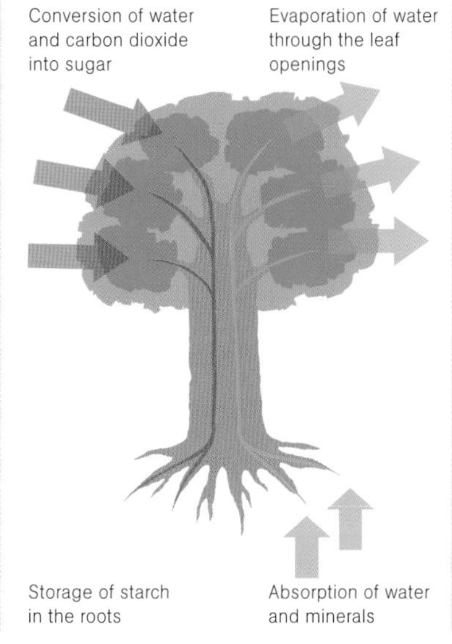

A plant's two transport systems: the xylem for water and minerals, and the phloem for organic nutrients.

practice

PHLOEM RESEARCH

Aphids feed on the concentrated sugar solutions in the phloem by inserting their sharp mouthpieces into sieve tubes. Biologists use this ability of aphids to study the content of the phloem. Using a laser, they detach the aphid's mouthpiece, collecting the phloem liquid in tiny tubes. They also analyze the honeydew fluid emitted by aphids. Scientists have discovered that the phloem carries messenger substances, hormones, viruses, and nucleic acids.

Much of the phloem liquid taken up by aphids from sieve tubes is secreted again as honeydew.

see also: Cycles of matter, pp. 278–279

secondary plant compounds

Secondary plant compounds or secondary metabolites belong to a group of compounds with varying chemical properties. They are nonessential to a plant's basic metabolic functions.

Secondary compounds are the metabolic products that accumulate in the plants tissues and are not directly essential to the plants survival. More than 30,000 secondary metabolites have already been discovered and, most likely, there are many more yet to be found. Many secondary compounds are still unknown because only about 15 to 20 percent of all plant species have been tested. Secondary compounds are often only present in very small concentrations, which makes detection even more difficult. Humans are able to take advantage of many of the properties of these secondary compounds in everyday life. Previously, it was thought that secondary plant compounds were merely waste products that needed to be flushed out of the plant's metabolic cycle. Today, researchers have found that these compounds play an important role in plant life. Many protect plants from herbivores by adding a bitter taste or making a plant poisonous. They may also act as a signal for pollinators, protect the plant from aggressive UV radiation by the sun, prevent evaporation or fight pathogenic microorganisms.

Hemlock is so poisonous that it was used to execute convicted prisoners in ancient Greece.

issues to solve

ANY secondary ant compounds cur in drugs such s cocaine or mariana. Recreational ug abuse poses a rious and unlved problem for ciety as a whole.

Typical secondary compounds

Alkaloids are typical secondary plant compounds produced by many different plant species, e.g., the nightshade and poppy families. Many alkaloids are toxic, but may be used in small quantities for therapeutic purposes. Familiar alkaloids include caffeine, cocaine, nicotine, and strychnine.

Essential oils are another type of secondary compounds. These are aromatic, volatile, oily compounds. Their original purpose is to attract pollinating insects, but they may also be used in medicine. For example, a chamomile vapor bath can be used to treat respiratory diseases.

Some secondary compounds are used to dye cloth. Indigo, for example, is a dark blue dye extracted from *Indigofera* species. This dye is now usually made synthetically.

in focus

MEDICAL PLANTS

Herbal remedies often derive their properties from secondary compounds and have been used for thousands of years and, despite a drop in usage due to modern medicines, they have recently gained popularity as it is believed that they produce fewer unwanted side effects. For example, cardiac glycosides from the foxglove species are known active medicinal plant substances. The active substance in aspirin, acetylsalicylic acid, originally came from willow, but is now produced synthetically.

Natural remedies in a traditional Chinese pharmacy.

In the centuries-old textile traditions throughout West Africa, clothes dyed with indigo signified wealth.

SEEDLESS PLANTS

Mosses, ferns, as well as many algae species do not produce flowers and are therefore unable to create seeds. Instead, they reproduce by means of spores. Many seedless plants display so-called alternate generations, where both types of reproduction—sexual and asexual—alternate from generation to generation.

algae

Algae are eukaryotic organisms that normally live in water and practice photosynthesis. The best known species are brown, red, and green algae. Reproduction in all three groups can be either sexual or asexual.

Algae do not form a true family group; however, they can be categorized according to their physical and chemical composition as well as the reserve substances and photosynthetic pigments they produce.

Multicellular, primarily marine brown algae can become rather large, as for example the giant bladder kelp that can reach a length of up to 197 feet (60 m) and grow up to 19.7 inches (50 cm) a day. At locations with nutrient-rich, relatively cold ocean currents, such as along the North American Pacific coast, this kelp forms huge underwater forests, so-called kelp forests, which provide a valuable habitat for numerous animals. On the other hand, smaller brown algae species are found in nearly all of the world's oceans, where they normally grow attached to the ocean substrate by means of an attachment organ. The brown or olive green coloration, displayed by many of the approximately 2,000 species, comes from dark fucoxanthin color pigments that cover the green chlorophyll.

Most of the 4,000 to 4,500 single- or multicelled red algae species can also be found in the sea. Special pigments enable them to survive down to a depth of 590 feet (180 m). Apart from chlorophyll, they also carry the pigments phycocyanin and phycoerythrin; the latter is responsible for the red coloration. With the aid of these pigments, shortwave sunlight penetrating to deep-water layers can still be used for photosynthesis.

The diverse group of green algae can be divided into single-celled and filamentous species; these can be branched or unbranched. There are also species with a leaflike structure. Colony formation has also been observed in some species. Green algae occur primarily in freshwater and their pigment content is similar to that of higher plants. Almost all forms have chloroplasts containing chlorophyll, which gives these algae their green coloration.

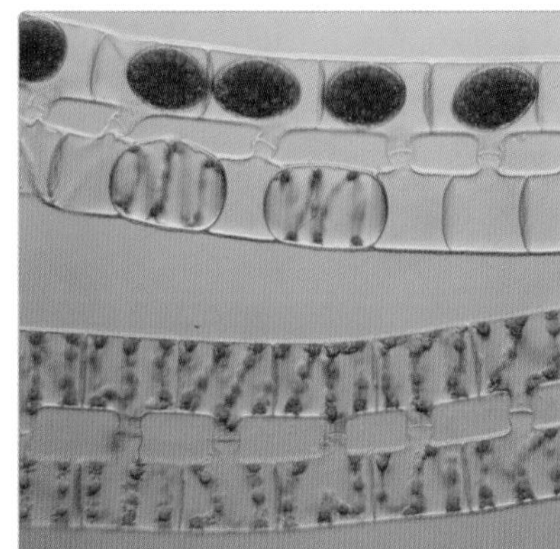

Spirogyra are a filamentous green algae species that forms in eutrophic freshwater. The lower cells contain chloroplasts that are arranged in spirals.

in focus

UTILIZATION OF ALGAE

Algae harvested at natural locations are mainly used for the production of iodine, bromine, vitamins, mineral substances, and proteins. Furthermore, they can function as binding and thickening agents in food production and livestock feed supplements. In biotechnical facilities in Israel, Japan, Australia, and some other countries, fast-growing unicellular algae are cultivated as food supplements and as ingredients for cosmetics.

In Japan some brown algae species are used as the basis for soups or in pickled form for snacks.

Kelp forests throughout the temperate and polar coastal oceans provide one of Earth's most productive ecosystems for a multitude of marine organisms.

see also: Cells, pp. 148–149

mosses

Mosses are small plants that can be found nearly everywhere on Earth. They are especially common in damp, shady environments.

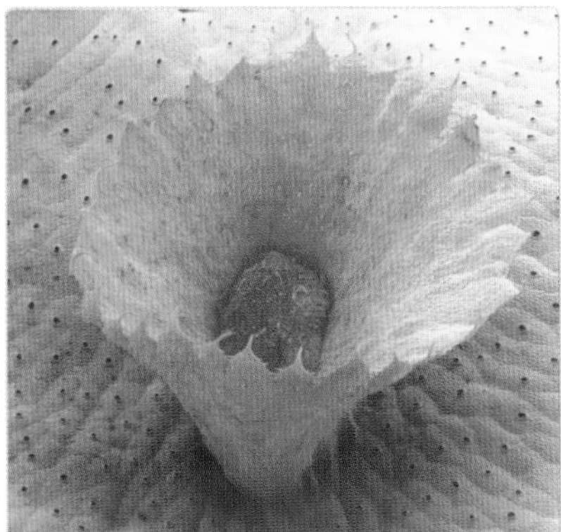

Liverworts: Vegetative reproduction occurs through special reproductive buds.

Based on differences in their anatomy, mosses can be divided into two large groups: liverworts and leafy mosses. The liverworts (*Hepaticae*) include some 10,000 species, which can again be divided into two groups: the thallose liverworts, in which the vegetative body has a lobed shape, and the leafy liverworts, which have small vertical or horizontal stems and small leaves without a central vein. The leafy mosses (*Bryopsida*) include approximately 15,000 species, which always possess a vegetative body divided into small stems and leaves, usually with a central vein.

basics

GETATIVE REPRO-CTION Many moss reproduce not ly sexually, but so asexually. In the ecies *Marchantia lymorpha,* for exnple, buds develop m individual surce cells in small productive vessels.

Mosses are ordinarily 0.4 to 4 inches (1–10 cm) tall, although some species can grow much larger.

The highly complicated sexual reproduction cycle of mosses depends on the production of spores. The cycle begins when a spore germinates, producing a tiny chain of cells called a protonema. Through budding, the protonema develops into a green moss plant (gametophyte), which attaches to the ground using fine cell threads called rhizoids. Rhizoids are not true roots, but are more comparable to the root hairs of higher plants. The gametophyte then develops female and male sex organs, called archegonia and antheridia, respectively. These may be found on different plants or on different branches of the same individual plant.

The archegonia are bottle-shaped vessels that hold the developing egg cells after fertilization by sperm cells from an antheridium. This can only occur in the presence of water (for instance, from rain), because the tailed sperm cells must actively swim to the egg. The fertilized egg cell develops into a sporophyte, which grows on top of the gametophyte, depending on it for water and nutrients. The sporophyte takes the form of a capsule on a stem, and the spores develop within the capsule, beginning a new cycle. Thus an alternation of generations between the sexual (gametophyte) and spore-producing (sporophyte) phases is characteristic of this reproductive process.

in focus

PEAT MOSSES

Peat mosses (*Sphagnum* sp.) play a key role in the development of peat bogs. Unlike most other plants, they are able to thrive in environments with stagnant water and low nutrient levels. Peat mosses can store large amounts of water, and their tips are constantly growing while their lower parts die off, leading to the accumulation of a thickening deposit of dead plant material. The upper layer is further thickened by the living peat mosses, which can often reach 20 times their dry weight by absorbing large amounts of water. Because the dead plant material decays very slowly in the resulting acidic and low-oxygen environment, a massive peat deposit can form over time.

A peat-cutter at work: Peat bogs take thousands of years to form and are important and delicate ecosystems.

Mosses developed about 400 million years ago from intertidal green algae. Moisture-retentive surfaces are hospitable to the growth of mosses. The smallest cracks are sufficient for mosses to grow on rocks.

whisk ferns, club mosses, and horsetails

The spore-producing vascular plants, Pteridophyta, includes the so-called true ferns, whisk ferns, club mosses, and horsetails. Their life cycle involves spore-based reproduction and the alternation of generations.

In ferns, the sporophyte is an independent plant, representing the dominant of the two reproductive phases known as the alternation of generations. In the true ferns, club mosses, and horsetails, the sporophyte consists of a central shoot, leaves, and roots, while among the whisk ferns, no true roots are present.

Whisk ferns are a simple, vascular plant with branching stems that contain scalelike structures rather than leaves.

Whisk ferns

Whisk ferns, found in the tropics or subtropics, consist of only one order (Psilotales) with two small families. Most whisk ferns possess only tiny, scalelike leaves, so that photosynthesis occurs predominantly in the central shoots. Their short-stemmed, rounded spore capsules (sporangia) are prominent, forming on the upper parts of the shoots, which usually have forking branches.

Club mosses

Club mosses, which enjoyed a golden age some 300 million years ago in the Upper Carboniferous period, include just over a thousand species. They have small or very narrow leaves (microphylls), usually in a spiral arrangement around the central shoot. The sporangia are typically gathered in thick bundles on the tips of the shoots.

Club mosses can be found almost all over the world. Most grow on the ground—sometimes producing yard-long creeping shoots—but other species are epiphytes, growing on other plants. Since club moss spores will often not germinate for several years, and then take up to 15 years to reach full maturity, many species are rare and are protected in several countries.

Horsetails

Horsetails also reached the high point of their development in the Carboniferous period. At that time the group had more species such as woody horsetails, which could reach a height of 98.4 feet (30 m). Today, only one genus remains with a few herbaceous species that form nonbranching fertile sprouts from a long stem emerging in springtime. These die off after producing spores. In contrast, the leafy summer sprouts that appear later are sterile, serving only to carry out metabolic processes and store energy.

The peak of the horsetails diversity was during the Carboniferous period. Only one genus of horsetails has survived into modern times.

THE RESURRECTION PLANT

in focus

An unusual plant from this group is *Selaginella lepydophylla*, a species of club moss that is often called the "resurrection plant." While most of its relatives grow in moist tropical forests, these plants can be found in extremely dry areas, such as parts of California and Mexico. In times of drought, they curl up into a ball and can survive for years in a dormant state. When rain falls, the stems rapidly absorb moisture from the air and spread out green shoots again.

One species of club moss has evolved a strategy to survive for long periods of drought.

see also: Ecosystems, pp. 80–87

ferns

The class of true ferns, with some 12,000 species, is the largest group within the spore-producing vascular plants. Its members are distributed worldwide, with the greatest diversity of species found in tropical regions.

Most ferns are herbaceous (nonwoody) inhabitants of shady forests, where they usually grow on the ground. There are also epiphytic species (see in focus), aquatic ferns, and some treelike representatives (the families Dicksoniaceae and Cyatheaceae), which can reach a height of 33 feet (11 m) under favorable conditions. All ferns exhibit a heteromorphic alternation of generations—that is, they alternate regularly between sexual and asexual generations that differ in appearance. Within this cycle, the plant we recognize as a fern is the sporophyte, or the individual on which the spores form. The spores are usually found in brown-colored spore capsules (sporangia). These are grouped together in small clumps (sori) on the underside of the fern fronds, which can be seen easily with the naked eye. They may also be covered by a protective film of tissue (indusium). As they ripen, the spores break free from the plant and are spread by the wind. After germination they form a small structure called a prothallium, which is usually soft and short-lived. This is the fern's gametophyte generation. The female and male sex organs (archegonia and antheridia) develop on the prothallium, and under damp conditions, tailed sperm from the antheridia are able to swim to the archegonia and fertilize the egg cells there, from which new ferns (the sporophyte generation) then develop.

■ Spori are made up of the spore capsules (sporangia), which are distributed in various patterns and shapes on the underside of fern fronds.

In many ferns, almost the entire visible part of the plant consists of leaves or fronds, which are often intricately feathered. They usually start out tightly rolled up in a "fiddlehead" shape.

This is the case, for example, with the common bracken *(Pteridium aquilinum)*, which can grow to a height of about seven feet (two m) or more under favorable conditions. The fronds emerge from a rhizome underground, which may reach a length of up to 164 feet (50 m). Aquatic fern leaves often have air-filled chambers, giving the plant buoyancy. Some of their leaves have evolved into rootlike structures.

THE REMAINS of ferns, club mosses, and whisk ferns from the Carboniferous period provide us with most our current reserves of hard coal today.

basics

EPIPHYTIC FERNS

The epiphytic species of ferns (those living on other plants) have—in addition to spore-bearing fronds—sterile leaves that function as nutrient collectors. These "nest fronds" lie flat or form open funnels for collecting water and falling plant material. This material breaks down and is metabolized by the fern. Nest fronds also form valuable humus as the old leaves decay, and their mat-like layers store moisture.

■ Epiphytic ferns live on other plants and must trap moisure and nutrients in their fronds.

in focus

■ Tree ferns tend to thrive in tropical to subtropical environments and can grow very tall in these warm, humid conditions.

gymnosperms | angiosperms | pollination | habitat specialists

SEED PLANTS

Seed-bearing plants (spermatophytes) include all plants whose structures can be divided into roots, stems, and leaves. Gymnosperms, in which the ovules are not enclosed within an ovary, differ from angiosperms (Magnoliaphyta; formerly Angiospermae), or flowering plants, in which the ovules lie within a ripened ovary, or fruit.

gymnosperms

Gymnosperms are a group of spermatophyte seed-bearing plants. This group includes only about 650 recent species, most of which belong to the class of conifers (Pinopsida).

Conifers are mainly evergreen trees (along with some shrubs), almost all with needle- or scale-shaped leaves (for protection against drying out in winter, among other functions). The male microsporangia and female gametophytes typically develop in separate cone-shaped structures. Conifers are nearly always pollinated by the wind; since no ovary is formed, the pollen grains fall directly onto the micropyle, the small channel that allows the pollen tube to enter the ovule for fertilization. Conifers are widespread in temperate and subarctic regions, where they are sometimes the dominant vegetation. An example is the taiga or boreal forest, which covers an area of some 5.4 million square miles (14 million km^2) in the Northern Hemisphere.

Some species from this group can become extremely large and ancient, notably the giant sequoias *(Sequoiadendrum giganteum)*, with extremely impressive specimens in the northwestern United States. Among the largest is "General Sherman"; estimated to be some 2,500 years old, it has reached a height of more than 272 feet (83 m), a diameter of about 36 feet (11 m) at the base, and a total mass of 1,500 tons (1,361 mT). The true Methuselah of trees, however, is the bristlecone pine *(Pinus longaeva)*, found in the mountains of eastern California. Only about 33 feet (10 m) tall, these trees have highly gnarled, bare-looking trunks. Only a few branches with green needles show that they are still alive. Experts estimate that some of these pines sprouted over 4,600 years ago, a time when humans just began a settled agricultural lifestyle.

Other divisions among the gymnosperms—which cannot be considered a group with a unified heritage—are the Ginkgophyta, with only one living species, quite recent in its development *(Ginkgo biloba)*; the Cycadophyta, similar in appearance to palms; and the Gnetophyta, which include the unusual Welwitschia *(Welwitschia mirabilis)*.

Giant sequoias, a species of redwood found in western California, are the world's largest trees.

The Welwitschia plant has an extremely long life span. The largest of these specimens is 2,000 years old and covers an area of 646 square feet (69 m^2).

The taiga, or boreal forest, is the world's largest terrestrial biome and covers most of inland Alaska, Canada, Sweden, Finland, Norway, and Russia. This biome is characterized by coniferous forests. With the exception of the tundra, the taiga is the coldest biome on Earth, yet it supports a varied and flourishing plant population.

see also: Classification of living things, pp. 168–169

angiosperms

Angiosperms (Magnoliophyta or Angiospermae) are divided into two categories: dicotyledons, which germinate with two cotyledons (seed leaves), and monocotyledons, which produce only one cotyledon.

The estimated number of angiosperm species is between 250,000 and 300,000. The ovules of angiosperms are not exposed on the carpels as seen in gymnosperms. Instead, they are enclosed inside an ovary. The ovary consists of several carpels attached to each other. Once the seeds are ripe, they are released from the ovary, which coincides with fruit growth. The cotyledons that are used as a characteristic for classification are already present in the seed and are the first part of the new germinating plant. They have a different appearance compared to the leaves that form later. Almost all of our cultivated plants belong to the angiosperms except for coniferous trees. The earliest angiosperms appeared during the Cretaceous about 110 million years ago.

The orchid family is one of the largest families of flowering plants, adding around 800 new species to its list every year. These sweet-smelling flowers are frequently used in perfumes.

Dicotyledons

Dicotyledons (Magnoliopsida or Dicotyledonae) include about three-quarters of the known Magnoliophyta species. Most have two cotyledons (rarely more) and their leaves usually have a distinctive stalk and midrib with lateral veins. Stipules are also common and the flowers are normally characterized by five whorls. These plants predominantly show secondary growth like trees that increase in thickness. In fact, most trees and shrubs belong to this group. Their vascular tissue is normally in ring form and a cambium (a layer of active growth with new cells adding to the thickness of the plant shoot) is always present.

Monocotyledons

The smaller group of monocotyledons (Liliopsida or Monocotyledonae), include approximately 50,000 species, most of which are herbaceous. They are characterized by having only one seed leaf, irregularly placed vascular tissue, and no cambium. Their leaves normally lack a stalk and their structure is less complex compared to the dicotyledons; the lack of stipules and roughly parallel veins are typical of these species. Liliopsida include grasses, orchids, lilies, irises, and palm trees.

in focus

INSECTIVORES

There are several plant species which use animals as additional nitrogen sources. Such plants often live in habitats low in nutrients, for example, moors and bogs. They are equipped with various mechanisms to attract and catch prey, mainly insects. These techniques can include sticky traps (e.g., sundew), jaw traps (e.g., Venus flytrap), and pitfall traps (e.g., pitcher plants). Once a prey has been caught, its body is dissolved by special enzymes.

The Venus flytrap is one of the only plant species capable of quick movements to catch prey.

With its giant leaves, the parasitic *Rafflesia* flower can weigh up to 22 pounds (10 kg) and is said to smell like rotting meat.

see also: Cycles of matter, pp. 278–279

pollination, fertilization, fruit, and seed dispersal

Seed-bearing plants produce seeds for reproduction. These consist of a plant embryo and nutritious tissue enclosed in a protective seed coat.

During pollination, male pollen is transferred from pollen sacs onto the stigma of a female flower. Gymnosperms (p. 188) such as spruce trees usually have flowers of separate sexes, so they are either female or male. Together they make up a cone; their pollen is dispersed by the wind.

Angiosperms typically have hermaphrodite flowers. They contain female reproductive organs, carpels, as well as male organs, stamen. Pollination mechanisms are not limited to relatively random wind or water dispersal; there are many other, sometimes highly specialized mechanisms. Pollen is most often carried by insects or other small animals which receive a "treat" in the form of nutritious pollen or sweet nectar from the female flower. Many plant species attract certain animal species, such as birds and bats, using color or scent. Flowers are also constructed to fit their corresponding pollinator. Some common types are composite, bell-shaped, and funnel-lipped flowers.

A dandelion dispersing seeds into the air.

Fruits and seeds

In angiosperms, pollen sticks to the stigma of the female flower and grows a pollen tube. Then the male gamete travels through the tube into the female ovary. This is where the actual fertilization takes place by the union of the ovule and gamete. Male and female genomes combine and a fruit develops within the first few weeks following fertilization. The flower disintegrates and the ovary increases in size and becomes fleshy. A seed begins to grow from the endosperm nucleus. An embryo plant is waiting inside the seed to create the next generation. The seeds are enclosed by the fruit until they reach maturity. The fruit is formed for optimal dispersal of the seeds. The fruits of the greater burdock are sticky and carry hooks to attach to animals themselves, which will then disperse the seeds. Maple and dandelion seeds are equipped with feathers, wings, or parachutes for wind dispersal.

Bees can remember colors; they prefer yellow and blue flowers that are rich in nectar.

basics

CROSS-POLLINATION increases genetic diversity and chances for better environmental adaptation.

Most plants have anatomical or biochemical mechanisms that prevent self-pollination.

Unlike gymnosperms the ovules of angiosperms are enclosed in a protected chamber of the ovary.

Clematis' seed tufts have a complicated flight apparatus that allows the seeds to be carried far away from the mother plant by the wind.

in focus

POLLINATION BY MAMMALS

While we usually think of insects as pollinators, bison also do the job.

Not only insects, but mammals can also help plants disperse pollen. The Australian honey possum has specialized in eating the nectar and pollen of banksia flowers by licking it off with its very long, bristly tongue. A large amount of pollen attaches to its fur while the animal is climbing the large and rigid flowers. The pollen is wiped off again, when the possum visits another plant. Many tropical plants are pollinated by mammals such as bats or flying foxes. Some of these plants include the baobab and wild banana trees.

habitat specialists

Plants use specific strategies to adapt to their environment. Climate conditions and the availability of water and nutrients, all affect their structures.

High temperatures and low precipitation force cacti and other succulent plants to collect and store water in order to survive long dry periods.

Essential dissolved minerals for growth and metabolism include nitrogen, phosphorous, calcium, magnesium, and iron. Potassium in particular is required to control the microscopic, pore-like openings of the stomata. These minerals are generally taken up by the roots, in the form of ions. In many cases plants have a symbiotic relationship with bacteria or fungi in the soil, increasing the roots available surface area and in turn providing the bacteria or fungi with sugar, a product of photosynthesis. Nitrogen fixing bacteria are also important in helping plants obtain molecular nitrogen.

Adaptation to temperature and moisture conditions

Optimal levels of resources for growth are not always available, therefore plants have adapted to survive a wide range of environmental conditions. They have many different strategies related to coping with water loss or excess in various extremes. In hot, dry environments plants reduce evaporation; mosses and flowering desert plants dry up and retreat into a dormant state and succulent plants store large amounts of water in their shoots and leaves, which they release only sparingly during droughts. Others minimize evaporation by having fewer or smaller leaves, or by growing deep root systems that spread out to absorb as much water as possible.

In moist environments, plants increase evaporation of water from their leaves (transpiration) by developing broad, thin leaves with minimal or no waxy coating. In variable climates that include periods of frost, the majority of plants shed their leaves in winter or produce a natural antifreeze substance that prevents ice crystals from forming in their cells. Aquatic plants, on the other hand, absorb water and dissolved minerals over their entire surfaces and therefore only have weak roots. Conversely, plants adapted to low light conditions tend to have large leaves with high chlorophyll concentrations in order to photosynthesize more efficiently.

in focus

SALT PLANTS

Salt plants populate salty habitats, either dry or flooded, often near a salt lake or the ocean. Some species may grow in either fresh or saltwater. Others prefer saltless ground but have been banished by competing plants. Yet others not only tolerate, but thrive under saline conditions. Salt-excluding species prevent salts from entering their tissues through reverse osmosis.

Salt-excreting species are able to take in highly saline water through their pores, and excrete excess salts through specialized glands in their leaves.

Mangroves are able to cope with salinities of 0–90 ppt.

Mountain flowers like the alpine rose are capable of thriving above the timberline. During the colder months photosynthesis is performed with the minimum of light left underneath a dense and heavy snow cover.

see also: Photosynthesis, p. 181

the fungi kingdom

Taxonomically, fungi used to be included within the plant kingdom. Today, modern analytical methods have determined that fungi are neither plants nor animals, but form an independent group within the eukaryotes.

structure | uses | dangers

FUNGI

Fungi are eukaryotic organisms. Among them are unicellular forms, such as yeasts; but most consist of mycelium, which is the vegetative part of a fungus made up of a mass of branching, threadlike cells called hyphae. From a human perspective, many fungi are very useful; we are indebted to them not only for items like bread, beer, and wine, but also for the first antibiotic. Fungi play an essential role in the material cycle of nature. However, many species can cause dangerous diseases.

Most of the edible fungi known to us have a finely branching network of threadlike cells, the mycelium, which is usually hidden, and the fruiting body that we perceive as the actual fungus. The mycelium consists of individual, filamentous hyphae, which are generally subdivided into individual compartments by so-called septa.

Growth and reproduction

Growth of the hyphae occurs at the tip and frequently there is branching, so that gradually a dense network of hyphae develops. The cell wall of fungi is made up primarily of chitin. Reproduction of fungi occurs via spores that are formed asexually or sexually. Asexual spores (conidia) are often budded off from spore carriers, while special fruiting bodies are formed during sexual reproduction. Depending on the species, the spores simply fall out of the fruiting body and are distributed by the wind, or sometimes they are actively catapulted out. Yeasts reproduce principally by means of cell division. Based on particular characteristics, fungi are placed into several groups: Basidiomycete (such as pileate mushrooms), Ascomycete (e.g., true yeasts, sac-, or cup fungi), Zygomycete (e.g., black bread mold), Glomeromycete (certain Mykorrhiza fungi), and the Chytridiomycete, which, among others, include a number of plant pathogens, causing infectious diseases.

Fungi species exist in a variety of distinctive shapes and sometimes striking colors.

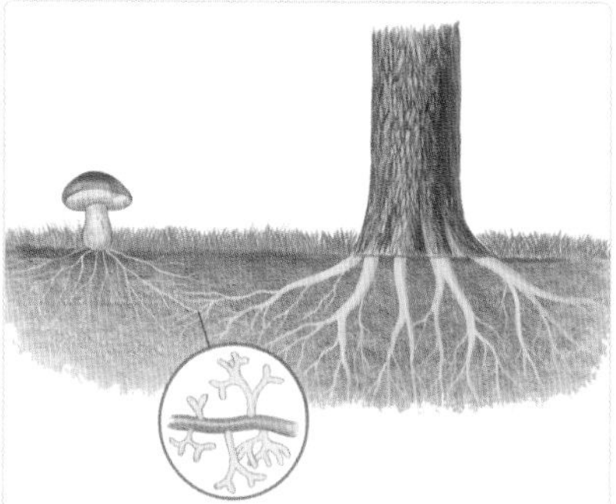

The threadlike mycelium, often hidden beneath soil or vegetation, absorbs nurients.

in focus

ANCIENT ORGANISM

Fungi can grow to significant sizes. Investigations in the United States revealed that the hyphae from a single specimen Basidiomycete *(Armillaria ostoyae)* covered an area of about 3.5 square miles (9 km^2) and was estimated to weigh approximately 600 tons. This places it among the largest and longest lived organisms on the Earth. Such growth would presumably require a period of more than 2,000 years.

A fungi specimen may be largely hidden underground and cover large areas.

Fungi are found in many different moist environments and play an important role in the recycling of decomposing materials. Some also form parasitic or symbiotic relationships with living plants.

useful and harmful fungi

Many fungi are considered useful due to the valuable services their cultivation can provide for humanity, while others are considered harmful because they may inflict significant harvest losses or cause disease.

With the aid of photosynthesis, plants produce an annual biomass (the total mass of organic material) of several billion tons. Over time, this biomass needs to be broken down again so that all the nutrients locked up in the material can be made available for use by other organisms and returned into the cycle of nature.

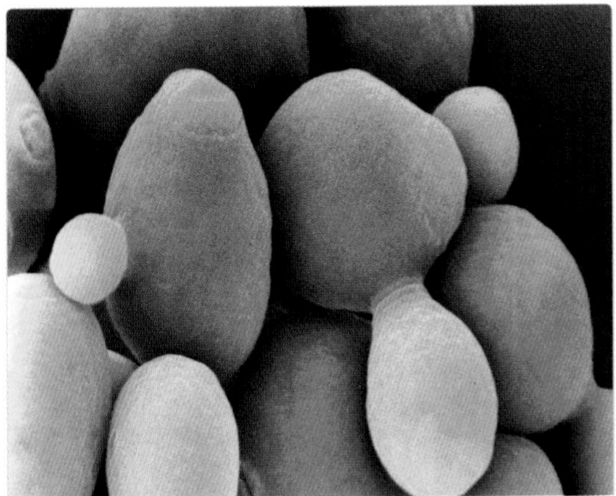

Yeast is a single-celled fungus that has been used throughout human history in the production of bread, wine, and beer.

Without this decomposition process, life on Earth would quickly come to a standstill. Fungi play a very significant role in this recycling, particularly in the breakdown of wood and other plant materials.

Fungi in the service of humanity

Humans have consumed fungi since ancient times, and fatal accidents have often occurred due to the potent toxins found in some species. On the other hand, the production of bread, wine, and beer—depicted in tomb art dating as far back as ancient Egypt—would not have been possible without the use of fungi.

Moreover, there are now a number of commercial products that are produced on a large scale with the aid of fungi, such as citric acid or vitamin B2, and of course antibiotics, such as penicillin.

Disease-causing fungi

There are a number of fungi that can be dangerous for humans; these include *Cryptococcus neoformans*, which can cause meningitis and is especially dangerous to people with a weakened immune system. If left untreated, this disease is nearly always fatal. Less dangerous, but still unpleasant, are fungal infections of the skin, such as foot fungus (athlete's foot).

Fungi that can trigger diseases among plants, such as loose smut (*Ustilago avenae*), used to cause harvest losses of up to 90 percent in the giant monocultures of North America. Coffee rust *(Hemileia vastatrix)* was the main cause behind the demise of the entire coffee production industry in Sri Lanka during the 19th century, with the result that there was a complete conversion to the cultivation of tea instead.

milestones

THE DISCOVERY OF PENICILLIN

In 1928, the bacteriologist Alexander Fleming made a discovery that proved to be one of the most significant in the history of medicine. He observed that a bacterial culture, contaminated by a fungus, had formed a narrow zone around it without bacteria. His assumption that the fungus was releasing a substance that inhibited growth proved correct, and this metabolic substance, penicillin, introduced the age of antibiotics.

An incidental discovery by Alexander Fleming in his laboratory changed medicine.

practice

SEARCH FOR TRUFFLES

Among humans, the most desired of the edible fungi are truffles, which are considered a delicacy and can be very expensive. The high cost of these fungi is mainly due to the fact that they grow under the soil surface and therefore can be difficult to find. However, animals with sensitive noses can detect the strong odor of truffles. Therefore, truffle collectors use pigs or specially trained sniffer dogs to detect these valuable delicacies hidden in forests.

People put the highly sensitive noses of pigs to use to help them locate buried truffles.

Although a number of fungi have been beneficial in human history, others may contain dangerous toxins that are poisonous to humans.

see also: Medicinal drugs and cosmetics, p. 305

BIOLOGY

ANIMALS

Animals are multicellular organisms that—in contrast to plants—must take in nourishment in the form of organic substances. These substances are generally broken down and absorbed in special cavities within the body (digestive systems). Animals can be divided into two large groups: invertebrates and vertebrates. The first group, composed of animals without a vertebral column, are sometimes referred to as lower animals. They include by far the largest number of species, even the highly diverse group of insects, along with others such as worms, bivalves, and gastropods. The vertebrates, on the other hand, include more highly developed animals, from fish to amphibians, reptiles, birds, and mammals, and thus also human beings.

sponges | cnidarians | worms | insects | crustaceans | mollusks

INVERTEBRATES

Invertebrates are multicellular organisms that lack a backbone. They include about 95 percent of all known animal species. The bodies of some invertebrates are completely exposed, while others are protected by an outer skeleton or a shell. They range from simple organisms such as sponges to the highly developed spiders or insects.

sponges

Sponges (phylum Porifera) are multicellular organisms with a very simple body—without organs, true muscles, or nerve cells. Adult sponges are sessile, that is, they are attached to a surface. The larvae can move freely.

At first sight, sponges may look like plants. They come in various shapes, and may look like crusts, shrubs, nets, barrels, or mushrooms. It was not until the 19th century that they were recognized as invertebrate animals. Now there are about 5,000 known species of sponge. Most live in the ocean, where they can survive down to depths of almost 20,000 feet (6,000 m), but there are also about 120 freshwater varieties. Different species of sponge can vary considerably in size. Some measure only a few fractions of an inch in diameter, while others can grow as large as over six feet (two m).

Structure

Although sponges have various basic body types, or construction plans, most sponges have an outer layer of surface cells (pinacocytes), as well as an inner layer of flagellated cells (choanocytes). Between the two layers, there are numerous mobile cells, similar to amoebae, and supporting skeletal structures made of calcium carbonate, silicic acid, or spongin.

Feeding and reproduction

Pores lead to the exterior of the sponge via channels originating in chambers lined with choanocytes. The flagellae that line the pores ensure a constant flow of water bringing food particles (detritus, plankton, and bacteria) into the interior of the sponge for the choanocytes or amoeboid cells to digest. Particles which cannot be digested flow into the central cavity and water flows out of a wide opening on top, the osculum, washing these particles back outside. The sponge is able to push water through the osculum at such speeds that only unused water will be sucked back in. The water flowing through the sponge delivers not only food particles but also oxygen for respiration.

Sponges may reproduce sexually or asexually, in which case they undergo division or budding. During sexual reproduction, a fertilized egg cell goes through several larval stages. After the planktonic phase, in which they float through the water, they attach to a surface and form a new sponge. During periods of low temperatures, sponges form gemmules that remain dormant, only turning into new sponges when temperatures rise.

Yellow sponges pump huge quantities of water through their hollow bodies while feeding.

practice

SHOWER SPONGES

Sponges have been used for body care since ancient times. These organisms are used due to their ability to absorb up to ten times their own weight and their elastic fibers, which allow water to be easily released by squeezing the sponge. Sponge divers go deep into the ocean to collect live sponges that are then gutted and sold as shower sponges. Today, however, most shower sponges are synthetic.

Shower sponges are harvested mainly in the Mediterranean, Caribbean, and Red Seas.

The *Latrunculiidae*, a branching sponge, lives in the Red Sea. Its sexual reproduction is hermaphroditic and viviparous.

see also: Coral reefs: underwater forests, p. 102

cnidarians

Common examples of the phylum Cnidaria include corals and jellyfish. Organisms live either individually or in colonies. Their tubular or bell-shaped bodies are armed with stinging cells.

The bodies of cnidarians primarily consist of two cell layers lining the mesoglea, which is a noncellular layer in between. These layers enclose the gastrovascular cavity, where digestion occurs. An opening links this hollow space to the outside and allows food particles to enter and waste products to exit. Cnidarians may occur as more or less stationary polyps or as medusae, or "jellyfish," which can swim freely.

Both types of cnidarians undergo cyclic changes in appearance from one generation to the next. For example, true jellyfish, such as the familiar bell-shaped jellyfish, produce eggs that develop into larvae. After a period of swarming, they will attach to a suitable surface and grow into a polyp. After reaching a certain size, the polyp will pinch off radial discs that then develop into medusae. Not all cnidarians reproduce sexually; some reproduce asexually, usually by budding. There are about 9,000 species, varying in size from a small fraction of an inch (a few millimeters) up to about six feet (two m).

Jellyfish may be carried by the tide or may swim with a backstroke created by contracting their bell.

Coral

Coral belong to the taxonomic class of the Anthozoa. The life cycle of these barrel-shaped polyps does not include a generation as medusae—instead, the larvae grow into new polyps straightaway. Coral usually have a basal plate that firmly attaches to the ground. Tentacles with toxic nematocysts surround the mouth to catch prey items and attack other predators.

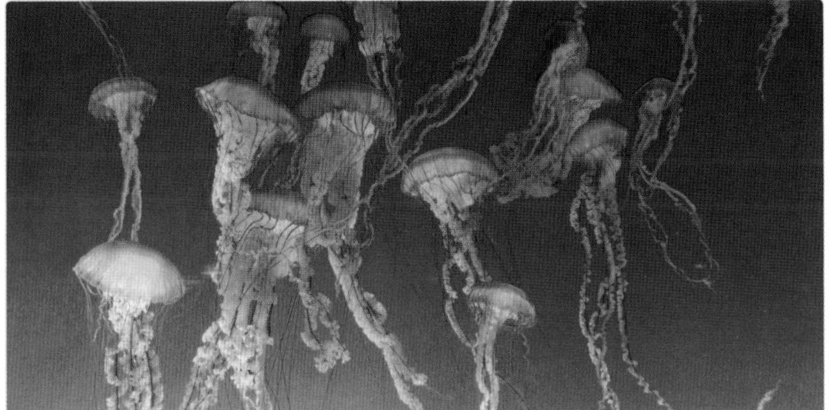

The poisons of jellyfish may act on the nervous system (neurotoxins) and paralyze the victim. Some species are extremely poisonous to humans, while others cause only mild skin irritations or burns.

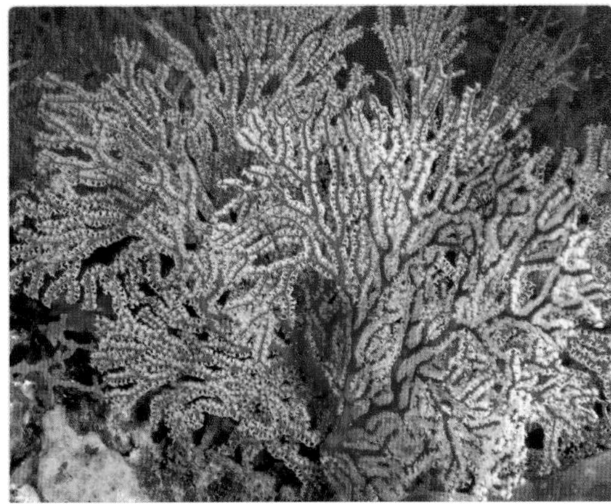

Coral are filter feeders. They eat plankton and other microorganisms carried past by the ocean current.

Jellyfish

Almost all larger species of jellyfish belong to the taxonomic class Scyphozoa. Among them is the giant jellyfish *Nemopilema nomurai*, which measures up to about six feet (two m), not including the long tentacles. This giant medusa weighs as much as 440 pounds (200 kg). The jellyfish bell consists of up to 99 percent water. Elastic tentacles with stinging cells hang from the edge of the bell. Inside the transparent body, the often colorful gonads are quite visible. In contrast, the corresponding polyps are relatively small.

issues to solve

CORAL REEFS are threatened with extinction. Greenhouse gas emissions cause carbon dioxide to dissolve in the ocean, increasing its acidity. Coral cannot survive in acidic water.

in focus

MINIATURE HARPOONS

Cnidarians have specialized cells for predation or protection. The most simple version of these stinging cells consists of a double-walled bladder filled with toxic and stinging substances and covered by a lid. The cavity also contains a threadlike stinging trigger coiled up in a spiral form and equipped with stiletto or spike bristles at the tip. Touching the tip releases the venom through the bristles, projecting it into the prey at high speed.

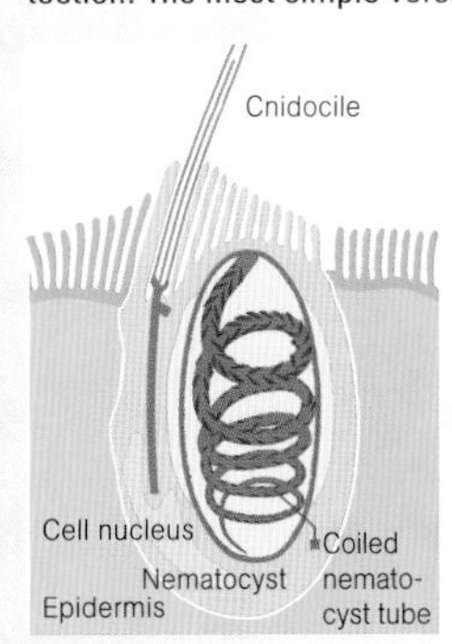

The sting of a cnidarian can be uncoiled at a very high speed.

worms

Worms include members of several different phyla, including flatworms (Platyhelminthes), roundworms (Nemathelminthes), and segmented worms (Annelida).

Most flatworms have a flat, sometimes paper-thin, body structure, filled with loose connective tissue. They lack advanced breathing organs and blood vessels and the digestive system ends blind without an anus. The nervous system spreads through the body like a mesh, with a single large nerve center, or ganglion. The reproductive organs may be quite complicated and in many species they take up most of the body cavity. Some flatworms live independently and others are parasitic. The most common flatworms include tapeworms, planarians, and leeches (see in focus).

Polychaeta, a class of annelid worms, live in the ocean and are remarkable for their variety of forms and feeding habits. Some are hunters, some scavengers, and others filter the surrounding water for food.

Roundworms

Roundworms are mostly round and often threadlike without segmentation. A layer of muscle below the skin allows for snakelike movements. The body cavity is usually filled with fluid. The most commonly known roundworms are the nematodes, which differ from flatworms by having a digestive tract with two openings: a mouth and an anus. Their reproductive organs are less complex than those of flatworms. Aquatic, terrestrial, and parasitic lifestyles are all common.

Some species of roundworms may even target people, such as the *Ascaris lumbricoides*, which can reach about 16 inches (40 cm) in length. Humans can also be infested by any roundworms, such as the *Trichinella spiralis*, carried by their pets.

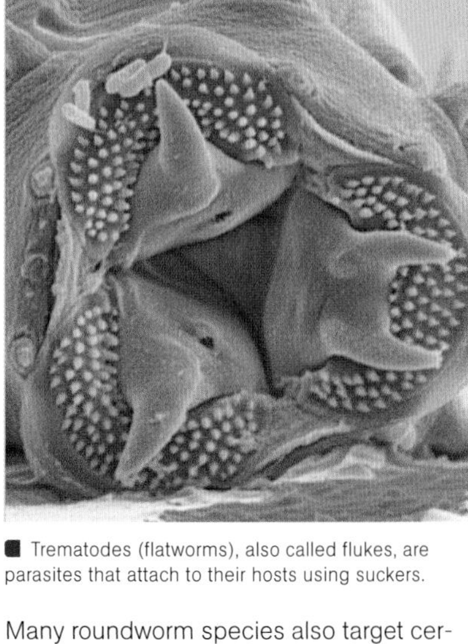

Trematodes (flatworms), also called flukes, are parasites that attach to their hosts using suckers.

Many roundworm species also target certain types of plants, such as potatoes, and can seriously damage crops.

Segmented worms

Segmented worms belong to the phylum Annelida. Their bodies are divided into segments, allowing them to have a closed circulatory system and a highly developed nervous system with nerve cords running the length of the body and branching out in each segment. Many species have advanced eyes and sensory organs. Common annelids are earthworms (*Lumbricus terrestris*) and lugworms (*Arenicola marina*).

> basics
>
> **PARASITIC NEMATODES** (roundworms) in whales may reach a length of up to about 30 feet (9 m).
>
> **GARDENS** can be home to as many as 37 earthworms per square foot (400 per m²).

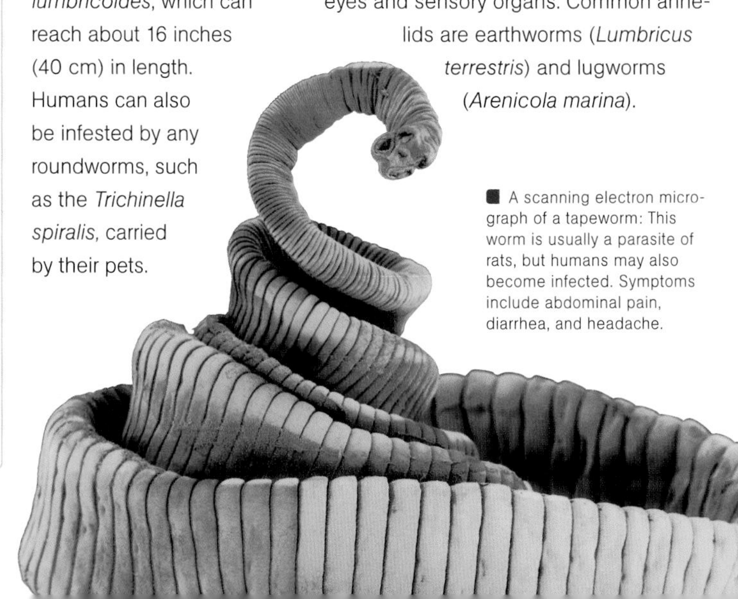

A scanning electron micrograph of a tapeworm: This worm is usually a parasite of rats, but humans may also become infected. Symptoms include abdominal pain, diarrhea, and headache.

> in focus
>
> ### SCHISTOSOMIASIS
>
> Schistosomiasis or bilharzia is a parasitic disease common in subtropical and tropical regions caused by flatworms of the genus *Schistosoma*. Its larvae enter the human body through the skin to mature. Female adult schistosomes then lay eggs that leave the body via urine or feces. If infected people do not receive medical attention, chronic infections of the liver, intestines, and bladder will develop. About 200 to 400 million people worldwide are estimated to suffer from this parasitic infection.
>
> 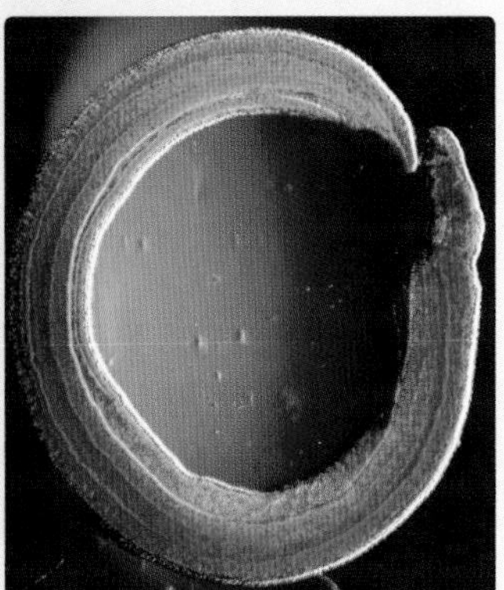
>
> Theodor Bilharz declared the cause of schistosomiasis for the first time in 1851.

insects

Insects (phylum Arthropoda) are by far the most diverse group of animals. More than a million species have been described to date, and there are probably many more unknown insects still to be discovered.

All insects are relatively small—even the largest species only reaches a length of about 12 inches (30 cm). They have conquered almost all terrestrial and freshwater habitats. All have a segmented hard outer protective layer, or exoskeleton, made primarily of chitin. Adult insects have three sections: a head, a thorax, and an abdomen. The head has a pair of antennae for feeling and smelling, as well as eyes and mouthparts, which may differ considerably between species and be adapted for chewing or piercing-sucking. Three pairs of legs and sometimes wings are attached to the thorax. The abdomen contains the major part of the digestive and blood circulatory systems as well as reproductive organs. Insects have a life cycle with several larval stages until the animal reaches reproductive age. Many species, including butterflies and beetles, change their entire appearance (holometabolism or complete metamorphosis) during this process. A pupal stage, during which the insect does not grow, comes before the final developmental step. If young or juvenile insects do not change considerably in appearance when they mature, the metamorphosis is considered incomplete (hemimetabolism). These larvae mature into adults increasing gradually in size without a pupal stage. Insects are the only invertebrates able to actively fly, contributing to their vast spread across various habitats. Their wings consist of an extremely thin membrane that is supported by a network of veins. Winged insects usually have two pairs of wings, but several species, especially parasites, have reduced wings. The latter are "secondarily wingless," as they have evolved from winged insects. Some insects never had any wings; they are "primarily wingless," as for example springtails.

Adult
Egg
Pupa
Larva

The metamorphosis of insects is controlled by hormones. The larvae shed their outer shell at regular intervals before the pupal stage.

basics

THE BOMBARDIER BEETLE sprays an acidic liquid when attacked. Two highly reactive chemicals are stored separately and then mixed inside the beetle's body.

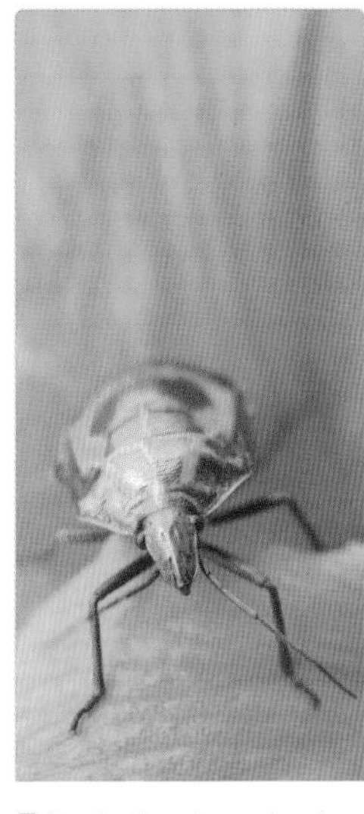

Beetles have two pairs of wings. One pair covers the other pair and protects the lower body.

SOCIAL INSECTS

practice

Some insect species, mainly ants, termites, and some bees and wasps, are social insects and live in large colonies to care for the next generation. In such insect colonies, the labor is strictly governed by social rules. Sterile workers take over several tasks, for example, gathering food, protecting the nest, or caring for the young. The queen alone is responsible for reproduction.

One method of bee communication is known as the waggle dance.

Dragonflies can hover in the air and change direction abruptly due to their ability to move their two pairs of wings independently.

crustaceans and spiders

Crustaceans and spiders, as well as the ancient scorpions, belong to the phylum Arthropoda or "arthropods." As close relatives to insects, they share many anatomical features.

There are about 38,000 species of crustaceans, most of which live in water. Crustaceans live in diverse environments and may look very different from each other. Tiny clam shrimps (Conchostraca) with a two-part shell that are only 0.02 inch (five mm) in size are as much part of the crustaceans as the American lobster (*Homarus americanus*), which has an elongated body and huge claws and can be longer than three feet (one m). Many crustaceans have two pairs of antennae and biramous legs, that is, legs that branch out into two segments. Crustaceans are equipped with gills, making them fully adapted to aquatic life. The bodies of spiders consist of two major parts: the cephalothorax at the front, which combines the head (cephalon) and midsection (thorax), as well as the abdomen at the rear. The front section contains the brain and stomach, eight legs, one pair of chelicerae or jaw-like structures, one pair of pedipalps (legs that can feel), and eyes attached to the front section.

Christmas Island red crabs migrate in huge groups to the coast, where they mate and lay eggs in the ocean. During this migration they may cover whole streets, and many crabs are killed or injured during the journey.

The abdomen includes the heart, part of the digestive tract, and the spinnerets, as well as the reproductive and respiratory organs. Many spiders build webs in order to capture their prey and produce venom for hunting and defense. The majority of spider venoms are actually not dangerous to humans.

TICKS AS DISEASE CARRIERS

in focus

Ticks of the genus *Ixodes* belong to the class of Arachnida, along with spiders and scorpions, and act as carriers for various diseases. One of these is Lyme disease, caused by a bacteria called *Borrelia burgdorferi*. Lyme disease may result in arthritis and infections of the heart and nervous system. In Europe, ticks may also carry a virus that causes the life threatening European early summer meningoencephalitis disease.

Ticks are found primarily in high grasses, scrub, and undergrowth areas.

Spiderwebs are made of long-chain protein molecules, giving them tensile strength and elasticity.

Scorpions still have the original form of spiders and there are approximately 2,000 known species in existence. They have large pinchers at the end of their pedipals and a very flexible tail. The tail is made up of five abdominal segments and is often armed with a thick venom gland with a sting at the tip. The venom of some species may be deadly to humans. Due to their nocturnal lifestyle and a preference for warmer temperatures, scorpions live mainly in tropical and subtropical regions, including deserts. Scorpions are carnivorous and hunt insects using their claws.

basics

THE AUSTRALIAN FUNNEL WEB spider belongs to the few species that may threaten human life. When males are searching for mates they may enter homes and gardens, and defend themselves aggressively if disturbed.

After birth, the baby scorpions climb onto the back of their mother. They remain there for protection until their first molting.

mollusks

Mollusks (phylum Mollusca) are a diverse group of animals that include snails, mussels, clams, and cephalopods. Mollusks may live either in saltwater, freshwater, or on land.

Cuttlefish are generally found near the seafloor. They have two relatively short tentacles.

All mollusks have a unique structure called the radula, a rasping organ similar to a tongue for feeding, though many species have only a reduced or embryonic radula. A rigid outer shell of calcium carbonate is also common to many mollusks. The shell protects and supports the soft body tissue.

The endangered giant clam, found in the Indian and Pacific Oceans, can grow to 4.5 feet (140 cm).

Snails

Snails (class Gastropoda) have a head with two tentacles and a muscular foot, flattened for crawling. The inner organs are located in the inner body cavity; in snails with shells, these organs are inside the spiral-shaped shell. When threatened, the animal can fully retreat inside its shell. In slugs, the intestines are located in the foot. There are two types of snails, classified according to how they breathe: gilled snails with gills and pulmonate snails with lungs.

Cephalopods

Just like snails, cephalopods (class Cephalopoda) also have a distinct head section. Attached tentacles are covered with many suckers. They often have highly developed eyes and beak-shaped jaws. Species living in light surroundings will often eject a dark tinted substance when threatened, so as to hide in the cloud of ink. Others can adapt quickly to the surroundings by changing color so they disappear against the background. Common animals of this class are squid, cuttlefish, octopus, and the beautiful pearly nautilus.

in focus

SHIPWORMS

The shipworm, which can grow to eight inches (20 cm), may look like a worm due to its elongated shape and soft tissue, but it is actually a clam with a reduced shell. The tiny shells are razor sharp allowing these animals to bore into wood. During the times when most ships were constructed with wood, the shipworm was especially unpopular for the damage it caused.

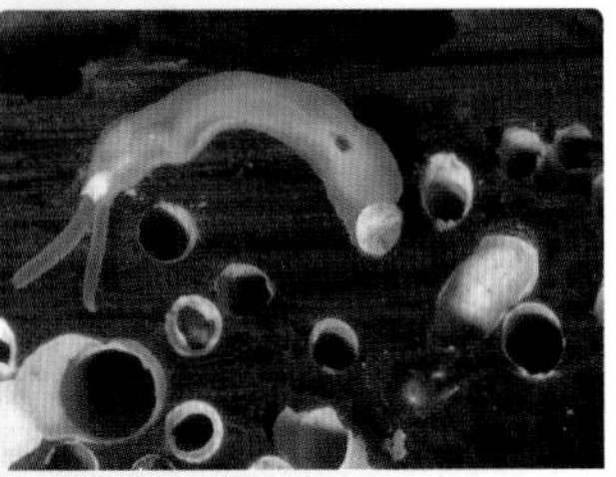

Piers or coastal reinforcements made of wood are still subject to damage caused by shipworms.

Mussels and clams

Exclusively aquatic mussels and clams (class Bivalvia) have a two-part shell that can be opened and closed by strong muscles. Some species are fixed in one place, while others can crawl or swim by using their shells to create a propelling effect. Mussels and clams are filter feeders, eating plankton and detritus from water that they suck in. Marine species usually reproduce through external fertilization.

basics

MANY SNAILS ARE HERMAPHRODITES, which means that they have both male and female germ cells. Before the actual act of mating, individuals of some species will shoot a tiny "Cupid's arrow" into the foot of their partner. The arrow introduces a slimy secretion, making the other snail more receptive to the sperm of the "shooter." Successful shooters often have twice as many offspring as those who miss the target.

Snails have one or two pairs of sensitive feelers. The eyes are located at the end of one pair.

see also: Loss of species diversity, p. 133

VERTEBRATES

Vertebrates include many of the best known species; however, they make up only a small percentage of all animals. All vertebrates have a spinal column. This is the central element of the body and connects all parts of the skeleton together. Fish, amphibians, reptiles, birds, and mammals are all vertebrates.

fish

Fish are the oldest and most varied group of vertebrates with over 30,000 different species. About 60 percent of fish species live in saltwater; the rest live in freshwater.

Fish exist in all aquatic habitats and are cold-blooded animals that breathe through gills. The majority have bony structures but a few—such as sharks and rays—have flexible cartilaginous skeletons and usually live in oceans. Bony fish inhabit environments from freshwater lakes and rivers to the ocean depths as deep as 13,000 feet (4,000 m). Some migrate between fresh- and saltwater habitats.

■ Rays have a flattened body and large pectoral fins, which they use to swim. Some species live on the ocean floor, others swim freely.

Anatomy of a fish

Like all vertebrates, a fish's body is built around its backbone. Support is provided by ribs and free "fish bones." Their body shapes vary greatly and are adapted to their particular habitat. Some bottom-dwelling species, like the flounder, are flat, whereas the pike and other swift hunters have torpedo-shaped bodies. Fish swim by flexing their bodies and pushing forward with their tail fins, using the paired pectoral and pelvic fins to maneuver. The dorsal, anal, and caudal (tail) fins help to stabilize them. The lateral line on their bodies has a row of organs to sense water currents and detect the positions of objects such as the seafloor and other creatures. Most have a thin skin covered with protective scales, but some bony fish have skin embedded with small, overlapping bony plates.

■ These Japanese koi carps, like about 40 percent of all known species of fish, live in freshwater such as rivers, lakes, and ponds.

Nutrition and reproduction

Fish feed off plants, plankton, or other fish—the position of their mouths is a good indication of their diet. Fish tend to reproduce outside their bodies. The female deposits her eggs in the water while a nearby male releases his sperm at the same time. A few fish, including the guppy and some sharks, give birth to fully formed young. The male fertilizes the eggs inside the female using an organ called a gonopodium and the eggs remain inside the mother until they are ready or almost ready to hatch, a process called ovoviviparity. After gestation, their offspring can usually swim within 24 hours.

in focus

GILLS

Aquatic animals have gills for breathing. Due to the slow dispersion of water, the thin gill filaments need to be flushed through with water all the time. Tiny cell projections (cilia) are used for ventilation to achieve flushing of the gills. Sharks and rays have a pumping mechanism for this purpose. Gills are also used for salt regulation, excretion of waste products, and, in some species, to take in food particles.

■ Gills contain filaments that expand in water but collapse if the animal leaves the water.

amphibians

Amphibians are descendants of the first vertebrates to move out of water onto land. Their lives still reflect that heritage: they spend their early life (larval phase) in water and the rest on land.

Most amphibians begin their lives in the water, spend most of their time near the water, and return to it to lay their eggs.

Amphibians include both tailed animals and tailless species and can be divided into two groups: Caudata and Anura. The Caudata group, which includes newts and salamanders, has over 400 species living in the Northern Hemisphere and the American tropics. Animals of this group have four legs, elongated bodies, a long tail, and often large eyes. Frogs and toads belong to the Anura group and are native everywhere outside the polar regions. They usually have powerful hind legs for hopping and jumping. Tree frogs have sticky toe pads, making them exceptional climbers.

Between water and land

Amphibians show their kinship to the first bony fish that used strengthened fins to climb onto land by beginning their lives as finned swimmers with gills. However, by adulthood they are air-breathing creatures with legs and are fully adapted to life on land. They are cold-blooded, so assume the surrounding temperature. Many spend winters snuggled into holes in the ground or protected by piles of leaves. Their skin is highly sensitive and must not dry out, as it supplements their lungs by absorbing both air and moisture. Some amphibians have skins with glands that secrete toxic or bad-tasting fluids to protect them from predators.

Why frogs croak

Frogs have an acute sense of hearing and strong voices. Male frogs have developed mating calls (croaks) that female frogs find alluring. Salamanders, on the other hand, are quiet and attract partners with special scents and bright colors. Amphibians can have highly complex mating rituals. Male pond frogs, for instance, release sperm to fertilize eggs laid as a cluster in the water. Female crested newts take up a packet of sperm for internal fertilization. Some amphibian clusters can contain over 10,000 eggs, whereas sometimes, only one egg is laid. In rare cases amphibian species even bear live young. Only a few amphibians breed on land; these lay eggs in rotting leaves or tree cavities.

Tree frogs like the red-eyed tree frog are exceptionally good climbers. They can cling firmly to a surface using their limbs and the moist skin on their belly. They also have toe pads which use capillarity to grip tight to surfaces.

basics

AMPHIBIAN NUMBERS are declining. Weed-killers and insecticides kill their food sources and are also absorbed through their skins. Others die on roads as they travel (often several miles) between their habitats.

in focus

THE AMPHIBIAN LIFE CYCLE

Spring is mating time for amphibians and egg-laying follows, usually in a pond. After emerging from the eggs, the larvae metamorphose into adults, often within a few weeks. Frogs undergo a particularly dramatic change from tadpoles to adults. They lose their tails and gills and develop powerful legs. Tadpoles are primarily vegetarians, but adult frogs eat insects and other animals. Salamanders feed on insect larvae and remain carnivorous throughout their lives.

The male midwife toad carries the eggs on his back until shortly before they hatch.

The conspicuous color of the red eft warns potential predators of the poison it can secrete from glands in its skin.

reptiles

Reptiles include snakes, lizards, turtles, crocodiles, and the extinct dinosaurs. They have been able to successfully live on land because of several key adaptations.

Crocodiles cannot chew. Instead, they tear off chunks of flesh by spinning their own bodies while gripping their prey.

Reptiles are well suited to life on land. Scales prevent their skin from drying out, their lungs are strong and efficient, and their eggs are encased in sturdy shells. Reptiles are cold-blooded, so they must regulate their body temperature by seeking shade or sunbathing. They are efficient users of food and can survive in nutrient-poor regions such as deserts.

REPTILES date from 300 million years ago, when three evolutionary lines diverged: One became turtles and tortoises, another dinosaurs, lizards, snakes, crocodiles, and birds. The third became mammals.

basics

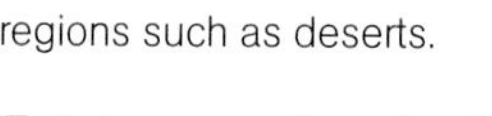

Snakes grow continuously and they shed their skin by detaching the outer layer.

Lizards

Lizards are the largest and most diverse species group of reptiles. Most are very small. Typically, they bury or cover their eggs and hibernate during cold weather. Except for the legless lizards (such as glass lizards), they have four legs and can be visually striking.

Turtles and tortoises

Turtles and tortoises have changed little in 150 million years of existence due to the protection offered by their shells. Most tortoises and freshwater turtles can withdraw their heads and legs into their shells, but sea turtles cannot. All are omnivorous and lay their eggs on land. Most are threatened with extinction due to human encroachment and habitat destruction.

The back and belly shells of turtles and tortoises are covered with skin and bone plates.

Crocodiles and alligators

The largest living reptiles are crocodiles and alligators. Limited to the warm regions of the world, they spend much of their lives in and around water. They often float just below the surface, breathing though their vertical nostrils. When they spot a potential meal, they lunge and pull their victim under water.

Snakes

Snakes have a keen sense of smell and their ability to sense vibrations and temperature variations helps them to hunt effectively. Venomous snakes use a pair of hollow fangs to inject a paralyzing or deadly nerve venom into their prey. Constrictors can unhinge their jaws in order to swallow a large prey.

Chameleons change color—often dramatically—and their protruding eyes allow them to see nearly 360°.

see also: Evolutionary trees, p. 169

aves

The ability to fly sets birds apart from other vertebrates. They are remarkably adapted for flight and display impressive capabilities in the air, often flying at speeds of 15–50 mph (24–80 km/h).

Birds of prey have a highly developed sense of sight, which they use to hunt their food.

Birds are distiguished by being the only vertebrates with feathers. Furthermore, not only can birds fly, but they also have excellent vision. Sight is especially highly developed in birds of prey, which must spot small animals from great heights. Beaks come in many forms and are adapted to eating habits. Raptors, for example, have sharp-edged beaks, woodpeckers have heavily reinforced beaks, and hummingbirds have tube-shaped beaks. Birds are warm-blooded with efficient hearts, lungs, and circulation. Their feathers reflect the lifestyle and behavior of the bird. In cold climates, birds have thick layers of feathers and insulating down. Many male birds have colorful plumage to attract mates, whereas those with dull-colored feathers use them as camouflage. When food supplies dwindle in the fall, many birds migrate to warmer areas, with some species traveling up to 6,200 miles (10,000 km) to reach their winter homes. About 8,000 species of birds are known to exist. A few, such as the African ostrich and the New Zealand kiwi, have lost the ability to fly. About 60 percent of all birds are passeriformes, or perching birds, including sparrows, swallows, and finches.

FLYING

in focus

Humans have been intent upon imitating the flight of birds for centuries, but have been unable to duplicate their efficiency. Birds use their powerful chest muscles to flap their wings to gain aerodynamic lift and hold their wings out to ride air currents. Birds weigh very little for their size—most weighing no more than 33 pounds (15 kg)—due to their extremely light, hollow bones, which also play a role in breathing.

Primary remiges
Greater primary coverts
Secondary remiges
Secondary remiges
Tertiary remiges
Shoulder feathers

The long flight feathers form the hydrofoil of the wing, which enables the bird to gain lift.

Mating and reproduction

Birds are territorial. They maintain and defend an area to mate and nest. Songbirds usually attract mates with vocal signals while others use visual signals, such as attractive plumage. Some birds, such as raptors, owls, nuthatches, and penguins, mate for life, whereas others have partner relationships only for the period of egg incubation. Some species use a nest for multiple years and can make them highly elaborate. A few—such as Antarctic penguins, which incubate their eggs standing up—do not build nests at all.

All birds lay eggs but the number varies from one to about 20. Most eggs are white, though brown, spotted, and pastel colors are not unusual. Parent birds take turns sitting on the eggs to maintain a warm temperature for the incubation period (11 days to 20 weeks). Once hatched, many chicks are fed by the parents until they learn to fly. Parasitic breeders (such as the cuckoo) lay their eggs in the nests of other birds, who consequently become unwitting foster parents.

Toucans are found in tropical and subtropical forests and are known by their colorful beaks.

Colibris fly at very high frequencies so they need a high-energy diet such as nectar.

Some birds, such as ostriches, have lost their ability to fly through the course of their evolution.

see also: Mesozoic era, p. 160

features | diversity

MAMMALS

Mammals make up a class of vertebrate animals. Recent mammalian species are divided into three subclasses: the higher mammals; the marsupials, for example, the koala, kangaroo, or oppossum, which live in Australia and America; and the egg-laying monotremes, such as the duck-billed platypus and the spiny anteater.

mammals: common features

Mammals have spread successfully over the entire world, and have even conquered extreme habitats such as deserts and the Arctic. In spite of all external diversities, they have numerous characteristics in common.

Mammals are the most highly developed vertebrate animals. Their constant body temperature grants them a life largely independent of ambient temperatures. Depending on the species, the body temperature of mammals lies between 96.8 and 102.2°F (36 and 39°C). This is facilitated by a high food intake—in comparison to reptiles, for example—and an increased rate of metabolism. In addition, they have temperature insulation, such as thick fur, to protect sensitive body organs. Mammals, such as whales or humans, where hair has been secondarily regressed, have an insulating layer of fat. A long digestive tract, dentition with four various types of teeth, and a powerful chewing musculature allow for the utilization of diverse food components. The mode of nutrition in mammals ranges from pure plant, insect, and meat eaters to omnivores.

Lung respiration, typical of mammals, together with an effective circulatory system, supplies the body cells efficiently with oxygen. The heart consists of two atria and two ventricles, which are completely separated from each other. This prevents the mixing of oxygen-rich and oxygen-deficient blood. The red blood corpuscles do not have cell nuclei (p. 148). This makes their shape more flexible and enables them to pass through the narrowest of capillaries.

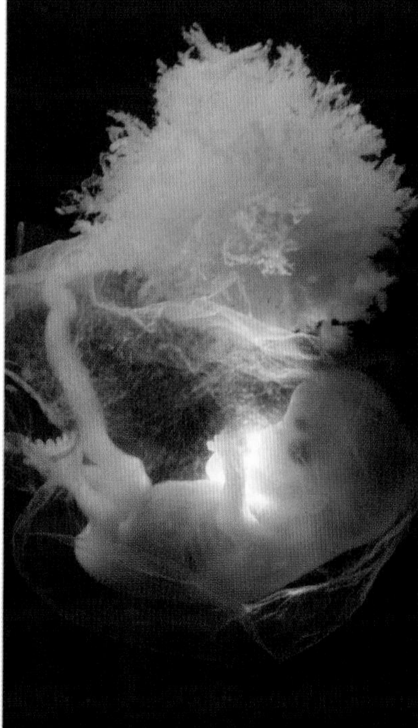

The placenta is a type of tissue that nourishes the fetus in the female body via the umbilical cord.

Progeny

All mammals, except for the egg-laying monotremes, give birth to live young. During the pregnancy in higher mammals, the fetus is linked to the circulatory system of the female and is supplied with nutrients via the placenta and through the umbilical cord. The exchange of oxygen and carbon dioxide also takes place via this route.

Following the birth, the progeny is raised with a special dietary substance: the fat- and nutrient-rich mother's milk. Milk is given off from special breast glands, which facilitate quick growth of the newborn. Breast-feeding also leads to the formation of a strong mother-child bond that is the basis of close social structures.

Polar bears inhabit northern polar regions. There they survive and breed in an icy environment.

in focus

MARINE MAMMALS IN DANGER

Whales and dolphins are not only endangered by commercial whaling activities: An estimated 60,000 animals die each year in enormous drift nets of the general fishing industry. Furthermore, there is overfishing of the seas, which takes food away from large marine mammals. Increasing pollution of the oceans by environmental toxins further restricts the habitat of these animals. "Noise pollution" from military and commercial shipping interferes with the orientation capabilities of whales and may cause them to become deaf.

Due to disorientation, more and more whales are stranded along coasts, where they die a painful death.

see also: Loss of species diversity, p. 133

mammals: diversity

From the fingernail-size Etruscan shrew to the almost a hundred-foot-long blue whale, the mere size differences among mammals are remarkable. Diverse modes of life have created a diversity of forms and functions.

Distinct adaptations to a particular way of life are, for example, revealed by the various shapes of limbs. The aquatic whales are fishlike in form. Bats in flight display a leathery skin extending between large fingers. Life beneath the earth produces the cylindrical, compact body of the mole, where the forelimbs have changed into short, powerful digging tools.

The gray-brown Etruscan shrew is one of the smallest mammals in the world. It is an endangered species.

The outer body envelope of mammals gives an idea of their diversity. This ranges from the enormously thick fur of polar bears, the spines of hedgehogs and porcupines to the bony plates of armadillos.

When danger approaches armadillos can protect themselves by quickly rolling into an armored ball.

Color of skin and hair is often adapted to habitat conditions, as for example, with snow hares. Skunks, on the other hand, give off strongly contrasting optical warning signals. With whales and sea cows, the furry coat has almost completely regressed.

With respect to diet, there are pure plant and meat eaters and species that will feed on anything. Meat-eaters have a short digestive tract, while plant-feeders break down and metabolize their hard-to-digest food in a longer digestive tract or, like ruminants, in a multichambered stomach.

Sensory diversity

The sense of smell, which is important for finding food, for territorial behavior, and for the recognition of other members of the same species, is particularly well developed in nose-oriented animals such as dogs or horses. In these animals the olfactory membrane is often extended onto the outside of the nostrils. Eye-oriented animals, such as cats and humans, have a highly developed visual faculty. Many diurnal animals are able to recognize colors, and nocturnal animals are able to enhance their visual acuity with a special reflection layer in their eyes. Cats also use extremely sensitive tactile hairs for orientation. Toothed whales and bats can orient themselves by means of echolocation, and the tip of an elephant's trunk has delicate sensors for mechanical stimuli.

in focus

SURVIVAL IN THE DESERT

The humps in camels serve as fat and energy reservoirs. These animals can take in enormous volumes of water within a few minutes, since the oval shape of their red blood cells prevents possible bursting. Special rectal cells remove liquid from fecal matter almost completely; the kidney and bladder condense urine to a point where only a single drop is released. The nostrils can be closed off, and special reticulating tissue in the nose cools the blood supply flowing to the heat-sensitive brain and eye cells.

The dromedary is perfectly adapted for survival in dry and arid regions of extreme heat.

basics

MAMMAL DENTITION consists of four distinct types of teeth: incisors, canines, and two types of molar teeth. Most species initially have a set of milk teeth, so-called deciduous teeth, that fall out and are replaced by permanent teeth during the second dentition.

Armadillos are prolific diggers. They use their sharp claws to dig for food, such as grubs, and to dig dens.

see also: Instinctive behavior, p. 261

monotremes and marsupials

Monotremes and marsupials have characteristics not seen in other mammals. Monotremes lay eggs and marsupials give birth to live young, which then climb into their mother's pouch to nurse and mature.

The monotremes and the marsupials are among the surviving original mammals.

Monotremes

Monotremes are very unusual mammals. Two to four weeks after mating, the female lays one or two soft-shelled eggs through a cloaca, a cavity into which the intestinal and urinary canals open. The eggs are then incubated for up to ten days before they hatch. After hatching, the babies feed on their mother's milk—like all other mammals.

The male platypus has poisonous spurs on its back legs. It is the only poisonous mammal. The poison is thought to be used when fighting over fertile females.

The platypus has a duck-like bill protruding from a broad, flat jaw strengthened by keratin plates. Its flat, beaver-like tail and webbed front paws allow it to hunt small creatures in the water. It lives in complex burrows in the riverbanks of Australia. Another monotreme, the echidna, lives only on land. Echidnas have strong claws to open insect nests. They use their long sticky tongues and tubular snouts to eat the insects and in case of danger they raise their spikes. All monotremes are loners, nocturnal, or active in the dawn hours.

Marsupials

The marsupials include some 270 mammals native to Australia, Papua New Guinea, and the Americas. Marsupials on the northern continents are often similar to the eutheria (placental) mammals, but in the south they are radically different.

After birth, the tiny baby has to climb from the birth channel to the mother's pouch using only its senses of smell and touch. Once there, it fastens onto one of the mother's mammary glands. The baby remains inside the pouch, suckling milk until it is large enough to leave. In the case of the red kangaroo, this suckling stage can last 235 days. The number of young born per birth varies among species up to a dozen babies, according to the degree and duration of protection they receive in the pouch. If the stay in the pouch is short or less protected—that is, if babies are more likely to die—more offspring will be produced to compensate for a higher rate of mortality.

The common feature of marsupials is a pouch in which mothers carry their young.

Koalas feed almost exclusively on eucalyptus leaves. The leaves contain a poison, which the koalas are able to tolerate to a certain extent. The concentration of poison is lower in older leaves.

see also: Ethology, pp. 260–261

rodents and lagomorphs

Although they can appear similar, rodents and lagomorphs developed independently of one another. They can be easily distinguished by the different incisor teeth on their upper jaw.

Rodents and lagomorphs inhabit the entire planet with the exception of Antarctica. Rodents mostly feed on leaves, seeds, fruits, roots, and tubers; a few also feed on insects or other invertebrates. Their teeth are supported by strong chewing muscles, which also give them their chubby-faced appearance. Many rodents—including marmots and house mice—live sociably in groups. Active at all times, they can be found in habitats ranging from water to high mountain tops. The synanthropic species, such as the house mouse, have adapted their habits to human environments. Beavers and South American capybaras inhabit wetlands and are good swimmers. The African spring hare is an outstanding jumper due to its kangaroo-like hind legs. Squirrels' tails help them balance when they climb.

Squirrels are excellent climbers and jumpers. They use their tails to steer and balance. Squirrels are solitary animals and generally interact only to mate.

The lagomorphs include both long-eared animals, such as hares and rabbits, and small-eared animals, such as the North American pika. Most lagomorphs have split upper lips and skin folds (harelips) and, when they open and shut their nostrils, their faces twitch. Their touch and hearing are well developed and aided by sensory hairs around their noses and ears. Their long, hair-covered hind legs enable them to run quickly—hares can achieve speeds of up to 50 miles per hour (80 km/h), helping them escape from predators. Their favored habitats are open fields, grasslands, plateaus, and semideserts. Lagomorphs are plant-eaters. To help them digest plant cellulose, they often excrete soft feces pellets that they then ingest a second time. Their young are raised in burrows or nests. Due to their short lifespan, lagomorphs have several litters of about nine each year. This high birth frequency compensates for the fragility of the young, which are often devoured by predatory animals and birds. The fur of wild hares and similar animals is almost always reddish or gray-brown for maximum camouflage.

Porcupines have long spines, which are actually adapted hairs. If the animal is threatened, it raises its spines to defend itself.

During the mating season from January to September, male hares fight with each other over females.

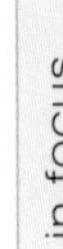

DIFFERENT TEETH

Rodents and lagomorphs can be distinguished by their teeth. Rodents have a pair of chisel-like incisors at the front of their upper and lower jaws. These teeth grow continuously and the chisel shape is formed through constant wear. Behind the incisors is a distinctive gap before the molars further back in the jaw. In contrast, all lagomorphs have two pairs of small "peg" teeth that sit behind the large incisors on the upper jaw. Because of this, lagomorphs are well adapted for a gnawing diet.

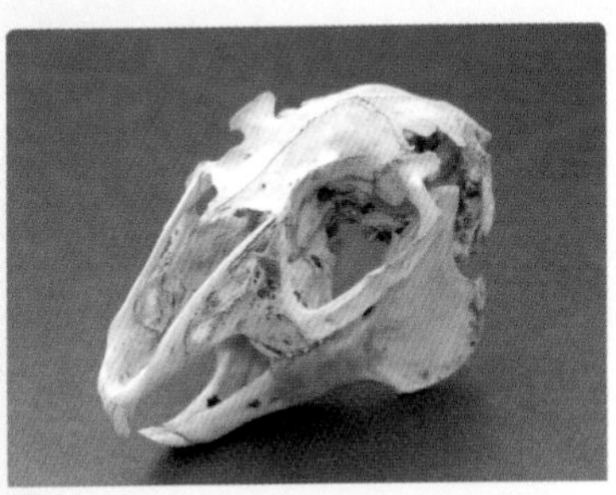

Rodents lack canine and premolar teeth, leaving a gap between their incisors and molars.

ungulates

Ungulates are hoofed mammals that have phalanges, or toes, that are encased within a hard covering. Ungulates include horses, cattle, sheep, goats, and many other animals, and almost all are herbivores.

Goats are most common in mountainous regions in Asia, Europe, and northern Africa. Their strong limbs and wide hips allow them to move easily over the rough terrain.

Ungulates include the largest, most impressive, and most commonly found land mammals. They can be identified by their toes (phalanges) that are covered by hooves. Except for the omnivorous pigs and the insect-eating anteater, all ungulates are plant-eaters. The group includes such domesticated animals as horses, cattle, sheep, goats, and pigs. Wild ungulates include elephants, rhinos, giraffes, and the sea cow, an ungulate that has returned to the water. Originally, ungulates were found on all the continents of the Earth except Australia.

Although they are vegetarians, ungulates cannot digest plants on their own as they lack the enzyme that ensures the complete digestion of cellulose. Thus, these animals require the presence of symbiotic bacteria, yeasts, and protozoa in their digestive systems. Many ungulates, including deer and cattle, are ruminants. Ruminants improve their breakdown of cellulose by repeatedly chewing predigested cuds.

Characteristic features

Most hoofed animals have long legs that helped them flee from predators in their original grassland and savanna habitats. Some ungulates (often males but some females, too) have developed other attributes such as horns (on cattle, for example), antlers (deer), or tusks (pigs).

Rather than use these aggressively, hoofed animals usually use these features to intimidate and assess the strength of their opponents. Hoofed animals are classified within two broad groups: the even-toed and the odd-toed. The even-toed ungulates include giraffes and hippos, whereas the odd-toed encompass the horses, rhinos, and tapirs. During their existence, all ungulates have slowly reduced their number of phalanges, so that in the case of modern horses there is only one hoof (toe) per leg.

Rhinoceroses are usually timid, but can become aggressive if approached.

in focus

ANTLERS

Male deer use their antlers for fighting rivals. Every year, they grow a set of horns from two points on the forehead called pedicles. During their growth, the bony parts are covered by a thick skin (velvet) with short hair, which supplies nutrients. At the beginning of the rut, male deer rub off the velvet on strong tree branches. During the fall or winter, a thin layer of bone between the antlers and pedicles dissolves and the deer shed their antlers.

Only male deer have antlers. They can grow as large as almost seven feet (two m) wide.

basics

DOMESTICATION first developed when early people began to farm. Hoofed animals were bred as transport and draft animals, as well as readily available sources of milk, wool, and meat.

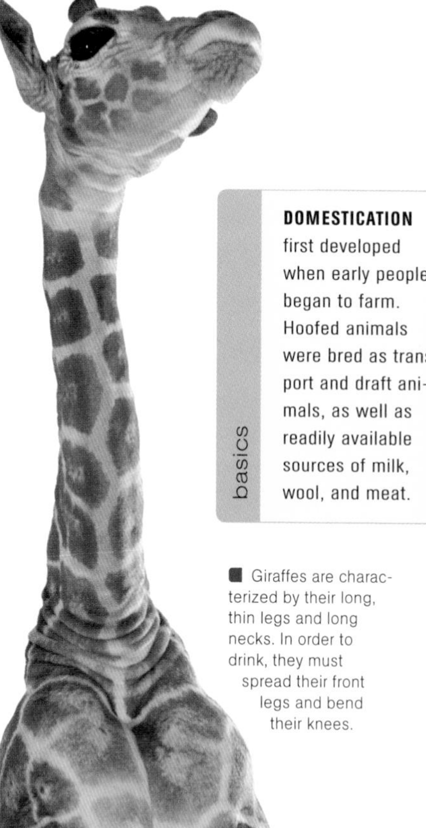

Giraffes are characterized by their long, thin legs and long necks. In order to drink, they must spread their front legs and bend their knees.

see also: Ecosystems, pp. 80–87

elephants and sea cows

Elephants and manatees are closely related to each other. The elephant is the largest mammal living on land, while sea cows (manatees and dugongs) have adapted themselves to living only in water.

The greatest danger today for sea cows does not come from hunters, but rather the pollution and human invasion of their habitats. For example, they can be seriously injured by motorboats.

Three species of elephant exist today: the African elephant, the forest elephant, and the Asian elephant. They live in grasslands, savannas, mountainous areas, and forests. An elephant's trunk is an elongated nose, with the nostrils located high up on the skull. Extremely strong trunk muscles allow it to be used to dig watering holes, tear bark off trees, intimidate enemies, and for defense. Because elephants' skin is sensitive and very thin in some places, especially near the stomach and behind the ears, the animals bathe in water and mud to cool and care for their skin. Since they do not possess sweat glands, excess heat is released through their large ears. Their pillar legs end in vestigial hooves with tissue padding. When walking, they make very little noise, despite their size. Their heavy tusks are transformed upper incisors that can grow to up to 11.5 feet (3.5 m) long.

The legendary memory of the elephant has been partly confirmed: even 70-year-olds can find their way back to the watering holes of their youth. Sea cows (manatees and dugongs), named for their herbivorous diet, are closely related to elephants—although there are few visual similarities. Only the short tusks of the dugong give a clue to their elephant relatives. They are fully adapted to aquatic life. Instead of forelegs they have flippers; their hind legs have receded, and the base of their body tapers to a rounded paddle-shaped tail. The body of a mature sea cow is around 13 feet (four m) long and weighs up to 1,320 pounds (600 kg). They swim slowly or drift in the water and can dive for up to 20 minutes. Sea cows live either alone or in small groups. A sea cow is born after a gestation period of 12 to 14 months and mothers suckle the young, only underwater. Manatees are found in the shallow waters of coastal areas and bays of tropical seas, and at the mouths of rivers.

issues to solve

THE IVORY TRADE and the poaching of elephants has nearly resulted in their extinction. A ban on trade since 1989 has helped little, due to smuggling and inefficient checks.

Elephant tusks are the world's main source of ivory. Tusks are sometimes removed to protect the animals from poachers.

Female elephants live in variously sized family groups with their offspring. The calves are protected not only by their own mothers, but also by the other members of the group.

sea mammals

The planet's largest living things are the Cetacea, which include whales, dolphins, and porpoises. Cetacea and other sea mammals have exceptional sensory abilities and impressive aquatic skills.

Whales and dolphins live in all the oceans on Earth, but despite their distribution, many cetaceans are threatened by extinction. There are two groups: the toothed whales (Odontoceti) and the toothless, or baleen, whales (Mysticeti). The baleens (including blue and humpback whales) get their name from the comblike keratinous plates (baleens) they have in place of teeth. These baleens are used to filter plankton or krill from the water. The toothed whales include dolphins, pot whales, killer whales (orcas), and narwhals. They feed mostly on fish or squid.

Hearing is the most important cetacean sense. It is very likely that marine mammals communicate with each other by means of sonar emitted frequencies from 5 to 280,000 Hz.

Cetaceans are hairless, warm-blooded creatures. Their layers of blubber (fat) help maintain their body temperature in the water. Their forelimbs have evolved into flippers used for steering, while their hind limbs have disappeared. The tail, known as a fluke, is their propellant. They are conscious breathers, meaning that breathing is not an automatic function but one actively controlled by the animal. They can dive deeply and stay below for long periods. Whales are adapted to holding their breath, some species for 90 minutes or more.

Newborn baby whales must adapt quickly. Their initial breath comes when they reach the surface for the first time. A blue whale calf suckles underwater for about seven months and grows 110 pounds (50 kg) heavier and 1.8 inches (4.5 cm) longer every day.

Bull walrus, which can be recognized by their strong imposing tusks, can weigh up to 4,500 pounds (2,000 kg). They spend much of their lives on the sea ice, are social animals, and feed primarily on mollusks.

Dolphins

Dolphins have a larger brain than humans and a fourth brain ventricle that allows them to remain mobile 24 hours a day. They relax by switching off half their brain for short periods. The other half takes over the duties of watching for danger and controlling their swimming and breathing.

Pinnipeds

Seals and walrus are members of the Pinneped family, meaning "fin-footed mammals." Seals can swim as fast as 28 miles per hour (35 km/h), but they spend much of their time on land and thus favor coastal habitats. They have dense coats of body hair, which has led to them being relentlessly hunted by humans. They give birth and fight on land. In the water they use their back flippers and tail like a rudder. Seals feed primarily on fish, while the leopard seal also hunts penguins.

Humpback whales can migrate up to 15,500 miles (25,000 km) each year. They only feed in summer and live off their fat reserves in winter.

basics

"WHALE SONG" is the name given to the acoustic communication between whales. Tones and tone sequences are produced, which can sometimes be heard hundreds of miles away.

THE EYES OF SEALS are relatively large and, due to an increased number of rods in the retina, are suited to the low light conditions underwater.

insectivores

Insectivores are a small and nonuniform order of about 450 animal species. Hedgehogs, shrews, and moles are the most familiar insectivores.

Some insectivores belong to the smallest mammals on Earth, including the Etruscan shrew (*Suncus etruscus*), which reaches a length of just over an inch (35 mm) without its tail and weighs only 0.07 ounce (two g). Apart from hedgehogs (Erinaceidae), shrews (Soricidae), and moles (Talpidae), there are two more families: the solenodons (Solenodontidae) from the Antilles and the tenrecs (Tenrecidae), which mainly live in Madagascar.

Hedgehogs and gymnures

Hedgehogs and gymnures only live in Europe, Asia, and Africa. Gymnures have thick, hairy fur, and usually a long tail, while hedgehogs have pointy spines, which have evolved from hair. They are connected through a ring muscle that covers the entire body surface and allows the animal to roll up like a ball with spines when threatened. Hedgehogs mainly eat insects and worms, but many also eat fruit and even carrion. They search for food when it gets dark.

The pointy spines of hedgehogs have evolved from hair to serve as a deterrent to and protection against predators.

Shrews

Species that belong to the family Soricidae may look like mice but they are not closely related to these rodents. They are very common in Europe, North and Central America, Asia, and major parts of Africa. Many species are active during the day and they are usually solitary. They mark their own territory and mainly feed on insects and insect larvae. Some species living near water bodies are excellent swimmers and divers, such as the European water shrew (*Neomys fodiens*) or the North American water shrew (*Sorex parva*).

Moles primarily eat earthworms and other small invertebrates. Their saliva contains a paralyzing toxin, so they are able to store their still living prey for later consumption.

Moles

Several European, Asian, and North American mole species spend a major part of their lives underground. Other species stay above ground, such as the American shrew mole (*Neurotrichus gibbsii*), while other moles have adapted very well to either an aquatic or semi-aquatic life. Species digging and living in underground path systems have cylindrical bodies with short extremities and front feet adapted for burrowing. Most moles cannot see well—they have probably evolved to lose their sight as it was not needed for life underground. Instead, they have highly developed touch and smell.

basics

HEDGEHOGS roll up and raise their long and sharp spines when threatened. This protects their bodies well from attackers, but does the exact opposite in the middle of the road. Every day, countless hedgehogs die from road traffic.

in focus

SHREW CARAVAN

When shrew mothers are disturbed, they often move their offspring to a new hiding place. Initially they carry the young with their mouths. Once they have grown bigger, some species will lead their offspring to a safer place by lining them up one after the other. Each one bites and holds onto the tail of the shrew in front, creating a caravan. This avoids losing offspring during the hasty move.

Shrews are fiercely territorial and dig burrows to catch prey or hide from predators.

see also: Ecosystems, pp. 276–277

predators

Predators are the hunters among the mammals. Found throughout the world, they have highly diverse lifestyles and widely varying methods of capturing their prey.

Predators have diverse lifestyles. Their most significant shared trait is daggerlike canine teeth and pairs of shearing carnassial cheek teeth that help them seize and tear into their prey. Faster predators can increase their hunting speed by using only part of their feet. Martens, for example, utilize only half of their foot.

The dentition of a lion features extremely elongated canine teeth for the killing and devouring of prey, such as gnus, zebras, and antelopes.

Others, such as bears, can increase speed by using the entire foot. Many animals use their excellent sense of smell to find their prey. Cats use their superior vision, which is functional even in semi-darkness. Social organization within animal groups relates to hunting and feeding. Wolves and lions live and hunt in packs. Others, including martens, leopards, and brown bears, hunt alone.

Friend or foe?

From surviving rock and cave art, we know that humans have had a complex and ancient relationship with predators. Humans and predators often hunted the same prey, and humans have always hunted predators—especially wild cats and bears—for their pelts. For a time, conservation resulted in the reemergence of wolves, bears, and lynx in central Europe, which saw the reawakening of age-old fears in humans of these predators.

Today, the creation of agricultural land from predator habitats has led to increased predator-human contacts while predator numbers are declining. Ecologically, these predators are important; they help to control wild herds of herbivores, such as deer, and keep those populations from exceeding the level that can be supported by the available food supply.

HUNTING BEHAVIOR

Hunting behavior is innate and further refined through learning. Each species has a characteristic strategy. Wolves and jackals are coursing predators; they chase their prey in packs until the victim tires, then they surround the exhausted prey while several pack members attack it simultaneously. Wolves coordinate their hunting through body movements, ear positioning, and vocalization, and group hierarchy determines which pack member has the privilege of eating first. In contrast, cats are stealth hunters. Whether they are small domesticated cats or large lions, felines sneak up on their prey with quiet steps and then pounce, using their claws to bring their victim down. A bite to the throat or back of the neck usually kills the victim.

Cats ambush or closely sneak up on their prey.

practice

ANATOMICALLY, giant pandas are classified as predators because recent genetic studies have shown they are bears, albeit bears that feed almost exclusively on bamboo.

basics

During their upstream spawning journey, salmon are a welcome prey for brown bears.

night fliers

The order Chiroptera belongs to the class of mammals and includes about 900 species of bats and flying foxes. These night fliers are the only mammals that can actively fly.

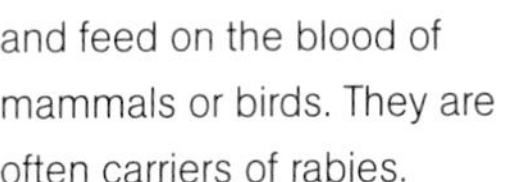

Townsend's big-eared bats *(Plecotus townsendii)* are the largest and best known family of bats, sometimes simply called common bats.

Bats are widespread around the globe, with the exception of very cold regions, while flying foxes only live in tropical or subtropical areas. Both groups of animals have unique front limbs that are very different from those of other mammals. Four of the five digits are significantly elongated and connected with a thin flying membrane that grows out of the lateral parts of the body. Although the majority of night fliers feeds on insects, some mainly eat fruit. Larger species may also occasionally hunt small vertebrates such as mice, birds, or frogs.

Gray-headed flying foxes *(Pteropus poliocephalus)* resting in Queensland, Australia.

Bats

Most bats are active at dawn or night; during daytime, they usually hide in a safe place. Though in general bats have poor vision, it is only a myth that they are blind. Their reproductive rate is low: many species only have one offspring a year. However, some bats can reach an age of up to 30 years. Species living in the temperate zone tend to hibernate. Before winter, these animals seek protected roosts in caves, tree holes, or buildings where thousands of bats may hibernate together. Bats are very social animals and also live in groups for the rest of the year. The females of some species even gather for collective birthing. The true vampire bats (family Desmondotidae) live in the tropical or subtropical regions of Central and South America and feed on the blood of mammals or birds. They are often carriers of rabies.

Flying foxes

The smaller group of flying foxes, also known as fruit bats, includes about 200 species. They were named after the fox-like shape of their heads. In contrast to bats, they have excellent vision. Only a few rousette fruit bats (*Rousettus*) are equipped with comparatively primitive echolocation. These animals exclusively feed on plants, mainly fruit, nectar, and pollen. They often pollinate the flowers while feeding. Flying foxes only live in warmer regions where hibernation is unnecessary.

basics

THE SMALLEST night flier, the bumblebee bat, is only about 1 inch (3 cm) long; the largest, the golden-capped flying fox, has a wingspan of over 6 feet (2 m).

A portrait of the hoary bat (*Lasiurus cinereus*): Hoary bats live a generally solitary life.

ECHOLOCATION

Most bats, especially insectivorous bat species, use echolocation (a biological sonar) for navigating and hunting at night. Bats emit ultrasounds of frequencies between 10 and 200 kHz. Humans are unable to hear these sounds. Potential prey items or obstacles bounce back these sound waves for the animals to detect with their extremely sensitive hearing. The waves are then converted by the brain into an image of the environment. This allows bats to hunt even in complete darkness.

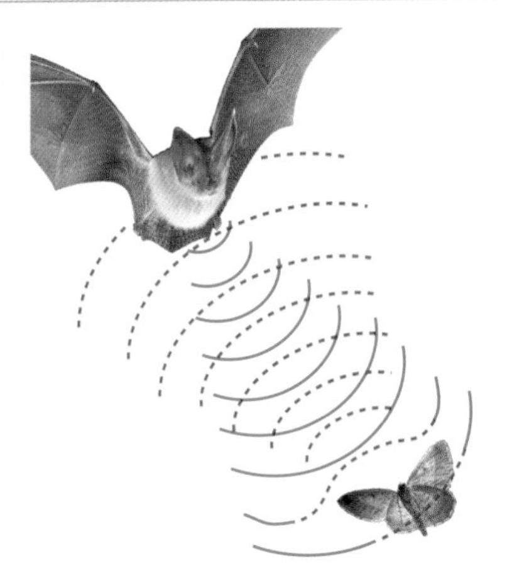

A bat produces ultrasound through its larynx and emits it through the nose or mouth.

primates

Primates, in comparison to most other animals, have a large brain with a strikingly enlarged cerebrum. This gives them the capacity to learn and to display complex social behavior.

Nonhuman primates live in the tropical and subtropical forests and savannas of Central and South America, Africa, and Asia.

Most primates are herbivores and live in trees, but some have adapted to life on the ground. They have relatively large brains and enhanced capacity for learning, tool use, and complex behaviors. Their large, forward-facing eyes let them see sharp detail in three dimensions. They can perceive contrast and color. Their supple hands and feet are adapted to feeling, gripping, and holding.

Communication and social life

Primates have highly developed social lives. They usually live in a family group or a harem—with one male and up to nine females—and they communicate through sounds, gestures, and facial expressions. Excreted chemicals (pheromones) can transmit messages, such as danger warnings or a desire for sex. Jane Goodall and other researchers have studied the behavior of chimpanzees, investigating their problem-solving skills and their ability to communicate with humans using symbols and sign language. Modern research divides primates into Strepsirrhini ("wet noses") and Haplorrhini ("dry noses"). Strepsirrhini tend to be nocturnal, with a smaller body size and keener sense of smell. Haplorrhini are mostly diurnal—active in the daytime—with some exceptions, such as the tarsiers. Strepsirrhini include lemurs, lorises, and galagos. Lemurs only live on the Comoro Islands and Madagascar. Lorises and galagos resemble sloths. Their powerful hands give them a secure grip on tree branches in Africa and South Asia. The Haplorrhini live in the tropical and subtropical zones of the Americas (capuchin monkeys and marmosets), in Africa (gibbons and great apes), and in South Asia (tarsiers). New World monkeys are found in the Americas, whereas Old World monkeys inhabit parts of Africa and Asia.

Primates' teeth are adapted to allow them to feed on both plants and meat.

Primates take intensive care of their young, allowing them plenty of time for learning and play.

GENEALOGY OF PRIMATES

in focus

The ancestors of most modern primates emerged as early as 55 million years ago. In the 19th century Charles Darwin, among others, suggested that humans and primates were closely related. Although some still dispute this theory for ideological reasons, multiple types of scientific studies clearly show a close biological kinship between humans and the great apes (such as chimpanzees, gorillas, and orangutans).

The first primates were tree-dwelling insect eaters. They emerged 65 million years ago.

Mutual or social grooming is sometimes called the social cement of the primate world. It bonds family groups together and is used to maintain or threaten social hierarchies and for reconciliation after a conflict.

see also: Evolution, pp. 140–145

humans

Modern humans, *Homo sapiens,* are distinguished from other living things by several physical characteristics and by the many intellectual and cultural achievements of several millennia.

No other living creature looks, lives, or moves like a human being. The evolution of humans has been supported by changes in the skeleton in response to changing conditions and demands.

Today, the pelvis is broad and tipped forward. The human backbone forms an S-shape, which absorbs vertical forces and allows for a wide range of upper torso movements. The big toe does not grip like those of the great apes; it lies parallel to the foot, which makes it easier to balance and move standing up. The arching bones in the sole of the foot provide cushioning during upright walking. The hands have changed dramatically, as they are no longer required to assist movement. With an opposable thumb and a rotating forearm, the human hand is an ideal tool for exploration, gripping, and manipulation.

Over the course of human development, the skull expanded and the face flattened, ridges over the eyes disappeared, and the nose and chin became more prominent. Body hair dramatically reduced in volume and coverage, and the teeth became smaller and more suited to a varied diet. The evolution of a moveable tongue, arched palate, and appropriately positioned larynx were critical to enabling the development of spoken language.

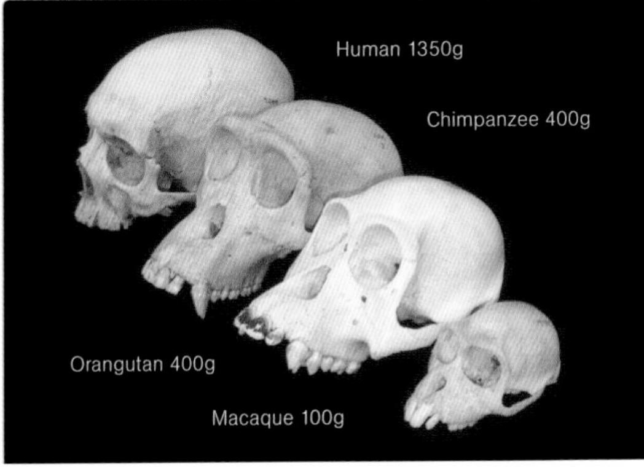

The *Homo sapiens* skull expanded in all directions, allowing for an enlarged brain behind a high forehead.

Prehistoric art: The human hand is capable of fine movements and extended usage.

TOOLS

in focus

Among the first tools used by early humans were stone axes, which appeared some 1.5 million years ago. With a rounded side for holding and a sharp edge chipped away on both sides to form a blade, it could be used for cutting, chopping, or scraping. This basic model was then further developed and refined.

Four flint tools from the Upper Paleolithic period, 35,000–10,000 B.C.

Intellectual and cultural development

Human physical and intellectual developments are closely related. Childhood lasts longer for humans than for chimpanzees, partly because humans are born at an earlier stage of development. Because of this, a significant part of brain development occurs after birth. In addition, the human brain has enhanced folds, which make efficient use of the space within the cranium. The larger brain has led to greater intelligence, an aptitude for language, and an ability to learn and engage in complex social behaviors—all crucial preconditions for the development of complicated languages and the emergence of human culture.

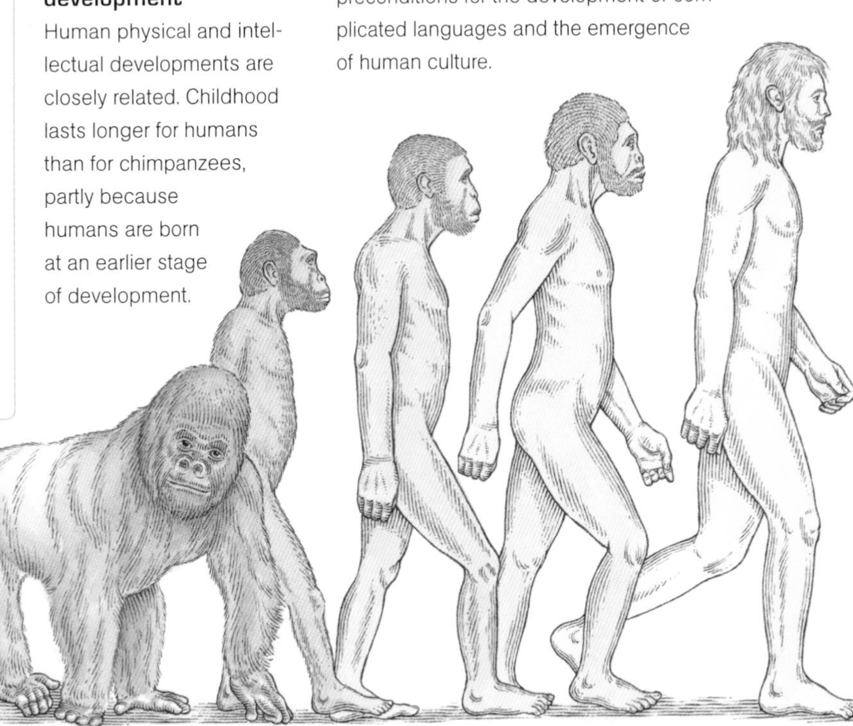

Unlike an arching backbone, the human spine is optimized for upright walking, forming an S-shape.

see also: Homo sapiens, p. 165

BIOLOGY

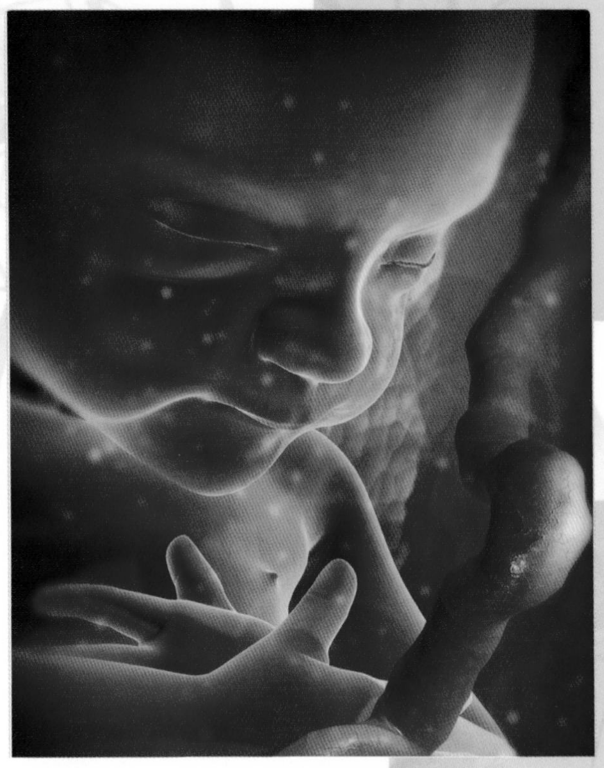

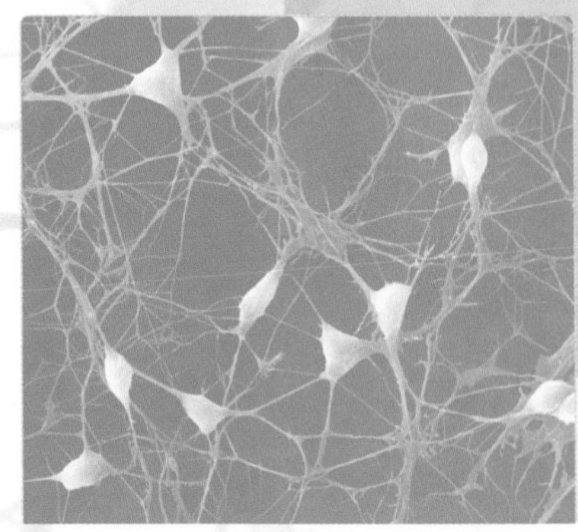

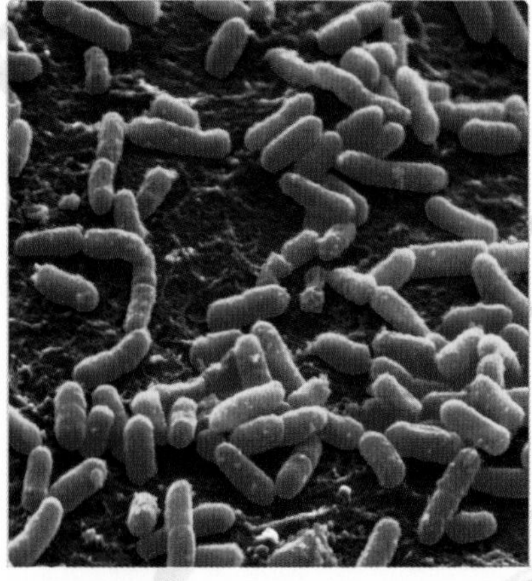

HUMANS

From a biological point of view human beings are grouped with mammals, more specifically with the great apes. However, humans hold a special place because of their highly developed brain. The human brain made it possible to cultivate complex languages and thus pass along new knowledge from generation to generation. The ability to solve practical problems—analyzing factors in the environment and then altering them to achieve a desired goal—also distinguishes human beings from other living things. In addition to their highly developed brain, humans have other special anatomical features, for instance, the intricate larynx that allows them to form the great variety of sounds involved in spoken language.

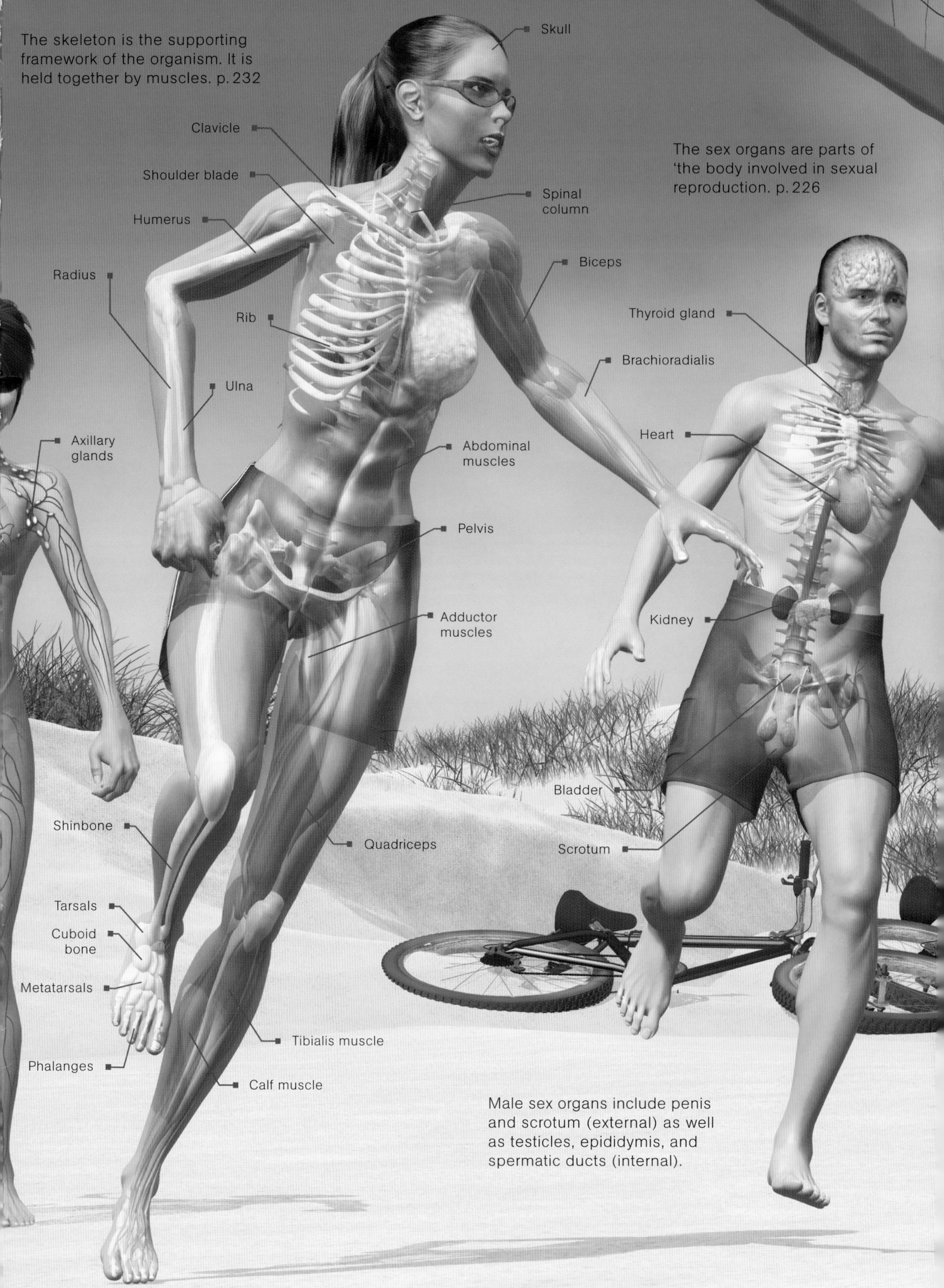

The skeleton is the supporting framework of the organism. It is held together by muscles. p. 232
Skull
Clavicle
Shoulder blade
Humerus
Radius
Rib
Ulna
Axillary glands
Spinal column
Biceps
Brachioradialis
Abdominal muscles
Pelvis
Adductor muscles
Quadriceps
Shinbone
Tarsals
Cuboid bone
Metatarsals
Phalanges
Tibialis muscle
Calf muscle
The sex organs are parts of 'the body involved in sexual reproduction. p. 226
Thyroid gland
Heart
Kidney
Bladder
Scrotum
Male sex organs include penis and scrotum (external) as well as testicles, epididymis, and spermatic ducts (internal).

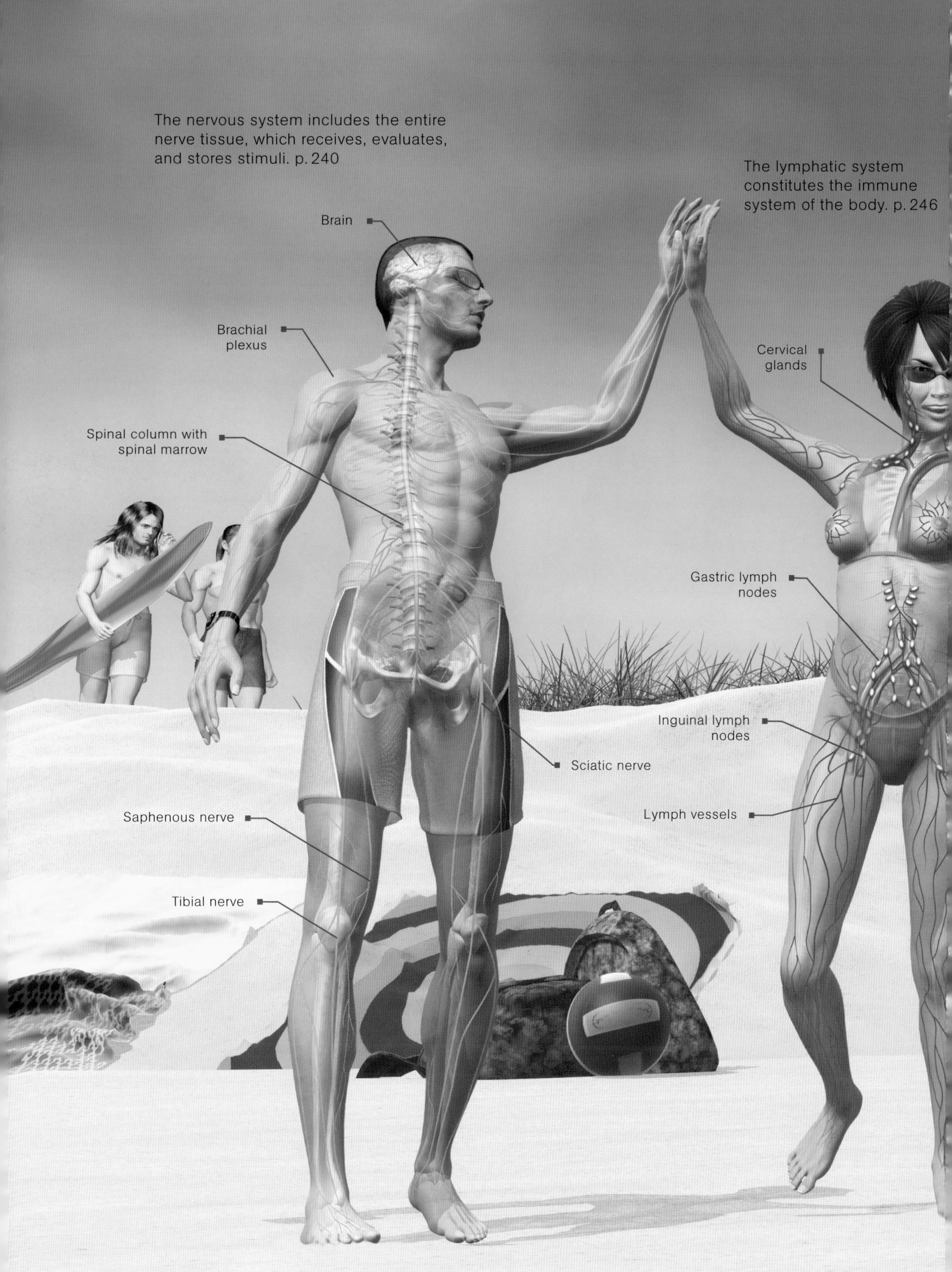
The nervous system includes the entire nerve tissue, which receives, evaluates, and stores stimuli. p. 240
The lymphatic system constitutes the immune system of the body. p. 246
Brain
Brachial plexus
Spinal column with spinal marrow
Sciatic nerve
Saphenous nerve
Tibial nerve
Cervical glands
Gastric lymph nodes
Inguinal lymph nodes
Lymph vessels

Parietal lobe (yellow) – sensory region
Frontal lobe (red) – motor region
Temporal lobe (green) – hearing region
Occipital lobe (blue) – sight region

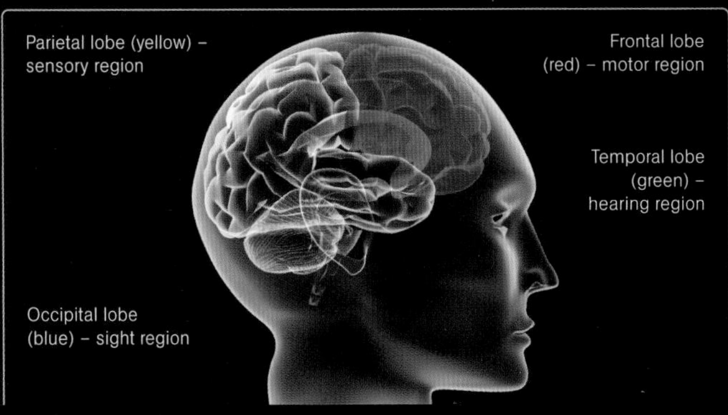

The cortex can be classified in four gross topographical regions called lobes, each with a special function. Also shown: cerebellum (light purple) and brain stem (dark purple)

HOSPITAL
SAG T11
SAG SE
10:13
SCAN 3
SCTIME 2:37
205*256
NSA 2 M
TR 409
TE 15
FOV 230
THK 5.0/ 0.5
SLICE 8/ 15
ANT 31.6
LEFT -3.9
CAUD -10.9 A
ANGLE AP 2
ANGLE CC 2
P W 889
L 559

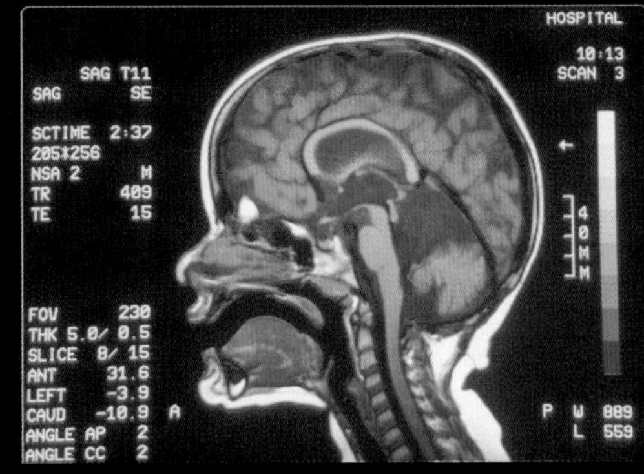

Brain research uses different tomographical methods to better understand the human brain.Also

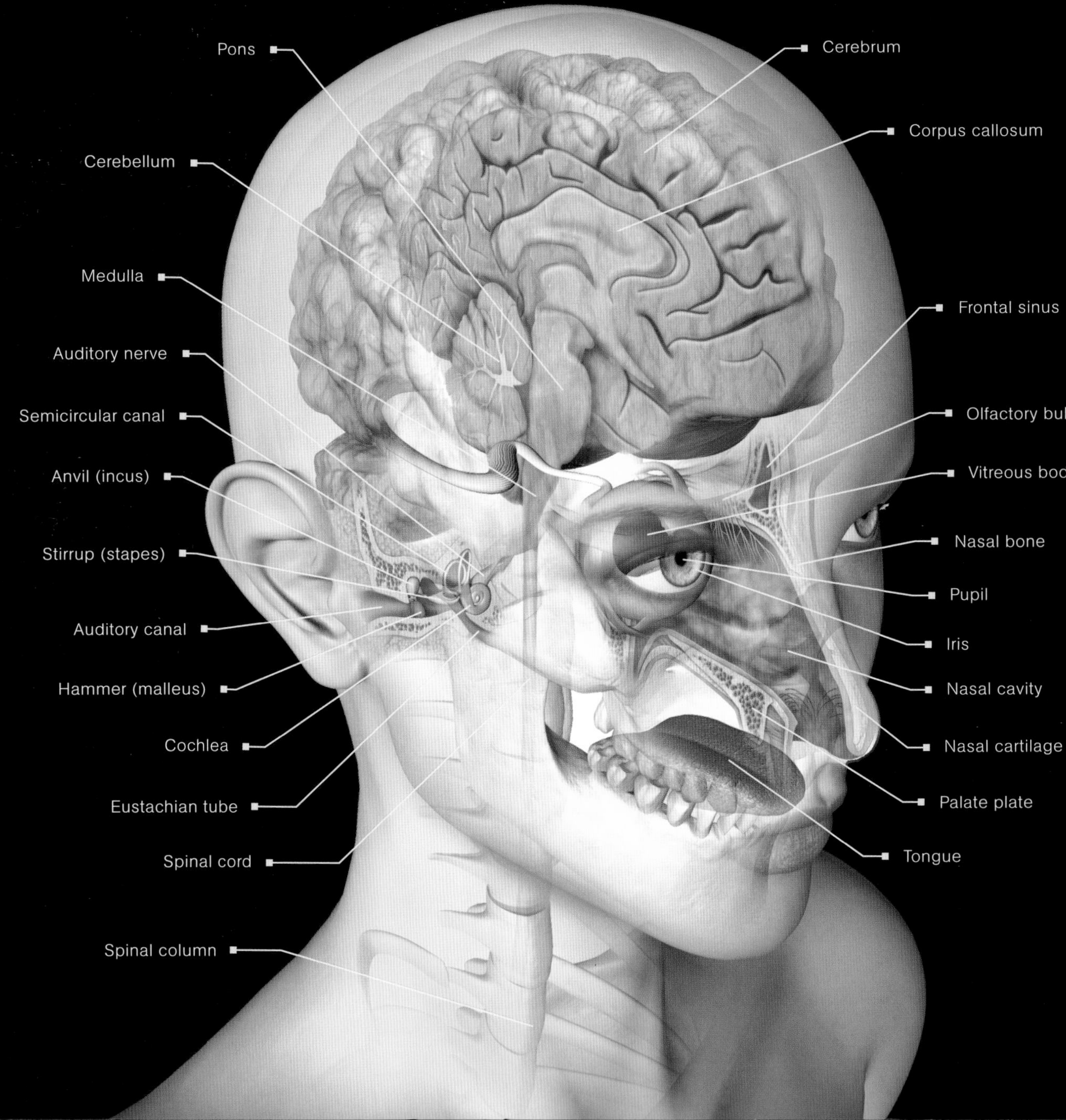

Pons
Cerebrum
Corpus callosum
Cerebellum
Medulla
Frontal sinus
Auditory nerve
Semicircular canal
Olfactory bul
Anvil (incus)
Vitreous boc
Stirrup (stapes)
Nasal bone
Pupil
Auditory canal
Iris
Hammer (malleus)
Nasal cavity
Cochlea
Nasal cartilage
Eustachian tube
Palate plate
Tongue
Spinal cord
Spinal column

Using our sense of smell can trigger particular memories and emotions.

Regular exercise can increase muscle size, strength, and endurance.

Hands are our main tools for physically manipulating our environment.

HUMAN BEINGS

The human body is an exquisitely complex mechanism. Its tissues and organ systems constantly work together to carry out the functions necessary to support life. The body's form and function are the result of a long evolutionary process, which is still ongoing today.

We perceive our environment through our sense organs. Stimuli reaching our eyes, ears, nose, skin, and tongue are translated into electrical nerve impulses and sent along to the brain. Far more flexible and powerful than any artificial intelligence system, the human brain processes a vast range of stimuli, organizes our thoughts and behavior, and coordinates our bodily functions. Thanks to increasingly refined imaging technologies, scientists can now observe the activity in various brain regions during dreams, intense emotions, and sensations of pain. An intricate network of nerves branching out from the brain and spinal cord coordinates our muscles, joints, and bones to produce both consciously and unconsciously directed movements, such as breathing, facial expressions, or sports activity. The musculoskeletal system gives the body form and stability, while also helping to protect the sensitive internal organs.

Alongside the nerves, other systems serve to ensure balance and harmony among different parts of the body. An arsenal of hormones and messenger substances coordinates the complex processes of growth, reproduction, and everyday metabolism. All of this requires energy, which is provided by our food and made available by the digestive system. Nutritional substances enter the bloodstream over the convoluted surface of the small intestine. The blood supplies all of the body's cells with energy from food, while also transporting life-giving oxygen, which we breathe in through our lungs. The blood and lymphatic vessels also help the body dispose of carbon dioxide and other waste products. Through the lungs, liver, kidneys, skin, and intestine, they are metabolized and excreted from the body.

Improved living conditions and the enormous advances in medical science have led to significant extensions of the human lifespan in the industrialized world. This progress has come at a price, however; as the body ages, its genetic and biochemical repair systems function less effectively, and tissues and organs begin to deteriorate. "Diseases of civilization," such as an unhealthy diet, obesity, lack of exercise, and substance abuse, also prevent the body from functioning optimally. Among the elderly in wealthy countries, numerous medical conditions, such as diabetes, arthritis, circulatory problems, and even cancer, are appearing more frequently.

Digestion is the process by which the body breaks down food into absorbable components. p. 236
Oral cavity
Esophagus
Liver
Stomach
Large intestine
Small intestine
Aorta
Heart
Femoral vein
The circulatory system is the largest transport system of the body. p. 234

The female sex organs include labia and clitoris (external) as well as vagina, uterus, fallopian tubes, and ovaries (internal).
Subclavian artery
Lung
Breast
Uterus
Fallopian tube
Ovary
Vagina

DEVELOPMENT

Human development begins when two gametes from the parents fuse to form the zygote. A new individual develops from the zygote over the course of about nine months. After puberty, humans are sexually mature and can reproduce. As the body ages, the efficiency and function of the organs gradually diminishes.

reproductive organs

Human reproduction is governed by a complex system of chemicals known as hormones. Patterns of behavior and specific male and female characteristics stimulate sexual arousal and lead to fertilization.

Human reproduction is ensured by the sex drive, which can be stimulated by, among other things, sex hormones, specific male or female sexual characteristics, and behavior.

Male reproductive process

Most adult men produce over 250 million sperm cells daily. These cells carry the genetic materials that can fertilize a human egg. In humans—and most other mammals—sperm is produced in testes located outside the body in the scrotum. Normal body temperature, 98.6°F (37°C), is too warm for sperm production, but the slightly cooler temperature in the scrotum is ideal. The sperm ripen in a long tube called the epididymis. When ejaculation occurs, the sperm exit the body by passing through the vas deferens and the urethra within the erect penis. The sperm cells are deposited inside the vagina during sex.

Female reproductive process

The female reproductive organs include two ovaries and fallopian tubes, the uterus, vagina, clitoris, and labia. Before birth, about 400,000 eggs develop in the ovaries. These eggs periodically mature after a woman reaches puberty. Multiple eggs, each covered with several layers of follicle cells for nourishment, develop during the female's monthly cycle, but generally only one grows to full size and is released into the fallopian tube during ovulation. If sperm enters the fallopian tube and fertilizes the egg, a pregnancy can occur. As the egg cell travels to the uterus, the remaining follicle forms the corpus luteum and secretes estrogen and progesterone. These prepare the walls of the uterus either for implantation of a fertilized egg or for breakdown and menstruation.

in focus

MENSTRUAL CYCLE

A woman's menstrual cycle lasts about 28 days. During the first five days of the cycle the lining of the uterus is shed via bleeding (menstruation) if fertilization has not occurred. The follicular phase lasts from the fifth to the 14th day, during which the pituitary gland releases a follicle-stimulating hormone (FSH) to ripen the follicle around an egg. Around the 14th day, the follicle produces estrogen to increase a luteinizing hormone (LH) that triggers the egg's release (ovulation). As the egg travels to the uterus, the woman's body temperature rises. From the 14th to the 28th day, the follicle left by the egg becomes the corpus luteum. It releases progesterone and prepares the lining of the uterus for either a fertilized egg or menstruation.

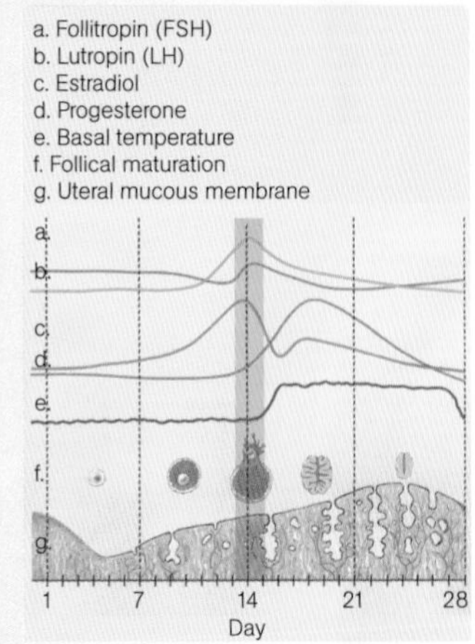

Monthly bleeding does not usually occur during pregnancy.

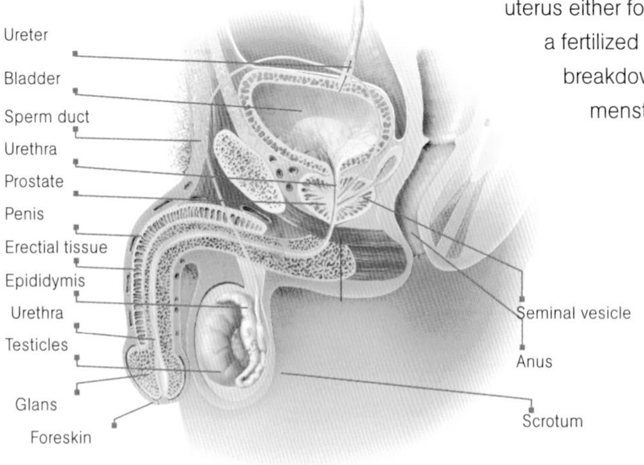

The semen consists of mature sperm cells (spermatoza) and seminal fluid secreted by the gonads, or sex glands.

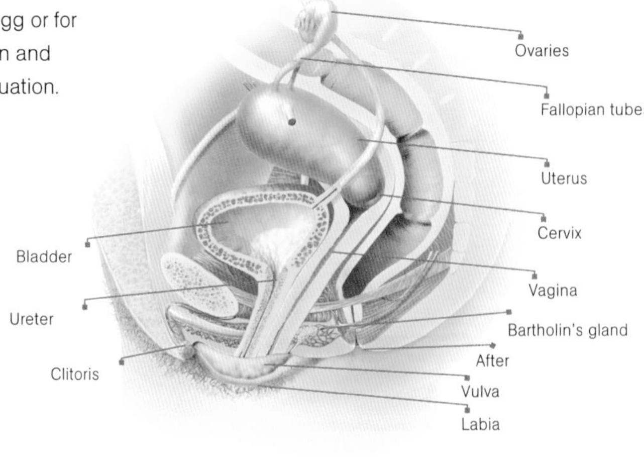

The uterus is a muscular organ. During pregnancy it expands enormously to provide sufficient space for the growing fetus.

see also: Human beings, pp. 220–225

reproduction

When humans reproduce, two differentiated germ cells (egg and sperm) merge. A new person develops from the fertilized egg—a process that takes approximately nine months.

Before the germ cells merge, the genetic material—which in normal cells occurs as a set of two—is reduced to a single set of chromosomes. This reduction happens during meiosis. During fertilization, one haploid germ cell from the mother combines with one from the father. Subsequently, the now fertilized egg cell (zygote) has a double set of genetic material: one set from the mother and one from the father.

IN VITRO FERTILIZATION

practice

If couples cannot conceive a child naturally, they may choose in vitro fertilization as an alternative. This process involves fertilizing a mature egg cell in a glass container. After a few days, the developing embryo is implanted into the uterus of the woman.

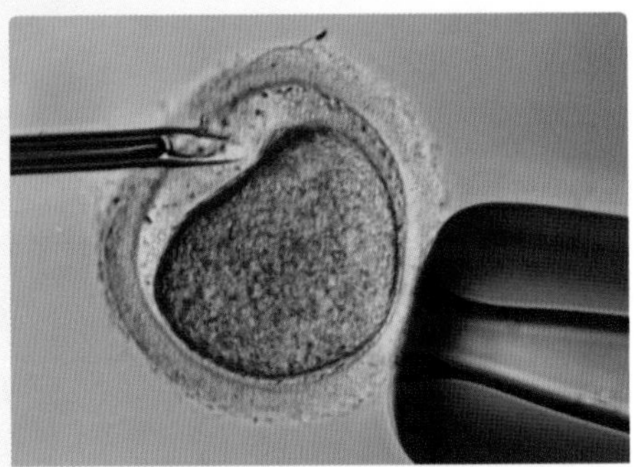

In vitro fertilization of a human egg cell.

Production of sperm

Males normally produce millions of sperm cells every day. But factors such as stress, illness, or toxins, for example nicotine, have a negative effect on this process. Male sperm cells develop in the seminiferous tubule of the testicles. They are then channeled into the epididymis for maturation and storage. Sperm cells measure about 60 micrometers and consist of a small head with a nucleus and a tail, or flagella, which propels the sperm forward.

Sperm cells are stored in a white and sticky liquid that is produced in the epididymis and take about three months to fully develop and mature. They are then ready to be released by muscle contractions leading to an ejaculation during an orgasm. Afterward, the sperm cells actively make their way through the uterus and into the fallopian tubes of the woman. Initially, there are several million sperm cells, but this number decreases rapidly and, finally, only one single sperm cell will be able to fertilize the female egg cell, unless nonidentical twins are conceived (see in focus).

Production of egg cells

Every four weeks, several egg cells mature within the female ovary. Usually one of them is released into the fallopian tubes during ovulation while the others degenerate. Egg cells are much larger than sperm cells. This is because they not only contain genetic material, but they also supply the nutrients necessary for an embryo to develop, if the egg is fertilized. Before fertilization, the mobile sperm cells are drawn to the egg cell by chemical attractants at increasing concentrations the closer to the egg cell they get. After fertilization, a fertilization membrane is formed and surrounds the egg to prevent further sperm cells from entry while the female and male nuclei merge. This concludes fertilization and begins the stage of embryonic development.

With a size of 150 μm, egg cells are the largest cells in the human body. They attract sperm cells with chemical attractants.

TWINS

in focus

If two eggs are released from the ovaries and fertilized by two different sperm cells, nonidentical twins (also called fraternal, or dizygotic, twins) develop. Just like siblings born from separate pregnancies, they each carry different genetic material. Sometimes, a fertilized egg divides into two embryos, which then develop separately. This results in monozygotic, or identical, twins. Each twin carries exactly the same genetic material. Around one of every 85 births is a twin pregnancy. Higher multiple births are even rarer.

Identical twins look very much the same while nonidentical twins have the same similarities and dissimilarities as normal siblings.

see also: Embryogenesis, p. 228

embryogenesis

A developing child goes through several phases as an embryo. Embryogenesis begins with the first cell divisions while the zygote travels to the uterus where the embryonic and fetal phases then occur.

Once an egg cell has been fertilized, it travels as a zygote through the fallopian tube into the uterus. At this time, the cell has already divided several times. The resulting cell cluster is similar to a mulberry *(Morus spp.)* and, hence, called a morula. The morula differentiates into an outer layer of cells which functions as nutritional tissue and an inner cell cluster from which the embryo develops. The inside of the morula fills with fluid, creating a germinal vesicle or blastula, which then plants itself into the uterine lining or endometrium. This process is called implantation or nidation. The placenta develops during this phase and creates a life-supporting connection between mother and child.

Embryonic phase

Within only a few days after implantation, two adjacent layers of cells develop from the interior cell cluster. These will later develop into three germ layers: ectoderm, endoderm, and mesoderm. The organ systems will emerge from these layers. The nervous system, the skin, and the mammary and sweat glands, for example, develop from the ectoderm, and the thyroid gland, liver, and pancreas develop from the endoderm. Starting in week four, the spine, heart, and eyes begin to appear. The body shape becomes visible after about eight weeks when the transition to the fetal phase happens.

basics

THE BABY'S DUE DATE is said to be 38 weeks after conception or 40 weeks after the first day of the last menstrual period. But 85 percent are born a few days earlier or later.

Fetal phase

The fetal phase is characterized by major growth of the unborn child. At the beginning of month five, the fetus weighs about 1.6 pounds (700 g) and noticeably moves around inside the amniotic sac. It can already move its fingers, toes, and mouth. The unborn child may be able to survive outside the womb from week 28. However, this very much depends on the individual progress made in developing the respiratory and nervous system. At birth, about 266 days following fertilization of the egg cell, a newborn weighs on average about 7.7 pounds (3,500 g) and measures almost 20 inches (about 50 cm).

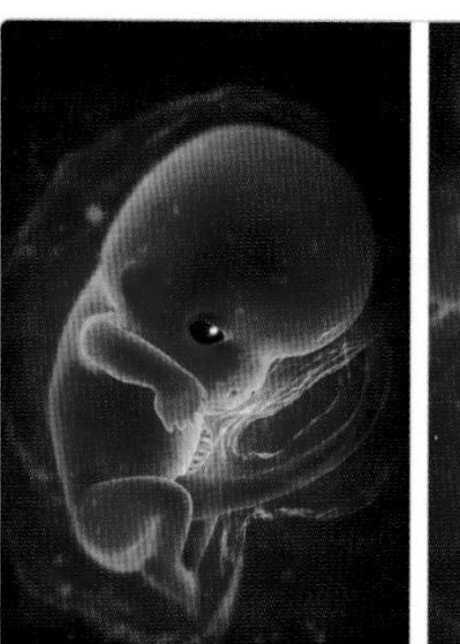
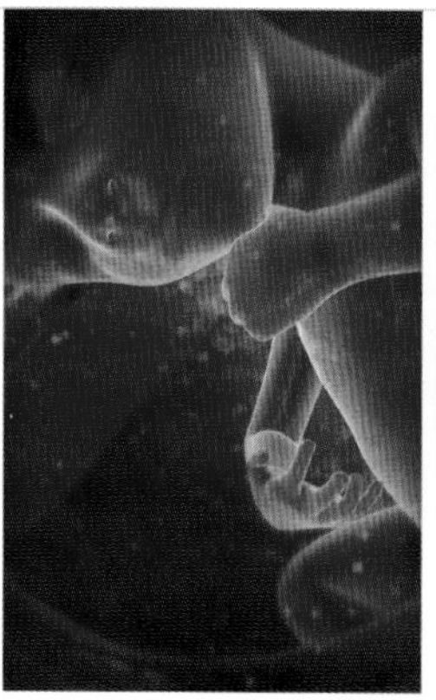
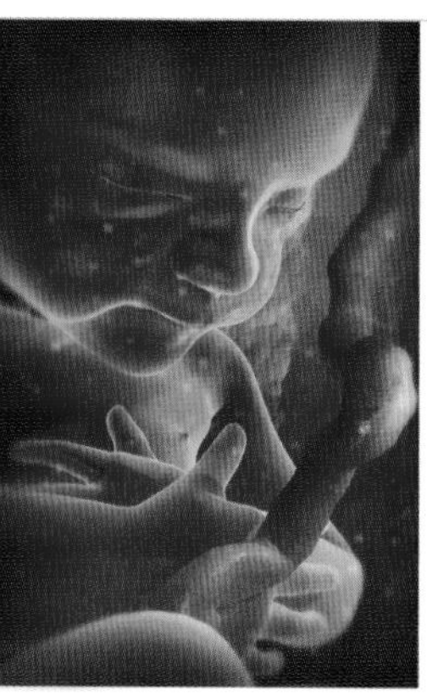
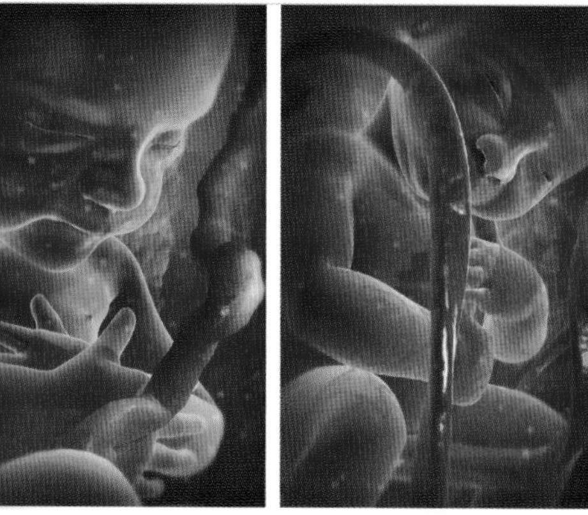

A human embryo takes about nine months to develop. During this time, the mother provides oxygen and essential nutrients through the umbilical cord. At birth, this cord is about 20 inches (50 cm) long and just under an inch (two cm) thick.

practice

AMNIOCENTESIS

An amniocentesis is a prenatal test often recommended for older pregnant women, whose children tend to experience a slightly elevated risk of chromosome defects. During this test, a sample of amniotic fluid is taken from the uterus and necrotic cells from the fluid are tested for potential defects. Before the amniotic sac is punctured through the abdominal wall with a local anesthesia, the precise position of the fetus is determined by ultrasound to prevent injury to the unborn child.

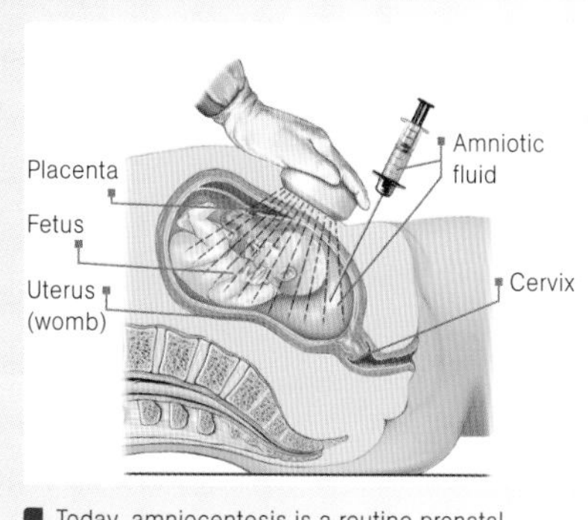

Today, amniocentesis is a routine prenatal diagnostic tool.

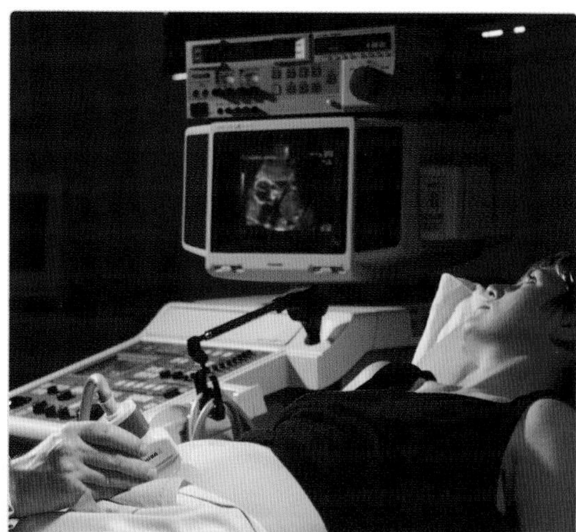

The size and position of the embryo as well as potential problems can be detected using an ultrasound scanner.

see also: Stem cell research, p. 257

puberty and aging

Puberty and aging are both phases of transition in the human life cycle: during puberty, the body reaches sexual maturity, while aging leads to a gradual loss of life functions.

■ Arguments between young people and their parents are common during puberty. Teenagers want to be treated like adults and have more responsibility, while parents want to continue to protect their children.

Girls will normally experience puberty, and the associated changes in their body, mind, and emotional state, at the age of ten to eleven. Boys can expect this to happen about a year later. During this phase in life, the sexual organs mature and begin to produce sperm or egg cells. Girls also experience their first menstruation. Secondary sex organs and characteristics develop (including breasts, pubic and underarm hair, facial hair, and voice changes). These physical changes are triggered by hormone signals from the pituitary gland (hypophysis), which tell the body to produce more sex hormones.

Changes become evident not only on the physical level, but also on an emotional level. Many adolescents look for new social ties outside the family; mood swings, the feeling of being misunderstood, and insecurity are all normal and common during this time of sexual orientation and beginning sex drive.

basics

IFE EXPECTANCY as more than dou- led in the last 150 ears in many in- lustrialized coun- ries. In the U.S., t is 77 years: in Japan, it is even igher than 80.

Aging

Aging is a biological process that begins at birth and gradually proceeds until death. Why people age is still unknown, and many researchers continue to study the phenomenon and test various theories of its possible causes. Scientists have found that cells can only divide a certain number of times before they die. This leads to the conclusion that cell death and aging are genetically predetermined or programmed. According to another theory, cells may also be damaged—and thus unable to properly reproduce and replicate themselves—by so-called free radicals, which are highly reactive oxygen compound ions. The body's defense mechanism is able to render these radicals harmless, but the limits to the effectiveness of this defense and repair system are thought to be genetically predefined. In addition, various other biochemical and physiological processes, psychological factors, and, of course, the individual lifestyle of every person significantly affect the very complicated process of aging. As an essential part of human society, aging also reflects cultural and societal conventions.

issues to solve

INCREASED LIFE EXPECTANCY and aging are causing problems in many Western countries. For example, the high percentage of older people in the total population has increased the proportion of people affected by dementia (estimated to be almost 25 million worldwide). Another unsolved problem is the increasing social isolation of the elderly.

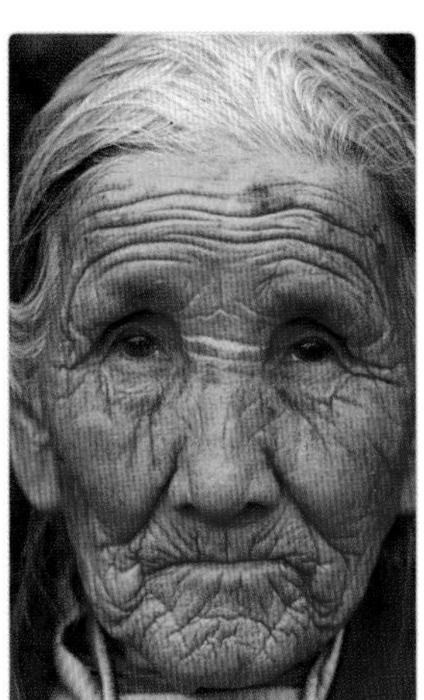

■ Skin elasticity decreases with age and wrinkles develop.

in focus

MENOPAUSE AND MIDLIFE CRISIS

Women between the age of 45 and 55 will normally experience a period of change called menopause (climacterium). This phase marks the end of a woman's fertile phase in life and frequently causes hormonal imbalances resulting in discomfort due to hot flashes and sleep disorders. Men can experience this phase as a so-called midlife crisis.

■ Potential signs of a midlife crisis are conflicts with the respective partner, a slowdown in career drive, or general unhappiness combined with mood swings.

tissues | organs | respiration | lungs

ANATOMY

The form, location, and structure of cells, tissues, organs, and whole body parts is nearly identical in all human beings. The spine and the bones give the body its stability, muscles and joints provide for its flexibility, and the blood circulation supplies all the tissues and internal organs with vital substances.

tissues and organs

There are specialized cells within the human body that organize themselves into tissues and organs. These cells are replaced continuously in a process that slows down with age.

Tissues are groups of cells with a unified structure and function held together by fibers or an outer body of cells. Organs consist of four basic types of cell tissues: epithelial, connective, nerve, and muscular.

Epithelial, connective, and nerve tissue

Epithelial tissues obstruct infectious organisms, protect against injury, and prevent loss of fluids. They are closely layered together. They cover body surfaces and organs and line cavities. Epithelial cells lining the lungs and intestines are in a single layer. Inside the nose, they are in multiple layers and combined with nasal hairs. The epithelial tissues of the skin can regrow rapidly to repair cuts and injuries.

Connective tissues support and bind other tissues. They can be liquid (blood), jellylike (tendons), and rigid (cartilage and bone). The loose connective (areolar) tissues are the most common in humans. These tissues, which connect the skin and organs and hold them in place, are highly flexible and tear-resistant. Dense connective tissue is found in the ligaments and tendons, which attach muscles to bones. Cartilage and bone constitute special connective tissues that support the body. Bone tissue is hardened by deposits of calcium phosphate, but is not brittle, and serves to protect the body as well as provide support. Nerve tissues in the spine, brain, and nervous system send electrical and chemical signals throughout the body. These tissues include nerve cells and the surrounding neuroglial cells.

Connective tissue cells typically lie relatively far apart and are embedded in a matrix.

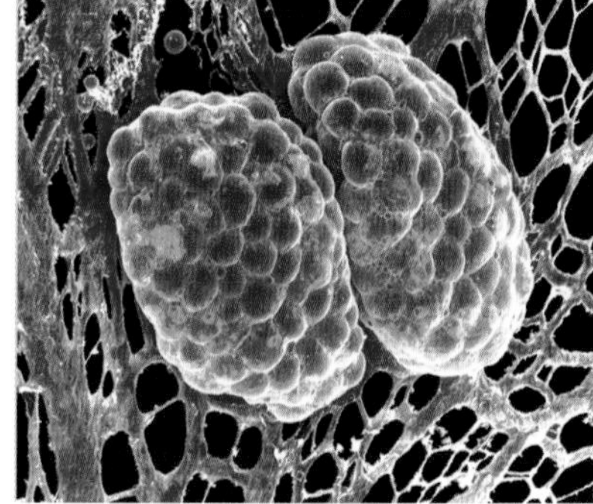

Fatty tissues are a form of connective tissue and are found in various parts of the body.

Muscle tissue

The elongated cells of muscle tissue can contract in response to a nerve impulse to cause movement. There are about 650 muscles in the body, making this the most common human tissue by volume. The muscles can be divided into three types: skeletal, smooth, and cardiac. Skeletal muscles are attached to bones by tendons and permit the body to move. Smooth muscles contract involuntarily and are found in the walls of the digestive tract, internal organs, and blood vessels. While they contract more slowly than skeletal muscles, the contraction persists for a longer period of time. Cardiac muscle, also involuntary, is resistant to fatigue allowing for continuous, rhythmic heart contractions.

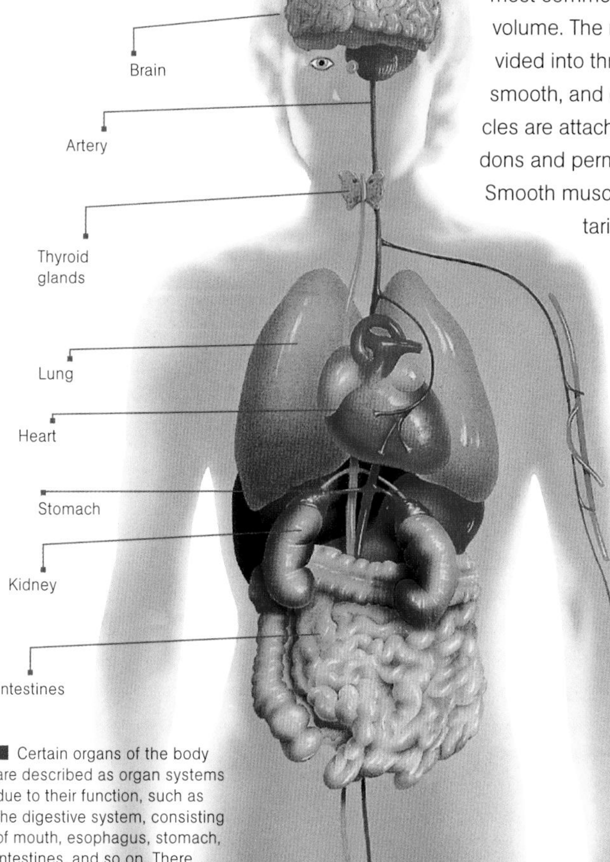

Certain organs of the body are described as organ systems due to their function, such as the digestive system, consisting of mouth, esophagus, stomach, intestines, and so on. There are also interactions and interdependencies between the organ systems.

see also: Human beings, p. 220–225

respiration and the lungs

The lungs are vital organs in the human body that provide oxygen required for most metabolic processes. Lungs also discharge carbon dioxide (CO_2), a metabolic waste product, which would otherwise poison the body.

The lungs are made up of two main components: the left and the right lungs. They are protected by the rib cage and flank the diaphragm and abdominal cavity. The right lung is divided into three lobes, whereas the left one is slightly smaller as it is divided only into two lobes. The heart is located between the lungs. This is where the large bronchial tubes (branches of the wind pipe or trachea) and blood vessels enter the lungs. The lungs are lined by a layer of skin called costal pleura, which is always covered by a thin film of liquid. This allows for movement of the lungs while they still fit closely against the chest wall. While breathing, the lungs are consequently able to slide along the walls of the chest with minimal resistance and obstruction.

Lungs are one of the human body's largest and most vital organs.

Gas exchange

Inhaled air travels along the wind pipe into the lungs. The wind pipe is a flexible tube that is kept open by the U-shaped cartilage of the trachea. In the upper part of the chest this tube splits into two bronchial tubes, which branch out further inside the lungs. They get smaller as they split so that air can reach all parts of the lungs. The bronchiole end in small sacs that are arranged similarly to grapes, called alveoli. There are perhaps several hundred million alveoli, and they are surrounded by a fine mesh of blood vessels, where gas exchange occurs. This is possible due to the extremely thin gas-permeable wall separating the inhaled air inside the alveoli and the blood inside the vessels. The oxygen concentration inside the alveoli is high while it is low in the blood, which ensures that oxygen can diffuse from the alveoli into the capillaries. The exchange of carbon dioxide functions in the same way: the carbon dioxide concentration in the blood is high and low in the alveoli. This gas exchange is effective due to the large, approximately 1,080-square-foot (100-m^2) total surface of the human lungs.

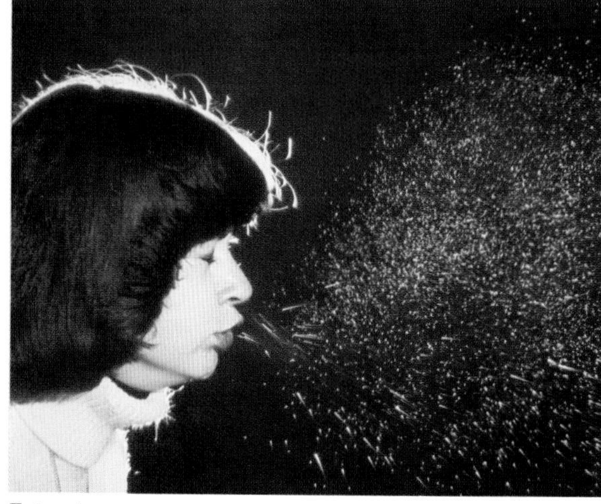

Sneezing—a sudden, forceful, and involuntary expulsion of air—is caused by an irritation of the mucous membranes of the nose or throat.

basics

SMOKER'S LUNG is a term for a respiratory disease mostly affecting smokers. One effect of this disease is a disrupted gas exchange of the breathing air and the blood due to the tar in tobacco.

practice

RESPIRATION AND EXERCISE

After two minutes of exercise the body responds by supplying muscles with oxygen. It adjusts the frequency and depth of breaths according to the activity level to ensure sufficient oxygen intake and carbon dioxide output. At rest adults breathe in and out about 16 to 18 times a minute, which passes about two gallons (10 l) of air through the lungs. During exercise, however, this amount can be much higher. In this case, up to 15 gallons (60 l) of air pass through the lungs per minute since a rising energy demand means a rising oxygen demand.

Proper exhalation during physical exercise, such as swimming, helps to increase the body's maximum lung capacity.

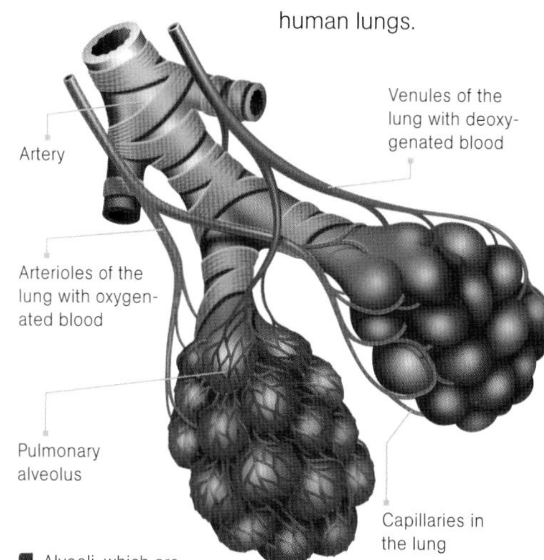

Alveoli, which are particular to mammalian lungs, appear at the end of bronchial tubes and are the site of gas exchanges in the lungs.

skeletal system: bones

Humans have a bony skeleton with movable joints. Several key adaptations provide strength and flexibility and enable human beings to walk upright.

The adult human skeleton has about 206 individual bones and several important functions. It supports the body and protects the inner organs. For example, the skull encases the brain and protects it from bumps and minor injuries; the rib cage protects the heart and lungs. Using the joints in the skeleton, muscles attached to individual bones and cartilage can contract and relax to move the body. The main body axis, or

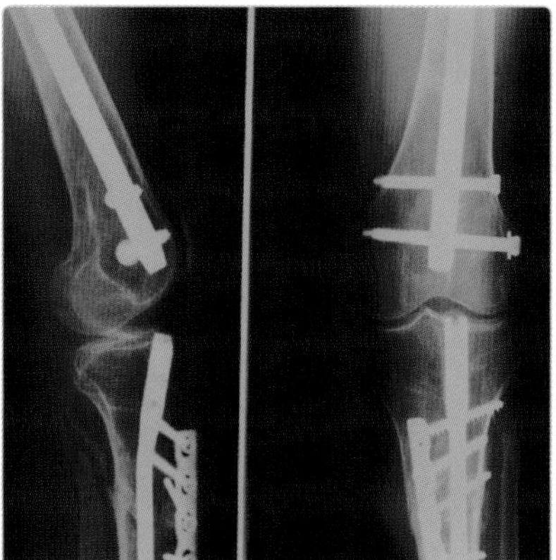

The x-rays show fractures that have been treated by the insertion of metal plates and screws.

axial skeleton, consists of the skull and vertebral column. Attached to the vertebral column are the arm and leg bones (the appendicular skeleton) and the shoulder and hip (pelvic) girdles. The vertebral column, with its S-curve structure, can absorb a great deal of sudden and heavy force while providing the support required for an upright posture.

Vertebral column

The human backbone, or vertebral column, consists of 33 vertebrae, each separated by spinal discs. The discs are made of cartilage and have a jellylike center allowing them to absorb shocks during walking, running, jumping, and other physical activity. The spinal cord—central axis of the nervous system—runs through a canal within the vertebral column. The spinal cord distributes signals between the brain and the individual organs in the body. An injury to the spinal cord, such as a broken vertebra or a ruptured disc, can damage the spinal cord and cause temporary or permanent paralysis.

Bones

The bones of humans and other mammals consist primarily of calcium phosphate. They are covered with a thin layer of connective tissue called the periosteum and have a compact outer layer and spongy inner scaffolding. Bone marrow is distributed on the inner scaffolding and laced with blood vessels. Red and white blood cells and platelets are formed in the marrow. The long bones in the thighs, shins, and arms differ from the flat bones that form the skull and ribs. However, all bones constantly reconstruct themselves. Thus, they can repair themselves after a break.

The human skeleton contains four main types of bones: long, short, flat, and irregular.

milestones

X-RAY TECHNOLOGY

While experimenting with cathode rays, W. C. Röntgen discovered an unknown form of radiation and named them x-rays. This revolutionized medical diagnostics and its potential in human medicine. Today, x-rays are a standard practice in medical diagnostics. Mammography and computerized tomography (CT) are specialized methods. Other scientific areas also take advantage of x-ray technology. Archaeology and art history, for example, rely on this technique to analyze specimens.

Wilhelm Conrad Röntgen (1845–1923) discovered x-rays and won the 1901 Nobel Prize in physics.

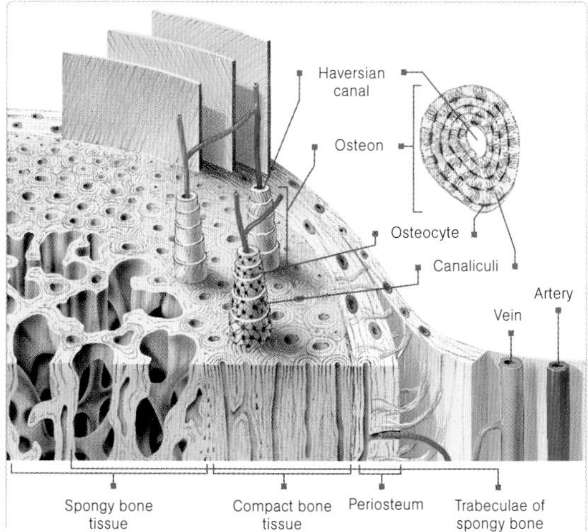

Bones are made from an active matrix of protein threads, which is continually replaced and rebuilt throughout the person's lifetime.

joints, muscles, and tendons

Coordinated movements are made possible by joints and flexible tendons connecting the rigid skeletal bones of the musculoskeletal system. Muscles are essential for the movement of limbs as well as organ functions.

Joints are a flexible connection between two bones. Normally, the end of a bone fits like a ball into the bowl-shaped depression of another bone or socket. Both joint surfaces are lined with soft and elastic cartilage that provides cushioning for the joint. The joint is surrounded by a capsule of connective tissue. The articular capsule forms a closed joint cavity filled with a viscous joint liquid called synovial fluid, which serves as lubricant. There are different types of joints characterized by their shape and direction of movement. These are mainly hinge joints like the elbow, ball and socket joints like the hip, and saddle type joints like the basal thumb joint.

Muscles and tendons

Muscles produce power for active movements by contracting. They are also responsible for maintaining body tension for an upright position, for example, while sitting or standing. Language would not be possible without muscles and even a smile requires more than 200 small muscles. Each muscle is made up of numerous bundles of individual muscle fibers embedded into the connective tissue. Striated muscle such as skeletal muscle or cardiac muscle is characterized by a high degree of alignment of contractile fibers. This muscle type is able to contract and expand much more rapidly and with greater force compared to smooth muscles (responsible for functions such as intestinal action or blood vessel tension), which are slower and show little fatigue. In contrast to skeletal muscle, which can be controlled by will, this muscle type cannot be influenced voluntarily as it is regulated by hormones rather than direct nervous impulses.

Tendons are firm and tight bundles of connective tissue (collagen or elastin fibers) that connect parts of the skeleton, provide elasticity, and transmit muscle force onto the skeletal bones. Tendons, despite their resistance to pulling force, are prone to painful sports injuries; they cannot endure gravitational forces or pressure for an extended period of time. Sudden muscle contractions or extreme loads may result in a full rupture of the tendon.

basics

A BONE FRACTURE requires immediate medical attention. The bone is straightened and fixed into position using plaster or traction bandages. Fractured bones take weeks to heal, sometimes up to a year.

Peroneus brevis muscle
Superior extensor retinaculum
Inferior extensor retinaculum
Achilles tendon
Calcaneus bone
Extensor digitorum longus tendons
Phalange bones

■ The foot is a complex system of bones, joints, muscles, and tendons.

■ Ruptured tendons (e.g., the Achilles tendon) are among the most common injuries during exercise.

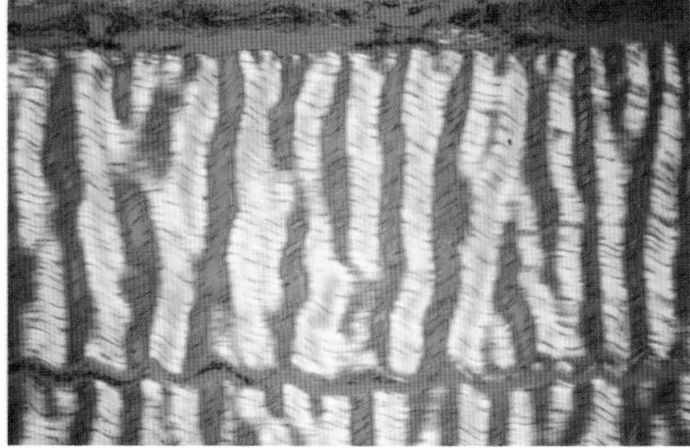

■ The longitudinal cut of the tendon shows the fibrous connective tissue that is capable of withstanding tension.

practice

PROSTHESES

Sometimes people lose limbs in serious accidents. Joints may be damaged through overuse so that they need to be replaced. In these cases, replica body parts are implanted into the body or attached externally as replacements (prostheses). The most obvious are artificial limbs: prostheses of arms or legs. Today, arm and leg prostheses have been developed that are controlled by microprocessors and allow complex movements and even sports.

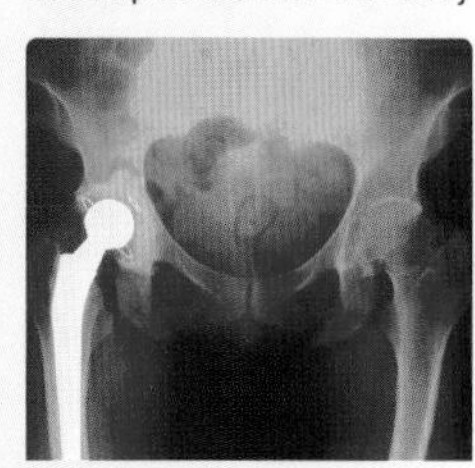

■ Endoprostheses are not externally visible. Examples are artificial hip joint (pictured) or knee joint replacement.

heart and circulation

The heart is the organ that distributes blood around the body through vessels such as arteries, veins, and capillaries.

The coronary blood vessels supply the heart muscle with oxygen-rich blood.

The human body's blood is circulated by one central organ: the heart. A healthy heart sends blood, oxygen, and various nutrients contained in the blood to all the organs and tissues. Made up almost entirely of muscle, an adult heart only weighs about nine ounces (300 g). The human heart consists of two halves that are divided by a wall. Each half is divided into subchambers: the upper atria and the lower ventricles. Sinus nodes, which act as the heart's own nervous system, cause the heart's chambers to contract and relax in a steady rhythm. Contractions pump the blood through the blood vessels that carry it to every part of the body.

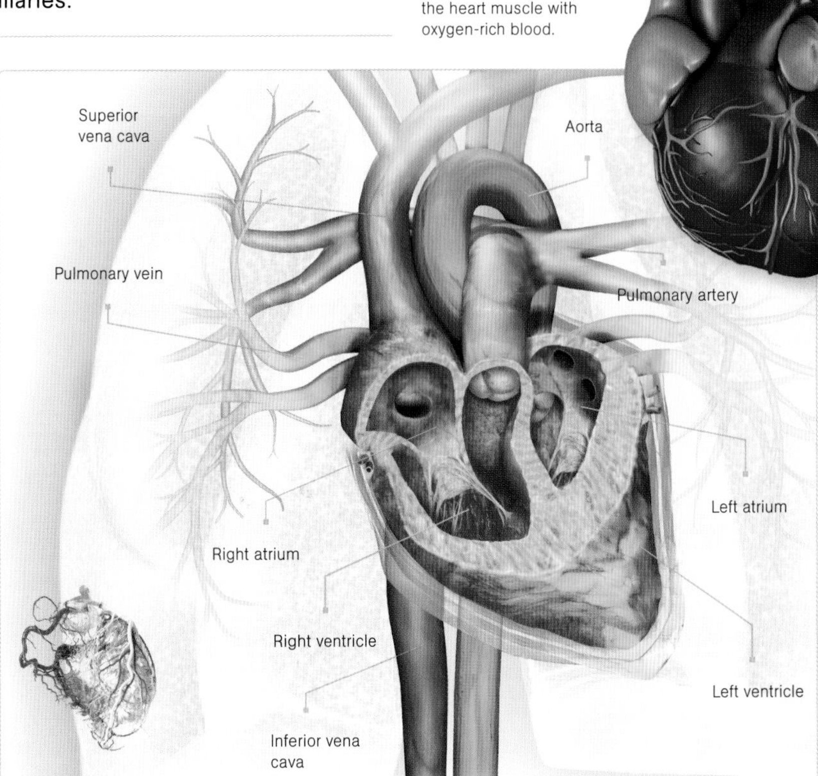

It is impossible for human beings and higher developed animals to live without a heart. It is one of the first organs created during embryonic development.

Blood circulation

Because all the blood in the human body is contained within vessels, the human circulatory system is called a "closed" system. The heart's left side pumps oxygen-rich blood throughout the body in arteries. As the oxygenated red blood cells circulate through the body's tissues, they release oxygen and absorb carbon dioxide. When the blood returns through the veins to the heart, the heart's right side pumps it to the lungs to be reoxygenated, after which the blood returns to the left side of the heart, continuing the cycle. Although blood vessels are not the cause of circulation, the layers of muscle that comprise them regulate the quantity of blood flow.

Diseases

Heart attacks, strokes, and other diseases of the cardiovascular system are common causes of death in industrialized parts of the world. Both heart attacks and strokes are caused by narrowed or blocked blood vessels clotted by blood cells or protein fibers. The heart can sustain irreversible damage if it is deprived of blood for more than 20 minutes—as can happen during a heart attack—much as nervous tissue may die when blood vessels in the brain are obstructed.

in focus

HEART TRANSPLANTATION

In 1967, Christiaan Barnard and his team performed the first successful heart transplant in South Africa. During the five-hour operation, the patient received a donor heart. In order to prevent the donor heart from being rejected as a foreign organ, his immune system was suppressed. The patient succumbed to pneumonia 18 days following the surgery as a consequence.

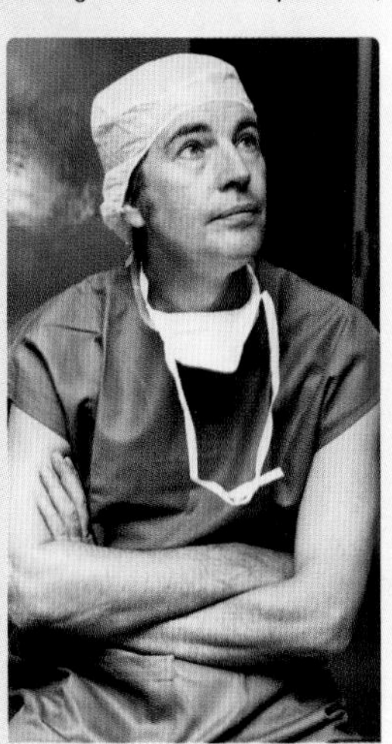

The accomplishments of Christiaan Barnard (1922–2001) in the field of cardiac surgery represent a significant breakthrough in life-extending surgery.

practice

PACEMAKER

Cardiac arrhythmia, a condition that causes the heart to beat erratically, sometimes requires more than just medication. A pacemaker, a small, battery-operated device designed to stabilize the heart's beat, can be implanted in the patient's chest and an electrode carrying an electric pulse from the pacemaker's battery can then be pushed through a vein into the heart. Today, thousands of people living in industrialized nations rely on pacemakers to regulate their heartbeats.

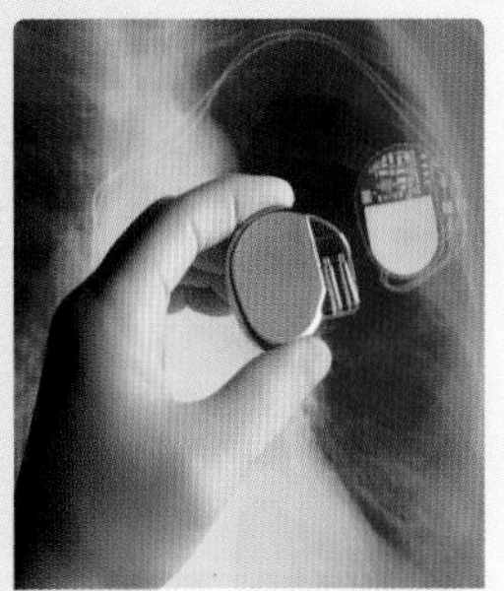

The lifespan of modern pacemakers lies between 5 and 12 years; on average 8 years.

see also: Human beings, pp. 220–225

blood

Blood is a bodily fluid, red in color, which circulates in the blood vessels and carries out important functions within the body. These include oxygen transport, heat regulation, signal transmission, and defense against illness, among other tasks.

In humans, blood cells make up about 45 percent of the blood by volume. The rest consists of plasma—a watery solution that makes the blood fluid—in which various substances are dissolved, primarily proteins but also electrolytes, carbohydrates, fats, and hormones. There are three different types of blood cells: the red blood cells (erythrocytes), white blood cells (leukocytes), and blood platelets (thrombocytes).

Red blood cells contain the substance hemoglobin, which gives blood its color. The hemoglobin allows the blood to transport oxygen from the lungs to tissues throughout the body and return carbon dioxide from the tissues to the lungs.

Since red blood cells are flexible and can change their shape as needed, they can pass through even the smallest blood vessels. The job of the white blood cells is to identify and destroy harmful organisms.

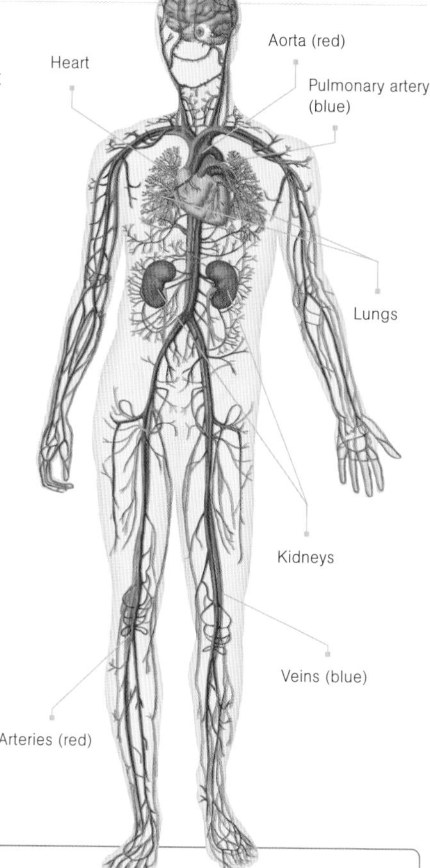

The main components of the human circulatory system are the heart, blood, and blood vessels.

BLOOD TYPES

practice

An individual's blood type, or blood group, is based on the characteristics of his or her red blood cells. Red blood cells can display specific structures, called antigens, on their surfaces. Various systems are used to divide individuals into blood groups. The best-known is the ABO system, which distinguishes four basic blood types: A, B, AB, and O. This information is especially vital for patients undergoing blood transfusions, since only blood from a donor with the same type as the recipient—except for Type O, which is universal—can be used. Otherwise, the blood cells will clump together, and life-threatening complications can arise.

Organ recipient	Organ donor: O	A	B	AB
O	♡	✕	✕	✕
A	♡	♡	✕	✕
B	♡	✕	♡	✕
AB	♡	♡	♡	♡

In 1901, the Viennese medical scientist Karl Landsteiner discovered the ABO blood system during his experiments.

basics

BLOOD VOLUME Adult humans have some 4–6 quarts or liters of blood, a total of 6–8% of their body weight. Men often have about one liter more than women.

BLOOD LOSS The loss of 15 to 20% of blood volume—about one quart or liter—can be life-threatening.

BLOOD TYPE DISTRIBUTION The proportion of the various blood types varies greatly in different parts of the world. For instance, in the United States, the majority of people have Type O blood, whereas Type B is more common in some parts of Asia, and Type A in much of Europe.

White blood cells can actively leave the bloodstream, for instance, in order to combat bacteria in bodily tissues and fight infections.

Blood platelets play a key role in blood clotting (hemostasis). When a blood vessel is damaged, the platelets are activated immediately. They help close the wound and limit blood loss.

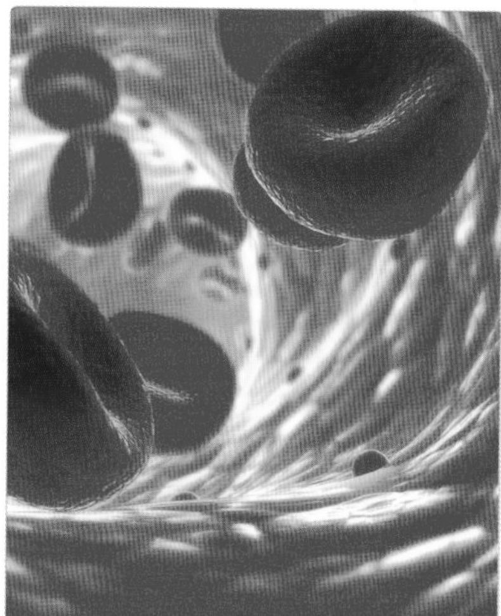

Red blood cells are responsible for transporting oxygen and carbon dioxide within the body.

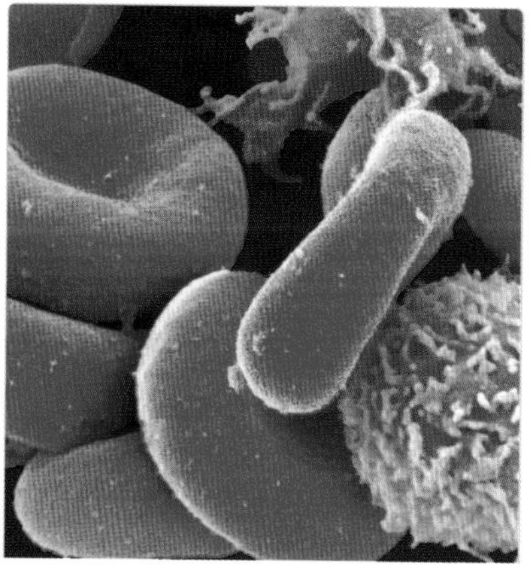

Blood cells are generated in the bone marrow.

nutrition | digestion | water | hormones

METABOLISM AND HORMONES

The human body regularly takes in nutritional material from outside, transports it and processes it, and eliminates waste and foreign material. Hormones play an important role in most metabolic processes.

nutrition

Plants produce their own nutrients; however, humans do not. Instead, we must consume food and liquids to obtain the energy and water we need to sustain our bodies.

Food is required as an energy source to fuel processes such as moving, thinking, blood distribution, and breathing. It also sustains the growth and repair of cells and tissues. Food energy is used by the metabolism. This refers to the processes in which an organism takes in, transports, and chemically transforms nutrients. It also includes the disposal of waste products.

Human metabolism requires carbohydrates, fat, protein, vitamins, and minerals. These must be taken in as part of the diet.

What happens to our food?

A bite of food follows a journey that starts in the mouth and continues through the esophagus, stomach, small and large intestines, and ends at the rectum and anus. During this process, each organ of the digestive system fulfills very specific functions. Other glands and organs, such as the salivary glands, pancreas, gall bladder, and liver, also contribute to digestion.

Path through mouth, stomach, and small intestine

Inside the mouth, the teeth break up food mechanically and combine it with saliva from the salivary glands. Enzymes in saliva break down carbohydrates into smaller sugar molecules. Swallowing pushes the food mixture down the esophagus into the stomach. There it combines with digestive fluids, including hydrochloric acid. The stomach mixes the food with a churning motion to break down cell structures. Enzymes also split proteins into smaller molecules. Two to six hours after a meal, the stomach empties into the small intestine, which is about 16 feet (five m) long, and where most digestion takes place. Bile from the gall bladder dissolves fats, while enzymes from the pancreas digest fats, proteins, and carbohydrates.

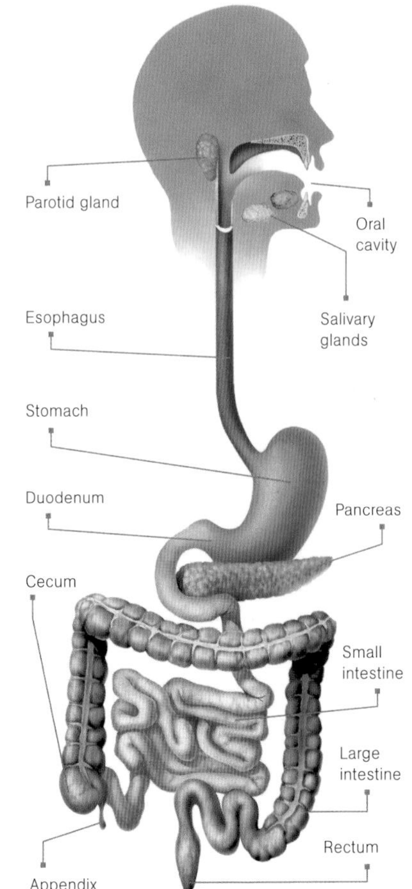

The digestion of food and nutrients takes place in the digestion track, assisted by various enzymes.

Diabetes

Diabetes is a disease that weakens the cells' ability to use sugar by disrupting the absorption of sugar from the bloodstream—a process carried out by insulin produced in the pancreas. Diabetes can be present from birth, result from poor nutrition, or come from metabolic problems related to aging.

in focus

CHEWING FOOD

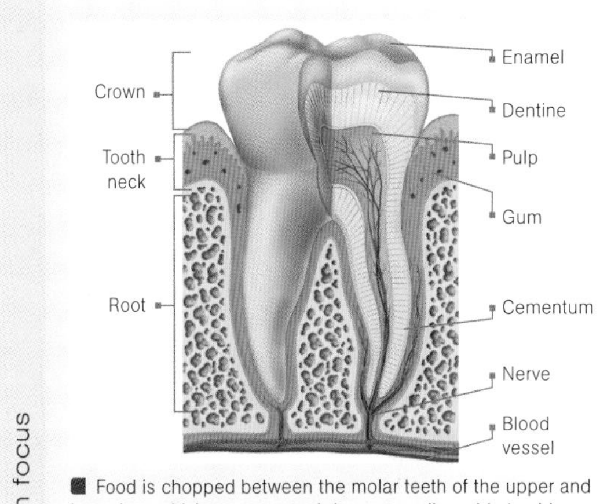

Food is chopped between the molar teeth of the upper and lower jaw, which move up and down as well as side to side.

During the chewing process, food comes into contact with saliva and the enzymes in the saliva begin the digestion process. A tooth consists of a tooth crown, neck, and root. The crown is coated with tooth enamel, the toughest and most durable substance in the entire human body. Once damaged or decayed, tooth enamel cannot be regenerated. The pulp is located inside of the tooth and is extremely sensitive to pain due to its high concentration of blood vessels and nerves.

see also: Human beings, pp. 220–225

digestion

The utilization of food mainly takes place in the small intestine. The required nutrients are then distributed through the body. Nutrition should be as varied as possible in order to assure an optimal supply of all "fuels."

After nutrients from the stomach have entered the small intestine, the breakdown of relatively large food molecules into smaller units continues. This takes place principally under the influence of various enzymes that break up proteins, fats, nucleic acids, and large sugar molecules such as starch into smaller building blocks. These nutrients leave the small intestine and are resorbed into the bloodstream, which enables the body to utilize them as energy suppliers for body functions. To facilitate this the wall of the small intestine consists of numerous intestinal villi, which enlarge the surface area of the small intestine to about 2,152 square feet (200 m^2).

■ *E. coli* is a bacterium that lives in the intestinal tracts of humans and animals. Though usually not harmful some strains are deadly.

In each intestinal villus there is a network of tiny blood vessels as well as a central lymph vessel. Amino acids and small sugar molecules enter the small blood vessels, which join into a portal vein that transports the nutrients from the intestine to the liver. Fats are broken down into glycerin and fatty acids, which are surrounded by special proteins. As tiny balls they then enter the lymph vessel from where they than reach the bloodstream.

Function of large intestine

The 4.2-foot (1.3-m)-long large intestine follows the small intestine. The remaining material yet to be digested from the digestive tract moves through the large intestine in about 12 to 24 hours. During this process the material becomes firmer, since water given off into the digestive tract, as a base substance in various digestive juices, is now being removed again. In total about 99 percent of the water added during the digestive process is resorbed by the small and large intestine. The rest is excreted as feces.

Essential nutrients

In addition to the "fuels" (such as fats, carbohydrates, proteins) required by the human body for conversion into energy and as raw material for many biosynthetic processes, there are also a number of important nutrients that must be provided in completed form. These include some amino acids, vitamins, trace elements (for example, potassium and phosphorous), and fatty acids, such as linolenic acid. If these nutrients are not provided to the body in sufficient quantities, malnutrition can occur, which can cause dangerous physical and mental effects.

■ The food guide pyramid is a nutritional guide that is used around the globe to illustrate the components of a healthy diet.

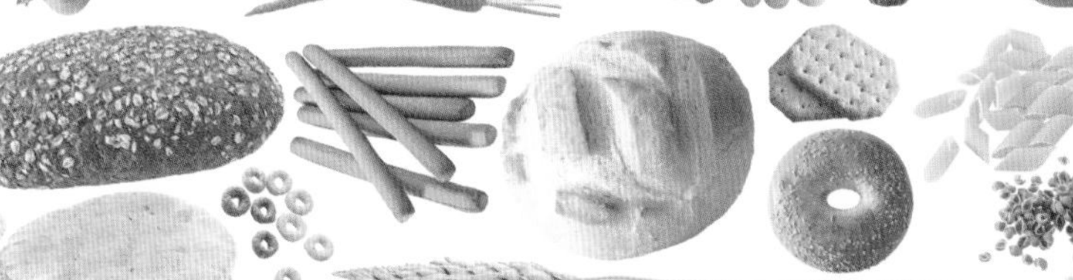

practice

PROBIOTIC FOODS

Probiotic yogurts belong to a new generation of food products, sometimes called "functional food." These products are supplemented with ingredients that are claimed to have health benefits. Probiotics are microorganisms that improve the natural intestinal flora and strengthen the immune system. Probiotics should not, however, be confused with medications.

■ The most popular "functional foods" are yogurts with probiotic bacterial cultures.

basics

DIABETES affects over 180 million people worldwide. It disrupts the absorption of sugar from the bloodstream—a process normally carried out by insulin produced in the pancreas. Diabetes can be controlled by diet, oral medication, or self-administered injections of artificial insulin.

water and the kidneys

Water in the body serves as a means of transportation and a dissolving agent, and it helps to regulate body temperature. The kidneys play a major role in maintaining the water balance in the body and removing toxins.

The human body contains between 65 and 75 percent water, depending on the sex and age of the person. The amount of water taken in, produced, and lost remains constant (water balance). Liquids are mainly taken in by drinking and, to a certain extent, by food intake; water is lost through the kidneys (urination), the sweat glands, and a small amount through respiration. If water is lost at a higher rate, for example, through heavy sweating, more water must be drunk in order to prevent dehydration. The consequences of dehydration can be serious and include fainting and dizziness, vision loss, extreme sleepiness, decreased urine output, and if left untreated, death. Symptoms usually begin if approximately 2 percent of the body's water is lost; losses of over 15 percent are usually fatal.

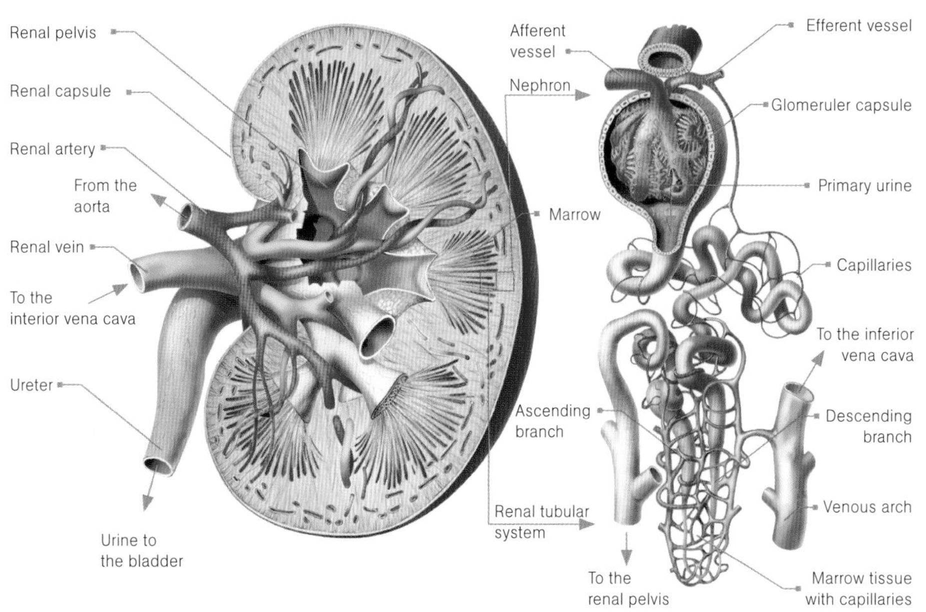

The human kidneys consist of functional subunits called nephrons.

SPORTS AND HYDRATION

practice

It is essential for everybody to drink enough and stay well hydrated. This is especially important during exercise, when the body loses a lot of liquids through sweating. If the lost water is not replaced in time, the blood viscosity, or blood thickness, is reduced. This means that the blood cannot flow quickly enough to provide the muscle cells with enough oxygen and nutrients. This may result in dizziness, vomiting, and cramping of the muscles.

Water lost through heavy sweating must be quickly replaced to keep the body functioning.

Kidneys

The kidneys are important organs that regulate the water cycle and remove waste products from the human body. These waste products usually reach the kidneys via the blood flow through the liver. After they pass the kidneys, they leave the body through urination.

Kidneys come in pairs and look like approximately four-inch (ten-cm)-long beans. They are located alongside the 12th vertebra of the chest, on either side of the spine. The organs are composed of a renal capsule that surrounds the renal cortex and medulla. The medulla contains 16 to 20 renal pyramids, whose tips point inward and have many openings for excreting urine. The opposite ends of the pyramids extend from the medulla into the renal cortex and contain about a million nephrons, which are responsible for urine production. Each nephron consists of several renal corpuscles that are surrounded by capillary blood vessels and are connected to a convoluted renal tubule that forms a U-shape in the medulla. The renal corpuscles take in waste products from the blood and transfer these into the tubules to produce primary urine. While passing through the tubules, essential substances and, moreover, most of the liquid is taken out again, which results in concentrated urine being excreted via the ureter.

issues to solve

KIDNEY TRANSPLANTS are relatively successful in case of kidney failure. The greatest problem is the limited number of people willing to donate their organs after they have died.

Drinking tea can contribute to our daily requirement to drink approximately 3.2 pints (1.5 l) of fluids.

hormones

Hormones are information carriers in the human body. Instead of the electrical impulses used by the nervous system, the hormone system uses chemical messenger substances and is, therefore, much slower.

Hormones are signaling molecules, responsible for regulating bodily functions, sending messages, and acting on organs and tissues. Moreover, hormones regulate functions as disparate as mood, growth, reproduction, and metabolism. The majority of hormones are generated in the endocrine glands, such as the thyroid gland, hypophysis (pituitary gland), or the pancreas. In addition, some tissue cells can also produce hormones, such as the cells of the stomach lining. The glands usually release hormones into the bloodstream, where they travel through the body to act as required. The hypophysis is of special importance, as the hormones of the hypophysis regulate the activities of other endocrine glands. Hormones may be a range of molecular types, including peptide hormones such as insulin, steroid hormones such as estrogen, and hormones composed of amino acids such as adrenaline.

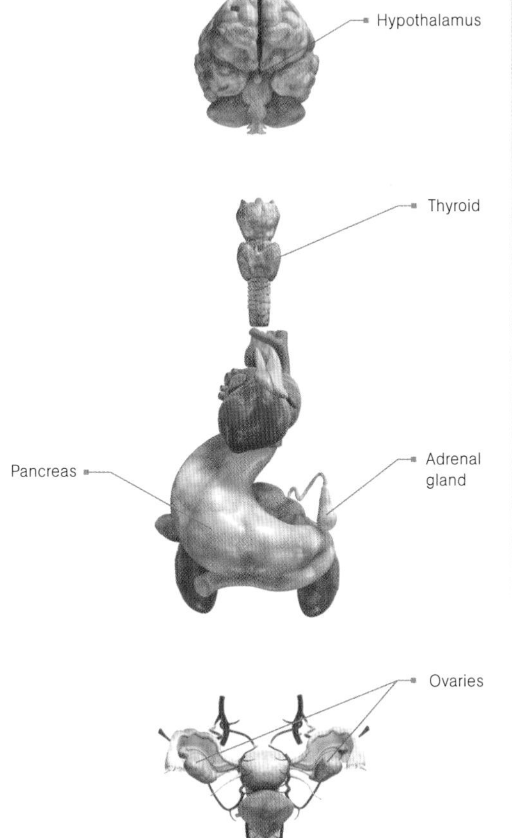

Hormones produced and secreted by glands like the pancreas and the hypothalamus are involved in almost all processes in the body.

Interaction of the hormones and nervous system

The hormone system and the nervous system are responsible for relaying information within the body and triggering reactions. The nervous system is mainly responsible for quick reactions, with transfer speeds of up to 268 miles per hour (120 m/s). The hormone system is designed for more long-term effects, with transfer speeds of about 0.2 inch per second (five mm/s). Some nerve cells can produce hormones. Humans have such neurosecretory cells in the hypothalamus, where hormonal and neural mechanisms are regulated and coordinated.

How hormones work

The way hormones work is best explained by an example: regulation of the blood sugar level. After eating a meal rich in carbohydrates, insulin is released by the pancreas in response to high blood sugar levels. Insulin causes the liver and muscles to take in glucose and store it as glycogen, which then lowers the blood sugar level. Glucagon, another hormone released by the pancreas, has the opposite effect. Released in response to low blood sugar levels, it increases the blood sugar level by accelerating breakdown of the sugar glycogen to glucose in the liver and muscles, as occurs during heavy exercise.

Endorphines are produced and released by the body after physical exertion, excitement, and orgasm. They are sometimes called "natural pain killers" and result in the phenomenon known as "runner's high."

basics

HUMAN GROWTH HORMONES are sometimes used for doping to increase an athlete's performance. This practice is dangerous as well as unethical: it may result in death due to an increased risk of heart attacks, diabetes, and perhaps also cancer.

practice

HORMONAL BIRTH CONTROL

When using hormonal birth control pills, the hypophysis is tricked into thinking that the woman taking the pills is pregnant. This is achieved through a combination of various hormones, primarily estrogen and progesterone. The hypophysis will then reduce production of the luteinizing hormone (LH), which usually triggers ovulation during the normal female cycle. If ovulation does not occur, pregnancy is impossible.

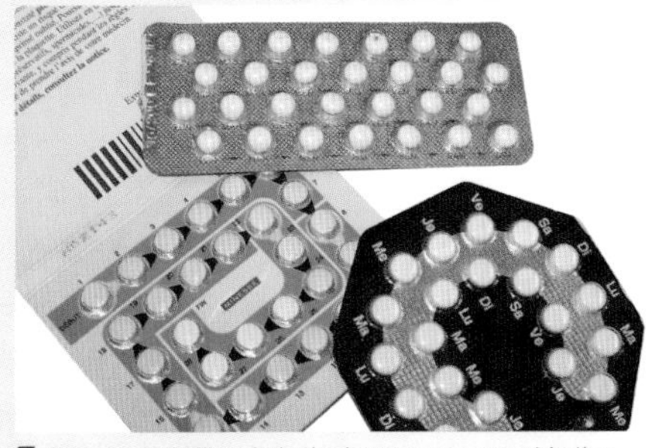

Birth control pills mimic the hormones present in the body during pregnancy.

see also: Puberty and aging, p. 229

nerve cells and signal transfer

Stimuli picked up from the environment are transferred to the brain through nerve cells. This information is then processed by the brain and, if required, a reaction (for example, a muscle contraction) is triggered via nerves.

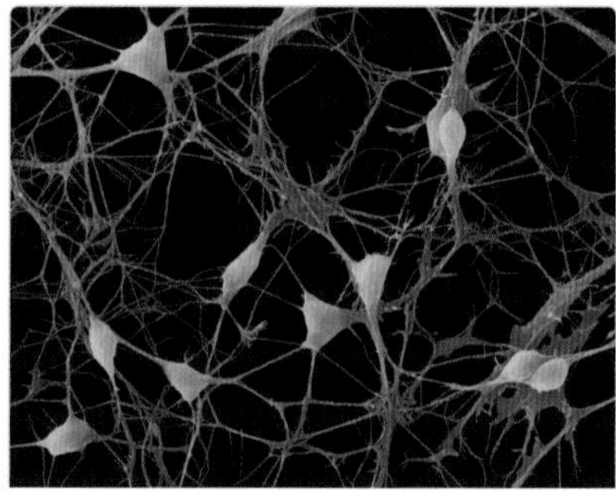

Cerebral cortex neurons: Neurons have a large cell body with several long, extending projections.

nerve cells | signal transfer | brain | spinal cord

NERVOUS SYSTEM

The human nervous system does not only control the vital processes of the body, it is also a mediator to the world around us, as it can pick up, evaluate, and store stimuli. It can be subdivided into the central nervous system (CNS), including spinal cord and brain, and the peripheral nervous system (PNS) with all nerve fibers of the body.

The functional and structural units of the nervous system are the nerve cells (neurons). The human body contains about 100 billion of these. They consist of a cell body (soma), where several short, usually finely branched processes, called dendrites, emerge. There is also a long, thin projection, referred to as axon or neuraxon. The latter functions in impulse transmission to other cells, and it can be surrounded by a myelin sheath that provides insulation and nutrition. The neurons are in contact with other cells via synapses. However, there is no direct contact, because the cells are kept apart by a two-nanometer-wide gap (synaptic gap). This prevents the action potential from a stimulated cell from crossing over directly to a cell that is at rest. Chemical carrier substances (transmitters) are used for the transfer, meaning that a conversion from electrical to chemical signals occurs.

Resting and action potential

The pickup and transfer of stimuli is brought about by changes in the membrane potential, for example, the unequal distribution of electrical charges within and outside of the nerve cell. Prerequisites for this are the selective permeability of the membrane and the activity of membrane-localized ion pumps that assure an unequal distribution of potassium (K^+), sodium (Na^+), chloride ions (Cl^-), and large protein anions (A^-). This way the concentration of K^+ and A^- ions within the neuron is high, while there are more Na^+ and Cl^--ions present on the outside. Although there is a concentration gradient, only potassium ions diffuse through the membrane to the outside in unexcited cells, that is, until the increasingly positive charge on the outside and negative charge on the inside no longer permit this. With a reverse potential of -70 to -90 mV (inside) equilibrium is reached. This state is referred to as a resting potential. When a nerve cell is stimulated, certain pores within the membrane change their conformation so that, for two milliseconds, these become permeable for Na^+ ions and the cell interior becomes temporarily positive relative to the exterior medium (depolarization). K^+ ions will then flow out again and so compensate for the Na^{+-} diffusion potential. During the Na^+ inflow and the K^+ outflow the potential reaches a peak value of +30 mV and a nerve impulse occurs. Brief changes of membrane potential is called action potential.

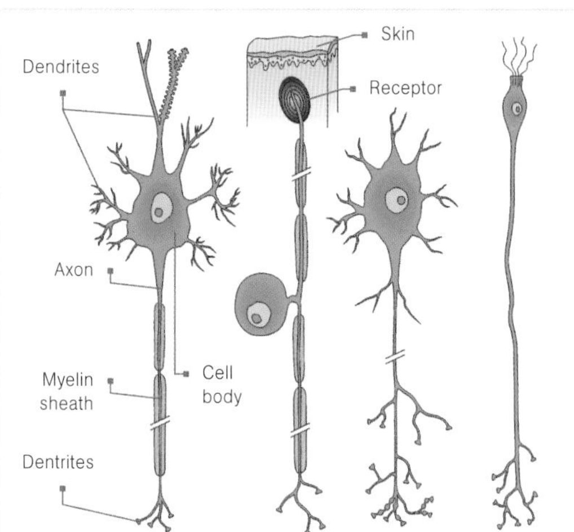

Various nerve cells (L to R): cerebral cortex; spinal ganglion; vegetative nervous system; sensory cell of a smelling mucous membrane.

The human nervous system controls our motor skills. Major damage to it can lead to paralysis.

basics

NERVE CELL length varies from a micron to more than 3 ft (1 m). A neuron can have up to 10,000 synapses. Transfer speed of nerve impulses can be up to 393 ft/s.

see also: Mathematics in the 21st century, p. 421

brain and spinal cord

The brain is an incredibly complex organ. It coordinates all incoming information. The spinal marrow receives and sends messages from the body to the brain.

Neurobiologists have long sought to understand the human brain. This high-performance organ processes all the sensory information arriving from the nervous system. The brain's tissue consists of neurons and glial cells. The skull bones, inner membranes, and cranial fluid all protect the brain.

Structure of the brain

The brain's symmetrical halves are connected by nerve fibers. The left half, or hemisphere, houses the language centers and processes analytical thinking. The right hemisphere processes intuitive and visual input. The brain's outer layer, the cerebral cortex, is about as thick as a little finger and has the highest concentration of neurons. Its surface area, and thus performance, is increased by having many folds. This outer layer is part of the cerebrum, the center of consciousness, perception, thought, emotion, and action. It is organized into lobes with specialized areas. Another section is the interbrain—the interface between the sensory organs and the cerebrum—which filters out unnecessary input to protect the brain from overload. It also regulates the body's fluid levels, temperature, and circadian rhythm. Some interbrain areas produce hormones to regulate important body functions. Beneath the interbrain, the midbrain—the brain's internal switching station—transfers information to other brain areas.

The cerebellum, at the back of the head, coordinates body movements. Along with the inner ear, it maintains the body's balance. Connected to the cerebellum is the medulla oblongata, which controls reflexes like swallowing and vomiting, and helps regulate the heart, breathing, and circulation.

The spinal marrow is part of the central nervous system and begins at the brain stem. From here it runs through the vertebral column to the lower back. Bunches of nerve fibers leave the spinal marrow at regular intervals and merge into spinal nerves. The nerve cells of the spinal marrow transmit signals from the body and the central nervous system.

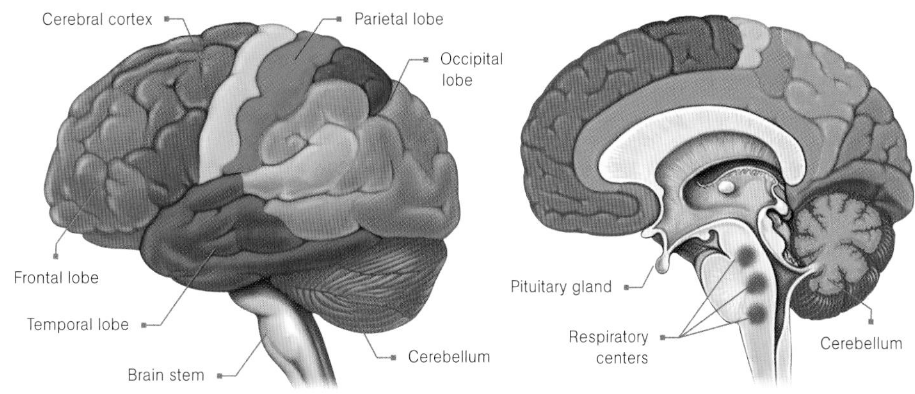

The cerebral cortex's deeply folded gray matter is organized according to function. The motor cortex controls the skeletal muscles. The sensory cortex processes senses such as sight and hearing. Association areas coordinate data; all relevant data is needed for full performance.

in focus: PAIN

Pain is a warning signal of the body. Nervous receptor systems such as numerous trigger points in the skin are stimulated and pass the information to the brain and spinal cord. This triggers reflexes to protect the body from the pain. This could be the quick pull back of a burnt hand from a hot stove. Complex vertebrates are most likely to have similar pain systems that are essential for survival.

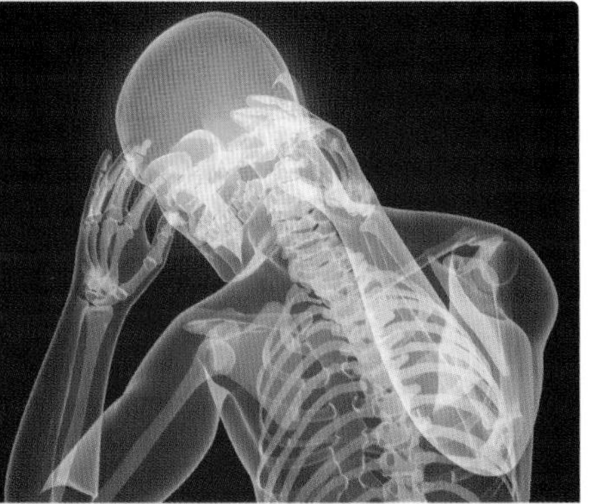

Pain may induce a range of behavioral changes designed to mitigate damage to the body.

basics

SEX DIFFERENCE: An adult female brain weighs 2.6 lb (1.2 kg); an adult male's brain weighs 2.8 lb (1.3 kg). However, female brains have more convolutions (folds).

THE BRAIN makes up 2 percent of total body weight, but it uses 20 percent of the body's blood volume.

The marrow consists of gray and white matter. The gray matter at the center is made of neurons and looks like a butterfly. This is coated by the nerve fibers of the white matter.

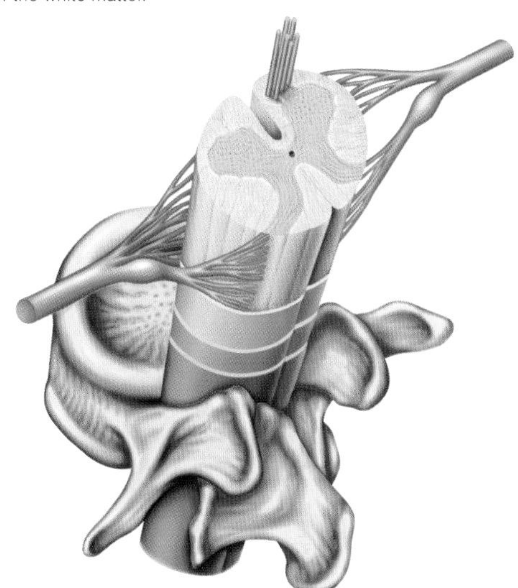

the eye

The human eye is an especially important organ, as it uses light information to provide us with essential details about our environment. Our eyes register the shape and color of objects and detect movement.

The human eye consists of a hollow spherical cavity with three segments. It is surrounded by an outer protective layer, the sclera, which turns into the strongly refractive cornea at the front of the eye ball, and covers the retina and vascular choroid coat. The front section of the eye contains a liquid-filled, and therefore refractive, anterior chamber; the iris and elastic pupil sit behind. The iris focuses light waves reflected from an object onto the sensory cells of the retina as a true, but smaller and upside down, picture. Here, the light waves pass through the large vitreous chamber filled with a gel called vitreous humor that, together with the sclera, gives shape to the eye. Light hits the retina, where sensory cells turn it into nerve impulses that are then channeled to the visual center of the brain through the optical nerve. The eye gets its color from pigments that are embedded in the iris.

eyes | ears | nose | tongue | skin

SENSORY ORGANS

Our sensory organs provide us with information about our environment. The five main senses are sight, hearing, smell, taste, and touch. Nowadays, it is also common to include the receptors that are located inside an organism or inside organs (proprioceptors) on the list of senses. The proprioceptors relay stimulation from within the body.

Accommodation and adaptation

The focus of the human eye can change by adjusting the curvature of the lens. The lens becomes flatter during distant viewing and thicker and more curved during close-up viewing. This is made possible because the lens is elastic. The ability to adjust the eye to be able to view objects at various distances is called accommodation. Moreover, the eye can also adjust to various levels of brightness. This process is called adaptation. In bright conditions, the pupil becomes smaller due to contractions of the iris muscles; in the dark, the pupil enlarges to allow more light to hit the retina.

Spatial vision is facilitated by both eyes working together with the visual cortex of the brain. The projections in each eye are slightly offset due to the distance between the eyes. Therefore, a different picture is projected in each eye. Both projections are simultaneously passed on to the visual cortex where a spatial picture with depth is produced, which takes into account the positioning of the eye and accommodation.

COLOR VISION

The retina contains two different photoreceptors: rods and cones. The very sensitive rods are responsible for distinguishing different levels of brightness while the cones are responsible for color vision. The cones are divided into three types with different ranges of absorption for green, red, and blue light; all other colors are calculated in the brain. The cones do not work in low light and with our rods we can only distinguish levels of gray. The human eye can see light radiation of wavelengths between 380 and 780 nanometers.

practice

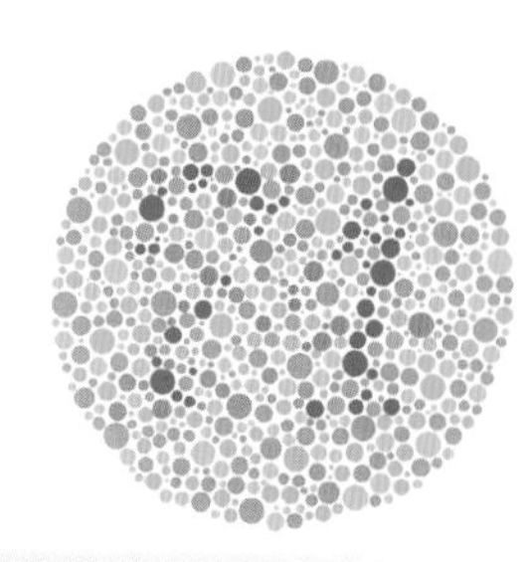

Color blindness is caused by the malfunction of at least one photo receptor.

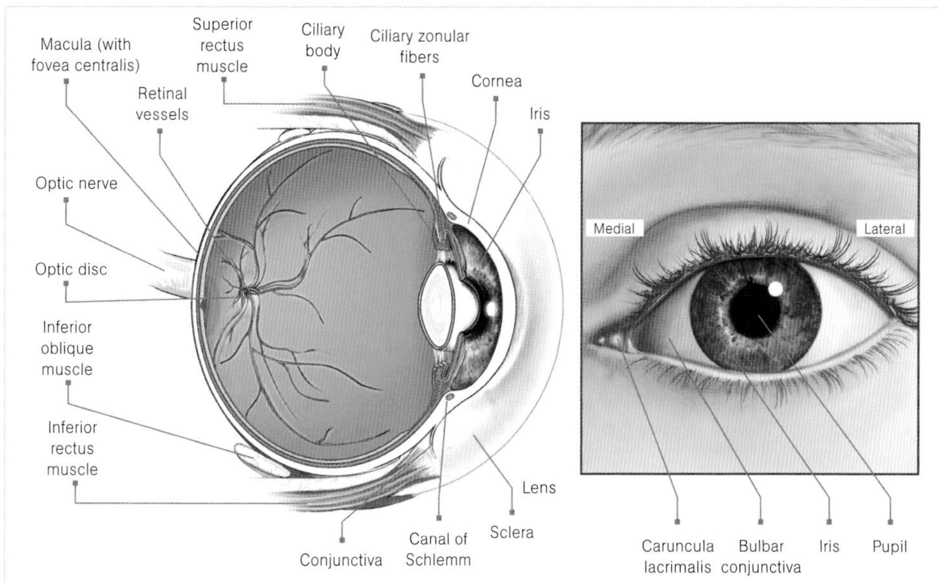

Sight is usually the most dominant of the human senses. Approximately 70 to 80 percent of our perceptions are influenced by what we see.

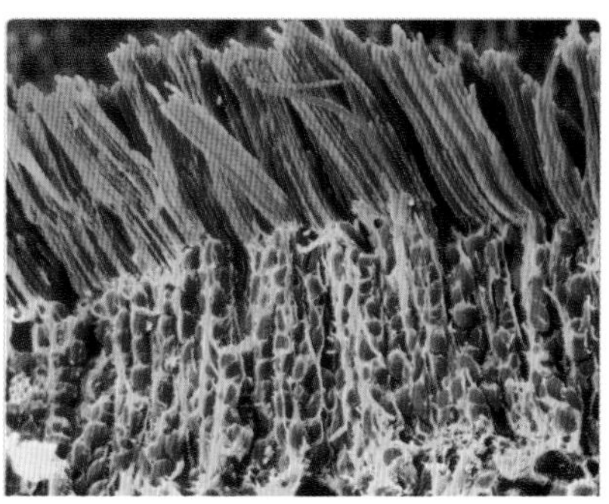

Rods and cones are special cells in the retina that contain optical pigments.

the ear

Animals often require good hearing for hunting or defense behavior. Humans mostly use hearing for communication among each other.

The human ear can be divided into three sections: the outer, middle, and inner ear. The outer ear consists of the auricle, the auditory canal lined with fine hairs and the tympanic membrane, or eardrum, which separates the outer ear from the middle ear. Specialized glands in the auditory canal produce earwax to protect the ear from dirt and dust. The auditory ossicles—the malleus (hammer), incus (anvil), and stapes (stirrup)—are located inside the middle ear and are the smallest bones in the human body. They connect to each other and to the tympanic membrane. The eustachian tube runs between the middle ear and the throat. Here, pressure differences between the middle ear and the atmosphere are equalized. The inner ear is filled with lymph and contains the cochlea, the actual hearing organ, which contains auditory sensory cells and semicircular canals. These form the vestibular apparatus, or balance system. When sound waves reach the ear, they cause the tympanic membrane to vibrate. The auditory ossicles amplify the sound waves and transfer them to the inner ear via the elliptical window, a membrane between the middle and inner ear. The vibrations are transferred to the liquid inside and are then registered by the tiny sensory hairs of the cochlea.

HEARING LOSS DUE TO NOISE

in focus

The human ear is very sensitive to excessive noise. Loss of hearing may already occur through a continuous noise impact of 85 decibels (dB). Traffic creates a noise level of about 80 dB and a jackhammer creates a noise level of about 110 dB. Some music concerts subject visitors to a sound level of up to 120 dB. Both short-term high noise impacts and long-term impacts result in loss of hearing.

Regular use of portable music players may result in irreversible hearing loss.

basics

PERFECT PITCH or absolute pitch is the ability to recognize tones solely by their sound. Statistics show that only one in 10,000 people have this ability.

Sense of balance

The vestibular organs responsible for balance are filled with liquid and are located in the inner ear. The three semicircular canals, and the utricle and saccule of the vestibule, together respond to movement and position. The semicircular canals and vestibule have sensory hair cells embedded into a gelatinous layer with tiny calcareous ear crystals (otoconia). During movement, the gelatinous mass shifts, bending the hair cells. The semicircular canals simultaneously register the rotational movement of the head in three dimensions. The saccule and utricle register linear changes in position within space. The combined sense of rotation and position results in the sense of balance.

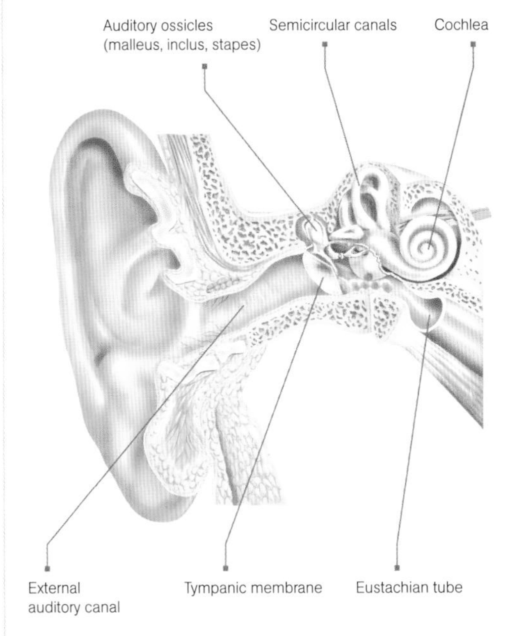

Hearing is the first sense to develop. Our ears can detect sound waves from 20 to 20,000 hertz.

People orient themselves using their sense of balance. This sense detects direction (above and below), angles of tilt, and the turning of the head in all directions.

nose and tongue

The senses of smell and taste are closely related to each other. This becomes obvious when a person has a cold; even the tastiest food seems bland due to an otherwise healthy, but now diminished sense of smell.

We can only sense smells that have evaporated and can be inhaled with the air that we breathe; therefore, the substances have to be in a gaseous state.

The olfactory epithelium inside the superior nasal concha is responsible for taking in smells. It consists of palisade-like supporting cells and threadlike olfactory cells in between. These turn into nerve fibers at the base of the epithelium. On top of the olfactory cells there are small cilia that act as receptors for gaseous fragrance compounds and generate a nervous impulse that is transmitted, via the olfactory bulb, to the appropriate area of the brain. The outside layer of the olfactory epithelium is covered by a thin film of liquid produced by specialized mucous glands, which are located between the supporting and olfactory cells. This mucous layer prevents the cells that are constantly exposed to air flow from drying out. The olfactory epithelium is, however, located slightly off the main stream of breathing air so that we begin sniffing as soon as we cannot clearly identify a certain scent. This narrows the lower part of the nose and the intermittent breathing draws in more air closer to the olfactory tissue located inside the nasal cavity.

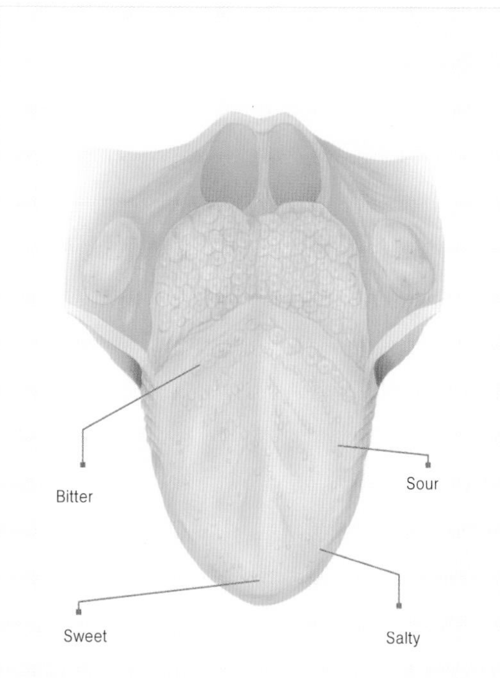

The human tongue possesses different types of taste buds that provide information about four taste sensations.

see also: Human beings, pp. 220–225

Sense of taste

One of the most important aspects of taste is the ability to differentiate edible from inedible or poisonous substances. This is possible through numerous taste buds, which are situated on the edges of small buds called papillae. They are predominantly located on the mucous membranes of the tongue and partly on those of the mouth. These taste buds contain receptor cells that are surrounded by supporting cells. Nerve fibers attach to the bottom of the receptor cells and transmit impulses of taste to the brain. The taste buds also contain basal cells which are responsible for the continuous reformation of new receptor cells. This is necessary due to their short life span of only about ten days. Despite having several thousands taste buds, we can only differentiate between four basic tastes: sour, sweet, salty, and bitter.

basics

BODY ODOR is as unique as a fingerprint. Personal scent—the result of sweat, body fluids, and broken down protein—may play an important part in how much we like or dislike another person's "smell."

in focus

BITTER AFTERTASTE

Taking in certain bitter substances, such as medicine, results in the familiar bitter aftertaste on the tongue. This is due to a delay in the flavor getting washed out of the dents lining the papillae where the taste buds are located. This is done by salivary glands at the base of the papillae, which are constantly producing saliva to remove all the old flavors and clear the taste buds for new impulses.

Besides bitter taste, sour taste sensations are also perceived as rather sharp and unpleasant.

skin

Tactile senses allow for the conscious sensation of stimuli such as touch, pressure, temperature, pain, and vibration. The skin is not equally receptive to stimuli, but sensitive at certain points containing sensory receptors.

Tactile senses can be divided into the sense of touch, the sensation of hot or cold temperature, and the sense of pain. The

An elephant's skin can be one inch (2.54 cm) thick on some parts of its body, yet it is highly sensitive.

receptors responsible for the sense of touch (mechanical receptors) are distributed across the body at different densities. The fingertips or lips, for example, have a lot of tactile receptor points, while the back, the upper arms, and thighs have only very few. Mechanical receptors can be divided into various types: Merkel's tactile disks and Meissner's corpuscles are stimulated when the skin changes shape; the Pacinian corpuscles, which are especially plentiful at the fingertips, react to pressure and vibration.

The human heat and cold receptors (thermoreceptors) are not only responsible for the reception and transmission of stimuli, but also play an important role in thermoregulation. They are very common in the face and especially in the region of the mouth where they form an almost continuous area. Since receptor cells for heat sensation are stimulated by high temperatures and receptor cells for cold sensation are stimulated by low temperatures, most people perceive water of a certain temperature as warmer or colder depending on which temperature the skin was exposed to prior to contact with the water.

Sense of pain

The human body is equipped with nociceptors, sensory receptors that can sense pain. These branched-out nerve ends inside the skin transmit pain stimuli to the central nervous system. If tissue is damaged, for example, due to bruising, cuts, or burns, the affected cells release messenger substances that trigger certain reactions in the nociceptors.

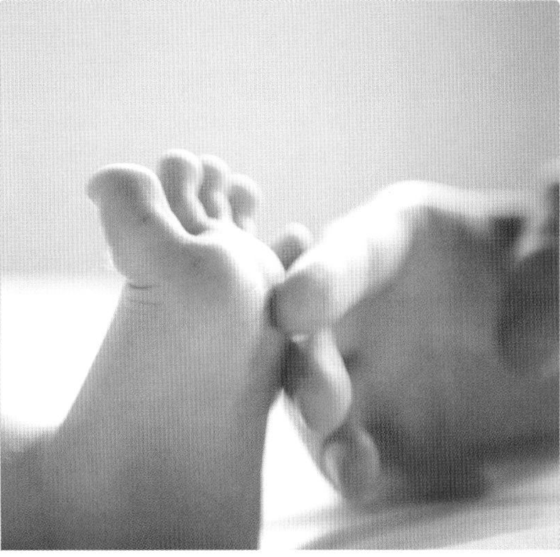

The skin is our body's largest organ and acts as a protective barrier. It is particularly sensitive in babies.

basics

MUTATION PREVENTS PAIN Just one single mutation can be enough to render a person extensively insensitive to pain. Six children from three families in northern Pakistan are affected by this mutation. Their lack of a normal sensation of pain has its downsides, as an important warning system of the body is not functioning. Several bone fractures remained unnoticed by the children.

in focus

BRAILLE ALPHABET

Although the tactile sense is usually the least developed sense in humans, it can be trained very effectively to perform at higher levels. Visually impaired people do this when they learn the Braille system. This way of writing was developed in 1825 by Louis Braille (1809–1852) who was already fully blind as a small child. He invented a system of letters composed of six raised dots which can be read by touching the symbols with the fingertips.

The Braille system has been adapted to suit different languages, including Chinese.

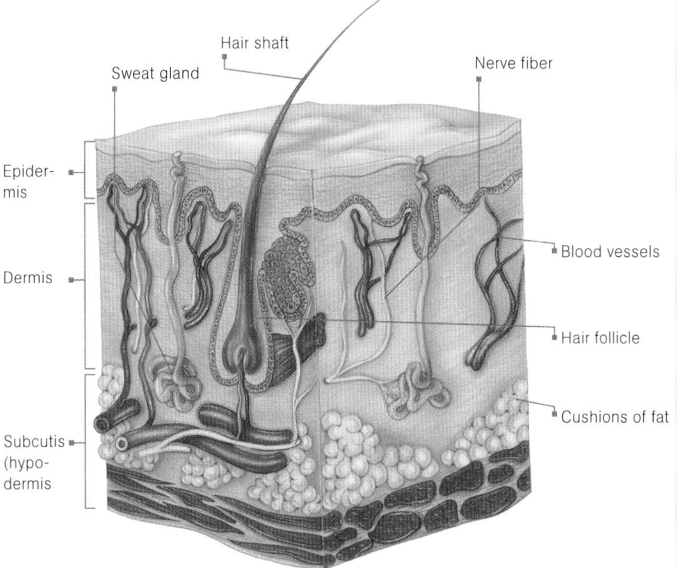

Skin is made up of three layers: the epidermis, dermis, and the subcutaneous tissue, which contain many specialized cells and structures.

see also: Nerve cells and signal transfer, p. 240

cell types | lymphatic system | immuse response

IMMUNE SYSTEM

The human immune system is responsible for protecting the body from pathogens. There are two types of immune systems: the innate nonspecific resistance fights germs by means of low pH values (stomach) or macrophages (leuocytes in the blood and tissue); the specific resistance is an acquired resistance to specific pathogens.

cell types and the lymphatic system

On a cellular level the immune system is supported by three main cell types, three groups of serum proteins (humoral defense system), and the lymphatic system with its lymphatic organs.

Leuocytes (white blood cells) play a major role in attacking pathogens. These mobile cells originate from the stem cells of the bone marrow and can pass through the walls of blood vessels. Leuocytes are mostly divided into granulocytes and monocytes (they develop into macrophages) for nonspecific immune responses as well as lymphocytes. The former are responsible for destroying germs by phagocytosis (ingestion). The latter can be grouped into B- and T-lymphocytes which

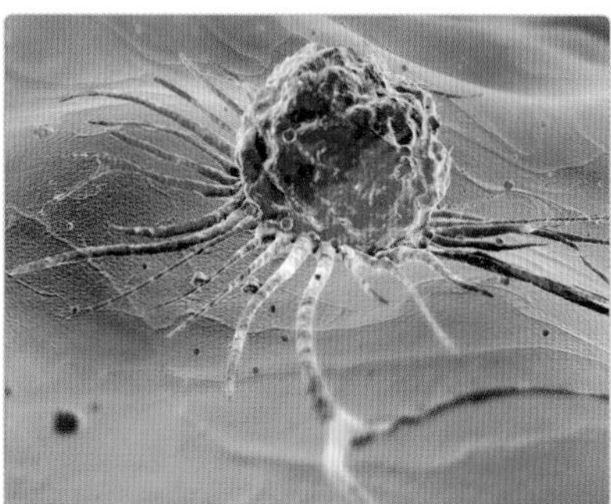

Microscopic view of a stem cell with processes. Almost all multicellular organisms have stem cells.

are of varying life spans. T-lymphocytes may reach 500 days old and serve as memory cells, as they can "remember" pathogens after a defeated infection.

Three types of serum proteins ensure immunity: antibodies (immunoglobulins), cytokines, and complement proteins. Antibodies are produced by B-lymphocytes and consist of a constant region and a variable region. Antibodies of the same type have identical constant regions. The variable region recognizes antigens, for example, certain areas on the surface of bacteria. The constant region determines functional properties of the antibody. Cytokines are messenger substances produced and released by immune and nonimmune cells. They function to stimulate and coordinate immune responses. The complement system is also involved in cell destruction and consists of several plasma proteins, which form a network with more than 30 different components.

The lymphatic system

The lymphatic system consists of lymphatic vessels as well as primary (for example, bone marrow and thymus) and secondary lymphatic organs (including the spleen and lymph nodes). The lymphatic organs produce and store cells of the immune system and regulate lymphatic vessels (via lymph nodes) and blood circulation (via the spleen). A dense network of lymphatic vessels traverses the entire body. Its function is the discharge of lymph fluid.

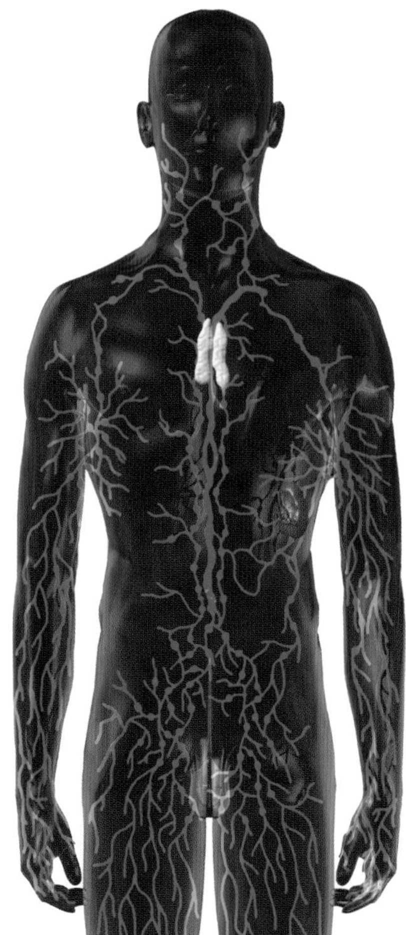

The lymphatic system is an important and major part of the immune system.

in focus

THE GOOD SIDE OF TONSILS

Normally we take little notice of our palatine tonsils, except when they are sore. They have a good side that may be less obvious to us, but they are still important to our immune system. In concert with the pharyngeal and lingual tonsils they recognize pathogens entering through the mouth and nose and help initiate an immune response. A few decades ago it was still common practice to remove the tonsils for preventative reasons. Today a removal of the tonsils is viewed with much more caution.

When tonsils become inflamed, they may need to be removed through surgery.

immune response to disease

Harmful bacteria, viruses, and other materials are constantly present in the environment and can cause illness and disease. To protect itself, the human body uses its immune system.

The role of the immune system is to recognize defective and foreign cells and combat them using both specific and nonspecific defensive strategies. The first line of defense is a series of nonspecific innate (passive) defenses. For example, the skin and mucous membrane coatings of the digestive, respiratory, and genital tracts have chemical defenses against most bacteria, viruses, and parasites. The skin's dead surface cells and the oils produced by the sebaceous glands form an effective barrier against many dangerous microbes. Other harmful bacteria are stopped by the enzymes and antibacterial proteins secreted by the mucous membranes. The specific defense mechanisms (the active or acquired response) of the immune system are triggered when special cells, lymphocytes called B-cells and T-cells, recognize individual antigens on the surface of the invading cells as foreign. When infection does occur, antibodies are formed to mount a defense targeted to the specific pathogen. The invading cells are attacked and consumed by white blood cells called macrophages. Inflammation and fever associated with infections increase the production and release of the defending macrophages, thus hastening the body's recovery.

Sneezing is a natural reflex that acts to expel dust and other foreign matter from the nose.

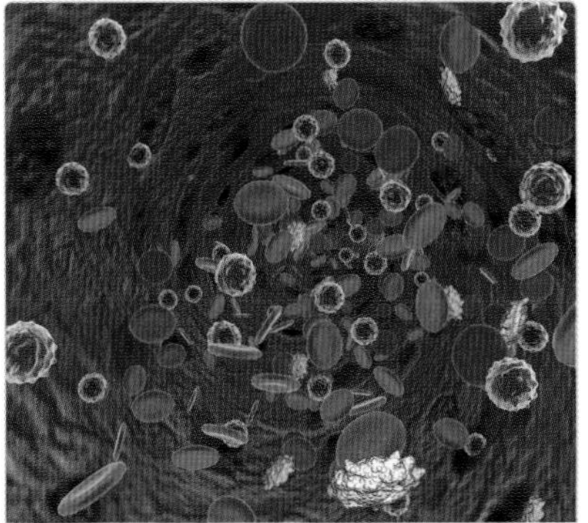

The body's defense cells circulate in the blood and lymphatic system and are also found in body tissues.

in focus

AIDS

AIDS (Acquired Immune Deficiency Syndrome) is a disease caused by the human immune deficiency virus (HIV). HIV is spread in body fluids such as blood, sperm, vaginal secretions, and mother's breast milk. Researchers are hunting for a cure; none has yet been found. Preventing infection, especially through protected sex, is the best defense against AIDS.

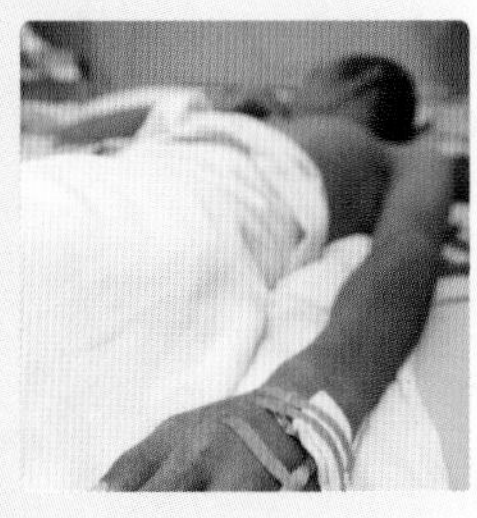

The immune system is attacked until tumors form and infections become common.

Primary infection

A primary infection occurs when an infectious organism enters a body for the first time. In response, the pathogen is isolated by an immune cell and broken into pieces displayed on the immune cell surface. This stimulates the production of specific defensive cells. The defensive cells try to consume the invaders through phagocytosis or kill them with superoxides. The immune system learns to recognize a pathogen by the chemicals it produces or the protein markers (antigens) on its surface. The immune system then produces antibodies that cling to the invaders. After a primary infection, antibodies and memory cells remain to help the immune system recognize and efficiently combat a new infection by the same pathogen.

basics

VACCINATION provides protection against specific diseases. In active immunization, weakened infectious cells are used to trigger the production of antibodies. With passive immunization, a serum with the needed antibodies is administered.

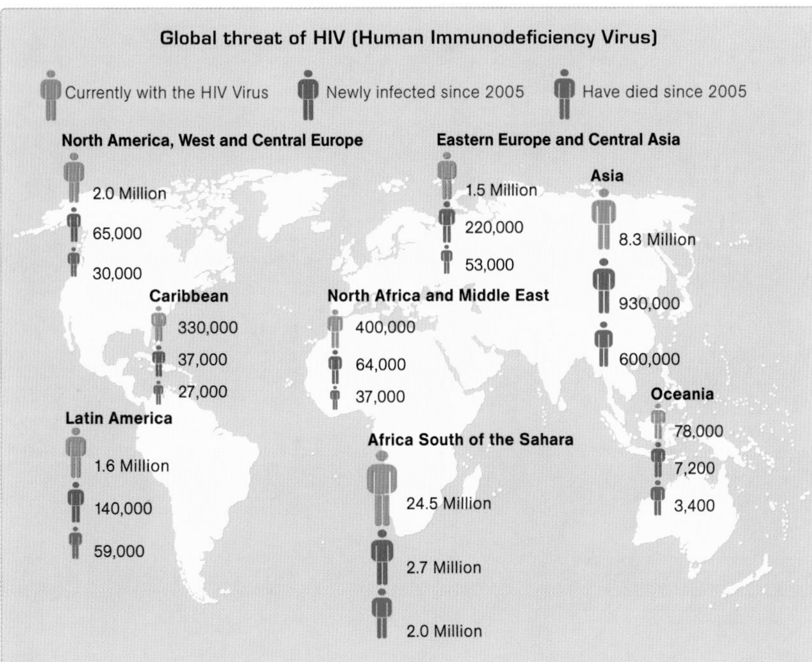

New HIV infections remain high even in countries where the transmission of the virus is well understood.

BIOLOGY

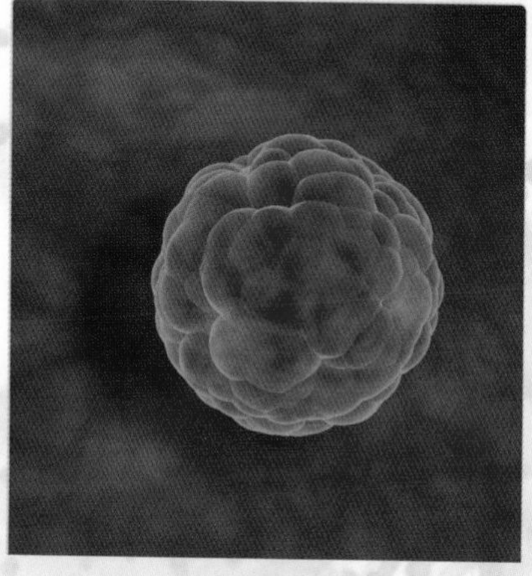

GENETICS AND HEREDITY

After Gregor Mendel made the first fundamental discoveries about the inheritance of genetic information some 150 years ago, many further questions about genome functioning and structure could be answered. Within a relatively short period of time humans have gained much valuable knowledge about fundamental molecular processes involved in gene expression, mutations, and genetic diseases.

Today scientists are able to target and manipulate certain genes of many organisms ranging from bacteria to humans. This ability provides new opportunities, for example the possibility of producing synthetic insulin using genetically modified microorganisms. However, it also bears risks and raises fundamental questions about the ethics of genetic manipulation.

GENETICS

The science of genetics examines the mechanisms governing the inheritance of traits by the offspring of living things. Within the field, a distinction is often made between general or classical genetics, which deals mainly with the formal aspects of heredity, and molecular genetics, which researches the underlying phenomena of inheritance on the molecular level.

genes

The carriers of genetic information are the genes. These are contained in the chromosomes: structures formed from DNA and proteins. The genetic information is represented by the order of nucleotides within the gene.

A gene is a segment of DNA (deoxyribonucleic acid) (p. 150) within a chromosome that encodes information for the synthesis of a particular protein. In eukaryotes (organisms with a true cell nucleus), most of the organism's genetic information is contained within the nucleus. In addition, the mitochondria (the cell's "power plants"), and chloroplasts in plants, also have their own small, separate sets of genes. In prokaryotes, which lack a cell nucleus, the DNA generally takes the form of a ring-shaped molecule within the cell's cytoplasm. Many bacteria also possess plasmids (smaller, nonchromosomal DNA molecules), which normally do not contain essential genetic information but may carry genes, such as those for toxin production, carbohydrate processing, or resistance against antibiotics. Species with these genetic additions often have competitive advantages over other bacteria. Taken together, all of the genetic information contained in an organism's genes constitute its genotype (the genetic constitution of an organism). The actual appearance of the organism—its phenotype—results from interactions between the genotype and internal and external environmental influences. Thus, organisms with identical genotypes do not necessarily have the same phenotype.

Organisms with double (diploid) sets of chromosomes, such as human beings, have two copies of each chromosome (homologous chromosomes). Therefore, each gene is represented twice. Different occurrences of specific genes are called alleles. Accordingly, humans have two alleles for each gene.

If two series of nucleotides are identical, the genes are called homozygous; conversely, if the pattern is different, they are heterozygous. Genes that occur in more than two different versions are multiple alleles. If an organism is heterozygous for a particular gene, its phenotype may be determined by only the dominant of the two alleles. The allele whose effect is overshadowed is recessive. If the phenotype is partially or equally influenced by both alleles, this is called intermediate inheritance, and the appearance of the heterozygous organism reflects a mixed form of the two homozygous phenotypes.

Genes are passed on through generations and give families a likeness to each other, such as similarities between a mother and daughter.

Ultraviolet light is used to read DNA-strand results at the Human Genome Lab.

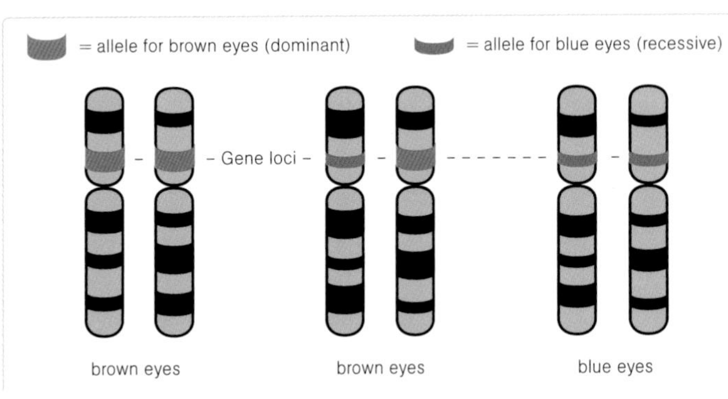

In a human with the alleles for both brown and blue eyes, the allele for brown eyes is more dominant, and, this phenotype, or physical appearance, will result.

see also: Nucleic acids: molecular building blocks, p. 298

hereditary rules

Classic genetics is the study of hereditary laws of individuals passing genes to their descendants. The first geneticist was a monk called Johann Gregor Mendel, who formulated the first genetic laws during the 19th century.

Initially, Mendel carried out his experiments using homozygous pea plants, which only differed in a few traits (for example, flower color). He crossed different varieties using artificial pollination, then carried out statistical analyses of his observations. His results showed regularities from which he deducted three laws, now known as Mendel's rules or Mendel's laws.

Inheritance of a trait

If two homozygous individuals of one species differing in one trait, for example, red and white flowers, are crossed, then the first generation (first filial generation F1) will produce offspring with the same expression of the trait (uniform). Therefore, Mendel's experiments only resulted in peas with red flowers. Alternating the parental sex during crossbreeding gave the same results with an even distribution of colors throughout the filial generation (reciprocity); for example, white female flowers and red male flowers, or vice versa. Mendel's first law is also referred to as the law of segregation, reflecting the concept of equal segregation. However, the second allele (white flowers) has not been lost but is passed on within the genotype of the F1 generation. The trait for flower color is present in the form of two alleles and the phenotype is determined by the dominant one. This is easy to prove by cross-breeding heterozygous individuals of the F1 generation. Their descendants (F2 generation) do not look alike, but their expressed characteristics differ with a ratio of 3:1 (dominant-recessive cross) or 1:2:1 (intermediate cross). Mendel's second law is also called the law of independent assortment.

Inheritance of multiple traits

If individuals differing in multiple traits (genes) are crossed, then each individual gene is passed on independently and recombines during gamete formation (the law of segregation and law of independent assortment both apply to each individual gene). However, gene pairs need to be located either on different chromosomes or far enough apart to allow free recombination. Mendel's third law, the law of dominance, is based on this concept.

GENETIC CONTROL ORGANISMS

practice

The fruit fly *Drosophila melanogaster* is an ideal species to study genetics. Gene mutations are easy to recognize in this species, because they manifest, for example, in a change in eye color or wing shape. Their genome also consists of only eight chromosomes which are extremely large and accessible in the salivary glands. Therefore, the genetic material is very easy to read.

Drosophila is one of the most examined animal species. It has been used since the early 20th century.

DISCOVERY OF HEREDITARY RULES

milestones

During his lifetime, the work of Austrian monk Johann Gregor Mendel (1822–1884) was largely ignored by the entire scientific community. Later, researchers made similar discoveries and his conclusions became the basis of an entirely new branch in sciences. Amazingly, it took Mendel only about 12 years to complete his revolutionary experiments. Afterward, his monastery appointed him as abbot: a position that required all of his energy and time.

Johann Gregor Mendel discovered the hereditary rules of inheritence in his experiments on pea plants.

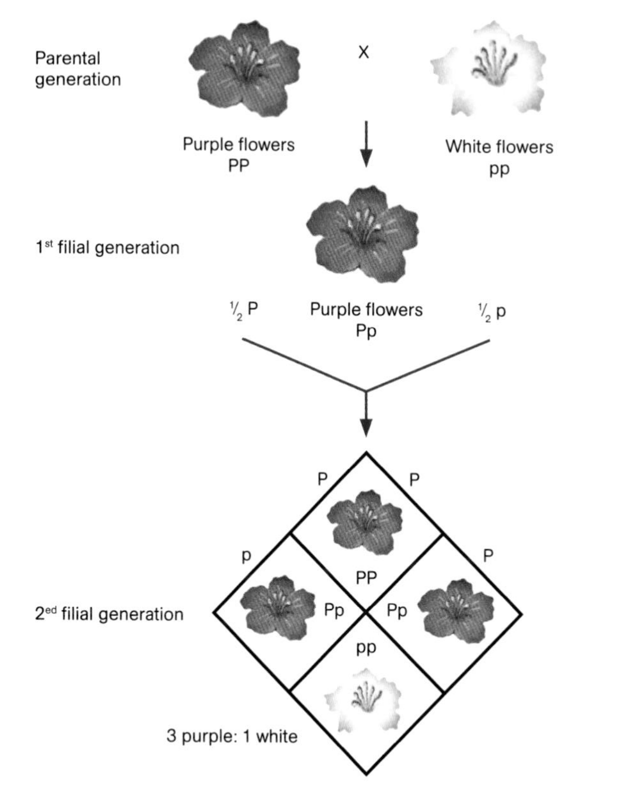

Certain traits, or genes, may not be expressed in the first generation but they are still passed on and appear in small ratios in later descendants.

feature characteristics: transcription

Following the one-gene-one-polypeptide hypothesis, one particular gene is always responsible for forming one polypeptide. Ribonucleic acid (RNA) plays an essential role in converting genetic information into proteins.

The genetic combination that produces red hair is relatively rare, as is the combination for green eyes.

RNA is found in the cell nucleus as well as in cytoplasm, the cell interior; in mitochondria, essentially the cell's power plant; ribosomes, sites of protein production; and chloroplasts, the plant organelles. It has a similar structure to DNA, but contains the sugar ribose (instead of deoxyribose), and the base uracil (instead of thymine), which can, however, also bond with adenine. RNA is normally single-stranded, but loops can develop within a strand due to base pairing. One can distinguish three different RNA types, which are based on their occurrence and function. Ribosomal RNA (rRNA) is—apart from proteins—the principal component of ribosomes. Transfer-RNA (tRNA) bonds amino acids and organic compounds with at least one amino and one carboxyl acid group. Then tRNA transports them to the ribosomes, where they are linked to a polypeptide chain with the aid of messenger-RNA (mRNA.)

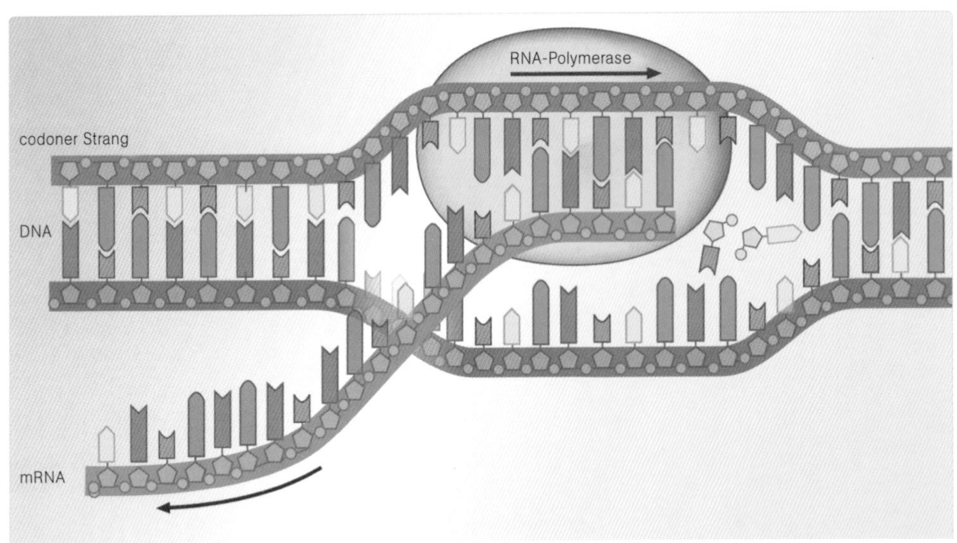

The RNA polymerase enzyme links corresponding free RNA nucleotides to the diverged DNA strand one sequence at a time, creating a growing, single strand of mRNA, which will separate at a certain point, allowing the DNA strands to merge again.

Transcription

The mRNA develops as a bit-by-bit copy of DNA in a process called transcription and subsequently brings the genetic information to the ribosomes where the transformation of information into proteins takes place (protein biosynthesis).

During transcription, the hydrogen bridges between the complementary DNA strands are split at those locations where genes are read and the single DNA strands diverge (initiation) so that the coded, genetic information can be transferred to mRNA. At a starting point of transcription (promoter), complementary base pairing begins between one of the DNA strands and free RNA nucleotides, linked by the enzyme RNA-polymerase, which moves along the DNA strand section-by-section, attaching one complementary nucleotide after another to the ribonucleotid chain (elongation).

DNA thus serves as a matrix for the synthesis of single mRNA strands. After reaching a certain frequency range (a terminator), the enzyme interrupts the synthesis. The completed RNA separates from the DNA, which now closes again, and carries the information to the ribosomes for biosynthesis.

in focus

PROTEINS

Proteins fulfill diverse tasks in all living beings. In protein biosynthesis, a sequence of DNA bases is transformed into a specific amino acid sequence in the protein molecule. Proteins are macromolecules composed of amino acids, which are arranged in chains and joined together by peptide bonds. These polypeptide chains (primary structure) can fold into spatial structures (secondary and tertiary structure).

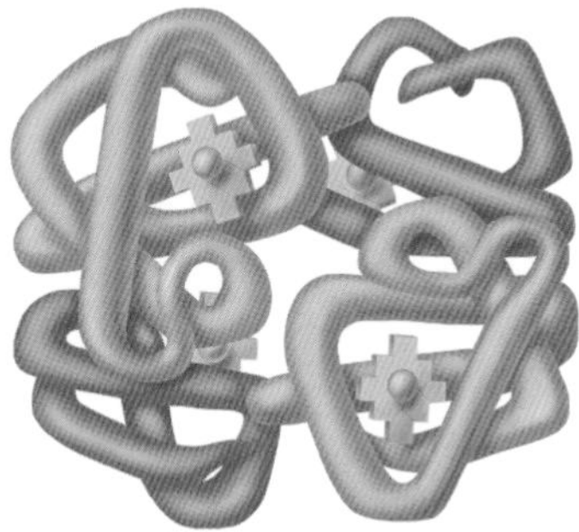

When a protein is made up of several chains, as for instance the blood pigment hemoglobin, this is referred to as a quaternary structure.

basics

SPLICING Since eukaryote DNA has coded and non-coded segments (introns and exons), the pre-mRNA formed by transcription initially also contains exons, which are spliced out before specific enzymes reunite the parts to form active mRNA again.

see also: Nucleic acids: molecular building blocks, p. 298

feature characteristics: translation

The genetic code is shared by all organisms and is made up of codons. Combinations of these instruct the formation of the amino acids that are used in protein biosynthesis.

The structure of DNA (p. 150) consists of nucleic acids with four different nitrogen bases. Of these, a unit of three successive bases determines an amino acid. These base triplets of DNA that code specific amino acids are called codons. Their totality forms the genetic code, and each of these codons corresponds to one codon of mRNA. All organisms use the same code; it is thus universal, although minor deviations are known to occur. Reading the codes takes place without overlapping; each nucleotide participates only on a single triplet and coding does not require an interval signal for the differentiation of triplets. The combination of four bases into sets of three theoretically offers 64 possibilities to code different amino acids. This is more than needed to represent the 20 amino acids that are required for protein synthesis. Therefore, there are several codon combinations for each amino acid.

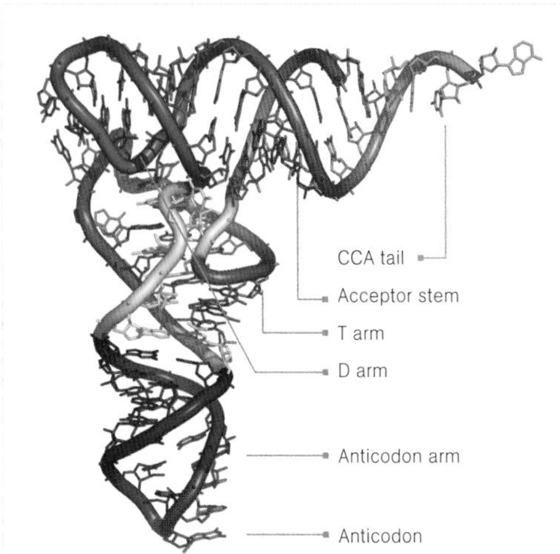

Transfer-RNA takes on the shape of a clover leaf if it is straightened out, but typically its structure is a folded L-shape.

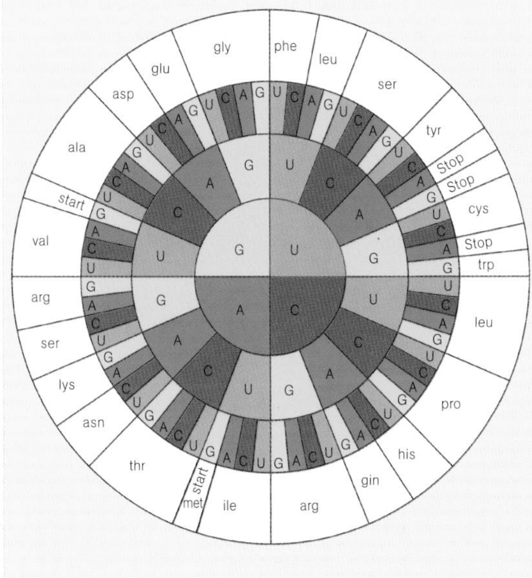

The four bases that make up the genetic code can form into 64 possible combinations of three letters known as codons.

Character expression: translation

During translation, the genetic information—now present in encoded form following transcription in to the mRNA sequence—is converted from amino acids for the formation of polypeptides (proteins). This takes place on ribosomes, tiny corpuscles located in the interior of all cells.

The amino acids required for protein biosynthesis are initially bonded inside the interior of a cell to a specific tRNA molecule, by using the nucleotides adenosintriphosphate (ATP) and with the aid of specific enzymes (aminoacyl-synthetase).

Due to base pairing, these tRNA molecules have a cloverleaf-like appearance. A specific base triplet (anticodon) is located in one of the loops, which can bond to the complementary codon of the mRNA. When tRNA molecules are loaded with the equivalent amino acids, they transport them to the mRNA located at the ribosomes. Base pairing occurs between the codon of the mRNA and anticodons of the tRNA molecule, and the amino acids are linked to each other through peptide bonds. After the mRNA has been completely read, the newly formed protein detaches from the ribosome.

Frequently, several ribosomes read the same mRNA (polysomes) and the information obtained is used repeatedly in this manner, before RNA-dismantling enzymes (ribonuclease) break them up.

The first codon is always for the amino acid methionine (starting codon). The end of the sequence is reached at one of the three stop-codons (UAG, UAA, and UGA), for which there is no loaded tRNA available.

in focus

RIBOSOMES

Living-beings with cell nucleus (eukaryotes) and those without (prokaryotes), have cell structures in which protein biosynthesis takes place. These so-called ribosomes are roundish particles consisting of two subunits, with a diameter of about 15 nm, made up of ribosomal RNA (rRNA) and proteins. The small subunit is responsible for recognizing mRNA, while free amino acids are linked to form a long chain with the aid of the large subunit. At the start of protein biosynthesis both parts of the ribosome come together and then they detach again after the protein has been completed.

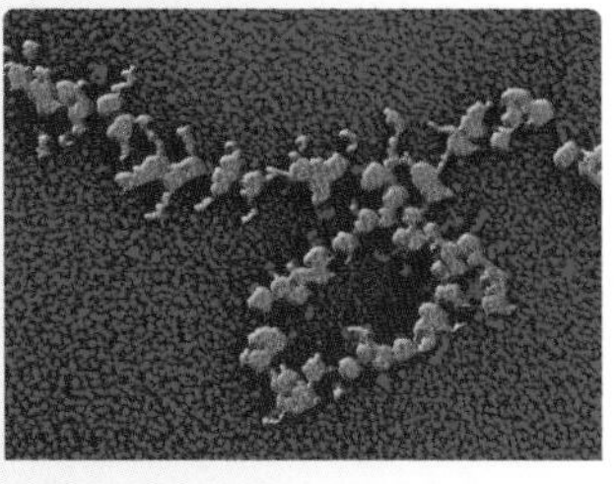

Protein biosynthesis occurs in cell structures called ribosomes

basics

ENZYMES are important catalysts in biochemical reactions, since they can lower the activation energy of these reactions. Most enzymes act very specifically, i.e., only in conjunction with certain substrates. Most enzymes are proteins, but there are also ribonucleic acids, so-called ribozymes, that have a catalytic effect.

mutations

Mutations may alter single genetic segments (gene mutation). Sometimes an entire chromosome is affected by a mutation (chromosome mutation) or even entire chromosome sets (genome mutation).

There are different types of mutations: a gene mutation alters a single gene, which results in the development of a new allele (gene expression); a point mutation affects only one base on a DNA-nucleotide chain. The deletion or addition of individual bases results in a frameshift mutation. In this case none of the nucleotide base triplets following the altered section of the nucleotide strand can be read correctly. All original information is lost. A base exchange may also result in no changes to the information. This may be the case if, for example, the altered triplet happens to encode the same amino acid as before. Since amino acids are mainly determined by the first and second position of a codon, point mutations of the third position on a triplet are usually less dramatic.

Chromosome mutations change the structure of individual chromosomes. This may be the result of chromosomes breaking up during division and fragments getting lost, known as deletion, or attaching to one of the sister chromatids in a process called duplication. Other potential mechanisms are an exchange of fragments between nonhomologous chromosomes (translocation) or the inverted reattachment of fragments onto the chromosome (inversion).

A genome mutation is an alteration in the number of chromosomes. Sometimes individual chromosome pairs or chromosomes do not separate during mitosis or meiosis, which is called a nondisjunction. This results in aneuploidy, a condition in which there is an increased or decreased number of chromosomes in the daughter cells. If one chromosome is missing, it is known as a monosomy and if one extra chromosome is present, it is a trisomy. A change that affects entire sets of chromosomes is called euploidy. Most organisms have a double set of chromosomes, which is known as diploidy; if instead three or more chromosome sets are present, then this is called polyploidy.

mutations | hereditary diseases

GENETICALLY INDUCED DISEASES

Mutations are random permanent alterations of genetic information. They may occur spontaneously or due to triggers such as ultraviolet light. In many cases a mutation does not affect an organism's functions. However, some mutations may result in serious disabilities or even death, as they may cause cancer or genetic diseases.

Albinism is a hereditary condition in animals (including humans) related to little or no production of the color pigment melanin in the skin, hair, and eyes.

Genetic mutations can result in unusual traits, such as crested growth patterns in cacti.

in focus

LACTOSE INTOLERANCE

Geneticists have found that earlier humans tolerated lactose only during childhood (lactose tolerance). About 9,000 years ago, a mutation emerged in humans of lighter skin that allowed them to tolerate lactose throughout their lives. Thus, many of their descendants are able to tolerate milk consumption after the age of weaning, while adults of Asian and African descent may suffer from digestive problems and discomfort.

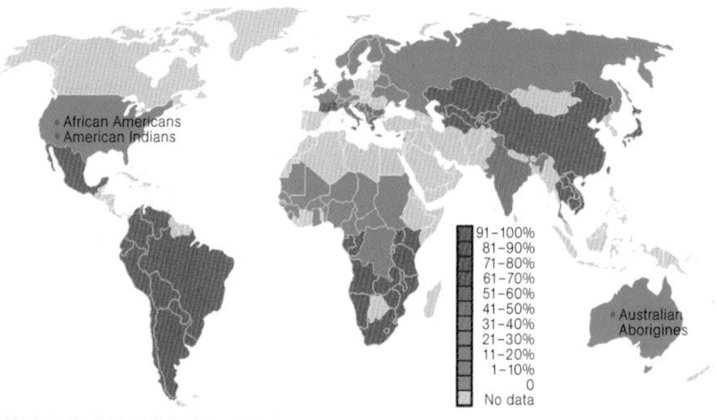

In parts of Asia and Africa, about 90 percent of the population is lactose intolerant; in Western Europe, Australia, and North America only 5 to 15 percent.

see also: Hereditary rules, p. 251

hereditary diseases

Hereditary diseases are caused by gene mutations that manifest themselves as a disease of an organism. During reproduction, the mutated genes are passed on to the progeny according to the Mendelian laws of inheritance.

Hereditary diseases, which are also known as genetic disorders, can be genetically passed on to new generations of offspring by autosomal or gonosomal chromosomes (gonosomes are the X and Y sex chromosomes, while autosomes make up the remainder); beyond that, it is possible to distinguish them by whether they are dominant or recessive hereditary diseases.

HEMOPHILIA

in focus

People who suffer from this incurable disease lack the blood-clotting substance that normally prevents blood loss due to an injury becoming too severe. Blood clotting for someone with hemophilia begins significantly later than it does in healthy humans, therefore even small wounds can lead to serious and potentially dangerous problems. This disease only affects men; women are only the carriers of this genetic disease. Nowadays there are possibilities to treat hemophilia selectively, for instance by administering cleaned blood-clotting factors.

In a person with hemophilia, wounds take longer to heal, resulting in continued blood loss over long time periods.

Autosomal-dominant hereditary diseases

An autosomal-dominant hereditary disease occurs when a certain gene shows a change, and this change—in spite of a normal second copy of the gene—leads to the development of the disease. For the children of a person affected by such a disease, there is 50 percent risk that they too will become ill; if both parents have the disease, the probability increases to 75 percent. With two damaged alleles (possible manifestation of the gene), an embryo normally dies prior to birth, so that most carriers of a dominant hereditary disease are heterozygous. A typical example of an autosomal-dominant hereditary disease is Marfan Syndrome, where a damaged allele leads to the formation of a defective structural protein, which in turn results in connective tissue weakness.

Autosomal-recessive hereditary diseases

For an autosomal-recessive disease to affect a person, there must be a change in both alleles of a particular gene. If only one allele is affected, the second allele can compensate for the consequences. Therefore when this disease occurs in a child from apparently healthy parents, both parents are heterozygous: carrying one mutated and one healthy copy of the gene. Examples of autosomal-recessive hereditary disease are albinism and sickle-cell anemia.

X-chromosomal hereditary diseases

For X-chromosomal hereditary diseases, the gene location for the affected protein is on the female sex chromosome. X-chromosomal-recessive hereditary diseases can occur in the male progeny of healthy parents, when the mother was heterozygous. A diseased father will pass the gene on to all of his daughters, but with a healthy homozygous mother, they will not get the disease, while their sons have a 50 percent risk of falling ill. For instance, hemophilia and red-green blindness are sex-related chromosomal hereditary processes, and are both passed on chromosomal-recessively. An example of an X-chromosomal-dominant hereditary disease is genetically related nyctalopia (night blindness).

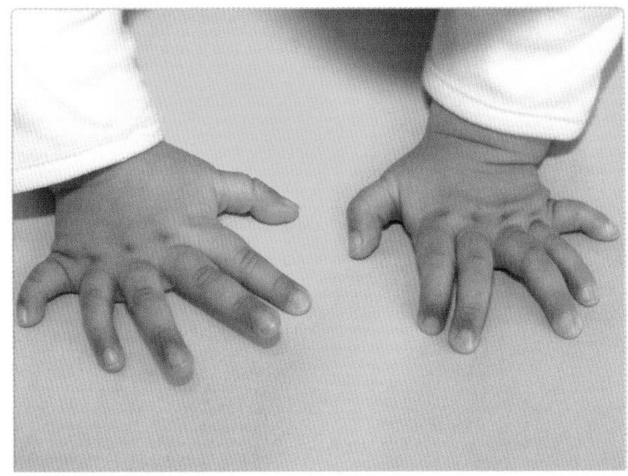

Polydactylism (extra fingers) is a mutation that can arise independently or can be a genetically inherited condition from the parents.

basics

CHROMOSOMAL ANOMALIES A deviating number of chromosomes (numerical aberration) or a differing chromosomal structure (structural aberration) can also cause hereditary disease. The most frequent numerical chromosomal aberration in humans is Trisomy 21, a defect that manifests itself in physical anomalies and mental retardation.

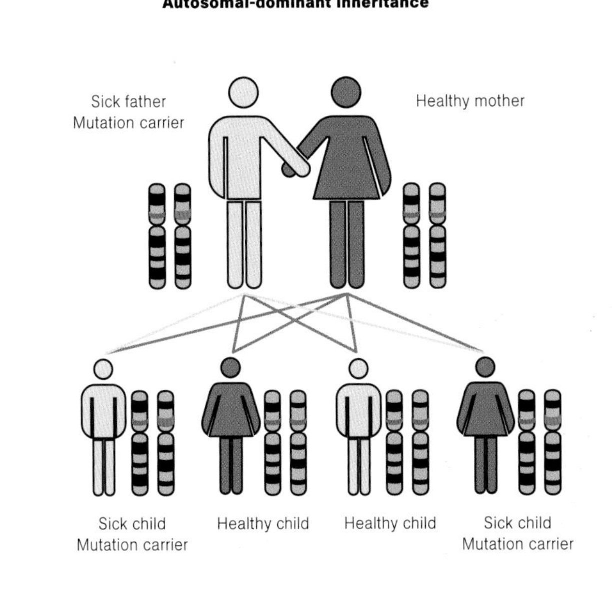

If only one parent is affected by an autosomal-dominant hereditary disease, the children have a 50 percent probability of being affected.

cloning

The age of modern gene technology began in the 1960s, with the discovery of the restriction enzyme, which enabled the selective production of DNA fragments. A number of other advances have furthered genetic technicians.

Cherry anemones *(Corynactis californica)* off Monterey, California, all of which are clones from a common ancestor.

cloning | stem cell research

GENE TECHNOLOGY

With modern gene technology it is now possible to selectively intervene in an organism's hereditary information or its biochemical control mechanisms. Genes and their regulators can be isolated and altered to produce particular genetic products, such as antibiotics, human insulin, or monoclonal antibodies in large quantities.

The cloning of genes—the integration of a particular DNA fragment in a transport medium (vector) for DNA and a subsequent transfer of the constructed fragment into a receiver cell—became possible when the first restriction enzyme (restriction endonuclease) was identified and described. These cut nucleic acids at very specific locations, so that afterward precisely defined partial fragments are available.

If restriction endonucleases cut both strands of a DNA double-strand with their bases offset, so-called sticky ends are formed (with a single strand overhanging the end sections). In the event of a straight cut, one talks about blunt ends. Subsequently, certain enzymes called ligases can rejoin the ends again. Restriction enzymes occur naturally in bacteria. Other important tools for geneticists are vectors (bacteriophages and plasmids), which are transport systems that can be used to funnel DNA fragments into a cell. Once these have been placed into a cell, the geneticist must ensure that it reproduces extrachromosomally, or that it becomes built into the genome.

Polymerase chain reaction

Another method commonly used is polymerase chain reaction (PCR) for the multiplication of specific DNA fragments, in which the target sequence is always doubled in successive cycles. A very small amount of DNA produces so much material within a short time that evidence of a particular DNA segment can be identified, for instance, by means of staining or agarose gel electrophoresis. In order to duplicate DNA, one utilizes primers (oligonucleotides), which attach themselves specifically to complementary sequences from the entire DNA in the sample. The primers are then lengthened by a DNA-dependent DNA-polymerase so that a copy is made. Subsequently, both strands are separated again by heat, so that a new cycle can take place.

basics

INSULIN SOURCES
While insulin for diabetic patients was formerly obtained from the pancreas of slaughtered animals, human insulin is now available for treatment, made from bacteria using new genetic-engineering methods.

Dolly the sheep was cloned at Scotland's Roslin Institute in July 5, 1996. As the first mammal to be cloned from an adult cell, Dolly was considered a milestone in the field of genetics. She died of a progressive lung disease in 2003 at the age of six.

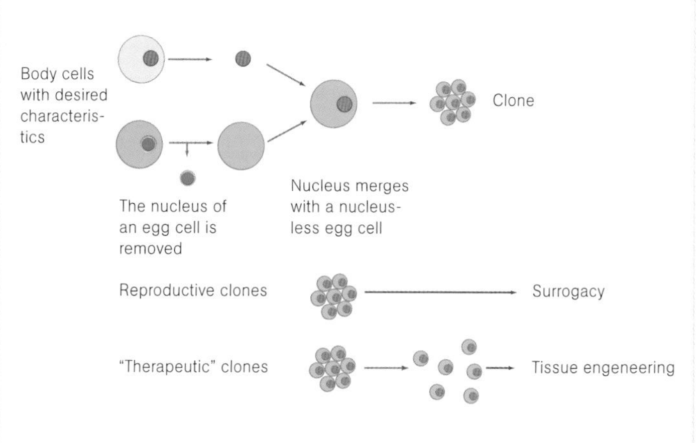

During the process of somatic cell nuclear transfer, the nucleus in an egg cell is replaced with another nucleus carrying the desired genetic information. In this way, dividing cells used for either reproductive or therapeutic means contain the target DNA.

see also: Genetically modified food, p. 344

stem cell research

Stem cells are undifferentiated body cells; they do not belong to cells of a definite cell type in an organism. They are divided into adult or somatic stem cells (originating from a fetus or adult) and embryonic stem cells.

Adult stem cells are responsible for producing replacement cells for renewing certain tissues. For example bone marrow stem cells can continuously restore blood components, while others can renew muscle or nerve cells. Adults have about 20 different kinds of stem cells.

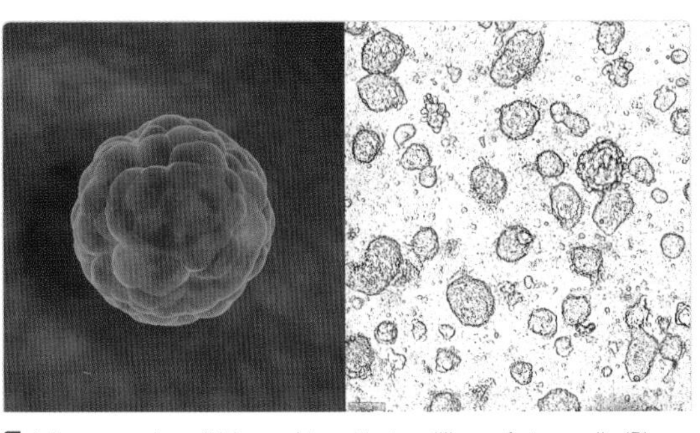

A human embryo (L) is used to cultivate millions of stem cells (R) that have the potential to turn into any cell in the human body.

Researchers hope to be able to apply this cell replacement ability to the therapeutic treatment of organ damage by taking a sample of a patients stem cells, allowing them to differentiate into specialized types in the lab, and reimplanting them into the body. Although adult stem cells have a much lower development potential than that of embryonic stem cells, their removal directly from the body, for example by biopsy, is not controversial and therefore most development and funding is directed into this branch of human stem cell research.

basics

RESEARCH RESTRICTIONS In some countries, for example, Germany and Ireland, it is illegal to use human embryos for stem cell production. Others allow their use but impose varying restrictions and conditions.

Embryonic stem cells

Embryonic stem cells are formed during the early phase of embryonic development. Cells of this type are able to divide infinitely. Theoretically they can develop into cells to fit any of the approximately 210 types of tissue found in the human body. This ability is called pluripotentiality. The possibility of producing embryonic stem cells, which can be multiplied in vitro almost infinitely, opens new doors for developmental research, especially in medicine.

In the future it may be possible to treat a heart attack victim by removing DNA from the patients body cell and transferring it into an egg cell without a nucleus. An embryo blastocyst, carrying the genetic information of the patient, could be grown in vitro. Embryonic stem cells could then be removed and treated in a way that would result in the development of heart muscle tissue, which could be used to replace damaged tissue. Rejection by the body would be unlikely as, after all, it consists of the bodies own tissue.

One day such therapeutic cloning could theoretically cure numerous diseases. However, this area of stem cell research is still very controversial.

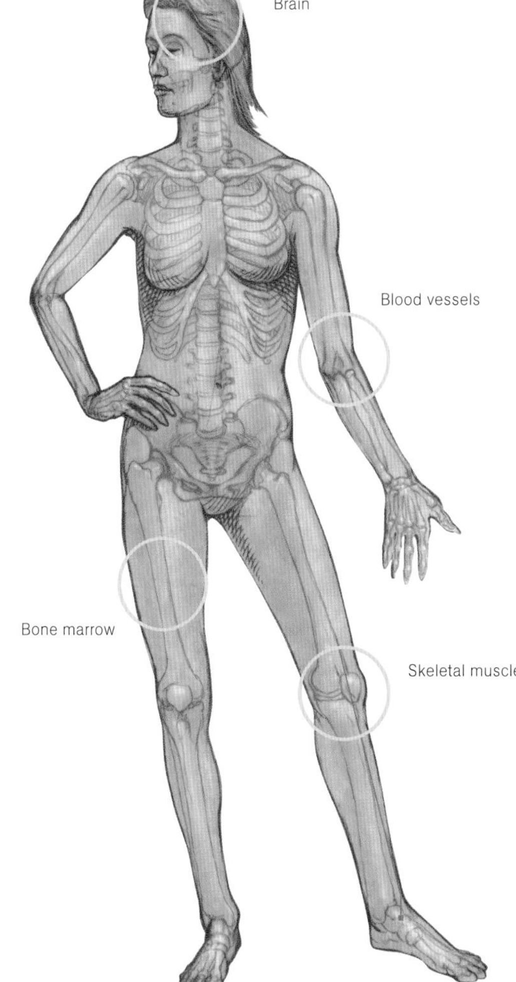

Stem cell research has the potential to lead to developments in repairing damaged tissues associated with health problems such as Parkinson's disease, heart disease, diabetes, and others.

in focus

ETHICAL DISPUTE

Some people are opposed to embryonic stem cell research, because although the embryos are at an extremely immature stage (an assemblage of cells called a blastocyst), they are destroyed during the stem-cell production process. At the center of this ethical discussion is the question of when an embryo begins to be a human being and should therefore be protected by human rights. Meanwhile researchers have been able to reprogram human skin cells so that their properties are almost equivalent to that of embryonic stem cells. This may offer a solution and render the discussion about ethical values irrelevant.

The president of the United States, George W. Bush, here at a press conference, addressed the controversial issues regarding human embyonic stem cell research.

BIOLOGY

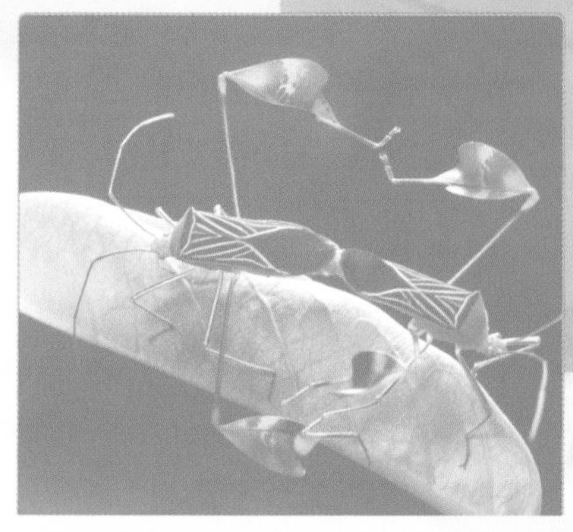

ETHOLOGY

Ethology is the study of animal or human behavior. A relatively new branch of biology, the field of ethology first had to concern itself with developing methods to attain objective and reproducible results. The foremost questions for ethologists today include determining which forms of behavior are genetically determined, or inborn, and which are learned as a result of individual experiences. Researchers are also attempting to discover the causes of behavior, what internal and external factors lead to certain behaviors, and what forms of social interaction have been developed by different species or social groups.

reactions | reflex | instinct

ETHOLOGY

Behavioral science is a branch of biology that studies the behavior of living organisms. Such behavior may include body posture, vocal sounds, changes in color, and the release of scent pheromones. Complex behavioral patterns can often be divided into macro (e.g., reproduction) and micro (e.g., courtship displays) behavioral categories.

what is behavior?

In classic behavioral science, a behavioral pattern is considered to be innate when it is essential for survival and already present at birth, as it is predetermined by the genetic make up of the organism.

A reflex is the simplest form of an innate behavior. It is a programmed reaction to an outside stimulus that is carried out unconsciously. For example, the eyelids close automatically as soon as a draft of air stimulates the surface of the eye and the pupils of a cat will contract as soon as it looks into bright light. These are reflexes which the organism does not have to learn; they are referred to as unconditioned reflexes (as opposed to conditioned reflexes, which are a result of learning). An unconditioned reflex is always an unconscious response, and therefore it is impossible to suppress it at will. Such a reflex always requires a stimulus that triggers a certain behavior. Many unconditioned reflexes exist in order to protect the organism, for example coughing, nausea, or the draw back reflex of the body part that was touching a hot object.

Anatomically, a reflex is based on a chain of stimulus and reaction, which is referred to as a reflex arc. A reflex arc begins at the receptor, passes through the central nervous system and from there into the organ that is to carry out a responsive action. A well-known example is the knee jerk or patellar reflex in humans, which is triggered by a light hit to the patellar tendon below the patella in the knee. This causes the thigh muscles to extend (stimulus), which excites the muscle spindles (receptors) and this excitement is then transmitted to the spinal cord (reflex center) via an afferent pathway (which conducts impluses toward the center). From there, it travels through a synapse and efferent neurons (which send impulses away from the center) back into the muscle. This triggers the immediate lifting of the lower leg. The reflex arc involved only runs through one switch point; therefore, it is referred to as a monosynaptic reflex. The knee jerk reflex is often used in medicine to test the function of the spinal cord and associated nerves. The real purpose of this reflex is to protect humans from injury when tripping. The lower leg moves forward quickly to a fall prevent a fall.

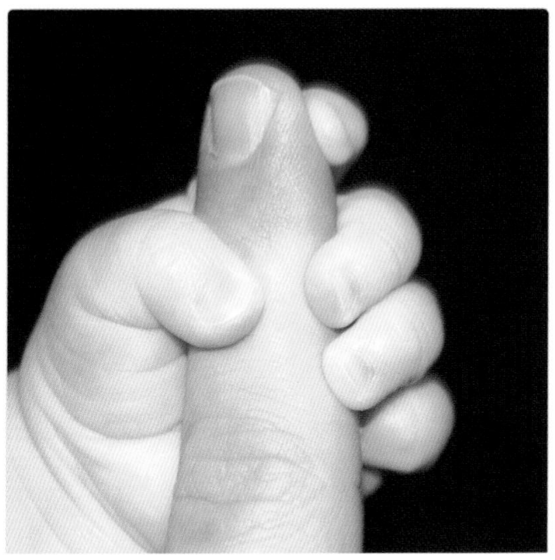

The palmar grasp reflex is a type of reflex exhibited by babies up to six months old. When an object is presented, a baby will close his/her fingers around it.

Web weaving is an intricate and innate behavioral pattern in spiders. Each web starts out with a single thread, which relatively quickly becomes an entire web. Through vibrations, spiders can sense prey trapped in their webs.

milestones

KASPAR HAUSER BEHAVIORAL STUDIES

Kaspar Hauser behavioral studies are carried out to see whether a certain behavior is innate or learnt. It involves raising animals without any social contact to suppress the development of normal behavior. Kaspar Hauser was an orphan found in Nuremberg in 1828 who claimed that he was held captive in a basement for his entire life.

Kaspar Hauser, found in 1828, showed signs of developmental retardation indicating a lack of childhood experiences.

see also: Homo sapiens, p. 165

instinctive behavior

Instinctive actions, or fixed action patterns, are similar to reflexes, rigid and irreversible but with a greater range and complexity. Examples are the nesting behavior of birds and the stockpiling behavior seen in hamsters.

In contrast to unconditioned reflexes the innate stimulus threshold, which triggers a stimulus of instinctive action, is variable. The process of a fixed action pattern can be divided into three phases. It begins with appetence behavior, that is the undirected search for a stimulus which may trigger an instinctive action (for example, undirected food searching during periods of starvation). The second phase is called taxis or directed appetence which is a targeted approach to the source of a stimulus (for example, smelling food). The third phase is the genetically coordinated fixed action pattern. It is a species-specific activity that is irreversible once it has been triggered. Following the previous food-related examples, this would be the intake of food.

Conditions for a fixed action pattern are a willingness to act (motivation), a sign stimulus, and an innate trigger mechanism. All three factors depend on the same motivation to carry out an action that requires a stimulus, but they can be triggered by various stimuli. This includes intrinsic factors such as hunger or hormones, as well as outside factors such as day length. Motivation decreases after successful completion of the fixed action pattern.

A stimulus that triggers instinctive behavior is called a sign stimulus; if it comes from an individual of the same species, it is called a releaser. In experiments dummies are used to determine which stimuli act like sign stimuli. Often, several sign stimuli or certain combinations of stimuli are more likely to trigger a response than one stimulus by itself. Dummies with artificially exaggerated sign stimuli may exceed the extent of the natural response (super-optimal sign stimulus). The nerve mechanism of the central nervous system that recognizes a sign stimulus and differentiates it from other stimuli is called the innate release mechanism. It triggers the appropriate behavior that corresponds to the sign stimulus.

basics

A DISPLACEMENT activity is the result of a conflict of instincts. The motivation to perform two, opposing actions is so equal that neither action is performed.

The common toad responds to the stimulus of a prey item with a series of linked reactions: orienting toward and stalking the prey; fixating its eyes on it; snapping; swallowing; and wiping its mouth with its forelimbs.

in focus

ACTION CHAINS

Fixed action patterns often do not occur in isolation, but rather in the form of a series of individual actions where one follows the next. A conspecific release of an instinctive action series by an animal is called an action chain. During the courtship display of sticklebacks, every action triggers the partner's next action.

Stickleback fish perform zigzag dancing rituals and use a system of interlocking releasers to orchestrate their mating.

Kittens like the one here push against their mothers' bellies with their paws in order to stimulate the following of milk.

see also: Types of cohabitation, pp. 274–275

courtship | social behavior | aggression | communication
circadian clock | conditioning

BEHAVIOR PATTERNS

Animals tend to react to certain stimuli with particular responses. These specific patterns regulate and simplify relations between animals of the same or different species. They include reproduction, care of young, communication, feeding, and defense, and may involve body movements and vocalizations.

courtship, mating, and care of the young

Higher animals have adopted distinct behaviors for courtship, mating, and rearing their young. These behaviors help them to find the best mate and produce healthy offspring.

Sexual reproduction, the procreation strategy of all higher animal groups, is maintained through courtship and mating rituals. Parents feed, protect and transmit skills and behaviors to their offspring. Behavior patterns form the foundation of family groups and other complex social structures, which may remain strong for life.

Searching for a partner

Each animal species uses unique strategies to search for and attract a potential partner. A male's display helps females recognize his potential, desirability, and suitability for mating. The female usually has the labor-intensive job of raising the young, so she selects a partner whose offspring will justify this investment. Courtship may involve visual stimuli, such as a rooster's comb, deer's antlers, or the bright colors of dragonflies. Other courtship devices include auditory cues, such as birdsong or specific odor-producing substances like pheromones, especially popular with insects.

Mating and care of the young

Mating ensures the successful transfer of the sperm cell to the egg cell. Fertilization occurs internally or externally, where sperm from the male fertilizes the female's egg cells. These mature into embryos inside the mother's body, in the case of most mammals, or in eggs laid by the mother, as for example in the case of reptiles and birds.

Nearly all invertebrates, amphibians, and reptiles reduce childcare to selecting or creating a place to deposit their eggs. This is often a hole or nest that protects the eggs from predators and the environment. Laying many eggs helps ensure that at least some offspring will survive. The young of many other animals are helpless and highly dependent on their parents. Particularly in the case of mammals, where care of the young is even more demanding and can last several years or more.

Courtship behavior among humans is complex, with many differing social commitments and rules.

in focus

BEGGING BEHAVIOR

Begging is a behavior mostly seen in helpless young animals. The purpose is to receive food or water from parents or social partners. Songbirds show a distinct begging behavior when young. They react to movements, elongating their necks and opening their beaks as wide as they can. The throat pattern becomes visible triggering feeding behavior by the parent birds.

Chicks begging for food: without parental feeding they would soon perish.

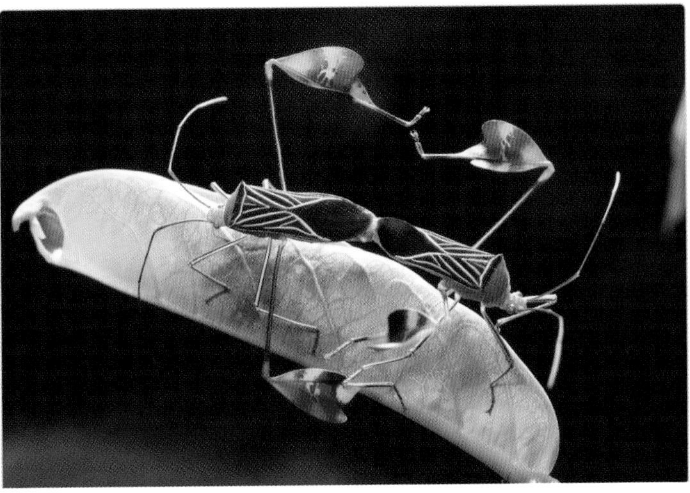

The mating methods of insects and beetles can sometimes be very unconventional, difficult, or even dangerous.

see also: Learned behavior, p. 267

feeding and social behavior

Behavior patterns facilitate group interaction by regulating individual rank, feeding priority, territorial boundaries, cooperative activities, and many other aspects of community life.

Among members of the same species, social interactions are simplified by means of specific behavior patterns that regulate their relationships. For example, a recognized system of rank and feeding priority helps to prevents disputes from flaring up over every meal, which would waste valuable time and energy.

basics

ANINE PACKS The ack is organized ccording to a famy structure consiting of parents and ffspring. Sexually ature wolves leave e family in search new territory.

Eating hierarchies

Individual animals signal their rank using body language and vocalizations, which are recognized by other group members. High-ranking animals display behaviors such as baring their teeth and hissing (lions), or pecking aggressively (chickens).

Low-ranking animals make submissive gestures, such as lowering the head and averting their gaze. Conflict is necessary only when the hierarchy is undergoing significant change. Even then, conflicts usually take the form of symbolic battles, fought without bloodshed.

A wolf pack consuming a shared meal: As a wolf pack will often leave almost no trace of an animal behind, feeding hierarchies become very important in sharing out what is often a limited food source.

Many patterns of group behavior favor the high-ranking animals at the expense of others. For instance, an alpha male may receive the largest and most nutritious portion of the prey. Generally, he is also the only male that reproduces with several females, to pass on the optimum genes to the next generation. Other behavior patterns clearly burden the individual while benefiting the group. For example, when a marmot whistles to warn the group of an approaching predator, it attracts the predator's attention. Cooperative behavior is also seen in group hunting, when all members benefit from the joint effort.

Learning within the group

Social behavior is only partly instinctive; it is also strongly influenced by watching and imitating parents and other adults. This becomes obvious in studies of animals raised in isolation, which show unusual behavioral disturbances when they lack the opportunity to learn social behavior patterns. Also, twins raised separately share the same genes, but behave differently.

During their first weeks of life, penguin hatchlings are highly dependent on their parents.

in focus

PLAYING

The behavior of play is mostly seen in juvenile mammals, including humans, and is an important part of normal behavioral development. During play, animals and humans try out initially random actions and observe how their environment reacts. They consequently learn to differentiate between random and non-random effects. The facial expressions and body language observed during this behavior signals "play time." This allows for a safe environment to practice and optimize hunting and fighting behavior that later becomes important as essential but dangerous survival tactics.

Through play, offspring learn behaviors that they will need as full-grown animals.

agonistic behavior

The powerful clash of antlers or the sight of bared, sharp teeth: these displays of aggression are an important part of the animal kingdom. A complex interplay between aggression and submission keeps animal life in balance.

The ethological term agonistic behavior includes any social behavior related to fighting and covers all types of behavior in response to negative influences from other organisms, including aggression and subordinance. Agonistic behavior assists animals in securing essential needs such as habitat and territory, food, and sexual partners. It includes two opposite components: aggressive and defensive behavior. The former can involve displaying, threatening, or attacking. Typical defensive behavioral patterns include placation, submission, and escape behavior.

Young Rocky Mountain male elk spar in Yellowstone National Forest, Wyoming. Their antlers provide a means of defense, as does a powerful front-leg kick. If threatened they will respond with gutteral grunts and posturing.

Inter- and intraspecific aggression

Aggression is the willingess to threaten or attack in response to external cues. Intraspecific aggression is targeted at individuals within the same species (conspecifics) while interspecific aggression targets individuals of other species. Aggression between different species includes predation or antipredator tactics. Intraspecific aggression occurs to secure subsistence needs such as food.

ANIMAL RANKING

in focus

Many vertebrate groups are organized by rank, determined by fights among rivals. This system ensures that the group is guided by a strong and experienced lead animal. Rank order largely prevents intraspecific fights for food, mating partners, and so on. Animals of higher rank often have a better chance of reproduction than low-ranked animals. Rankings are usually dynamic and reorganized after regular fights for position.

Within a pride, or a group of lions, the most experienced and strongest lions lead.

Aggression control

Fights are usually preceded by displays and threatening behavior. A common tactic to appear more threatening is to make the body seem

Male peafowls, also called peacocks, display their extravagant tail plumage as part of courtship.

larger by increasing its visual outline, for example, by straightening the body, raising the tail, and bristling the fur. This can be accompanied by acoustic threats, such as growling. If threats escalate to a conflict situation, fight or flight responses will result in either an attack or escape outcome. However, threatening displays are usually sufficient to diffuse the situation. A stronger animal is less likely to fight if the other animal displays submissive behavior (for example, lowering the tail). Pacification behavior may also inhibit the victor from killing, which is important for a species as a whole. Serious fights may lead to grave injuries or death, while ritual fights, usually preceded by extended threat displays, tend to cause little or no injury. The latter have fixed rules and are usually carried out to secure or improve an animal's social rank. Serious injuries in this case are rare.

basics

TERRITORIAL behavior is a form of aggression through which territory is defended against conspecifics. Boundaries are set by marks, or acoustic or optical signals.

Cobras can enlarge their hoods in order to appear bigger and thus intimidate an enemy.

see also: Competition, p. 273

communication

Communication plays an important role in social interactions between animals. A variety of signs are used to exchange messages, such as optical and acoustic signals.

Signs for intraspecific communication frequently involve optical signals. Gestures, facial expressions, body coloration, and posture are all indicators of either aggressive or submissive intentions. The advantage of this type of communication is that it is quick and direct; the disadvantage is that these signals usually only work in close proximity of each other and under certain circumstances (for example, during daylight).

> **NSECT COMMUNI-ATION** Certain ants re equipped with p to 15 pheromone lands to communi-ate with one an-ther. Honey bees ven dance to ommunicate a find.
>
> basics

Acoustic signals are vocals or noises of any other kind. Often they are used to set the boundaries of a territory (for instance, the knocking of a woodpecker). Other typical acoustic signals include vocal communication, for example, between ducklings and their parents, or croaking in frogs. Chemical signals are often used as scent marks to set the boundaries of a territory. They may also serve as path markers for ants or as pheromones to find mating partners (see in focus). Tactile signals—that is communication via body contact—is mainly observed in mammals (for example, social grooming among primates).

Subtle forms of visual communication such as the curve of a lip can send strong signals between chimpanzees. Some of their communicative behavior share many similarities with human language.

A golden-fronted woodpecker: Bird communication includes vocalizations such as songs and calls.

Ritualization

Ritualization is a fixed action pattern that has been changed during the course of evolution so that conspecifics may communicate with each other more explicitly and therefore more efficiently. Courtship dances in many bird species are examples of such modified behavioral patterns where sections from other action chains, such as grooming of feathers or nesting behavior, have been incorporated. Roosters, for example, paw the ground during courtship like hens attracting their chicks.

Body parts that are involved in certain communication signals are often particularly conspicuous and colored to emphasize ritualized movements. For example, the male fiddler crab has asymmetric pincers. One of them is significantly larger and its function is to attract females that are ready for mating. Ritualized behavior, in general, also functions as an aggression control mechanism.

PHEROMONES IN SILK MOTHS

Chemical signals used for intraspecific communication in many animal species require such sensitive sensory organs that they are superior to any technological system built by humans. The antennae of the male silk moth *(Bombyx mori)* are a typical example. They are able to detect pheromones produced by the females, even at tiny concentrations (about one molecule pheromone per 1,016 molecules of air), from up to several miles away.

A silk moth's sense of smell responds to various kinds of chemicals from other insects.

in focus

see also: Types of cohabitation, pp. 274–275

the biological clock

The timing of many behavioral patterns is determined by certain biological rhythms, which are usually based on endogenous circadian clocks as well as certain outside factors called timers.

It is most likely that all animals follow a daily routine governed by a biological clock, which determines, for example, when an individual gets tired and when it will wake up. The suprachiasmatic nucleus (SCN) is an 800-micrometer small structure situated behind the eyes in mammals, exactly where the visual nerves cross. Here certain light-sensitive sensory cells of the eye feed the SCN with information about the environment. The release of hormones then triggers certain responses. Light is an important timer of the daily rhythm, but there are other environmental conditions, such as temperature, which fluctuate and may influence the biological clock. There are also other cycles apart from the circadian rhythm: for example, the circannual rhythm, which is also referred to as the inner calendar. This system regulates migrating behavior in birds as well as reproductive behavior in many animals.

Navigation in long-distance migratory birds is based on various senses. Birds can also detect and utilize the Earth's magnetic fields, use the sun as a compass, as well as make mental maps of an area for orientation.

in focus

THE HAMSTER'S BIOLOGICAL CLOCK

Since genes that regulate the biological clock are activated or inhibited in a 24-hour rhythm, how does this affect hibernation? Researchers have found that genes responsible for maintaining the 24-hour cycle are turned off during hamster hibernation, so that the corresponding hormones, which ensure sleeping and waking up at certain times, are not produced at all. The hamster's regular daily rhythms return with the onset of spring.

The tiny hamster's biological cycles alter over the winter to include periods of hibernation.

Animal migration

Animal migration is the long-distance movement to predetermined locations.

Daylight has a significant effect on the circadian rhythm of animals, plants, fungi, and cyanobacteria.

Many different groups of animals migrate regularly and for a variety of reasons. Some fish migrate for spawning, such as the European eel *(Anguilla anguilla)*, which spends most of its lifetime in European rivers. This animal travels thousands of miles into the Sargasso Sea for spawning. The young eels hatch here and, within a period of three years, move back into fresh waters. They find exactly the same rivers, in which their parents reached maturity. Salmon have a similar life cycle, except that they grow up in the ocean and move into the rivers for spawning.

Ungulates from the Serengeti migrate every year in search of food. Thousands of wildebeests, zebras, gazelles, and other mammals pass through the treeless plains to find rich pastures to feed on. Bird migration is the most well-known and easily recognized form of migration. At the beginning of winter, many bird species leave their nesting grounds to fly to warmer regions. These journeys often cover thousands of miles and require precise and intricate navigational techniques.

issues to solve

BIRD FLU Since migratory birds are susceptible to avian flu, there are concerns that the virus could spread via their droppings, even if the birds never touch land.

see also: Energy flow and food chains, p. 277

learned behavior

Learning is the capacity to store individual experiences in long-term memory and use them to adapt to new situations. The ability to learn significantly enhances an individual's chances of survival in a changing environment. Learned behavior includes imprinting and conditioning.

Imprinting is a basic and necessary learning experience in early childhood and in young animals. It is generally irreversible, and occurs only once during an individual's development, during a short window of time called the critical period—in birds, for instance, this is shortly after hatching. Imprinting leads to a specific reaction in the individual, such as a gosling following its mother. In object imprinting, an object or another animal provides the stimulus for a particular reaction, while in behavioral imprinting, it is a pattern of action.

■ Using animals in circuses is somewhat controversial due to potentially cruel training methods.

Conditioning

Conditioning is a learning process in which certain behaviors are associated with specific stimuli. For instance, a puff of air (an unconditioned stimulus) naturally stimulates the eyeblink reflex (an unconditioned response). When a musical tone is played (a neutral stimulus), there is no eyeblink. However, if this neutral stimulus always comes just before a puff of air, after a while the musical tone alone will induce the eyeblink reflex. This new stimulus is called a conditioned stimulus, and the reaction to it a conditioned response. Learning by developing conditioned responses is known as classical conditioning.

milestones

MAN AND GOOSE

The Austrian zoologist Konrad Lorenz (1903–1989), who won the Nobel Prize in 1973 for his work on comparative behavior, is considered one of the founders of ethology. The concept of imprinting is linked to Lorenz's name as he was able to show that goslings will accept anything as their mother as long as it is nearby shortly after they hatch and makes the right sounds.

■ Konrad Lorenz's imprinted goose, Martina, followed him around for her entire life.

A learning process that brings together two ordinarily independent factors is called conditioned association. For example, many birds over the course of their lives learn that certain materials are especially favorable for nest building, and so search for these. Thus, based on positive experiences, a previously neutral stimulus comes to evoke an associated response. Animals also learn from negative experiences. If an originally neutral stimulus is linked to a frightening or painful occurrence, the animal will begin to avoid that stimulus; this is called conditioned aversion.

If an animal is always rewarded immediately after exhibiting a particular behavior, the behavior becomes associated with the reward; the behavior acts as a tool through which the desired result is achieved. This is referred to as operant conditioning.

basics

IMITATION Primates learn through imitation. Young chimps eat the same fruits as their mothers, who reinforce this by offering or taking away certain foods.

■ Pigeons have learned to adapt themselves extremely well to the urban landscape. They have learned to congregate around places, often tourists areas in big cities, where people feed them bread crumbs.

see also: Human environmental impact, pp. 280–281

BIOLOGY

ECOLOGY

The term ecology was first introduced in 1866 by German biologist Ernst Haeckel. However, the study of the relationships of living things with each other and their environments has only gained broader public attention as more and more people have become aware of the rising degradation of the environment. They hope to gain new insights from this branch of science to help solve the emerging problems. The field of ecology is often subdivided according to the size of the systems being studied, such as ecophysiology, population ecology, and ecosystem ecology. A further branch is human ecology, which studies the complex interactions between human beings and their environments.

abiotic factors | light | temperature | water

ECOLOGY OF THE INDIVIDUAL

Ecology is a branch of biology that deals with the interrelationship of organisms among themselves and with their biotic (animate) and abiotic (inanimate) environment. The study area of ecology is the ecosystem, which is the unity made up of the environment (biotope) and the living ecological community (biocenosis).

abiotic factors: light

Within the interrelations of an organism to its environment, one distinguishes between influences from the inanimate environment (abiotic environmental factors) and effects due to other organisms (biotic factors).

Abiotic factors are nonliving physical and chemical factors that affect the ability of an organism to reproduce and survive, for example temperature, water, and light. Biotic (or animate) factors, on the other hand, include predator-prey relationships, competition, symbiosis, and parasitism, among others. Depending on the type, individual environmental factors can have rather variable effects on the existence of an organism. Graphically, this influence is usually illustrated with the aid of a tolerance curve. The section along the curve most favorable for an organism is called the optimum; the threshold values, within which an organism can exist, are called minimum and maximum. A tolerance curve also shows a section where an organism can still survive but is unable to reproduce (pessimum). The tolerance range between minimum and maximum indicates the ecological potential of an organism for the respective environmental factor that is being considered. Species that have a narrow tolerance range are referred to as being stenoecious; species with a broad ecological potential are euryoecious. The most favorable biotope is where the preference ranges intersect those environmental factors that are important for an organism. This then limits the number of individuals of a species due to unfavorable environmental conditions (sphere of influence of environmental factors). For instance, water is usually the limiting factor for desert plants, while on the other hand, temperature and the availability of nutrient salts make growth for these plants still possible.

Tolerance curves help illustrate the variable effects of individual environmental factors, such as temperature and water, on the existence of an organism.

The sun is an important abiotic factor for this basking crocodile in Lake Baringo, Kenya.

Light as ecological factor

The intensity and duration of solar radiation have a direct influence on plants and animals. Plants require light as an energy source for photosynthesis; sun-adapted plants (heliophytes) require a lot of sun, while plants in shaded areas are better able to use any existing light. Light also plays an important role for many animals, for example, to initiate reproductive cycles or stimulate migratory movement in birds.

basics

INDICATOR PLANTS
Plants can act as indicators of particular environmental conditions, for example, stinging nettle is an indicator plant for nitrogen-rich soil.

Heather (*Calluna vulgaris*) is a perennial shrub that grows on well-drained and acidic soils, in open sunny areas or in moderate shade. It is widely found in Europe.

see also: Biocenosis and biotope, p. 276

abiotic factors: temperature and water

Apart from the equatorial region where there are constantly high temperatures throughout the year, plants and animals are often exposed to daily rhythms and temperature variations to which they need to adapt.

Among animals, metabolically controlled heat regulation exists only in those species which are warm-blooded (homoiothermic), in other words, species that are able to maintain a largely constant body temperature. Therefore, throughout their evolution, these animals were able to penetrate into permanently cold biotopes. In cold-blooded (poikilothermic) animals, such as insects, amphibians, or reptiles, body temperature is determined by the ambient temperature, which necessitates strong dependence on temperature stability. When temperature falls below a critical value there is initially a reversible torpor; nevertheless, cold-induced death can occur if temperature continues to fall. Conversely, if temperature exceeds a critical value, this will lead to an irreversible heat rigor, leading to heat-induced death due to the protein coagulation. If animals have a broad temperature range they are eurythermic; if the temperature range is rather narrow, they are stenothermal species. Certain conformities (climatic rules) can be observed in the relationship between temperature and the appearance of an animal. According to Bergmann's Rule, homeothermic animals of a species and of related species grow larger in cold climates than they do in warmer regions (relative to volume, larger animals have a smaller surface area, which is advantageous for heat balance). Allen's Rule (proportional rule) stipulates that individuals of a species (and related species) in colder climatic zones have smaller body appendages (e.g., ears) than those in warmer regions, in order to conserve heat.

The fennec fox, with its distinctly large ears, is a small fox species found in the Sahara of North Africa. Their large ears help dissipate heat and keep them cool much in the same way large elephant ears do.

HIBERNATION

in focus

For many warm-blooded animals, the cold season is associated with increased heat requirements, but because of reduced food availability, the needed energy is often insufficient. Therefore, insect-eating bats, for example, go into hibernation. During this dormancy period, the body temperature can drop below 41°F (5°C), which then leads to a drastic slowdown of the animals metabolism, heart rate, and respiration.

The emergence of groundhogs from hibernation is celebrated in North America.

basics

RGT RULE How strongly temperature can affect an organism's life can be estimated by the RGT (reaction-speed-temperature) rule, which states that a temperature increase of 50°F (10°C) doubles the reaction speeds of metabolic processes.

Water as an ecological factor

Animals in regions with few open water areas usually attempt to keep their water loss as low as possible, for example by means of a nocturnal or subterranean mode of life. Plants in regions with low precipitation also have special adaptations to restrict water loss. Many xerophytes (dry plants) have protective layers and hidden stomata to defend them against drying out.

A rounded body shape, thick fur, and small ears are features that help the arctic fox survive cold cimates.

growth | regulation | competition

POPULATIONS

The totality of individuals of a species that form a reproductive community within a delineated space is called a population. Each population has a uniform gene pool, which can be different from the gene pool of another population of the same species, due to different climatic zones. Population size is subject to constant changes.

growth and regulation

Factors that determine growth rate (the difference between birth and death rates) include biotic (animate) factors and abiotic (inanimate) factors, external and internal, and density-dependent and density-independent factors.

If all essential resources are present in sufficient quantities for the growth of a population, there will be—after an initial phase—exponential growth. This is because the progeny of a generation will produce new progeny, while the death rate will remain stable. Yet, normally the growth of a population is under the influence of various environmental factors, especially by limitations of food sources. Therefore, after a reproductive period with exponential growth, there will be a stationary phase during which developing and dying organisms balance each other out, until the die-off phase sets in, when more organisms die than new ones are born. Such growth sequences can readily be observed in a laboratory where bacterial culture can grow on a well-defined nutrient medium. The density of a population is often the regulatory mechanism within a population, in other words the number of individuals of a species that inhabit a certain area per unit area. For instance, strongly growing populations quickly use up all food sources, resulting in a decline of the growth rate. Other density-dependent factors can also influence population growth, for example intraspecific competition, enemies (parasites or predators), infectious diseases, as well as social stress. Furthermore, density-independent factors, such as interspecific competition or gradual climatic influences and their consequences, usually have additional effects on population density. Among animals, the dominance of one species can also lead to a predator-prey relationship. In such cases, the dominant species (predator) uses the individuals of the inferior species (prey) as food, so that there is a dependency between predator and prey in respect to population density.

Nevertheless, the total inventory of the prey population remains largely unaffected, provided available food resources of the respective biotope permit this.

basics

REPRODUCTIVE SELF-RESTRAINT
Some animals have developed regulatory mechanisms to prevent over-population. For example, birds of prey may only feed the first-born chick and let the others starve.

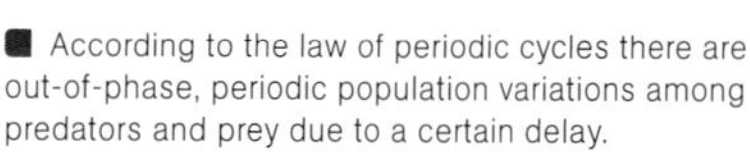

According to the law of periodic cycles there are out-of-phase, periodic population variations among predators and prey due to a certain delay.

There are various reasons for a decline in flamingo populations: susceptibility of chicks to avian predators (an estimated 30 percent of chicks die within the first year); egg removal by the local human population; flooding and hurricanes, which have caused the desertion of entire nest sites; and disruptions from tourists in motorboats.

see also: Feeding and social behavior, p. 263

competition

The intra- and interspecific competition—that is, the contest among organisms for limited resources, such as food and living space—are significant biotic environmental factors that often have a large influence on biomass.

With intraspecific competition, the individuals of a species are in direct competition for biotic and abiotic resources, while with interspecific competition there are two or more species that have similar requirements. Within this context one refers to the totality of all biotic and abiotic environmental factors important for a species as the ecological niche of that species. This term, however, does not describe space but rather respective interactions. The occupation of various ecological niches within the same ecosystem is regulated by interspecific competition: the competitive behavior increases the closer the resemblance between two species.

> basics
>
> **BIOLOGICAL BALANCE** Mutual competition, recruitment and departure of individuals, and the dependency relationship among species leads to the establishment of a dynamic ecological equilibrium.

In order to avoid such competitive situations if possible, many species have developed special adaptations to their environments. An example of this is niche occupancy, which is an effective method to avoid interspecific competition and therefore facilitates the coexistence of many species within the same biotope. Niche occupancy can occur through several different factors, for instance through the variation of main activity periods (day and night activity), diverse food particle sizes, searching for food at various sites, different temperature optima, distinct times for reproduction and for brood care, and so on. Occasionally, niche occupancy can occur within a species that lowers intraspecific competition (intraspecific competition widens the niche, inter-specific reduces the scope of competition). According to the competitive exclusion principle, there can never be two species with identical ecological niches in the same environment.

However, it is possible that different, geographically separated species can occupy comparable ecological niches. The development of similar forms, organs, and modes of life in this case is called convergence (for example, the similar body forms of fish and aquatic marine mammals, such as dolphins).

A spotted hyena and white-backed vultures duel over a carcass. During dry spells or periods when food is hard to come by, animals must be extra aggressive when competing for their meals.

Bindweed wrapping around a cornstalk: Not only animals, but also plants compete with one another.

> in focus
>
> ### NICHE OCCUPANCY
>
> A good example of niche occupancy is food acquisition among ducks. The gray goose grazes on land for terrestrial plants; the teal searches for tiny plant food along the water surface; mallard ducks, northern pintails, and mute swans dabble for water plants, worms, and small crustaceans; while the Goosander hunts for free-swimming food organisms. Varying neck lengths allow species to reach different depths.

Ducks and geese are able to share a pond habitat since all acquire food differently.

see also: Agonistic behavior, p. 264

mutual benefit

In symbiosis, species that are different but adapted to each other live together. Here, one distinguishes between spatial relationships (endo- and ectosymbiosis) or according to the degree of the reciprocal dependency.

Endosymbiosis is a symbiotic relationship where one of the partners (symbionts) lives in the body of the other. An example of such a close relationship is found among lichen, where fungi live in association with green algae or cyanobacteria. The shape and structure of the lichen aggregation is created by fungi (mycobiont); the algae or bacteria (photobionts) are usually located in the upper part of the thallus (vegetative body), so that they obtain sufficient sunlight for photosynthesis. The advantage for the photobiont is, among others, protection against desiccation. On the other hand, the mycobionts receive a large part of their required nutrients from their active (photosynthesizing) partners. Consequently, lichen can grow even in extreme locations. This relationship, where both partners are completely dependent on each other, is called mutualism.

In ectosymbiosis, the symbionts do not live inside the body of their partner. An example of this is the symbiosis of clown fish and sea anemones. The latter protect the fish with their stinging tentacles and, at the same time, profit from food remnants generated by clown fish feeding among their tentacles. Moreover, the fish often defend their symbiotic partner against predators. Frequently, symbiotic organisms are subdivided based on the type of their dependency. Commensalism, for example, occurs when only one partner derives benefit from the association. This is the case when, for instance, animals benefit from the protection of a partner, without damaging or being of use to that partner. The burrows of large marine worms or shellfish very often contain "uninvited guests," not only for protection but also because they utilize food remnants or waste products.

Protocooperation occurs when two organisms enter into only loose relationships, which could benefit both, but are not imperative because the participants are also able to live alone. An example of this is the already mentioned partnership between clown fish and sea anemones.

mutual benefit | parasitism

TYPES OF COHABITATION

A particular form of cohabitation between different species, advantageous for both partners, is symbiosis (also referred to as mutualism). This is in contrast to other parasitic relationships where organisms feed partially or completely at the expense of others, while they live temporarily or permanently on or inside their body.

Clown fish and sea anemones famously form a near-perfect, mutually beneficial relationship.

Oxpecker birds and hippos have developed a symbiotic relationship, in which the birds prey on parasites on the hippo's body. This means that the birds are well fed and the hippopotamus is rid of potentially harmful ticks.

in focus

MYCORRHIZA

Mycorrhiza is an important type of cohabitation between the roots of higher plants and fungi. There is a difference between endotrophic mycorrhiza, where fungal hyphae penetrate the root cells, and ectotrophic mycorrhiza, where the roots are enclosed by an envelope of fungal hyphae that penetrate only into the interspaces between cells (extracellular) of root bark.

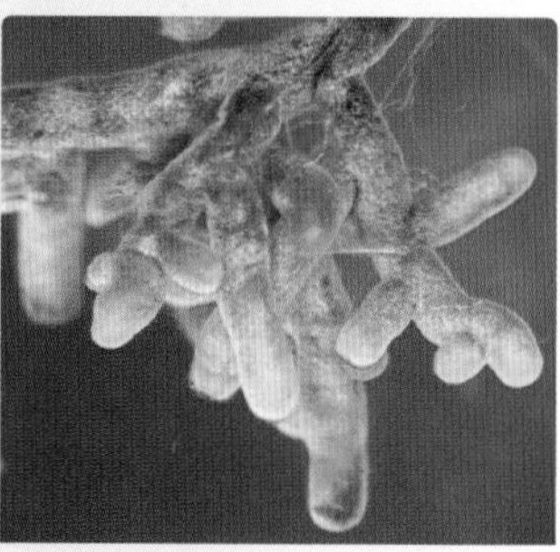

About 95 percent of all higher plants live in a symbiotic relationship.

parasitism

Parasites are organisms that withdraw nutrients from other species (hosts). This may damage the host but it will not, at least during the early stage, kill it. Parasites are mostly specialized for a specific host (host specificity).

Parasites are intricate organisms who have complex relationships to their hosts. Some parasites (phytoparasites) are parasitic on plants, while others (zooparasites) use animals as hosts. There is a futher separation between ectoparasites (internal) and endoparasites (external) among animal parasites, as well as between full and partial plant parasites.

The Oriental rat flea is a rodent parasite and can transmit the typhus disease.

SAPROPHYTES are organisms that obtain their nutrients from dead and decaying organic plant or animal matter. Typical saprophytes include many fungi and bacteria, and, among plants, the rare ghost plant (or indian pipe).

basics

Animal ecto- and endoparasites

Ectoparasites, which live outside the host's body, frequently feed on the blood of the host, for instance, fleas, lice, or ticks. Some of these (such as the head louse in humans) are host specific, while ticks do not have a particular host specificity. These parasites often display special adaptations, for example, lice are not only wingless, but they also have a flattened body and special oral tools adapted to their mode of feeding. Temporary parasites, such as mosquitoes, carry out nonpermanent attacks. Animal endoparasites, on the other hand, include all parasites that live inside a host. They are found most commonly in the digestive tract and in the blood; however, they can also occur in muscle tissue. Endoparasites show even greater adaptations to their particular mode of life. For instance tapeworms do not have a digestive tract and absorb the food digested by the host directly along their body surface.

Partial and full plant parasites

Among plant parasites, partial parasites (hemiparasites) differ from full parasites (holoparasites). The latter no longer use photosynthesis because they remove all of their required nutrients from a host plant. An example is the toothwort (*Lathraea*), which forms a long, heavily branched subterranean root system to which numerous suction organs (haustoria) are attached. They are used by the parasite to tap into the roots of other plants, specifically trees like alders and beeches. Partial parasites, however, still photosynthesize, but they remove water and nutrient salts from their host plants. A typical semiparasite is the evergreen mistletoe, which lives on trees and is normally not a threat to its host.

Mistletoe are semiparasitic plants that grow attached to trees and shrubs, though they generally do not damage them.

in focus

NEST (BROOD) PARASITISM

In nest parasitism, animals take advantage of the brood behavior of another species, whereby the progeny of the host species is adversely affected in favor of their own progeny. This form of parasitism is found among birds and insects. For instance larvae of the ichneumon wasp parasitize the bodies of certain caterpillars. However, the most well-known example is that of the cuckoo, which lays its eggs in the nests of other birds. These birds then raise the cuckoo's chicks as one of their own. After hatching, the young cuckoo usually throws the eggs or the chicks of the host parents out of the nest.

In many cases the parasitic cuckoo nestling is raised by significantly smaller host parents.

biocenosis and biotope

The totality of living beings, including animals, plants, and microorganisms of an ecosystem is referred to as biocenosis; the spatially delineable environment with its typical, ambient conditions is called the biotope.

Black bears have an extremely varied diet that includes fish, honey, berries, nuts, insects, and carrion.

A biocenosis, or ecological community, is characterized by the number, abundance (average number of individuals of a species in an area), and areal distribution of species. The species inhabit different ecological niches, so species diversity is primarily dependent upon the number of ecological niches in an ecosystem; the higher that number is the more species can avoid competition with other species. In a biocenosis, there is generally constancy in specimen density within local species. In an ideal case, natural material cycles are stable, so that a dynamic equilibrium is established. If this is disturbed, the overall character of the ecosystem changes and it can be partly or entirely destroyed. Moreover, since ecosystems are interrelated, the equilibrium of adjacent ecosystems can also be affected. Biotopes, the delineated environments of a biocenosis, display a more or less uniform state or condition, and can be distinguished from each other. Similarly, a habitat refers to the location where a particular species occurs.

biocenosis | biotope | energy flow | food chains

ECOSYSTEMS

Ecosystems are dynamic systems of inanimate (abiotic) and animate (biotic) components. They are interconnected through material and energy cycles. Abiotic factors include air, soil, climate, and food requirements; the biotic factors are plants, animals, fungi, and bacteria. All ecosystems on Earth make up the biosphere.

Aquatic ecosystems, such as lakes, are varied and delicately balanced.

Energy flow within ecosystems

The organisms in an ecosystem can be categorized according to their function into producers, consumers, and destruents (decomposers). The producers, or manufacturers, include all organisms that are capable of photosynthesizing. These are principally plants, but also include autotrophic bacteria. They synthesize organic compounds (biomass) from inorganic substances, which are then used as food by all heterotrophic organisms within the ecosystem. The consumers, or users, include all plant-eaters that are directly reliant upon the photosynthetic capability of producers (primary consumers), as well as small and large meat-eaters (secondary and tertiary consumers). The destruents include bacteria and fungi, which reduce dead organic substances to water and carbon dioxide, as well as into minerals. These are then available again as nutrients.

basics

SUCCESSION AND CLIMAX The stable, final state of an ecosystem is called the climax condition. Succession is the state of changes in the ecosystem that can lead to a climax community.

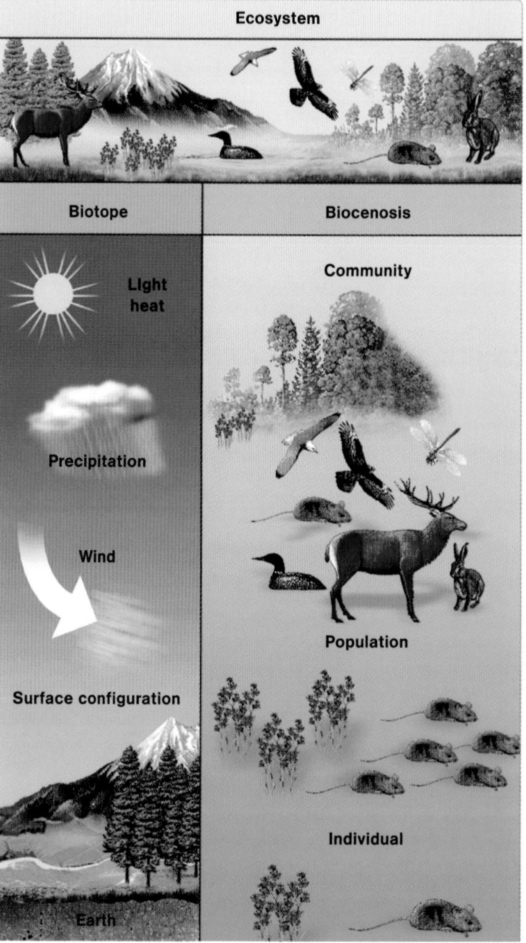

Ecological systems are made up of biocenoses and biotopes, between which occur complex and reciprocal effects.

see also: Environmental protection, pp. 128–137

energy flow and food chains

Food chains, or food networks, describe the feeding relationships between species within a certain ecosystem. The system illustrates the intricate interactions between organisms in terms of consumption and production.

The organisms of an ecosystem that are linked to each other by production and consumption of biomass form food chains that are generally interrelated within often very complex food networks. Simplified, trophic (food) relationships in all ecosystems can be depicted as follows: The sun is always the energy source. Its energy transforms autotrophic organisms, that is, mainly plants and cyanobacteria (primary producers) into chemical energy with the aid of photosynthesis. This energy is stored in the form of energy-rich substances within their cells. Subsequently, plant-eating animals feed on the producers. These are the consumers of the first order, which modify the energy taken in and store it in their cells. The next link in the chain are the meat-eaters, which feed on the plant-eaters and which are then designated as consumers of the second order. They are often eaten by other species (consumers of the second order, and so on). Meat-eating animals that no longer have any natural enemies make up the end of the food chain (apex, or top-level predators). Upon their death other organisms, such as the decomposers, begin to break down these terminal links of the food chain. Decomposers are mainly bacteria and fungi, which disintegrate these dead organisms into low-molecular components. Primary producers then utilize these particles so that the top predators' organic substances are returned to the perpetual cycle of nature.

Under natural conditions, such pure food chains can exist, however, only in biotopes that are deficient in species. In most ecosystems there is a net of interrelated food chains, called a trophic web. Typical of such trophic webs are numerous cross-connections between different organisms. This arises principally from the fact that animals usually do not feed on a single species alone, but rather have a far wider and richer food spectrum.

Energy moves from lower to higher tropic levels: plants are eaten by herbivores, who are consumed by carnivores, who, in turn, are eaten by top-level predators.

basics

INTERFERENCE If a food chain is interrupted, e.g., by pesticides entering water, the subsequent links in that chain also perish. Broader food webs are less sensitive.

in focus

FOOD PYRAMID

Within food chains, the weight of consumers increases only by about one-tenth of the amount of food consumed, since most of the food serves to generate energy and not to increase mass. Part of the biomass is also lost as heat and via excretion. Thus, food chains rarely have more than five links, otherwise there would be little left for terminal consumers such as humans.

A bird of prey devouring mice: The loss from one food level to the next is about 90%.

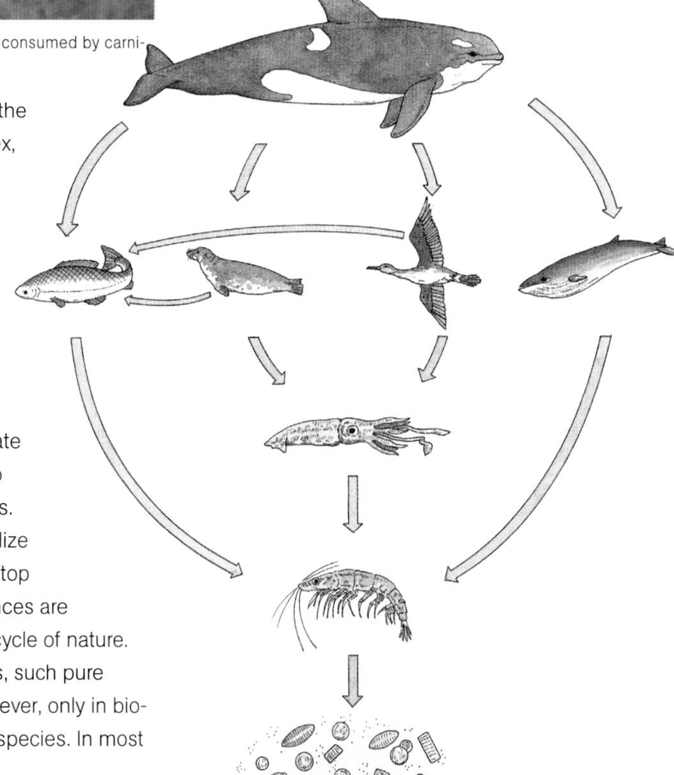

Biomass is steadily reduced from one level up to the next in a trophic pyramid.

see also: Feeding and social behavior, p. 263

carbon cycle | nitrogen cycle | phosphorus cycle | water cycle

CYCLE OF MATTER

Material cycling, as recurring processes, enable a return of matter into food chains. Thus, an ecosystem in terms of its turn over can be independent of an external supply of matter, except for the (required) energy supply, e.g., through solar radiation. Important material cycles include the carbon, nitrogen, phosphorous, and hydrogen cycles.

carbon cycle and nitrogen cycle

Carbon plays a particularly important role in ecosystems, as it participates in fundamental biological processes. Nitrogen is also an essential substance as it is required for the synthesis of proteins and nucleic acids.

All organic compounds contain carbon. The carbon cycle is a complex series of processes through which all of the carbon atoms in existence rotate. The very same carbons in our bodies today have been used in countless other molecules since the beginning of time. The carbon cycle starts out when autotrophs (organisms that can use inorganic material for nutrients, such as green plants) pick up carbon from the air in the form of carbon dioxide. During photosynthesis, this is reduced to carbohydrates (sugars), which can serve as food for plant eaters. Subsequently, carbon is oxidized back to CO_2 through respiration and fermentation. It is then returned to the air, so that the cycle is closed and balanced out again normally. In recent decades, however, the carbon dioxide content in the atmosphere has steadily risen, especially due to the increased use of fossil fuels and other anthropogenic (or human-made) interventions into the global ecosystem. This has raised concerns about excessive carbon dioxide levels.

Carbon is released through the combustion of organic material

The decay of animal and plant matter produces carbon

Forests absorb the majority of the planet's carbon (above ground)

Almost every organism on Earth needs carbon to survive. Unlike energy, carbon is constantly cycled and reused, as the Earth only has a fixed amount of carbon. Consequently, the carbon cycle can be seen as one of the most fundamental forms of recycling.

The nitrogen cycle

Nitrogen is an essential substance for all organisms because it is required for the synthesis of proteins and nucleic acids. So that nitrogen is available in sufficient amounts, individual forms of nitrogen are constantly being provided by a type of cyclic replenishment.

The nitrogen-fixing bacteria rhizobium form nodules on soybean roots and convert nitrogen into nitrogen compounds.

When nitrogen compounds from the excretions of animals, animal cadavers, plant remnants, and so forth reach the soil, they will eventually be broken down (proteolysis). Nitrogen is then released in the form of ammonia (NH_3). Subsequently, bacteria oxidize this ammonia (through a process called nitrification) into nitrate (NO_3^-). This newly created substance is then absorbed by green plants and built into organic compounds (proteins and nucleic acids) in a process known as nitrogen assimilation.

Plant proteins and nucleic acids are then utilized by plant-eating animals for the development of their own proteins, because they cannot absorb inorganic nitrogen. When plants and animals die, these nitrogen compounds are returned to the soil and the nitrogen cycle begins all over again.

basics

DENITRIFICATION

The opposite process of nitrogen fixation is denitrification, where bacteria take in nitrate (nitrate respiration). This releases elementary nitrogen.

see also: The climate system, pp. 118–127

phosphorus cycle and water cycle

Phosphorus plays an important role for all organisms in many processes involving energy transfer and energy maintenance. In addition, water is an indispensable component in all activities of life's processes.

Phosphorus, which occurs in nature almost exclusively in the form of organic compounds or phosphate, participates in various metabolic reactions. It is also an indispensable component of nucleic acids and enzymes, as well as the building block for bones and teeth. Phosphorus enters the phosphorus cycle exclusively in the form of phosphate ions (PO_4^{3-}). These originate principally from the effects of weathering and the breakdown of phosphorus containing rocks, as well as from phosphorus waste. Some of this phosphate is then taken up by plants and is eventually returned to the soil by the so-called phytophagous food chain (through grain cereals and their consumption by animals and humans). At that point it is converted back again to phosphate by bacteria, and it can then reenter the phosphorus cycle. Another part of the phosphate is washed out through groundwater, streams, and rivers into lakes and the ocean, where most of it is taken up by phytoplankton and then flows back into the food chain. These final phosphorus links are then often returned to the mainland, for instance through fishing activities, so that the cycle is consequently closed once more.

basics

GUANO The phosphorus and nitrogen-rich excrement deposits of sea birds are referred to as guano and used to be a popular fertilizer, mainly during the 19th century.

The water cycle

The constant movement of water associated with changes in aggregate conditions, between the oceans, the atmosphere, and the mainland is referred to as the water cycle. The cycle commences with evaporated water (water vapor) rising from water surfaces (especially from the oceans but also from lakes and rivers), from the soil (evaporation), and from the plant cover (transpiration) entering the atmosphere.

in focus

EUTROPHICATION

If suddenly large amounts of phosphate are added to a lake or a pond, for example, via wastewater, it can destroy the fragile ecological balance. Due to rapid algae growth (a sudden increase in food material), animals will quckly reproduce. This leads to a significant increase in oxygen consumption, so that in an extreme case a lack of oxygen may make the waters uninhabitable.

A pond overrun with algae: Algae is sometimes used to capture fertilizers and waste.

The driving force for the formation of water vapor is solar energy. The water vapor drifts away due to air currents, and following condensation and the formation of clouds it is returned to Earth in the form of rain or snow. Approximately 80 percent of this precipitation falls over the oceans. Part of the remaining 20 percet evaporates from the mainland and is returned directly back to the atmosphere. Some of the water also sinks into the ground and so supplies the groundwater. This groundwater later returns to the surface through springs or runs off the land surface into rivers and then back into the ocean. There the water cycle begins again.

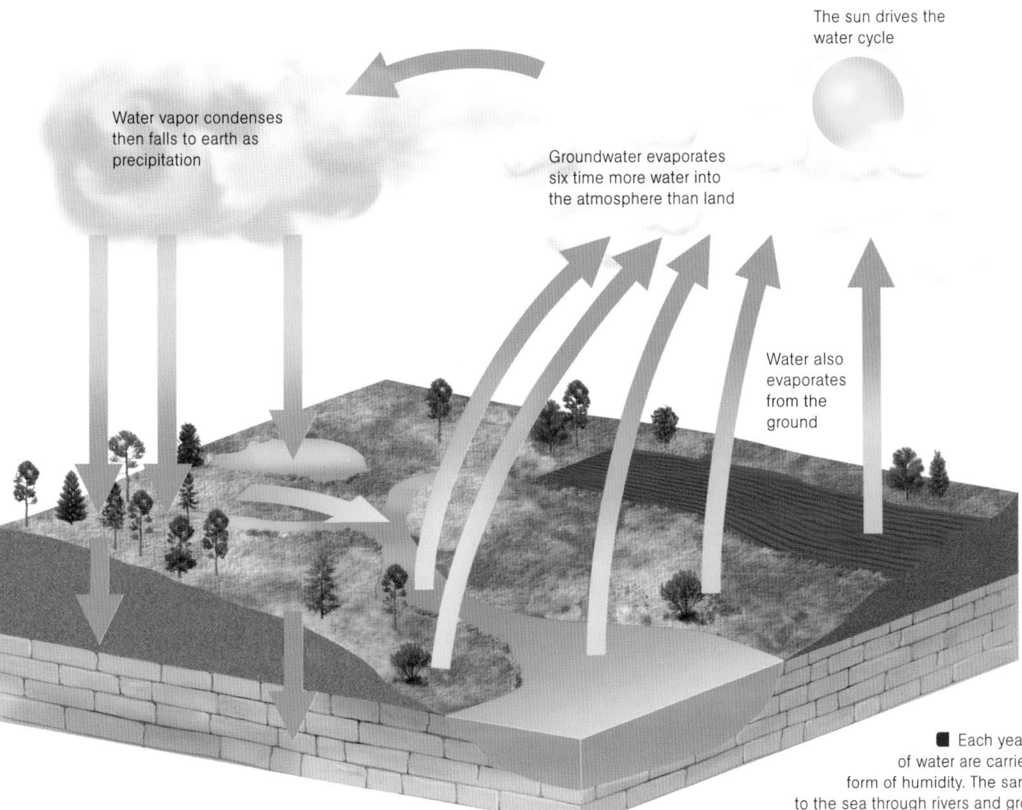

Each year 9,500 cubic miles (39,700 km^3) of water are carried from the oceans to land in the form of humidity. The same amount subsequently returns to the sea through rivers and groundwater channels.

see also: Global warming, p. 126

agriculture | cities

HUMAN ENVIRONMENTAL IMPACT

Humans have changed their environments according to their needs from time immemorial. A typical change to natural ecosystems is the clear-cutting of forests for food production. Many natural ecosystems are also destroyed due to increasing urban development.

agricultural ecosystems

Agricultural ecosystems, or cultivation ecosystems, are created by humans for agricultural use—that is, mainly to produce food and other biological raw materials. These human-made ecosystems come to replace natural ones.

Agricultural ecosystems need to be cultivated and, therefore, controlled and monitored. For example, the level of nutrients needs to be topped up regularly by fertilizers, as the removal of the crop disrupts the material cycle. Without fertilization, an agricultural ecosystem breaks down within a short period of time and sooner or later converts back to an environment that is governed by natural processes. To ensure economically viable agricultural areas, it is often perceived necessary to clear cultivated areas of all noneconomic species, which results in monocultures of a single crop species being sold. These monocultures are very different from the natural environment. They reduce the diversity of plants and animals to a minimum and shorten food chains.

Native animal and plant species are eliminated not only due to intensive cultivation but also due to the use of insecticides, fungicides, and herbicides used to protect crop plants. Populations of other organisms, especially so-called weeds and pest species, often grow significantly as a result (see in focus). They may even invade the area from other regions. Ecosystems adjacent to intensively cultivated areas are also often negatively impacted. For instance, fertilizers and pesticides may be washed from a field into creeks and rivers. In this manner, toxic substances may accumulate in the food chain. In an attempt to tackle these problems—and also to recognize the growing awareness of these concerns among consumers—some agricultural products are now being produced in more ecologically sound ways. These involve recognizing natural material cycles and cutting down on the use of synthetic fertilizers and pesticides as much as possible, even if the crop yield may be lower. The more naturally an agricultural area is maintained, the greater its ability to self-regulate. This is a typical characteristic of natural ecosystems.

Planting corn in burnt soil near a national park in the Ivory Coast contributes to deforestation.

Apple pests can cause great damage to an orchard. Apple growers are thus looking at low-cost and environmentally friendly ways to quell these pests: for instance, by introducing insects that are predators of the pests.

in focus

PESTS ON THE INCREASE

In order to maximize the use of agricultural areas, most are cultivated as intensively as possible by planting vast areas with only one type of crop. However, such monocultures are ideal environments for pest species, which may populate these areas very quickly. These pest species are usually killed by the use of insecticides, herbicides, or fungicides, which is certainly effective in the short term. However, not only do these toxic substances accumulate in the food crops, but the pest species also develop resistances against such pesticides within only a few generations. Pest control then becomes a real challenge. Therefore, biological methods are becoming more and more popular as long-term solutions for pest control.

A farmer sprays pesticides on his farm in Chiang Mai, Thailand.

see also: Environmental concerns in everyday life, p. 136

cities

Until well into the 18th century, most people were living in agricultural areas. This changed with industrialization, when more and more people migrated into the cities—rather unnatural ecosystems.

Urban environments were built by humans according to their ideas and requirements. They are influenced by abiotic factors such as climate, water, light, and heat, and have little in common with natural ecosystems. Most urban areas are covered by buildings, roads, and footpaths. The pavement seals soil in and thus prevents precipitation from permeating the ground; this consequently disturbs the natural material exchange between the air and soil. Unpaved areas are much dryer compared to the surrounding areas outside of the city. This is in large part due to the effects of targeted drainage. The ground also suffers from other factors such as road salt and industrial waste.

issues to solve

UBURBAN RECRE-TIONAL AREAS

owadays even the reas surrounding lany cities are hanging and losing leir natural charac-eristics as they are ntensively used and nodified for recre-tional purposes. hese ecosystems equire many pro-ective measures to nsure that they do ot disappear.

The urban climate is also different from the climate of surrounding areas. The many buildings store significant quantities of heat during the day and release it during the night. Fossil fuel is constantly burnt by heaters, power plants, and industrial facilities. High building densities in cities also prevent the exchange of air masses, due to the lack of wind and evaporative cooling, which are themselves the results of a lack of vegetation. In addition, frequent smog days are the result of particles and fumes from exhaust pipes in the air, preventing the release of heat energy into the atmosphere. This is why the annual average temperature is at least one degree higher in cities than in surrounding areas. There are also fewer days of freezing and the humidity can be up to ten percent lower. Moreover, biodiversity is significantly lower in urban areas compared to natural ecosystems, as living conditions are comparatively unfavorable for most plants and animals. Nevertheless, urban residential areas offer many opportunities for small habitats. These are mainly areas such as gaps in the pavement of sidewalks or mulched areas around urban trees with limited opportunities to grow due to landscaping maintenance. Even large cities are frequently visited by several invading species. For example, common pigeons *(Columba livia)* or the common kestrel *(Falco tinnunculus)*, which would originally nest on rocky walls, find suitable nesting areas in tall city buildings. They now live year-round in many of our cities.

in focus

LICHEN AS A BIOLOGICAL INDICATOR

In some cities, lichen is used as a biological indicator to test air quality. Lichen is very sensitive to pollutants as well as changes in climate. If fruticose lichen still grows in an urban area, it is said to have clean air. If lichen mainly consists of foliose lichen, then the air is considered slightly polluted (50–70 μg of sulfur dioxide/m^3). The air is considered to be polluted (70–100 μg sulfur dioxide/m^3) if crustose lichen are dominant or heavily polluted (more than 100 μg sulfur dioxide/m^3) if only few crustose lichen are growing in an area.

Lichen, which are found on rocks and trees, often look like peeling paint.

Dallas is the third largest city in the state of Texas and the ninth largest in the United States. It has the greatest inland metropolitan area in the country.

CHEMISTRY

CHEONG HING
STORE LTD.
COSMETICS
PERFUMERIES
KENWOOD
TSANG FOOK PIANO CO

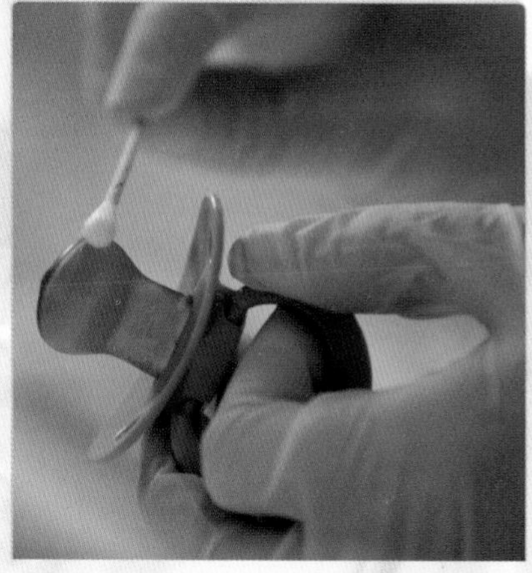

INORGANIC CHEMISTRY

In science and in everyday life, chemistry appears as a world of substances: whether pure materials or mixtures; elements or complex molecules; stable matter or reactive compounds. Chemistry describes a microcosm that is constantly in motion and forming new combinations, whether in the natural environment, industrial laboratories, or our own households. Today's chemists work not only in the laboratory but also in front of the computer, where they simulate new combinations, test their theories, and try out potential applications.

Detecting tiny amounts of harmful substances—such as in food, the human bloodstream, or the environment—is only one of the many important applications in the field of chemistry. With nanotechnology, for instance, scientists are taking decisive steps into the territory of atoms and molecules, gaining opportunities to develop completely new and "intelligent" materials.

elements | compounds | periodic table | chemical bonds

MATTER

The ancient Greek philosophers believed that all matter consisted of four elements: fire, water, air, and earth. It was not until modern times that researchers succeeded in identifying the building blocks of matter. Through this they were able to give order to the various materials in the world and explain chemical changes between substances.

from matter into elements

Compounds, mixtures, and elements differ by the ways in which they are separated into component parts. An element consists of atoms and cannot be broken down by physical or chemical means.

Our environment consists of chemicals and over 30 million are listed in the Chemical Abstracts database. Almost 12 million of these are commercially viable and we constantly develop new ones. Worldwide, 400,000 new chemicals are created each year. Chemicals may be elements, compounds, or mixtures.

Elements

Elements have only one type of atom and cannot be broken down by chemical or mechanical means though they can be broken using particle accelerators. Examples of elements include hydrogen, oxygen, gold, and carbon. Carbon is one of the most common elements and is found in many varied forms. For example, both graphite and diamonds are forms of carbon even though they look very different. The known elements are listed in the periodic table of elements.

Compounds

Compounds are combinations of elements that have bonded together to form molecules. They cannot be broken down by mechanical means though they can be broken into their components by other means. For example, water will separate into oxygen and hydrogen when an electric current is applied. Compounds can appear very different from their components. For example, a molecule of common salt (NaCl) consists of one atom of sodium (Na) and one atom of chlorine (Cl). In its pure form sodium is soft and silvery white. Chlorine in its pure form is usually a pale green gas, highly toxic, and water reactive.

■ Gold is very soft. For use in jewelry, an alloy (a type of mixture) is made, usually with silver and copper.

Mixtures

Mixtures often have a uniform appearance but they can be broken into their components by mechanical means. Seawater is a mixture that includes the compounds water and salt. The alloys brass and bronze are mixtures of metals with special characteristics. In other cases, such as emulsions, each type of liquid can be easily seen. Many oil-and-vinegar salad dressings are emulsions. There are several common ways to separate mixtures into their components. Filtration is one method, used when the particle sizes differ or can be made to differ with a solvent. Heating is another method. Wine is a common mixture that can be separated into its component parts using distillation, a heat process: Distillation of wines is used to either create a more alcoholic beverage such as brandy, or converting perished and surplus wine stock into bioethanol (biofuel) or industrial alcohols.

practice

DISTILLATION OF WINE

Wine has been produced since 6000 B.C. It is a mixture of ethanol, water, and other substances. The ethanol can be extracted by distilling a heat process. When wine is heated it starts to boil at approximately 172°F (78°C), the boiling point of ethanol. The ethanol turns into a gas and rises leaving the other materials behind in liquid form. If the fumes are collected in a cooler container, the ethanol condenses into a liquid. If you continue to heat the wine the boiling point will rise to 212°F (100°C), the boiling point of water.

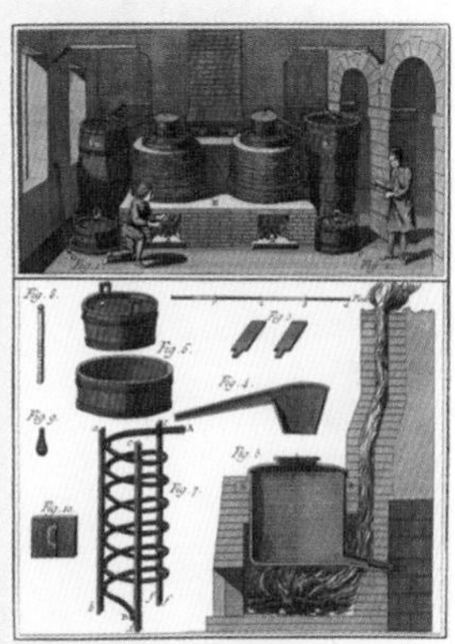

■ The retort and alembic have been used for distillation since A.D. 800 and are still in use today.

■ Some mines use cyanide to mine gold, creating a toxic compound called gold cyanide.

see also: Quarks, electrons, and co., p. 328

atoms: the building blocks of nature

The first step in the development of modern chemistry was the recognition that atoms existed. Today it is known that every element consists of a specific kind of atom.

At the start of the 19th century, John Dalton, a science teacher from Manchester, England, promoted the theory that all matter is composed of indivisible atoms. The atoms of any element are identical in their mass and chemical makeup. In the years that followed, atoms were assumed by scientists to be round elastic objects that were uniformly filled with matter. But later in that century this view was definitively refuted when physicist J. J. Thomson showed that negatively charged particles, electrons, can be separated from their atoms. Furthermore, Antoine Henri Becquerel (1852–1908) was actually able to observe natural radioactivity. Through these discoveries physicists began to understand that the atom could be split.

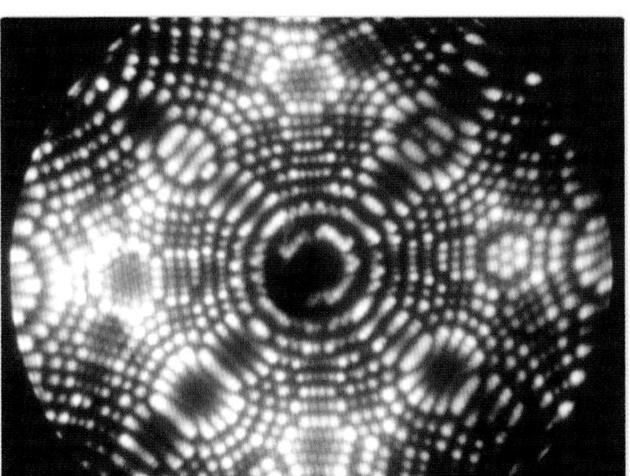

Every bright spot on this field ion microscope image of platinum is an individual atom.

Mainly empty space

At the start of the 20th century, Ernest Rutherford bombarded a thin sheet of gold leaf with alpha particles emitted from a radioactive element. Almost all of the particles penetrated the gold leaf with no deflection, but a few were deflected and some even bounced back. "*It was unbelievable, almost as if you were to fire a 38-cm artillery shell at a piece of tissue paper and have it come back and hit you,*" wrote Rutherford later. He concluded that the atoms in the gold leaf consisted mostly of empty space, but at the same time there was a mass in the center of these atoms that could deflect the incoming particles. This led Rutherford to the idea that atoms are constructed of a positively charged core and a negatively charged shell. This idea was further developed by Niels Bohr and others. According to them, the shell of an atom consists of electrons that circle a nucleus, which is made up of protons and neutrons. Electrons move in fixed orbits, much like the planets. The Bohr atom model is based on a bold hypothesis. From the perspective of classical physics the electrons that orbit the nucleus of the atom would release their energy in the form of radiation. If this were true the electrons would eventually crash into the nucleus of the atom. The physicist Niels Bohr advanced the proposition that this is exactly what electrons do not do. Starting from this assumption, he developed his own generalized and idealized model of the atom. Using this model, Bohr could explain the wavelength of the light radiated by hydrogen atoms that were in an excited state.

The particle bombardment of elements is used to deduce the atomic structure of atoms, due to their miniscule size and unpredictable nature.

in focus

WAVE MECHANICS AND ORBITALS

A problem for physicists from Bohr to Erwin S. Schrödinger was that it seemed that for half the time they would be applying the classical laws of physics, and the other half they would be applying the laws of quantum physics. Today, quantum wave mechanics posits that the position and momentum of an electron cannot be simultaneously determined exactly. From this perspective there are no more electron orbits. Instead, the movement of electrons is described and determined by a mathematical function: the wave function. The wave function value yields the probability of an electron being present within a given small volume. The region in which an electron may be found around an atom has a characteristic shape; this region is called an orbital.

The Rutherford model of the atom, although inaccurate, is widely recognized.

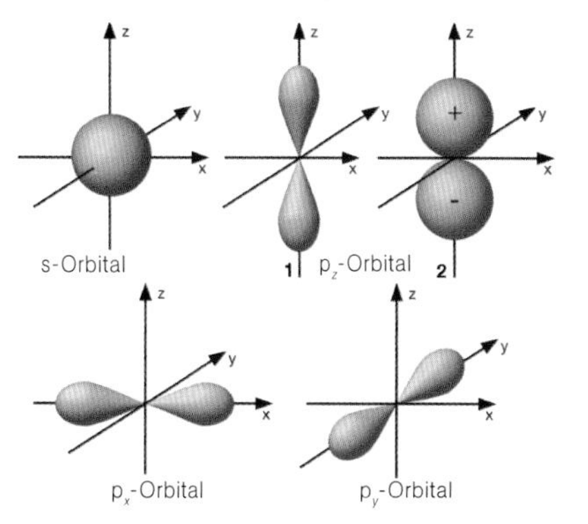

A computer-generated image of the atomic regions called orbitals.

see also: Nuclear power, p. 349

the periodic table

A copy of the periodic table hangs in every chemistry classroom around the world. Today it includes 111 confirmed elements, arranged in groups and periods according to their atomic number and properties.

Chemists had long speculated about an order for the elements when, in 1829, the German chemist Johan Döbereiner found several groups of three elements ("triads") with similar chemical properties and noticed that the atomic mass of the element in the middle is approximately the same as the average of the mass of the other elements.

19th-century discoveries

Many elements were discovered in the mid-19th century. In 1869 Russian scientist Dimitri Mendeleyev and German scientist Lothar Meyer independently developed a periodic table. Mendeleyev classified the elements in ascending atomic weight order and placed related elements under each other in vertical columns. He also left blank spaces under aluminum and silicon to accommodate elements yet to be found. About a century later, gallium and germanium were discovered and had the properties predicted by Mendeleyev. This contributed greatly to the scientific acceptance of the periodic table.

NEW ELEMENTS Today, scientists can use nuclear fusion to make elements with a short half-life. In the future, man-made elements may become more stable.

issues to solve

The periodic table today

The periodic table shows the elements with their chemical symbols arranged in rows in the order of their atomic number. The atomic number (also called the proton number) is the number of positive elementary particles (protons) in the nucleus of the element's atom. It uniquely identifies a chemical element.

In an atom of neutral charge, atomic number is equal to the number of electrons. Hydrogen atoms (symbol: H) have the simplest structure since they have only one proton. Uranium atoms contain 92 protons—these are the largest atoms found in nature.

The element's chemical properties are determined by the number of electrons in its outermost shell. Elements with similar properties are arranged in 18 vertical groups. For example, all the elements in the eighth main group (the "noble gases") have eight electrons in the outer shell—except for helium, which has two—and do not react with other elements. The horizontal rows are the periods (1–7).

The atomic mass, measured in unified mass units, is calculated as the combined mass of protons and neutrons, because the weight of electrons is so small as to be insignificant. The atomic mass differs between isotopes of one element.

Neon: elements in one group of the periodic table have similar properties. The gases of group eight, called the noble gases, are all stable and have low reactivity, due to their full valence (outer) shell of electrons.

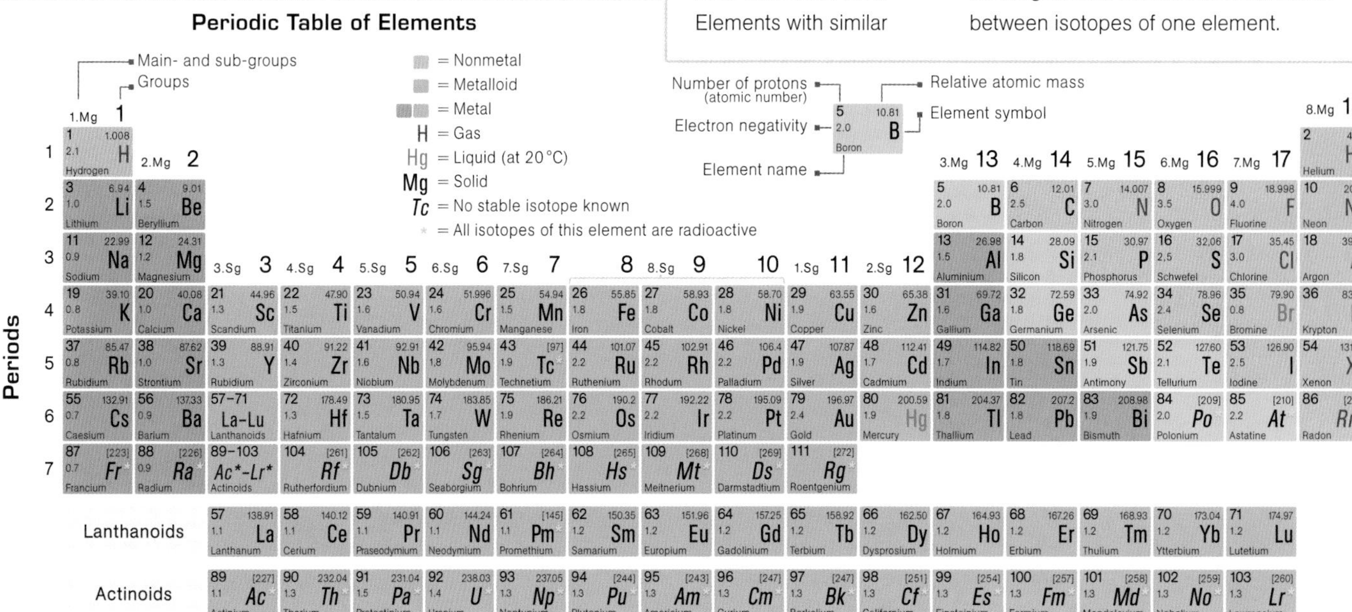

The periodic table of elements represents all elements in order of their atomic number. The elements in this chart are arranged horizontally in periods, and vertically in groups according to their chemical properties. Within the main groups further differentiation is made based on the elements' properties (such as electrical conductivity). Elements within subgroups are more similar to each other than those in other groups. A further group of elements (the g-series) has been suggested as theoretically possible.

the power of chemical bonds

Chemical bonds between atoms are formed when atoms come into contact with each other and their electron shells interact. The bonds may be covalent, ionic, or metallic.

Atoms bond upon contact with each other. For example, two hydrogen atoms that are far apart do not exert any force on each other. This changes when they make contact and the laws of electrostatics come into play. The negatively charged electron shell of one atom attracts the positively charged nucleus of the other. The electron shells "coalesce" and a zone of negative charge is formed around the two nuclei.

Free electrons from outer shells of metal atoms

Metal ions

The outer (valence) electrons in a metal easily disassociate from the nucleus. Thus metallic bonds—where electrons are disassociated and free to move—cause metals to be ductile and malleable. The shape of the metal can be changed without the bonds breaking apart.

Attraction and repulsion

This process has limits. The closer two atoms come, the greater the force of repulsion. The forces of attraction and repulsion combine, so that some distance remains between the atoms and a molecule is formed.

Hydrogen chloride is formed by a covalent bond between chlorine and hydrogen. The bonding electrons are more strongly attracted to the chlorine than the hydrogen. Such simple bonds are called monomers. When there are two monomers, the compound is a dimer; with three monomers, the compound is a trimer. The electrical charge in hydrogen chloride molecules is unequal; it has two poles. This bonding is called polar covalent.

Electronegativity (EN) is a measure of how much an atom in a molecule attracts electrons in a bond. This determines whether the bonding electrons are completely transferred to an atom or shared between the two atoms. When the EN-difference of the atoms is greater than 1.8, an ionic bond is formed. When the EN-difference is very small, the bond is nonpolar covalent. Sometimes, one atom completely acquires the electrons of another, which results in the generation of ions. The force of attraction between ions is very high due to their opposite charges, and they build a very stable ionic lattice structure.

Solid ionic compounds often form a lattice due to the attraction and repulsion forces between positive and negative charges. This makes them very strong, but brittle.

Metals, like ionic compounds, have a definite structure. The electrons are not bound to a particular atom. Instead, they form a sea of negative charge that binds the positive nuclei together. Most covalent molecules have only a few atoms. Carbon atoms are an exception. They are able to form bonds with four other atoms, leading to complex molecules. Carbon chemistry thus forms its own branch of chemistry: organic chemistry.

STRUCTURES OF DIFFERENTLY BONDED MATERIALS

in focus

In covalent bonds, electrons are shared by the bonding atoms, and a stable connection—a molecule—is formed. The resulting molecules may be small, medium-sized, or even "gigantic," depending on the number and type of atoms involved. Small compounds are generally volatile and gaseous (such as HCl, perfumes, and aromatic oils), with low melting and boiling points. They generally do not conduct electricity well. Large molecules can form three-dimensional structures and chains (such as polymers and proteins).

Solid ionic compounds are often powdery, salt-like substances, consisting of many small crystals (with a crystalline lattice structure). They have very high melting and boiling points. They are often soluble in water, where the positive cations and negative anions are good conductors of electricity.

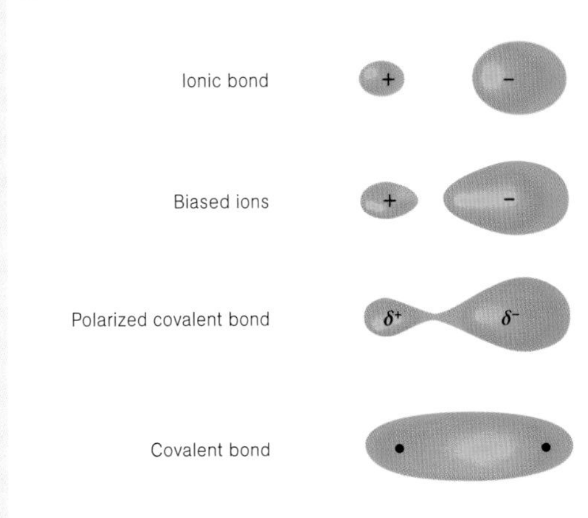

Sharing an electron pair produces a molecule, while the transfer of electrons results in two ions.

see also: Elementary particles, pp. 328–329

transformations | reactions | reactor | catalysts

MATERIAL IN FLUX

Chemical processes are everywhere. Engines transform gasoline into exhaust gases. Iron chains become rust. In an oven, a raw roast is transformed. In our bodies, our food is transformed into new materials. Industrially, large amounts of new substances are created by controlling chemical reactions.

new substances are formed

Chemical reactions are one way we form new substances, both willingly and unwillingly. The reactivity of different compounds can create surprising and useful combinations, or damage and breakdown vital componentry.

New substances are produced when chemical reactions occur. All of these chemical transformations involve either an increase or a decrease of energy. As an example, when 0.141 ounces (4 g) of zinc powder is mixed with 0.0755 ounces (2 g) of sulfur in a dry flask and the resulting mixture is touched with a red-hot wire, flames and white smoke appear. After cooling, a grayish white mixture remains. This reaction releases heat and is called exothermic. In endothermic reactions, however, energy is removed from the environment. The amount of energy needed to start a reaction (in this case from the red-hot wire) is called the activation energy. The reaction between zinc (Zn) and sulfur (S) can be described by the equation: $Zn+S \Rightarrow ZnS$. The start materials, or reactants, are to the left of the arrow. To the right of the arrow is the final product. While the zinc is reacting with the sulfur, it is also reacting with oxygen (O_2) from the air, creating zinc oxide: $2Zn+O_2 \Rightarrow 2ZnO$.

The coefficient two denotes the number of zinc atoms. The subscript $_2$ is used to denote an oxygen molecule consisting of two oxygen atoms and a zinc oxide molecule that has two oxygen atoms.

It is also possible to measure the result of this reaction in "moles." A mole consists of six times 10^{23} particles. The equation shows that two moles of zinc atoms react with one mole of oxygen molecules to produce two moles of zinc oxide molecules.

Redox reactions

Originally the term oxidation was used only to describe a reaction in which a substance combines with oxygen. Today every particle that transfers electrons to a reaction partner is said to be oxidized. For example, zinc is oxidized during a reaction with sulfur. At the same time, every particle that takes up electrons is said to be reduced. This is an oxidation-reduction, or redox, reaction.

Acid-base reactions

Acids dissolve metals and taste sour, while bases feel slippery and taste bitter. Acids turn litmus paper red, and bases turn it blue. Acids release protons (H+) to other ions or molecules, and bases accept protons.

If an acid reacts with a base, its protons are accepted by the base. The reaction between hydrochloric acid (HCl) and ammonia (NH_3) is an example of this kind of reaction, described by the equation $HCl+NH_3 \Rightarrow Cl^- + NH_4$. The superscript indicates that the chlorine becomes an ion with a single negative charge, and that the ammonia becomes an ion with a single positive charge.

■ Swimming pools must be kept sanitary by destroying bacteria and other organic materials. A bleach, calcium hypoclorite ($Ca(OCl)_2$), is often used. This dissolves in water to form hypochlorous acid (HClO), which acts as a powerful oxidising agent on organic material, killing most bacteria within milliseconds. The reaction is powerfully exothemic, producing heat.

PREVENTING RUST

■ Boats and ships are highly susceptible to rust due to constant exposure to salty water and harsh weather.

When iron and steel come in to contact with oxygen in a wet environment, a chemical reaction takes place. The metal is oxidized and an iron oxide compound called rust develops. Metals require protection against this reaction, passively with protective layers of paint or some noble metal or actively with metallic coating of a non-noble substance (zinc, aluminium, or magnesium alloy). Active corrosion protection is commonly used in shipbuilding where the coating dissolves, functioning as a sacrificial anode while the iron acts as a cathode.

practice

from test tube to reactor

Before you can produce large quantities of a substance, you must first learn what reactions are needed to generate the substance and how those reactions can be scaled up for industrial production.

At the end of the 18th century, the French Academy of Sciences offered a prize for the development of a method for creating washing soda (sodium carbonate) from common salt (sodium chloride). Sodium carbonate was used in large quantities as a washing agent and for the production of glass. The process developed a few years later by Nicolas Leblanc (1742–1806) was the beginning of the modern chemical industry, reducing the cost of producing soda to just one-ninth of the previous cost. However, waste products resulting from the Leblanc process were bad for people's health and a burden on the environment. Because of this, Leblanc's process was replaced. Current environmental concerns, along with efficiency are important reasons to develop new production methods for chemical substances.

Originally, sodium carbonate was also used to produce caustic soda (sodium hydroxide) and large amounts are still used worldwide—around 60 million tons yearly. Caustic soda is produced by the electrolysis of sodium chloride. Through this method, sodium chloride is dissolved in water to become negatively charged chloride ions and positively charged sodium ions. The chloride ions move in water to a graphite rod, which is positively charged and connected to a power supply of direct current. Chlorine is formed when the chloride ions transfer their electrons to the positively charged rod, called the "anode." The power supply pumps the electrons from the anode to a negatively charged steel sheet, called the "cathode." The electrons (e-) are transferred to water molecules (H_2O) so that hydrogen (H_2) and hydroxide ions (OH-) are formed according to the equation: $2\,H_2O + 2e^- = 2OH^- + H_2$. Caustic soda (sodium hydroxide in water) is formed from the hydroxide ions and the sodium ions. A technical difficulty lies in the fact that the cathode products of hydrogen and caustic soda have to be kept separate from the anode product of chloride in order to avoid undesired side reactions. This process is chlorine-alkali electrolysis and has been continually improved over time.

Reactions in micro-reactors, however, yield a substantially better product than reactions that take place in large agitating tubs. In addition, the integrity of the process is higher, since the reactions can be controlled in a targeted manner. The rate of producing chemicals can also be increased, meaning micro-reactors will be used more and more in industry.

milestones

HABER PROCESS FOR AMMONIUM

Ammonia is an important substance, and its derivatives are used in fertilizers and explosives. At first, the synthesis of ammonium from nitrogen and hydrogen failed, however, early in the 20th century, Fritz Haber (1868–1934) finally found that ammonium yield was increased if the reaction occurred under external pressure together with a catalyst to raise the rate of reaction.

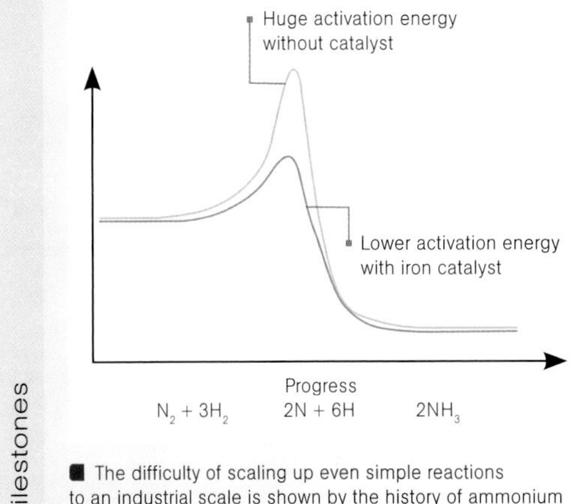

The difficulty of scaling up even simple reactions to an industrial scale is shown by the history of ammonium production.

Production of iron

An example of the difficult path between the discovery of a chemical reaction and its technical implementation is that of iron and steel. Iron has been extracted from iron ore in blast furnaces since the 14th century. Only in the 18th century, when coke was used to bind oxygen to the iron ores, did iron become an important material. New methods of making iron are still being invented, in order to control the carbon content in iron as economically as possible.

Electrolysis is performed on an industrial scale in order to satisfy demand.

in focus

CATALYSTS

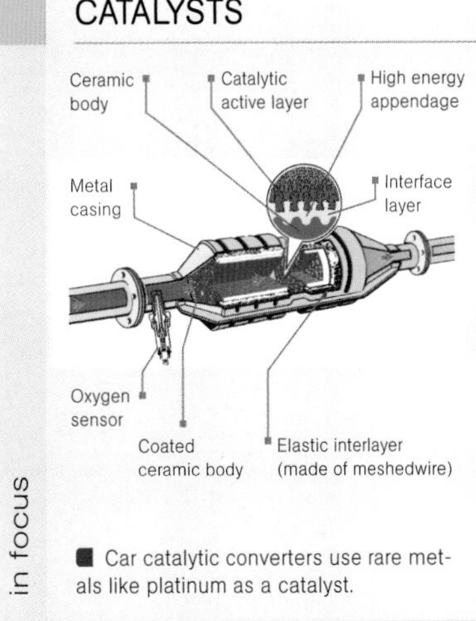

Car catalytic converters use rare metals like platinum as a catalyst.

Chemical reactions start only after the activation energy level is reached. Catalysts lower the energy level required for this process. One of the best known catalysts is used in automobiles where it helps convert poisonous exhaust gases into harmless gases. Catalysts are widely used in the chemical industry as they speed up industrial scale processes, even though they are not actually required for the reaction, saving on money and energy costs.

see also: Modern biotechnology, p. 303

with retorts and computers

Even though computers are a part of every chemistry lab, there are still researchers who synthesize new materials every day. The job of a professional chemist is more varied than many imagine.

biochemistry | research | analysis | computers | materials

THE WORK OF CHEMISTS

Chemistry has its origins in the world of alchemy. It includes the search for the chemical structure of even the minutest quantities of materials. From nanotechnology to macroscale production, chemists seek new materials and more useful products using both computers and traditional laboratory equipment.

If we create a picture of a chemist at work, most of us will probably think of people in white coats using test tubes and retorts (distilling glasses) to mix substances in a lab. The actual routine of chemists is very different from this, especially for industrial chemists. For industry, production procedures that have been discovered need to be applied on a large scale. In order to accomplish this task, industrial chemists work in pilot plants, are involved in control engineering, and perform numerous calculations.

The responsibilities of a chemist include participation in an additional interface between research and production. Chemists help client firms to market the products in an efficient and safe manner and advise them on how the product should be developed in the future, in order to promote further marketing success.

A few chemists working in universities, government agencies, and research departments of private industry fit the classic image of a chemist at work. They work in a "preparative" capacity; that is, they produce new substances with a large repertoire of methods and instruments. One of the most common of these methods is the conversion of raw materials or reactants into a solution. The solution is boiled and the vaporized materials that come from the retort are condensed in a vertical cooler, so that it flows back to the retort. Chemists call this procedure reverse flow injection. While the analytical chemistry is very much the core of the field of chemistry (analyzing compounds for their properties) there exists many other fields such as biochemistry that study the chemical reactions in living matter. Biochemists have an important role producing compounds that are used in pharmaceutical drugs. Theoretical physicists can often be found in the field of physics working on nanotechnology and astrochemistry. Chemistry is often thought of as the "central science," connecting all the branches of science together.

Centrifuges are used throughout the field of chemistry to separate substances by sedimentation.

practice

COMPUTER MODELING

For chemists, computers are as important as retorts. With them, theoretical chemists can predict the structure and properties of molecules and, for example, simulate the shape of the molecules of a nonexistent drug and deduce its properties. Only after simulation shows the feasibility of a new drug will a company invest the money needed to synthesize the substance and bring it to market. Others are using computers to search for non-medical substances used in industrial processes like electrolysis or welding.

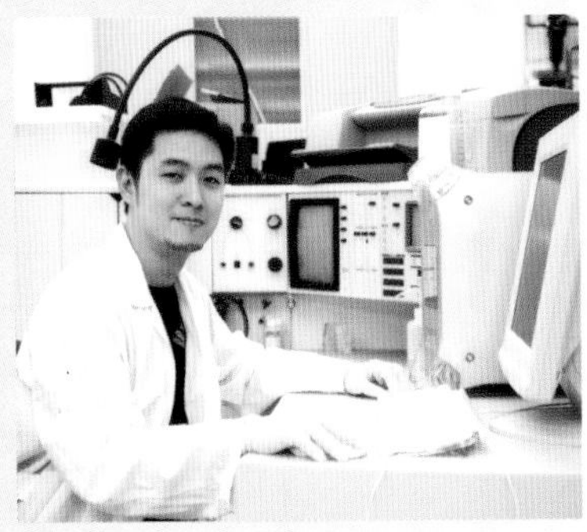

Computers are important tools in modeling the actions of a molecule, compound, or reaction.

Chemists often work on the atomic scale, building materials from molecules and atoms.

searching for clues

Analytical chemists often use detective skills as well as an ability to apply multiple methods of analysis to reach their goal: a better understanding of what lies inside molecules.

Analytical chemists can tackle problems as varied as confirming the age of a piece of art or detecting hazardous substances in air or food, or examining the quality of crystals to be used in microchips.

Blood and urine samples that have been taken from athletes are tested for performance-enhancing drugs. A higher than usual concentration of testosterone (pictured) may indicate the use of anabolic steroids.

Progress through analytics

The 19th-century German chemist Carl Remigius Fresenius observed, *"It can be easily proved that all great advances in chemistry are more or less directly related to new or improved analytical methods."* Since chemical reactions must be identified before they can be improved, his statement remains valid. Only when we understand the exact structure and composition of a chemical can improvements be made in medicine, solar cells, aircraft engines, or other products.

Solving crimes in the lab

Today people can be identified and in the case of a criminal be convicted with only a tiny amount of sperm or skin. The evidence is in the cells. The genetic material—such as DNA—in even a tiny sample can be duplicated with enzymes, using the technique called polymerase chain reaction (PCR). Next, the DNA is analyzed and fragments are separated onto a gel based on their sizes. Ultimately, the fragments are displayed in the form of a bar code, very similar to the codes on today's retail products. Except for identical twins (who have the same genetic makeup), each person's gene code is unique and can be used to identify them.

Doping analysis

Traces of drugs remain in urine longer than in blood. However, it is not always possible to find an athlete guilty of doping using only evidence found in urine. The analysts must also determine the quantity of the substance. For example, the consumption of a massive amount of caffeine can be determined and prohibited, but an athlete with caffeine levels expected from normal consumption of drinks like coffee, tea, or cola is not considered to be doped.

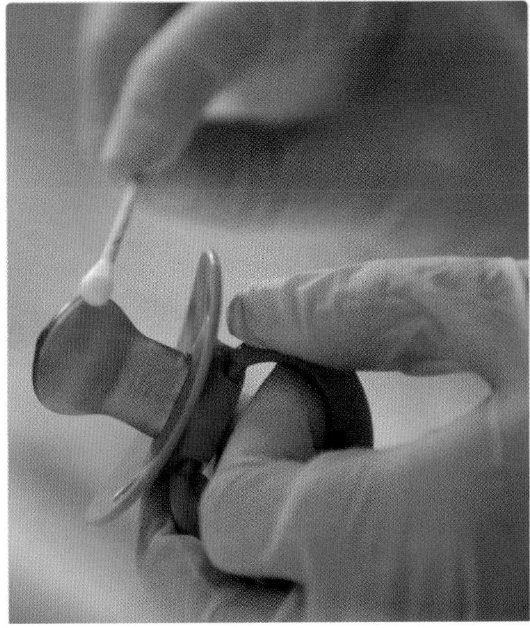

DNA-analysis can determine the identity of a person, e.g., in criminal investigation, clarify questions of kinship relations, and examine the probability of contracting certain diseases.

Human bodies produce many substances at known and established levels. Medical tests are used to establish whether an abnormally high amount of a particular substance is present.

practice

TOOLS OF ANALYSIS

For analysis, a substance is usually separated from other materials. Often, high-performance liquid chromatography (HPLC) is used for this purpose. A mass spectrometer may then be used to identify the separated molecules. This equipment can detect a billionth of a gram of a substance in a kilogram of material, roughly the equivalent to detecting one lump of sugar in a swimming pool. To test very high concentrations of a substance, chemists may turn to a nuclear magnetic resonance (NMR) spectrometer.

Because any one particular analytical method can provide only part of a total answer, analysts often use a combination of testing methods before answering a question.

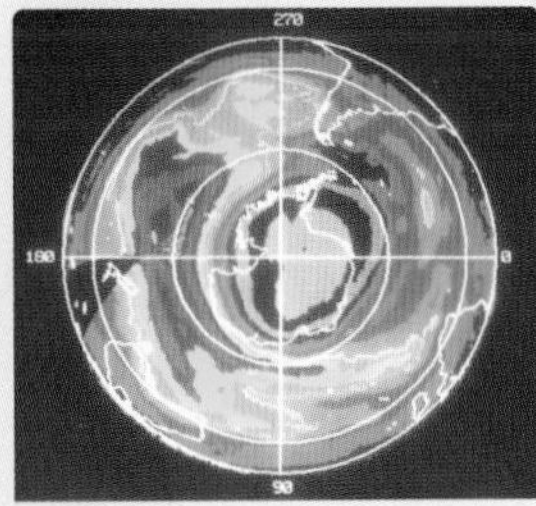

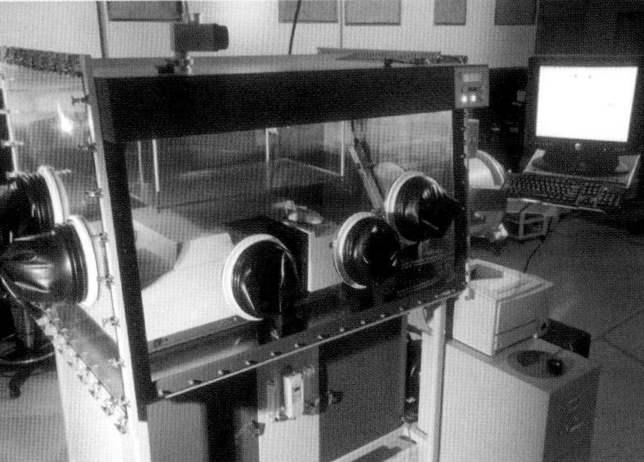

A Total Ozone Mapping Spectrometer (TOMS) image showing ozone depletion over Antarctic and a spectrometer used in the optics industry.

semiconductors

Temperature and the ability of certain materials to conduct electricity are related. This was discovered in the 19th century and roughly a century later became the basis of information technology.

Computers, cell phones, digital cameras, and medical technologies, such as computed tomography and artificial pacemakers, depend on semiconductors; these

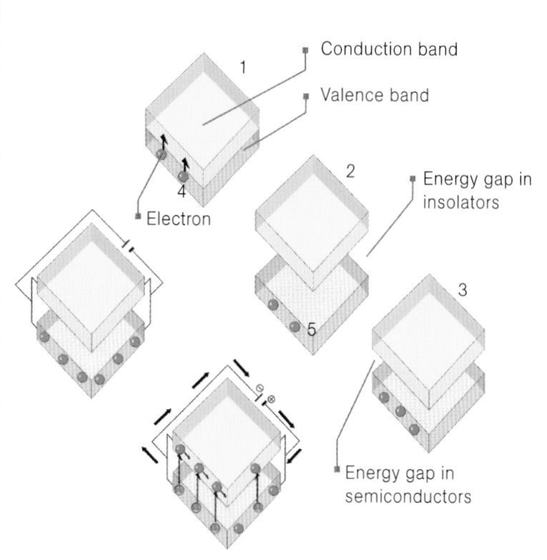

The electronic band structure of semiconductors and insulators.

may be made of elemental materials and vary in their band structure. At 68°F (20°C) semiconductors conduct electricity better than insulators but not as well as metals. Temperature is an important factor. For most semiconductors, the ability to conduct goes up with rising heat.

When we hear the term "semiconductor," we usually think of silicon, one of the most important semiconductors. However, it is just one of 600 known inorganic semiconductors. Other important ones are the III-V semiconductors, such as gallium arsenide.

> **basics**
>
> **AMONG CHEMICAL** elements there are solid substances that conduct electricity (conductors) and those which do not (insulators). However, with semiconductors, conductivity can be influenced by energy through heat, light, and an applied electrical current.

The perfect crystal

Most semiconductors are close to perfect crystals. A bar of manufactured semiconductor material is melted piece by piece, then, during slow cooling, the atoms arrange themselves in a crystalline structure. To produce a very thin layer of crystal, the gaseous semiconductor is condensed on a cold surface. Today, great importance is placed on the methods used to crystallize semiconductor materials.

Desired donor atoms

The conductivity of a semiconductor can be changed by adding impurities in a process called doping. Donor atoms are introduced into the crystalline structure by firing an ion beam into the material. The higher the energy with which the ions collide with the semiconductor, the deeper they penetrate.

Semiconductor applications

Semiconductors are critical in microtechnology and information technology. They are used to convert light to electrical energy and electrical energy to light. Many calculators are powered by solar cells thatconvert light into electricity. Light-emitting diodes are used in many electronic displays. While electrons in free atoms have energy levels, electrons in solids form energy bands. This band structure explains the properties of semiconductors. Between the bands of different energy there are gaps. In a semiconductor the highest energy level occupied with electrons is the valence band while the lowest unoccupied energy level is the conduction band. In insulators electrons cannot cross the band gap or, if they do, they will be captured again by the huge attraction of the atomic nuclei, and electric current cannot flow. In doped semiconductors electrons build an additional narrow band that allows them to easily reach the conduction band and contribute to conductivity.

Solar panels situated in a desert, a location that receives large amounts of solar radiation.

> **practice**
>
> **PRODUCING COMPUTER MICROCHIPS**
>
> 
>
> To build microchips, semiconductors must be doped and tiny circuits built. The semiconductor is covered with a photoresistant varnish. A "mask" is then reproduced on the varnish using UV radiation or electron beams. A chemical bath dissolves the varnish in specified areas. Acids then etch the circuits. This work must be done in clean rooms because if a dust particle settles upon the surface of a microchip circuit, the microchip will not work properly.
>
> In a clean room, the use of many normal items, such as paper and fabrics, are avoided.

see also: Computer components, p. 379

materials of tomorrow

Tomorrow's materials will be lighter, more stable, more heatproof, and more "intelligent." New nanomaterials with microscopic structures open up new options while improvements continue for steel and ceramics.

Steel contains iron and varying alloying elements. A alloying element or elements change the character of the steel. For example steel with a 15 percent manganese content enriched with a 3 percent aluminum and silicon content does not rupture, even at tensions as high as 1,100 megapascals. That is the equivalent to the weight of ten bull elephants on an area the size of a postage stamp. Conventional steel can resist only 700 megapascals. Another type of steel can be stretched lengthwise by around 90 percent without rupturing. When this type of steel is used in cars, the behavior of the vehicles in a crash is much improved. Furthermore, this particular steel can reduce the weight of an auto body by around 20 percent, helping to improve gas mileage and performance. Steel of this type is difficult to produce but, in the future, most automobiles will benefit from its use.

In the future, "memory metals" could regain their original shape after an accident.

New Boeing airliners are making extensive use of metal with shape memory in order to reduce engine noise.

see also: Industrial engineering, p. 375

HIGH-PERFORMANCE CERAMICS

practice

Modern high-performance ceramics are very different from the fragile earthenware of ancient times. Ceramics can be made breakproof by adding carbon fibers. Fitted on the nose of a space shuttle, these reinforced ceramics protect the vehicle from the enormous heat produced during reentry into the atmosphere in the form of ceramic tiles. A similar ceramic is used for rotor disks in brakes because of its ability to resist wear and corrosion.

The reentry temperature of the space shuttle is so high that its tiles regularly need replacing.

Substances with memory

There has been an accident and a fender was dented, wouldn't it be good if the dent could just disappear? That could be a reality if the fender were made of a shape memory material. These advanced materials "remember" their original shape and return to it when heat is applied. So the dent from the accident could be repaired by simply applying heat, a task that can be done at home without requiring a trip to an auto body shop.

Metals that remember shape

We already have plastics with shape memory. However, there are also metals with the same characteristic. A special nickel-titanium alloy allowed the European research satellite ENVISAT-1 to open its "eyes" after it reached its orbit in 2002. Shape memory metals are also used in medicine. For example, wire nettings called stents are placed in coronary arteries that are constricted by disease. When exposed to the warmth of blood, the stents expand and support the arteries thereby improving blood flow. Many physicians and other researchers are experimenting with additional materials and transplants to improve the lives of their patients, bringing the materials of tomorrow to the hospitals of today.

CHEMISTRY

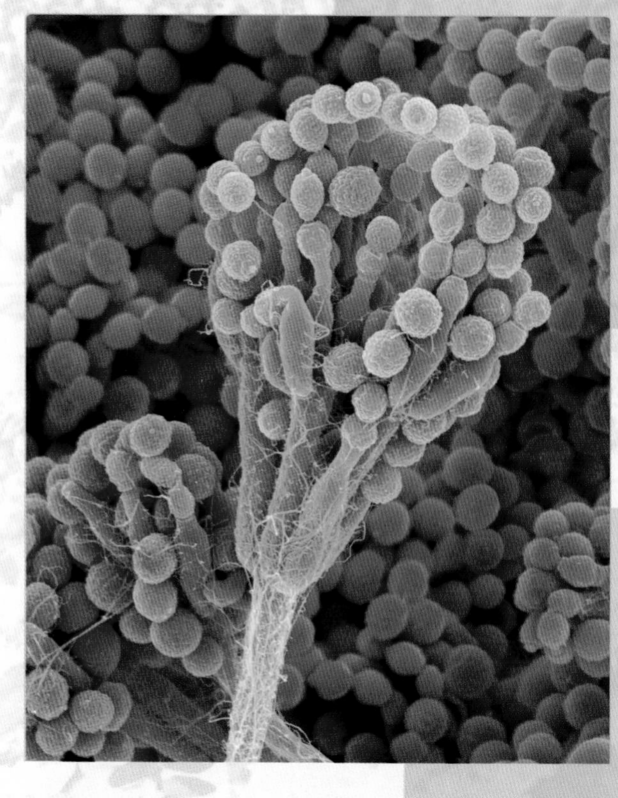

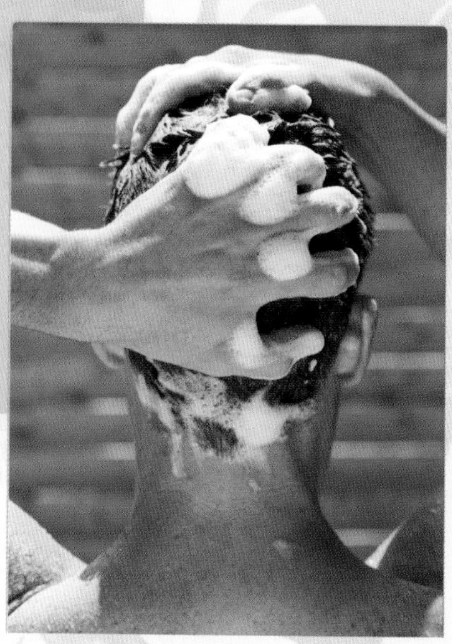

ORGANIC AND BIOCHEMISTRY

Molecules are particles composed of two or more atoms bonded together. They can include only a single element or a combination of elements. Every living thing is made of molecules, most of which contain the element carbon. With its endless variety of combinational possibilities, carbon forms the basic building block of organic chemistry.

In order to survive, all living things must take up and give off various substances. Normally our food contains all of the essential nutrients necessary for life. When our biochemical processes become unbalanced, however, medicines and supplements can help repair damage or replace what is missing. Thus, organic chemistry also focuses on the development of useful drugs and other substances that do not occur in nature. Many of the products of chemistry and biotechnology have become part of everyday life in the modern world. As economic factors they carry great weight—yet they also pose the risk of unforeseeable consequences for human beings and the environment.

carbon: an extremely versatile element

More than 200 years ago, chemists coined the term "organic chemistry." At that time it was postulated that living matter is fundamentally different from dead matter. This proved to be an error.

Up to the middle of the 19th century, the conviction prevailed that natural substances comprised elements other than those known at the time, principally discovered by the mining industry.

organic chemistry | hydrocarbons | compounds | aromatics

CARBON

Carbon has an unusual characteristic: It is the only element that can form chains and rings on its own, using single or multiple bonds. The compounds it forms with hydrogen are the basis of modern organic chemistry. Its original task, interaction with living matter, has nowadays been taken over by biochemistry, which investigates four classes of substances: carbohydrates, nucleic acids, lipids, and amino acids. This then provides the foundation of biotechnology.

Nitrogen
Oxygen
Hydrogen
Carbon

Carbon is the 13th most common element (by weight) on Earth.

However, with a pioneering experiment the German chemist Friedrich Wöhler (1800–1882) refuted this assumption. He synthesized the organic molecule urea (carbamide) from the inorganic substance ammonium cyanate. With that, he showed that organic substances are composed of already discovered elements. Consequently the separation of organic and inorganic chemistry has lost its purpose; however, it is still being maintained till today because the element carbon is unique in that it can form many millions of compounds on its own. In simple terms one now refers to organic chemistry as the chemistry of hydrocarbons and their derivatives.

The largest part of hydrocarbons comes from crude oil and natural gas, which are refined by petrochemistry. Hydrocarbons play a significant role in fuels (e.g., gasoline), as a solvent and as chemical raw materials: for instance, plastics. It would be difficult to imagine life without hydrocarbons; without them automobiles would not run and there would not be any modern textiles or plastics.

Pure hydrocarbons are chemical compounds made up of the elements of carbon (C) and hydrogen (H). For that, carbon atoms are linked to each other by means of covalent bonds (electron-pair bonding) in linear or branched chains or rings (cyclic hydrocarbons). Hydrocarbons are classified as saturated alkanes (paraffins)—the carbon atoms are linked to each other by simple bonds—and unsaturated compounds, alkenes (olefins) with at least one double bond, and alkynes, which have at least a triple bond. For instance, a simple cyclic hydrocarbon is benzole (benzene), which consists of a ring of six carbon and six hydrogen atoms with three double bonds. Compounds derived from benzole are re- erred to as aromatics, also known as aromatic compounds.

basics

MORE THAN 99% of carbon occurs in the form of minerals and 0.03% only as organic compounds. Of that tiny fraction about two-thirds are found in the form of fossil fuels.

milestones

FIRST SYNTHESIS OF AN ORGANIC SUBSTANCE

The German chemist Friedrich Wöhler isolated the elements aluminum, beryllium, silicon, and boron; and in 1828 he was first to demonstrate that organic substances contain inorganic elements. Subsequently he and the chemist Justus Liebig developed the "radical" theory and identified a multitude of organic-chemical compounds.

Friedrich Wöhler was a highly influential scientist of his time, contributing to the founding of modern biochemistry.

Hydrocarbons such as oil or natural gas provide for large amounts of the world's electric generation.

see also: Biotechnology, pp. 302–303

carbohydrates: the world of sugars

In our culture sugar belongs to the essential staple foods. Chemically, the category of sugars consists of sweet-tasting organic compounds of carbon, oxygen, and hydrogen.

Sugars are also referred to as hydrates of carbon, or carbohydrates. These hydrates are compounds that are bonded to water by electrostatic forces. Chemically, sugars are polyvalent alcohols, since sugar molecules contain one or several hydroxyl groups. The simplest carbohydrate is glycoaldehyde. The most important natural sugars are pentose and hexose sugars. These include ribose, glucose, and fructose. These sugars are made up of a single sugar molecule, and therefore called monosaccharides. However, this group was named after saccharose, that is made up of two sugar molecules: glucose and fructose. It, like lactose (milk sugar, made up of glucose and galactose), is a disaccharide. Apart from monosaccharides and disaccharides, there are also polysaccharides that consist of many thousands of sugar molecules. The best known example is starch, which serves as energy storage for plants such as potatoes and corn (maize). The photosynthetic activities of plants produce glucose and bond up to a thousand glucose molecules together into large, branched structures that are stored in the form of starch grains. Cellulose, a component in the cell walls of plants, is also a polysaccharide made up of glucose units. Next to their roles as a supplier and store of energy, many carbohydrates also serve as structural elements. The element that most frequently occurs in the human body is ribose. After giving off one oxygen atom (deoxidation), it is used in the human genome as deoxyribonucleic acid (DNA). Also, one finds oxyribonucleic acid (RNA) functioning as amino acid carriers and information messengers in protein biosynthesis (messenger RNA), as well as in ribosomes. Nearly all cells of the human body have carbohydrates and their derivatives on their surface areas. The best example is the structures on red blood cells. Most tissue cells have sugar molecules, so that certain cells of the immune system can distinguish between the one's own tissue and foreign tissue.

In Europe, sugar (sucrose) extraction takes place exclusively through the cultivation of sugar beets.

STRUCTURAL DETERMINATION OF SUGAR

milestones

German chemist Emil Hermann Fischer (1852–1919) discovered the spatial structure of saccharides and the chemistry of carbohydrates. He received the Nobel Prize in chemistry in 1902 for his work on carbohydrates. After getting cancer, probably resulting from his work with toxic phenyl hydrazine, one of his discoveries, Fischer took his own life in 1919.

Emil Hermann Fischer is considered the founder of classic organic chemistry.

Sugar beets can be grown commercially in a wide range of climates.

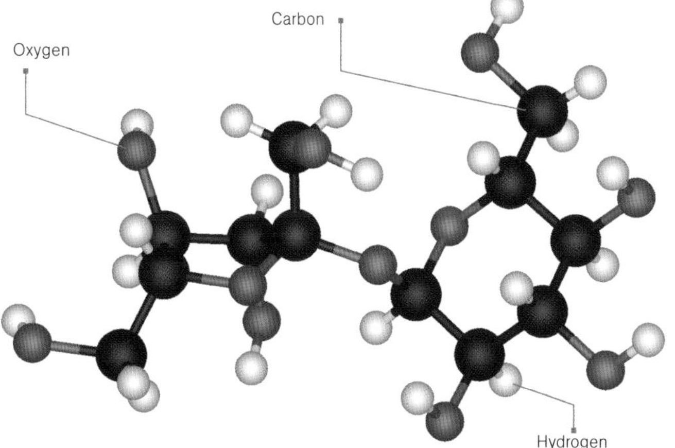

A three-dimensional model of a sucrose molecule.

see also: Food technology, pp. 340–341

nucleic acids: molecular building blocks

Our sex, the color of our eyes, and all other characteristics are stored in our hereditary material. This task is handled by four molecules: adenine, cytosine, guanine, and thymine.

Up to about 150 years ago, scientists believed that hereditary information was stored in proteins. It was not until 1869 that the Swiss physiologist Friedrich Miescher (1844–1895) discovered a whitish substance in white blood cell nuclei. Since it differed from proteins, he referred to it instead as nuclein. Twenty years later the biochemist Albrecht Kossel (1853–1927) wrote: "*While digesting nucleins with diluted acids I have discovered a new base of general occurrence. This base, for which I suggest the name 'Adenine,' was prepared from the pancreatic glands of cattle.*" After this discovery, Kossel isolated another three molecules and verified they were related to the pentose carbohydrate ribose. Building on this, his student, Hermann Steudel, discovered that each carbohydrate molecule bonded to a molecule of phosphoric acid. This meant that at the turn of the 20th century it was understood how the building blocks of hereditary material—the mononucleotides—were structured. Each nucleotide possesses a phosphate remnant to which carbohydrate bonds. The carbohydrate is a deoxyribose (a ribose lacking an oxygen atom). One of the four bases—adenine, cytosine, guanine, or thymine—bonds to the deoxyribose. Due to the groundbreaking research of the Canadian bacteriologist Oswald Avery (1877–1955), It was discovered that hereditary characteristics are transferred from bacterial cells by deoxyribonucleic acid (DNA). Yet a mystery remained as to the connection between individual components of mononucleotides, and how they formed DNA. The Austrian-American biochemist Erwin Chargaff (1905–2002) discovered in 1952 that the bases adenine and thymine, as well as cytosine and guanine occurred in equal proportions. Afterward the English biochemist Rosalind Franklin (1920–1958) took ground-breaking x-rays of crystallized DNA. The American biochemist James Watson and his British colleague Francis Crick used Chargaff's and Pauling's discoveries as well as Franklin's x-rays, and resolved the spatial structure of DNA, solving the puzzle of DNA and establishing the basis of modern gene technology.

■ Rosalind Franklin's famous x-ray images of the structure of DNA were the first time the building blocks of life had ever been witnessed.

basics

EVERY HUMAN has in his or her DNA certain unique traits. These can be used to prove, through evidence, a suspect's guilt in many cases of violent crimes.

milestones

DISCOVERY OF THE STRUCTURE OF DNA

The biochemists James Dewey Watson and Francis Harry Compton Crick, and biophysicist Maurice Hugh Frederick Wilkins, were awarded the Nobel Prize, for discovering the molecular structure of nucleic acids. Rosalind Elsie Franklin, who contributed significantly to the discovery, had passed away four years earlier of cancer.

■ James Watson and Francis Crick, contributed to the discovery of DNA, during a period of great interest in genetics.

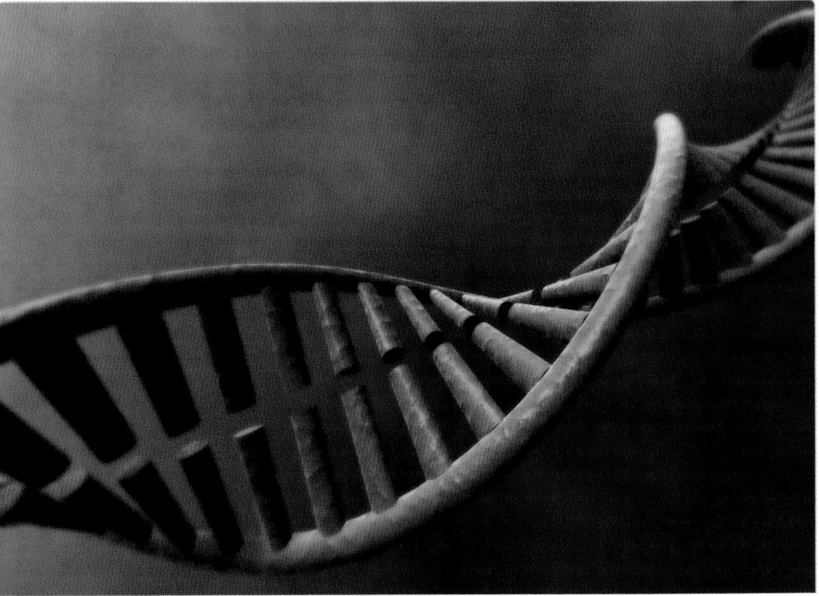

■ The DNA double helix holds the genetic instructions for development and growth for all living organisms.

see also: Genetics, pp. 250–253

lipids and fatty acids

Lipids and fatty acids are essential for the nutrition and health of all living organisms. Butter or margarine, candle wax, and olive oil consist mostly of these water-insoluble substances.

Plants and animals produce these water-insoluble or hydrophobic (derived from the Greek word meaning "water repelling") substances that serve as structural elements in cell membranes, energy reservoirs, or as messenger substances. Lipids are divided into various categories: fatty acids, triacylglycerides (fats and oils), waxes, phospholipids, sphingolipids, lipopolysaccharides, and isoprenoids (such as steroids and carotenoids). Fatty acids are unbranched hydrocarbons that have carbonic acid at their end. The simplest fatty acid, butanoic acid, contains a chain of four carbon atoms and is a monocarbonic acid. The carbon atoms are connected by simple or multiple bonds. The former are said to be saturated, while those with at least one double bond are called unsaturated fatty acids. Fats, such as butter, contain triacylglycerides, the largest group of dietary lipids. These consist of a sugar alcohol, glycerin, and three molecules of fatty acids. The principal components of animal membranes are phospholipids. They are distinguished from triacylglycerides in that the glycerin molecule is esterized with two long fatty acids only.

Olive oil is mainly composed of oelic acid and linoleic acid that have positive effects for the health.

There is a phosphoric acid molecule at the third carbon atom of glycerin, which forms the hydrophilic (derived from the Greek word meaning "water loving") or polar head of the phospholipid. Yet, the fatty acids are hydrophobic and form the nonpolar tail. In contrast, steroids such as cholesterol are structured differently. These lipids have four rings and one hydrocarbon chain. Cholesterins are located in outer membranes and assure that these become more "rigid"—they increase the melting point of a membrane. Therefore fish have very little cholesterol, otherwise their membranes would become rigid and they would die. Apart from these simple lipids, there are also complex lipids in nature. These are made up—just like lipoproteins—of several components. They consist of several lipids and a single protein and transport cholesterols to and from individual cells.

basics

LIPIDS function as:

- energy suppliers
- energy storage
- membrane building blocks
- signal molecules
- hormones
- fat-soluble vitamins
- pigments

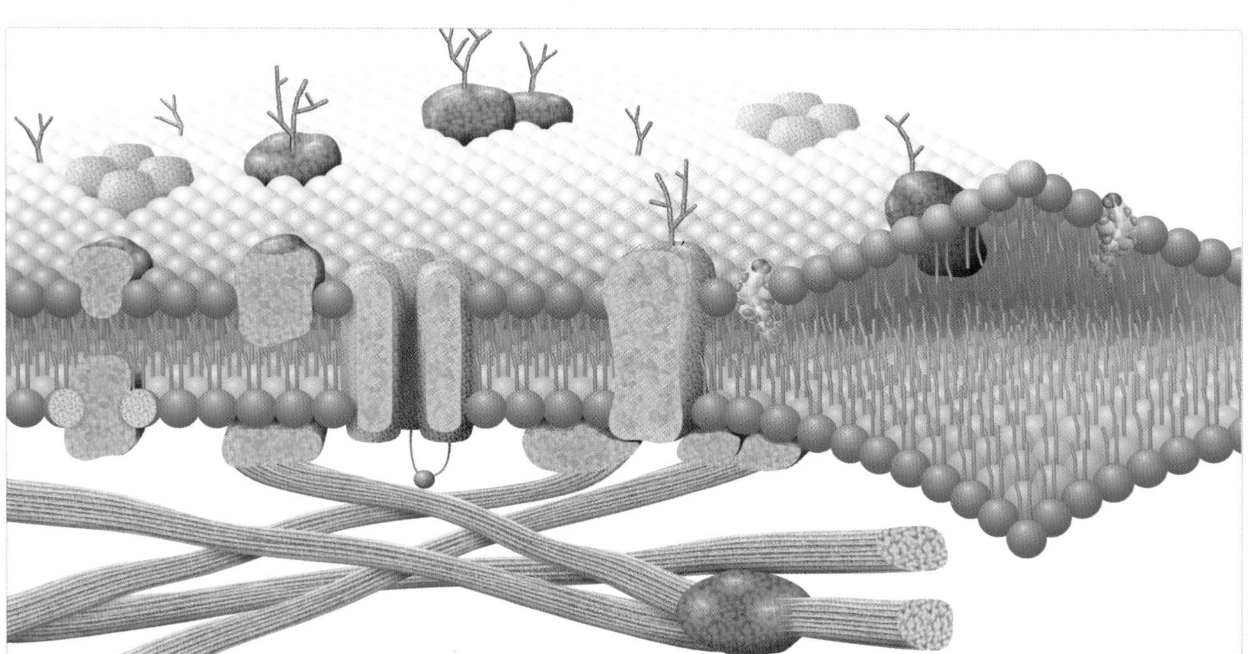

Animal cell membranes consist of a lipid double layer. The hydrophobic "tails" of the lipids will settle so that the hydrophilic "heads" are pointing toward the outside.

see also: Metabolism and hormones, pp. 236–239

amino acids

Hair and fingernails are external and visible structures made of proteins, which play an important role in almost all biological processes that take place in the human body.

As early as 1836, the Swedish chemist Jöns Jacob Berzelius (1779–1848) coined the term proteins (from the Greek *proteios*, meaning "primary"). Proteins are the most important components of all living cells. They consist of individual building blocks, the amino acids, which are arranged in long chains. Amino acids, in turn, are carbonic acids, which have an amino group at the first carbon atom next to the carbonic acid. Also located at the first carbon atom is a side chain that is specific for each amino acid. These side chains are straight, circular, or branched hydrocarbons. The human body requires 20 different amino acids, but the body itself can only produce some of these; the remainder will need to come from food, meaning a balanced diet is very important.

basics

THE HUMAN BODY can produce only part of the amino acids it requires. Those that have to be taken in with food are referred to as "essential" (required for survival) amino acids. "Conditionally essential" amino acids are amino acids that are sometimes essential: for example, for people with a specific disease or nutritional need.

The composition of individual amino acids to proteins (protein biosynthesis) takes place along the ribosomes. The sequence of individual amino acids relative to each other is determined by the sequence of DNA. During protein biosynthesis, amino acids are linked via peptide bonds.

Structural features

These large proteins consist of many thousands of amino acids that are bonded to each other, as in a chain of pearls. Such a long chain forms spatial structures. The primary structure refers to the sequence of individual amino acids. Its side chains are distributed irregularly in space, or they are aligned in a regular pattern; this arrangement is the secondary structure. This includes a spiral, the α-helix, and an almost parallel arrangement, the β-sheet (also β-pleated sheet), where certain segments of the chain run parallel to each other. Some proteins are rolled up like a ball of wool. In such an arrangement, it can happen that, for instance, the first amino acid of the chain (primary structure) is located in close proximity to an amino acid that is actually in position 800. The spatial relationship of amino acids, which are in part within the primary structure but located far apart, is referred to as the tertiary structure.

Hemoglobin is a protein in red blood cells, which transports oxygen. This protein consists of four individual proteins: the subunits. How these subunits are spatially oriented in relation to each other is, in turn, described by the quaternary structure. In some subunits, proteins are linked by a covalent bond via two sulfur atoms that form a so-called disulfide bridge. Proteins take over various functions in the human body: they transport oxygen (hemoglobin) and other substances, they move muscles (myosin), serve as mechanical support (collagens), and attack foreign substances (antibodies). As hormones, proteins transmit signals, bond hormones (receptors), and accelerate reactions (enzymes).

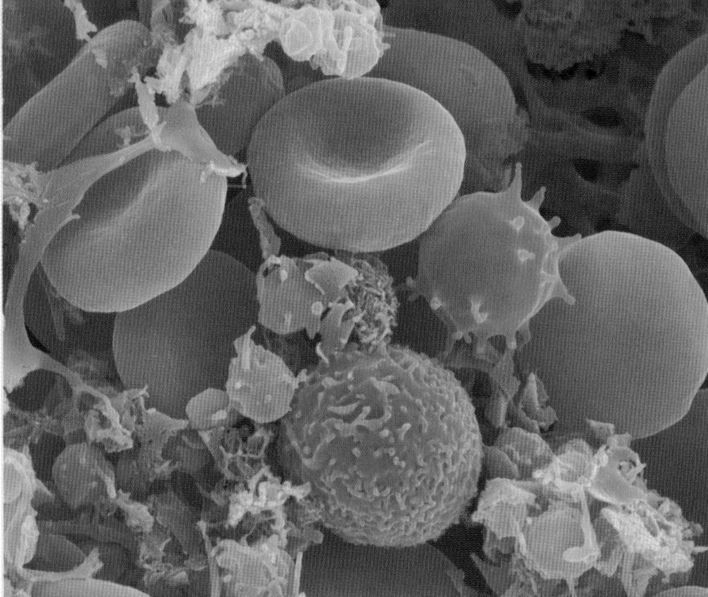

Hemoglobin is the iron-containing protein in the red blood cells and is responsible for transporting oxygen throughout the body.

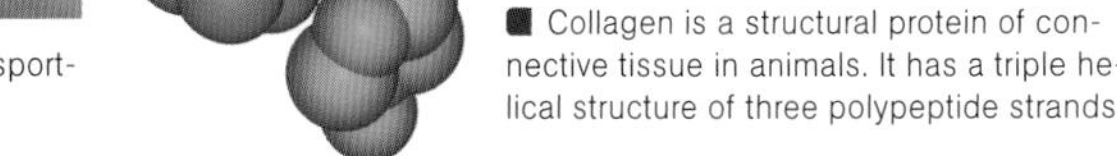

Collagen is a structural protein of connective tissue in animals. It has a triple helical structure of three polypeptide strands.

enzymes: active catalysts

The most important role in the metabolism of living organisms is played by enzymes, which catalyze billions of metabolic reactions every second.

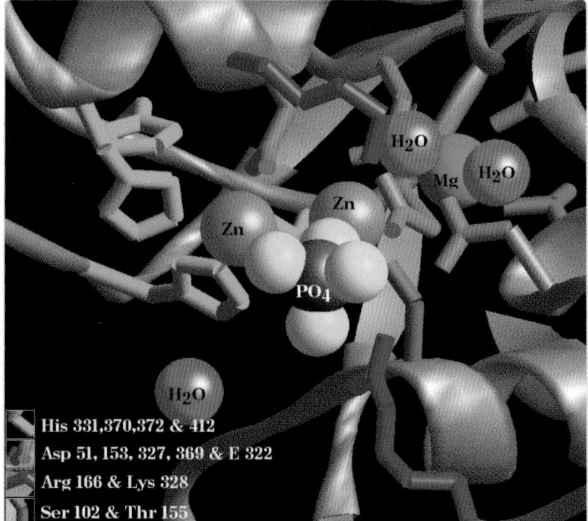

An illustration of the active center of an enzyme together with a substrate molecule.

In 1878, the German physiologist Wilhelm Kühne (1837–1900) introduced the term "enzyme." At that time, the word "ferment" was being used. Alcoholic fermentation had been known for centuries, but it was not until 1897 that the German chemist Eduard Buchner (1860–1917) discovered that enzymes are also effective outside of living cells. Most enzymes are composed of proteins, and some of ribonucleic acids. Many enzymes are made up of two parts that form together the holoenzyme: a protein foundation structure (apoenzyme) and a nonproteinaceous component (coenzyme or prosthetic group). Coenzymes are substances that transfer or remove, respectively, certain molecules or atoms to or from the substrate.

Enzymes act as catalysts (Greek: *catalysis,* meaning "decomposition and dissolution") and accelerate metabolic reactions by lowering the activation energy. For that, they bond substrates to their active center, where the prosthetic group is located. The active center bonds only with specific substrates of an enzyme-substrate-complex (key-keyhole principle). However, while an enzyme only ever bonds with a specific substrate, cells need entire cascades of enzymes; for instance, in glycolysis.

Cells possess numerous possibilities to regulate the activity of their enzymes. Such control mechanisms are particularly important when certain reactions are only needed for a short period of time. In a reaction sequence, frequently the enzyme that catalyzes the first step is finally suppressed by the end product. This feedback inhibition prevents cells from producing an excess of the final product. There are principally two different ways this occurs. During the process of competing inhibition, one molecule (the inhibitor) bonds to the active center, so that the actual substrate can no longer bond to that site. On the other hand, during noncompetitive inhibition, the inhibitor has its own bonding site away from the active center and so assures, by means of a spatial, structural change, that the substrate does not bind to the active center.

All enzymes are placed into various groups according to their different tasks. For instance, the digestive enzyme trypsin breaks up proteins taken up by the small intestine. These proteins are called proteases. On the other hand, nucleases break up DNA or RNA, lipases reduce triacylglycerides to fatty acids and glycerin, and glycosidases liberate the monosaccharides from di- and polysaccharides.

milestones

THE FIRST ISOLATED ENZYME

The American biochemist James Batcheller Sumner (1887–1955) worked with the enzyme urease, taken from soybeans (this enzyme catalyzes the breakdown of urea into ammonia and carbon dioxide). Within this work, he was the first researcher in 1926 to succeed in isolating and crystallizing an enzyme.

At the time, this was such a radical development it was ignored and ridiculed; however, in 1946 he was awarded one half of the Nobel Prize in chemistry—the other half of the prize went to two fellow Americans, John Howard Northrop (1891–1987) and Wendell Meredith Stanley (1904–1971), for their scientific research on the isolation and crystallization of enzymes and viral proteins.

James Sumner, codiscoverer of enzyme isolation.

The process of cheese production of this utilizes rennet, a substance whose active enzyme (rennin) functions to coagulate milk. It occurs naturally in the stomachs of calves and other cud-chewing animals.

origin | development | modern

BIOTECHNOLOGY

Biotechnology is a rather young field of technological applications, yet some of its methods are much older. As early as the fifth millennium B.C., the inhabitants of Mesopotamia utilized the capability of certain microorganisms to brew beer from cereal grains, and the ancient Egyptians also knew several types of wine as early as 3900 B.C.

origin and development

In 1992, the United Nations defined in its Convention on Biological Diversity that biotechnology uses biological systems in order to produce certain products, such as antibiotics or foodstuffs.

For a long time, microbiological processes have been used to purposefully manipulate the genetic material of plants and animals. However, this was done mainly intuitively, without a clear understanding of the scientific ideas behind it. Since the Neolithic period, humans have used artificial selection and crossbreeding to rear the plants most useful to them; they cultivated beer and wine for more than 5,000 years and soon thereafter began making sourdough bread.

Apart from alcoholic fermentation, it was also known during antiquity how to cultivate acetic and lactic acid bacteria for the purpose of preserving milk.

In contrast to modern biotechnology, this "old" biotechnology took place outside the laboratory.

In the course of the 20th century, modern biotechnology achieved several profound breakthroughs. Significant steps include the discovery of penicillin by Alexander Fleming in 1928 and the clarification of the molecular structure of DNA by Watson and Crick in 1953 (p. 150), the introduction of the methods of gene technology since 1973, and finally the decoding of human hereditary material a few years ago (2001).

In the meantime, biotechnology has become one of the key technologies of the 21st century. According to an estimate by a British consulting firm, there were more than 2,000 companies (with about 60,000 employees) existing in Europe in 2004, which were developing exclusively biotechnological products or procedures.

Biotechnology principally uses methods of genetic engineering (gene technology) and of biochemistry. In the process of this type of work, microorganisms are changed so that they take on desired characteristics, in order to produce certain products, such as insulin. Apart from microorganisms, modern biotechnology also modifies enzymes and even plants—for instance, to produce larger yields or to become more resistant to disease.

Penicillin was the first antibiotic to be discovered by Scottish biologist Alexander Fleming.

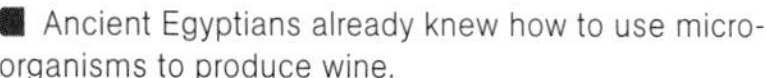

Ancient Egyptians already knew how to use microorganisms to produce wine.

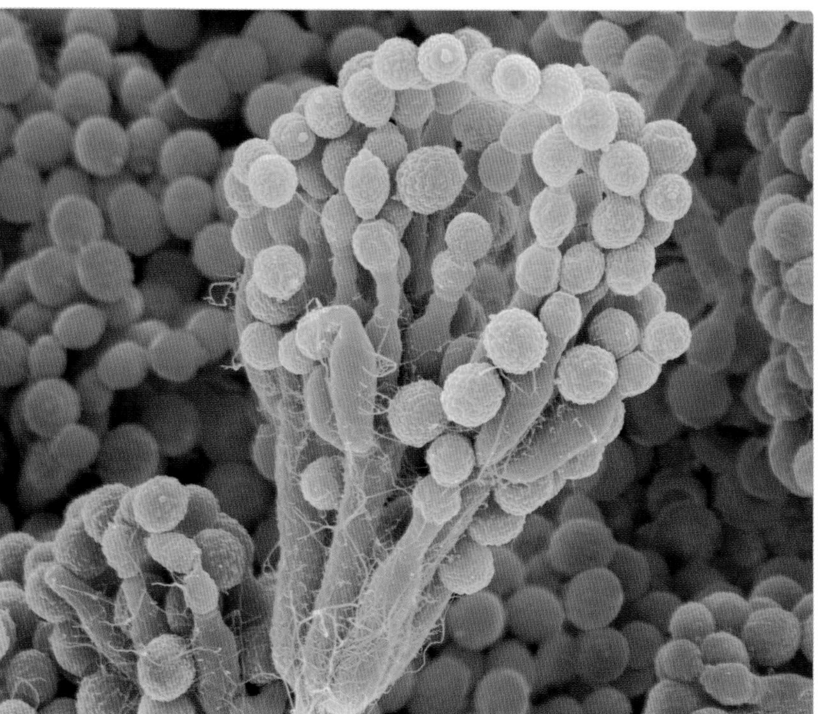

Reproductive structures of *Penicillium chrysogenum* (also known as *Penicillium notatum*). This common household mold is used in the production of both cheese and the antibiotic penicillin.

see also: Gene technology, pp. 256–257

modern biotechnology

The fields of application in modern biotechnology are diverse, but they fall into several main branches. Although these branches cannot be delineated precisely, a color-coded system can be used for easy recognition.

The color-coding system has not (yet) been officially adopted, but can aid in clarifying the different fields of application.

Green biotechnology deals with agriculture, particularly genetically modifying commercially used plants, like cereal grains or maize (corn), to produce greater yields or become resistant to harmful organisms. This branch also deals with the further development of future food products to have new, additional characteristics: for instance, potatoes that will be able to produce surface molecules of certain disease carriers, in order to "immunize" humans against these diseases.

Therapy and diagnostics of diseases and the production of new medications belong to the red biotechnology. This branch not only develops new medications—for instance, human insulin—but also aids burn victims who need skin transplants. Tiny skin sections removed from the patient can be grown artificially in a laboratory into larger skin sections. When the patient recieves a skin transplant consisting exclusively of its own bodycells (a so-called autotransplant), rejection by the body is unlikely. Growing new tissue (tissue engineering), however, is not restricted to skin. It is also being tested for other tissue and even for organs. Another objective of red biotechnology is the selective use of medications or active components on diseased tissue, for example drug-delivery systems. For that purpose, "transporters" such as lipoproteins or viruses are specifically modified to transport active components to certain cells, where they are then released.

White technology optimizes industrial processes for manufacturing certain products. Plant raw materials, such as starch, oils, or cellulose, are modified in order to produce fibers, chemicals, or plastics with improved characteristics. The extraction of bioalcohol (ethanol) from plants as alternative fuel for motor cars is one of its objectives. A classic application of white biotechnology is the optimization of certain enzymes used in detergents. Detergents that use enzymes (proteases) to breakdown and remove substances such as blood or milk products are normally active only in a temperature range from around 68 to 149°F (20 to 65°C). If optimum efficiency is at about 140°F (60°C), there is only limited activity at 86°F (30°C), and they can be virtually inactive after only a short period at 203°F (95°C). Therefore attempts are made to improve the optimum

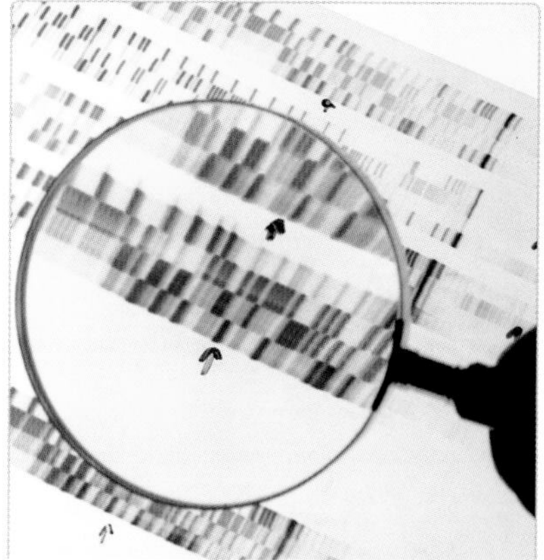

DNA analysis begins by extracting DNA from the cells in a sample of blood or other fluid or tissue.

temperature of these enzymes, in order to conserve energy through lower washing temperatures.

For more than a hundred years, wastewater from cities and municipalities has not only been mechanically filtered, but biologically cleaned also. Gray biotechnology attempts to improve the conditions of the processes involved. It also deals with the decontamination of soils that may have been polluted by a chemical factory or a waste dump.

Blue technology deals with microorganisms in the world's ocean; it studies deep-sea bacteria that live on hot vents (so-called black smokers), because these particular bacteria remain active at very high temperatures. Less clearly delineated is brown biotechnology, which focuses on environmental technology, and yellow biotechnology, which concentrates on activities involving foodstuff and raw materials associated with it.

Biotechnology is used in the production of grains, such as sorghum, which can be fermented in ethanol processing plants optimum to produce ethanol for use as a biofuel.

see also: The work of chemists, p. 290

EVERYDAY MATTER

For millennia humans have been governed by simple truths. Food will spoil and be consumed by pests while people will die and suffer from disease. It is a natural part of the human condition to strive to control the forces of nature with technology, so for that reason we have developed pesticides to protect our crops, and medicines to cure and fight disease. These improvements benefit our surroundings as we try to enhance the quality of our lives.

chemistry and nutrition

Artificial fertilizers and pesticides are necessary to produce enough food for the Earth's growing population. We also need better distribution of food and manufactured preservatives to reduce food waste.

The use of artificial fertilizers began in 1850. Today, a farmer can produce five times as much corn (maize) on an acre of land as his counterpart could have in the mid-19th century. This extraordinary revolution started when the German chemist Justus von Liebig discovered that plants extract nutrients from the soil and that these nutrients are then unavailable for use by future plants. To maintain agricultural output, the extracted nutrients—especially nitrogen, phosphorus, and potassium—must be returned to the soil. This is the role of fertilizers. Today, chemically generated fertilizers are designed for each crop based on its specific nutritional requirements. For example, the fertilizers used for corn are different from those used for wheat or soybeans. Modern fertilizers include calcium, magnesium, sulfur, and sometimes trace elements that play an important role in nutrition for specific plants.

Pest control

Fertilizers are not solely responsible for the huge increase in agricultural production. Advances in pest and disease control have also played an important role. Starting in the second half of the 19th century, farmers turned to chemical compounds to fight crop-threatening organisms. For example, Victorian era farmers used the "Bordeaux mixture" (burnt lime in a copper sulfate solution) to fight fungal diseases.

In the 1940s, it was discovered that chlorinated hydrocarbons and organic phosphorus compounds could efficiently kill insects. Herbicides were found that suppressed the weeds and grasses, which were consuming light, water, and nutrients that would otherwise be available to crops.

New pesticides are constantly under development. This is necessary because fungi and insects gradually become resistant or insensitive to the old ones. In addition, the chemical controls currently being developed are becoming increasingly efficient. Years ago, a farmer might have used 11 pounds (5 kg) of pesticides on an acre of crops, while today a farmer can achieve the same amount of pest control with only 3.5 ounces (100 g) of chemicals.

Chemical repellents applied to ripening sunflowers can help reduce blackbird damage, which is a chronic problem from seed formation to harvest.

Preservatives

In earlier times, when fungi and rot often spoiled food, famine was common. Today, substances are used to reduce the growth of fungi and repel insects, enabling food to be stored for a longer time. Commonly used preservatives include sorbic acid, nitrites, vitamin C, and vitamin E, which prevent the multiplication of pathogens. However, there is debate today about the possible negative effects of such additives.

Drying is the normal means of preservation for cereal grains such as rice, wheat, barley, millet and rye.

practice

HOW PESTICIDES WORK

Pesticides are chemical or biological agents that are used alone or as a mixture against organisms considered pests in agriculture and forestry. They range from protective products to substances for eradication of pest organisms in the widest sense. The best-known of these are insecticides against insects, herbicides for weed control, and fungicides against fungi and fungal diseases. Many pesticides are also toxic to humans; therefore, protective suits are usually worn during large-scale agricultural applications. Chemical residue from pesticides in food items is particularly significant as a health concern for humans. For this reason, limits have been set for residual levels in both food for humans and animal feed, which must not be exceeded.

High grasshopper populations can completely destroy fields and gardens.

see also: Food preservation, p. 345

medicinal drugs and cosmetics

Today's increasingly innovative medication plays a decisive role in preventing and treating diseases. More than 80 percent of the active ingredients in modern pharmaceuticals are generated chemically.

Although medicinal drugs do not guarantee eternal health, we still owe much to their role in our lives. Vaccines have eradicated polio in most of the world and people who are infected with HIV can delay the onset of AIDS by using medication. With

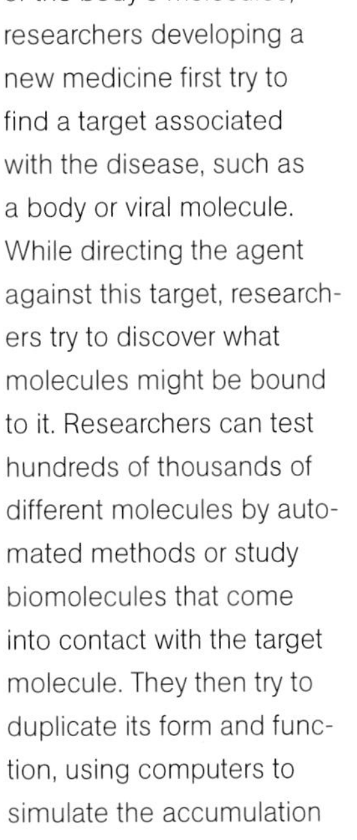

In China, both modern and traditional researchers and drug makers are competing to develop an effective treatment for AIDS.

the aid of drugs, gastrointestinal ulcers are cured within a week, making surgery unnecessary and certain pharmaceutical agents called statins lower the cholesterol levels of people who are suffering from cardiovascular disease or considered at risk. Pharmaceutical companies have suggested that even the most advanced stages of cancer may someday be treatable by drug therapies.

The search for active ingredients

Every drug's abilityto help the body depends upon its active ingredient, such as a plant extract (whether of natural origin or generated genetically)—although active ingredients are increasingly chemical in nature.

Because diseases are usually caused by invading viruses or by faulty interactions of the body's molecules, researchers developing a new medicine first try to find a target associated with the disease, such as a body or viral molecule. While directing the agent against this target, researchers try to discover what molecules might be bound to it. Researchers can test hundreds of thousands of different molecules by automated methods or study biomolecules that come into contact with the target molecule. They then try to duplicate its form and function, using computers to simulate the accumulation of molecules in the target. When scientists have found the appropriate molecules, they conduct tests, for example to discover whether possible agents bind only to the target without affecting similar biomolecules (something that can cause a variety of undesired side-effects).

Extensive tests

Candidates for active ingredients are tested on cell cultures and animals so researchers can analyze how they change in the organisms, how they and the products of their decomposition are distributed, their levels of toxicity, and whether they damage an organisms' genetic makeup. About three out of every four agents are rejected during these trials. The remainder of these are used in clinical trials on human beings that test first the drug's safety and then its effectiveness. Only one out of every 5,000 active ingredients discovered reaches the market. About 25 active ingredients are certified worldwide each year. The development of a new medicine takes an average of 12 years and costs about $800 million.

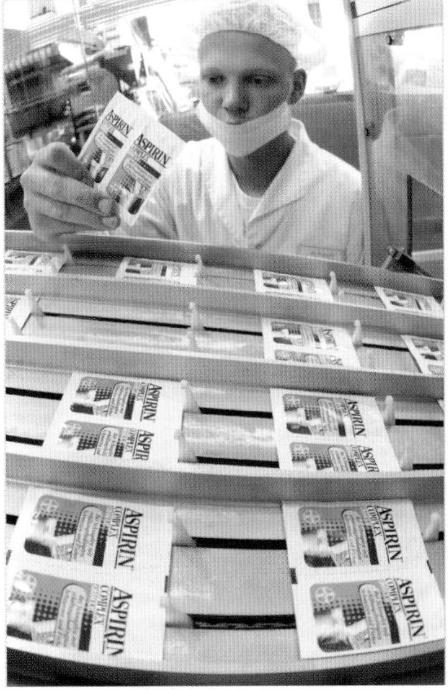

Medicinal ingredients must be intensively clinically tested before being tested on humans.

VACCINE CONCERNS

At the center of the vaccination controversy often lie questions of effectiveness, the possibility of adverse side-effects and the age at which to recieve vaccinations.

issues to solve

HOW SHAMPOO WORKS

The principal task of a shampoo is to clean the hair. For that purpose, shampoo contains surface-active substances, so-called surfactants, that become attached to the water-insoluble dirt and fat deposits in hair. Subsequently, they are washed away again with water. However, fat removal from hair must not be too radical (normal soap is therefore unsuitable), because fatty substances in hair also have a protective function. Beyond that, shampoos may also contain ingredients for various other effects, such as anti-dandruff agents, fat replacement substances, enhancers, protein hydrolysates, and perfume oils.

Sodium laureate sulfate makes shampoo foam.

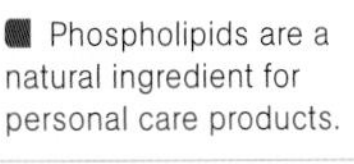

Phospholipids are a natural ingredient for personal care products.

in focus

plastics

Plastics are cheap and easy to process. The almost endless list of plastic products includes upholstered furniture, lacquers and paint, DVDs, casings of household appliances, and air bags in cars.

Nylon replaced silk in parachutes after World War II.

Due to their low cost, plastics are often used for single use products; however, because many are not biodegradable, they fill trash dumps and create a pervasive litter problem. Society needs to utilize more recycled and biodegradable plastics to avoid future enviromental problems.

In many ways, plastics have influenced the very fabric of modern society, from the decrease in glass bottles, wine corks, and Bakelite to the their role in the empowerment of women through Tupperware parties in the 1950s.

Composition

Plastics are built up from thousands of small groups of atoms bound together into a large macromolecule. They are known as polymers: a term derived from the Greek words *poly* ("multiple") and *meros* ("parts"). Thermoplastics soften and become moldable when they are warmed, while thermosetting plastics do not soften when heated; they change color or generally decompose. The most important thermosetting plastics are synthetic resins used for lacquers. Elastomers can be deformed at room temperature using pressure or tension but will return to their original shape once the pressure or tension is released. Foam materials and polyesters are elastomers.

Light and moldable

Plastics have significant advantages over materials such as wood or metal. They are light in weight and resist damage from weather and chemicals. They are excellent insulators for heat and electrical currents. Furthermore, they can be shaped easily. One widely used shaping technique is injection molding. In this process, thermoplastic granules are melted and then injected into the hollow mold. The mold determines the form and surface of the finished product. Many plastic bottles are made using injection molding. Because of their wide use in short-lived products, plastics are a growing disposal problem. The initial advantage of chemical stability later becomes a disadvantage, since they decompose very slowly in trash dumps and cluttered roadways, where they end up. Thus the issue of plastic recycling is growing in importance.

In the future?

PPV foils made from advanced plastics show promise as a new and flexible light source. Electronic paper made from a thin and flexible screen of plastic foils could replace ink and paper, using electronic ink for text that changes upon the influence of an electric current.

issues to solve

PVC (POLYVINYL CHLORIDE) is commonly used in many applications. However, it is very costly to recycle and is often included in ordinary rubbish leading to many countries developing detailed recycling systems to deal with plastics.

Plastics are often shredded and then reused as road ballast and building materials.

basics

MOST PLASTICS melt when heated. This is because their threadlike molecular chains are held together only by weak molecular forces. Nonplastics such as rubber can withstand higher temperatures.

in focus

POLYCARBONATE

Polycarbonate CDs, CD-ROMs, and DVDs are used today in numbers so large it is difficult to imagine. The 35 billion discs produced in 2001 would create a tower about 19,000 miles (30,000 km) high if the discs were stacked on top of one another. Bottles, MP3 players, and sunglasses are also made of this useful thermoplastic. Polycarbonates are widely used because they are transparent, inherently stable, mechanically firm, resistant to light, and easy to clean.

Plastic bottles are constructed during the process of injection molding.

nanotechnology

Products made of nanomaterials are already being produced around the world. It is expected that nanoproducts of every shape and sort will be used more and more in our everyday lives in the years to come.

A nanometer is to a meter (equal to 3.2 feet) what the diameter of a hazelnut is to that of the Earth. Those who work with structures less than a hundred nanometers in size work in the world of nanotechnology, where classical principles of physics and chemistry do not always apply. Today, research is primarily focused on nanostructured surfaces, nanoparticles, and the mixture of nanoparticles with materials such as plastic.

The creation of nanoparticles

Nanoparticles can be created by reducing structures and objects, a method that the semiconductor industry currently uses to miniaturize microchips. Objects can also be built by manipulating individual atoms and molecules. Scientists use the tip of a scanning, tunneling microscope to move atoms to and fro, merging them into larger entities if needed. Because this procedure is time-consuming and costly, researchers are hunting for ways to get atoms and molecules to arrange themselves.

From sun protection to computers

Nanoparticles have already been added to many common substances, helping sunscreens deflect UV radiation, for example, and windshields to reflect heat and light; wafer-thin nanocoatings keep automobile paint and plastic glasses from scratching and mirrors from misting up. Nanostructured substances prevent dirt from gathering on self-cleaning bathtubs and roofing tiles.

Many nanotechnological discoveries will soon come into play in many areas of our lives. It is hoped that the surfaces of nanoparticles can be coated with biological matter that will allow nanocontainers loaded with poisons to attach themselves to cancerous cells. Nanostructures may also be able to suppress the formation of blood vessels that supply tumors. Carbon nanotubes (atom-thick graphite sheets rolled into seamless cylinders) can be used to manufacture transistors and simple logical circuits that may soon replace silicon transistors, the miniaturization of which has reached its physical limits. Researchers today are investigating higher performance batteries, fuel cells, energy converters, and forgery-protected documents.

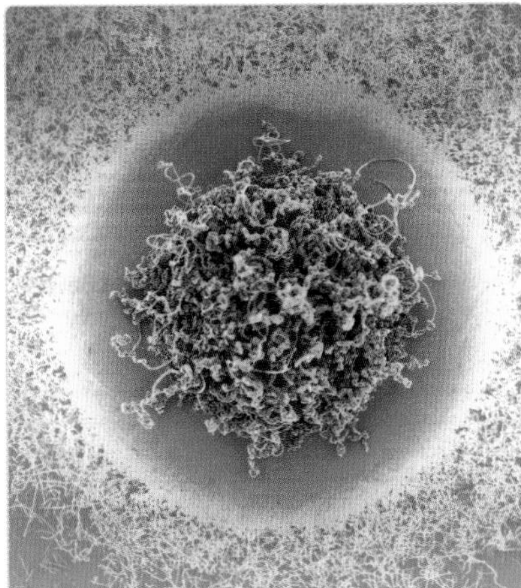

Carbon nanotube structures are regarded as having endless practical applications in modern life.

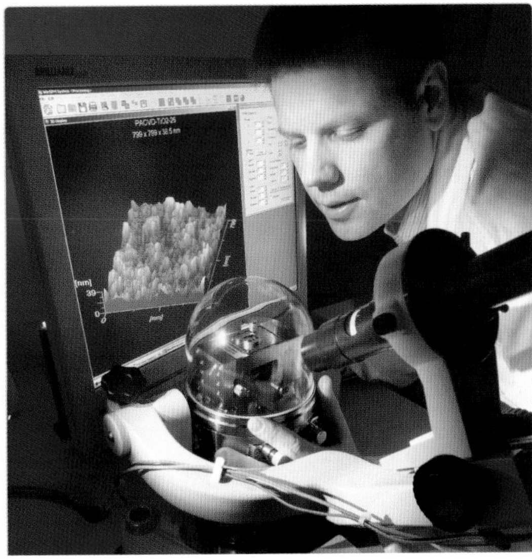

Materials act very differently at the nanoscale; however, due to their minute dimensions, nanomaterials are hard to study.

Properties of nanomaterials

Materials often change when shrunk to nanoscale. Gold, which has a distinctly yellow color and very low chemical reactivity, turns red at the nanoscale and acts as a catalyst, accelerating reactions between other substances. This is because the proportion of its surface area to its volume changes, a fundamental principle of nanotechnology. The greater the surface area, the greater the possibility of chemical and physical exchange with the environment.

practice

CARBON NANOTUBES

Nanotubes are tiny tubes with a diameter of only a few nanometers and a length that can be up to one millimeter. By comparison, a human hair is 50,000 times thicker. Particularly well researched are carbon nanotubes (CNTs), but there are also nanotubes made of other substances. Carbon nanotubes are built up from one or several graphite layers that are rolled cylindrically. They can be single- or multi-walled, and depending upon production conditions, bundles or threads of nanotubes can also occur. Nanotubes conduct heat very well and, according to their structure, they can be good semiconductors or even superconductors at certain temperatures. Moreover, they do not tear and they are extremely stable against chemical and thermal influences.

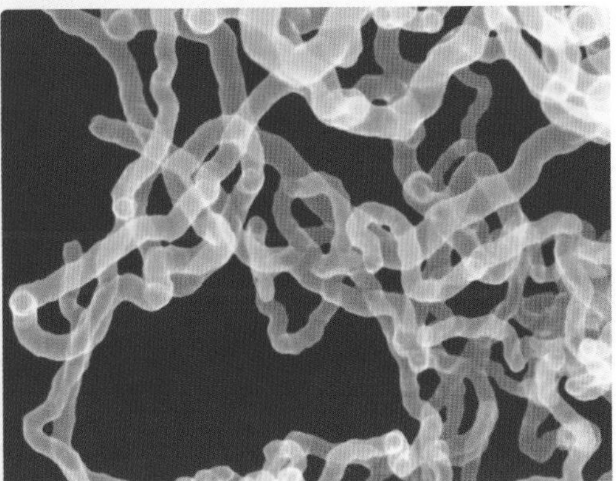

It is hoped that nanotubes can improve computer chip efficiency, as standard technology is unable to deliver new breakthroughs.

see also: Materials of tomorrow, p. 293

chemistry | production chain | accidents | pollution

ECONOMY AND ECOLOGY

No part of the economy can manage without the products of the chemical industry. The industry generates billions of dollars in revenue; however, we should not ignore the possible danger from their products and installations.

economy

The chemical industry and its many products are involved in every branch of industrial production from automobile assembly lines to building constructions, employing millions of people worldwide.

The chemical industry constantly researches and develops new materials of every description to help businesses deliver new and improved products to consumer and industrial markets. New materials help increase the durability of consumer goods such as shoes, clothing, and sports equipment. Such materials can also give capital goods—such as industrial robots, heavy machines, and electronic parts—longer lifespans. The advent of new and innovative synthetics has allowed lighter, more comfortable automobiles to be manufactured, and wood floors and furniture no longer release hazardous substances now that manufacturers are able to use new, solvent-free adhesives to assemble them. New materials with brand-new optical properties allow large quantities of data to be exchanged at high speeds using fiber-optic cables. Advanced chemical innovations help reduce the many expenses associated with production processes. The actual contributed commercial value of chemical products is relatively minor in comparison to the total value of most end products of which they are a part, but their absence would put the economy of the entire industrialized world in a serious bind.

Adhesives are constantly developed with ever stronger adhesion and reduced environmental impact.

Chemistry in the global market

The United States makes more chemicals than any other country in the world. In 2005, American chemical companies produced chemicals worth approximately $593 billion and employed about 880,000 people. In addition, many jobs in ancillary industries directly depend upon the chemical industry, and consumer spending on chemical-related products creates new jobs in many other fields. Although the U.S. dominates the international chemical market, Germany, the Netherlands, the UK, and Japan are also major players in global chemical production.

Processes in the benzene production chain

From raw materials to end products, the chemical industry is its own biggest customer. The production of chemicals takes place across many stages, each of which requires processing that involves a variety of other chemicals. One major production chain begins with chlorine-alkali electrolysis, and another chain begins with the production of ammonia (p. 289). The initial product of yet another important production chain is crude oil.

Benzene, for example, is converted into cumene before substances can be developed from it that are used to produce lacquers and plastics. Even at its basic initial stage, benzene can also be converted into ethylbenzene, cyclohexane, aniline, and chlorobenzene.

practice

SULFURIC ACID

Sulfuric acid (H_2SO_4) belongs to the most important inorganic acid, and it is produced in huge quantities by the chemical industry. However, only a small portion of it is used directly as sulfuric acid; most of it is utilized as a base ingredient in the manufacture of secondary products. Sulfuric acid is particularly important for the production of fertilizers (e.g., ammonium sulfate) and other inorganic acids, and for paints and cosmetics. The sulfuric acid production of a particular country can be used to gauge its level of development and the significance of its chemical industry.

Sulfuric acid is a highly corrosive substance, and care must be taken in its storage.

Chemical factories and refineries are a part of the landscape of industrialized nations.

see also: From test tube to reactor, p. 289

accidents

Many chemicals have dark sides. Both smog and polluted waters are among the numerous side effects of chemical production, especially in countries that do not have stringent environmental regulations.

Even the staunchest believers in the benefits of applied chemistry are willing to admit that when it comes to chemical plants, safety is not guaranteed. It is widely accepted that as global conditions continue to change, humankind's reliance on these techniques will only increase, making Earth's situation even more precarious. Chemical plants sometimes catch fire or explode, and still more accidents have occurred while transporting chemicals. The most serious incident to date occurred in 1984, when 40 tons of methyl isocyanate leaked from a tank of gas in the Indian city of Bhopal. The resulting chemical cloud killed at least 2,000 people, injured more than 100,000, and exposed over 500,000 to harmful gases.

Burning chemical fires bring a secondary hazard, often extending unseen beyond the flames and smoke. The cost of incurable degenerative disorders and birth defects is not known.

Side effects

Chemical production brings with it negative and long-term consequences. In Asia and South America, regulations (such as those concerning the desulfurization of gaseous waste) are less stringent than those in Europe and North America. But when weather conditions keep gaseous waste—caused by industrial production and emitted from vehicles—from escaping into the higher atmosphere, smog forms. "Summer smog" is formed by emissions from solvents that release volatile compounds. Solar radiation causes these compounds to react with the nitric oxide in vehicle exhaust, forming ozone and other pollutants that affect breathing, irritate mucous membranes, and lead to circulatory disorders.

The Bhopal disaster is a focus for protests over the alleged lack of responsibility and compensation.

Long-term consequences

If environmental pollution is not detected in time, the effects may not be known until years after the damage has already been done, making countermeasures remedial instead of preventative. Such environmental problems include the CFC (chlorofluorocarbon)-produced hole in the ozone layer, damage done to forests by acid rain, and global pollution of the environment by DDT (dichloro-diphenyl-trichloroethane).

Poisonous dioxins

Dioxins are produced by such things as garbage incineration, paper manufacturing, forest fires, and diesel engines. They quickly spread in the environment where they can be found in small quantities in foods and humans. As extremely toxic substances, dioxins can increase the likelihood of cancer. In 1976, a cloud released from a chemical factory near the Italian town of Seveso killed many animals and caused chloracne in about 200 people. Dioxins comprise about 210 compounds of a similar chemical structure with varying toxicity.

in focus

DDT (DICHLORO-DIPHENYL-TRICHLOROETHANE)

Greenpeace is an active campaigner against the use of DDT, which is still in use in parts of the developing world.

This insecticide proved to be an effective tool against mosquitoes and malaria, which was significantly reduced in the middle of the 20th century. It was realized only later that DDT, sprayed in vast quantities by farmers, wound up heavily concentrated in the food chain. This caused an array of health issues, such as cancers and asthma, as well as child development problems. It also affected wildlife by thinning the eggshells of birds. DDT was gradually prohibited in all industrial countries.

PHYSICS AND TECHNOLOGY

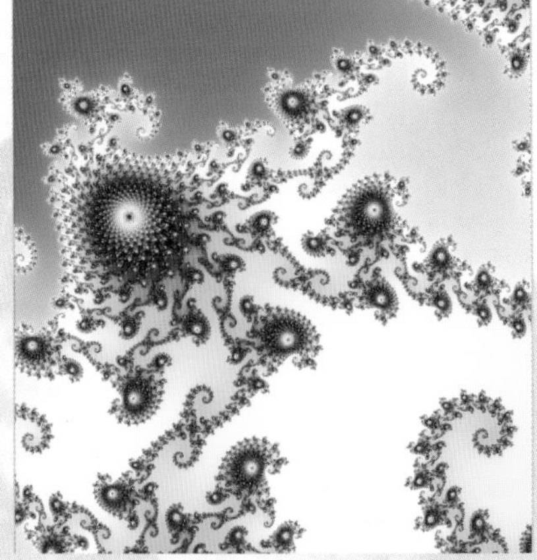

PHYSICS

Physicists study natural forces, the characteristics of matter, and how physical laws interact with one another. Trying to explain natural phenomena seems almost impossible; after all, they range from the smallest particles to the universe as a whole. Nevertheless, they can be described using terms such as distance, time, speed, acceleration, mass, and charge. Most natural phenomena are extremely complex processes, which is why their physical laws are mainly detected through targeted experiments. Ultimately, the laws are generalized using mathematics as a language. Insights into the physics of natural phenomena collected in this manner can then be applied to other natural science fields, as well as engineering disciplines.

energy and matter

Every natural process is an exchange or conversion of energy. Mass and energy are equivalent due to the fact that mass is a characteristic of matter and a form of energy.

types | quantity | forces | interactions

ENERGY

Even in ancient times or earlier, humans have been wondering about the matter of the world. The idea of four basic elements was widespread until the Middle Ages. These elements were fire, water, earth, and air. Modern physics poses the same question about nature and its subject matter in many different new ways.

Generations of physicists have been studying the connection between force, energy, and motion for centuries. Finally, in the 19th century, a concept was formulated to answer this equation, a term derived from the Greek word *ergon*. The idea originated from the question of how machines accomplish mechanical work (moving a body through force). Scientists today define energy as the ability of a machine, a living organism, or any other "system" to do work. A car, for example, filled with fuel as energy can transport persons and goods; the gravitational energy of the Earth causes water to flow toward the ocean, as well as apples to fall off trees and skiers to speed down a trail with hardly any effort at all. There are many different types of energy: kinetic energy, thermal energy, (bio-) chemical energy, radiant energy, atomic or nuclear energy, electric and magnetic energy, and gravitational energy. All planets, rocks, molecules, and living organisms in the universe have one or another type of energy acting upon them, frequently even several at a time. All processes happening in the universe involve an exchange or a transfer of energy from one form into another.

Physical quantities

In physics, physical quantities are defined and often represent certain characteristics that can be given in measurable units. Energy, the most important quantity, has the unit "joule" (J; after James P. Joule, 1818–1889). Mass is measured in pounds, time and space in seconds (s) and feet (ft), and speeds in feet per second (f/s). Force (p. 313) has the unit Newton (N; after Isaac Newton, 1643–1727). We calculate newtons with the formula $1 N = 1 kg \times m/s^2$ and joules with $1 J = 1 N \times m$. Electrical units are voltage (volt, V), amperage (ampere, A), or resistance (Ohm, Ω).

Albert Einstein, who won the 1921 Nobel Prize in physics, revolutionized this field when he was able to explain the association between mass and energy in a simple way with his concept of mass-energy equivalence formula ($E = mc^2$).

Nature has the ability to free energy tied up in the weather systems in many spectacular ways.

The cause of all life on Earth is the release of nuclear energy from the interior of the sun.

in focus

CONSERVATION OF ENERGY

The concept of energy is part of the most important law in physics: the law of conservation of energy. This law states that the sum of all types of energy in the entire universe always remains the same. The same is true for all so-called closed systems such as machines, animals, and atoms that do not exchange any energy or matter with their environment.

Kinetic energy is converted to heat due to conservation of energy, as it can only change forms and never be destroyed.

see also: Dark matter and dark energy, p. 23

forces, fields, interactions

The idea of energy may be useful in theoretical physics, but it is not a very tangible concept. Here, an older term may be more useful that is used both in physics and in everyday language: force.

Physicists consider everything that triggers a change in speed as a force. Figure skaters use muscle force to push their bodies up from the ground and then are pulled back onto the ice by gravitational force.

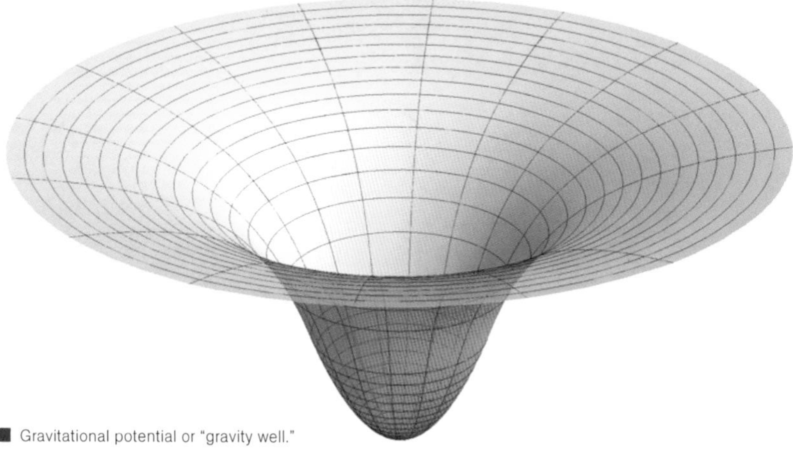

Gravitational potential or "gravity well."

They skate elegantly by taking advantage of the interaction between gravitational force, the inner forces of the ice, and the ice skates. Forces also interact between the sun and the planets of the universe and also between an electron and an atomic nucleus. All known forces, such as muscle force, explosive force, friction force, buoyancy, and magnetism, can actually be derived from the four fundamental forces: gravitational, electromagnetic (p. 322), and two different kinds of nuclear forces (p. 329), which act over comparatively short distances. Gravitation is the only one of the four fundamental forces that affects all objects in the universe, as gravitation is a direct result of the spatial and temporal structure (p. 326–327). In contrast, the electromagnetic interactions only affect electrically charged particles (electrons) or objects (magnets).

Gravitation is the weakest of all the four forces; however, it is the one that we feel the effects of constantly.

Fields

Before discussing topics such as heat, work, or tsunamis, we need to first consider the third concept that, besides energy and force, fields exist to describe physical processes. Electric, magnetic, and gravitational fields are everywhere. They summarize in easily understandable ways how forces influence a particle at a certain point. The relationship between energy, force, and a field can best be explained by an example: An electron with the electric charge "E" travels at high speed through a four-inch (ten-cm)-long cable toward the positive pole of a battery. During this time, the electron reacts to an electric field (E), which keeps getting stronger and accelerates the electron with a force $F = e \times E$. The amount of electric energy converted into kinetic energy of the electron is the product of the force and the distance over which the force was active; F multiplied by four inches (F x 10 cm).

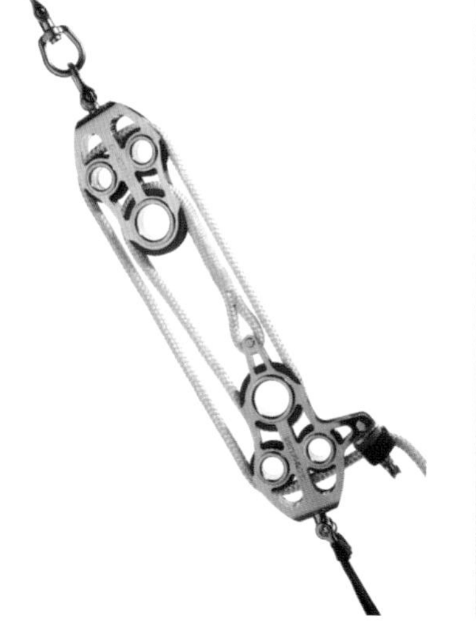

A pulley block reduces the force required to pull an object, but the energy required remains the same.

practice

FRICTION

Friction force is a less talked about force in physics. Rather than a fundamental concept, friction force is rather a collective term for processes where (technically usable) energy is converted into heat or lost in a different way. Without friction, a car engine would only need to run at the start, electricity could be stored forever, and spaceships would not require a heat shield during re-entry into the atmosphere.

If friction between car tires and the road is too low, dangerous skidding occurs.

rest and motion

The names of Johannes Kepler, Galileo Galilei, and Isaac Newton represent a time of development and revolutionary science emerging from a collection of inventions and experiences: the science of motion and its causes.

rest | motion | gravity | velocity | momentum | mass

MECHANICS

Mechanics is the oldest subject area in physics. Even early humans accomplished challenging tasks using levers and simple tools during the Stone Age. The ancient Greeks turned the development of machines into an art, the *mechanike techne*. The Romans derived the word *machina* from this Greek term, and thus modern scientists began using the terms "mechanics" and "technology."

One of the most important realizations in mechanics is that each body has its own center of gravity. It does not matter if the body is a planet, a human, or a crystal. Its movement through time and space can be described as if the entire mass was concentrated in this one center of gravity (as long as the air resistance is ignored). Grasping this abstract idea played a major role in advancing human intellect. Because of it, scientists were able to study motion by itself and to transfer findings to any applicable example.

Newton's cradle: a device that operates using the principle of the conservation of momentum.

Velocity and acceleration

The study of an object in motion is called kinematics (from the Greek word *kinein:* "to move") and is a subject area in mechanics. One of the most important laws of kinematics states that velocity (v) is the quotient of the distance traveled (s) and the travel time (t): $v = s/t$. Its unit is miles per second (km/s).

In order to calculate the acceleration (a), the difference in velocity is divided by the travel time: $a = v/t$. This calculation gets more complicated when the value and direction of the kinematic quantities change during the process that is being investigated. Certain mathematical methods are required: differential and integral calculus. These methods were eventually developed by scholars such as the Englishman Isaac Newton and the German Gottfried Wilhelm Leibniz.

Impact and momentum

The mathematical description of the planetary orbit of a cannon ball was a tremendous success of mechanics that gave it credence as a science. However, the causes of motion were described by physicists. This is how the concept of force (p. 313) was developed. An object, such as a ball, is accelerated by a greater force the smaller the mass of the object.

Let us consider the product of mass and velocity: momentum. Every impact changes the momentum exactly by the degree of the force acting on the object. Momentum also has a law of conservation of momentum, similar to the law of conservation of energy (p. 312). A few mathematical conversions produce another conserved quantity for rotating objects, the angular momentum. The laws of conservation are an important tool that is used to formulate and solve equations for every particle collision, every pirouette on ice, and so on, without having to consider the details of these processes.

The space shuttle accelerates in eight minutes from a stand to an orbital speed of about 17,398 mph.

The angular momentum of a figure skater does not change during a pirouette.

see also: Practical application, p. 409

of apples and planets

Prehistoric people already knew that objects have a certain weight. The novel idea is that the weight of, for example, a milk bag is actually the reciprocal attraction between the milk and the Earth.

Mass is not only a factor in Newton's principle of action. It is also the critical quantity of gravitational force FG, also referred to as gravity or gravitational attraction. A reciprocal attraction of two masses only depends on the product of their masses and the distance r between the two: $FG = G \times m^1 \times m^2/r^2$. The constant of gravitation G is a quantity that remains a mysterious constant of nature. Its value is always the same in the entire universe and at all times, just like the electrical charge of an electron. To date, nobody knows why it is always the same.

Liquid mechanics

Archimedes' discovery of buoyant force is a story that is just as fascinating as the .legend about Isaac Newton's apple (see milestones): While he was bathing he discovered that a crown, which he was commissioned to check for its gold content, became lighter once he put it in the water. He was so excited about his finding that he ran naked through town and shouted "Eureka" ("I found it!"). Ever since, the mechanics of liquids and gases have become an important scientific discipline.

Research areas include the investigation of the air resistance of flying objects or vehicles (known from the cW-value). Scientists of this discipline also investigate the behavior of liquids in pipelines as well as oceans and the atmosphere. This subject area is called hydromechanics. It is not a basic science but mathematically very challenging. This is why the largest and fastest (for its time) computer in the world was constructed in 2002 for the purpose of hydrodynamic calculations on climate. This Japanese computer was called Earth Simulator.

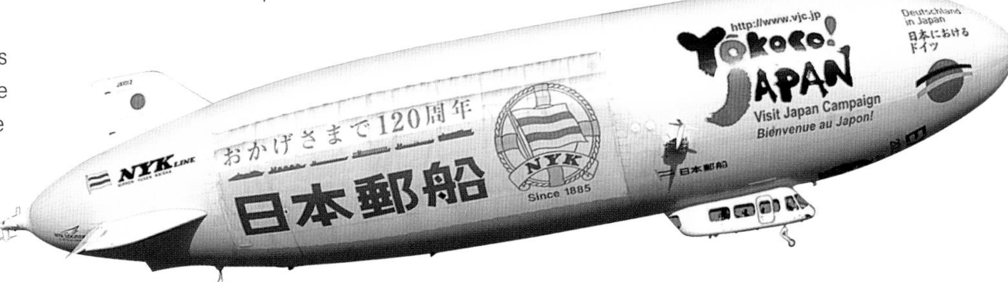

A zeppelin hovers due to its hydrodynamic buoyant force.

NEWTON'S APPLE

Legend has it that Isaac Newton, while resting under a tree, observed how an apple fell from a branch to the ground. He was struck by the revolutionary idea that the apple was drawn to the Earth by the same gravitational force that existed between the planets and the sun. Thus the classic theory of gravitational force was born. Newton recognized the mass of an object (the "heavy," as he called it) as a source of the ever present force of gravity.

This is how Isaac Newton is said to have come up with the general law of gravitation.

milestones

Mechanics: an outdated discipline?

Mechanics in the classical sense of Galilei and Newton's work has not been pursued as a novel research area for centuries. New discoveries are unlikely. Moreover, some of the basic theories of classical mechanics were negated by Albert Einstein's theory of relativity (p. 330) and quantum mechanics (p. 326), which was developed only shortly after. And even today's scope of mechanics has been questioned by the new chaos theory (p. 334).

Many problems cannot be solved or their solutions are "chaotic." Given these doubts, is mechanics still a relevant discipline? Absolutely. The theory of relativity only needs to be considered for extremely exceptional velocities and quantum mechanics differs from the principles of classical mechanics only when considering submicroscopic elementary particles (p. 328). The original laws of (hydro-) mechanics apply to everyone who wants to build high-rises, send satellites into space, or build wind turbines to use the power of wind. Today's civilization depends on technology that is based on the art of engineering and advanced knowledge of mechanics—for throughout the course of history, mechanics has been the basis for new inventions.

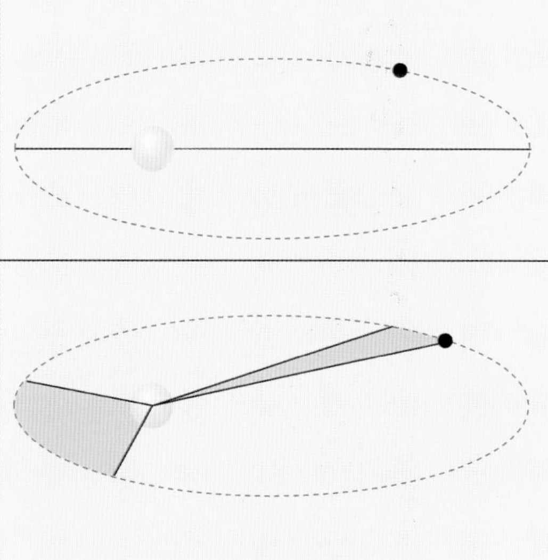

Johannes Kepler's laws of planetary motion can be derived from the law of gravitation.

resonance | vibration | diffraction | interference

VIBRATIONS AND WAVES

Most areas in physics investigate well-defined issues—for example, electromagnetics or thermodynamics. The study of vibrations is different: Practically any vibration or oscillation can alternate between a high or low intensity. This explains why this area of research is so interesting and relevant to the entire field of natural sciences.

vibrating systems

Just as the Beach Boys sang *Good Vibrations* in 1966, vibrations can be passed on in physics as well—sometimes resulting in harmony and sometimes in a catastrophe.

Physical vibration can be much more than a sweep of a pendulum or a child on a swing. For example, atoms of a carbon dioxide molecule are swinging against each other continuously. Vibrations inside the Earth tell us much about its structure, but they can also cause damage through earthquakes. The "inner clock" of humans, animals, and plants runs on periodically recurring chemical reactions. The rise and fall of stock prices as well as poll ratings of politicians are also familiar examples. Regardless of how diverse the swinging or vibrating systems in nature and society may be, the same physical and mathematical methods of description apply to all of them.

■ The Foucault pendulum always swings in the same direction. What moves is the Earth beneath it.

■ Tacoma bridge vibrating violently due to hurricane winds causing a violent oscillating motion.

Resonance: from sound to catastrophe

As soon as we hear a rhythmic sound nearby, we will often begin to unconsciously move along with it. Not only humans can be made to swing or vibrate, the same can be done to a tuning fork or a blade of grass. The scientific expression for this phenomenon is "resonance" (from Latin *resonare*: "resonate").

Not every sound causes an object to resonate. A soprano singer has to hit the note precisely so that it will break glass. We have to push our children on a swing in precisely the right rhythm that allows the swing to go even higher. Why? Because of the so-called natural or free vibrations of the glass or the swing. Each system has certain rhythms (scientifically, this is referred to as frequencies). If such a free vibration acts upon the object from the outside, then it will vibrate more and more—in the worst case, it may even vibrate until it is mechanically destroyed.

Taipei 101

The Taipei Financial Center, called Taipei 101, is a mega-high-rise building in the Taiwanese capital Taipei, which is situated in a high-risk area for earthquakes and hurricanes. To protect the building from suffering damage, the architects and engineers designing the building decided to place a huge ball weighing several tons in one of the top floors. Vibrations of the building caused by waves from earthquakes or hurricanes are dampened by the ball's easily induced movement.

in focus

THE SINE FUNCTION

The study of vibration is relevant to mathematics, physics, and mechanics. The simple form of vibration with a constant frequency and constant maximum deflection is described by the mathematical sine function. The plotted function looks like a curve with gentle waves—its Latin name *sinus* is derived from this shape.

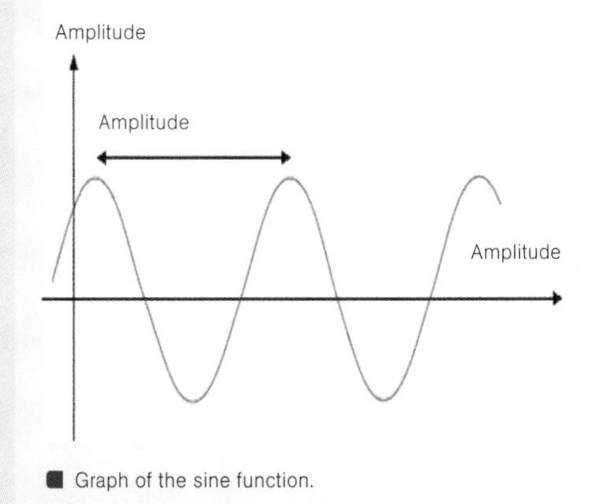

■ Graph of the sine function.

see also: Earthquakes, pp. 68–71

waves everywhere

Waves are familiar in everyday life, but they are a more difficult concept to grasp than vibration. Wherever there are vibrations, waves can also occur. A medium such as a gas or liquid is required to transmit acoustic or water waves, while electromagnetic waves of light and radio require no medium.

Small children experiment with waves in the bathtub and older children play in the waves at the beach. Our two most important senses, hearing (p. 314) and sight (p. 318), use waves: acoustic waves or sound and light. Radio waves, infrared light, x-rays, and light are all electromagnetic waves with different frequencies. The creators of quantum physics (p. 322) were able to show that even electrons, atoms, and all materials have wave properties.

A vibration can be visually observed through watching the waves that develop in a standing body of water or oil.

Origin and expansion

The origin of a wave is always a vibration: for example, the string of a violin being played. First this will cause a small volume of air to vibrate, which, in turn, causes its surrounding environment to vibrate and so on. Thus the state of motion vibration expands in all directions through the medium of air, which is receptive to vibration. If the vibrations continue, the room will be filled over time by a typical wave pattern, which changes periodically at every point. If the original vibration was of short duration—for example, when a rock is thrown into the water—only a shortwave train is generated, which naturally dissipates and is eliminated due to friction.

Diffraction and interference

Two of the many wave phenomena are especially interesting: diffraction and interference. Both are based on Huygens' Principle, named after Dutch scientist Christiaan Huygens (1629–1695). According to this principle, a circular or spherical wave can originate from any point of a wave pattern with exactly the same frequency and wavelength. So if a wave is traveling along a wall toward a corner, it can expand from this corner in the form of a spherical wave, as if it were suddenly set free. This diffraction teaches us a great deal about the properties of waves.

Constructive interference is where two waves superimpose in such a way that the crest of one wave will always be on top of the crest of the other wave. Their impacts combined result in especially strong local vibrations. In contrast, during destructive interference a crest is always on top of a trough. If both wave patterns have the same height, then the interfering waves cancel each other out entirely.

During supersonic flight, jet aircraft create pressure waves both in front and behind. The sonic boom is created when the highly compressed forward wave suddenly increases in pressure then is equalized by the rear.

in focus

TSUNAMI

What was different about the monster wave on December 26, 2004, that resulted in a catastrophe? With normal water waves, only the surface of the water body is in motion; however, a tsunami brings not only the surface into motion but lifts and then drops an entire water column in the ocean. This column may be several miles tall. The size of such a wave makes it tremendously powerful, and if it hits a coastline, the entire wave energy is concentrated onto a very flat layer of water. This causes the water to move differently than in the open sea. It moves quickly and far with a high amount of destructive power.

The tsunami catastrophe of December 26, 2004, resulted in hundreds of thousands of deaths.

waves | noise | ultrasound | sonar

ACOUSTICS

Acoustics is the study of sound waves traveling through air, liquids, and solid objects. Just as hearing is one of our most relied upon senses, speech is the basis of the growth of human society. Acoustics could be said to have contributed to art, science, and society.

waves in the air

Among the varieties of waves occurring in nature, sound waves are especially familiar to us, since we possess a sense organ—the ear—that can pick up an astonishing amount of information from them.

Waves are vibrations in motion (p. 316). What is actually vibrating, though, as a sound wave travels? It is the atoms and molecules of the medium through which the sound is moving. Vibrating around a resting point, they cause their neighbors to vibrate as well. These vibrations—and thus sound—can move through nearly anything: a piano string, a container of air, the ocean, and even our Milky Way galaxy. Sound waves in the air have a special characteristic: air molecules vibrate only in the direction in which the wave is traveling. This is called a longitudinal wave. In contrast, waves moving through solid objects, such as the Earth, can also have vibrations perpendicular to the direction of travel (transverse waves).

Sound, music, noise

The concepts of wave physics find their counterparts in acoustics: for example, a sound wave of a particular frequency is perceived as a specific tone. The amount of energy carried by a sound wave is registered by the listener as loudness or volume. Whether a sound is heard as music, speech, noise, or static depends on the various sine waves that combine to make up the sound. The term "tone color" is used to describe this phenomenon. Sine waves with frequencies related to each other in simple ratios—such as 1:2, 2:3, or 3:5—are perceived as harmonious. Frequency relationships such as 4:17 or 97:111, on the other hand, are less pleasant to the ear. If many different frequencies with similar energy levels are emitted at the same time, we hear static or "white noise." The speed of sound depends on the substance through which it is moving, but it is usually greater than 621 miles per hour (1,000 km/h). This is still much lower than the speed of light, however, as is evident when we see a flash of lightning long before the sound of the thunder arrives.

practice

CREATING TONES WITH MUSICAL INSTRUMENTS

In stringed musical instruments (such as the violin, guitar, or piano), sounds are produced when metal strings are set in motion, vibrating separately from each other. The frequency of the vibration—and thus the particular note produced—depends on the length of the string. With brass and woodwind instruments, on the other hand, the musician's blowing action causes vibrations in a column of air inside the instrument. By pressing different valves, the player can change the length of the air passages and thus the pitch of the note.

Violins create sound through friction caused by the movement of the bow upon the violin string.

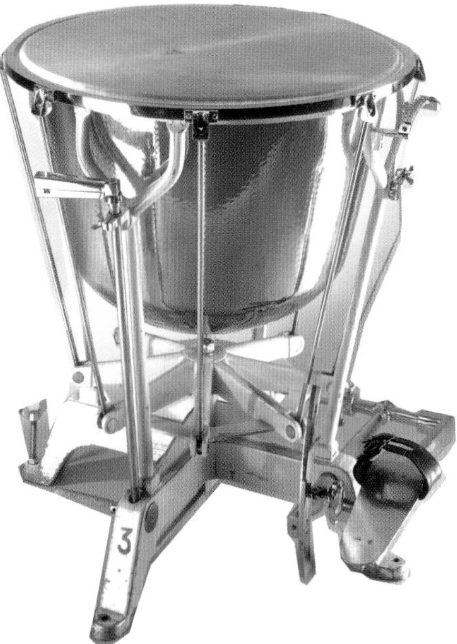

With percussion instruments such as the bass drum, the sound is highly variable and distinctive.

Cosmic microwave background radiation provides evidence for the big bang theory. Soundwaves moving through space shortly after the big bang affected the way galaxies are clustered today.

ultrasound

The use of ultrasound in medical exams—for instance, during pregnancy—has become routine. However, ultrasound is also very useful in technological applications, such as depth measurement.

The vibrations of a sound wave in air can have a very wide range of frequencies, from less than one per second (below one hertz or Hz) to 1,000 times a second (one kilohertz or kHz) and in certain cases even more. The musical notes with the lowest frequencies—that is, those with the deepest pitch—vibrate at about 20 Hz, while the highest notes audible to people reach around 15–20 kHz (although the upper limits of hearing decrease with age). All sounds with frequencies over this limit are called ultrasound ("ultra" means "beyond" in Latin).

Ultrasound in nature

Although people cannot hear ultrasound frequencies, some animals can. Dogs can perceive tones reaching up to some 40–50 kHz, like those produced by special dog whistles. Bats and dolphins have an even higher hearing limit: well over 100 kHz. There is a practical reason for this; because high-frequency ultrasound waves move in precise lines, like light waves, bats and dolphins can use them for orientation, sending out ultrasound signals just as a ship's sonar system does. From the echoes, the animals' auditory systems can determine the position, speed, and even the shape of their prey (or enemies). Another variation can be found among the elephants, who are able to produce infrasound tones with frequencies of below 20 Hz. Spreading out in all directions over a very broad area, the sounds serve as a kind of "elephant cell phone network."

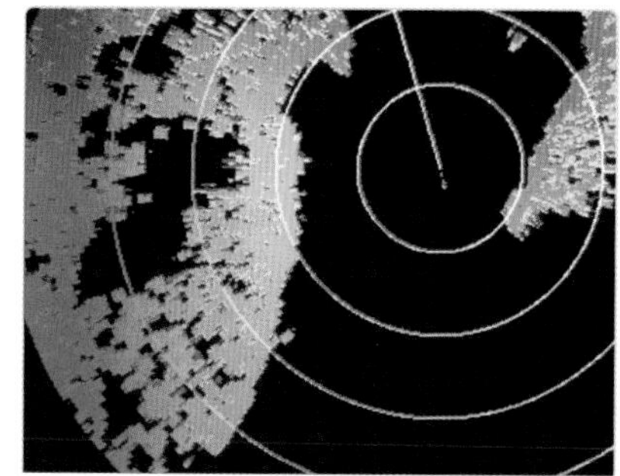

Sonar (sound navigation and ranging) uses underwater sound propagation to communicate, navigate, and locate other vessels.

Dolphins orient themselves using ultrasound, emitting calls and then interpreting the echoes. Their teeth are arranged in a way that works as an array or antenna to receive the incoming sound, and the lower jaw transmits the waves to the middle ear.

basics

INFRASOUND Tones below 20 Hz. Pressure waves in air with frequencies lower than 0.1 Hz are no longer considered to be sound.

AUDIBLE SOUND The range of frequencies that can be heard by young, healthy humans stretches from 20–20,000 Hz.

ULTRASOUND Frequencies over the upper limit of human hearing (20 kHz). Sounds up to several hundred KHz can be perceived by some species of animals.

HYPERSOUND Sounds above 1 GHz (1 billion hertz). As the frequency increases further, hypersound is no longer considered a wave, according to the laws of quantum mechanics, but as so-called quasiparticles.

practice

HOW IS AN ULTRASOUND PICTURE PRODUCED?

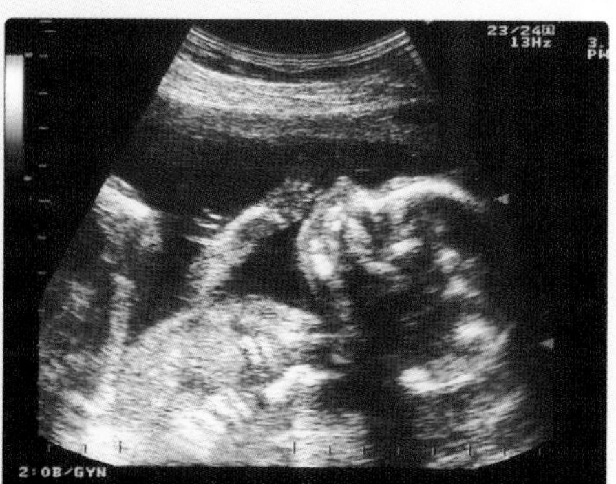

Bones, muscles, and other tissues produce different ultrasound echo patterns. Displayed as a picture, they provide a sonogram—the "first snapshot" of a baby in the womb.

Depending on the frequency used and the type of body tissue being examined, ultrasound waves penetrate to different depths within the human body. Ultrasound echo patterns can produce a complex and informative picture of bodily structures, which an experienced doctor can use to draw many conclusions about the patient's health. One difficulty arises from the fact that ultrasound cannot enter the body directly from the air. For this reason, special transmitter heads and gels are used to transmit the waves.

Healing and cleaning

Among humans, ultrasound is not only useful for ship navigation: obstetricians and many other medical specialists use ultrasound to examine the human body in a relatively non-intrusive manner. Ultrasound cleaning devices use the energy produced by these waves to loosen dirt particles. In medicine, doctors use the same principle to break up gall bladder or kidney stones.

see also: Sea mammals, p. 212

temperature | motion | radiation | entropy

THERMODYNAMICS

Warmth and cold are qualities that can be directly perceived by people. The physical principles underlying these phenomena long remained unclear and still remain difficult to understand. Surprising connections between heat energy, statistics, and disorder are revealed by the basic laws of thermodynamics.

heat and temperature

Two objects combined behave like a body with the sum of both masses. However, if we pour cool water at 50°F (10°C) into warm water at 86°F (30°C), we obtain not 50°F + 86°F = 104°F (40°C), but water at 68°F (20°C).

The reason for this is that temperature and heat are phenomena of "many-particle" systems, or large assemblies of particles. They derive from the random movements carried out by all atoms or molecules in a substance. Named after their discoverer, the botanist Robert Brown (1773–1858), these vibrating, rotating, or lateral movements are called Brownian motion. This motion can be modeled and predicted using the laws of stochastic processes and probability theory (p. 417). In concrete terms, the temperature of a gas, for example, is the average random motion energy of the particles within it. A high temperature means that the particles are rapidly moving, vibrating, and rotating. Slower motion corresponds to lower temperatures. Another term for temperature is "thermal energy" or "heat energy." The term "heat" alone, on the other hand, describes the energy that passes from one substance to another. For example, a carpeted floor feels "warmer" than a tile floor, since carpet drawsless heat energy out of our feet than ceramic tiles do at the same temperature.

Perpetual motion?

Even before people recognized what heat is, physicists were familiar with the so-called first law of thermodynamics. According to this law, the sum of mechanical work done and heat gained (or lost) remains constant. This is a direct consequence of the law of conservation of energy. Anyone who is skeptical about it may try—like unsuccessful inventors of earlier generations—to build a perpetual motion machine, a closed system-based device that theoretically performs work without the addition of extra energy. The law of conservation of energy dooms these attempts to failure; however, that does not stop many people attempting to do it anyway.

The water-screw perpetual-motion machine: Water falls onto a paddle wheel, which then turns a screw that returns the water to its previous height.

The physical unit of temperature is the kelvin (not "kelvin degree"). Absolute zero lies at 0K, while water freezes at 273.15K and boils at 373.15K. Yet most technical applications use the Celsius scale, in which water freezes at 0°C. On the Celsius scale, nominally negative temperatures can occur.

issues to solve

ASTRONOMERS have measured the prevailing temperature in space: 3K (-454°F). However, using special cooling devices—for instance, laser beams that slow atomic motion—scientists have reached even lower temperatures. The record achieved by experimentation lies at less than a billionth of a kelvin. This may be the lowest temperature in the universe.

The coldest point in the universe: Using the unusual states of matter generated by Bose-Einstein condensates, scientists have cooled rubidium atoms to 170 nanokelvins (nK), very close to absolute zero.

in focus

HEAT RADIATION

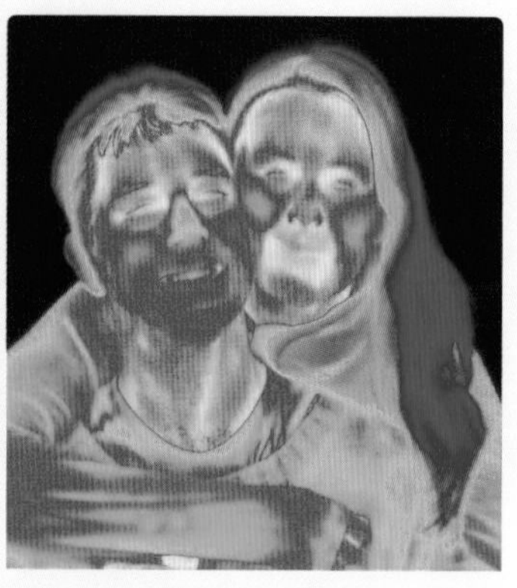

Every object constantly sends out electromagnetic radiation (p. 321), at frequencies and wavelengths that depend on its temperature. At room temperature, this heat radiation lies in the infrared zone and is thus invisible to the human eye. Snakes can perceive it, however, and it can be made visible using infrared cameras or night-vision devices. The heat radiation of objects at 5500 kelvin, on the other hand, is bright-yellow visible light. This is the temperature of the sun's surface.

A thermographic (infrared) camera image of a couple embracing.

order and disorder

When we pour milk into a cup of coffee, it mixes forming a light brown liquid. The opposite—the spontaneous separation of coffee and milk—has never been observed, although it is theoretically possible.

Mixing of liquids: You would need to do this every second for approximately one million years to finally view spontaneous separation.

The spontaneous separation of a mixed liquid is possible, although it does not seem at all obvious, according to the laws of thermodynamics. The concept of thermodynamics includes a condition or value called entropy (often identified with "disorder"), which was first identified in the field of chemistry. Entropy helps balance the energy equations for certain chemical reactions: for instance, a gas has higher entropy than a liquid; warm objects have higher entropy than cold ones; and the entropy of a mixture is higher than that of an "orderly" system, which is separated into components.

The key to understanding entropy was discovered by Ludwig Boltzmann (1844–1906). His realization was that entropy describes the probability of a particular state within a system. Thus, there may be only one way to sort the papers on a desk by size and content, but there are countless ways to scatter them in unorganized piles. Similarly, milk and coffee molecules can be distributed in many more ways as a mixture than if they were neatly divided into layers. For this reason, unorganized conditions are generally more likely. In fact, the number of atoms in a cup of coffee is so unimaginably large that the probability of spontaneous separation is less than 1 in 10^{-20}.

The second law

Our knowledge about entropy is summarized in the second law of thermodynamics: entropy (or "disorder") in a closed system never decreases. If we look at the universe as a whole, then total entropy must always increase over time. This means that sometime in the far future, structures such as galaxies, planetary systems, and even chromosomes can be expected to break up. The thermodynamically stable final condition would be a uniformly mixed cloud of assorted particles at a constant temperature, permeating the entire universe. Many multiples of the current age of the universe will have to pass before this state, called "heat death," is reached, a state where all energy will be evenly distributed throughout the universe, a universe that will be barely warmer than absolute zero.

in focus

LIFE BRINGS ORDER

Living things are highly intricate, orderly structures, whose complexity has continually increased through evolution. We live—viewed on a cosmic scale—in a paradise of order. The Earth has been warmed by the sun for billions of years and we are far from reaching thermal equilibrium. The complex self-organizing system called life has plenty of time to avoid the inevitability of entropy and heat death.

A spider's web: a complex ordered structure indicative of the order brought through life.

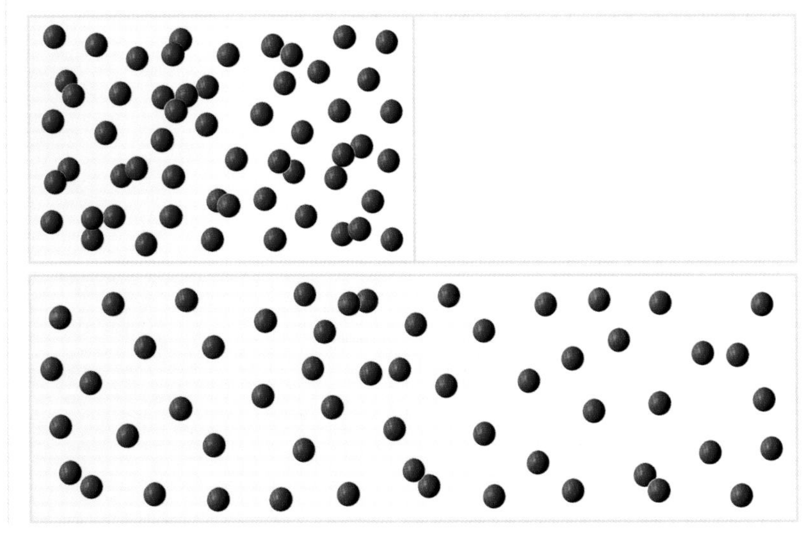

Systems theorists and complexity theorists study life's ability to continually increase complexity against the trend toward disorder, although they have not yet truly solved the puzzle (p. 335).

flow | watts | volts | induction | fields

ELECTROMAGNETISM

Gravity—the fact that all objects attract each other in proportion to their mass—is familiar to everyone. In fact, life without it can hardly be imagined. Electricity and magnetism are different; even through the early 19th-century phenomena such as naturally magnetic minerals were scarcely known and not understood.

electricity

Although we take the use of electricity as a power source for granted today, most people know little about its underlying physical principles. The key to understanding is the concept of electrical charge.

While the "source" of gravitation is mass, electricity arises from two kinds of charges: positive and negative. Opposing charges attract each other while similar ones repel. If a positive and negative charge of the same strength are combined, they will outwardly appear neutral. In everyday life, electricity is scarcely noticed because almost all carriers of charge on Earth—and elsewhere in the universe—have found an opposite partner that counterbalances their effect.

The flow of electricity

Because of their strong force of attraction, it is easy to accelerate negatively charged electrons (p. 324) toward the positive pole of a battery. Even over large distances, the electrons race through metals and other electrical conductors. This flow of electrical charge is called current. The unit of electrical current strength is the ampere (A). In a current of one ampere, a charge of one coulomb (C) is transported each second; one C is the charge of 6.24 × 1,018 electrons (not a round number for historical reasons).

Watts and volts

Many electrical phenomena can be better understood if we think of the electrons in a wire as water running through a system of pipes. The downward tilt of the pipes corresponds to the electrical potential, or tension, which is measured in volts (V), referred to as "voltage." The greater the difference in potential, the more electrical energy the electrons gain by flowing downward. The electrical power of lamps or electric motors equals the energy used divided by the time taken; its unit is the watt (W) and is referred to as "wattage." The electrical energy consumed by a device such as a lamp is the current strength multiplied by the potential; in our analogy, the work done by a waterwheel depends on its height and the amount of water flowing per second.

Electricity pylons traverse the countryside, an ever present reminder of society's need for electricity.

basics

A BIG DIFFERENCE

The electrical force of repulsion between two protons—or hydrogen nuclei—is some 10^{36} (an undecillion) times stronger than their gravitational attraction. Because of this, for almost all processes involving atoms and molecules, and thus all chemical reactions, electricity (along with magnetism) is the decisive force.

in focus

AMBER

Amber is mineralized resin from ancient trees. Thanks to its especially high electrical resistance it serves as a good insulator. The electrons within it are bound particularly tightly to their atoms. By rubbing, however, it is possible to transfer electrons onto a piece of amber, thus giving it an electrical charge. This effect was already familiar in the ancient world; in fact, it gave electricity its name: the Greek word for amber is "electron."

Amber was known about since the fourth century B.C.

In this example, the first barrel is full of water and has a high potential flow (voltage) but the tap is closed. Once the valve is opened and the water flows, the potential energy also drops and the flow rate is reduced. Therefore voltage drops but the wattage remains constant.

see also: Energy technology, pp. 346–353

mysterious magnetism

The magnetic compass is believed to have been used in China as early as the second millennium B.C. In ancient Greece people knew that stones found near the city of Magnesia attracted bits of iron. Yet explanations for these phenomena did not come until many centuries later.

Like electricity, magnetism has two opposing forms, here called north and south poles. Again, similar poles repel each other, while opposite ones attract. A special attribute of magnetism is seen in the fact that if a bar magnet is cut across the middle, both pieces again have north and south poles. If these pieces are further divided, the same thing happens. This can be explained by the theory of elementary magnets, which states that magnets consist of a huge number of submicroscopic magnetic particles, all oriented in the same direction so that their effects are combined. In the 20th century, researchers succeeded in identifying these particles as magnetized atoms.

The magnetic field of the Earth is not static, we can tell from geological records that the magnetic poles drift, sometimes rarely even swapping locations.

Electromagnetism

The similarities between electrical and magnetic phenomena caused early 19th-century physicists to speculate about a common cause for the two. They were encouraged by the discovery that electrical currents and magnetic fields influence each other.

In the 1860s these efforts culminated in James C. Maxwell's theory of electromagnetism. In short, it states that: First, every electrical charge creates an electrical field, which in turn can affect other charged objects. Second, moving electrical charges create a magnetic field. This is true both for an electron traveling around a nucleus and an electric current flowing through a coil of wire. Finally, in both electric and magnetic fields, wavelike disturbances spread outward at the speed of light. Their value in a particular medium (glass, water, vacuum, etc.) can be directly deduced from the medium's electrical and magnetic properties.

An electromagnet can lift many tons of steel scrap metal, vehicles, and cargo containers.

practice

INDUCTION AND INDUCTION LOOPS

Changing magnetic fields induce an electrical current. This discovery was a milestone on the path toward electromagnetic theory. It also has many technological applications, for instance in the generator and electric motor, as well as the transformer, which is used to alter electrical potential. Induction loops can be used to count vehicles and automatically change traffic signals. Microphones also use the principle of induction. In recent years the inductive transfer of electrical energy has become especially important since it is used to recharge pacemaker batteries without an operation.

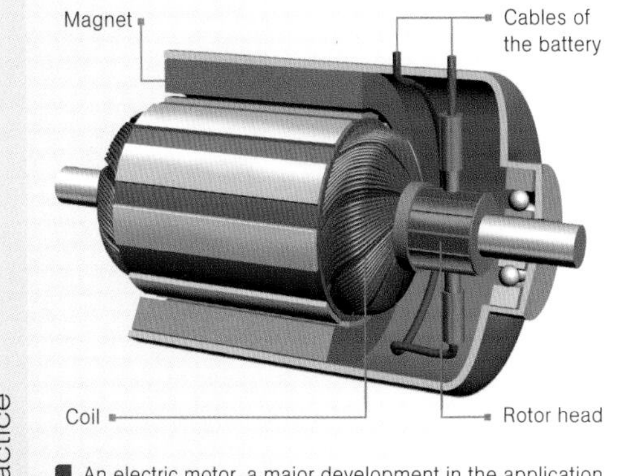

An electric motor, a major development in the application of electromagnetism, is used in countless modern devices.

issues to solve

THE SEARCH FOR MAGNETIC MONOPOLES Particle physics theory requires that a hypothetical particle with a singular magnetic pole, or monopole, exist. Instead, all magnetic materials or particles always have both north and south poles. According to speculative theories attempting to unify all of the fundamental forces, however, these magnetic monopoles could have been present shortly after the big bang. In complex experiments, scientists are currently attempting to demonstrate their existence—so far without success.

The third principle is especially significant for physicists, since light is also an electromagnetic wave, and therefore optics (p. 324) is a specialized branch of electromagnetism.

light | prisms | precision optics | lasers

OPTICS

Up to a few years ago, optics were considered one of the areas of physics that promised few innovations: telescopes and lasers have been known for some time. All this changed with the arrival of photonics. A Nobel Prize in 2005 in the area of optics, together with a predicted computer revolution, gave rise to great expectations.

light

Without light, life on Earth as we know it would be impossible. The principal source of natural light on Earth is the sun. Yet physics deals not only with light visible to man, it also deals with electromagnetic radiation.

It is the energy of sunlight that makes our planet habitable. We value a well-lit enviroment, and the basis of our nutrition is the ability of plants to produce organic substances from light, carbon dioxide, and water. Futhermore, we as humans require sunlight to produce vitamin D, a lack of which leads to liver, kidney, and bone disorders. Sunlight is also essential for our mental balance: many people become depressed during winter or when deprived of light. For a long time, there were two competing answers to the question as to the nature of light. According to Isaac Newton (1643–1727), light consisted of tiny particles that move from the light source through space in a straight line. On the other hand, Christiaan Huygens (1629–1695) described light as being wavelike.

During the 19th century, this wave model seemed to persevere; for instance, Thomas Young (1773–1829) showed in 1803 that light could overlay and interfere, a phenomenon that occurs only with waves (p. 317). James C. Maxwell (1831–1879) was even able to derive light waves from his electromagnetic basic equations.

However, the turning point came at the beginning of the 20th century: atomic experiments proved that light is always given off in individual particles called photons. If such a photon encounters an electron, it behaves just like a particle with impulse and speed.

Both explanatory models were therefore shown to be correct. Thus quantum mechanics (p. 326) explains that light as well as matter have wave and particle characteristics. In many cases, one of the two characteristics outweighs the other, and sometimes both models have to be applied. On the basis of this work, French physicist Louis de Broglie (1892–1987) developed an entirely new academic field called wave mechanics, which combined the studies of both light and matter.

A prism splits a white light beam into its constituent spectral colors.

The visible spectrum of electromagnetic radiation

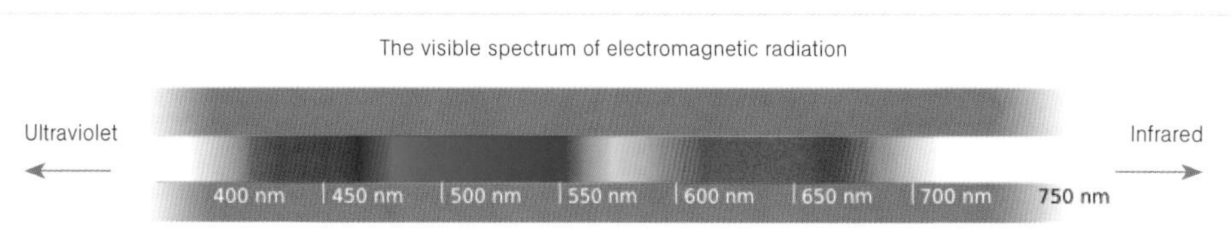

The electromagnetic spectrum extends from radio waves, through microwaves and infrared light to visible and UV light. Increasing energy leads to x- and gamma ray radiation. With decreasing wavelength, the transported energy grows and the particle characteristics of radiation increases. The human eye can only perceive a very small portion of electromagnetic radiation with a wavelength between approximately 400 nanometers (blue-violet light) and 700 nanometers (red light).

in focus

GOETHE'S THEORY OF COLOR

Johann Wolfgang Goethe (1749–1832) was not only a poet, writer, thinker, and politician, he also had interests in history and natural sciences. In 1792 he published his *Contributions to Optics* and in 1810 his *Theory of Colors*. Although in some ways flawed, his theories about color perception are still considered groundbreaking.

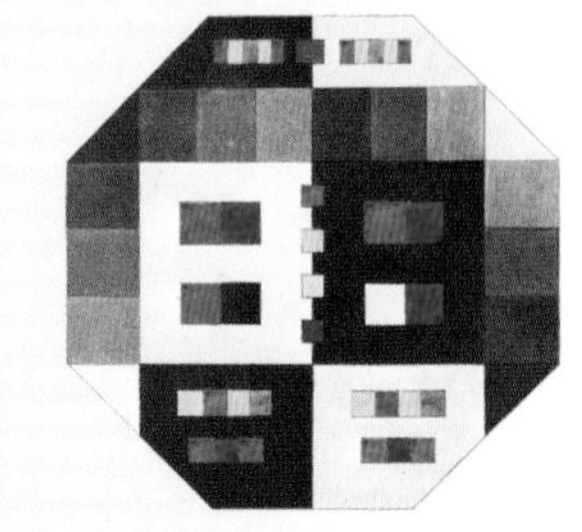

In *Theory of Colors*, Goethe graphically describes the process of color refraction.

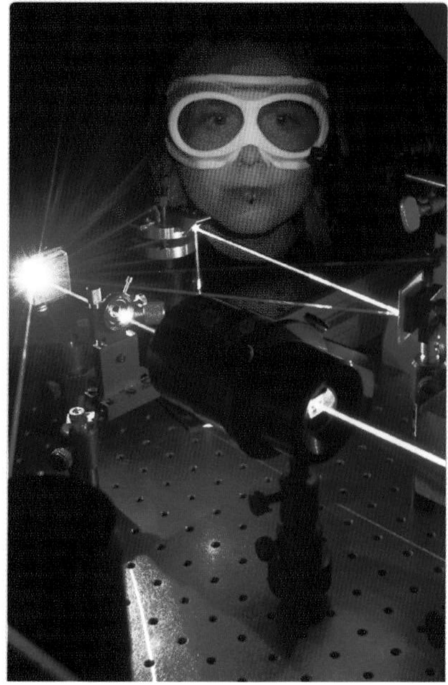

Lasers are used for many industrial operations, such as the examination of semiconductor samples.

see also: Theory of relativity, pp. 330–331

from microscope to optoelectronics

The first spectacles were made from polished semiprecious stones, as early as the late Middle Ages. Yet it was the invention of the microscope and the telescope that laid the foundation for precision optics.

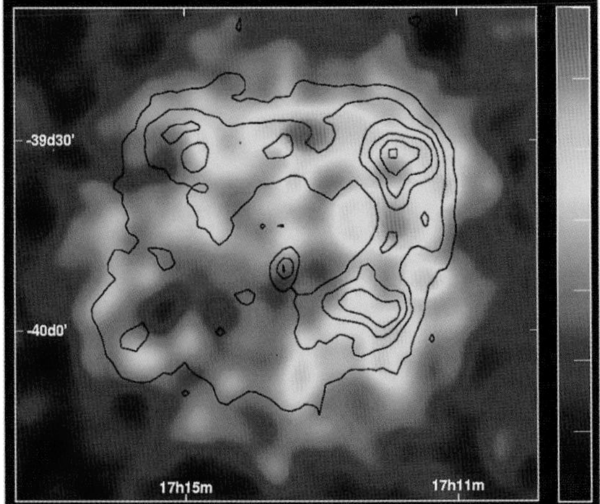

UV (ultraviolet) optics have found use in space-based telescopes, used to measure the massive amounts of UV radiation emitted by the sun and nearby galaxies.

Telescopes and microscopes both take advantage of the fact that light can be deviated from its linear propagation path when it is passed through suitably polished glass lenses. For instance, a condensing lens concentrates parallel light rays into a point: the focal point. Two such lenses—placed in line at the proper distance apart—form an object enlarged on the retina of an observer. When used on a telescope and in a microscope these are called objective and ocular (in Latin *oculus:* "eye") lenses.

Nowadays optical equipment fulfills diverse tasks in measuring technology, biological and medical research, and in astronomy. From radio telescopes and UV detectors to x-ray microscopes, all parts of the visible and invisible spectrum are being utilized and studied. Microscopes have even been developed for observing the particle radiation from electrons and neutrons.

Calculating with light

Computers are sometimes referred to as electronic brains because information can be transmitted, manipulated, or stored in the form of electrons. However, electrical resistance limits current flow and creates undesirable waste heat, the smaller the circuits become. Therefore increasingly powerful cooling needs to be built into modern computers. Moreover, electrons influence each other because of their electric charge, meaning there is a limit to how small we can make standard computer chips.

A way out of this troublesome dilemma would be with photons: that is with light particles. They take up the same space in almost unlimited numbers, all traveling at the speed of light. That is the reason why so much data can be sent simultaneously via fiber optics at such a high speed.

It is hoped that significant advances can be achieved within the new field of pure optical circuit elements with so-called nonlinear crystals. The way to an advanced computer made of light or "photonic brain" may possibly not be very far, running at speeds that will make modern supercomputers look seriously outdated.

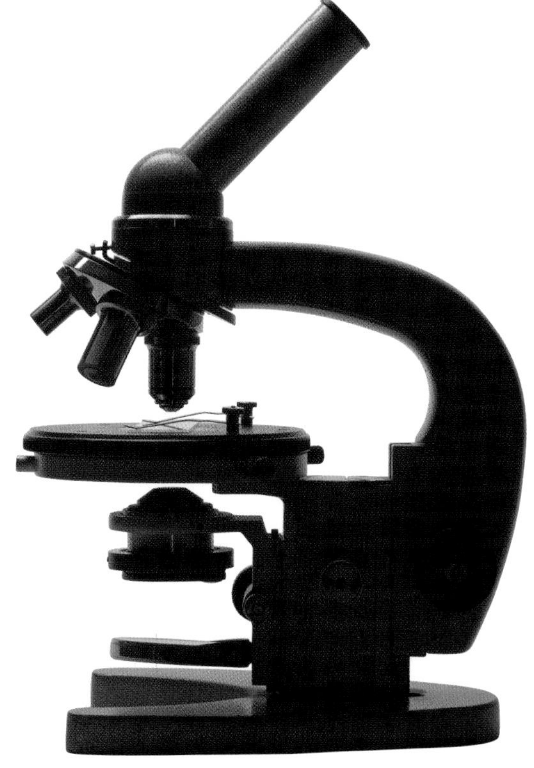

A laboratory telescope is a high precision optical device used to observe that which is not perceivable to the human eye.

practice

LASER

Normal light, such as that generated by the sun, is a mixture of many different waveforms with different directions, energies, and so on. Lasers, however, behave differently. With a pulsing amplification process, a large number of identical waves are being generated, where every wave crest matches the adjacent wave crest, and every wave trough matches the next wave trough. An extremely bundled and straight beam is created. In everyday life, we encounter lasers in CD and DVD players and supermarket checkouts, often as unseen but vital components.

An industrial laser cuts steel into precise shapes.

see also: Integral calculus, p. 415

quanta waves | neutrinos | clocks | superconductors

QUANTUM MECHANICS

Quantum physics investigates the behavior and conformity of matter in the atomic and subatomic realm. The basis of this physics theory—with the physicist Max Planck (1858–1947) as its originator—was largely established during the first half of the 20th century. Quantum mechanics is regarded as one of the most important foundations of modern physics.

the new physics

In 1890 almost all mysteries in physics appeared to be solved. In fact, the future Nobel Prize winner Max Planck was advised against studying physics, ostensibly because there was nothing new to be discovered.

Although physics was believed to be almost completely understood, there were unanswered questions relating to the frequency distribution of thermal radiation given off by an object. Although James C. Maxwell had resolved the basic equations of the nature of electromagnetic radiation (p. 323), these failed to describe this behavior. In 1900, Max Planck succeeded after he hypothesized that light and thermal radiation is given off in individual portions only. He referred to these light packages as "quanta," without being aware that he had given a name to a revolution in physics.

In 1905 Albert Einstein explained the "photoelectric effect" with the aid of Planck's light quanta, for which he was awarded the Nobel Prize in 1921. Shortly thereafter, it was realized that an atom could only be stable when its electrons moved in "quantized orbits" around the nucleus. With Planck's quantum hypothesis, it was possible to achieve what previously seemed impossible: to predict the behavior of light, electrons, atoms, and molecules.

Probability waves

The basic assumptions of the quantum theory have radical consequences: it is generally impossible to indicate precisely the energy or location of a particle; one can only make probability statements. These probabilities are given by means of wave functions, and are able to resolve the wave-particle controversy of the light theory: the light quanta are particles, but in order to be able to indicate the probability of their position, one has to consider the nature of their wave behavior.

The Atomium in Brussels displays the positions of iron atoms within a crystal.

The quantum vacuum

Further developments in quantum mechanics, such as quantum field theory and quantum electrodynamics, revealed that there was no absolute vacuum. Within the realm of the quantum theory, particles and photons are constantly created out of nothing and then immediately disintegrate. These "ghost particles" or neutrinos have tiny, but measurable effects, which—for instance—could provide problems in the future for those who build microchips.

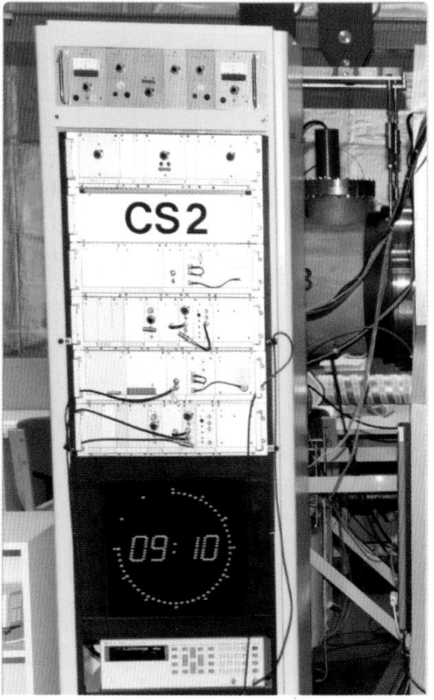

An atomic clock measures the resonance of atoms, creating highly accurate time signals.

in focus

DOUBLE SLIT EXPERIMENT

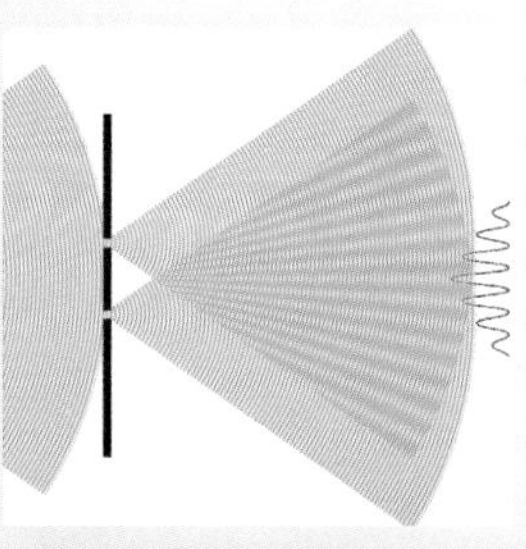

A bathtub is quite suitable for experiments with waves. When a piece of wood is dipped into water at regular intervals, this will create moving wave fronts. When these waves encounter a wall with two slots, an interference pattern will develop on the other side (p. 317). In 1961, such a double split experiment was used to discover the probability waves of electrons and in 2002, physicists selected this experiment as the most beautiful in physics.

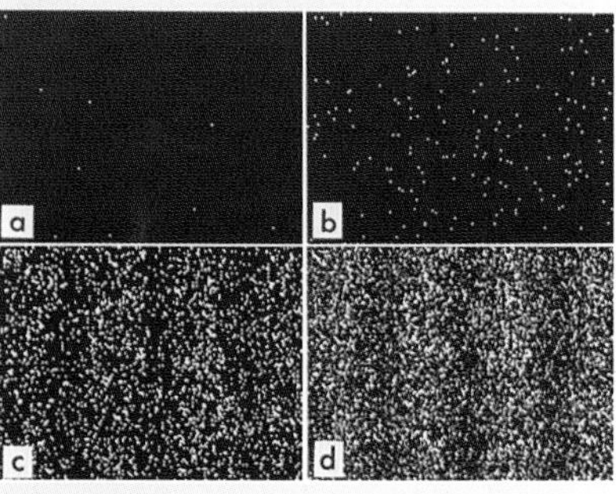

Behind a wall with a two slits, waves form an interference pattern.

see also: Matter, pp. 284–285

quantum effects in our daily lives

Quantum physical effects play an essential role in modern technology: examples are lasers, electron microscopes, atomic clocks, and superconductors. They all promise very new and exciting commercial applications.

The magnetic field of a superconducting continuous current keeps the superconductor suspended in the field of another magnet.

Radio-controlled clocks are popular in the home because they always give the correct time. Where do they get their time signal and why is it so precise? Many universities and research groups operate these clocks, transmitting signals over the air that are used to correctly set the time. The signal used is generated by an atomic clock with a precision that is a direct consequence of quantum laws. These laws fix the permitted orbits (or better, the quantum conditions) of the electrons of an atom. This means that there is a fixed energy difference between two conditions of an electron, which is the same throughout the universe for all atoms of an element. This energy difference corresponds to a frequency, and since this frequency is the reciprocal value of time, an exact measurement of the frequency provides a correspondingly exact time signal. Nowadays the best atomic clocks reach precisions of $1{:}10^{15}$ and better.

The behavior of LEDs (light-emitting diodes) can best be explained through quantum mechanics.

see also: Semiconductors, p. 292

Superconductivity

When electric current flows through a wire, that wire warms up. The reason is the electric resistance of the conductor (in this case, the wire). Even with good quality copper wires, this cannot be avoided, or can it? Here a quantum effect is at work, which would be impossible to describe through conventional physics. Quantum physics, however, shows that at very low temperatures, electrons form pairs that suffer less interference from the actions of atoms and impurities in the semiconductor and so exchange less energy with the atoms of the wire. This leads to an almost lossless conductivity in the material. Many metals display such superconductivity, or lack of resistance, at temperatures of a few kelvins (a few degrees above absolute zero).

Ceramics

The principles of superconductivity discovered in 1911 by Heike Kamerlingh Onnes, (1853–1926), only became really interesting when the first ceramic conductors were discovered in 1987. These conductors displayed an interesting property: upon being cooled, using liquid nitrogen, to a temperature of 100K (-279.76°F) they would become superconductive. Modern high temperature superconductors (HTS or High-T_c) can be cooled with much less effort and their use promises higher degrees of efficiency in energy-transfer cables, transformers, and motors.

issues to solve

COMMERCIAL APPLICATIONS are impeded by the fact that high temperature semiconductor ceramic, in contrast to metal, are brittle and difficult to manufacture into wires. However the integration of a 393-foot-(120-m) long cable into the electricity grid is currently being tested in Detroit.

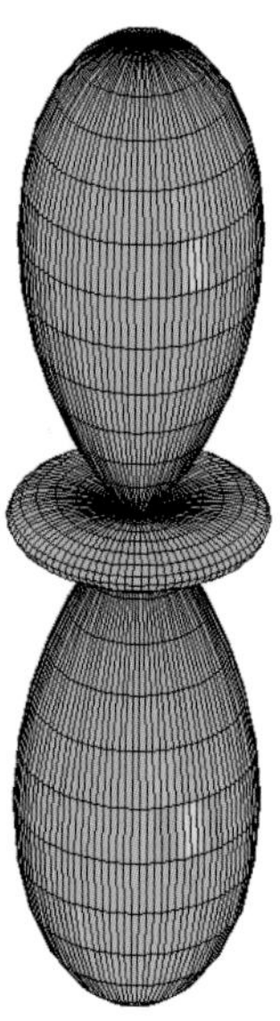

The regions of an atom where electrons are found can be calculated, giving a result that is known as the "atomic orbital."

quarks | electrons | neutrinos | forces | interactions

ELEMENTARY PARTICLES

Even in ancient Greece, philosophers were already debating whether the world is made up of indivisible atoms. In the 19th and early 20th centuries, science seemed to have answered this question, thanks to modern atomic theory and the discovery of most of the chemical elements. Today, however, we know that even the particles within atomic nuclei have their own inner structure. Quantum physics explores the characteristics of the most basic particles currently known in the universe.

quarks, electrons, and co.

The elementary particles known today are unimaginably small—billions of times smaller than an atom. They are categorized according to qualities such as mass, charge, and interactions.

To better understand the structure of matter, think of the human body: it consists of organs such as the heart, liver, and bones, which are made up of countless cells. These cells are composed of molecules, which in turn are formed from atoms. The atoms, for their part, contain a small, relatively heavy nucleus and a shell of light electrons. While the electrons are indeed basic particles, according to current knowledge, the atomic nucleus has a complex inner structure. Each nucleus contains positively charged protons and electrically neutral particles called neutrons, which are nearly equal in mass. Protons and neutrons, in turn, have their own inner configuration; each contains three "quarks." These particles—squeezed together within an unimaginably small space—can only be described with the help of quantum physics and relativity theory (pp. 326, 330). The energy of their mutual attraction is so great that particles within protons and neutrons are constantly forming out of the vacuum and disappearing again.

A basic principle of quantum physics is that all objects possess both wavelike and particle-like properties. Thus even electrical and nuclear forces operate in particle-like ways, through force-carrying particles. One variety of these is already familiar: the photon (p. 324). This quantum (indivisible unit of energy) of electromagnetic radiation is the carrier for all electrical and magnetic interactions. In addition, the electron has two heavier siblings: muons (μ) and tau (τ) particles. The so-called up and down quarks, which are found within protons and neutrons in atomic nuclei, also each have two larger relatives. In addition, the electrons μ and τ each have partners called neutrinos, which are electrically neutral and almost—but not completely—without mass.

All known matter in the universe is composed of these particles: quarks, electrons, and neutrinos.

A new element Roentgenium (Rg) was discovered by a German team 1994. It was discovered by experimenting with cold fusion.

basics

IN THE LATE 1990s, astronomers discovered that we know significantly less about the universe than previously thought: observable matter and energy (quarks, electrons, and force particles) account for only 4% of the total matter and energy in the universe. 23% is composed of unknown mass, detected through its gravitational influence (dark matter), and nearly three-quarters of the energy is so-called dark energy, about which we know practically nothing.

The Large Hadron Collider is a particle accelerator built at CERN.

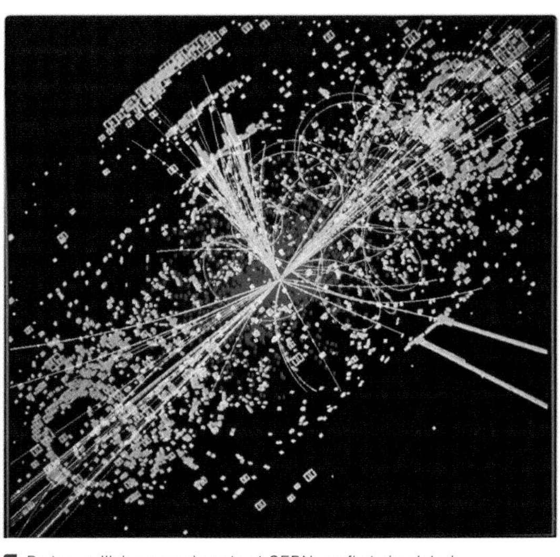

Proton-collision experiments at CERN are first simulated.

see also: Matter, pp. 284–285

fundamental forces

In the 19th century, James C. Maxwell unified the theories of electromagnetism and optics. Today quantum field theory brings together three of the four fundamental forces of nature, excluding only gravitation.

In principle, gravitation and the electromagnetic force operate over any distance, but their influence decreases in proportion to the square of the distance of separation. That is all the two forces have in common, however. Currently gravitation can only be described using the general theory of

The inner view of the JET (Joint European Torus) fusion reactor device.

relativity (Albert Einstein, 1915), but this theory breaks down at the level of elementary particles. On the other hand, quantum electrodynamics (QED), a description of electromagnetism using quantum field theory, explains all of the interactions within atomic electron shells as well as among atoms and molecules, with great precision.

The strong and weak interactions

The discovery of radioactivity by Antoine Henri Becquerel with husband and wife team Pierre and Marie Curie brought two additional natural forces to the attention of physicists. One is even more powerful than electromagnetic interactions and is thus called the strong interaction. Its force particles are called gluons. Only quarks are affected by this force, since they are the only particles (besides gluons themselves) to possess a so-called color charge. In analogy to visual color, there are three kinds of color charge, and three quarks of different colors produce a color-neutral proton. The strong interaction is so well shielded that it only affects particles at distances smaller than 33 to 50 feet (10 to15 m). The weak interaction or weak nuclear force (100 billion times weaker than electromagnetism, but still 1,025 times stronger than gravitation) is involved in processes such as radioactive beta decay. All known particles are affected by this interaction. Its basic effect is the transformation of quarks into electrons or neutrinos, and

Unimaginable amounts of energy are released through the explosion of even a single atomic or hydrogen bomb.

vice versa. Here, too, the distance over which the force operates is extremely small. The weak and strong interactions are the causes of radioactivity—that is, the transformation of atomic nuclei accompanied by the emission of various kinds of radiation. Radioactive alpha decay is a special kind of nuclear fission, with two protons and two neutrons (an alpha particle) splitting off from the nucleus together. In beta decay, electrons are emitted as beta radiation, while the emissions of gamma decay are extremely high-energy protons (gamma rays). Other decay processes within large nuclei, such as their fission into two fragments, produce energy, as does the fusion of two lighter, weakly bound nuclei.

milestones

THE "NOBEL FAMILY" CURIE

Marie Curie received two Nobel Prizes. The first, shared with Antoine Henri Becquerel and her husband Pierre Curie, was in the field of physics for the study of radioctivity, in which she played a key role in the discovery of forces within the nucleus. The second was awarded to her for her work in chemistry. Her daughter Irène Joliot and son-in-law Frédéric also received Nobel awards, making them a true "Nobel family."

Marie Curie (1867–1934).

see also: Theory of relativity, pp. 330–331

everything is relative

Even today, most people assume that time and space are absolute, unchanging points of reference. Yet since the publication of Albert Einstein's special theory of relativity in 1905, we have known that this is not true.

In his renowned theory of relativity, the German-born Nobel Prize-winning theoretical physicist Albert Einstein (1879–1955) discarded the now discredited theory of ether and came to a logical conclusion: the speed of light remains constant, regardless of a light source's motion. Instead, space and time are altered in accordance with the object's movement. For instance, as a rocket accelerates, time passes more slowly for the rocket, and its length shortens. Conversely, from the rocket's point of view, clocks on Earth advance more quickly, and the planet takes on more of an egg shape (the Earth's radius shrinks in the direction of the rocket's movement). But which version is really true? According to the principle of relativity, both are true, since space and time are always relative to an object's motion.

time | space | light | mass | gravity

THEORY OF RELATIVITY

"Relativistic" physics began with a failed experiment: in the late 19th century, people believed that waves were disturbances in a medium called the ether. This was discredited by the groundbreaking work of Albert Einstein.

The speed of light

Not only is the speed of light always and everywhere the same, it is also the absolute maximum speed limit, as unbreachable as the low temperature limit of absolute zero (p. 320). What happens if we take an electron that is already moving very rapidly and try to accelerate it further? It becomes more massive, and thus, because of its Newtonian inertia (p. 314), it opposes further acceleration with ever stronger resistance. In the most extreme case, its mass can become infinitely large. The precise speed of light can only be achieved by particles with no mass, such as photons (p. 329). However, these particles can never stand still; outside of a lab the speed of light is unvarying.

Mass and energy

In this context, a clear description of the properties of mass is important. On the one hand, it is a measure of inertia. On the other hand, it continuously increases as the motion energy of a body rises. This means, however, that mass is actually a form of energy. In 1905 Einstein was able to calculate the total energy E of a mass m, using the speed of light c, resulting in the ground-breaking equation $E = mc^2$. This formula, the mass-energy equivalence equation—surely the most famous in physics—summarizes one of the most important insights in the history of relativistic physics in just a few symbols.

THE TWIN PARADOX

in focus

If one twin sister travels through space at a high speed while the other remains on Earth, time passes more slowly for the traveling twin, and upon returning, she is younger than her sister. This seeming paradox has been confirmed both theoretically and experimentally (with elementary particles). It only appears paradoxical, however; it would contradict natural laws only if time were absolute, flowing at the same rate in all systems. This is precisely what does not occur.

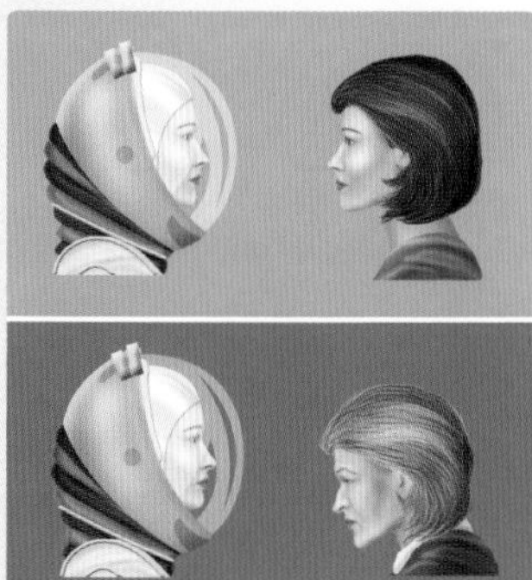

■ The younger twin greets her aged sister.

ALBERT EINSTEIN: THE FATHER OF MODERN PHYSICS

milestones

Albert Einstein (1879–1955), one of the greatest physicists of all time, transformed our understanding of time and space. His world renowned theory of relativity and the mass-energy equivalence equation are his most well-known works; however, he published works on fields as diverse as probability and quantum theory. In 1921 he was awarded the Nobel Prize in physics for his discovery of the photoelectric effect.

■ Albert Einstein left the world an enduring legacy of scientific accomplishments.

■ The most famous formula in the world, the mass-energy equivalence equation, has been displayed on skyscrapers, ships, T-shirts, and countless books and magazines.

see also: The universe, pp. 20–23

space, time, and mass

Mass also has a third quality: it is the cause or "charge" of gravity. Einstein explained how this gravitational mass relates to inertial mass in his general theory of relativity (1919), 14 years after the special theory of relativity.

Einstein gave a radical answer to the question of the relationship between gravitational and inertial mass: they are identical! Since inertial mass is inseparably related to the changing nature of space and time (inertia increases as the length shortens), this is also the case for gravitational mass. As a result, gravitation—the attraction between masses—is directly linked to space-time. Thus the fundamental statement of the general theory of relativity can be understood as follows: gravity is a distortion in space-time. The more massive a body is, the more it warps space and time, which is expressed in the accelerated movement of other nearby bodies.

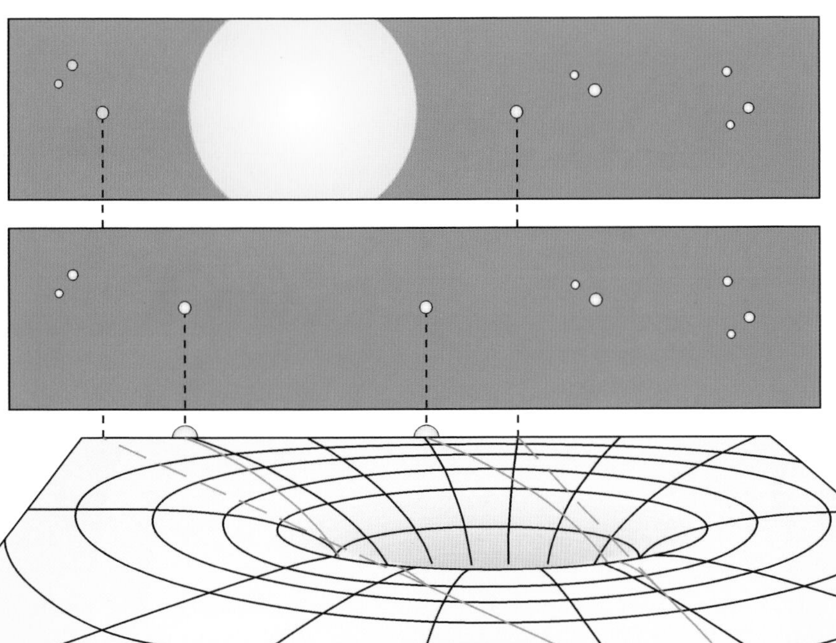

Curved space-time: An observer looking at a light source would discover that the light was affected by the presence of the gravity well of celestial bodies.

Einstein's triumph

Mathematically, general relativity is significantly more complicated than special relativity. Nevertheless, it was confirmed in a spectacular manner in 1919. A solar eclipse made it possible to show that stars in the vicinity of the sun appeared in slightly different positions, since their light was displaced by the sun's gravity. The observed displacement agreed with Einstein's predictions, and he became a celebrity overnight. Few people truly understood the meaning of Einstein's ideas—aside from "$E = mc^2$" and "everything is relative"—but in combination with his striking personality, they made him world famous. Being Jewish and a pacifist as well as a modern scientist, however, he was targeted for persecution by the Nazis. In December 1932, he emigrated to the United States, not visiting Germany again until 1952.

"You have reached your destination" (thanks to general relativity)

Neither special nor general relativity is significant at the speeds achieved by cars or airplanes. Atomic nuclei in particle accelerators reach slightly relativistic speeds, but few experiments deal directly with these effects. However, one technology dependent on relativity has already become a fixture of modern life: satellite navigation with systems such as GPS, GLONASS, or Galileo. Microwave signals are sent to the system's satellites, and their transit time is used to calculate the distance to the satellite and thus the device's own position. This would be a simple trilateration calculation—a routine task in measurement technology—if the Earth's mass did not produce a tiny distortion in space. This alters the time taken by the signal so significantly that without corrections based on general relativity, accurate and modern GPS (Global Positioning System) equipment would be a mere pipe dream.

THE SEARCH FOR GRAVITY WELLS

practice

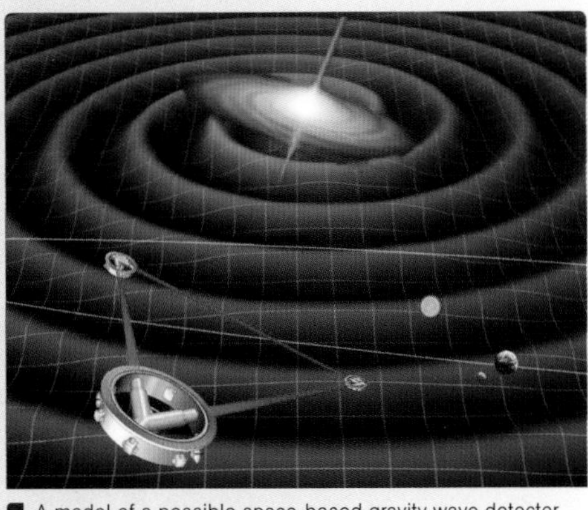

A model of a possible space-based gravity wave detector.

From a mathematical point of view, Albert Einstein's equations resemble more complex versions of the mathematician James C. Maxwell's electrodynamic equations (p. 323). In general relativity, moving disturbances of space and time—gravity waves—are possible. These waves cause unimaginably small variations in length around a trillionth of a billionth of a meter (10^{-21} m). A gravity well is the pull of gravity that a large body in space exerts.

the great unification

In the early 20th century, quantum mechanics (p. 316) and relativity theory (p. 330) exposed the limits of classical physics. More than this, they raised fundamental new questions.

theory | particles | supersymmetry | dark matter | black holes

THEORY OF EVERYTHING

According to the medieval philosopher William of Ockham, we should use as few theoretical assumptions as possible in our attempts to describe natural processes.

First, the atom was replaced as the basic building block of nature by an increasingly diverse collection of elementary particles. Unfortunately, however, the two main pillars of the modern physical view of the world—quantum mechanics and relativity theory—showed themselves to be fundamentally incompatible. Curved space-time cannot be integrated into quantum theory, therefore the greatest current challenge for modern physics is to achieve a unified description of all four fundamental natural forces or interactions. This attempt has become known as the search for a "theory of everything," a term that was first used with an ironic undertone.

Encouraging observations

While the first half of the 20th century was a time of discovery and experimentation, the second half of the century brought the first steps toward the reconciliation of the many theories utilized by the field of physics. First, the theory of quarks brought order into the "particle zoo" when it was discoved that, except for electrons and neutrinos, nearly all known particles—even those within atomic nuclei—turned out to be combinations of two or three quarks. Then, experiments showed that at rising energy levels, the differences between electromagnetism and the strong and weak nuclear forces begin to disappear. A temperature was even identified at which the three forces behave identically—although this temperature is so high that it cannot be reproduced in any laboratory.

The third success came with the unification of quantum mechanics, special relativity, and electrodynamics, into quantum electrodynamics (QED). Mathematically, this theory is closely related to the quantum theory of the strong interaction or color force, quantum chromodynamics (QCD).

In 1950, Werner Karl Heisenberg believed he had found the theory of everything. However, it unfortunately failed to stand up under testing.

The pioneer mystery

Even today, mysteries still appear in unexpected places. Space probes were launched in the direction of Jupiter and Saturn during the 1970s as part of the Pioneer and Voyager programs. They have since left the solar system, but their positions are still being tracked. In recent years, however, it has become clear that their paths cannot be described accurately either by known classical celestial mechanics or by relativity theory.

milestones

THE ELECTROWEAK FORCE

Born in 1933, Steven Weinberg is one of the most prominent U.S. physicists of his generation. His book *The First Three Minutes,* describing the time just after the big bang, was an international bestseller. In his most important contribution to physics, he combined electromagnetism and the weak nuclear force into the "electroweak force," showing that both the forces arose from an original electroweak force existing shortly after the big bang.

Steven Weinberg, the 1979 Nobel Prize winner in physics.

Pioneers 10 and 11 are still traveling beyond our solar system.

see also: Pure mathematics, p. 418

supersymmetry

Encouraged by the successful "electroweak" unification, researchers next attempted to add the strong interaction or color force. This has proved much more difficult.

The most promising attempt to find a theory of everything has been the concept of supersymmetry and superpartners. Symmetries—such as those of before and after, right and left, electricity and magnetism, QED and QCD—have always fascinated physicists. The current standard model of the elementary particles has a one-sided and asymmetrical aspect. The quantum mechanical spin of electrons, neutrinos, and quarks has a value of a half, while that of energy particles, such as photons and gluons, is one. As early as 1973, the hypothesis was formulated that these elementary particles have "superpartners" with different spin values. What advantages would be gained from this expansion in particles? The first is mathematical: the superpartners would prevent the appearance of undesirable infinite values in various equations, which then require "fine-tuning." In addition, the interactions between superpartners are closely related to the characteristics of space and time—a promising point of connection with general relativity theory. Ultimately, it was demonstrated in 1975 that, except for supersymmetry, there can be no other undiscovered symmetry in the universe.

Bringing dark matter to light

Supersymmetry could also solve the mystery of dark matter (p. 329). It seems evident that superpartners could only transform into other superpartners; otherwise, scientists would have detected the products of their decay long ago. Thus, there must be at least one smallest supersymmetrical particle that is nearly or completely stable. This would be an ideal candidate for dark matter: a non-observable particle with a large mass, present throughout the universe.

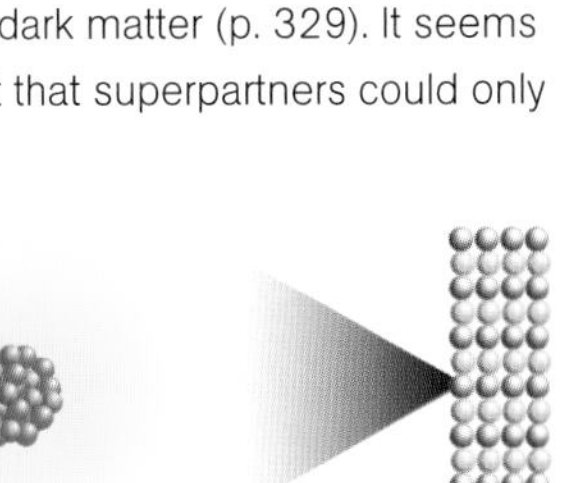

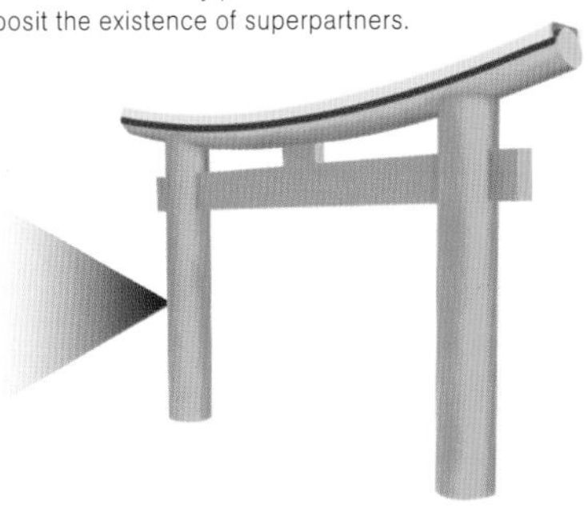

From macro to micro, the expected symmetrical state of elementary particles led scientists to posit the existence of superpartners.

Black holes

Stephen Hawking discovered an interesting link between quantum and gravitational theory. He studied the way virtual particles arising from fluctuations in the quantum vacuum (p. 326) behave in the vicinity of black holes. The gravitation of black holes is so strong that particle-antiparticle pairs are ripped apart. While one is swallowed up by the black hole, the other may escape and become an elementary particle. Thus, black holes emit a weak particle stream known as "Hawking radiation."

milestones

HAWKING RADIATION

British theoretical physicist Stephen Hawking (b. 1942) is, like Isaac Newton before him, the Lucasian Professor of Mathematics at Cambridge University. In 1974, he provided the theoretical argument for the existence of Hawking radiation. There is much debate over this exotic radiation and many experiments are under way to vindicate or disprove this theory.

Stephen Hawking is a renowned scientist and author of the bestselling book *A Brief History of Time.*

issues to solve

FINE-TUNING presents a problem in physics. Whenever theoretical models do not match up with the results, fine-tuning of the equations is required in order to match the observations. However, this does not mean that the equation was wrong, but rather that there is an unexplained effect upon the results. One possible explanation is the anthropic principle, stating that the results are being changed merely from being observed.

in focus

STRING THEORY

String theory interprets elementary particles as states of vibration in unimaginably small strings. This concept is automatically supersymmetrical and can also be applied to space and time (and thus to gravity). However, string theory only works in a system of at least ten spatial dimensions leading to new conclusions about the nature of gravity.

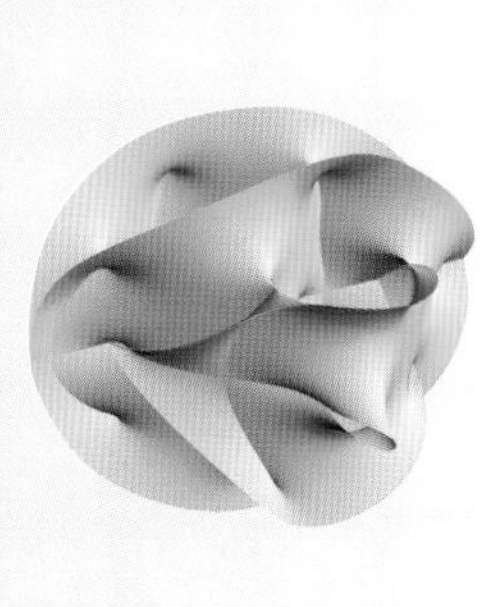

Many physicists see string theory as a key component within a theory of everything.

see also: Dark matter and dark energy, p. 23

determinism | probability | fractals | applications

CHAOS IN THEORY AND PRACTICE

In ancient Greece, the term "chaos" was used to describe the original state of the world that had yet to be shaped. Today the term is used in physics with a meaning that differs greatly from the original term.

order from disorder

A model developed by the meteorologist Edward N. Lorenz (1917–2008) may be the best way to illustrate the idea of "deterministic chaos," which is the subject of investigation in physical chaos theory research.

Edward Lorenz presented the image of a butterfly flying around in the Amazon rain forest. Its flapping wings result in tiny air movements which, in turn, would interact with slightly larger air masses and, then again, with even larger masses of air. This is how, in theory, a little butterfly could eventually cause a hurricane in New York or China. This concept is called the butterfly effect: small triggers creating large effects.

During the past decade, several systems were identified that show how tiny changes of the original setup could, over time, lead to completely different results: the weather, snooker, avalanches, the heartbeat, or a series of connected pendulums. Physicists and mathematicians are very interested in these systems because they are based on seemingly chaotic mathematical laws.

Kepler's laws of planetary motion were the basis for Newton to develop laws in mechanics. The laws of both scientists are slightly flawed as they only apply to one system at a time, either a sun and a planet or a planet and a moon. Generally the three-body problem cannot be solved. If this was also not enough, given adequate time, most orbits are actually not stable but distinctly chaotic. However, due to the large mass of the sun and relatively long distances between the orbiting bodies, "enough time" can be defined for our planetary system as a billion years. Therefore we do not need to worry about potential collisions anytime soon. However, in the early days of the solar system, unstable orbits were more common. Some findings indicate that the outer planets must have swapped positions with one another at one point, and that the Earth's moon most likely originates from a collision between the Earth and a protoplanet the size of Mars.

Although classical physics is deterministic, there exists no computer powerful enough to correctly model and predict the movement of a flag flapping in the wind.

CHAOS IN THE KITCHEN

practice

By making flaky pastry at home, you can carry out a basic experiment of chaos research. After rolling and folding the dough 25 times, each layer has reduced its thickness to one atom and the exact position of a dough particle cannot be easily predicted. Its path is as complex as the fractal images that were first created by the mathematician Benoit Mandelbrot (p. 335).

Flaky pastry is a chaotic system that everyone can make at home.

Chaos and population

A practical use of chaos theory is in predicting population growth in locations with limited resources or space, which commonly exhibit chaotic behavior. This natural growth and decline of species populations is especially interesting when applied to human communities.

The weather is a highly complex, nonlinear system, which makes weather forecasting very difficult.

TURBULENCE

in focus

"Turbulence" may be a familiar term from everyday usage. In physics, this term has a special meaning. For instance, in a flowing liquid that is mixed by numerous and turbulent whirls of varying sizes, dyed droplets of liquid cannot be recognized after a short period of time mixing. Neighboring droplets quickly reemerge in impossible to predict locations. This is where the study and research of chaos theory is used.

Turbulant eddies swirling in cigarette smoke.

see also: Mechanics, pp. 314–315

fractals in nature and technology

The behavior of chaotic systems seems as if no rules applied. However, in reality, when an abstract approach is used, it becomes clear that it is based on hidden deterministic rules that explain the inherent order in chaos.

Physicists like to use diagrams such as the distance-over-time diagram in mechanics or the phase space chart where the location and velocity (or rather the momentum) of one or several particles can be mathematically deduced.

This is exactly where chaotic deterministic systems come into play. A phase space chart gives a complex and structured image of the process of motion. Sometimes, paths of chaotic particles will group along a bent curve without ever actually reaching it or they may fill certain areas but never touch other areas. Mathematical analysis shows that these shapes cannot even be categorized into a dimension such as a line, area, or body. For example, a circle with an infinite number of small holes has an infinitely small area. Overall it may look like a messy pile of lines, but there is much more to it. Such mathematical objects have a fractal dimension (for example, 2.3 or 1 and 3/7) that can be calculated. A typical trait of chaotic systems is that their trajectory charts form infinitely complex fractal patterns.

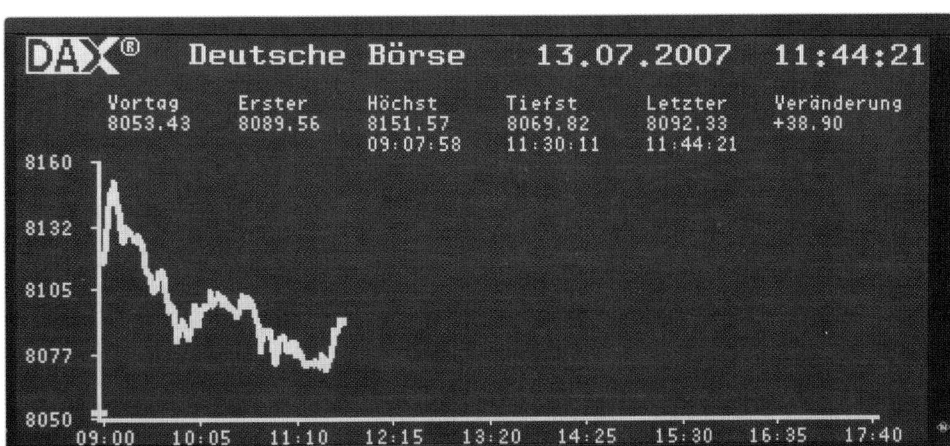

■ Fractals in the stock market: The rise and flow of stocks and shares show a fractal pattern; it is hard to see the difference between a chart of ten years, a month, or even an hour. They all share the same pattern of sharp increases and decreases.

Fractals or fractal objects occur in nature. For example, fern leaves or cauliflower, especially the Romanesco variety, grow little copies of themselves on their fern fronds or stalks, which, in turn, grow even smaller copies of themselves and so on. Another good example is how the British coast continually gets more complex the closer you look at it. This is because each magnification adds more detail than the last. Mathematician Benoit Mandelbrot (b. 1924), known as the father of fractal geometry, calculated that the west coast of Great Britain has the fractal dimension of 1.25. The stock market, in comparison, is an example of temporal fractals. At first sight, it is impossible to distinguish a ten-year chart from a chart of a month, an hour, or even a minute. They all show a similar pattern of sharp spikes and lows.

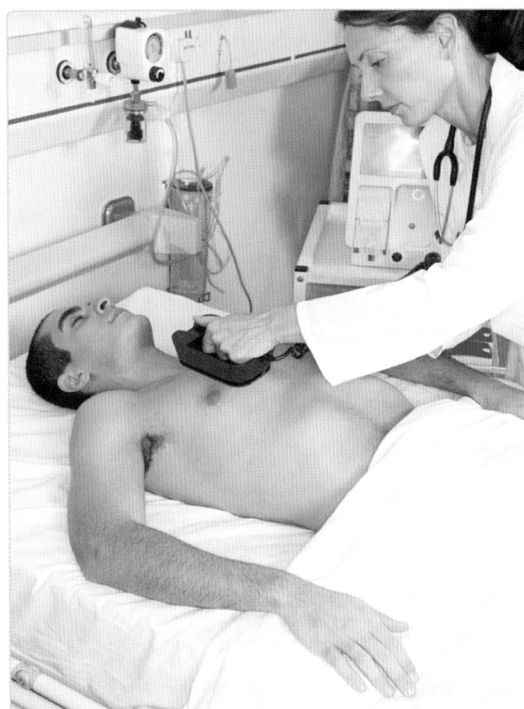

■ Medical physicists have been able to show that cardiac arrests are of chaotic nature, and carry out research to build better defibrillators.

in focus

FRACTALS

Chaotic systems may provide interesting trajectory charts, but perhaps the most beautiful fractal images are designed mathematically. The instantly recognizable appleman set, for instance, is created by repeatedly applying a function to itself. Changing the original parameters projects new patterns onto the computer screen. This concept has many applications. Computer-generated movies, for example, produce trees and other objects using the same technique; the more fractals are used the more natural an object looks.

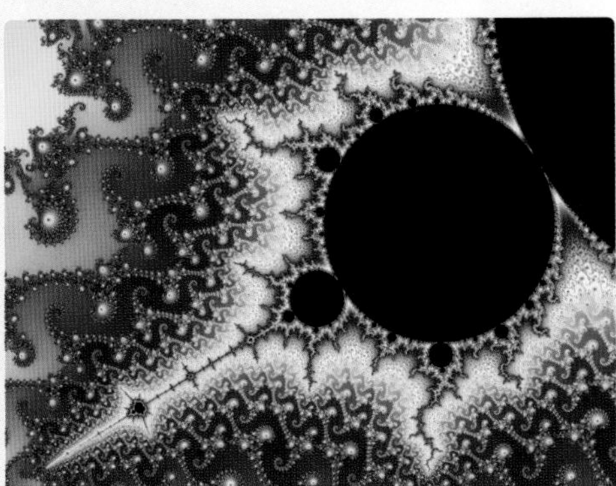

■ Benoit Mandelbrot's appleman fractal set.

see also: Vectors, p. 413

models and fluid mechanics

The study of the flow of liquids and gases has many modern applications. It is used to improve the aerodynamic qualities of cars and planes, model the behavior of liquids in a pipe, and predict the actions of people in a crowd.

In hydrodynamics, flow patterns in pipelines or water bodies are modeled using differential equations, the same equations used by astrophysicists when describing the motion of stars within a galaxy or the path of a galaxy in a large accumulation of galaxies. When solving these equations, it is irrelevant what objects are being transported, water molecules, stars, or entire galaxies. Similar flow patterns may emerge from modeling a water pipe or an accumulation of galaxies. This shows one of the strengths of the science of abstract physics as it simplifies a normally complicated thought or natural process into an abstract model that can be applied to many different situations.

mechanics | traffic | medicine | computing

NEW ISSUES IN PHYSICS

Robert B. Laughlin, born in 1950 and Nobel Prize winner in physics, published his book *A Different Universe: Reinventing Physics from the Bottom Down* in 2005. In his book, he wants physics to turn its approach upside down. He believes that physicists should stop searching for the fundamental formula of the universe and its smallest particles. Instead, they should focus their attention on the complex and interdisciplinary phenomena seen in nature. Models could be effectively applied to these phenomena enriching the scope of physics and taking it to the next level.

Traffic and flow

Let's consider an example: the physics of traffic. This is a young discipline that emerged in the 1990s applying hydrodynamic laws to solve problems of traffic. Here, cars take over the role of specific molecules and assumptions are made regarding their motion patterns. There are, for example, certain probabilities as to how often a car will change lanes, what the preferred speed is, how fast its driver can react, and so on. The equations for flow patterns are modified using these variables in order to solve problems like why some traffic jams happen for no obvious reason. Physicists were also able to prove the positive effects of speed limits on the flow of traffic (and on the number of accidents).

These traffic models can also be applied to pedestrians. Newly constructed soccer stadiums are now optimized using traffic simulations that help identify the ideal locations for escape routes and emergency exits. Access from Mecca, towards the Jamarat Bridge, where the religious stoning-of-the-devil ritual is performed, has been optimized by German traffic physicist Dirk Helbing and others. They succeeded in preventing the repeat of numerous deaths in 2004 and 2006 after the holy site was reconstructed under their guidance saving hundreds of lives a year.

Studying the effect of airflow on a cyclist in a wind tunnel.

The Kabba in the Al-Masjid al-Haram mosque in Mecca: Fluid mechanics improved the flow of pilgrims as many were being crushed in the massive crowds generated during the Hajj.

the environment, humans, and the brain

Three examples of new interdisciplinary research areas that emerged from the traditional disciplines are environmental physics, medical physics, and the research of neural networks.

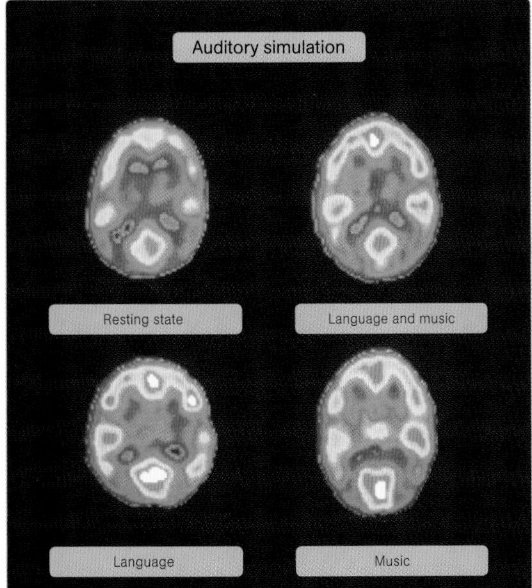

Modern imaging techniques can identify which regions of the brain are active while the test subject is solving a math problem.

In 2007, climate change became one of the most discussed issues worldwide. Climate researchers have been warning about the consequences of rising CO_2 levels and the destruction of the ozone layer since the 1970s. Environmental physicists play an important role in the research community and they have been investigating environmental issues using physical methods while theoretical physicists use physical models to solve problems in meteorology, soil science, oceanography, and glaciology. Environmental physicists, once looked down on by their colleagues researching quarks and black holes, have now become an established part of the scientific community.

Medical physics

At first sight, topics such as the human metabolism, immune system, or even emotions do not seem to apply to physics but more and more physicists are involved in medical research. Not only do they bring in their knowledge in all technical aspects of development and maintenance of medical instruments, but more importantly, their expertise applies to the areas of radiology and medical imaging. A physicist's knowledge about the use of (and protection from) radioactive substances is valued in radiology. The scope for physicists in medical imaging extends from the x-ray machine to the ultrasound, magnetic resonance imaging (MRI) to positron emission tomography (PET), and many more methods for viewing a patient's body literally inside out, and even inside the brain.

Medical advances in science can replace limbs with new albeit artificial limbs.

Neural networks

In computer sciences and physics, a neural network is a simulation of artificial nerve cells or neurons connected in a way that resembles the circuits of the brain as closely as possible in order to study the mechanisms of thought. A distant goal is the creation of an artificial intelligence (AI) construct within a computer. So far, research has concentrated on the myriad differences between microprocessors and the brain: microprocessors with their super fast connections between millions of transistors each and, on the other hand, the brain with its slow networks of billions of interconnected nerve cells. When it comes to performing raw calculations, humans cannot keep up with modern computers. However, simple tasks such as the recognition of patterns or the identification of faces and voices can only be accomplished by highly complex artificial neural networks. Outside the realms of science fiction, we are still far away from developing true artificial intelligence; however, the many popular machine translation services on the Internet use cutting-edge artificial intelligence research in a very real and tangible manner.

CAREERS IN PHYSICS

in focus

If you dream about finding the fundamental formula of the universe and receiving the Nobel Prize then you probably should consider a career in physics. This diverse field includes academic professors and particle, theoretical, and quantum physicists, all dedicated to the challenging task of unraveling the mysteries of the world and the universe around us.

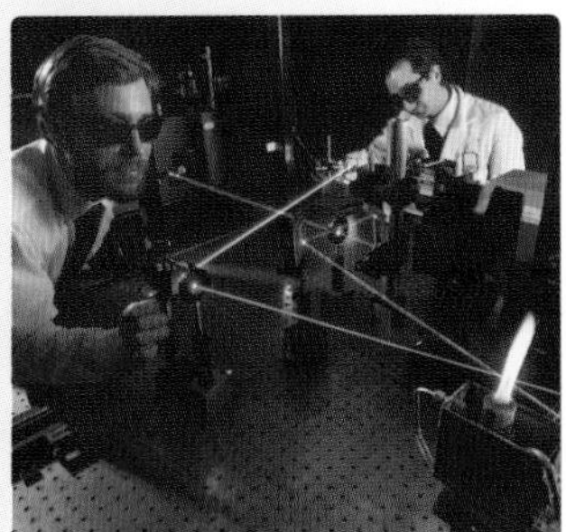

Physicists at work in a highly demanding field that requires comprehensive knowledge of mathematics and science.

see also: Human beings, p. 220

PHYSICS AND TECHNOLOGY

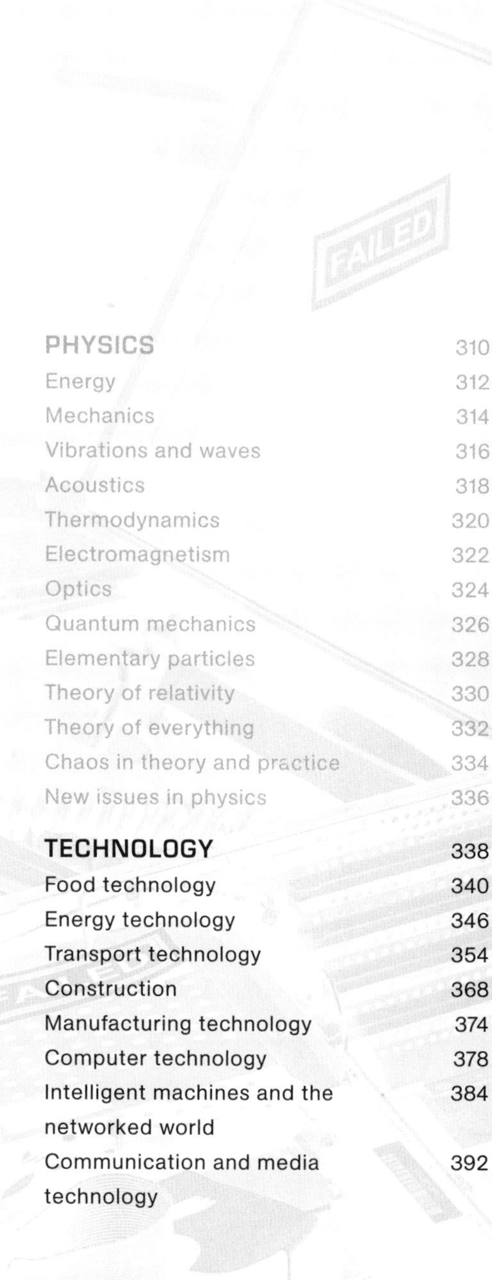

TECHNOLOGY

Humans research natural forces not only for the sake of new discoveries but also to use the findings to their advantage. Nowadays manual labor has largely been replaced by machines; in some cases machines are completely taking over certain processes. For example, machines produce food and energy, as well as transport passengers and goods by land, sea, air, and even space. Others simplify administration, planning, or organizational tasks; for instance, computers support office work, education, and research. The global connection between humans and machines is made possible by computer networks that allow for the merging of old and new ways of communication. Finally, computers can even be programmed to simulate the intelligent performances of living organisms and transfer this artificial intelligence into motion carried out by robots.

agriculture | animal husbandry | fisheries | gm food | preservation

FOOD TECHNOLOGY

Food technology is important as only a few natural products are used for direct human consumption. A wide range of technological applications are used to process most agricultural commodities. Food technology also plays an important role in food conservation.

agriculture

Agriculture has undergone tremendous development from its beginnings to the present day. Farms range in size from small self-subsistence operations to large, highly developed companies with massive machinery.

Today, due to structural reforms, individual farms operate on very large tracts of agricultural land.

Agriculture has played a decisive role in human history. When people began to domesticate animals and plant crops, they became sedentary. Hunter-gatherers became farmers. The systematic breeding of farm animals and cultivation of crops made it possible to feed more people—not just the farmers. Other occupations developed, along with a system for the division of labor. This development formed the basis for the beginning of our civilization.

Because they pollinate agricultural crops, bees are both directly and indirectly responsible for supplying one-third of the food consumed by humans. A mass mortality incident involving bee colonies in the U.S. could result in a food crisis.

Still, to this day, securing our food supply remains a very important task. About a third of working people worldwide are involved in agriculture.

Chemical fertilizers

As they grow, plants remove nutrients from the soil. Chemical fertilizers provide a method for adjusting the nutrient imbalance in the fields. Before fertilizers were available, fields used on an ongoing basis had to be cultivated according to a crop-rotation cycle, including a fallow period. In fact, crop rotation has been a feature of agricultural management in Europe since the Middle Ages. Fields are sown in a three-year cycle, usually starting with winter grain, then summer grain, followed by a fallow year. Crop rotation is beneficial because it avoids the buildup of pathogens and pests that often occurs when one species is continuously cropped. With chemical fertilizers, crop rotation is no longer necessary—a development that led to monocultural production in many places (that is, the same types of plants were cultivated for many years). This has had negative effects on the environment and reduced biodiversity.

Green revolution

Since the early 1960s, the use of chemical fertilizers and plant treatment agents (for controlling weeds and pests) has resulted in an enormous increase in production per area cultivated, effectively transforming agriculture. Efforts to use modern agricultural technology to secure a food supply over the long term for fast growing populations, particularly in Asia, Africa, and Latin America, received solid political support. This international commitment became known as the "green revolution." The fear that the Earth could no longer supply its inhabitants with food at some point seemed to disappear. In America, increased agricultural production meant that by the end of the 20th century, a single agricultural worker could feed 130 people. Just a century earlier this ratio was only 1 to 2.5. However, critics of the green revolution point out the increased use of pesticides and decreased biodiversity as causes of concern.

in focus

ECOLOGICAL AGRICULTURE

Pesticides, chemical fertilizers, growth stimulants, and genetic engineering are avoided in ecological agriculture. Production methods are environmentally sound and emphasize ecological aspects and the protection of the environment. In many countries, foods labeled as organic must meet strict legal requirements. In the U.S., the EU, and Japan, farms that use ecological methods must be licensed and monitored regularly.

On-site farm store on an organic farm: The food sold here is organically grown, that is, using no synthetic fertilizers or pesticides, or genetically modified organisms.

agricultural machinery and technology

Machines are used to accomplish almost all agricultural tasks from working the soil (tilling) to sowing and harvesting. These implements, either self-propelled or pulled by another vehicle, have greatly increased crop yields.

In recent decades structural reforms have increased the size of farms cultivated by individual agricultural concerns. These reforms have also led to a steady increase in the automation of agricultural processes; more and more machines are used today in order to carry out the work quickly and effectively.

Mills are the interfaces where natural grain products meet food processing technology. Sacks of flour are filled here and sent on their way for further processing.

Procuring agricultural machinery often involves a huge investment that small and midsize operators cannot afford. For this reason small farms join forces in machinery cooperatives to share the machinery. Sometimes external companies are contracted to perform particular tasks. The most important agricultural machinery include tractors, harvesters, plows, harrows, sowing machines, planters, fertilizer applicators, and crop sprayers for applying plant treatment agents. In fact, major trade fairs for agricultural technology and machinery take place regularly around the world.

Automation on the farm

Technology continues to gain significance in agriculture, and not only in the cultivation of agricultural land. Machines are also used in other areas of agriculture: for example, in animal husbandry (p. 342). Computers control feed distribution in the animals' stalls, and milking machines are used in dairy operations.

Further processing

After the harvest, most raw agricultural products undergo further processing. For example, grain is dried and corn is milled for use as animal feed and fermented into silage. Yet many steps of these processes are not carried out on the farms themselves, but rather in "downstream" companies. Grain is delivered to a mill where it is ground into flour. Livestock sold to a slaughterhouse is processed into meat and meat products. Some agricultural commodities are used immediately in the food industry or in other trade businesses.

The use of efficient machines, chemical fertilizers, and pest management treatments has more than doubled the size of the area used for grain production worldwide.

in focus

ENVIRONMENTAL EFFECTS

Using chemicals in the form of agricultural fertilizers and pesticides has had serious consequences for the environment. The application of nitrate and phosphate in fertilizers greatly impacts rivers and lakes and has disturbed the natural biological balance. As a result bodies of waters can "collapse." A few of the pesticide treatments (such as DDT) had such a negative effect on the environment and the food chain that their application has been prohibited.

Pest control from the air: Crop dusters specially approved for agricultural use have huge chemical tanks and application equipment.

see also: Environmental exploitation, pp. 130–133

animal husbandry

Animal husbandry includes the keeping, feeding, and care of animals. It is not only an agricultural practice, it also extends to the keeping of pets and zoo animals.

Wild animals were first domesticated around 10,000 years ago when aurochs (forerunners of modern cattle), wild pigs, and wolves began to be bred by early farmers. Most breeds of farm animals today have been domesticated. However wild animals are also sometimes kept in enclosures for the production of venison and other game products. Animals are kept primarily for economic purposes: to be sold to provide food and raw materials. In some countries, however, animals are also used to provide transportation.

Today animals are milked almost exclusively with milking machines. Some milking machines are permanently installed (so-called "milking parlors") and others are mobile for use in tie stalls or in pastures.

A free range environment is considered by many people to be more appropriate for hens than the conditions on a factory farm. Cage-free hens have more freedom of movement.

Factory farming practices

To increase productivity and decrease production costs, animals are often kept in the smallest amount of space possible. For example, some egg-laying hens are kept in cage arrangements called "laying batteries." These factory farming methods have reduced the price of meat and meat products, making them readily available and affordable for people in industrialized countries.

Meat

Keeping farm animals for their meat is known as producing "feeder animals." The animals most widely used for meat include beef cattle, hogs, and poultry. Special feed mixes and hormones can be used to hasten the growth of farm animals and thus get them ready for sale sooner. Hormone use, however, is now prohibited in many countries due to health risks for consumers.

In feedlot operations, production costs are often kept as low as possible by keeping feeder pigs together in pens where they do not have enough room.

Milk and eggs

Animal husbandry is also used to produce milk and eggs. Dairy cows are generally used for milk production in Europe and North America, but sheep, goats, and water buffalo are also kept for this purpose. Dairy cattle are kept in either tie stalls or pens where they enjoy relatively free movement.

in focus

ECOLOGICAL AGRICULTURE AND ANIMAL-APPROPRIATE HUSBANDRY PRACTICES

Husbandry practices based on the animals' natural living conditions are known as animal-appropriate husbandry practices. Here, the animals maintain their innate behavioral patterns, which is not possible under factory farming conditions. Animal-appropriate husbandry practices include adequate living space and natural or near natural feed. In economic terms, using these practices to produce agricultural commodities takes more effort and is more expensive. Collecting eggs takes longer, milking requires more work, and the animals gain weight more slowly because they are free to move about. However, the animals are healthier and there are far fewer negative environmental impacts. Due to this, animal-appropriate practices are most often applied in organic and ecological agriculture.

Animal-appropriate husbandry practices involve animals being kept in healthier and more natural conditions.

see also: Environmental concerns in everyday life, pp. 136–137

fisheries

The fishing industry is concerned with the catching and breeding of all animals that live in the water. The main products of the industry are human and animal food.

Both the oceans and inland waters provide rich sources of food. Fresh, edible fish is stored at temperatures between 0 to 35°F (0 to 2°C) and then sold for consumption.

Fish are not the only targets of the fishing industry: in addition, invertebrates such as mollusks and crabs can be caught. This may include bivalves, such as oysters or mussels, or large crustaceans, such as lobsters, crawfish, prawns, and shrimp. Because fish and other sea creatures live in the vast oceans, it is difficult for humans to have an influence on their migratory movements. For this reason, a wide variety of fishing methods have been developed throughout history and across the world. In the Mesolithic period, humans had already begun to exploit the water as a food source. They caught fish with hooks, nets, or gill nets. Modern fishing methods range from a hook on a line to fishing vehicles equipped with high technology.

Aquaculture includes fish farming, the fastest growing food production sector. Cultured fish account for almost one-fifth of today's worldwide fish production.

Fishing is among the most significant methods for using the ocean's resources. However, sustainable use is only possible if populations are allowed to renew themselves.

Fishing methods

Some traditional fishing methods in use for many centuries include gill nets and weirs. Gill nets are fine mesh nets that are fastened to the ocean floor and positioned vertically; weirs have a barrel or funnel shape and are made of wicker, net, or wire mesh. Purse seines are nets for catching schools of fish not far below the surface of the water (such as mackerel, tuna, salmon, or herring). Drift nets are attached to buoys. These nets, sometimes miles long, catch fish and other living creatures at random, which is why this method is controversial and has been banned in many places. Trawling makes use of sack-shaped nets to catch specific schools of fish. Fishing boats drag these nets through the water at various depths. If they are dragged along the ocean floor, however, they destroy almost every living creature there. Longlines are simply fishing lines (with hooks and bait)—except that they are up to 60 miles (100 km) long and have thousands of hooks attached.

in focus

FISHING UNDER CRITIQUE

A fishing operation usually catches not only the desired fish, but also other fish, birds, and mammals that are killed in the process (by-catch). The by-catch varies in size and scope depending on the fishing method used. To reduce the by-catch—especially of rare or endangered animals—more selective fishing methods are increasingly demanded. In addition, due to decreasing fish populations, there are now international agreements regarding overfishing. Fishery protection zones, where fishing is limited, have also been established.

Each year millions of tons of tuna are caught; the by-catch, however, often includes dolphins.

basics

HOW DO WE KNOW IF A FISH IS STILL FRESH? A chef uses the following criteria to determine if a recently caught fish is still fresh: its eyes should be clear, the gills a shiny, dark red, the fish should not smell like "fish," and the flesh should give a little but still feel elastic when pressed lightly.

genetically modified food

Through selective breeding, people have long exercised indirect influence over the genetic makeup of plants and animals. Today, however, genetic technologies make it possible to directly manipulate an organism's genes.

Because of consumers' concerns, more and more countries are establishing labeling regulations and guidelines on traceability.

Food may come into contact with genetic technologies in various ways. In the simplest case, a genetically engineered natural product is eaten directly as food. However, many genetically modified (GM) plants undergo further processing after harvest and are only used as edible ingredients. For instance, starch can be derived from GM corn, and GM soybeans are used to make soybean oil or soy meal. Many additives are produced with the help of genetically engineered microorganisms (such as yeast and bacteria), and thus GM products enter the food supply in a roundabout manner. Indirect contact with genetically modified organisms (GMO) can also occur; for example, if livestock consume feed made from GM plants. Technology does not have the capability for the detection of GM components in the resulting meat products.

The approval process

A GM plant may produce specific proteins or other substances that have never before been part of the human food supply. These substances potentially have negative health effects: for instance, they may act as toxins or allergens. Accordingly an extensive testing and approval process is mandated for foods of this kind, including individual ingredients and additives, as well as whole foods. Approval is granted only if the food is shown to be safe in accordance with scientific standards. This is accomplished by comparing the GM product with its traditional counterpart.

Criticism of GM food and genetic technology

The long-term effects of the use of GM plants on people and the environment have not yet been adequately determined. This is the main focus of protests against the use of genetic technology in agriculture. Since GM plants are grown in open fields, their release is something that cannot be undone. They may cross (hybridize) with other plants or crowd out traditional plant species—with unforeseeable consequences in areas such as species diversity. Despite attempts to evaluate risks to human health, it is not possible to test every component of a GM plant. Furthermore, since legal requirements and corresponding oversight systems have not been instituted in every country, there is a risk that field testing or agricultural use of GM plants could get out of control. Objections have also been raised against the patenting of GM organisms, which gives companies ownership rights to seeds and plants. This practice may give rise to monopolies in the food industry, with incalculable costs, especially for impoverished or developing countries.

issues to solve

THE CULTIVATION of genetically modified crops has expanded to more than 100 million acres worldwide. However, the long-term effects on people, animals, and the enviroment have not yet been determined scientifically.

in focus

GENETIC ENGINEERING

The tobacco plant can be easily genetically modified and cultivated at low costs.

Genes contain the biological information of heredity; that is, they are responsible for the inherited characteristics of an organism. Through genetic engineering, these characteristics can be transferred directly to another organism when new genes are inserted into a living cell. Genetically modified organisms often exhibit characteristics that are not found in nature. For instance, bacteria can be altered so that they produce new substances, while plants can be made resistant to pests or diseases. Genes can even be transferred among unrelated species—a feat that is impossible using traditional breeding methods.

The most commonly available genetically modified foods since the 1990s are soybeans and corn. In 2005 the total surface area of land cultivated with genetically modified organisms had grown to approximately 200 million acres, of which more than 50 percent were in the U.S.

see also: Genes, p. 250

food preservation

Most fresh foods spoil soon after production or harvest. Because of this, people have sought ways to preserve perishable foods since the beginning of human history.

The changes we observe in natural products as they decay are mainly the work of bacteria, yeast, or mold. These microorganisms draw nourishment from the food and then multiply. In so doing, they change the characteristics of the food: for instance, it may become mushy, foul smelling, or moldy. In most cases, people view these changes as undesirable and attempt to stop or at least slow them. Numerous preservation methods have been developed to suppress the activity of microorganisms in foodstuffs. In general, these can be divided into physical, biological, and chemical techniques. Some processes, such as sterilization, all microorganisms in a substance are destroyed. Other methods (such as the pasteurization and refrigeration of milk) merely slow their growth, so that the food retains its quality for a certain length of time. Radical methods that destroy all microorganisms may be highly effective, but they are not always the best solution, as they can negatively affect the food's taste, nutritional value and texture.

PASTEURIZATION

milestones

Louis Pasteur discovered that most food-borne microorganisms can be destroyed by briefly heating the food to 140–158°F (60–70°C). This method, which became known as pasteurization, is still commonly used to preserve liquid foods. For instance, milk is heated to about 161–167°F (72–75°C) for 15 to 40 seconds. Yet the results are not completely sterile, and consequently, they remain fresh only for a limited time.

Louis Pasteur, French chemist (1822–1895).

Physical methods

One of the oldest food conservation techniques is drying. Dried foods can be safely stored for long periods, since the lack of water prevents microorganisms from reproducing. The oldest and simplest method is air drying. Food in canning jars or metal cans is preserved through the use of heat, which destroys all of the microorganisms within the food. However, the longer a food is heated, and the higher the temperature used, the more the quality of the food will suffer. Accordingly, gentler techniques have been developed, such as pasteurization (see milestones). With refrigeration and freezing, the growth of bacteria and other organisms is sharply reduced, while many frozen foods remain quite similar to the fresh product after thawing. Microorganisms can also be destroyed with ionizing radiation (irradiation).

Most types of fish and meat can be preserved by drying or may be both preserved and flavored through the use of smoke, typically in a smokehouse. The combination of heat, to dry the food without cooking it, and the addition of aromatic hydrocarbons from the smoke preserves the food.

The production of all alcoholic beverages employs ethanol fermentation by yeast. Wines and brandies are produced by fermentation of the natural sugars present in fruit, whereas beers and whiskeys are produced from grain. The fermentation must take place in a vessel that is arranged to allow carbon dioxide to escape, but that prevents outside air from coming in.

Biological, chemical, and other techniques

In some foods, the natural changes caused by microorganisms are considered desirable. For example, if fresh milk is left out in the open, certain types of bacteria will feed on its sugars, producing lactic acid and suppressing the growth of other unwanted microorganisms (and converting the milk to yogurt). In alcoholic fermentation, yeast populations expand and give off alcohol, which has a preservative effect. Chemical preservatives can also be used to prevent bacterial growth. The use of nitrites to preserve meat products is widespread. Other well-known chemical preservatives include benzoic acid and sorbic acid. Food can also be preserved by smoking or salting it, or by adding sugar.

basics

FOOD PRESERVATION TECHNIQUES

PHYSICAL methods: water removal (drying), heating, refrigeration, freezing, irradiation

BIOLOGICAL methods: fermentation (e.g., alcoholic)

CHEMICAL methods: addition of chemical preservatives

OTHER methods: smoking, salting, addition of sugar

Canning involves cooking fruits or vegetables, sealing them in sterile cans or jars, and boiling the containers to kill or weaken any remaining bacteria as a form of pasteurization.

energy production, transport, and storage

A modern communications-based society with high mobility depends upon universally available energy. A wide range of technology is used to fulfill this demand.

production | consumption | fossil fuels | nuclear | solar | wind | water | biomass

ENERGY TECHNOLOGY

Since the discovery of fire, techniques for producing and consuming energy have been growing more efficient and complex. Early sources of energy included the sun, wood, and water power. During the industrial revolution, coal and electricity were added, followed by nuclear energy in the 1950s. Today, the dwindling of fossil fuel reserves is spurring the development of new technology that uses renewable energy sources. The main uses of energy include heating, transportation, producing electricity, and operating machinery.

Energy can be present in a multitude of forms. For example, it can be stored chemically in gasoline and then made available for use through the process of combustion. A pendulum already set in motion has kinetic energy, while a tensed spring has potential energy. Electricity is considered one of the most useful forms of energy, since it can be transported efficiently and is used for a wide variety of applications.

Energy production and use

From the perspective of physics, all technological devices that produce or use energy can be viewed as energy transformers.

High-tension cables are mounted on isolators made from glass, ceramic, or plastic. Their ribbed shape allows problematic moisture to escape easily.

Energy production refers to the process of converting energy from a natural resource, such as coal, into a form that can be easily used by the general public, such as electricity. The resulting energy can then be employed for a particular application, such as to start up and run a computer.

Technological devices that are commonly used to produce energy include generators, thermocouples (for the production of electricity), and furnaces (for the production of heat). Devices that use energy include electric motors, lights, and ovens.

Transporting energy

Energy must be transported to make it available where it is most needed. To transport electricity, copper cables in the form of high-tension wires are used. When electricity flows through a conductor (in this case a wire), the resistance of the conductor leads to a loss in energy. The wire itself "uses" electricity by converting some of the electrical energy to heat. Therefore, in order to minimize these losses, energy is transformed into a different state. Low voltage electricity is converted to a higher voltage and lower current, so that power can be efficiently transported to the desired destination.

Energy loss during transformation

In whichever form energy is being transported, the second law of thermodynamics states that energy transformations usually involve a loss of some of the energy. For instance, kinetic energy can be transported using gears, belts, or chains. However, at the same time a portion of the kinetic energy is converted through friction into unwanted heat energy. Radiation energy is lost through dissipation; even a pendulum releases energy in the form of heat as it is slowed by friction and air resistance.

Energy storage

Apart from making transportation easier, storing energy can also be useful for other reasons, such as saving energy for use whenever and wherever it is needed. Electricity can be stored electrochemically in batteries or dry cells, or as a charge within a capacitor. Transformation into and back from other forms of energy offers a multitude of additional possibilities. For example, if energy is used to lift water with a pump, the result is stored potential energy in the form of increased water pressure.

practice

DRY CELLS AND BATTERIES

Dry cells and batteries store chemical energy. The once popular low-performing nickel-cadmium batteries are now widely banned for environmental reasons. High-performance lithium-ion batteries power laptops and cell phones are powerful mobile energy storage units. Because they are economically and environmentally friendly, rechargeable batteries are popular. They must be recharged with special devices, or they can overload, catch fire, or even explode.

Batteries harness the energy released by a reaction between chemical components.

see also: Semiconductors, p. 292

energy consumption

There is only a limited supply of most technologically useful carriers of energy, such as coal and natural gas. Reducing energy consumption is thus a key challenge for engineers, architects, and transportation planners.

The difference between sensible and wasteful energy use is clearly evident in the case of lighting technology. Ordinary light-bulbs produce light when electricity flows through a metal filament; the filament's resistance produces heat and it glows. Most of the energy it releases, however, is in the form of invisible ultraviolet radiation. The visible light emitted by a standard bulb represents only one to two percent of the electrical energy it consumes. Fluorescent tubes and compact fluorescent lightbulbs, on the other hand, transform 10 percent of the electrical energy they use into visible light. In these bulbs, electricity stimulates the electrons in a gas, raising their energy levels and causing the gas atoms to emit ultraviolet radiation. A layer of fluorescent material within the bulb then converts these ultraviolet emissions to visible light.

Even more efficient are light-emitting diodes (LEDs). Like computer processors, LEDs are constructed from semiconducting materials. When an electrical current is applied to an LED, electrons within the semiconductor fall to a lower energy level and emit visible light. Due to this direct transformation of electricity into light, LEDs achieve efficiency levels of up to 20 percent.

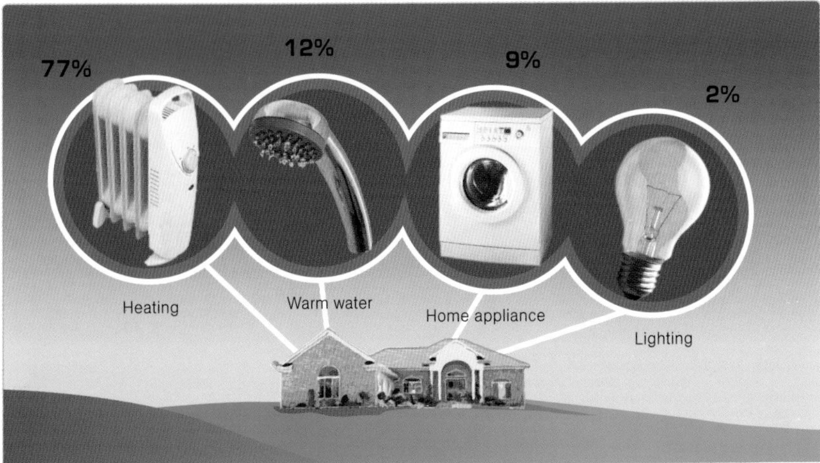

■ The heating and cooling system is the largest energy user in the typical home, followed by water heating, electrical appliances, and lighting.

■ Computers use increasing amounts of electricity, especially when they are left running day and night.

"Passive houses"

Many people are careful to buy energy-efficient appliances and use compact fluorescent bulbs. However, the most significant factor in home energy consumption for most households is heating (and air-conditioning in warm climates). Fortunately, heating is also the area of household technology with the greatest potential for savings, above all through enhanced insulation. State-of-the-art insulation, used and often required in new construction, significantly reduces heating and cooling costs in comparison with older homes. Especially in Europe, a standard known as the "passive house" calls for such efficient insulation that the warmth generated from cooking, electrical appliances, and the inhabitants' bodies is sufficient to create a comfortable indoor environment. A separate heating system is not necessary: only an electric ventilation system and a water heater are needed.

in focus

"ENERGY HOGS"

Electric heating systems: When power plants generate electricity using a heat source (such as coal), large amounts of energy are wasted. Using the resulting electricity to produce heat for a home—again with substantial energy losses—can be extremely inefficient.
Sport utility vehicles: These popular vehicles combine the off-road and towing abilities of a pickup truck, while still being able to carry numerous passengers. SUV's attract extensive criticism for their low fuel efficiency, due to being classed as a "light truck" and so not having to conform to the stricter fuel standards for smaller passenger vehicles.

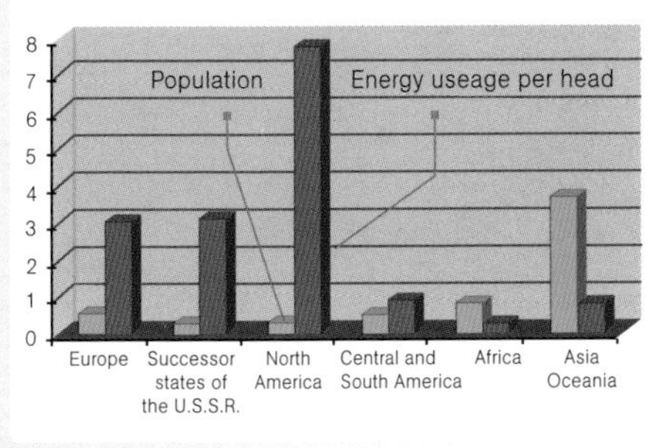

■ Energy usage per head in various countries.

basics

HOW MUCH ENERGY DO THEY USE?

The following list shows the approximate power consumption of various devices and vehicles per hour:

- Lightbulb: 60 watts; equally bright compact fluorescent bulb: 11 watts; LED: 5–10 watts.
- Handheld hair dryer: 1,000–2,000 watts (1–2 kW)
- Electric stove: 2–10 kW
- Gas heating for a single-family home (older construction): approximately 10–30 kW
- Compact car: 30 kW
- High-performance sports car: 200–500 kW
- Intercity bullet train: approximately 3,000–8,000 kW
- Oil tanker: up to 100,000 kW

see also: Energy-efficient construction, p. 371

fossil fuels: coal, oil, and natural gas

Fossil fuels provide the bulk of the energy that powers the industrialized and developing world. As demand for fossil fuels increases, the technical undertakings needed to satisfy it must be constantly adapted.

Oil and natural gas pipelines in extreme climates must be well protected against corrosion.

Oil is the most important fuel, heat source, and raw material in the chemical industry, but before it can be efficiently utilized, it must be refined. Refineries first remove sulfur from the oil. The resulting oil is then separated through distillation into its lighter and heavier components: gasoline, diesel, heating oil, and tar. To improve engine performance and reduce wear, various substances are added to gasoline.

Coal, on the other hand, often needs only a mechanical cleaning before it can be used. With so-called "cracking" techniques, it is also possible to produce liquid fuel from coal, for instance, to power vehicles. However, this process is technically complex and not yet cost-effective.

Natural gas contains 85–98 percent methane, along with other hydrocarbons, carbon dioxide, and sometimes helium. For easier transport and for use as a fuel, it is compressed and partly liquefied. The heat produced during this compression process can be used for many purposes, including to heat homes or swimming pools.

To reach the coal, gigantic excavators often have to remove layers of earth as thick as 300 feet (91 m).

Energy production

Chemical energy stored in gasoline and other fuel is released and used in a multitude of ways. Many power plants produce energy using combustion engines. The fuel's energy is converted to heat, which turns water into steam, which then exerts pressure to turn a turbine. The turbine drives a generator to produce electricity through the principle of induction (p. 323). Thus, the energy from coal, oil, or natural gas must undergo numerous transformations before it can be used in the form of electricity. During electricity production, excess heat is also given off, which can be transferred to other locations, for example, to heat buildings. This process of power-heat coupling, or cogeneration, makes better use of the same amount of fuel. Increased efficiency is important as combustion releases large amounts of CO_2 into the atmosphere, a major contributor to the greenhouse effect.

EXTRACTION OF OIL AND NATURAL GAS

Tens of thousands of production facilities worldwide tap the Earth's reserves of oil and natural gas on land and at sea. Producing countries estimate their combined reserves at approximately 1.2 billion barrels of oil (160 billion tons). Based on this figure the supply should last for another 40 to 50 years. Oil fields lying deep under the surface of the Earth or the sea are mined using pipes equipped with diamond-tipped drill heads. At first, the pressure exerted on the deposits from the Earth is often enough to bring the oil to the surface. Later, pumps or injections of natural gas or water may be needed, followed by complicated special procedures as the reserve is depleted. Even with the latest technology, an oil field cannot be fully exploited: about 20–40 percent of the oil remains in the ground. Tanker ships or pipelines transport the oil and natural gas to refineries that process the materials to produce usable substances.

Rigs obtain oil and natural gas through holes drilled in the ocean bed. These structures may be attached to the ocean floor, consist of an artificial island, or be a floating platform.

practice

see also: Fuel cells and biomass, p. 353

nuclear power

In the early days of nuclear technology, hopes were raised of unlimited energy. It did not take long for the potential dangers to be recognized, although nuclear power still supplies a portion of the world's electricity.

Nuclear technology works by splitting atomic nuclei (fission) to release vast amounts of energy. To begin the process, neutrons bombard fissionable material such as uranium or plutonium. When a neutron hits an atom's nucleus at just the perfect speed, the nucleus splits apart. This produces smaller atoms, more free neutrons, and energy. The free neutrons then split other nuclei, initializing a chain reaction. A nuclear reactor's core contains fuel rods of fissionable material bundled together. During controlled fission, moderators such as water or graphite control the neutrons' speed and control rods slow or stop the chain reaction when necessary. The heat from fission is then collected by a coolant substance and used to drive turbines and generators.

Pressurized water reactors are the most common type of power-producing reactor. They use superheated water as coolant and neutron moderator. The primary coolant loop is kept under high pressure to prevent the water from boiling.

Types of reactors

Reactors differ according to the fission process used and the materials used as rods, coolant, and moderators. Boiling water reactors have one cooling cycle; the water that cools the reactor also drives the turbines as steam. Pressurized water reactors have separate water circulation systems. "Fast breeder" reactors, with liquid sodium coolant, "breed" new fissionable material, extracting more energy from a given amount of uranium. Pebble-bed reactors, on the other hand, use helium coolant, and spheres of fissionable material instead of rods. There are various risks associated with nuclear technology and different hazards are produced by each type of reactor. In boiling water reactors, defective turbine housings can leak radioactive water into the environment from the primary circulation system. In pebble-bed reactors, helium at over 1832°F (1000°C) can contact water, abruptly vaporize, and subsequently explode. Even properly functioning reactors produce radioactive waste that remains dangerous to plant life, animals, and humans for thousands of years. Despite the multitude of risks, nuclear power will continue to be used and innovations in safety mechanisms and more efficient utilization of fissionable material will constantly be discovered.

Antinuclear activists block the transportation of radioactive waste with a dummy transport container.

 see also: Energy and matter, p. 312

REACTOR SAFETY

in focus

Nuclear reactors pose several different kinds of risks, with the "worst case scenario" causing destruction similar to that of an atomic bomb. In addition to the direct devastation of pressure waves and fire, lasting radioactive contamination of the surrounding area is a serious threat. Smaller-scale accidents may expose workers to harmful or fatal doses of radiation, and ongoing leaks of radioactivity into the environment may occur. Finally, radioactive material may be diverted and used to construct "dirty bombs," which even small groups can use to threaten entire metropolitan areas.

At least in Western Europe, numerous measures have been put in place to increase the safety levels of nuclear plants. However, nuclear power remains a risky technology. To date, no insurance company has been willing to protect the operators of a nuclear plant against the consequences of a meltdown: the risks are ultimately carried by society itself.

Since the 1986 accident in the Chernobyl nuclear plant, the surrounding land has remained radioactively contaminated for miles around.

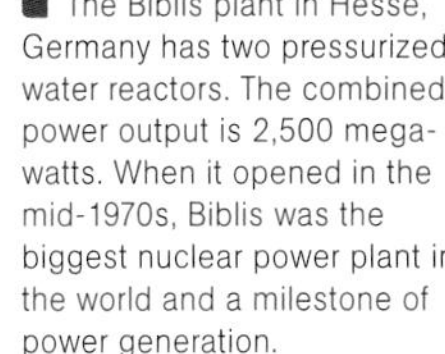

The Biblis plant in Hesse, Germany has two pressurized water reactors. The combined power output is 2,500 megawatts. When it opened in the mid-1970s, Biblis was the biggest nuclear power plant in the world and a milestone of power generation.

solar technology

Radiation from the sun can be transformed into useful energy without the release of harmful emissions. Much research is now being invested in developing solar technology.

The sun can be compared to a gigantic nuclear reactor. Inside the sun, atomic nuclei fuse together, releasing huge amounts of energy that then reach the Earth in the form of solar radiation. The aim of solar technology is to harness this energy and transform it so that it is useful to people. This goal is not pointless, since if the sun's entire energy output could be put to use, around 2,500 times more energy would be available than the current worldwide demand.

The term solar technology includes various systems. Photovoltaics, for example, use the same principle as light-emitting diodes, only in the opposite direction. Absorbed light is converted into electricity using silicon-based semiconductors. Commonly used solar cells are capable of converting 8 to 16 percent of the light they capture into electricity. Solar thermal devices, however, use the sun's energy to heat water, which is then stored in "solar batteries" (insulated water tanks). Other technologies using solar power include solar-chemical systems, solar chimneys, and photochemical installations. The great advantages of solar technologies are their low-cost and emission-free operation.

There are, however, also significant disadvantages. The life span of solar cells is only about 20 to 30 years. Moreover, their production is costly and often dependent on government subsidies. Many countries offer private rebate incentives for the use of solar power, meaning solar cells can pay for themselves in five to ten years.

THE SOLAR ENERGY CYCLE

in focus

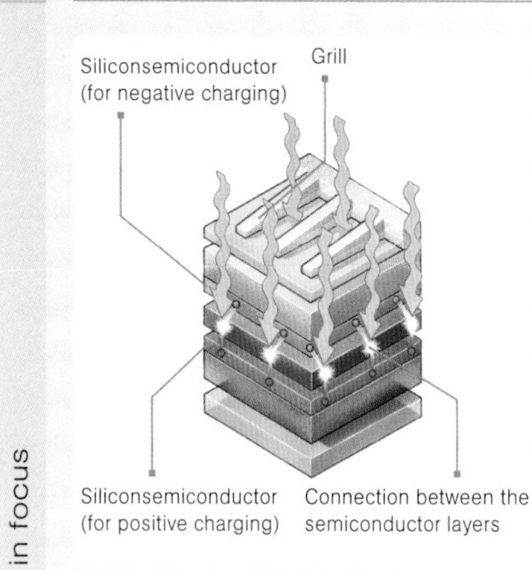

Most homes with solar panels feed the power produced into the public electricity grid. Electrical devices in the home draw their electricity, as usual, from the electric utility. This makes sense when the solar power produced has a greater value than the cost of power from the local utility. In self-sufficient "island systems," solar power is stored directly and in batteries.

When sunlight falls on a photovoltaic cell, the photons activate the electrons in the cell. These migrate through metal contacts between the two semiconductor layers, following the internal electric field. The circuit is closed once an appliance is attached to the solar cell.

Solar parks consist of millions of solar cells connected together in large modules and directed toward the sun. Large solar parks are constructed so that they are able to orient themselves toward the sunlight and thus attain maximum energy efficiency.

Future research and development

In the future, mirrors and magnifying lenses will be used to increase the efficiency of solar cells by supplying them with more light ("concentrator cells"). In addition, multiple layers of semiconductors are expected to allow solar cells to exploit a wider spectrum of light. Using an extremely thin coating of silicon for the cells may also significantly reduce production costs.

basics

SOLAR DEVICES are classified according to their power level. They may use several dozen watts (as in parking meters), a few kilowatts (residential systems), or several megawatts (large "solar parks" or arrays of solarpanels).

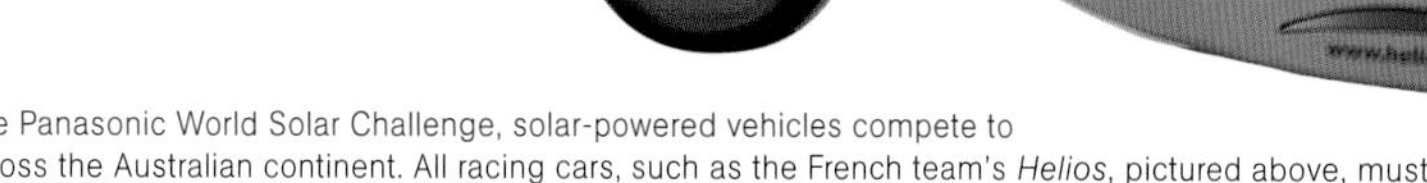

In the Panasonic World Solar Challenge, solar-powered vehicles compete to race across the Australian continent. All racing cars, such as the French team's *Helios*, pictured above, must cover the 1,865-mile (3,000-km) stretch between Darwin and Adelaide using only solar power.

see also: Energy-efficient construction, p. 371

wind energy

Not every good idea is a new one, as witnessed in the case of renewable energy technologies. Sailboats and windmills have had a long tradition of utilizing wind power. But what happens when the wind stops blowing?

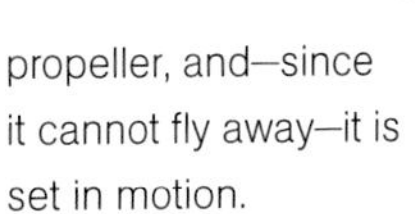

Exposed mountainous areas (such as Germany's Harz Mountains or the central Spanish highlands) typically experience much more wind than protected inland valleys.

Currents in the gaseous layer surrounding our planet—that is, the air—are perceived as wind. As a fundamental principle, air moves from high-pressure zones to areas where the pressure is lower. These pressure gradients, in turn, are caused by temperature differences—which result from variations from place to place in the amount of the sun's radiation reaching the surface. In other words: wind energy is solar energy. It is one of the oldest forms of energy used by humans.

The propeller principle

What distinguishes a modern wind turbine from a traditional windmill? At a glance there are two obvious differences: a wind turbine is significantly taller, and it has only three blades or less. The height of modern wind power systems is easy to explain, since average wind strength steadily increases with altitude. The shape of the blades, meanwhile, can be explained through comparison with airplane propellers. Since propellers are used to move the plane forward, they are constructed for the efficient conversion of power into movement. A wind turbine simply reverses this process: the wind pushes on the propeller, and—since it cannot fly away—it is set in motion.

Using wind energy

The wind's strength can vary from place to place and time to time. Because of this, the successful use of wind

A windmill undergoes a maintenance check on Mykonos island in Greece: A wind turbine differs from a windmill in that the mechanical energy it produces is converted into electricity, rather than used directly.

power requires either the addition of backup energy sources—to make up for power shortfalls during periods of low wind—or effective power-storage technology. Storage options include large, nearly friction-free flywheels and pump-based storage systems. With the latter, for instance, part of the generated wind power is used to pump water up to a higher level, where it is stored in a tank or reservoir. In times of insufficient wind, the water is released to flow downhill, turning a turbine and generating electricity.

in focus

OFFSHORE WIND PARKS

During the past 15 years, some 18,000 wind turbines have been installed in Germany, most of them on the north German plains. Unfortunately, however, most of the country's favorable inland locations have been utilized. Further expansion potential is offered by offshore wind parks in the North and Baltic Seas. Because the wind is significantly stronger there and blows consistently almost all year, these offshore locations promise a substantially more efficient "wind harvest." Although questions of conservation and impact on wildlife remain problematic, these must be weighed against the potential environmental damage caused by coal-fired or nuclear power plants with similar levels of energy output. Plans for further development in the U.S. and Canada are already in the works.

Germany's Borkum West II is set to be one of the world's largest offshore wind parks when it is completed in 2009.

see also: Wind, storms, and anomalies, p. 117

water power and geothermal energy

As fossil fuel resources are used up and concern about their environmental effects increases, alternative forms of energy are gaining prominence: water and geothermal energy are two of the most important.

Energy in the natural world can take many forms—from heat deep within Earth to the kinetic energy of ocean waves, rivers, and wind. To convert this energy into usable forms, various processes are used to produce electricity, heat, or fuel. Some processes have proven cost-effective and are already in use. Others are still in the testing phase or exist only on paper. The search for renewable energy is driven by the need to reduce carbon dioxide emissions and to get independent of fossil fuels.

Bathers in Iceland enjoy the hot spring lakes in front of a thermal power plant. Virtually all of the country's electricity and heating comes from hydroelectric power and geothermal water reserves.

Water power

Most hydropower plants use generators to convert kinetic energy into electricity, such as low-pressure hydropower plants are installed in rivers. High-pressure power plants, located in dams, transform the potential energy of water in the reservoir into kinetic energy. Innovative new tidal power stations use the movement of seawater to produce an air current, which drives a propeller to produce electricity. Experimental osmotic devices create pressure using various combinations of freshwater and saltwater, and wave generators use the motion of waves to produce electricity by induction.

issues to solve

COMBINATION POWER PLANTS If 20 to 30 facilities are managed together, a constant minimum supply of electricity can be ensured, even when the wind does not blow, or tides undergo periodic cycles. This may be the solution to one major obstacle of renewable energy.

Geothermal energy

Geothermal technology makes use of the high temperatures found deep in the Earth. Hot water is brought up from a drilled borehole and then utilized directly for heating, or transferred to a heat pump to produce electricity. Through a second borehole, an equal amount of cold water is returned to the ground. A variation is the "hot dry rock" method, used in places with high temperatures and without natural water supplies. Water is first pumped into the ground to be heated before being returned to the surface. Geothermal power is considered to be one of the best potential renewable energy sources. At this stage, however, the technology is still in its early phases.

in focus

SEAFLOW TURBINES

A pilot program of seaflow turbines has been in operation off the British coast since 2002. Ocean currents from the tides cause huge rotor blades to turn, and this energy is then transferred to a generator. A current of some eight feet (2.5 m) per second is enough to produce 350 kilowatts of relatively constant electrical power—totally independent of the weather. The company responsible for the project envisions installations with a capacity of one megawatt each, consisting of twin-rotor towers. These facilities would be capable of supplying 40 percent of Great Britain's electricity needs.

Seaflow turbines (here raised out of the water for maintenance), have 33-foot-(10-m)-long blades.

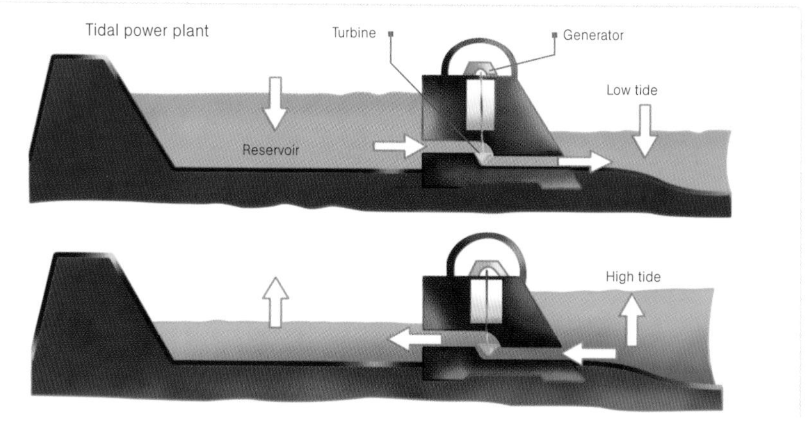

Barrage tidal power:as the tide rises, energy is generated through turbines in the sluice gates. The sluice gates are closed, then the basin is again emptied through the turbines after the tide falls.

see also: Ecologically friendly consumption, p. 137

fuel cells and biomass

Fuel cells are energy converters that obtain electrical energy from hydrogen or hydrogen-containing hydrocarbon. On the other hand, biomass is a regenerative energy carrier—fuel obtained from renewable resources.

At a Volkswagen plant, a low-temperature fuel cell is given the final check prior to start-up, in order to ensure safety and efficiency.

The fuel cell is the big sister of the alkali cell (domestic type of battery), that is, an electrochemical energy converter. The main difference between dry cell batteries and rechargeable (storage) batteries on one side and fuel cells on the other is as follows: While household batteries and large storage batteries store electrical energy when charged, which is given off again later on, fuel cells—like combustion engines—are supplied constantly with fuel. For that purpose, hydrogen or a hydrocarbon-containing gas such as methane, are used. In the first case, the waste gas is water, and in the second case, it is carbon dioxide. This process generates distinctly less waste heat than combustion engines, hence it is referred to as a "cold burn." Although this efficient technology is exacting—requiring extreme precision—and not yet sufficiently refined for all applications, its potential is impressive. The spectrum of possible applications extends from decentralized, small power plants to vehicle propulsion to micro-fuel cells for notebook computers and camping cookware equipment.

Biomass utilization

The very first energy carrier was wood. The use of fire can even be considered the starting point of human history as such. Especially since the start of industrialization, fossil fuels such as coal, crude oil, and natural gas have increasingly displaced wood. However, from an ecological point of view, wood has a distinct advantage: along with straw and other so-called biomass components, wood releases only as much climatically harmful carbon dioxide (CO_2) as plants have removed from the atmosphere during their growth period. Within this context, burning biomass is "CO_2-neutral." Meanwhile, in view of rising crude-oil prices, the cultivation of energy plants has become an interesting economic question. Fuel for wood-chip or pellet heating has a similar cost factor as heating oil, and in a few years the costs related to such wood fuels will presumably be distinctly less. Additionally, there are still problems with biomass burning due to a high micro-dust component. It is also feared that competition will arise between the cultivation of food products and fuel plants, which can presumably only be defused by intelligent and sound agricultural politics.

basics

BIOMASS During photosynthesis, plants (along with some bacteria) convert carbon dioxide (CO_2) and water (H_2O) into biomass (for example, hydrocarbons, $Cx(H_2O)y$) and gaseous oxygen (O_2). Here, sunlight serves as an energy source. During combustion (or digestion), this energy is released again. Therefore biomass is, in effect, stored solar energy.

in focus

ALTERNATIVE ENERGY PRODUCTION TECHNOLOGIES

The scarcer and more expensive fossil fuels become, and the faster global warming occurs, the more alternative energy technologies are going to be used. Wind generators, solar parks, and high-pressure hydropower plants are being further refined and deployed on a larger scale than ever before. The use of tidal currents to drive turbines has so far only been used in prototype form. Solar cells are also relatively widely used, yet the amount of solar power being utilized today is still far behind that of water or wind.

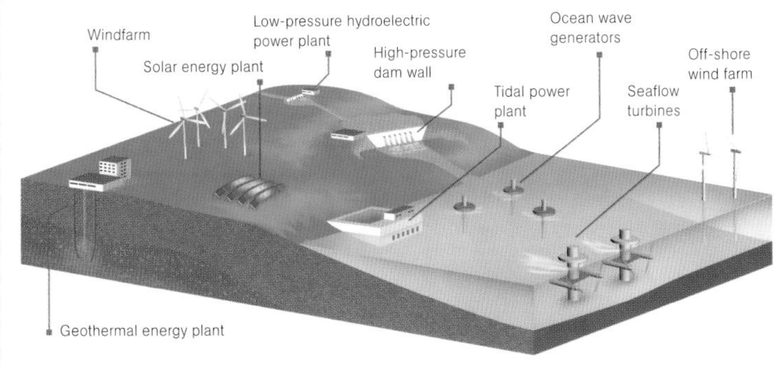

Newer alternative energy technologies, such as tidal turbines, are being further explored.

Oil palm seeds can be used as biomass. But mass cultivation can become a threat to the rain forest.

TRANSPORT TECHNOLOGY

In the beginning of the 19th century, the steam engine became a great success. It revolutionized travel by enabling speeds seven times faster than a horse-drawn carriage. This revolution in speed continues today. New challenges include improving vehicle safety, using alternative means of energy in automobile engineering, and integrating information and communication technology.

automobiles: motor and body

By making transport power and mobility available to the individual, the invention of the automobile has had a profound influence on personal and social life, especially since it became affordable to the middle class.

The motor is the heart of the automobile, and its function is to produce kinetic energy by conversion of thermal energy and either electric or chemical energy into movement. Higher oil prices and stricter emission standards, together with a rising customer demand have all inspired the global automobile industry to begin exploring alternative energy sources. These include biodiesel fuels (containing corn and soybean by-products), electric engines, and hybrid engines (utilizing both fuel and electricity).

Transmission and chassis

Between the motor and the steering wheel is the power transmission, which includes the clutch, manual transmission, power divider, drive shaft, and differential gear. These work in conjunction in order to relay, distribute, and regulate the energy produced by the engine. This energy is transmitted through the transaxle to the chassis, which is subsequently responsible for transmitting that energy to the roadway through the tires, suspension, brakes, and steering mechanisms.

The number of cars worldwide was estimated at 750 million in 2006: 1 car for every 9 people.

Body

The body, or exterior, of the automobile provides visual appeal, but more importantly it is responsible for the safety of its passengers. Automobile bodies made of a single piece of metal, called unitized bodies, are common. Recent improvements include a skeletal structure made of hollow sections, allowing the use of lighter materials while still retaining strength and rigidity. Aluminum, magnesium, and synthetic materials, in addition to steel, are now frequently used.

Automotive electronics

The electronic system of an automobile includes the ignition system, battery, starter, safety features, and security system. It also offers modern convenience through optional comfort systems such as heated seats and digital accessories.

in focus

THE PETROL MOTOR

In gasoline engines, the fuel is finely sprayed and mixed with air in the carburetor, then forced into the cylinder and compressed by the piston. A spark plug ignites the mixture, causing an explosion that drives the piston outward. In four-stroke motors, the piston injects the fuel and discharges spent gases. Petrol motors allow for quicker cycling, but are less efficient.

In a four-stroke engine, a piston completes four movements per cycle.

Assembly line in a Toyota factory: The production system developed by the Japanese firm in the 1950s influenced car producers around the world. In 2007, Toyota was the second most profitable firm after General Motors.

see also: Fuel cells and biomass, p. 353

automobiles: safety

Modern automobiles offer comfort, speed, and a visually appealing design. In addition to being a status symbol, however, they must also meet increasing road safety standards.

In recent years, the number of vehicles on the road has increased, while the number of accidents that result in serious injury or death has decreased. Multiple improvements in the safety systems of automobiles have contributed to this decline.

Appropriately sized children's seats and fitted seat belts give children the best chance of surviving the accelerating forces of an accident.

ELECTRONIC STABILITY CONTROL

practice

Electronic Stability Control (ESC) is an active safety system. Sensors constantly evaluate each wheel and a microcomputer evaluates the data from the sensors. In the case of a loss of traction (skidding), the ESC takes control. Each wheel is able to brake independently of the others until stable traction is established again. During normal braking, the ESC locks the wheels, but can carry out full braking automatically if there is an emergency.

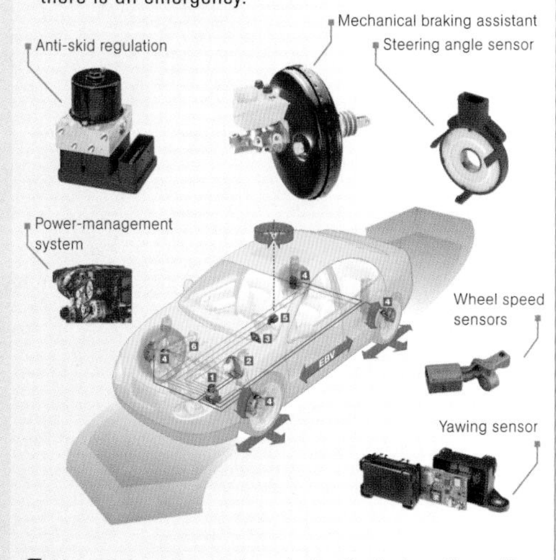

Manual steering controls are hydraulically strengthened by the ESC system.

Passive safety systems

Passive safety systems minimize the consequences of accidents for drivers and passengers. The most obvious of these systems are safety belts, headrests, and airbags. The "crunch zone" is another passive safety system that involves the point of impact (usually the bumpers) deforming softly and absorbing the energy of the collision—lessening the effect on the passengers inside the vehicle. Innovative materials, including shatter-resistant glass for the windshield and carbon fibers in the body structure, also help to reduce the risk of injury during an accident.

Active safety systems

Active safety systems, including braking aids, steering aids, and warning systems, are also used to help avoid accidents. Electronic systems, including vehicle tracking systems, are being combined with passive safety systems and other active safety systems to improve passenger safety.

Navigation and global positioning

There are many assisting systems that help the driver navigate the car. The electronic vehicle tracking system helps a driver when driving in reverse by issuing a warning signal when the vehicle is too close to an object behind it—some cars even include rearview cameras. Cruise control systems help a driver maintain a selected speed, which also improves fuel economy. More recently, global positioning systems (GPS) are becoming common in vehicles. These systems help the driver by displaying the vehicle's position, using satellite technology, and can give turn-by-turn directions to a destination selected by the user.

In an accident, the head is cushioned from impact by the airbag. Some luxury cars also have side impact airbags in the door.

basics

SEAT BELTS save lives. Since they were made compulsory in the 1970s, deaths and serious injuries from accidents are far less common.

CHILD SEATS protect children from injury and are mandatory in many countries. Baby carriers must be specially mounted and not used where there is an air bag.

The crunch zone effectively lengthens the braking distance. The impact of the collision is thereby lessened.

alternative propulsion systems

Gasoline and other combustible chemical fuels contain huge amounts of energy per pint or pound. However, the burning gases released into the air contribute to global warming, while dwindling oil reserves drive gasoline prices ever higher. The need for alternative fuels is urgent.

An increase in heavy traffic contributes to escalating environmental damage. Reduced-emission and emission-free vehicles, such as electric and solar cars, are being improved to combat the rise in air pollution.

One alternative to combustion engines is the electric motor. Contrary to popular belief, this is not a new idea; in fact, electric cars already existed in the 1830s, decades before the inventions of Gottlieb Daimler and Karl Benz. The first automobile to exceed 60 miles per hour (100 km/h) was propelled by electricity. However, because of its constant breakdowns, it was nicknamed *"la jamais contente"* (French for "she who is never satisfied"). Unfortunately due to technical limitations and high costs, electric vehicle technology was neglected in favor of the combustion engine. Electric motors, nevertheless, have considerable advantages in principle: they are highly efficient (wasting very little energy), lightweight, and they produce no exhaust during their operation. Furthermore, they can transfer power effectively to the drivetrain at almost any speed, and they can use the kinetic energy of braking to generate additional electricity.

On the other hand, electric cars also have significant drawbacks to consider. In terms of energy supply, batteries can only store relatively little energy per unit of weight, which explains why ordinary car batteries are so heavy. Even today a limited range of operation is the main argument against the purchase of a fully electric car.

in focus

HYBRID VEHICLES

There is a type of vehicle that avoids many difficulties facing its counterparts, since it uses at least two energy sources, with two energy storage systems built into the car. Hybrid vehicles use one-third less fuel than comparable models with traditional gas or diesel engines. They achieve this by using both a combustion engine and an electric motor. The two systems complement each other, allowing each to be used optimally. Moreover, like trains and streetcars, these vehicles generate additional electricity during the process of braking.

A hybrid drive combines the advantages of electrical motors and combustion engines.

Transportation is one of the leading sources of greenhouse gas emissions in the U.S.

Solarmobiles

It is not quite true that an electric car produces no emissions; the electricity used to power it must be produced somehow. If it comes from a coal-fired or natural-gas power plant, the pollution is merely shifted from the car's exhaust pipe to the plant's smokestacks. The situation is different for solar cars (or electric cars charged using solar power): these so-called solarmobiles are truly emission-free. In principle they can also get by without storing electricity in batteries. Unfortunately, however, today's solar cells are still unable to provide a sufficient flow of power to transport several people at a reasonable speed.

A Ford Think City electric car at the first U.S. alternative fuel station in San Diego.

new developments

With the advance of globalization, the automobile market has undergone rapid changes. New market participants have appeared, offering innovative concepts to satisfy growing demands, while the more established manufacturers struggle to meet new global challenges.

This century will see the rise of new markets and developments: China has long been known as a nation of bicyclists, and the first Chinese automobile factory was only constructed in 1953. The country mainly produced trucks until the 1990s, but an increasing number of cars have been built since that time, at first mainly in cooperation with Western firms. Today the Chinese automobile industry is expanding by 15 percent per year, that is, even faster than the overall economy in this booming country. By 2020 yearly production is expected to rise to at least 20 million vehicles—significantly more than (and in some cases double) the yearly production of most large automakers, including their foreign subsidiaries.

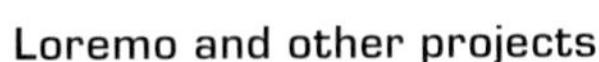

Loremo and other projects

The Loremo AG is not a multinational corporation but an innovative start-up from Germany. The name "Loremo" stands for "Low (air) Resistance Mobile." The sporty-looking prototype is said to get 78 miles to a gallon of diesel fuel (3 l/100 km); a variant with an electric motor is also planned. Although series production will not begin until 2009/2010, the first display models have attracted great attention in the automotive industry.

Even more visionary are vehicles such as the OneCat, a French project featuring a propulsion system based on pressurized air. Also called a gas-expansion motor, this system needs only compressed air as a fuel. A series of unsuccessful trials in recent years have made the OneCat's future appear questionable. However, significantly more progress has been made on propulsion systems using fuel cells, with the first prototypes already appearing in the 1990s (p. 353).

■ The Loremo: This light-weight vehicle is set to be one of the most environmentally friendly diesel-fueled vehicles developed.

milestones

THE WANKEL ENGINE

The propulsion system invented by Felix Wankel (1902–1988) does away with traditional pistons and cylinders. Instead, fuel is burned within the chambers of a housing containing a three-sided, oval-like rotor, causing it to turn. This rotation is transferred directly to the drivetrain. In practice, however, this impressive idea has not conquered the automobile market; because of its higher fuel consumption and some technical issues. Its success remains confined to various niche markets. Nontheless, the Wankel engine design is not yet obsolete: Mazda introduced a hydrogen-powered Wankel engine in 2006 that does not emit any carbon dioxide.

■ Instead of cylindrical pistons, a Wankel engine uses a housing containing three rotating chambers that pass through a cycle of fuel intake, combustion, and exhaust.

in focus

TATA MOTORS

India—the second country to reach one billion in population—also has an automobile industry, which is rapidly developing into a global player. Most notably, the Tata Motors company made headlines worldwide in early 2008. This subsidiary of the diversified Indian firm Tata Sons Ltd. announced the introduction of the world's least expensive production car, the "Nano," offering a four-door model for around $2,500. At the same time, the company presented a takeover bid for the traditional British brands Jaguar and Land Rover.

■ Built by Indian carmaker Tata Motors, the four-door Tata Nano, introduced at the New Delhi Auto Show in January 2008, features a two-cylinder engine mounted in the rear of the car.

see also: Fuel cells and biomass, p. 353

bicycles and motorbikes

Two-wheelers, whether powered by muscles or motors, can be used as a means of transportation, sports equipment, and even toys. Today, they are more popular than ever.

High-capacity street motorcycles achieve speeds of up to 186 mph (300 km/h). Due to their lower air resistance and friction, they are more efficient than four-wheel vehicles.

Centrifugal force

While a bicycle or motorcycle may seem easier to steer, the forces of physics affect two-wheel vehicles. Centripetal forces act perpendicular to any body that is changing its direction of motion. During a right turn, the rider experiences a force to the left and leans into the curve in order to balance the acting forces.

In the typical motorcycle gearbox, the gears are constantly engaged or meshed together.

Horsepower

The method of propulsion differs between bicycles and motorcycles. The bicycle rider applies pressure to the pedal, rotating the front sprocket wheel, which in turn moves the chain, transferring the circular motion to the rear sprocket wheel and the rear axle. In the case of the motorcycle, a two- or four-stroke combustion engine powers the circular motion of the front wheel, while a chain or wheel drive transmits the power through either a transmission or gearbox to the rear wheel and axle.

The gearbox

The gearbox controls the number of revolutions per minute in an engine by pairing up the large and small gears in different combinations. The gears are controlled by the clutch, which is usually made up of two friction disks. Once the clutch is released, the disks separate from each other, which consequently allows the gears to be shifted.

Innovations in two-wheelers

Innovations in two-wheelers include new frame materials, tougher body materials, upgraded braking systems, improved suspensions, and the addition of computer systems. These innovations are not only stylish, but they also improve security and comfort for the rider.

basics

GEARLESS HUB DYNAMOS used to generate electricity are highly efficient (70%), low maintenance, and reliable, even in snow.

HYDRAULIC RIM BRAKES have advantages including efficiency, ease of control, and endurance. They are also maintenance-free.

in focus

GYROSCOPIC EFFECT

From a physics standpoint, a wheel works like a toy top. The gyroscopic effect stabilizes the horizontal rotational axis once the wheels are moving fast enough. Once the wheel slows down, it can start to wobble, until the gyroscopic effect is no longer strong enough to create stabilization. While the gyroscopic effect contributes to stabilizing a two-wheeler, its effect is not as great as the centripetal forces that help drive a two-wheeler.

Technological innovations and new materials like shaped aluminum or carbon fiber help to reduce weight and improve the aerodynamic efficiency of the rider's position.

railways

Surface and underground railroads utilize more than 621,373 miles (one million km) of track worldwide. Trams, used for passenger traffic and transporting goods, are an alternate form of rail travel.

In California, San Francisco's famous trolley cars are powered by underground pull cables.

The TGV—Train à Grande Vitesse—reaches speeds of 357 mph (575 km/h) using two electro-motor engines.

Vehicles that ride on rails are more economical and can carry a greater payload than road vehicles. Their greatest disadvantage is that they are confined to the railway system and cannot reach some areas.

While there are various methods used to construct railways and tracks, they all follow the same basic pattern. Construction begins with a bed of crushed stone or concrete dampers in those sections where trains will operate at high speeds. Next, beams of wood or pre-stressed concrete are placed at regular intervals on the bed. The last step is to fasten the steel rails to the beams. The train cars are held to the rails by flanges on the wheels. The wheel-track connection is stabilized by the double frustrum formed by the joined right and left wheels.

International train service has been simplified by the extensive standardization of track width (gauge). Where different standards exist, as is the case near the border of Poland and Belarus, passengers and goods must change trains, or the entire car is transferred to a different chassis.

Propulsion

Trains are pulled or pushed along the tracks by locomotives. Locomotives can be powered in many ways, including diesel engines, steam engines, and electric power. The diesel locomotives fall into three categories: diesel only, diesel-electric, and diesel-hydraulic. Electric locomotives acquire electricity through a bus bar or overhead contact line.

Current supply

Electric cars can be powered by either alternating current or direct current. Underground railways and trams are the primary users of direct current. Although heavy transformers are needed, most large electric railways use alternating current. The heavy transformers are used to convert industrial network direct current to alternating current with a smaller frequency.

basics

THE LONGEST RAILWAY stretch in the world is the Trans-Siberian Railway at 5,771 miles (9,288 km).

THE HIGHEST RAILWAY runs from China to Tibet. In parts it is as high as 16,640 feet (5,072 m).

THE DEEPEST RAILWAY in the world is in Jordan, where one underground railway runs at 820 feet (250 m) below sea level.

MAGNETIC LEVITATION TRAIN

practice

As the name suggests, maglev, or magnetic levitation trains, are propelled by magnets: alternating fields moving ahead of the vehicle accelerate or slow it as needed. Since a cushion of air replaces the friction between track and wheels, significantly higher speeds are possible. Maglev trains can also climb hills more easily. One problem, however, is the current lack of infrastructure, which is much more expensive to lay than ordinary train lines. Finally, because of its intensive energy use, this technology is not feasible for large-scale freight transportation. Development of maglev trains by the Transrapid consortium began in 1969, but to date the only passenger track in the world is a short stretch in Shanghai.

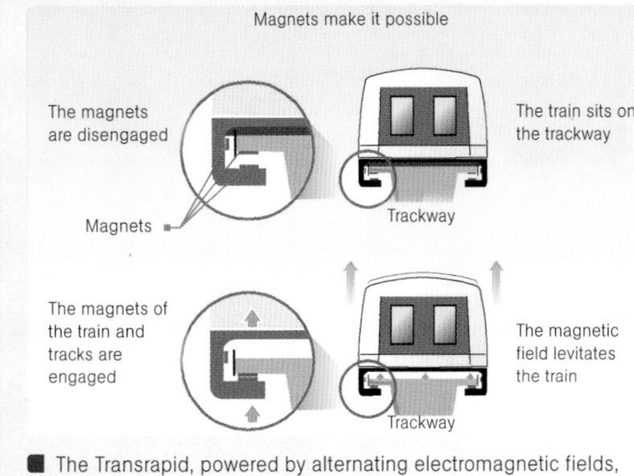

The Transrapid, powered by alternating electromagnetic fields, uses a magnetic suspension system to hover above the rails.

motor ships

Using ships to transport goods by sea is more popular than ever. Technical innovations to improve security and profitability are constantly in demand.

Globalization has led to an increase in the quantity of goods exchanged between continents. For large quantities and heavy individual items, shipping by sea is still the most economical method of transport. Popular in the first millennium B.C., sea transport has remained popular to this day and registered steady annual growth rates for many years.

basics

HOVERCRAFT traverse both land and water. They move by floating on a downward stream of air produced by fans, and thus appear to "hover."

DRIVE GENERATORS powered by water currents produce little pollution and are important in inland shipping.

Ship construction

As shipping companies demand faster, safer cargo and passenger ships, shipyards in Europe, North America, and Asia are eager to respond. Modern ship builders face the same basic problems as ship builders faced 2,000 years ago. The shape of the hull below the waterline governs the speed and stability of the ship. Container ships carry dry goods and manufactured products in truck-size containers. They are wide and relatively flat, making them very large and stable, but slow. Powered by 12- or 14-cylinder diesel motors capable of 10,000 horsepower each, they can reach speeds of around 25 knots (29 mph or 46.3 km/h). Built on the same principles, bulk carriers transport unpackaged cargo, such as coal and wheat, while tankers transport bulk liquids, such as petroleum and chlorine.

Speedboats, designed to maximize speed rather than load capacity, are built differently than cargo ships. Speedboat hulls are designed to transverse water with minimum water displacement. At high speeds, speedboats actually lift out of the water. Motorboats can be powered by an inboard motor, an outboard motor, or a hybrid of the two.

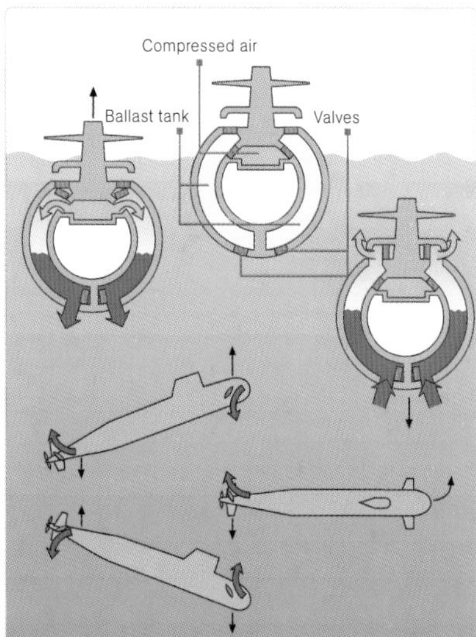

Submarines dive and rise by filling large tanks with seawater. To dive, the tanks are opened. Air escapes from the top, while water flows in from the bottom. Once the overall density of the submarine is equal to the water around it, it has neutral buoyancy and will remain at that depth.

Maximum demand: Speed, mobility, low draft, and high capacity are all desirable traits for a cruise liner. These traits can be maximized by a slim hull design.

Engines

At the core of every ship is the engine. Gas turbines, electrical motors, and diesel motors are all used to power ships. The power of the engine is transferred through the drive shaft directly to the propellers. A gas turbine engine, used in aircraft carriers and nuclear submarines, has a high power-to-weight ratio, allowing for fast acceleration, jet propulsion and long distances between refueling.

in focus

SAFER SHIPPING

Supertankers and huge container ships often carry dangerous cargo, and their fuel (marine diesel) is an environmental hazard. For this reason, modern ships are built with double-hulled construction. Specially trained pilots board the ship for difficult maneuvers and harbor docking. A fully loaded container ship has an extremely long braking distance and a huge turning radius. An ongoing problem is the tendency to cut costs by eliminating prudent safety measures, flying "flags of convenience," and hiring inexperienced crew from low-wage countries. The threat of piracy also remains acute today, especially in Southeast Asia.

The *Colombo Express*, the world's largest container ship, can load up to 8,750 containers.

see also: Water power and geothermal energy, p. 352

sailboats

Sailing ships, boats, and recreational vehicles use wind energy, currents, and muscular strength for propulsion and direction.

After the industrial revolution, sailing ships were replaced by steamers because they were less dependent on the weather, required smaller crews, and could get out to sea more quickly. However, sailing remains a popular recreation.

■ The *Gorch Fock*, a school ship of the German Navy, was build in 1958. More than 11,000 cadets have been trained on the barque.

Physics of sailing

Wind energy supplies the power to move a sailing craft. Wind currents can hit from directly behind or from the sides, splitting into two currents that flow around the sail. The air flowing over the windward side of the sail moves faster than the air flowing over the leeward side, creating more pressure on the leeward side. This is the force that moves the craft. The angle of the sail relative to the wind's direction determines the drift; if the angle is wrong, the sails will flutter and there is no movement. Using a process called tacking, sailing vessels can move in a direction opposite the wind.

Heeling and stability

Heeling occurs when air currents hit the sides of the sails, causing the ship to tilt along its longitudinal axis. This can lead to capsizing. A boat's ability to resist heeling is directly dependent on its stability. In keeled ships, a large weight is used to counteract heeling. The undersides of these ships make up 30 to 50 percent of the overall weight. When the angle of tilt is increased, the keel generates a torque opposite to the tilt and keeps the ship upright. Small sailboats do not have a keel. Instead, they rely on a centerboard and wide-type hull for stability. These boats can easily roll back to an upright position. A small boat's crew can also contribute to stability by moving over to one side of the boat to help balance it.

■ Surfing was known to the ancient Polynesian culture, but windsurfing did not emerge as a sport until the latter half of the 20th century.

Wave and muscular power

Windsurfing and kite surfing are two popular non-motorized water sports that use wind power. The board and rider are propelled forward by wind caught in the sails or kites. The surfer uses his body weight and muscular power to determine the direction of movement, to ride waves, or perform tricks like aerial jumps and somersaults.

HYDRODYNAMICS

in focus

During motion in the water, a boat's hull produces bow and wake waves. A ship cannot move faster than the bow and wake waves. The maximum speed a boat can achieve is called the speed of the hull; the shape of the hull determines this speed. Longer ships create longer waves, which disperse rapidly, allowing the boat to reach higher speeds.

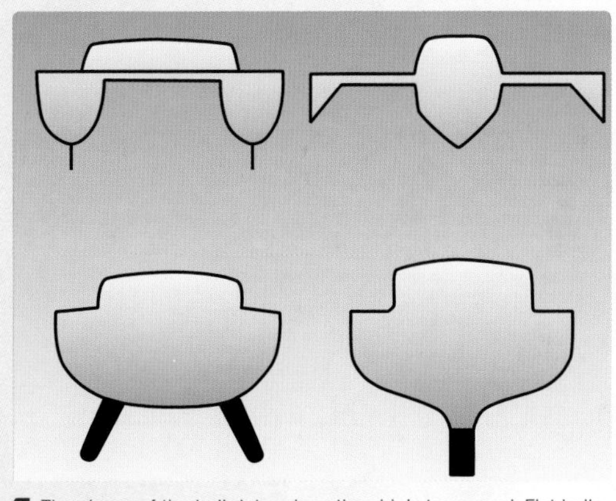

■ The shape of the hull determines the ship's top speed. Flat hulls can carry larger loads but make a ship slower.

Long, flat hulls with wide sterns produce lift forces. Caused by the dynamics of the heel and bow waves, these forces allow the boat to skim over the water surface, a process called planing. When planing, the boat can move faster than the hull speed because it is no longer being slowed down by wave friction. Instead, it is riding just on top of the surface, where there is far less friction.

airplanes: propulsion

The dream of flying finally became a reality with the first airplane built by Orville and Wilbur Wright in 1903. Since then, the significance of flight has increased rapidly due to advances in technology and the need to transport goods and travelers worldwide.

Since airplanes are heavier than air, forces that are greater than the weight of the plane must act on the aircraft for flight to become possible.

Liftoff

Four forces act on an airplane during flight: weight, lift, thrust, and drag. These forces counteract each other to provide optimal conditions for a plane to take off and fly. To overcome the weight of the plane, aircrafts generate lift. The airflow that is produced while a plane moves forward on the runway is split by its curved wings. The upper airflow travels over the top curved part of the wing, and the lower airflow passes under the flatter underside of the wing. Airflow velocity above the wing is greater than velocity on the lower side of the wing, causing pressure below the wing to be greater. The resulting lift is explained by Bernoulli's principle, which states that an increase in velocity occurs with a decrease in pressure. The magnitude of lift depends on the speed of airflow and the shape of the wing.

The force of air resistance opposing the motion of the plane is called drag. To overcome drag, the airplane produces thrust, or propulsion, from a force such as an engine. The plane requires an appropriate amount of thrust to counteract drag and remain in flight.

Piloting

In an action similar to steering, the plane is piloted with the help of a rudder, elevators, and ailerons. The pilot controls these devices through the control stick, column, wheel, or side stick, and pedals for the rudder. Movement around the longitudinal axis of the airplane, which is called rolling, is regulated by the ailerons. They are the flaps located on the farthest point of the wings and can be moved upward or downward independently of each other, causing the lift of the wings to change. The plane rolls toward the direction of the wing that has the aileron tilted upward. The pilot's control stick or column operates the ailerons.

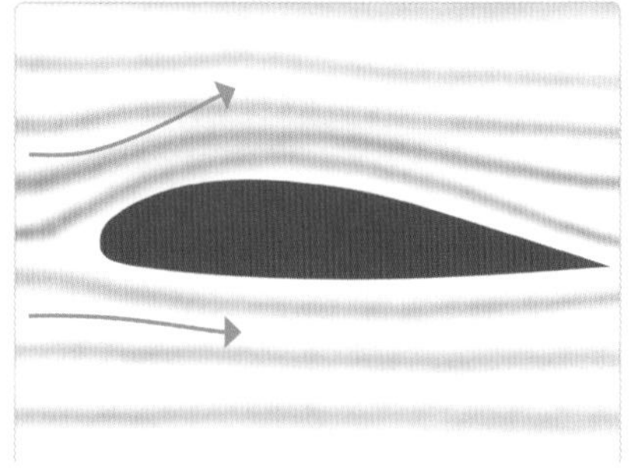

The airflow is split when it hits the wings. The upper airflow has a longer path to follow than the lower one. This creates a difference in pressure so that the airplane is lifted upward.

Airplane movement to the left and right around the vertical axis is controlled by the rudder. Similar to the structure of a boat, the airplane rudder is attached to the tail and lies perpendicular to the plane. It is used during the landing approach and balances out the undesired yaw, which is a side to side movement caused by the ailerons. The control pedals for the rudder also steer the wheels of the plane on the ground. Elevators are part of the horizontal tail of the airplane and rotate the airplane around its lateral axis; that is, elevators control the pitch of the plane. The control stick or wheel operates the elevators.

basics

AILERONS, RUDDERS, AND ELEVATORS enable a plane to rotate around its three axes.

AIR TRANSPORT will triple by the year 2015, according to estimates.

AIR TRAFFIC CONTROL (ATC) regulates aircraft flying throughout the majority of airspace.

practice

TURBOFANS

A turbofan functions according to the recoil principle. Air flowing into the fan is compressed in several stages and heated up to nearly 1112°F (600°C). The airstream is mixed with fuel in a combustion chamber and ignited. During the burn, the gases stream through a turbine. This generates the energy for the compressors. The gases are released through a nozzle, and the plane is pushed by the force of expulsion.

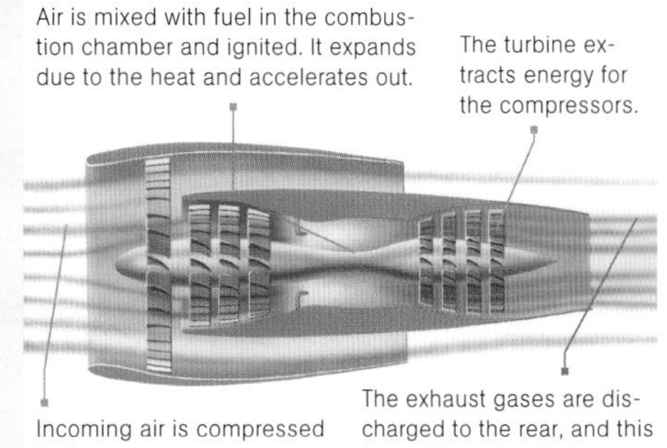

Turbofans are a kind of jet engine similar to a turbojet. They consist of a ducted fan that is powered by a smaller diameter turbojet engine built behind it.

see also: Human beings in space, p. 50

airplanes: speed

Military aircraft are highly specialized, and often pioneer new technologies before these are used for civilian purposes. Of particular importance are the speed and stealth abilities of these vehicles.

Naval aircraft carriers bring military aircraft to all parts of the world. Military aircraft have many purposes, including transport, reconnaissance, air warfare, and interception. Modern fighter aircraft may travel at supersonic speed—faster than the speed of sound in air. Supersonic speeds are measured in Mach numbers: the speed of the aircraft divided by the speed of sound in the air that the aircraft is moving in. Mach 1 is the speed of sound—about 745 miles per hour (1,200 km/h). Using afterburners and special high-performance jet engines, military jets can achieve speeds of around Mach 2; some reach Mach 3.

basics

BUSINESS FLIGHTS using supersonic vertical takeoff planes as air taxes are currently being developed.

As a plane moves through the air, it "pushes" the air in front of it, producing a front of compressed air. At speeds faster than the speed of sound in air, the sound waves cannot move past the front of compressed air created by the plane, and thus pressure builds. This pressure creates spherical shock waves at the tip and rear of the plane that spiral outward along the flight path. Although these waves are not themselves sound waves, when they reach the human ear, they are perceived as a single or double bang. Temperature variations, humidity, pollution, and winds can all have an effect on how a sonic boom is perceived on the ground.

The F-16 is a single-engined, multirole tactical aircraft with a thrust-to-weight ratio greater than one, thus it can climb and accelerate vertically.

milestones

THE CONCORDE

Civilian aviation experienced a rapid expansion after the Second World War, with larger and faster jets transporting ever more passengers. With the advent of the Concorde, even the sound barrier was no longer a limit. However, after a disastrous crash on July 25, 2000, and further declines in demand for the luxury service, operations of the Concorde ceased in 2003.

Air France and British Airways began offering supersonic transatlantic flights on the Concorde in 1976.

in focus

SPACESHIPONE

The first privately funded manned spaceflight took place in the summer of 2004. Motivated by a $10-million incentive from the Ansari X Prize (but incurring development costs of approximately $25 million), Scaled Composites built a suborbital space plane with a hybrid rocket engine. Boosted by a launch craft to a height of 50,000 feet (15 km)—it achieved an altitude of more than 328,000 feet (100 km) above the ground. The crew experienced some three-and-a-half minutes of weightlessness. SpaceShipOne is officially designated a glider, as most of its independent flight is actually unpowered.

SpaceShipOne achieved three space flights before being retired. It is now on display in the National Air and Space Museum in Washington D.C.

Camouflage

Military flights currently use acoustic and radar camouflage techniques. Stealth technology includes surfaces that deflect rather than reflect radar signals and the use of radar-absorbent materials, such as iron ball paints, which convert electromagnetic energy into heat. Active or optical camouflage is mostly still in the development stage. Active camouflage makes the camouflaged object appear not merely similar to its surroundings, but effectively invisible through the use of mimicry; secondly, active camouflage changes the appearance of the object as changes occur in the background. Strategies could involve coatings or panels capable of changing color or luminosity, or the use of organic light emitting diodes to create "invisibility cloaks."

see also: Ecologically friendly consumption, p. 137

light aircraft

Not every aircraft is as impressively huge as the A380 or Boeing's 787 Dreamliner. On the contrary, light aircrafts and gliders can be smaller than a car. They usually carry one or two people and often have no engine.

Gliders, paragliders, and hang gliders can fly for long periods of time without needing an onboard power source. Instead, they are designed to make maximum use of air currents. To gain height, gliders use thermals, which are rising columns of air that occur naturally. Thermals called slope winds form on the windward side of a mountain, and those known as lee waves form on the downwind side of a mountain. While people using hang gliders can often launch the glider from a high location, such as a cliff top, most other gliders require an external power source to get them off the ground and into the air. Two commonly used launching power sources are winches with stationary motors and powered airplanes, but some gliders do have their own engines. Gliders of all kinds seek an optimal relationship between the loss of altitude incurred and the distance that is covered. These aircraft should be light, so they can rise quickly in thermal columns. However, since the weight of a craft allows it to travel faster, they cannot be excessively light. Another design concern involves the ability to turn quickly, since this means that the aircraft can spend longer periods circling in a thermal. Furthermore, since gliders must sometimes land without an airfield (an "out landing"), they need to be easy to transport. Some gliders have water tanks in the wings that provide extra weight, allowing the pilot to adjust the craft's center of mass. The water, however, must be jettisoned prior to landing in order to reduce the stress on the frame.

Gliders, used for recreation and aerial photography, are lightweight and carry up to two pilots.

basics

FUEL SAVINGS of around 25 to 35 percent, longer ranges, and higher payloads are cause for flying wings to revolutionize aviation as an alternative to airplanes with tails.

INEXPERIENCE can lead to dangerous accidents. Also, a good general state of health is required, especially for hang gliders and paragliders.

see also: Atmosphere, pp. 112–113

in focus

HANG GLIDERS

A hang glider only consists of the utterly necessary parts of an airplane: wings and something for the pilot to hang on to. Today, a few navigation instruments, such as variometers or GPS devices, are also included. Pilots steer by shifting their weight. Because of this, pilots need physical strength and may suffer back problems.

Hang gliding is a breathtaking experience, but it requires a lot of training and can be quite strenuous.

Another category of especially small air vehicles is ultralight or microlight aircraft. In affluent countries, these now account for up to 20 percent of the civil aviation fleet. In the U.S., these small engine-powered machines for one or two people can be flown without any license (although good training is strongly advisable). They are limited in weight and speed and must not be flown at night or over populated areas. In Europe there are varying degrees of regulations, but some kind of license is always required.

Waterplanes do not require a landing strip. Many light aircraft are equipped for "out landings."

helicopters

Unlike any other aircraft, helicopters can hover, take off perpendicularly, fly slowly, and fly backward. Due to their flexibility, they have a wide variety of uses.

The versatility of the helicopter is due to the fact that helicopters do not have wings bolted to the aircraft. Thus, aerodynamics is a less important element of design than in airplanes, and helicopter construction can focus on optimizing load volumes, fixed weight, and stability.

Aerodynamics

Helicopter aerodynamics differ from that of a fixed-wing aircraft. Like an airplane, the helicopter needs to exert lift and thrust to take off, but the mechanics are different. Most helicopters are propelled by turbines that power the shaft on which the rotor blades are fixed. The blades spin on the top of the helicopter, and air flowing over the blade has a higher speed than air flowing underneath, thereby creating a difference in pressure and causing lift. The tail rotor, spinning sideways, provides thrust to move the helicopter forward, but also provides balance to the rotational force of the main rotor to prevent the fuselage from spinning in the opposite direction. A freewheel between the turbine and gears also greatly reduces the spinning velocity transfer from the rotor to the turbine. In some cases, two main rotors on top of the craft spin in opposite directions to balance the rotational forces and eliminate any need for a tail rotor.

Instead of wings, a helicopter has two different rotor systems: the top rotor lifts, and the tail rotor provides thrust and keeps the craft balanced.

Due to their special abilities, like taking off and landing vertically and hovering for longer periods, helicopters are often used for rescue.

Control

Helicopters change direction on three axes through rolling (tilting sideways), yawing (turning left or right), and pitching (tilting forward and back). The controls include a collective pitch control lever, a cyclic pitch control lever, and pedals for the tail rotor. In order to control ascent and descent, the pilot uses the collective pitch control lever to change the angle of the rotor blades. The steeper the angle, the greater the amount of lift. To control pitch, the pilot adjusts the angle of the main rotor blades at different points during the rotation. Pressing the cyclic pitch control lever forward causes the angle of the rear rotor blade to increase, which causes the helicopter to move forward. To control roll, the pilot steepens the angle of the right or the left rotor blade at the proper point of the rotation. Anti-torque pedals, corresponding to the rudder pedals on a fixed-wing aircraft, control the tail rotor, which manages helicopter movement along its normal axis (yaw). In the case of helicopters with two main rotors and no tail rotor, yaw control is possible through a cyclical blade adjustment of both the rotors.

SWASHPLATES

The central elements of steering control in helicopters are swashplates. Attached to the cyclic and collective control levers, they translate the pilot's steering commands into movement by controlling the main rotor blades. Two swashplates are located in the rotor shaft under the rotor head, and each has a different function: the upper swashplate rotates with the rotor, and the lower swashplate is tilted left, right, forward, or backward for cyclical blade control. The swashplates shift upward or downward along the rotor axis for collective blade control. The angle of attack is steered by the swashplates, and the control and driving rods.

Rotor blade
Variation of the angle of attack
Driving rod
Swashplates
Balance joints
Control rod

Rotor head with swashplates.

practice

see also: Wind energy, p. 351

rockets

A rocket engine functions according to the reaction principle, just as the jet engine of an aircraft does. However, rockets require far greater thrust than jet aircrafts and must also be able to function in airless outer space.

Rockets work according to the reaction principle, whereby the propulsion unit ejects a mass—for example, in the form of combustion gas—and rocket and expulsion mass repel each other, pushing the rocket forward. This force is referred to as thrust, which increases with exhaust speed and fuel throughput: that is the fuel mass used per second. A substantial difference, however, is that an aircraft engine uses atmospheric oxygen to burn fuel, while a rocket must carry oxygen with it for the chemical propellant to function in airless outer space. There are ways to create thrust. Chemical propellant engines use solid or liquid fuels, together with an oxidant such as liquid oxygen or an oxygen-containing chemical compound. In starting liquid-fuel rockets, fuel and oxidant are pumped into the combustion chamber and ignited. The exhaust gases generated escape at high speeds through the jet. Liquid hydrogen, hydrocarbons, or hydrazine, among others, are used as rocket fuel. With solid-fuel rockets, fuel and oxidant are mixed as solids in the combustion chamber. Hybrid propellants usually combine a solid fuel with a liquid oxidant.

THE STAGES PRINCIPLE

To propel a significant payload, such as a satellite, into Earth orbit, a process of staging must be employed. In using this principle, each rocket unit (stage) is built with its own engine and fuel tanks. The stages are cast off once their fuel has been used up, at which point the next stage is ignited. This has the effect of reducing the weight of the rocket, meaning that it gets faster acceleration from the thrust. The stage engines can be arranged one on top of another or side by side.

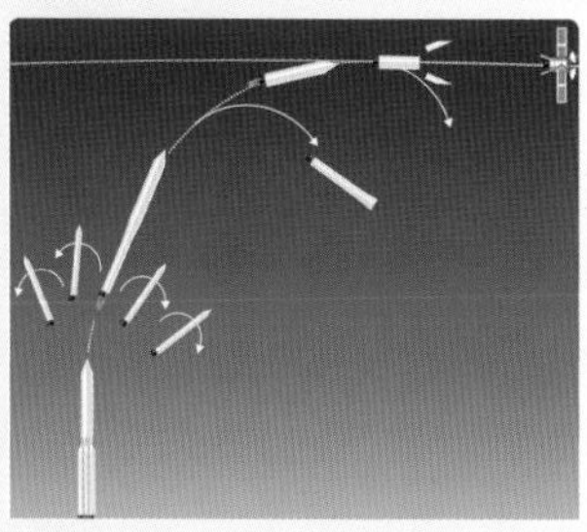

Multistage rockets can be used to increase efficiency and acceleration.

in focus

Cooling procedure

The combustion of rocket fuel creates extreme heat; therefore, the combustion chamber and the jet must be cooled. In large engines, liquid fuel is pumped through tubes alongside the combustion chamber and jet as coolant. Another method is to apply an ablative coating of material that conducts heat poorly and is gradually worn away. In solid-fuel propulsion units, the mixing ratio of fuel to oxidant along the periphery is specifically adjusted so that the burning temperature is low in that area.

Electric propulsion

Electric rocket engines use an ionizing gas or a metal that can readily be vaporized. This is accelerated by electric or magnetic forces and subsequently expelled. Since this type of engine does not produce sufficient thrust, it cannot be used for launches from Earth. Instead, electric engines are sometimes used for position adjustments of satellites or for operating space probes.

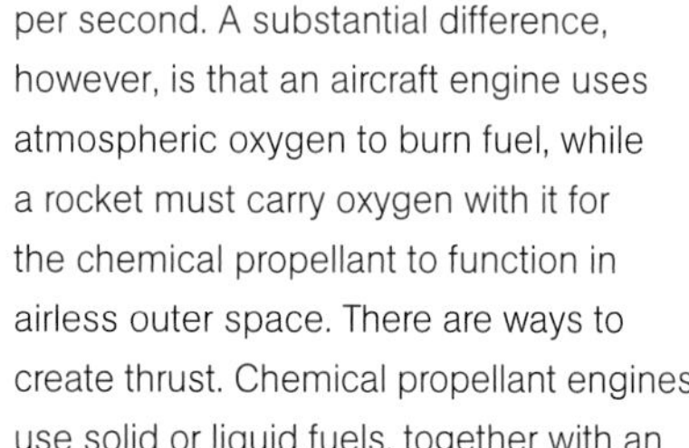

A Proton-M booster rocket being moved to its launch pad.

The ion thruster engine SMART 1 was used on a probe launched by the European Space Agency (ESA).

Solid fuel rocket boosters are used in staging: They detach from the launch vehicle once their propellent has been used up.

to outer space and back again

Carrying a space vehicle into Earth's orbit requires a vast amount of energy. On its return, the kinetic energy of the space vehicle is reduced through friction with the air in the Earth's atmosphere.

Satellites and astronauts are transported into space by means of a launch vehicle, but it is not enough just to carry the vehicle to the altitude of its intended orbit. Once there, it must be accelerated to the required orbital speed. For instance, for a low orbit at an altitude of 186 miles (300 km) this needs to be almost five miles per second (eight km/s).

The NASA space shuttle Columbia just after a launch: The space shuttle made many successful trips into space, but was destroyed on reentry into the Earth's atmosphere in 2003 on its 28th flight.

Examples of orbital maneuvers

If a space vehicle is to move from a lower to a higher orbit, its engine needs to "burn" for a specific period. A higher speed extends the orbit into an ellipse. At the highest point of the elliptical trajectory, the engine is started up again. The thrust must accelerate the space vehicle to exactly the speed required for the new orbit.

During a rendezvous maneuver, space vehicles are guided toward each other along their orbits. If need be, they can then be coupled mechanically. This facilitates servicing satellites in space and is also used to transport equipment and astronauts to the international space station.

During such a rendezvous, it is common for one space vehicle to be passive, while the other space vehicle approaches it. The movements of the pursuing vehicle must be closely controlled so that positions, speeds, alignments, and rotations of the vehicles correspond with each other. This complicated maneuver is monitored via radar by ground stations and the pursuing vehicle.

Atmosphere reentry

The vehicle's return to Earth is initiated in orbit by igniting the deceleration engine. Entering into the Earth's atmosphere is a critical maneuver because the spacecraft is moving at such great speed. Friction with the denser air in the atmosphere results in the extreme heating of the craft, while also braking the speed.

The heating and braking of the spacecraft are dependent on both the speed and the angle of reentry. Braking action can be severe for a manned capsule. This can be mitigated, however, by shaping the capsule to create a slight buoyancy effect. Although part of the frictional heat is dissipated by the air current, a heat shield is essential if the space vehicle is not to burn up on reentry.

So-called ablative cooling uses materials, which will melt off or vaporize and so remove heat in the process. For instance, the charcoaled ablation shield of the Apollo capsules maintained inside temperatures below 80.6°F (27°C), while outside temperature were in excess of 3632°F (2000°C). Another possibility is protective heat tiles, such as those that have been used on the space shuttle.

Finally, space capsules land by means of parachutes. On some capsules, ground impact is reduced by firing additional braking rockets when the capsule is close to the ground. The space shuttle has the advantage that it can land like an aircraft. The disadvantage is its greater starting weight and it is greater technical complexity.

The Canadarm is a robotic arm used on NASA space shuttles for maneuvering large payloads in orbit.

basics

IF THE STARTING location is close to the Equator, the rocket is already moving at the rotation speed of the Earth, and thus requires less energy for acceleration.

THE FUEL needed to overcome the gravitational force of the Earth and the resistance of the air can often make up about 90 percent of the starting weight of a rocket.

The Atmospheric Reentry Demonstrator (ARD) was tested in 1998 to analyze reentry technologies.

see also: Vectors, p. 413

building construction

Over centuries, master builders have passed on their experience with regional building materials and craft skills. They have left behind a diverse cultural heritage. More recently, architecture has become increasingly characterized by industrial materials, high tech, and formal standardization.

Concrete structures allow almost any kind of shape. Shell structures like the roofs of the opera house in Sydney are reinforced with steel rod mats and can span wide spaces without further support.

construction | skyscrapers | sustainability | energy-efficiency | infrastructure projects

CONSTRUCTION

The fundamental physical principle in structural engineering is the balance of forces. Building construction refers to architecture of any type, size, and shape. The spectrum ranges from private structural conversions or additions to the new construction of entire cities. Civil engineers, however, are involved with ground level or subterranean structures, which are mostly infrastructure projects such as roads, tunnels, and canals.

An adequately strong foundation is one of the most important prerequisites for a building project. Depending on the substrate's condition and the weight of the building, a load-bearing construction can rest either on a continuous bedplate or on individual point and strip foundations that extend down to frost-free level. Since the beginning of structural engineering, two fundamental construction principles can be distinguished: massive construction and skeleton construction. Archaic clay huts and pile dwellings conform to these principles, as well as complex systems of the modern building industry. Massive construction usually consists of a load-bearing structure of disk or cup-shaped elements. These can be of cemented bricks, of concrete poured on-site, or they can be assembled from prefabricated concrete components. Reinforced concrete is more loadable than brickwork, because of the embedded steel reinforcement. This method also offers more scope in terms of support spans and structural shapes. Lattice structures, made of rod-shaped steel, reinforced concrete, or timber, are characteristic of skeleton construction. The fields between pillars and beams can be filled in with wall or ceiling slabs. These take on a stiffening function, preventing the distortion of the carrying structure under its own load.

Completion works

After the building shell's completion, roof and windows are weatherproofed. Heating, plumbing, and electrics are installed. At the same time, stairs, wall partitioning, floors, doors, internal wall surfaces, and built-in furniture are finished. In many cases, security, insulation, and sound- and fireproofing need to be taken into consideration.

in focus

STATICS

Each building forms a static assembly of many different building components. These must be joined to each other tightly, but they also need to be sufficiently flexible to react to particular stresses. All components are dimensioned so that they are jointly able to absorb any pressure, tensile, and bending force that will arise, without deforming. Here the structural engineer must take into account traffic loads, the dead weight of the building, external pressure created by wind, water, snow, or ice, as well as special structural stresses caused by vehicle impact, earthquakes, or avalanches.

All loads acting upon a structure are dissipated into the ground via load-carrying structural components.

The lattice truss structure of Colossos in Germany is one of the world's largest wooden roller coasters. Wood is more elastic than steel and thus allows a very intense riding experience.

see also: Cities, p. 281

skyscrapers

Not only the public's fascination with great height has made skyscrapers popular—their design is also economical. However, since height comes with its own set of dangers, safety is an ongoing concern.

THE CHICAGO SCHOOL

milestones

Two structural engineering innovations—the fire-resistant steel skeleton and the electric elevator—made it possible to build the first tall office building in Chicago in 1885, the 138-foot (42 m) Home Insurance Building. The Chicago school, a loose grouping of architects and engineers, went on to develope the commercial skyscraper as a completely new type of building design.

Louis H. Sullivan (1856–1924) is considered the creator of the modern skyscraper.

The steel frame of a skyscraper is designed with economics in mind. The building's geometry and floor plan are designed to maximize the usable floor space, while minimizing construction costs. In the case of extremely tall buildings, horizontal wind forces pose a special challenge. As a building's height increases, the force of the wind increases exponentially. These forces must be precisely calculated to ensure sufficient strength and stability.

basics

SKYSCRAPERS have a system consisting of multiple coordinated elevators.

ELEVATORS in high-rises use geared traction machines with an electric motor to lift the cabins.

SAFETY SYSTEMS make elevators the safest means of mass transportation.

Steel frame construction

Skyscrapers are mostly built on a skeletal structure made of steel. A supporting framework consists of vertical columns and horizontal girder beams. The steel frame as a general rule consists of vertical, horizontal, and cross-sectional steel girders, to which precast reinforced concrete units and external wall panels are added.

Reinforcing systems

Vertical supports come in several types: steel frames, tube designs, concrete cores, and megastructures. In steel framed skyscrapers, vertical columns with immovable wall units are placed in the central portion of the building. In the case of tube supports, the building envelope is created as a tubular cross section. These are considered to be the safest skyscraper designs.

Skyscraper safety

Skyscrapers also need to withstand the challenges posed by natural catastrophes and fire. Elevators and emergency exits are constructed and operated in fire-resistant reinforced shells. There are air spaces between the floors to prevent the spread of fire. In earthquake-prone regions, skyscrapers are constructed on an elastic foundation that vibrates with the Earth without breaking. A new system detects oscillations and uses hydraulic cylinders controlled by a computer to adjust the position of the building.

A Chinese construction worker goes about his work on a skyscraper under construction in the mega-city of Wuhan, central China's most populous city.

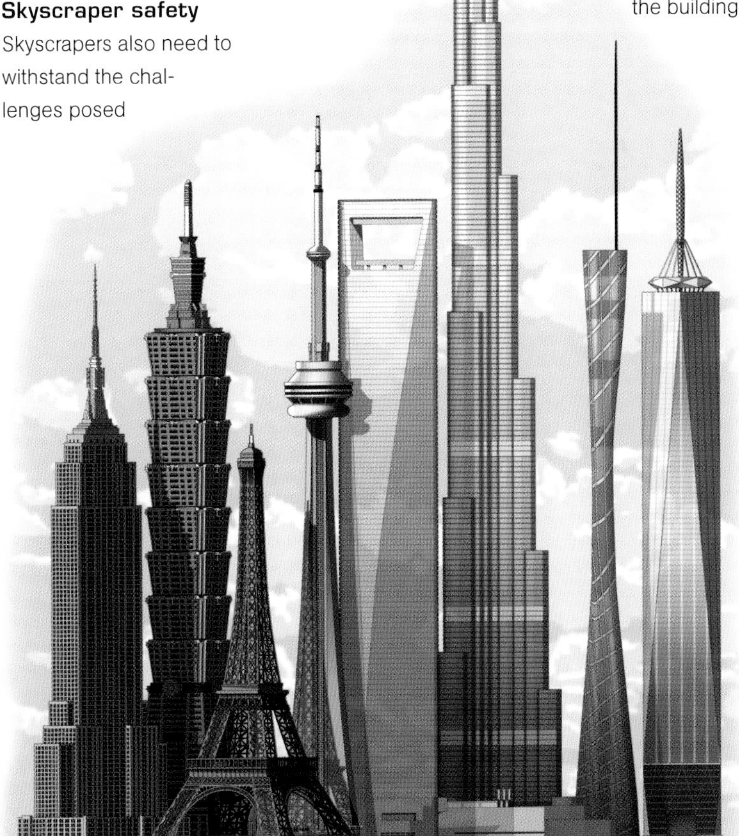

Reaching for the sky (from left to right): Empire State Building, Taipei 101, Eiffel Tower, CN Tower, Shanghai World Financial Center, Burj Dubai, Guangzhou Television and Sightseeing Tower, and Freedom Tower.

environmentally sustainable construction

Since the last decades, recognition of the limitations of the Earth's fossil resources has led to a new awareness among those who plan and implement both private and public construction projects.

The UK's largest eco-village in Beddington, London, allows for a lifestyle with no carbon emissions.

The efficient and environmentally sustainable use of land and locally available building materials has long been a necessitiy, especially in densely settled or resource-poor areas. These practices also helped safeguard the long-term supply of natural resources. However, after the far-reaching changes brought about by the industrial revolution, and in the face of economic crises such as that of the 1930s, the need became evident for new concepts for housing and construction, such as those developed by American architect Frank Lloyd Wright (1867–1959). With his economical, flat-roofed homes and his vision of including the natural world in his architectural creations—as with the house Fallingwater, built into a cliff above a waterfall—Wright became a forerunner of today's movement that demonstrates that environmentally conscious construction can also be aesthetically appealing.

The grass-covered roof of this "eco-villa" in Switzerland provides excellent insulation, while the pond serves as a natural habitat for local flora and fauna.

WOOD is a renewable resource. Trees bind CO_2 as they grow, removing it from the environment.

CLAY helps maintain a comfortable indoor environment. As building material, it is gaining renewed popularity.

STRAW can be used as insulating material in wood-frame buildings. Straw roofs provide excellent rain protection.

basics

New needs, new concepts

Today ecologically sustainable construction means, above all, considering both the present and future needs of human beings and the environment. Thus it seeks to avoid destroying natural habitats or imposing burdens on future generations. To accomplish this, builders must not only consider factors related to the construction site itself—such as the use of locally available, renewable building materials—but they must also ensure that the extraction, manufacturing, and transportation processes involved are environmentally sustainable, as for instance, minimizing emissions, waste generation, energy use, and water consumption.

Those involved in large construction projects, such as property owners and developers, regional and local governments, construction companies, architects, and engineers, must work together to ensure environmental sustainability. This is no less important for private households, which also constitute a significant part of the problem. In the United States, for example, approximately 30 percent of carbon monoxide emissions stem from environmentally unfriendly home airconditioning, heating, lighting, water, sewer, and waste disposal systems.

NATURAL CLIMATE CONTROL

in focus

A traditional element of Persian architecture still in use today is the windcatcher. This natural "air conditioning" system is both economical and sustainable. The height of the tower, which extends down into the building's lowest levels, is important because it takes advantage of the chimney effect. During the night, air cools the interior walls, which have stored daytime heat. The warmed air flows upward within the tower and escapes. The walls of the building also radiate heat during the night. Air channels that can be individually opened and closed optimize these processes.

Taking advantage of natural thermal currents, windcatchers provide effective cooling while ensuring sufficient ventilation.

see also: Ecologically friendly consumption, p. 137

energy-efficient construction

The oil crisis of 1973–1974 exposed the delicate balance between the rising demand for energy and the Earth's limited resources. Today energy-efficient construction is an essential, expanding area of building technology.

Land values can be a decisive factor in construction plans: high square-feet costs often lead developers to build taller structures for optimal economic returns. They then seek to balance above-average purchase, planning, and construction costs with reduced operational costs, although the building's occupants expect a comfortable indoor environment throughout the building at all times. The result is a constant artificial climate within the building, with correspondingly high levels of energy use. The main objective of energy-efficient construction is to reduce these costs. This can be achieved in two ways: firstly, by taking advantage of cost-free energy sources whenever possible, and secondly, by using energy more efficiently or recapturing it. For this purpose, the building must be designed with measures to reduce its energy budget, including the use of alternative energy sources. Another important consideration is minimizing energy loss: for instance, by optimizing insulation, using glass surfaces effectively, and installing efficient climate control, electrical, and water systems.

■ The rooftops of these houses are used to produce energy. Each roof surface is covered with photovoltaic modules.

Modern buildings often include automatic control systems for ventilation and sun protection, which provide a centrally managed, uniform climate within the whole building. For increased energy efficiency, a decentralized control system can also be added. Increasing numbers of buildings are incorporating high-tech materials and technological systems such as "smart" two-layer facades that react automatically to temperature and weather changes. Just as important, however, are in-house energy production systems, such as solar collectors, photovoltaic panels, geothermal heaters, and waterrecycling mechanisms. Measures such as these can be implemented both in new construction and in the renovation of older buildings, making them more energy efficient. Here, the ideal is the "energy-plus house," which operates so efficiently that it produces more electricity than it uses. The excess can then be diverted into the local electrical system.

Over the long term, however, a number of problems still remain to be solved before energy-efficient systems for homes and buildings become common place: for instance, the high cost involved in manufacturing, managing, and repairing such technologically complex systems.

HEAT PUMPS

practice

Heat pumps are capable of extracting heat from an area with a relatively low temperature, such as the ground, and moving it to an area of higher temperature, with the input of mechanical work: for instance, from a pump or compressor. This heat can then be used to warm a building, for example. Gases or liquids are used to help extract the heat. Compared to direct electrical heaters, heat pumps consume less electricity.

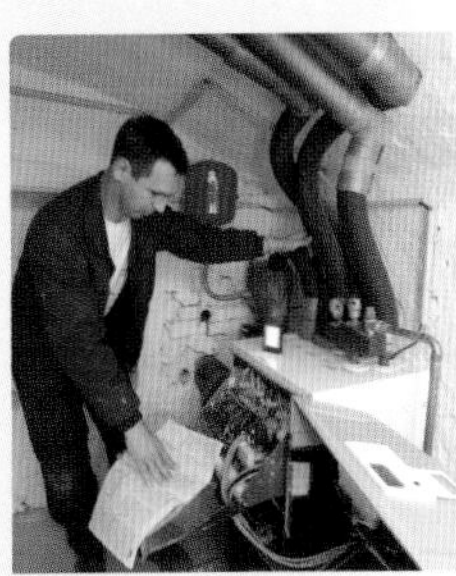

■ A heat pump in a single-family home, with control module, hot water, and buffer storage systems.

basics

CONVENTIONAL HOUSES require 25,360–95,100 Btu per square foot (80–300 kWh/m^2) for heating each year.

LOW-ENERGY HOUSES use less than 25,360 Btu per square feet (80 kWh/m^2).

PASSIVE-ENERGY HOUSES need less than 4,750 Btu per square feet (15 kWh/m^2).

THREE-LITER HOUSES require only 3 liters (about 3 quarts) of oil or natural gas per year per square yard.

ENERGY-PLUS HOUSES produce more energy than they use. The surplus can be sold to the local energy utility.

■ Like heliotropic plants, the apartments within this visionary eco-skyscraper can follow the sun or turn away from it.

roads and tunnels

Civil engineering includes the construction of roads, tunnels, canals, dams, and bridges. A deciding factor in design is the stability of the geological subsoil in the area.

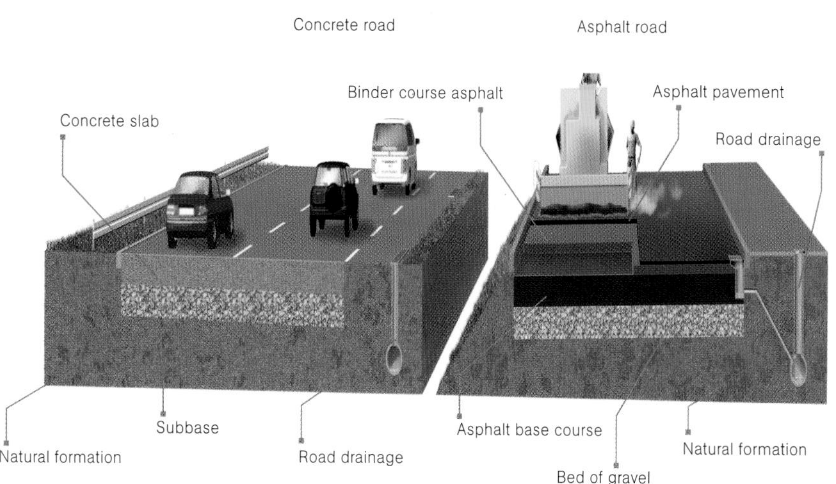

■ Roads must be sturdy enough to sustain consistent traffic flow. They are made of a superstructure, or top cover, to protect them from climatic damage, and a substructure, which often consists of compacted soil.

Roads have to be strong enough to withstand the load of heavy traffic for several decades. Roadways are subjected to compressing and shearing forces, as well as forces exerted by the turbulence of passing traffic. Additional forces are exerted on a roadway along curves, in high-speed sections, and in sections subjected to the sudden, jolting impact of braking vehicles. In addition to the stress of traffic, forces of nature, such as extreme temperature fluctuations, also trigger damage to roadways.

■ Road surfaces are under the most stress at traffic lights due to braking and accelerating.

Structure of roads

Roads are made of a superstructure and a substructure—both of which are designed with a view toward local climatic conditions. The superstructure consists of a surface course or top cover to protect the road construction and a lower base course to distribute the load. If required, a protection against frost is also used. The substructure includes an 8-to-32-inch (20-to-80-cm) underlying layer and an embankment, often consisting of compacted soil. The substructure can extend up to 6.6 feet (two m) below the roadway.

Slope of the road

All roads have a transverse slope to allow surface water to drain off. Where the road curves, this slope is increased to counteract the centrifugal force of the vehicles. A road's longitudinal (front-to-back) slope is determined by the structure of the surrounding terrain and must be designed to accommodate the driving speed of vehicles—the faster the traffic on the road, the flatter the road needs to be.

Tunnels

Tunnels provide shorter routes through uneven terrain, but careful geological analysis of the surrounding area is required before their construction. No matter how safe these structures may be, traffic accidents can never be entirely avoided. Sometimes an accident can escalate into a full-blown disaster, such as in the Mont Blanc tunnel in 1999, and at other times in the Tauern and Gotthard tunnels. Tunnel construction begins by loosening the rock mass with hammers, drills, or dynamite. The loosened rock is then removed with tunnel-boring machines. Modern tunnel construction uses concrete, steel arches, and other construction materials to prevent the collapse of any cleared cavity spaces. In addition, ventilation, water flow, and fire precautions must also be taken into account.

basics

ASPHALT is the carefully refined residue from the distillation process of selected crude oils. The primary use of asphalt is in road construction, where it is used as the binder for the aggregate particles.

WHISPER ASPHALT Whisper asphalt has over 20% air voids and a very smooth riding surface.

in focus

THE GOTTHARD BASE TUNNEL

When it opens in 2015, the Swiss-Italian railway tunnel will be the longest in the world. Its 95-mile (153-km) course was measured out assisted by GPS and laser. Many factors had to be considered in the structural planning, including varying rock layers, water pressure, and the shifting and growth of the Alps. The tunneling machines are often up to 1,300 feet (400 m) long.

■ The 30-foot (9-m)-wide tunneling shield "Sissy" spins as it drills through rock mass.

see also: Transport technology, pp. 354–359

bridges and dams

Bridges, dikes, and dams are among the most demanding challenges of structural engineering. They are often subject to extreme and variable forces from tides, winds, or moving masses of snow and earth.

Bridges are built to span rivers, valleys, and crossroads. All possess a horizontal span and a means of vertical support, such as vertical piers, arches, or towers called pylons. The design of a bridge must take into account the form of the surrounding terrain, the depth and solidity of the bedrock or soil for supporting piers or pylons, the load of traffic that the bridge will bear, climate conditions, and forces such as wind and water currents. Other factors include the bridge's construction costs and long-term maintenance costs, its overall appearance, and locally available building materials.

Arched bridges can be erected with steel as the main supporting material, but for massive bridges a pressure-resistant material such as concrete must be used. The arch can withstand tremendous pressure, and, as a structural component for building bridges, allows pressure to be distributed across its curve. Arched bridges made of steel arches can have a span of 1,640 feet (500 m).

Suspension bridges are constructed to span more than 2,620 feet (800 m). These have two high pylons with two horizontal steel supporting cables attached to them. Underneath, a concrete roadway is attached with vertical cables called hangers. The entire load and stress is transferred to the supporting cables via the hangers and directed to the pylons as a vertical force. Suspension bridges are susceptible to damage from vibrations triggered by wind.

Dams, dikes, and retaining walls

Canals and dams control and block the flow of rivers. Reservoir dams provide drinking water or power for industrial areas and protect against floods. Dikes and retaining walls are built mainly to protect against floods or landslides. Most artificial dams are embankment dams, made of compacted earth. Embankment dams use their weight to hold back the force of the water. They can be covered with concrete and filled with earth or rock. Retaining walls in hilly areas hold back water, earth, rocks, or snow. In Peru, the Inca built terraces on otherwise unusable slopes. They also built canals and aqueducts to irrigate the soil for agriculture. When a large dam is built, thousands of people may need to relocate from areas that will be flooded. Because of this, the construction of dams, canals, and dikes often has great environmental impact on a region.

in focus

MOVABLE BRIDGES

When ships and roadways cross, a movable bridge is often the most functional and economic solution, particularly in harbor cities visited by large ocean liners. Depending on the design of the bridge, the roadway, in parts or as one, may be raised at an angle, hoisted whole, or swung to the sides. Road traffic must wait until the ship has passed.

The central roadway of the Tower Bridge in London can be lifted almost vertically by electrical pumps.

issues to solve

HUGE DAMS like the Three Gorges Project in China or the High Aswan Dam in Egypt often have great environmental impact. When a large dam is built in a highly populated area, natural and cultural heritage may be lost.

The mile-long (1.6-km) Golden Gate Bridge stretches across San Francisco Bay. Only suspension bridges can be built to span such great lengths.

The Glen Canyon Dam on the Colorado River in Arizona: It provides energy and flood protection and is a source of drinking water.

see also: Civil engineering, p. 368

production | optimization | engineering | assembly lines

MANUFACTURING TECHNOLOGY

In this era of globalization, rarely is a product completely manufactured in a single location. Information networks are intricately tied with international production systems. Many simple components are manufactured in multiple locations around the world and then assembled in another location to make a final product, such as a computer, an automobile, or an airplane. Logistics, conservation of resources, and protection of the environment are all vital to the production process.

the production process

While the optimization of labor, automation, and computerization contributes greatly to production in modern factories, in many individual production still continues manually and on a smaller scale.

The steps of the production process are the same in any industry. Workers use tools and other mechanical equipment to shape and form individual elements in order to build a finished product. This finished product is put through quality control tests, transported, stored, and finally sold to consumers–local, abroad, or both. Throughout this process end-of-life products and waste materials must be dealt with through either recycling or reuse, in compilance with local regulations.

Process optimization

There are many factors that need to be considered to optimize the production process. The amount of working capital needed for wages and raw materials and the quantity and quality of goods to be produced are the first factors to be decided. To manufacture these goods, it is important to determine whether it is cost effective to make each component, or purchase it (make or buy). If buying components, timing the supply correctly can save storage costs (just-in-time inventory strategy). The manufacturing run is another important decision. Is it more cost effective to produce large quantities of the same product (mass production), smaller quantities of the same product (batch production), or unique items (custom-made production)? Cost-effective delivery requires access to roads, railways, and other transportation. Any technology used in production should be flexible enough to allow the company to manufacture and deliver a product to a customer's exact requirements. While the coordination of all the aspects of the production process is never easy, comparing processes that other companies use (process benchmarking) can help determine what will optimize efficiency within an operation.

It is fast and energy-efficient to transport finished goods by rail.

MANUFACTURING

practice

Hand-manufactured goods are often regarded as more expensive, higher quality, and desirable than mass-produced items.

When a product undergoes a manual process of production, even if machinery and assembly lines are used, it is manufactured (Latin: *manus* ("hand") and *factura* ("production").

To produce a large quantity of high-quality units within a short timeframe, the production process is subdivided into steps performed by different workers using different machines. Certain articles of a higher value, such as porcelain figures and musical instruments, are still manufactured entirely by hand.

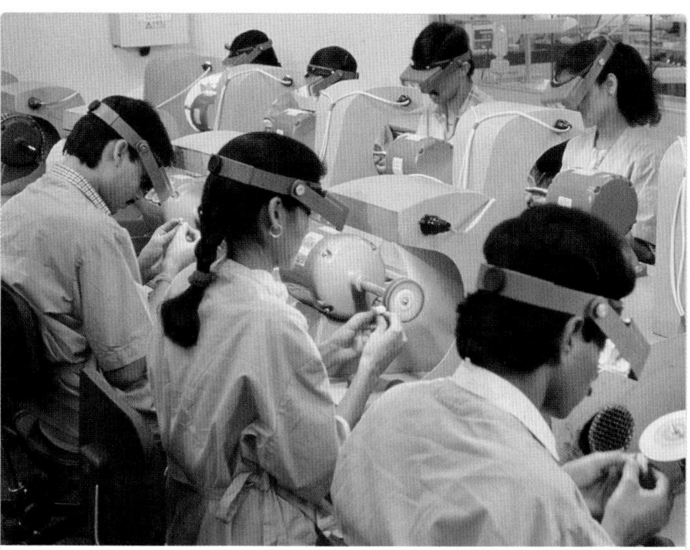

Quality control is used to ensure the quality of finished goods reaches the manafacturers and the customers expectations.

industrial engineering

The central phase of the production process is manufacturing. Manufacturing is the point at which products are produced in their final form from raw materials and/or component parts.

Up until the 18th century, all processing of raw materials was done by hand or with the aid of hand tools. The industrial revolution led to almost all manufacturing being performed by machine. Since the 1980s, the use of computerized controls and robots has maximized automation. Factories that used to rely on a large number of workers to complete the production process now use fully automated production lines that can be operated by an engineer and a small group of technicians.

Furnaces have been used for over 2,000 years to create iron and more recently steel, products that demand industrial infrastructure for efficient production.

Power supply

Large waterwheels once provided power for manufacturing, but the industrial revolution replaced them with the steam engine. At the beginning of the 20th century factories began to use electricity for power. Today most production lines are linked to a computerized control center through data cables.

Industrial processes

Many industrial processes are required to produce a finished product. Examples of reshaping and remolding processes include the melting, pouring, and bending of iron, as well as the heating and shaping of plastics and synthetic materials. Laser beams are utilized instead of files and saws to produce the required form and size. Bonding techniques, including rivets, welding, and adhesives, combine to assemble the individual components. Chemical processes can be useful in altering the surface of the product (coating), completely transforming the look of the product.

Industrial robots on a Nissan Motors production line.

Assembly lines

At the beginning of the 20th century the production process was revolutionized when Henry Ford introduced the concept of assembly line production in car manufacturing. In assembly line production, the production process is divided into multiple functional stages carried out by unskilled workers. Chutes, cranes, vehicles, and robotic arms are vital in the production process, in addition to the assembly lines. As computer-controlled robots take over the majority of mechanical activities, human interaction is limited to the planning stages, control of the processes, and periodic maintenance.

in focus

CNC MACHINES

Computerized Numerical Control (CNC) machines are controlled by a computer program that directs the machine to perform specific tasks with minimal effort from the worker. One example is a milling machine that is programmed to cut away excess metal, precisely fabricating components. CNC machines produce items faster and with more accuracy than traditional manual processes. In addition, these machines can automatically perform quality control checks.

Production of a single contact lens with a CNC lathe, a task that requires precision and predictable quality.

environmental protection and consumption

As resources become scarce and ecological destruction increases, methods to reduce consumption, and improve technology to protect the environment become increasingly necessary for ourselves and our children.

The high rate of consumption of raw materials, such as petroleum and metals, will cause a shortage of raw materials in the foreseeable future. In addition, the damage to the environment caused by waste materials will have negative consequences for future generations. So as to counteract this, we must adopt production methods that make economic use of raw materials and recycle waste products.

A field of rapeseed plants: In Europe this crop is popular in the refining of biodiesel, as it can be used as a replacement for diesel in automobiles without modification. Diesel mixed with biodiesel is very common and results in much less pollution than the pure fuel.

Waste management and cleansing

There are many environmental protection procedures that can be employed to lessen the harmful effects of the manufacturing process. Filters can be installed to reduce the amount of hazardous substances in liquid effluents and exhaust gases. Chemical and physical processes can reduce or eliminate toxic materials. Catalysts, substances that speed up reactions, can be used to reduce emissions. One example is the metal alloys in a vehicle's catalytic converter that reduce the emission of hazardous substances in the exhaust by converting nitrous oxide, hydrocarbons, and carbon monoxide into nontoxic carbon dioxide, nitrogen, and water. Catalysts can also be used to neutralize poisonous substances. Scrubber systems remove particulate matter and dangerous gases from industrial exhaust streams. One such technique uses limestone to change sulfur dioxide into gypsum that is recycled and used in drywall manufacturing. Environmental engineering is not implemented as often as it could be because it does increase production costs. Environmental protection processes can only reduce and slow down destruction, not eliminate it. The only way to achieve total environmental protection is to cease production.

Cotton production consumes a large amount of water, and has been a factor in the desertification of a number of inland areas.

Economic production

Economic production includes designing products and manufacturing processes that are meant to limit the need and use of raw materials. Recycled materials can be incorporated into the process to decrease the use of raw materials. Recycling leftover raw materials, waste products, and waste energy are other aspects of economic production.

in focus

TIMBER INDUSTRY

Wood is a renewable resource when managed correctly. Under a program of sustainable logging, harvesters take only the amount from a forest that can grow back over a certain time period. Clear-cutting, especially in nutrient-poor tropical zones, destroys a forest's potential for regeneration. This leads to erosion that brings about poor quality grasslands or even deserts. The newly barren land also loses its ability to absorb CO_2 from the air and thus can no longer counteract global warming.

Bulldozer in a rain forest.

recycling and disposal

Disposing of garbage and unwanted goods is expensive. Increases in waste, environmental pollution and shortages of raw materials have resulted in improvements in waste management and recycling.

Industrial countries produce more waste now than ever before. To protect against contamination and damage, almost all manufactured goods are sold in packaging. To lower production costs and reduce the environmental impact, many packages are designed to save space and weigh less during transport. In addition to the packaging creating more waste, the product itself will eventually break or become obsolete, turning into another piece of waste. To compound these environmental effects, many products are made of synthetic materials that either do not decompose, or emit poisonous substances into the environment during decomposition.

Waste disposal

We are developing new ways to counteract and limit the effects of hazardous waste in the environment. To protect groundwater, waste disposal sites are built on an impermeable base or artificial foundation. Incineration of waste is used to produce energy while reducing the waste to an environmentally safe form, although toxic gases are still likely to be released. To determine the best method for disposal, certain factors must be taken into consideration, including the type of waste, its potential for causing harm, and the costs incurred. Certain waste products, such as radioactive waste, cannot be disposed of in an environmentally friendly manner.

Recycling and downcycling

The process of manufacturing new products of value from waste or scrap is known as recycling. Refilling used ink cartridges is a classic example. Downcycling is the process of using waste to manufacture products of lesser value than the original, as for example, in using chopped up synthetic or plastic packaging for the production of concrete instead of just cement.

Because recycling procedures are labor-intensive and have high technical requirements, an effective conservation program requires genuine commitment and involvement from every member of society.

Waste electrical equipment is a growing problem. Components are often toxic and non-biodegradable.

RECYCLING OF GLASS

practice

The reuse and recycling of glass has a long history. Whole or broken glass containers may be melted down and either mixed with new raw materials or used as-is to form new products, depending on the quality level required. Sorting glass by type and color generally makes recycling easier. Most European countries currently recycle more than half of their used glass containers.

Glass can by recycled again and again.

basics

RECYCLING OF PAPER

While cardboard can be made from low-quality paper that has been recycled many times, high-quality recycled paper is used for newspapers, hygiene products, and packaging, Because paper fibers become shorter with every reuse, printing paper must be made using a substantial portion of fresh pulp.

Landfill waste disposal is no longer a viable option: not only is it environmentally destructive, but many heavily populated countries are simply running out of space for garbage.

the computer | components | input | output | data storage | programs

COMPUTER TECHNOLOGY

Digital computing has become an important aspect of everyday life since its beginning in 1940. Innovations led to mobile computing such as laptops, personal digital assistants (PDAs), and mobile telephones. The term "information age" has been coined to describe modern times, due to the central role of computers. Research and development goals are aimed at more compact, versatile, and universal computers, as well as improving processing speeds and reliability.

the ubiquitous computer

In 1977, Ken Olson, president of the Digital Equipment Corporation, is reported to have said: "There is no reason for any individual to have a computer in his home." Time has proved him wrong.

In 1941, German engineer Konrad Zuse built the first working programmable computer: the Z3. Weighing over one ton, the Z3 was used for calculations in aircraft design and used relays as switches. Two years later, the United States started building the ENIAC (Electronic Numerical Integrator and Computer) from which the term computer remains in use. The ENIAC, built for the military to compute projectile flight paths, used electron tubes as switches, making it the first electronically programmable computer.

Computer technology is used in almost every aspect of daily life. Today, it has become integral to everyday objects, from mundane kitchen appliances to life-saving pacemakers. The many and varied applications it is utilized for may be observed in cars, cell phones, watches and clocks, home entertainment, security systems, and hospitals.

Into the future

The computer industry plays a dynamic and important role in the world's economy. New technologies are constantly being developed, including supercomputers, microchips with extremely high computational power, and highly miniaturized computers. The use of decentralized computational power is on the rise. For example, NASA is taking advantage of grid computing by distributing data packets to private users whose unused computer processing power is then utilized to perform NASA's extensive calculations. A problem that would normally take one computer a significant amount of time to solve can be broken into smaller parts and distributed to thousands of computers for computation, before being returned and assembled again to reveal the answer.

in focus

MINIATURIZED MOBILE COMPUTERS

Laptop computers must be as small and as light as possible. To increase the efficiency of the battery, laptops are equipped with special mobile processors designed to perform at minimal power consumption. Expansion slots and ports allow easy attachment of external devices.

Compact components allow laptops to achieve small dimensions, although at a higher cost than desktop models.

The Universal Automatic Computer (UNIVAC) was the first computer designed for commercial use.

One of the fastest computers in the world is the JUGENE (Jülich Blue Gene) supercomputer in Germany.

see also: Artificial intelligence, p. 385

computer components

The hardware of a computer consists primarily of the processor, a control module, the bus, the memory and input and output units. In modern computers, the central processor is both the processor and control module.

Personal computers are now used mostly for office applications and personal activities, whereas large computers, such as mainframes, are used by corporations, government agencies, universities, and the military as high-performance central computers that can be simultaneously accessed by many users. Both types of computer are made up of the same key components.

A PC expansion board with capacitors, microprocessor, and USB connection ports.

Elementary components

A central element of most computers is a circuit board called the motherboard, which enables almost all of the computer's components to communicate with each other. The BIOS (Basic Input Output System) is found in a small chip in the motherboard. This program provides access to the computer's hardware components. The central functional unit, the processor chip, is also attached to the motherboard. The motherboard can generate substantial heat, so it must be cooled by a fan. The processor carries out most of the computing operations, and is therefore the most important factor in determining the speed of the system. When data must be made available to the processor quickly, it is loaded in the main memory. If the main memory is small, programs that need to process large volumes of data quickly (such as video programs or games) will run slowly. This problem can now be solved with high-end graphics cards. When paired with a high-performance three-dimensional microchip, they provide even real-time games with realistic, high-resolution images. Within the computer, data is exchanged through the bus, a connecting device with multiple conductors. The transfer rate within the system is heavily influenced by the amount of information that the bus can transfer per unit of time. With universal serial bus (USB) connections, many different devices can be connected to the computer and supplied with power. Modems and network cards (p. 383) are attached to the bus in order to connect the computer to the Internet or to other computers in a local area network (LAN).

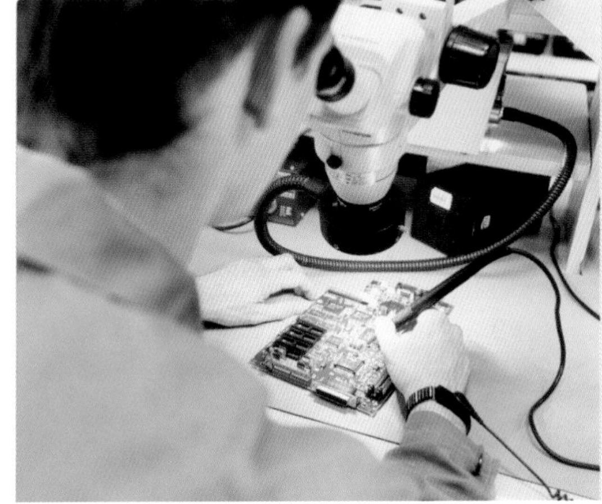

According to a 2004 study, the production of one computer requires more than 63 gallons (240 l) of fuel.

milestones

THE APPLE IMAC

Many computer manufacturers from the 1970s and 1980s, such as Atari, Commodore, or Schneider, have since disappeared from the market. Alongside Microsoft Windows-compatible PCs, the company Apple Computers, founded by Steve Jobs (also founder of Pixar and a leading Disney shareholder), has survived. Initially sold in 1998, the iMac was the first personal computer to combine all components in the monitor housing, and was designed for inbuilt, easy Internet access. The newest generation of iMacs has an improved performance and lost none of its style.

The iMac revolutionized the appearance of the PC and the fortunes of Apple Computers.

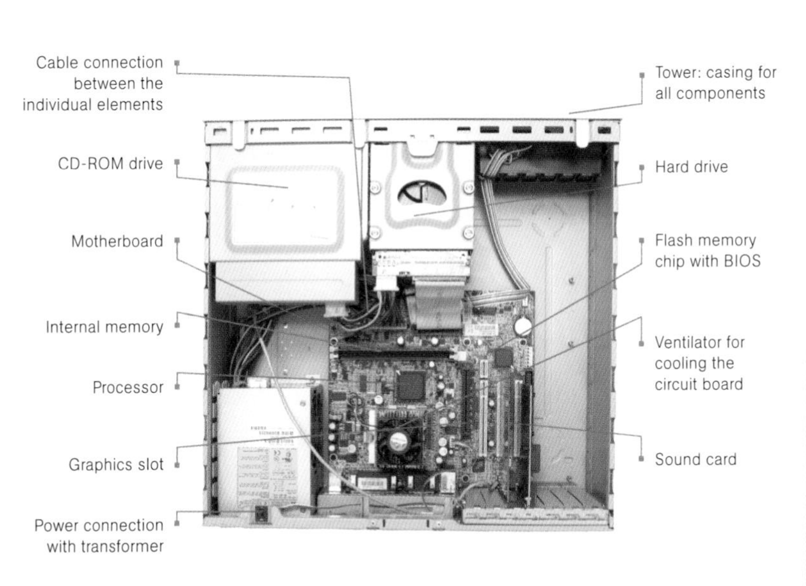

Different computer components require different types of connections. External components such as monitors or external drives also have specific requirements.

input and output devices

The performance and capabilities of the personal computer have been constantly expanded since its introduction. Internal and external modifications have equipped computers for new tasks, such as multimedia applications.

The first computers only had levers and buttons to input numbers and mathematical operations, and the output was displayed by control lights. After several years, the idea of a "monitor" was adapted from air traffic control stations. Devices called teleprinters are likely to have served as models for the computer keyboard. Further developments aimed at minimizing fatigue for the user, optimizing control, and simplifying operations allowed greater amounts of data and instructions to be entered as quickly and simply as possible.

Today, the workplaces of many people in industrialized nations are equipped with computers and peripheral devices, from monitors and keyboards to more specialized equipment.

Input and output devices

The most important input device is the keyboard, which enables users to type in characters and program the computer. Another optional input device is the mouse, or alternatively, touch pads and trackballs, which are used in the case of laptops and tamper-proof computer terminals. Public computers, such as ticket machines at railway stations, often use touch screens instead of a keyboard. Game consoles have specially developed game pads as input devices. Scanners are used to record optical information, while microphones record acoustic information. Microphones are also used to operate phonetically controlled computers. The most important output device is the monitor, which displays results for the user in graphical form. Editing text and pictures would be impossible without the monitor. The bulky cathode ray tubes of the past are increasingly being replaced with flat panel screens. Other output devices are speakers, which have been standard equipment since the 1990s, and printers. The Internet can be considered an output device that offers many possible connections.

Virtual worlds

Cyberspace or virtual reality scenarios represent nonexistent virtual worlds by means of computers. The goal is to use people's real world movements as data input for a computer program that can interpret and translate the actions into the virtual world. The input and output are intended to mimic reality as closely as possible.

in focus

THE MOUSE

Unlike the mechanical computer mouse, new generations of mice use an embedded light source—usually a light-emitting diode (LED)—to illuminate the surface below. Reflections are picked up by an optical sensor. The laser mouse is a further technological advance: using a laser diode instead of a standard LED, it perceives contrast even on very smooth surfaces.

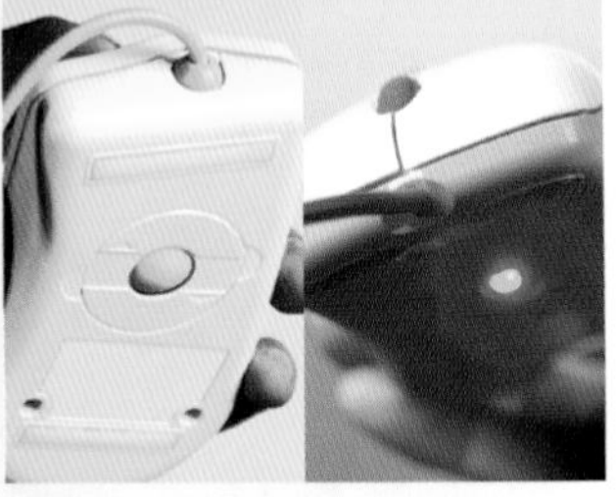

The LED of an optical mouse compared to the ball of a mechanical mouse.

practice

FLIGHT SIMULATOR

Normally, a user sits close to the computer and interacts with the screen, mouse, and keyboard using only a selected few of their senses. In simulators, however, computers generate virtual events in a pseudo-environment designed to be as realistic as possible. Flight simulators for pilot training use the operating instruments and display devices of a real cockpit to serve as input and output devices. This provides fight experience without the risk and can be a very effective learning tool.

Flight simulators were originally developed for training purposes for pilots and astronauts.

see also: Applications, p. 383

data storage devices

Secure data storage and strategies to achieve it have become increasingly important for both individuals and businesses. Technical innovations have brought improved storage media to a wide range of users.

Data storage must meet high technological standards. Increasingly, devices are required to store large amounts of data securely for long periods of time. The storage medium should be compact and transportable, allow quick access, and connect with many systems—all at an attractive price.

Portable music players with flash memory drives store thousands of songs in a very compact device.

Standard mobile storage media

Floppy disks and Zip disks have been largely replaced by compact discs with read-only memory (CD-ROMs). More efficient digital versatile discs (DVDs) have been available since 1996. Blu-ray discs (BD) are the latest generation, offering up to 50 GB of storage. Flash memory drives with a USB connection are double-sided devices with a faster access time. These drives can be connected to almost all computer systems without any special reading device. Flash memory can be electronically erased and reprogrammed. Flash drives may have a capacity of many gigabytes, and are small and simple to handle.

Storage media with special tasks

In the computer motherboard's primary storage or main memory, data is temporarily saved for immediate processing. Graphic and video cards also have their own main memory. BIOS is a kind of permanent memory in which the manufacturer stores important system data. This data cannot normally be modified by the user. Random-access memory (RAM) allows faster data transfers than storage devices such as the hard drive, however in most cases, the data is lost once the computer powers down. New forms of RAM may allow for powerless data retention.

New technologies

More efficient materials and novel ideas for data storage are now being researched. A team in Japan are investigating an optical storage medium using tiny plastic balls, 500 nanometers in diameter, with a fluorescent colorant. Each ball is modified during saving so that the colorant lights up during reading. Every ball represents one bit. This method would provide several times as much memory as a DVD in a comparable amount of space. Data has even been transferred onto DNA molecules and implanted into bacteria in the form of genetic information. The data remains unmodified and readable even after the bacteria has reproduced hundreds of times. It remains to be seen whether this ambitious research will ever result in a marketable product.

Flash drives utilize a type of writable RAM that stores data as charge in a capacitor.

All modern personal computers have a CD-ROM drive with which data can be read.

practice

PHYSICS OF DATA STORAGE

To store binary data, storage material needs two forms—to represent 0 and 1. A hard disk has magnetic units, polarized to represent 0 or 1. The smallest units of a USB memory stick are tiny transistors that conduct or block current. CD-ROMs have one-bit indentations, scanned by a laser. In magnetooptical disks, a laser changes the state of the magnetic material on the disk.

Rewritable compact discs, with a storage capacity of 650 MB, were first released in 1996. With suitable burners, CDs can be rewritten up to 1,000 times.

basics

BITS, the smallest memory unit, have one of two values (0 or 1), represented by positive and negative charges.

TO ENCODE 1 keyboard character, bits are grouped into a byte, usually in groups of 8. Since $2^8 = 256$, the 8 bits can have 256 possible combinations.

1 MEGABYTE usually equals 1 million bytes. However, since computer memory systems are based on powers of two, 1 megabyte $= 2^{20} =$ 1,048,576 bytes.

operating systems

The software that makes the use of a computer possible is called the operating system. It manages and controls the running of the computer—from starting and ending the programs to managing attached devices.

Computers can only perform the functions for which they are programmed. Programs are written in languages that use strict logic and formal commands, often based on mathematical functions composed of algorithms. Some common functions include logical sequences of events and their consequences, conditional queries, processing the size and position of objects, and access to external hardware like the hard disk or the Internet.

Viruses can infect and damage a computer system after entering through an email attachment, downloaded materials, or other source.

Assembler, interpreter, and programming languages

A processor receives instructions for computing and assigning memory space in an assembly language. These languages are easy to learn and are translated into computer code via the assembler software. A disadvantage is that every type of processor requires its own assembly language. All types of processors, however, can execute high-level programming languages. Compiler software translates these programs into assembly language.

High-level programming languages

High-level languages are primarily used in application programs. General-purpose languages are used for a variety of applications; domain-specific languages are suited for special applications only. Declarative languages define conditions to be satisfied by the program, and object-oriented languages execute commands for virtual objects. A few innovative languages do not use text commands. Instead, their functions are activated by mouse-clicking symbols displayed on the screen.

Examples of languages

C++ is a popular high-level language used for object-oriented programming, databases, and back-end applications.

The JavaScript language expands the scope of possibilities on the Internet. It can run programs that detect the user's hardware and adjust the display to be compatible with this hardware.

JavaScript is distinct from Java, another object-oriented language used to make applets for Web pages. HTML (Hypertext Markup Language) is not a programming language, but rather one for formatting Web pages (p. 388). Hypertext Preprocessor (PHP) is an increasingly popular language used to implement dynamic Web sites.

Competitors at a Programming Olympiad in 2000.

issues to solve

"TROJANS"—recalling the Greek myth—are seemingly harmless pieces of software that unload hidden programs to breach or even damage a user's computer system.

SPYWARE programs gain unauthorized access to personal data, passwords, and code words.

DIALER programs link the user to expensive dial-up numbers—without the consent of the user.

in focus

OPERATING SYSTEMS

A distinctive logo is a part of the marketing strategy of the larger software producers, from Windows, to Apple, to Linux.

Linux is an open source operating system, meaning the user has access to the source code in which programs are written. The user can modify it or write additional programs that can be appended to it. Linux programs are distributed free of cost. Firms and private users are increasingly deciding to use open source systems.

Microsoft Windows is the most used operating system in the world. It was introduced to the market in 1985 and its source code remains proprietary to this day.

Mac OS X is a Unix-based operating system that is extremely easy to use. It is more stable than Windows, and particularly popular in creative professions.

see also: Energy technology, pp. 346–347

applications

Applications are software programs whose functions are accessed directly by the user. Examples include word processing and spreadsheet programs as well as computer-aided design (CAD) programs and games.

Office programs include word processors, spreadsheets, presentation aids, and databases. Microsoft Office is the current market leader. OpenOffice.org is an increasingly popular "open source" program, meaning it is free to download and use and can be modified according to the user's needs.

Multimedia

Multimedia programs edit photos, videos, and sound files. They include animation tools, layout programs, and Web site design applications. They range from simple user-friendly freeware applications, open source image editing software such as "The GIMP" (GNU Image Manipulation Program) or expensive professional programs, which require extensive training in order to use. Multimedia editing programs differ with their approach to the compression of files, visual color depths and resolution.

Professional applications focus on editing large files, in high resolutions and file sizes, with no data loss through compression (Lossless Format).

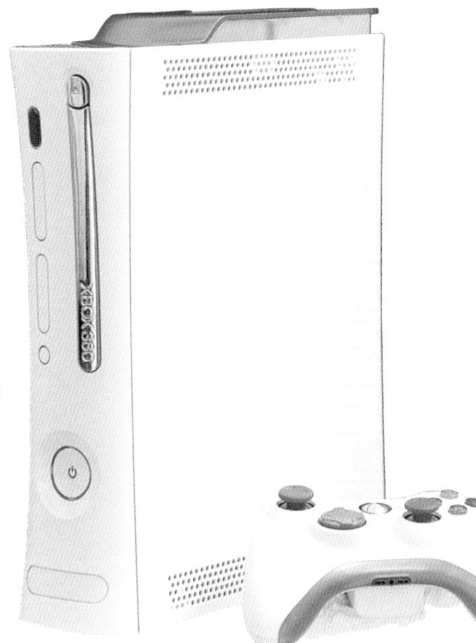

Stationary consoles, like the PlayStation, Wii, and Xbox, use a television as an output device.

Home user applications attempt to achieve a balance between quality and small file size by allowing some data loss through compression (Lossy Format).

Portable game systems first became popular in the 1980s and 1990s. Today, they are smaller than ever.

Computer games

The computer game industry has strongly influenced the development of modern computers. Computer game designers design games to run on the very latest computer hardware, which encourages customers to frequently upgrade. As a result, in the 1980s, computer gamers often shared a keyboard on a home computer while today, gamers enjoy multiplayer three-dimensional games on their own computers, connected via a local network or the Internet. This is all made possible by high-performance processors, memory, and display cards. Consoles, on the other hand, are specialized computer game hardware that is developed over a number of years. This means that although consoles are often less powerful than desktop computers, games run faster and with greater stability.

in focus

GAMEBOYS AND CONSOLES

Games can be played on computers, arcade machines, and dedicated video game machines called consoles. They can often use the same video and music formats as PCs and are increasingly compatible with them. Handheld consoles, like the PlayStation Portable and Nintendo DS have an integrated display. Some handhelds can be used to play games over the Internet and watch movies. Consoles are relatively affordable when compared to a computer, however, they are limited in the amount of peripherals that can be connected.

Newer game consoles are also able to connect to the Internet or a local network.

Since their beginnings in *Pong* (1972) or *Pac-Man* (1980), computer games have become a powerful creative and economic force, made possible by huge technical advances.

intelligent machines

Robots assist people by carrying out complex, monotonous, or dangerous work, but they still lack the judgment and versatility of human beings.

Robots are computer-controlled machines designed to complete specific tasks, such as industrial robots used for welding and assembly. However, their programming only allows them to complete particular tasks within a specific work environment. If there are any changes in the work procedures, they must be reprogrammed. This is often done using a "teach-in" method: a human operator takes the robot through the steps of the process, which are then stored by the robot. Robots that are autonomously mobile are more independent—they are able to walk, drive, swim, fly, and react to sudden events. They can currently handle work such as providing building security, serving meals, guiding visitors through an exhibit, or exploring distant planets.

INTELLIGENT MACHINES AND THE NETWORKED WORLD

Electronic networks allow both people and machines to communicate. An increasing number of machines are equipped with sophisticated artificial intelligence and can be programmed to handle many types of work, while computers may be utilized even within seemingly simple and everyday objects.

Robots designed to move across rough terrain are often modeled after insects—even simulating a biological nervous system with a computer program. According to the situation, each leg "decides" whether and how it should move.

A robot receives data about its environment and components through inbuilt sensors. For example, the distance to an obstacle can be determined by measuring the time taken for an ultrasound signal to bounce off the object and return, or detailed data about the environment may be provided by a camera. A voltmeter is used to measure the remaining charge in a battery, and an internal protractor is able to register the position of a robot's gripper arm. A robot can monitor its own location by using a GPS unit or with the help of special markers in its vicinity. Robots move using actuators, which may be legs, wheels, gripper arms, or tools—all driven by electric motors. The more complex the robot's equipment the greater skill it will have; however, programming it becomes increasingly difficult.

Programming a robot

A robot's computer program is designed to use sensory data to evaluate the environment and its own physical elements, as well as to account for the possibility of faulty data. For example, a robot may calculate the distance it has traveled from the number of rotations of its wheels, but even a simple malfunction, such as a slipping wheel, will produce errors. In this case, the robot could be designed to be able to recognize the problem and compensate using additional sensor data. Depending on the robot's mission, its program determines which actuators should be used and how. The robot must be able to combine different actions appropriately; for instance, it may move toward a target while simultaneously avoiding obstacles. To help with these complex actions, programmers use artificial intelligence technology (p. 385).

practice

THE ROBOCUP

This international competition presents a playful challenge, inspiring creativity among students and scientists and promoting innovative research. The vision is to create a team of fully autonomous humanoid robots capable of defeating the human World Cup soccer champions by 2050.

The RoboCup combines robotics and the world's most popular sport: soccer.

Industrial robots are programmed to carry out limited, specialized tasks, working rapidly, precisely, and indefatigably.

see also: New mathematics, pp. 418–421

artificial intelligence

Artificial intelligence (AI) is a specialized branch of computer science. All researchers attempt to imitate some of the abilities of living things, such as perception, learning, logical reasoning, and language using computers.

Individual biological nerve cells (neurons) can be simulated by computers and connected to each other in layers, forming a so-called neural network. Data entering the input layer is processed by the network and passed along to the output layer. The data processed in the outer layer depends on the pattern of connections between the individual neurons. During the learning phase, the connections are adjusted in accordance with sample data, so that afterwards the system can identify such items as handwritten letters, even if each letter is not always written in the same way.

Computer-based knowledge

Storing knowledge is a complicated process and provides a much greater challenge than simply recording ordinary data. Knowledge about a chair, for instance, goes far beyond simple information about its shape and location. In many contexts its function as a place to sit may be considerably more important. Beyond this basic function, a chair can also be used, for example, as a surface to rest objects or as a stepstool. There is also knowledge about events and actions, and even knowledge about knowledge itself—such as its scope, reliability, and origin. This "knowledge data" can be combined into statements. By using several statements, logical rules can be used to derive further statements. These methods are used by so-called expert systems to figure out, for instance, what has caused a machine to malfunction and even to plan its repair.

In some areas computers are superior to humans. In 2006, for example, the program Deep Fritz defeated reigning world chess champion Vladimir Kramnik.

THE TURING TEST

In the 1950s, mathematician Alan Turing devised a test to clarify the question of whether a computer can be regarded as intelligent. The structure of the test requires a human judge to engage in a natural language conversation on a computer. They must identify whether they are conversing with a human or a computer programmed with human-like responses. If the computer can be identified, it has failed the Turing test.

Alan Mathison Turing (1912–1954) explored the theoretical potential of computers.

practice

AI and robotics

The first robots were programmed according to the model "perceive ⟹ plan ⟹ act." The downside of this technique is the great deal of computational time required, and the slow reaction speed of the machines when unexpected events arise.

Later, researchers imitated the stimulus-response principle followed by living things, such as that used when we reflexively pull our fingers away from a hot stove. These complicated stimulus-response chains are used, for example, when we are climbing stairs while being simultaneously absorbed in a conversation. Today, both methods are combined for programming robots to complete complex missions.

Local kindergartners gather around a communication robot at a shopping mall in Fukuoka, western Japan.

see also: Mathematics in the 21st century, p. 421

computer networks

Our modern world has been shaped by linking computers and therefore people together. Networks allow data to flow from machine to machine—across a room or around the globe.

With only a little effort, computers can be linked together in small networks. In turn, these Local Area Networks (LANs) can form the links and nodes of broader or even global network structures. Printers and other devices can also be included in networks,

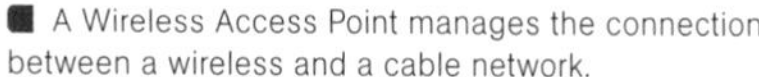

A Wireless Access Point manages the connection between a wireless and a cable network.

allowing multiple users to access them. Thus, computers can exchange information, a printer can be shared by several computers, and data of interest to all users can be stored on a common server. A server is a computer or program that offers services within a network. For instance, a mail server distributes personal messages (e-mails) to recipients.

Local area networks

The computers in a LAN are usually located in the same building and connected to the network via network cards. Applications pass data to the network card and receive external data from it. An operating system, such as Windows, Mac OS, or Linux, is required to organize this data exchange.

The network card feeds data into the network in the form of packets. Data packets are exchanged according to set technical rules, called protocols. Likewise, computer networks are constructed using particular technical standards. Ethernet is widely used for wired networks, while Wireless LAN (WLAN) is most common for wireless networks. The interface between a WLAN and a wired network is managed by a special device, a Wireless Access Point.

At LAN parties, guests link their computers together and compete in multiplayer games.

Broader networks

A Metropolitan Area Network (MAN) connects computers and entire LANs, typically within a city. It usually relies on fiber optic cables and radio links to enable rapid data exchange. A Wide Area Network (WAN) covers an even larger territory. WANs may be linked via satellite or undersea cable to a Global Area Network, or GAN.

DISTRIBUTED COMPUTING

practice

Scientific research often requires large amounts of computing power. Instead of using a supercomputer, the tasks can be split up and divided among many ordinary computers working at the same time. In the SETI@home project of the University of California at Berkeley, for example, Internet users download radiotelescope data to their home computers. When the computer is not in use, a screen saver software package searches the data for possible signals from alien civilizations.

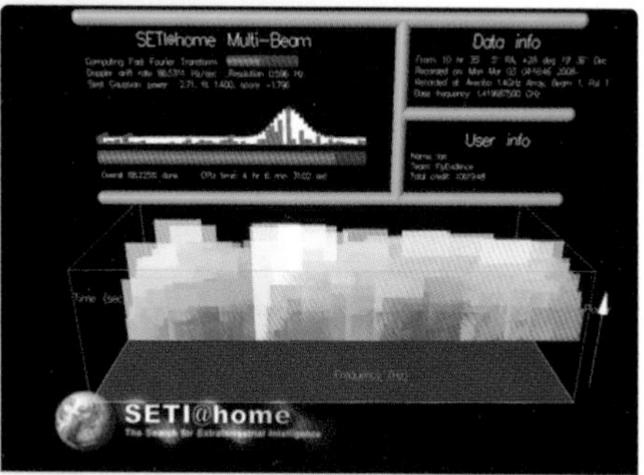

The SETI@home screen saver displays the progress of its data analysis.

All devices in a LAN exchange data over the ethernet, a cable network. Ethernet technology can also connect distant computers via fiber-optic cables or radio waves.

the internet

The word Internet derives from "Interconnected Network." Its advanced technology allows connected computers in large or small networks around the globe to exchange data. Countless applications rest on this foundation.

The Internet is a specialized network of connections and computers. It links smaller computer networks around the world, such as the local networks maintained by businesses or universities.

The dawn of the Internet age

The research that eventually gave rise to the Internet commenced in the United States during the 1950s. It was conducted by the Advanced Research Projects Agency (ARPA) of the U.S. Department of Defense. In 1969, ARPANET—an experimental computer network connecting four universities—was put in service. Additional networks were then gradually connected, and enhanced technologies and services such as electronic mail (e-mail) were introduced. Thanks to technology standardization and rising international interest, the Internet ultimately emerged.

■ Subnetworks are often connected by fiber-optic cables, the "backbone" of the Internet.

■ Internet cafés, also known as cybercafés, offer customers access to the worldwide information network.

Protocols for data transport

The transmission of data between computers is regulated by Internet Protocol (IP). The data is broken up and transmitted in smaller packets. Subnetworks within the Internet are connected through routers. Packets include the addresses of the sender and receiver. These IP addresses consist of a series of numbers separated by periods, such as 60.230.200.100. If a data packet is not addressed to the router's home network, it is passed along to other networks. There are numerous other protocols in addition to IP. For instance, the Transmission Control Protocol (TCP) ensures that the data reaches the right application program in the receiving computer.

Domain names

IP addresses, with their long strings of numbers, are not very user-friendly. As a result, the Domain Name System (DNS) was developed to assign a unique domain name to each IP address. Name servers translate between the two. When a user enters a domain name, the application program requests the corresponding IP address from the name server.

basics

THE INTERNET is often thought to be equivalent to its applications, such as the World Wide Web. In fact, it is the transport system for data that makes these applications possible.

ORGANIZERS OF THE INTERNET

in focus

What happens with the Internet is determined by numerous associations. The Internet Engineering Task Force (IETF) handles organizational and technical problems. It develops protocols, routing techniques, and security guidelines. Network Information Centers assign Internet addresses. The World Wide Web Consortium (W3C) focuses on the standards with which Web sites are programmed and transmitted.

■ Work groups of the IETF in Vancouver.

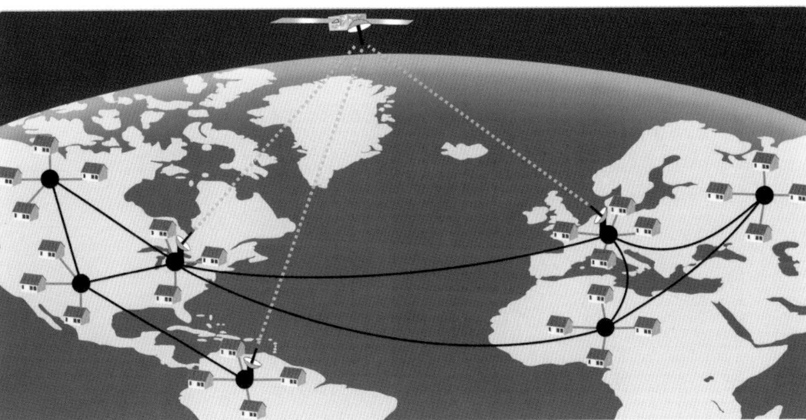

■ The Internet connects local networks around the globe. Internet providers offer access to individual users.

the world wide web

The idea of the World Wide Web revolutionized the Internet, turning it into an information center for all. In contrast to newspapers, radio, and television, users everywhere can contribute to its content and development.

The World Wide Web, or simply the Web, is a much-prized use of the Internet (p. 387). Internet users can navigate through the jungle of information with the help of a Web browser. This application calls up documents stored on Web servers over the Internet and then displays these documents as Web pages. Web pages can include text, photos, graphics, music, and videos. Hyperlinks, or links, are among the most important elements of a Web page. They point to other Web pages. When the user clicks on a link, the browser loads the corresponding document.

basics

THE INTERNET began in 1969 with four networked computers. By 1991 there were 375,000. Today it exceeds 400 million. The World Wide Web can be abbreviated to Web, WWW, or W3.

Structure of a Web page

The document used to create a Web page includes the page's content and also describes how the information should be structured and displayed. This structure is described with hypertext markup language (HTML). The browser interprets this language to display the information as intended. It is not difficult to learn how to design simple Web pages.

Browsers find documents on the Internet using the uniform resource locator (URL). For instance: http://www.example.com/many/examples/example1.html indicates that the Web page example1.html can be found in the directory /many/examples/. The Web server for this page has the domain name www.example.com. The page is requested by the browser according to a particular protocol (p. 387) and transmitted by the Web server using hypertext transfer protocol, HTTP. The URL can be typed directly into the browser, or it may be embedded in a document as a link.

HTML (hypertext markup language) is a text-based language used to control the content of Web pages. Special programs, called Web browsers, interpret these instructions to present Web pages to the user.

Information searches on the Web

Information on the Web can be found with the help of search engines and Web directories. When the user enters search terms, search engines list pages containing those terms. To do this, search engines review Web pages using special computer programs (called Web crawlers, spiders, or robots) to identify search terms and create a database. Web directories are collections of URLs, which are assembled and managed by people. They cover only limited subject areas, but are able to offer accurate descriptions of the Web pages they list.

milestones

THE FIRST WEB SITE

The revolution began at the European nuclear research center CERN in Geneva. In 1989, physicist Tim Berners-Lee suggested a new method for navigating through a sea of documents. A user-friendly information network would be constructed for the CERN scientists. Systems engineer Robert Cailliau helped implement the idea. The team developed a server, a browser, and hypertext to display documents. Today, the Internet is a global mass medium.

Sir Tim Berners-Lee, here with former UN Secretary General Kofi Annan, invented the WWW.

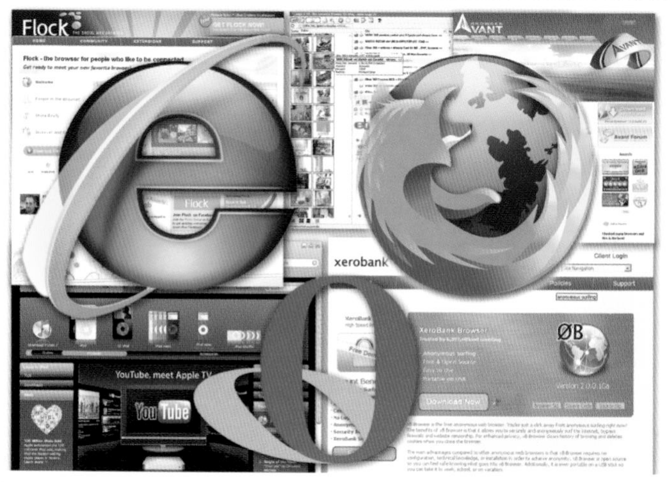

Web browsers are programs through which texts, pictures, and hyperlinks can be seen and understood. They are the user interface that makes searching the WWW possible.

see also: Operating systems, p. 382

the networked world

The Internet and World Wide Web connect computers, people, and ideas around the globe. The Internet's commerical opportunities were quickly discovered, and new forms of community emerged as well.

The Internet (p. 387) has changed society. There is no faster way to disseminate information and make it available worldwide. Opinions and ideas can be published on personal Web pages (p. 388) and blogs (online diaries) and new projects have been made possible by this interconnectivity. Wikipedia, for example, is an online encyclopedia created and constantly expanded by users around the world. Open-source software is developed in a similar manner. Groups focusing on political and social issues also use the Internet to organize. Because of this, many non-democratic governments restrict Internet use among their populations, blocking access to undesirable political or religious content.

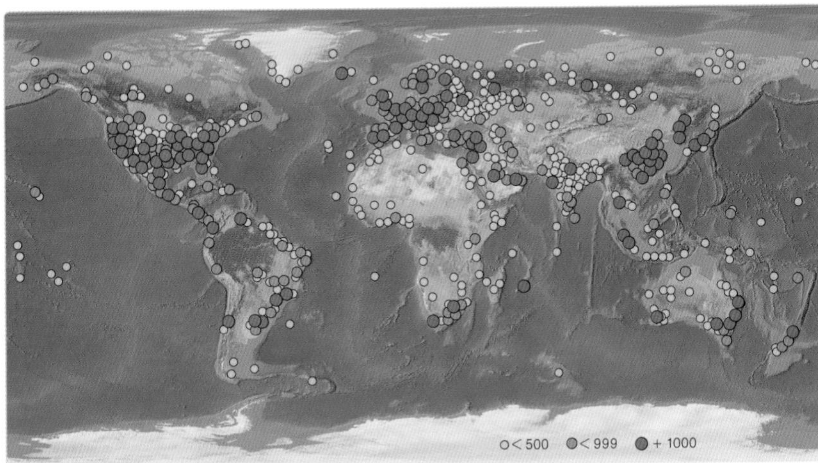

Internet world map 2007: This map shows the distribution of Internet routers worldwide. North America and Europe have 70 percent of the total, while Asia has the next largest share with 14 percent.

The work and business worlds

E-mail makes it easy to transmit documents and keep in touch with clients and colleagues around the world. In addition, specialists around the world can work together on the same project. Companies and tradespeople advertise goods and services online, and an enormous variety of products can be purchased directly on the Web, with payments made by credit card or through other payment systems. Encryption protects account information and passwords and makes online banking possible. Absolute security, however, is not assured—just as it is not when you pay by credit card in a store or conduct transactions with a bank teller.

The Internet has changed the way we work. New modes of work have developed; for example, teleworkers work at least partly from home.

Crime

Unauthorized users can gain access to sensitive information via the Internet, and data can be falsified or destroyed.

The operating system Linux is worked on by software developers around the world.

Servers containing privileged information are protected by firewalls; however, computer specialists can sometimes breach these by exploiting errors in the security system or using spy programs or Trojan horses. Such techniques are not only used by criminals and spies. In some countries, similar procedures are under consideration or in use by the police to search private computers. Critics point to the secrecy of these investigations in contrast to legally authorized searches.

in focus

THE INTERNET OF ORDINARY OBJECTS

Everyday objects could one day be equipped with tiny computers. In this vision, the refrigerator knows when to order more milk and the front door accepts packages when no one is home. These things could become reality through a combination of RFID (p. 391), advanced sensors, artificial intelligence, and nanotechnology.

An armband could call the ambulance after a fall or accident.

see also: The Internet, p. 387

biometric techniques

Biometric techniques focus on an individual's physical traits or behavioral patterns, instead of using a PIN code, password, magnetic card, or key. This allows for a quicker and more secure verification of people's identities.

■ Physical traits that tend not to change dramatically over time, such as facial or iris geometry, are used for biometric identification.

Biometric techniques are largely free from disadvantages such as forgotten or stolen passwords or lost keys. Physical characteristics used by biometric systems include fingerprints and geometry of the face, iris, or hand. Behavior-based identifiers include signatures, voices, and keystroke patterns. Personal traits can change over time, however, especially those related to behavior. A person's individual data is first entered into the biometric system in a process called enrollment: for example, a fingerprint scan or a photograph of his or her face. Depending on the system used, a data set or template is produced from these raw data, using special mathematical procedures. This template can be stored in the individual's personal ID card or passport or in a database, along with their identifying information to be accessed as needed. An individual's identity may be checked, for example, during an access-control procedure. The physical traits are captured again. If the two match, the system registers that the person's identity has been recognized.

Secure operation

Biometric systems are subject to minor or major errors in recognition, due to traits changing over time, such as age, illness, or the presence of foreign substances such as dirt. Errors can also arise, for example, from changes in hairstyles, contact lenses or when fingers are placed differently on the scanner. To compensate, traits are not expected to match exactly but just to display an adequate degree of similarity. Some biometric characteristics may be better suited to particular applications. Care must be taken that a biometric system does not accept counterfeited fingerprints or severed fingers. Due to cost reasons, techniques available to prevent this are more likely to be used only in high-security applications. In addition, several biometric systems may be combined, or they may be monitored on-site by security personnel.

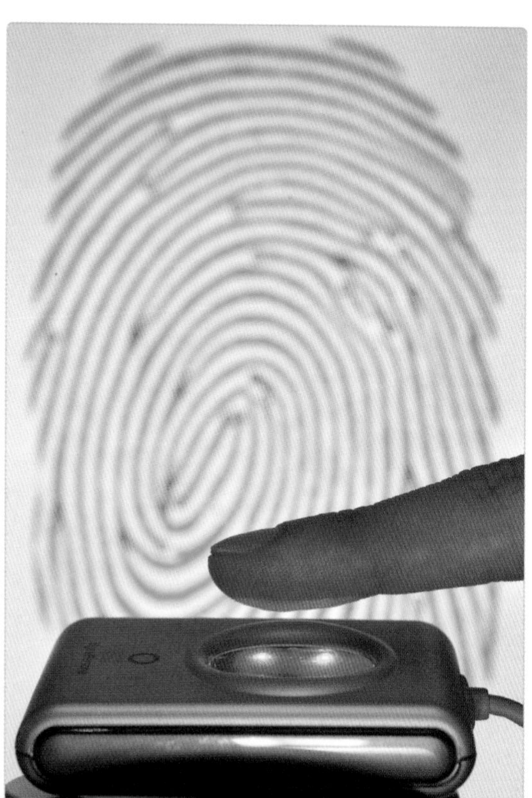

■ Portable fingerprint scanners provide useful biometric identity and security management.

practice

USE OF BIOMETRIC TECHNIQUES

In many computer systems, fingerprints are used to authorize individual users. Registered customers of some businesses can pay with the touch of a finger. The fingerprints of criminal suspects are compared by computer with prints left at the crime scene. Internal gates to control entry into critical areas—for instance, in a nuclear power plant—are monitored by various biometric systems as well as security personnel. Some nations' passports have embedded radio-readable RFID chips (p. 391) that store biometric data.

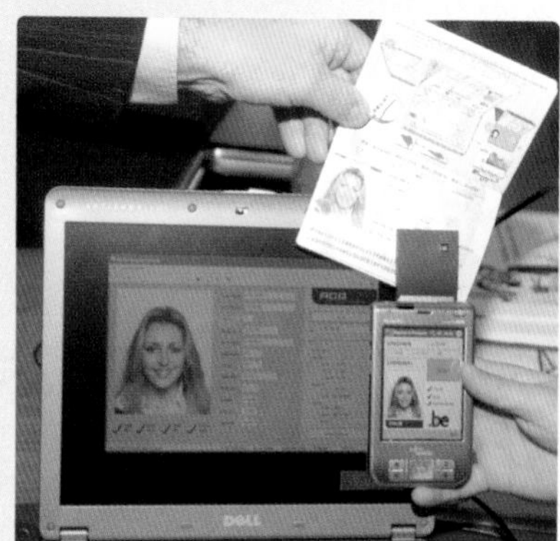

■ Digital storage of biometric traits can reduce the counterfeiting and misuse of passports.

see also: Computer technology, pp. 378–383

rfid: identification through radio waves

RFID technology is expected to gradually replace barcodes on consumer products. The system offers useful applications for business, industry, and everyday life. However, it also raises privacy concerns.

RFID, or radio frequency identification, is a technology used to identify objects, goods, and living things with the help of radio waves. A special RFID transponder or tag, which includes a tiny computer chip and miniature antenna, is attached to the target object. Data on the chip may include the tag's serial number as well as other information, such as the item number of a product or the access code for a building. An RFID tag transmits its data whenever it receives a signal from a special reading device. Most tags operate without their own power sources, receiving the energy they need from the reader's signal. A computer system or network with a database is a necessary component of an RFID system. The database may store detailed information, for example the price, expiration date, and location of items in the warehouse or store, in addition to the product code and serial number.

RFID reader devices may be stationary, portable, permanant, or subject to alteration. The tags may take the form of adhesive "smart labels," or transponders in a card-sized format may be used as customer loyalty cards, tickets, or entry passes. Sturdier transponders are used, for example, to track industrial parts during the manufacturing process. Transponders with a wide reception range, equipped with their own batteries, are attached to shipping containers. RFID tags are also noteworthy for the amount of data they can hold. More expensive tags can offer security measures to prevent unauthorized users from reading or changing the data. Some transponders function as sensors. Their response signals may vary depending on their position or temperature. RFID sensors can be used to monitor the temperature of frozen foods, for example, or to report wirelessly on a truck's tire pressure. RFID tags also make it harder to counterfeit products. They can be embedded in passports or other documents (p. 390), hindering fraud. RFID transponders in ID cards or keys can help control access to a building or vehicle. Similarly, they can monitor the time spent by a worker in a workplace.

■ RFID tags store information electronically and are replacing more traditional bar codes as product identifiers.

■ Depending on the technology used, the reception range for reading an RFID tag can range from anywhere between less than an inch to several feet.

■ RFID tags in tiny glass cylinders, the size of a grain of rice, can be implanted under the skin of pets, livestock, or people.

in focus

RFID PRIVACY RIGHTS

Because RFID tags can be surreptitiously attached and read, some people are concerned about the possibility that unauthorized persons may use them to monitor others or interfere with the right to anonymity. To safeguard privacy rights, numerous demands have been made. For example, consumers should be fully informed about RFID systems in use and must be allowed to destroy any product tags after purchase; furthermore, unauthorized tag reading and access to sensitive databases must be prohibited and information must only be used for its intended purpose.

■ Citizens' organizations and privacy activists see harmful potential in the misuse of RFID technology.

telephone networks | mobile telephone | video | radio | TV | film technology | printing

COMMUNICATION AND MEDIA TECHNOLOGY

Forms of media affect and saturate every aspect of our daily lives. Media spreads information and advertising messages, documents world affairs as well as private events, and serves recreational purposes. Modern technology has made this more convenient than ever.

fixed-line telephone networks

Thanks to the digitization of the telephone system, calling has become much more convenient in the last few years. Numerous new services have also become available.

Today, most fixed-line systems include fiber-optic network cables that link different individual cities or geographic areas. However, individual customer connections, also known as the "last mile," still usually consist of the original double copper wire that connects them to the telephone company's local exchange. When a user picks up or activates the telephone receiver, it forms an electrical circuit to the local exchange. Dialing a phone number sends two combined tones for each digit to the exchange office, initiating the process that connects the telephone call.

DIGITIZING VOICE SIGNALS

in focus

To digitize a voice signal, a converter samples the sound signal several thousand times per second. Each value is represented as a binary series of ones and zeros, easily processed by computers. In this way, many telephone conversations and accompanying data signals can be transmitted almost simultaneously over a single line.

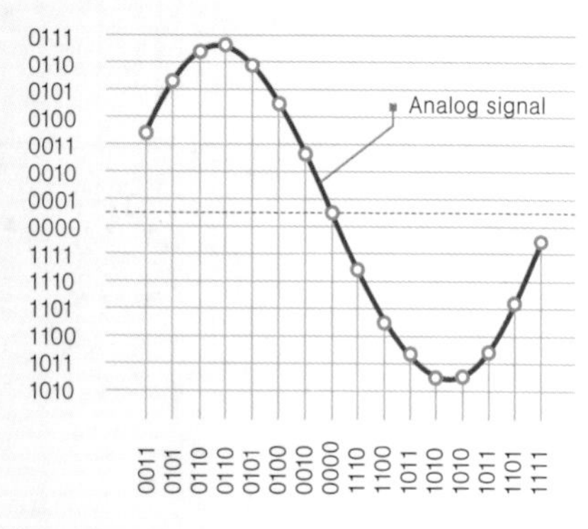

■ The most common method of encoding an analog voice signal and encoding it into a digital bit stream.

■ With the spread of cell phones (p. 395) in the industrialized nations, traditional telephones gained a new name: fixed-lines or landlines.

Telephone exchanges

Exchanges are the individual nodes in the telephone network. The local exchange is responsible for managing customers' connections. When a user makes a call, the local exchange is the first point of contact. The person being called is often not directly connected to the same local exchange, therefore the phone call may be routed over various higher-level centers to reach their local exchange.

Main centers function as connection points between distant networks, routing calls from one network to those of other domestic telephone providers. International phone calls are routed through international gateways.

With the introduction of computer technology into telephone networks, separate channels were established for voice and signal data. The data carried by the signal network includes the phone number being called and the local exchange of the call recipient, in addition to connection and disconnection information for the voice channel.

basics

WIRELESS TELEPHONES consist of a mobile unit and a base station. The mobile unit maintains a radio link with the base station, which is connected to the local telephone network.

EARLY FIXED-LINE telephone networks used only copper wires for all connections.

ISDN

ISDN, or Integrated Services Digital Network, is an international standard for digital communications networks. With traditional analog landlines, the voice signal is not digitized until it reaches the local exchange. ISDN digitizes the entire connection, starting at the telephone or other device. In addition, several channels can be used at the same time within one connection, allowing users to make phone calls, surf the Internet, or send a fax at the same exact time. ISDN also offers numerous convenient features for its users, such as caller identification, call forwarding, and the management of multiple callers simultaneously.

■ A workman prepares telephone switches for the installation of digital equipment at one of Moscow's automatic telephone exchanges.

connecting to the internet

For many years, telephone connections acted as bottlenecks, hindering the transmission of large amounts of data. The development of modern transmission technologies has significantly improved Internet connectivity.

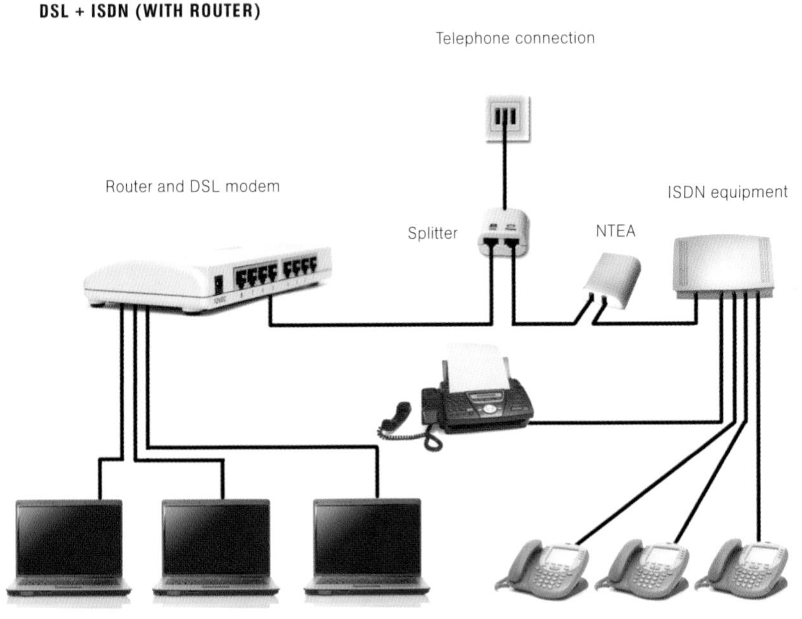

■ At the telephone connection, the splitter device divides the two types of signals, delivering voice signals to the telephones and data signals to the ADSL modem, which then prepares the data appropriately and sends it, for instance, to the network cards of attached computers.

Many communications networks, such as cell phone networks (p. 395), telephone landlines (p. 392), and cable TV networks, are connected to the Internet. The data packets sent and received are reformatted appropriately at the network gateway and Internet access points are made available by local providers. Special adapters allow notebook computers to access the Internet over a cell phone or wireless local computer network. For home computers and small networks, an ISDN card or ADSL modem may be used to transfer data between the computer and the telephone connection. The Internet access point is managed by the local exchange (p. 392).

■ Newer cell phones and other portable devices support direct connections to the Internet utilizing built-in wireless capabilities.

ADSL

Thanks to modern fiber-optic technology, the main arteries of the fixed-line telephone system allow increasingly rapid transmission. The copper wiring of connections for home telephones, however, was not designed for heavy-duty data exchange, although its possibilities have not yet been exhausted. The ADSL (Asymmetric Digital Subscriber Line) system significantly increases the performance of the connection wire by using a wider range of frequencies for signal transmission than was usual in the past, and enhancing the ratio of data to noise. In addition, the data are translated into signal series that allow for the correction of errors.

ADSL is especially suitable for applications using Internet browsers (p. 388). Internet users typically send only small amounts of data, such as the commands needed to call up a website, but may receive large amounts in the form of text, graphics, and video files. ADSL works asymmetrically: customers can receive more data per second than they can send. However, as the maximum data transmission rate declines in proportion to the length of the wire, the distance to the nearest local exchange should not be more than a few miles.

ADSL can be used along with traditional telephone services. The voice and data signals are transmitted together over the telephone lines, and are divided at the user's connection by a splitter device.

basics

IN COMPARISON with older connection methods such as dial-up modems or ISDN, broadband Internet connections offer significantly higher rates of data transmission.

ADSL IS only one of many forms of DSL technology.

practice

OPEN WIRELESS NETWORKS

Open wireless networks or "hotspots" are created and managed by private individuals on a noncommercial basis. An internet connection is shared publicly using a wireless network access point (p. 386), which employs no encryption or password security, enabling Internet access for anybody in the locality. Due to the lack of security, it is very unwise to access banking details or to access unencrypted Web sites using open wireless networks.

■ Many cafés offer open wireless Internet access to their customers.

internet telephone systems

A new alternative to traditional telephone networks has appeared in the last couple of years. It uses computers and computer networks to route voice calls over the Internet.

VoIP broadband cordless phones come with a built-in Voice over IP broadband service.

Telephone calling over computer networks (p.386) using the Internet (p.387) is called IP telephony, Internet telephony, or Voice over IP (VoIP). Networks transport not only ordinary data, but also carry a digitized voice signal (p.392) along with routing data to establish the connection. In IP calling, a voice signal is transformed into digital impulses and sent over the network in the form of data packets. At the receiving end, the packets are reassembled. The data transfer is regulated by the so-called IP protocol.

Calls can be made using special IP telephones or by computers equipped with special software. Ordinary telephones also have the capability to be connected to the system when they are fitted with IP adapters. Making phone calls over a computer network can often result in significant cost savings for the user, and even the ordinary telephone network can be accessed using VoIP gateways. A user's computer or IP telephone is identified with an IP address (p.387). One problem is that the caller's IP address must be known but this can be subject to frequent change.

The solution to this problem is that the VoIP system contacts a special server with their user identification. They then transfer the relevent IP address details and set up the connection. In this way users are not tied to a particular computer or phone line and can be reached at any location with an Internet connection.

Reception quality

At the receiving end, the data packets formed from the voice signal are assembled in a specific order.

However, when data packets arrive too late at the destination device, they can no longer be integrated into the voice stream, which leads to a perceived drop in the call quality. This transmitting of voice signals in data packets presents quite different challenges in comparison to those involved in the transfer of ordinary computer data, for instance the data transfer rate must be sufficiently high to ensure good quality.

basics

WITHIN ORGANIZATIONS and businesses, increasing numbers of IP telephone networks are being established.

IP CALLING may be limited to a network or location, but thanks to the Internet, it can also be used worldwide.

in focus

SECURITY ISSUES

Internet telephone systems allow an exchange of data, like any other Internet connection. This allows for unauthorized access to communications; for example, a call may be monitored or diverted. Although encryption is possible, it can reduce sound quality. As with e-mail, spam, or advertisements can also be received.

Often VoIP programs are used over wireless connections or in Internet cafés, leaving users open to password theft.

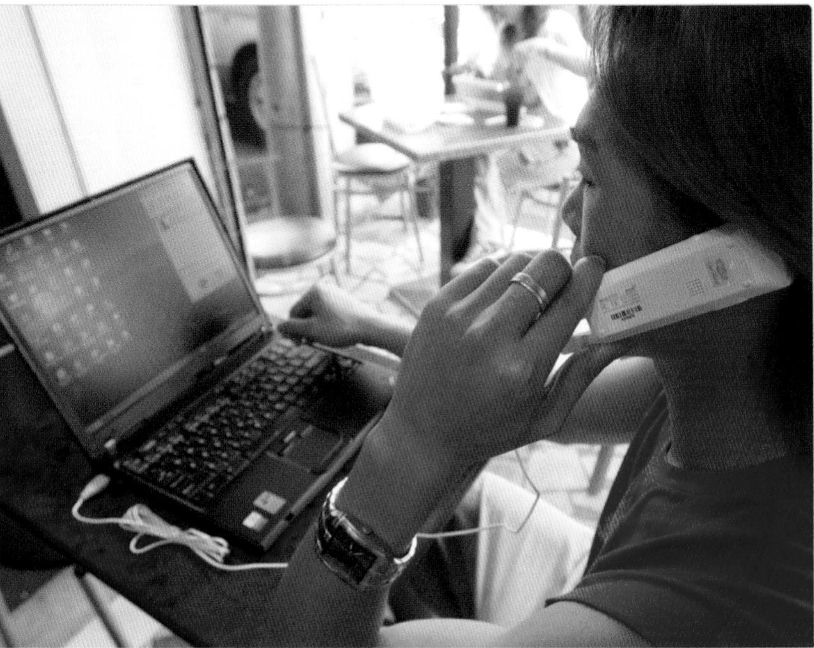

A person using a VoIP phone connected to a laptop. Computers, PC keyboards, and even cell phones are being produced with the capability to make VoIP calls already built in.

mobile telephone networks

Mobile telephone networks allow users to be reached at any time and in most parts of the world. Modern mobile devices are increasingly powerful and versatile and have become an indispensable modern device.

Mobile phones or cell phones are small, portable telephones, connected via radio to mobile telephone networks. A mobile network consists of fixed transmitting and receiving centers, called cell sites or base stations. Each base station is responsible for a specific area. Together, they create a network of cells providing full coverage. In city centers, base stations may be only a few hundred yards or meters apart, while in rural areas they might be separated by several miles or kilometers. Neighboring cells use different frequencies to avoid interference. Within each cell, a large number of users can place calls simultaneously. This means that their radio signals must be easily distinguishable. This is achieved, for instance, through the use of different frequencies, or through cooperatively timing transmission and reception.

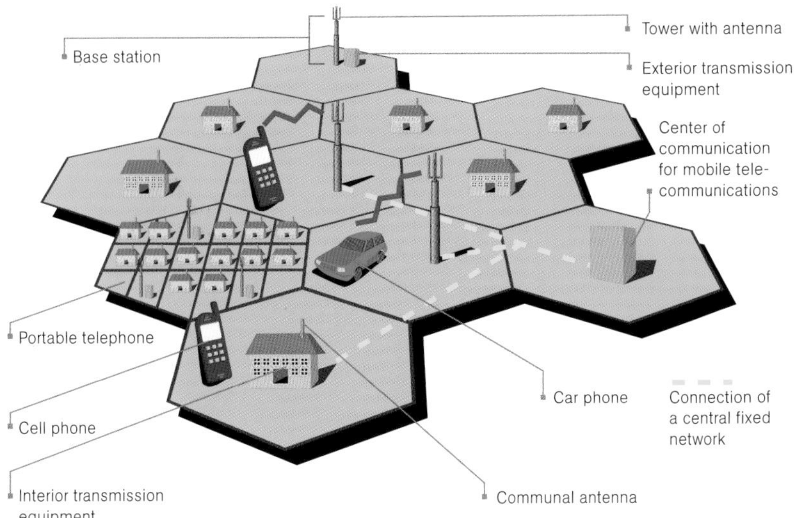

The base stations for individual cells are connected to each other by wires or two-way radio connections. They are also connected to higher-level switching stations.

ELECTROSMOG

This term refers to the environmental effects caused by electromagnetic fields. These are most notably produced by high-tension wires, electrical appliances, computers, radio towers, and mobile phone base stations. The fields can affect living things as well as electronic devices. The effect of low levels of electrosmog generated by personal electronics and its potential harmful effects is controversial. High numbers of people have claimed negative health effects; however, further studies by doctors have been inconclusive.

Crowded transmission masts can often be seen in urban areas.

in focus

Technical standards

The Global System for Mobile Communications, or GSM, is an international standard for digital mobile phone networks. It includes voice telephone systems and also the transmission of text messages. The GSM standard has gradually been expanded, with the addition of HSCSD, GPRS, and EDGE, among others. These standards and technology allow mobile and cell phones to carry out data transfer and offer Internet access.

Among the latest standards for mobile phones and cell phone networks is the Universal Mobile Telecommunications System, or UMTS, which allows even faster data transmission. Presently, GSM and UMTS networks are being operated in parallel, but UMTS will soon replace GSM.

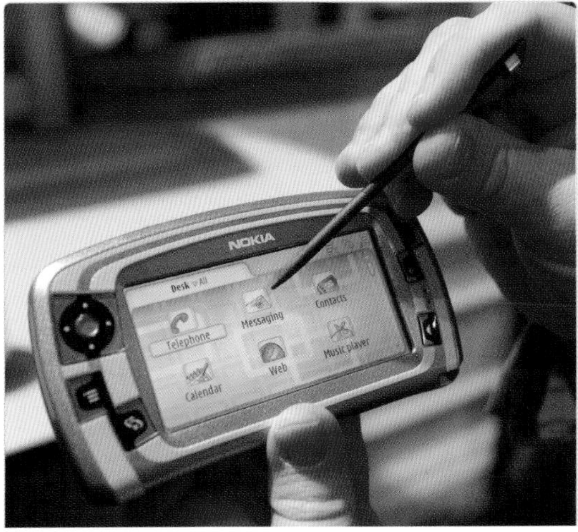

Smartphones combine the features of a personal digital assistant (PDA) with a cell phone.

basics

IN ADDITION to land-based cell phone systems, satellite networks also offer mobile calling services. Because of their higher calling costs and larger, more awkward phones, they are mainly used in places offering no other means of communication, such as unpopulated areas. Despite the cost, so-called Sat Phones remain popular with sailors, mountain climbers, and the military.

The universal mobile phone

Modern cell phones serve as portable offices, featuring diaries, word processing, e-mail and web browsing (p. 388). They can also receive radio and television broadcasts, function as digital cameras and with an optional GPS, operate similarly to an in-car satellite navigation system. Cell phones can also be used to buy items in special stores, pay for parking or purchase tickets.

audio technology

The conversion of sound waves into electrical signals opens up a wide range of possibilities for recording voice, music, and other sounds, and then storing, processing, and replaying them.

Sound waves are traveling vibrations in gases, liquids, or solids. Humans can generally perceive a range of some 16 to 20,000 hertz (cycles per second), although this varies according to age, hearing ability, and the volume of the sound.

Various microphones are needed for different applications, such as voice, music, or scientific sound analysis.

Microphones

To convert sound into electrical signals, microphones are used. Microphones can be categorized largely based on their sound fidelity, frequency range, and applications for which they are used. Various methods can be used to transform sound into electrical impulses. Most microphones have a thin elastic membrane that is vibrated by sound waves. In moving-coil microphones, the membrane transfers its motion to a coil within a magnetic field. This produces an alternating electrical current corresponding to the pattern of the sound. The resulting electrical signals are then amplified, so that they can be recorded and further processed.

Storage

Today voice and music are usually recorded and stored digitally. To do this, electrical signals coming from the microphone are sampled several thousand times per second. The measured values are then coded and stored for later processing on a computer.

Music can be stored in the form of MP3 files on a computer's hard drive or other storage medium. The huge volume of data is reduced using various techniques. For instance, only the components of the recording that humans can actually hear are stored. The compact disc (CD), on the other hand, is an optical storage medium. It consists of a plastic disc with a layer of aluminum and a transparent coating. The digitized sound signal is stored in a series of tiny indentations in the aluminum layer, laid down in spiral form around the disc. Playback devices use the beam of a semiconductor laser to read the data from the disc. The original electrical signal can then be recreated from the digital data, amplified and transformed back into sound by speakers.

Speakers

Speakers create sound waves from electrical signals. A common type is the dynamic speaker, which uses an electrical coil suspended within a cup-shaped permanent magnet. The coil is connected to a membrane. When a pattern of electrical signals flows through the coil, the membrane vibrates in response to the signal, producing sound. Larger membranes are better suited for emitting lower notes, while smaller membranes reproduce higher tones more effectively. Thus signals are routed to various sizes of speaker within a speaker box, depending on their frequency ranges.

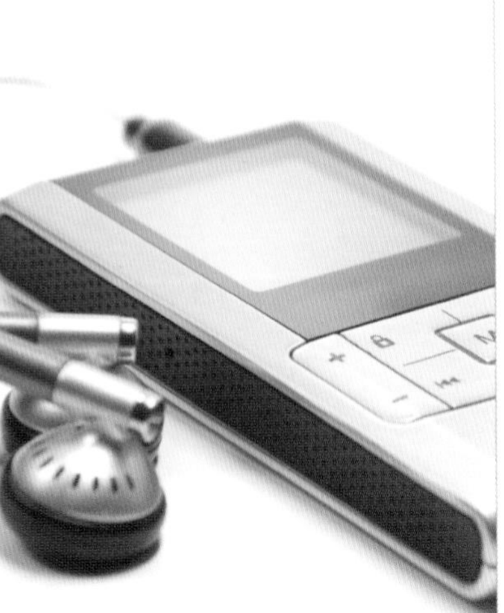

Modern digital music players can hold hundreds of almost CD-quality albums in less space than a CD case.

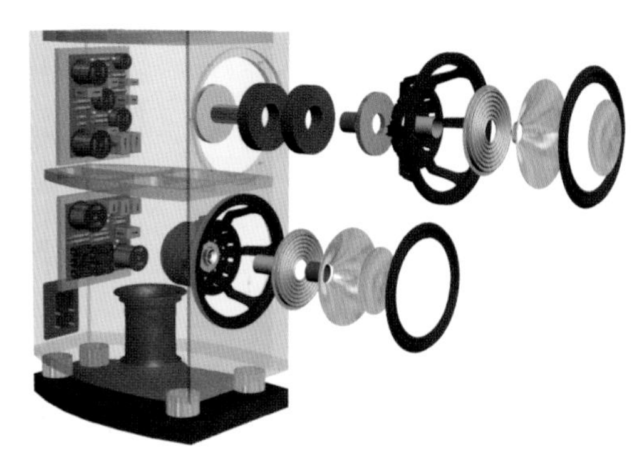

Multiple loudspeakers for various frequencies are assembled together in a box.

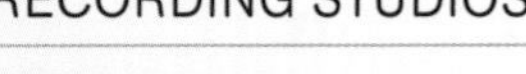

RECORDING STUDIOS

practice

In a recording studio, voices, music, and other sounds are recorded and processed for audio CDs, radio broadcasts, or movie sound-tracks. The facility may include a control room as well as the studio where the recordings are made. Soundproofing and other techniques are used to create excellent acoustic conditions. With the help of multi-track recording devices or computer-based systems, several different sound sources can be recorded separately.

At the mixing console, recordings can be combined or "mixed" as desired.

basics

SINE WAVES are perceived by the ear as "pure" tones. Increasing numbers of cycles per second (higher frequencies) produce a higher note. Noises are made up of a mixture of tones, resulting in irregular sound waves.

digital cameras

Traditional film cameras are rapidly being replaced by digital models. Instead of using photographic film, today they are equipped with high-resolution digital sensors.

A basic camera is an opaque-fronted box containing a photodetector on its back. Earlier, the sensor was a strip of light-sensitive photographic film; however, today an electronic chip is used. Light reaches this chip through optical lenses in the camera's opposite side.

A digital camera: The large screen enables a preview of previously taken photographs, an advantage not shared by older film cameras.

Shutter and aperture

The shutter of a digital camera regulates the sensor's exposure time. Usually open for just a fraction of a second, it must remain as such until the sensor has received sufficient light to produce an image. A short exposure time allows the photodetector to capture a clear image of a target, even if it is moving rapidly. Longer exposure times blur the movement, a desired effect in art photography. The amount of light reaching the sensor depends on the aperture, or circular opening within the lens. This device is similar to the iris in the human eye.

The shutter speed and aperture setting must be coordinated: if the exposure time is short, the aperture must open wide, and vice versa. The aperture setting also affects the sharpness of the picture. A small opening means increased sharpness, and a greater depth of field. A larger aperture setting reduces the depth of field. More modern cameras can automatically adjust these settings.

The photographic lens

The camera's lens assembly is focused to form the sharpest possible image on the sensor. It usually consists of several optical lenses, to compensate for color or image distortions. A standard lens captures a scene with almost no magnification, while a wide-angle lens has a short focal length, allowing the sensor to capture broader views, such as landscapes. The long focal length of telephoto lenses produces a narrow picture angle making distant objects that appear small fill the photo.

With a zoom lens the focal length can be changed, while the focus remains sharp, allowing the user to change the size of the image and adjust the viewing angle.

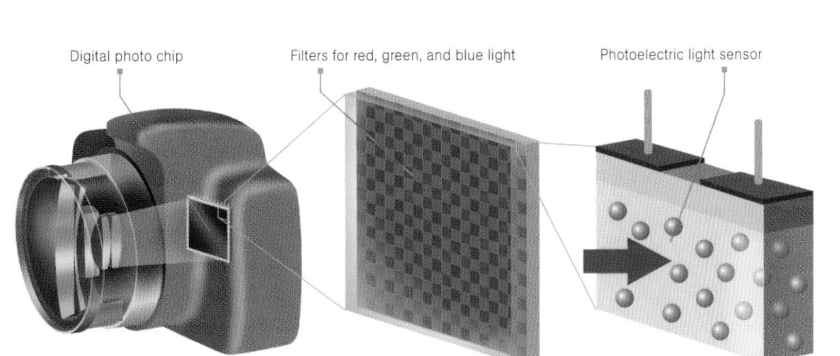

An example of the checkerboard layout of pixels, or "bayer filter array," used in the vast majority of modern digital cameras.

in focus

PHOTO DETECTORS

A digital camera's sensor contains tiny, light-sensitive semiconductor pixels arranged in a checkerboard pattern. As light enters the camera and falls upon each pixel, it releases a certain number of electrons depending on the light intensity. The values of the resulting electrical charges are then measured. As every pixel is also covered by a red, green, or blue filter, the light's color value can also be measured, producing a color photograph.

A charged-coupled device (CCD) from a digital camera is an analog shift register.

A photographic lens, mainly used in professional SLR cameras.

basics

THE NUMBER OF PIXELS used by modern digital cameras is more than sufficient. Thus, more important factors for picture quality include the size of the sensor and its photo elements, the quality of the lens, and the skill of the photographer.

video technology

Video technology provides the means to record, store, process, and view sequences of images. Using computers, both professionals and amateurs can easily work with their video material.

A professional video camera, as used by TV broadcasters and professionals around the world.

A video camera takes a series of individual pictures in rapid succession which are converted into electronic signals, so that they can be stored electronically or processed using a computer. The recording can also be viewed immediately on a monitor. When the images are presented to the eye in rapid succession, they create the perception of movement.

Video cameras with built-in storage are called camcorders. A microphone is included, to record sound and high-end cameras allow users to attach specialized microphones (p. 396) depending on their needs. The sound signal can be monitored during the recording process using earphones.

The most important part of a video camera is the photosensor (p. 397). Many newer models have separate sensors for each of the three primary colors: red, green, and blue. Electronic optic stabilizers are also used, to prevent blurring.

PLASMA SCREENS

in focus

Each pixel in a plasma screen has three tiny chambers. These contain a low-pressure mixture of the noble gases neon and xenon. Each chamber can be electronically "ignited," causing the gas to emit ultraviolet radiation. The radiation then strikes a light-emitting substance. Depending on the chamber, the substance shines red, green, or blue.

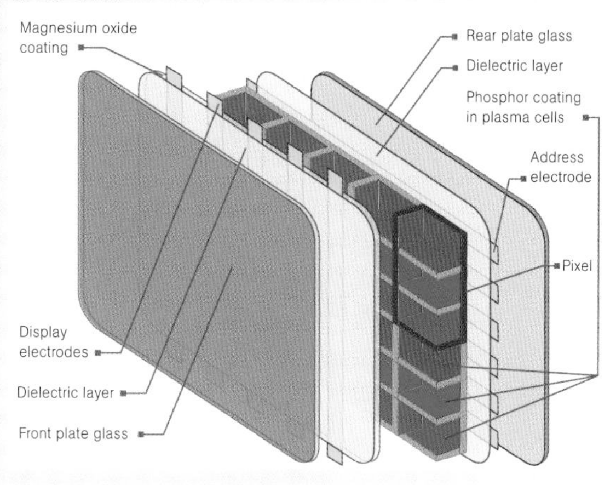

Structure of a plasma display: With modern LCD and plasma screens, the pixels are stimulated by electronic signals. Thus these screens can be quite flat and compact in shape.

TO PRODUCE COLOR effects, pixels within a screen use the "additive method," based on the primary colors red, green, and blue. The combination of red and green light produces yellow. Green and blue together make cyan, while red and blue produce magenta. Combining all three of the primary colors produces white light.

basics

Storage and processing

Modern camcorders store images digitally. Magnetic cassette tape has long been used for this purpose; however, modern camcorders now store recordings on built-in hard drives or less spacious recordable DVDs and memory cards. Camcorders using memory cards for storage can be very compact in size. In order to process a recording, it must be transferred onto a computer's hard drive, where editing software can be used to rearrange scenes, create transitions, and add special effects, titles, and sound.

Video screens

Video material is generally viewed on screens, such as televisions or computer monitors. On the screen, each individual picture is made up of tiny pixels. For color screens, each pixel includes three illuminated dots in the primary colors red, green, and blue. The brightness of the dots varies to produce a particular color that is perceived by the eye.

In traditional television screens, a stream of electrons stimulates individual pixels to produce light. They do, however, require a large vacuum-containing cathode tube, so they are increasingly being replaced by the more efficient plasma and TFT (thin-film transistor) flat-screens.

Video screens come in all shapes and sizes, and are often grouped together in broadcast and news offices displaying breaking news.

see also: Optics, pp. 324–325

radio technology

Wired connections are not always convenient for the transmission of news or other information. Instead, electromagnetic waves are used, traversing the world with an unseen plethora of radio transmissions.

Electromagnetic waves consist of modulating electrical and magnetic fields, sent and received using various kinds of transducers, commonly known as antennas.

A parabolic radio antenna: This sort of transmitter creates a highly focused and strong signal.

The frequency of an electromagnetic wave is its number of vibration cycles per second. The unit of frequency is hertz (Hz). Another characteristic of the wave is its wavelength, or the distance between two successive wave peaks. Waves in the range of millions of hertz (megahertz or MHz), are used for radio broadcasts, among other applications. They may have wavelengths of several yards. Microwaves, in contrast, have frequencies in the billions of hertz (gigahertz or GHz), and wavelengths of only a fraction of an inch. The size and shape of an antenna depends on the wavelength range for which it is designed. Thin metal rods are used to receive typical radio frequencies. For microwaves, on the other hand, parabolic antennas are used. The "bowl" of the antenna concentrates the waves at the parabola's focal point. There they are picked up by a detector and sent to the receiving device.

basics

THE ELECTROMAGNETIC spectrum is split into frequency ranges, or bands, which are used for different technological applications. Various bands are allocated for radio, aircraft communications, satellite television, and two-way radio links.

Broadcast methods

The use of electromagnetic waves to transmit relies on a principle called modulation: the signal is "imprinted" in some way on a carrier wave. The type of carrier wave used depends on the application; for instance, a VHF (very high frequency) radio station uses a carrier wave in the VHF band. Several types of modulation are used for various radio technologies. In the simplest case, the strength of the carrier wave varies according to the pattern of the sound signal (amplitude modulation or AM). The receiving device must be tuned to the desired carrier frequency. The waves received are then amplified, and a demodulator separates the variations from the carrier wave, reproducing the voice signal. Another modulation method (frequency modulation or FM) alters the frequency of the carrier wave according to the pattern of the transmitted signal.

Digitization

As with many other fields of electronics and telecommunications, radio technology is increasingly moving toward digitization. Voice and video signals are transformed into data (p. 394) that can be manipulated using computers. These data files are easier to process and manage; for instance, data errors caused by electromagnetic disturbances can be corrected even after many transformations and transmission over long distances, creating a much clearer signal. Digitization also allows various types of data and routing information to be combined. Therefore, in digital radio systems, supplemental text information can be transmitted along with the audio signal.

An amateur "ham radio" enthusiast: Although termed amateur, they are often highly skilled radio operators, frequently relied on in emergencies, and when standard communication systems break down.

in focus

PROPAGATION OF RADIO WAVES

In space, waves spread out in straight lines. On Earth, their movement depends on the wavelength. Longer waves act as "ground waves," spreading with the help of the Earth's electrically conductive surface. Shorter waves expand in all directions; however, they are directed back to the Earth's surface by the ionized upper atmosphere. Through a zigzagging process of reflection, they travel around the entire planet.

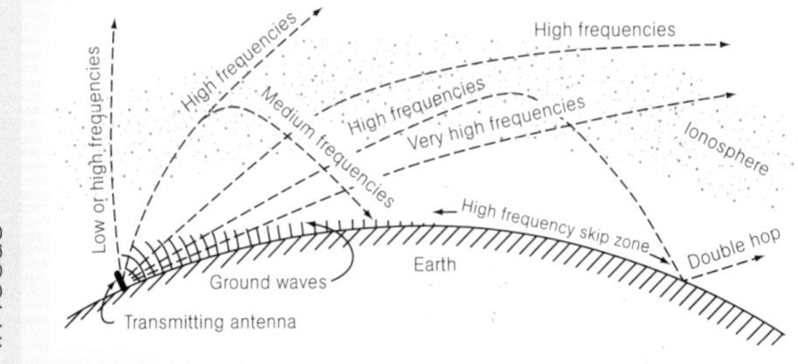

see also: Vibrations and waves, pp. 316–317

internet radio and tv

Today an enormous variety of movies, television programs, and radio broadcasts can be transmitted over the Internet and other networks, giving rise to a range of new applications, media, and services.

Internet, or Web radio, is traditional radio broadcasting transmitted over the Internet (p. 387). Many radio stations offer online programming, allowing access to material that aired previously. Stations can also reach listeners over the Internet who cannot receive over-the-air broadcasts. Many radio stations have emerged that offer exclusively Internet-based programming.

Internet-TV programs can be viewed using a computer, mobile phone, or a television equipped with a specialized add-on receiver.

Sending and receiving technology

Internet radio programs are transmitted as a constant data "stream." Unlike normal downloads, users need not download large data files. A stream is processed as it downloads, meaning users can begin listening to the radio program almost immediately. In fact, in the case of live programming, this is the only option. This type of data transfer is called streaming and can be received using streaming media computer software, as well as wirelessly connected mobile devices.

Internet TV and IPTV

Broadcasting movies and television programs over the Internet is called Internet TV or Web TV. This system is similar to Internet radio. However, because of the larger amounts of data, high-speed Internet connections are necessary for good quality viewing. Television broadcasts are also offered under a system called IPTV; these are often broadcast over dedicated networks using Internet standards.

Internet broadcasts are accessible to nearly everyone, often at no cost. However, since the Internet is an open network, broadcast quality can vary. IPTV, by nature of being offered mainly by providers connected to larger data networks, is usually of better quality.

UMTS or 3G capable phones and laptops allow Internet access at speeds capable of streaming media.

Podcasting

In podcasting, audio or video files are offered for downloading via the Internet, usually free of charge. An extremely wide range of content is offered through podcasting. Increasingly, professional producers are also getting involved; for instance podcasts may be offered to supplement the content of online magazines or political party Web sites.

DUE TO THE MODERN developments of the Internet, traditional media such as television, radio, and magazines are beginning to overlap and merge. This media convergence is, in turn, giving rise to new applications.

IN A VIDEO-on-demand system, users can choose films from an archive. The movies are transmitted over an Internet connection or other data network from special high-speed video servers maintained by the provider.

basics

Many podcasts and Internet radio streams are privately produced, allowing anyone to establish his or her own "broadcasting studio"; however, intellectual property rights to the broadcast material must be considered.

film technology

Producing a movie or documentary for the cinema or DVD is a highly complex undertaking, both technically and logistically. A film must go through numerous production stages before it can be distributed to the public.

An old style cinema: Before the advent of television, lavish movie halls were commonplace, and part of the popular culture of the age.

Traditional movie cameras shoot 24 frames per second on a light-sensitive film strip. The film is fed to the camera from a reel, rolling frame by frame in front of the lens where it is quickly exposed and rolled onto another spool.

The camera can also be set to expose more or less than 24 frames per second. When the resulting film is replayed at a normal speed, the action on the screen appears either faster (time-lapse photography) or slower (slow motion).

When the camera must be maneuvered during filming, a special steadicam can be used. With this system, the camera sits on a freely moving joint on an armature just above the center of gravity. The joint is attached to a iso-elastic arm which acts as a shock absorber. The monitor and batteries are attached lower on the armature, serving as a counterweight.

MODERN CINEMAS use complex interconnected sound systems with speakers on all sides of the theater. Special speakers provide low-frequency effects such as rumbling, booming, and vibration.

basics

Digital video cameras (p. 397) are replacing traditional film cameras. While both types of camera offer advantages and disadvantages, digital films are significantly easier to copy, distribute, and store than traditional film reels.

Movie sound tracks are often recorded separately from the film. To ensure that the sound and picture fit together, they are electronically synchronized. Voices, music, and background sounds can be externally recorded, allowing audio to be used that is unaffected by the acoustic conditions of the filming location.

Movie studios create many blockbuster films as part of a multibillion-dollar worldwide industry.

Special effects

Starting with a digital version of the film, special effects can be added. With blue-screen technology, the background is shot separately. The actors are then filmed in front of a blue background, a color that can easily be distinguished from skin tones. Using a computer the actors can then be superimposed onto the background. Backgrounds, scenes, or individual characters can also be created through computer animation. These can then be copied into the movie or pieced together to create a fully animated film.

FILM EDITING

During the editing process individual scenes are arranged for cinematographic effect in order to create the finished film. For Hollywood-type movies, the film reels are usually optically scanned and then stored digitally. Individual scenes can then be processed, graphically manipulated on a computer, and reassembled. The final result is copied onto a film reel or DVD.

Computers have given unprecedented flexibility to film editors and have become an essential tool in the industry.

practice

Desktop computer software can be used not only for the creation of special effects, but also for the modeling of real-world props.

printing technologies

The performance of today's printing presses would surely astonish Johannes Gutenberg, who invented printing with movable type in Europe. Modern methods have resulted in affordable printing on a large scale.

Although the rise of digital information and the spread of the Internet has challenged the position of the book in modern culture, paper printing is still a popular information medium. The classical methods still used for printing can be divided into three main groups: raised, flat, and gravure printing. In raised printing, the letters or images carrying the ink to be printed are higher than the surrounding surface, as with the movable type originally used by Gutenberg. The most important flat printing method (p. 403) is the type seen in books, which are often printed using offset printing techniques. In gravure printing, the image to be printed is represented as indentations in the printing plate. This method is used in lithography and linocuts, and is especially popular for applications such as magazines, newspapers, and other high-volume texts. Gravure printing offers an excellent color print quality and low per-page costs; however, the production of each printing plate is quite expensive.

Offset printing presses, a development of lithographic printing priciples using the interaction of water and oil, are commonly used in commercial mass printing, such as books.

On-demand publishing

Because the creation of printing plates is relatively complex, traditional book printing requires a minimum run of some thousand copies. For small presses and unknown authors, this often represents a substantial financial risk. Help for these publishers came in the form of "print on demand" (POD) technology, developed in the 1990s. Using a purely digital process, POD does not require the creation of printing plates, and books are printed only when an order is received. There is another side to the resulting POD boom, however: nearly any text can be published without previous quality screening or editing. Because of this, these publications now have a questionable reputation.

Glossy color magazines with large readerships are usually published using gravure printing.

milestones

JOHANNES GUTENBERG

Johannes Gensfleisch (1400–1468), also known as Gutenberg, did not invent book printing. Books were being produced in China as early as the 11th century and printing processes using stamps were already familiar in the ancient world. His innovation lay in introducing units of type that could be quickly rearranged and reused for each page. This transformed the book into a product that could be duplicated many times. The spread of the Bible—especially in Martin Luther's translation—led to far-reaching changes in European culture and society.

In the mid-15th century, Johannes Gutenberg invented book printing with movable type.

in focus

MASS-MARKET PRINTING

Modern technology has led to economic mass-printing techniques.

Many innovations have led to modern mass-printing technologies, including the invention of the steam-powered press (1810) and the rotary press (1845). Today, the manufacture of printing machines is a prominent branch of industry. The world leader in offset printing press production is the German company Heidelberger Druckmaschinen AG, with a market share of 40 percent. The market segment for consumer computer printers enjoys high sales figures; however, these printers are not suitable for mass-market printing.

newspapers and magazines

Newspapers and magazines should be up-to-date and as economical as possible, but are not generally intended for long-term use. Thus, different printing and production methods are used than for books.

Newspapers are printed in large, continuous rolls of paper using offset printing machines.

Dramatic news headlines on huge rotating rollers of printing machines often appear in old movies. These offset printers, employed by many newspapers, use a technique involving the interaction of water and oil. The image is first transferred to a printing plate in the form of water and oil-resistant areas. Printing ink is only taken up in the water-resistant areas. The plate is mounted on a drum and pressed to a blanket cylinder, which in turn prints the image on the paper. This indirect printing method is used to help protect the expensive printing plate. In book printing, so-called page signatures consist of 16 or 32 pages being processed at a time; however, newspaper printing uses continuous rolls of paper in a roll-to-roll or rotary printing process.

Stock photography agencies

Newspapers and magazines without illustrations are unattractive to many readers. However, few newspapers can afford to send their own photographers to the hot spots of international events. In turn, even professional photographers may lack the resources to successfully market their photos. The gap is filled by stock photography agencies. These agencies purchase the photo rights from photographers, paying either a flat rate or a portion of the proceeds from each photo. Publishers, newspapers, and advertising agencies can subscribe to the agencies' catalogs and search their databases for the right photo for nearly any news article or written editorial piece.

Behind the scenes at a newspaper office

What you see in today's newspaper was determined yesterday by news editors and department chiefs in an editors' conference. For important topics, the publisher and sometimes even the newspaper's owner may influence the story's presentation. Staff journalists research the topics selected, or assignments may be given to reporters and correspondents. Another common option is to purchase finished articles from a news agency. In-depth articles and series may be planned over a longer term or bought from freelance journalists. Photo editors are responsible for illustrations, which may come from the articles' authors or from photo agencies. Text, photos, and advertising are assembled by layout staff on a computer using page layout software before being sent on for proofreading. When the editor-in-chief approves the final edition, it is sent electronically to the printer and in many cases is also published on the Internet.

The influential *New York Times* newspaper (founded in 1851) earned the nickname "The Gray Lady" for its sober style and appearance.

basics

PHOTOCOPIERS AND laser or LED printers use the same basic technology: the image to be printed is transferred to a rotating drum using light from a laser or LED. Toner attaches to the electrically charged areas on the drum and paper is pulled over it, transferring the image to paper.

in focus

NEWSPAPER OF THE FUTURE

Increasing numbers of people are turning to their computers for fast and up-to-date news information. However, printed media is still popular and is likely to remain so into the future. Many newspapers now publish their information on the Internet, retaining the typical layout. The reader can either print the pages, or load the text on to a portable device to read as digital text. Researchers have also been experimenting with the concept of electronic paper for several years. This would allow digital text to be read off a screen as thin and flexible as paper.

Nowadays many traditional newspapers have their own Internet portal.

MATHEMATICS

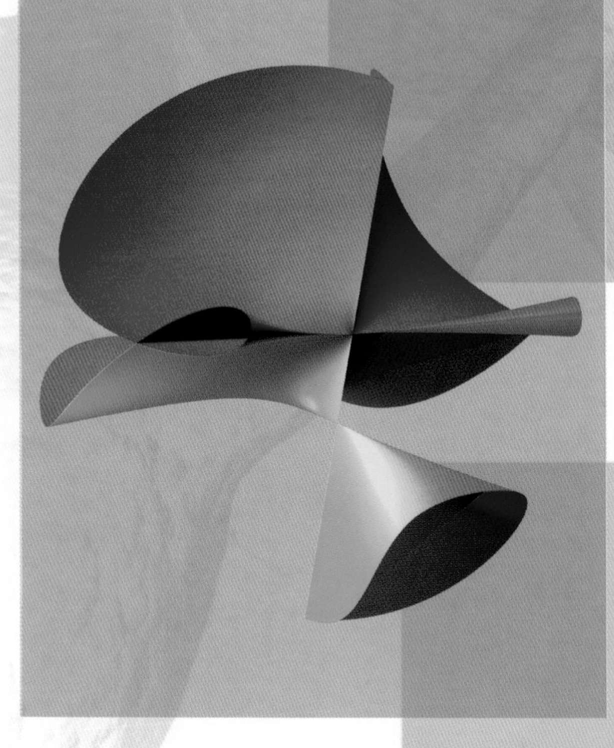

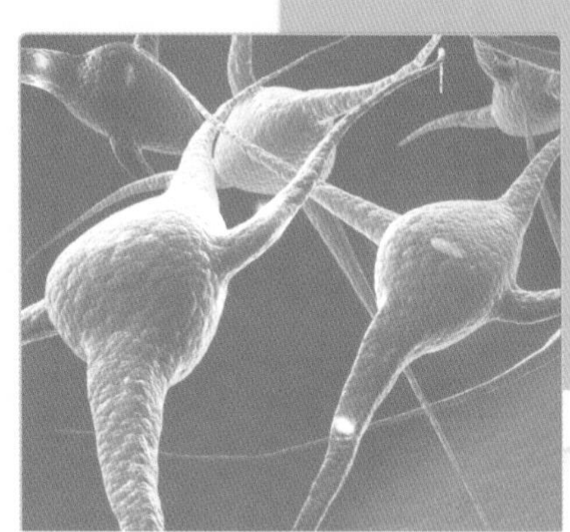

MATHEMATICS

Mathematics is a science based on abstract notions, which in turn are created by logical thought processes from concepts such as numbers, shapes, and structures. It is distinguished from other sciences by the validity of its findings. Yet, while all findings in the natural sciences must be confirmed by experimentation, only logical thinking proves mathematical statements. Mathematical theorems are based on strictly logical evidence, and are therefore final and generally valid truths. Moreover, although these results are pure thought processes, they find remarkable applicability in the natural sciences and in technology.

the first mathematicians

Civilizations in the ancient Middle East, China, India, and ancient America explored mathematical problems long before mathematics became a discipline in itself.

Early artwork depicting an Egyptian farmer during a land survey.

history | numeric systems | subject | practical application

HISTORY OF MATHEMATICS

Mathematics as a science was more than likely established by Pythagoras of ancient Greece. He and his teacher Thales of Miletus are considered two of the first philosophers. From its very beginning, mathematics was tied to philosophy. To this day, logic is still very much a part of both disciplines, mathematics and philosophy.

The first approaches to mathematics as we know it today date back about 5,000 years (before 3000 B.C.), from Mesopotamia. They generated the Sumerian, Babylonian, and Assyrian cultures. The most important mathematical scripts originate from the old Babylonian period (around 2000 B.C.). These included formulas for calculating areas, volume, approximate values for the number pi, calculation of √2 with an accuracy of six decimal digits and references to the Pythagorean theorem about a right-angled triangle.

The most important Egyptian sources that remain from the Middle Kingdom (around 2100 to around 1790 B.C.) are two papyrus manuscripts and one leather script proving that basic calculation methods, fractions, geometry, and the number pi were already known in ancient Egypt. Despite these advances, neither of the two cultures has passed down any mathematical proofs.

High point of mathematics

Mathematical historians concentrate mainly on ancient Greece, as the Romans were predominantly preoccupied with architecture and law. The Greek philosopher Thales of Miletus is considered the first mathematician who demonstrated by logical argumentation that his theorems were valid. He was born around 620 B.C. The Greeks have passed down formulas, laws, and rules about geometry which are still valid today. Not only did they prove their own formulas but also the ones previously stated by the Babylonians and Egyptians.

In contrast to the Babylonian, Egyptian, and Greek civilizations, the history of Chinese mathematics dates back to 4,000 years ago. *The Jiuzhang Suanshu* or *The Nine Chapters on the Mathematical Art* is the most famous Chinese textbook on mathematics of all time. It originates from about A.D. 100 and contains solutions to problems as well as algorithms. Additional texts about the proof of algorithms were written up to the 13th century A.D., the high point of mathematics in China.

Calculations and geometric tools were also necessary in ancient architecture.

milestones

EUCLID'S ELEMENTS

Euclid's manuscript *The Elements* is one of the most significant sources of Greek mathematics. It is the first systematic summary of fundamental geometry and arithmetics. Euclid derives the characteristics of geometric objects and natural numbers from axioms (basic statements that do not require proof) and covers geometric algebra, proportions, primes, divisibility of numbers, and the method of exhaustion.

Euclid of Alexandria, Greek mathematician, around 365 to 300 B.C.

see also: Construction, pp. 368–369

numeric systems

Humans created the concept of mathematical numbers to describe quantities such as the size of an animal herd with a limited amount of effort, and to put these quantities in order.

The development of numeric systems was preceded by a lengthy process. This is due to the level of abstraction required in associating the same terms with the same quantities of objects in order to find a generally binding description of quantities; for example not only three cups are associated with the number three, but also three spoons, three apples, and so on. Therefore, the number "three" is the result of an abstraction process, which took some time to establish in various cultures. Today still, some indigenous people only use small numbers, which they associate with particular, tangible objects.

INVENTION OF THE NUMBER ZERO

in focus

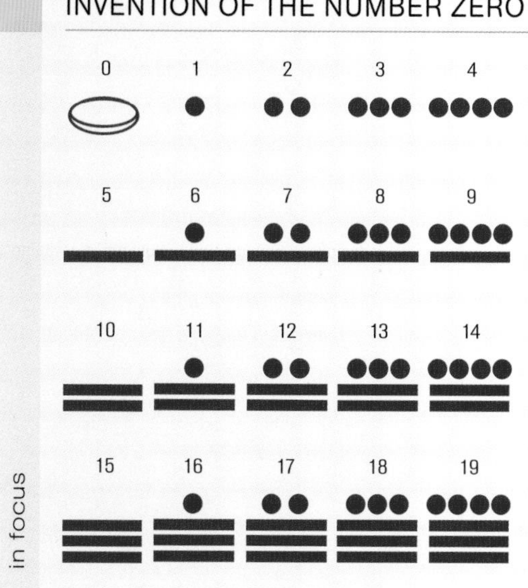

The number zero has been invented three times: by the Babylonians (around 500 B.C.), the Maya (around 50 B.C.) and by the Indians (around A.D. 500). A zero digit plays an important role in a positional system, as it allows differentiating, for example, between the numbers 309, 390, and 39. Before the invention of a placeholder character, the Babylonians would always leave the appropriate position empty which often resulted in misunderstandings and uncertainties. Leonardo Fibonacci introduced the zero in Europe (around 1200).

Numeric character of the Maya: the symbol for zero looks like a snail shell.

THE ISHANGO BONE is a notched bone that is at least 20,000 years old, found in Central Africa in 1950. Some researchers suspect the first traces of an arithmetic system, while others suggest the notches might represent a moon calendar.

issues to solve

Fingers and toes

Humans have always used their fingers for counting. This explains the different increments used in various numeric systems.

Some cultures used the five fingers of one hand or ten fingers of both hands, while others used the ten toes in addition to the fingers. The numeric system in ancient Egypt, for example, was based on increments of ten, while the Maya and Aztec counted in increments of 20. The French number *quatre-vingt* (four times 20) meaning 80, and a similar form used in Danish, are traces of this numeric system that uses increments of 20. The Babylonians were an exception: they used increments of 60. The

Calculation of the area of a lot in Umma, Mesopotamia, written in Sumerian cuneiform script, 2550 B.C.

graduation of hours into 60 minutes of 60 seconds each and angular units are still used.

Additive and positional system (place-value notation)

In an additive system, the value of a number is given by the sum of its digits, where the order of number signs does not affect the value. Such a system was used by the Romans among others, but large numbers are difficult to express and the system is completely unsuitable for multiplication, fractional arithmetics, and advanced mathematics. The positional system differs, as the value of individual digits depends on the position. This is seen in our decimal system (e.g., $1{,}243 = 1 \times 10^3 + 2 \times 10^2 + 4 \times 10^1 + 3$). The latter originates from India and was imported into Europe by Arabic mathematicians. The Babylonian sexagesimal system is based on the same principle (e.g., *243 = 2 x 602 + 4 x 601 + 3*).

European	0	1	2	3	4	5	6	7	8	9
Greek		α	β	γ	δ	ϵ	ς	ζ	η	θ
Chinese	○	一	二	三	四	五	六	七	八	九
Urdu Naqsh	ز	ز	ژ	س	ش	ص	ض	ط	ظ	ع
Devanagari (Hindi)	०	१	२	३	४	५	६	७	८	९
Roman	0	I	II	III	IV	V	VI	VII	VII	IX

X - ten, L - fifty, C - hundred, D - five hundred: Roman numerals are not suitable for arithmetic.

The decimal system with ten digits from zero to nine has been established throughout the world. However, its characterization varies.

see also: Humans, p. 217

the subject of mathematics

Mathematics is more than numbers and geometric figures, it is a logical science that deals with structure, space, quantity, and change. Mathematicians study the relationships between these concepts.

Mathematics is classified as a structural science and works on theories that it creates. A theory is a system of statements about an object. Mathematics takes the relationships between objects and generates a structure called a theorem. A theorem, while derived from basic statements or axioms that are taken to be self-evidently true, can only be proved by making further assumptions. The power of its theorems gives mathematics its importance. If a question is able be described mathematically, then it can also be solved mathematically.

Process of abstraction

Mathematics allows abstraction and disengagement from concrete objects, while still providing for retranslation into everyday situations. For example, through observation the average number of times that a certain number would be rolled on a die with every sixth throw can be determined. To save time, mathematics can be used to determine the chances of rolling say a "six" without even having to throw the dice.

Fields of application

Mathematics is composed of many different branches, some of which may overlap. Mathematicians segregate into two groups, pure mathematics and applied mathematics. These divisions are based more on what their goals are than what branch of mathematics they study. Pure mathematicians study mathematics in an abstract form, with little thought given to practical application. Applied mathematics, as the name implies, focuses on the ability to apply mathematical knowledge to solve real problems.

The major branches of mathematics include number theory, topology (an extension of geometry), numerical analysis, and discrete mathematics (dealing with finite countable structures), along with the disciplines normally taught in school, such as algebra and geometry. Numerical analysis and discrete mathematics are new fields of study that were developed in the 20th century with practical applications to sciences, business, and other domains in mind.

basics

AXIOMS are elementary statements that do not require any proof.

AN AXIOMATIC SYSTEM is a set of noncontradictory theorems of a mathematical theory.

MATHEMATICAL FACTS are derived from axioms, such as seen in classic geometry.

Sum	Expected results	abs. freq.	rel. freq.	P(X)
2	(1|1)	15	0.0375	1/36 = 0.028
3	(1|2); (2|1)	14	0.035	2/36 = 0.056
4	(1|3); (2|2); (3|1)	30	0.075	3/36 = 0.083
5	(1|4); (2|3); (3|2); (4|1)	45	0.1125	4/36 = 0.111
6	(1|5); (2|4); (3|3); (4|2); (5|1)	54	0.135	5/36 = 0.139
7	(1|6); (2|5); (3|4); (4|3);(5|2); (6|1)	58	0.145	6/36 = 0.167
8	(2|6); (3|5); (4|4) (5|3); (6|2)	51	0.1275	5/36 = 0.139
9	(3|6); (4|5); (5|4); (6|3)	62	0.155	4/36 = 0.111
10	(4|6); (5|5); (6|4)	38	0.095	3/36 = 0.083
11	(5|6); (6|5)	25	0.0625	2/36 = 0.056
12	(6| 6)	8	0.02	1/36 = 0.028

Procedure for calculating the probability (P(X)) of a certain sum occuring when two dice are thrown 400 times.

milestones

FIBONACCI NUMBERS

Leonardo da Pisa, also known as Fibonacci, was an arithmetician in Pisa and is considered the most influential mathematician of the Middle Ages. With his greatest work *Liber Abaci* he introduced Europe to Indian arithmetics and the Arabic number system we use today. In modern mathematics, his name is associated with a series of numbers. Attention was drawn to this series due to the famous rabbit problem, which was described in *Liber Abaci*. The number series begins with zero and one and each of the following numbers is the sum of the two previous Fibonacci numbers. Amazingly this number series reemerges in many other areas, for example the so-called golden section, Pascal's triangle, and the spiral-shaped alignment of leaves or seeds in many plants.

Leonardo da Pisa, also known as Fibonacci, was born around 1180.

In virtual table tennis, numerics and mathematics play an important role in data processing and calculations.

see also: Probability, pp. 416–417

practical application

Many scientific areas take advantage of applied mathematics. Modern technological advances, in particular, would not have been possible without mathematics.

Astronomy, physics, geodesy, and economics have always contributed to mathematical advances and, vice versa, mathematics has been the basis for advances in these scientific areas. Isaac Newton, for example, developed infinitesimal calculus (p. 414), which is the basis of analysis for mathematically describing the physical law that force equals rate of change in momentum. While studying heat propagation in solid objects, including constructional engineering, material physics, and design applications to test the practicality of a plan.

High-tech

Due to the continued development of numerical technology and computer performance, many aspects of society are now mathematized and computerized. Almost all high-tech applications are directly related to applied mathematics.

Electronic chips: Complex algorithms are used to optimize the arrangement of the wafer.

Wind tunnel testing: Complex algorithms are used to simulate and model aerodynamics.

Jean Baptiste Joseph Fourier also researched wave equations which describe the expansion of waves. Not only did he deduct these equations but he also found an approach for a solution, the so-called Fourier series, which is applied in many mathematical areas, for example, statistics. Examples include electronic chips, design of planes and high-speed trains, exploration of fossil fuel and natural gas sources, and other high-tech areas.

in focus

BOOLEAN ALGEBRA

Boolean algebra is named after the English mathematician George Boole (1815–1864). It is the basis of logical connectors. They are the calculation basis in dual arithmetics, which is based on the application of four arithmetic operations (addition, subtraction, multiplication, and division) using the dual-number system, that is the binary numbers zero and one. They connect two input values with one output value using three basic functions: AND gate, OR gate, and NOT gate.

The mathematician George Boole developed Boolean algebra.

Theories for the digital age

Mathematicians have also developed theories which were later applied in other areas. Complex numbers that were developed in the 16th century have now become the basis for mathematically describing electromagnetics, quantum mechanics, and so on. Another example is Boolean algebra, which is the basis for digital technology, control engineering for machines and plants and all computer programming languages. Even areas in architecture are based on mathematical principles

Quantities of goods and their associated sizes need to be calculated in order to coordinate their logistics.

geometry | arithmetic | numbers

CLASSICAL MATHEMATICS

The classical themes of mathematics include numbers, geometric forms, and calculations. Euclid of Alexandria, a Greek mathematician and one of the most famous communicators of classical geometry, discusses geometric figures in three-dimensional space and on a plane in *Elements*, his most popular work. Arithmetic (number theory), defined as the science of numbers, is another classical branch of mathematics and includes basic arithmetic operations and the divisibility of numbers, as well as other calculations.

geometry

Classical geometry, one of the first branches of mathematics, describes the relationship between geometric structures. Written by Euclid, the axiomatic system is the basis for classical geometry.

Originally, geometry dealt only with measurements and comparisons of an object using well-known standards, such as an angle, a length, or a surface, that were easy to reproduce.

Euclidian geometry

Euclidian geometry uses clear definitions of points, lines, and straight lines and can be employed without using common numbers. Basic operations, such as bisecting a line or an angle, are possible without knowing the size by using tools such as protractors and rulers, regardless of whether or not they possess scales. All basic operations can be reduced to these elementary steps: drawing a straight line between two given points, drawing a circle around a given point and then through a second given point, and duplicating a given length.

Geometry with numbers

Geometric constructions are the basis of arithmetic operations. For example, addition corresponds to the joining of two lengths, while a rectangle corresponds to multiplying two lengths. Using numbers for both sides of the rectangle, the surface area can be found by multiplying those numbers.

The Pythagorean theorem is fundamental to geometry. The length of the diagonal of a square can be found using the Pythagorean theorem: $c^2 = a^2 + b^2$, where c is the length of the diagonal.

Using 1 for the sides of the square (*a* and *b*) results in: $c^2 = 1^2 + 1^2 = 2$.

milestones

THE PYTHAGOREAN THEOREM

Pythagoras was born in 570 B.C. on the Greek island of Samos. His most famous work is the Pythagorean theorem, based on elements from Indian, Babylonian, and Egyptian mathematics. It provides a theoretical foundation for geometric constructions involving right angles. His theorem states that for any right triangle, the area of the square of the hypotenuse (the longest side, or *c*) is equal to the sum of the areas of the squares of the other two sides (*a* and *b*). Therefore, $a^2 + b^2 = c^2$.

Pythagoras of Samos (570–510 B.C.) is one of the founding fathers of mathematics.

practice

FROM ANGLES TO TIME

A circle can be divided into any number of parts. The Babylonians defined a circle as 360° because 360, when divided, always results in a whole number. Why? The number 60 has the most divisors of any number between 1 and 100. If 60° is divided twice, 15° is obtained. If you divide 360° by 15°, you arrive at 24. This is the basis for telling time. The rotation of the Earth is divided into 24 segments (hours). An hour is then divided into 60 minutes of 60 seconds each.

The angle of the sun's rays can be used to measure time.

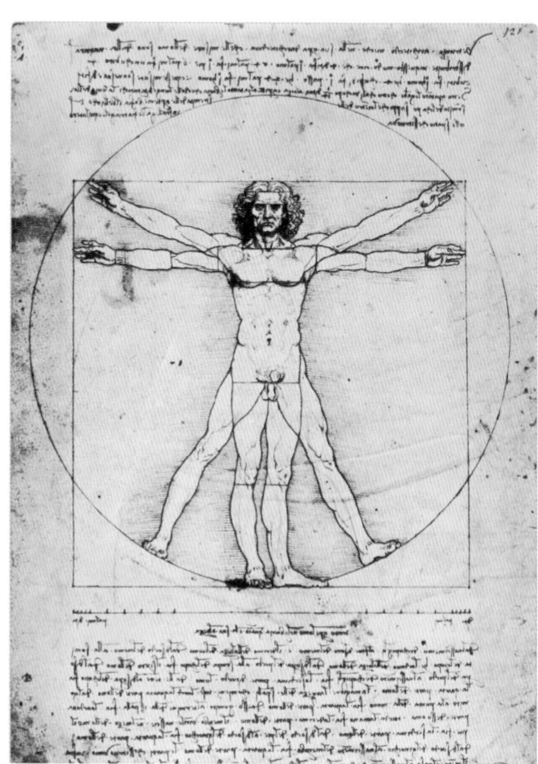

The Vitruvian Man, by Leonardo da Vinci, is a world-renowned painting that connects aesthetics and geometry.

see also: Humans, p. 217

arithmetic and numbers

In a strict sense, arithmetic, or number theory, primarily deals with calculations using numbers, leading to the examination and development of the principles of different types of numbers.

Arithmetic deals with calculations using basic rules derived from an intuitive and natural handling of objects. For example, it does not make a difference whether a person first adds two cows to four cows and then adds another three, or if he adds four cows to three and so on. This is called the associative law: $(a + b) + c = a + (b + c)$. It applies not only to natural numbers, but to all numbers in the number system.

Why are new numbers needed?

Generally, you cannot divide or subtract using only natural whole numbers. For example, dividing eight apples between three people, or distributing three apples among five people, is not possible using only whole numbers. These calculations can, however, be solved through the introduction of new kinds of numbers such as fractional and negative numbers. Eight apples divided by three people results in eight-thirds of an apple per person, and three subtracted by five is a negative two.

■ An abacas is an old mechanical calculating tool.

Number ranges

Natural numbers, N, allow addition and multiplication unconditionally. The set of integers, Z, includes negative whole numbers and allows for subtraction to occur unconditionally. The set of fractional numbers, Q+, allows for division. The set of rational numbers, Q, made up of N, Z, and Q+, allows all of the basic arithmetic operations to be carried out, except for dividing by zero.

The set of real numbers, *R*, includes irrational numbers and allows nonterminating nonrecurring decimal fractions, such as π and square roots, to be expressed. There are also the imaginary numbers, which are purely thought up. These may be meaningless in some contexts but not in others. Physics and engineering calculations are made easier through the application of imaginary numbers.

■ The structures of music and mathematics are closely linked and can be incorporated into each other.

milestones

RAMANUJAN'S NOTEBOOKS

Srinivasa Ramanujan (1887–1920) grew up in a poor family in a small South Indian town. Self-taught, he gained most of his early knowledge from two books, which included over 6,000 theorems. Later, he was able to prepare his own set of propositions which included insights into pi, prime numbers, and partition functions. He spent five years at Cambridge University's Trinity College, but due to health problems he returned to India where he died one year later. He left behind over 600 formulas, without proofs, in a notebook that caused a great sensation when it was rediscovered in 1976.

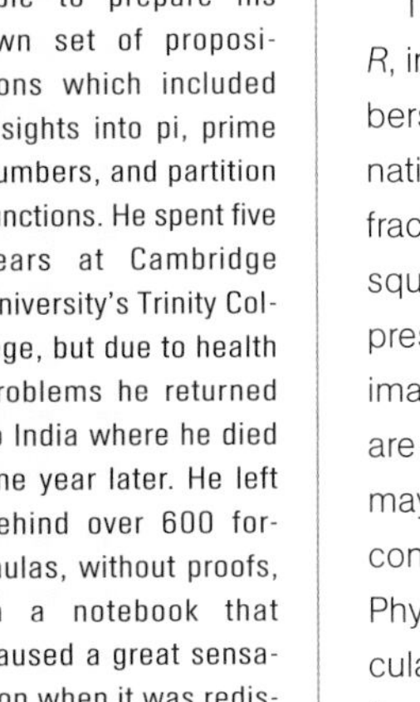

■ The exceptional mathematician Srinivasa Ramanujan, (1887–1920) died aged 33.

in focus

COMPUTING LAWS

Commutativity: The factors are exchangeable in multiplication, and the addends are exchangeable in addition.
Associativity: The order of addends or factors is immaterial, but the calculations that are within brackets are always done first.
Distributivity: When an equation uses both multiplication and addition, every addend in a bracket must be multiplied by the factor outside the bracket. Multiplication can take place after common factors are combined. This principle forms the basis of the binomial theorem.

■ The basic laws of arithmetic are a fundamental component of every school curriculum.

coordinate geometry

In coordinate geometry every point is allotted a fixed place in the coordinate system around a selected set of axes. A set of points can be represented by equations or using graphs.

If you are in a new city you can orient yourself using a fixed reference point and find out the direction and distance between this place and your next destination. In a coordinate system points are specified used to define the set of points in the line. By substituting a value for *x*, the value of *y* can be found, defining the coordinates of a point. Any coordinates of a point *(x, y)* on a straight line will satisfy the equation.

Coordinate geometry is put into practical use in GPS navigation systems in cars and other vehicles.

coordinate geometry | vectors

ANALYTIC GEOMETRY

All geometry problems are solved graphically. Using algebra and geometry together, the same problems can be solved with calculations using variables. Vectors are mathematical objects with length and direction. They can be represented by arrows placed on a coordinate system and denoted by letters set in boldface or with tiny arrows on top. They are used to describe physical quantities.

using a pair of numbers that describe the distance and direction from your fixed reference point, or origin. The origin is where two selected perpendicular axes intersect *(0, 0)*. The coordinates *(3, 4)* would therefore refer to a point three units to the right of and four units above the origin.

basics

A FUNCTION gives exactly one y-value in the output set for each x-value in the input set.

A LINEAR FUNCTION will produce a straight line if all the pairs *(x, y)* are plotted on a graph.

A point set

A straight line is an infinite number of points. The equation $y = 2x + 3$ describes a straight line and can be frequently used. Both geometry and algebra answer the question of how many points are common to two sets of points. Coordinate geometry is very important in mathematics and is a vital tool for physics; for instance it is possible to graphically depict movement of an object by choosing time as the independent variable.

Geometry and algebra

Geometry and algebra work well together. Algebra can describe and solve a geometric problem; geometry can be used to find the solution to an algebraic equation. In cases where an exact solution is impossible or too difficult to obtain, graphs are

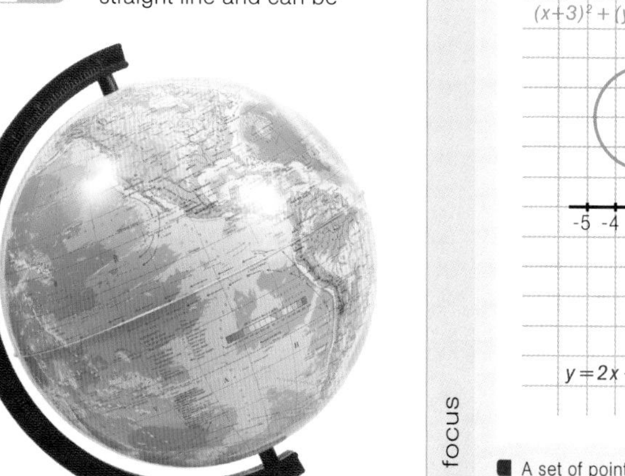

The geographic coordinates system divides the globe into latitude and longitude.

in focus

CARTESIAN COORDINATE SYSTEMS

A Cartesian or rectangular coordinate system includes two perpendicular axes that form a plane, as well as coordinate lines that are at equal distances from each other. A value along the horizontal axis is called an x-coordinate or abscissa, while a value on the vertical axis is called a y-coordinate or ordinate.

The point at which the axes cross, with the coordinates *(0, 0)*, is called the origin. This system was named after its inventor, the French philosopher René Descartes, who also investigated the fields of algebra and Euclidean geometry. The idea was also developed at the same time by Pierre de Fermat, although Fermat did not publish his findings. Today, the Cartesian system is the most widely used coordinate system, since it offers the most effective way to represent geometric concepts such as scalar products.

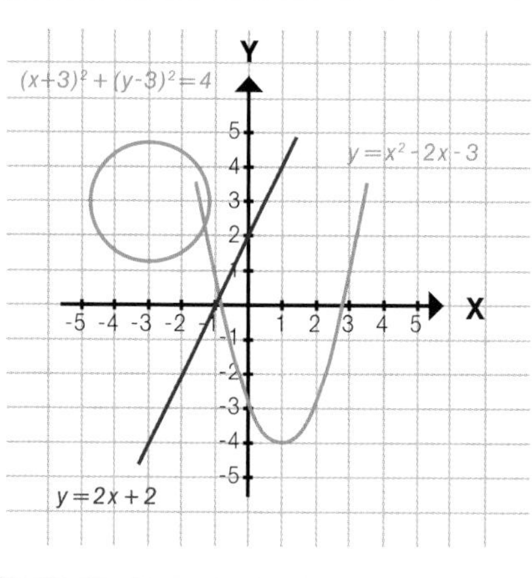

A set of points can be displayed as an equation or in graphic form within a coordinate system.

vectors

Vectors are used to describe movements in planes and space and are represented in a coordinate system by their components in *x*, *y*, and sometimes *z* directions.

A vector is often drawn as an arrow representing the movement of a point in space. The length (magnitude) of an arrow and its direction, make up a vector quantity. In geometry, all arrows with the same direction and length represent the same vector. A vector can be placed anywhere on a coordinate system and can also be defined by a pair or triplet of numbers, with the understanding that the arrow begins at the origin. A vector space is a set of vectors and various specific operations in the form of an algebraic equation. Because the structural content is independent of real quantities, vector calculations are useful in analytic geometry, physics, and engineering.

■ Adding the vectors to a diagram tells us the length of the arrow's direction.

Points and point sets

Vectors describe many different quantities. A velocity vector defines the direction and speed of an object. A position vector defines the location of a point relative to the origin of a coordinate system. The linear equation describing the set of points on a straight line is: $\vec{x} = \vec{a} + r\vec{b}$. To reach $\vec{x}$, start at $\vec{a}$ (the position vector) and travel a distance equal to r times $\vec{b}$, in the direction specified by $\vec{b}$ (the velocity vector). A plane can be represented by: $\vec{x} = \vec{a} + r\vec{b} + s\vec{c}$.

A second velocity vector $s\vec{c}$ represents a distance equal to s times $\vec{c}$, in the direction specified by $\vec{c}$, on a different plane.

Geometry without drawings

Descriptions of geometrical objects can be depicted using calculations instead of diagrams. Using calculations, common points of different sets of points and distance calculations can be found. In physics and computer graphics, complex problems in which vectors play a role are solved through pure calculation. Full-length animated movies also use vector graphics.

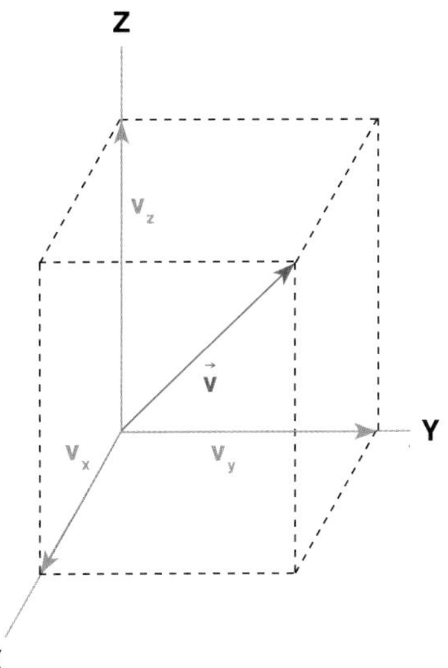

■ A vector in a three-dimensional space: Every point *V(x,y,z)* is defined by the vector: $\vec{v} = \begin{pmatrix} x \\ y \\ z \end{pmatrix}$ whose coordinates are *x*, *y*, and *z*.

in focus

COMPLEX NUMBERS AS VECTORS

Complex numbers expand the domain of real numbers by introducing a new value, *i*, defined as the square root of *-1*, so that $i^2 = -1$. Using this new number, the roots of negative as well as positive numbers can be calculated. Complex numbers are represented according to the formula $a + bi$, whereby *a* and *b* are real numbers and *i* is an imaginary unit. One convenient way to represent complex numbers is the complex plane (Argand diagram: a coordinate system in which the x-axis represents the real component, *a*, and the y-axis the imaginary value, *b*). Therefore every point on the plane corresponds to a unique complex number, making it comparable to a two-dimensional linear space or vector field.

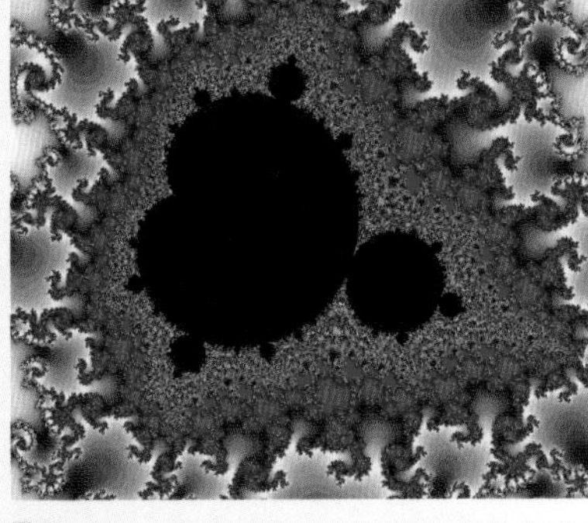

In accordance with their definition, the addition of complex numbers is comparable to that of vectors. However, their multiplication is different; multiplying two complex numbers produces a third vector of different magnitude and direction.

■ The Mandelbrot set plays an important role in chaos theory. This geometric figure is produced using a defined series of complex numbers.

practice

VECTORS IN PHYSICS

Variables with both a magnitude and a direction are known as vectors. They include magnetic and electrical field-strength speed, acceleration, and force. However, mass and energy are non directional and thus are considered scalar. A communication satellite in orbit around the Earth at a constant speed would keep moving in a straight line due to its inertia; however, the acceleration of gravity applies a force perpendicular to the direction of its movement, which affects its motion; thus only the direction of the speed vector changes.

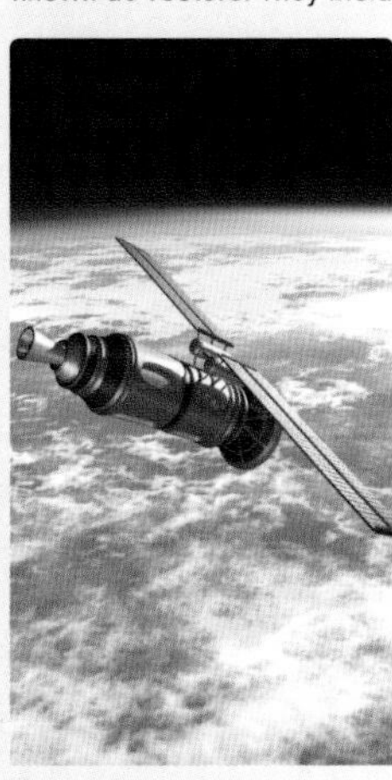

■ An orbiting communications satellite is influenced by the force of Earth's gravity.

see also: Exploring the universe, pp. 48–51

differential calculus | integral calculus

INFINITESIMAL CALCULUS

Infinitesimal calculus, at its simplest, is calculating with infinitely small or infinitely large numbers. In application, it focuses on how certain mathematical relationships change when a variable approaches a certain limit, and is part of analysis. Analysis, or the study of functions, also includes differential calculus and integral calculus, which look at functions in infinitesimal sections.

differential calculus

Differential calculus, founded in part by Sir Isaac Newton and Gottfried Wilhelm Leibniz, describes the behavior of a function in infinitely small parts.

The information available about, for instance, a hike up a mountain to a certain height above sea level is usually of interest to a hiker, but perhaps how steep the mountain is and how difficult the climb will be is of the most importance. Differential calculus focuses on this aspect: the steepness—or slope—of graphs of functions.

Slope, secant, difference quotient

The term slope describes the steepness of a straight line; a curved line has a constantly changing slope. A straight line that cuts the graph of a function *f(x)* at two points (x_1 and x_2) is called the secant, and the difference quotient $\frac{f(x2)-f(x1)}{x_2-x_1}$ is used to approximate the slope of a specific section of the curve.

Boundary value, tangent, derivative

To precisely measure the slope of a curve at a particular section, the secant should cut the curve to produce an infinitesimal section. Such an infinitesimal straight line is called the tangent (from the Latin *tangere* "to touch") and appears to touch the "back" of the graph of a function. The slope of the tangent line is the "limit" (from the Latin *limus* for "border") of the slope of the secant and is expressed mathematically as: $lim \frac{\Delta y}{\Delta x} = \frac{dy}{dx} = f'(x)$

This differntial calculation gives the slope of the curve at this point, called the derivative of the graph at point *x* and written as *f'(x)* or *df/dx*.

The flight path of a released object corresponds to the tangent of the circular orbit at the point of release.

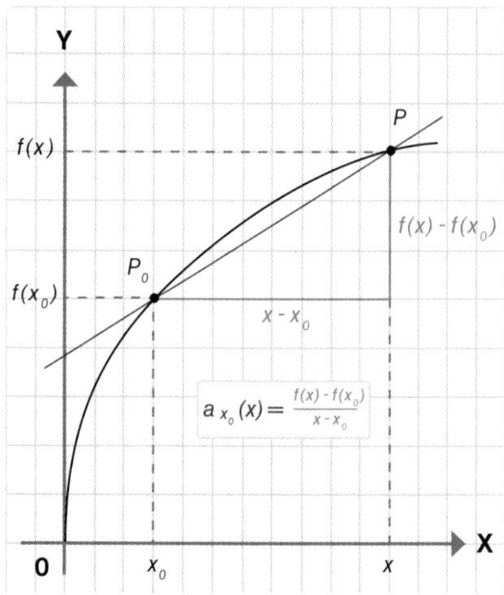

The secant cuts the graph of a function in two points so that the slope can be ascertained.

Differential calculus finds application in real processes by using the derivative to specify the rate of a process at different points. In business economics, marginal costs are determined by taking the derivative of the total costs as a function of the quantity produced. Companies are then able to determine how far they should be able to lower the cost of a product through infinitely small steps before profits are affected.

DEVELOPMENT OF CALCULUS

Calculus was independently developed by Gottfried Wilhelm Leibniz and Isaac Newton in the late 17th century. Leibniz based his theory on the use of geometric processes to solve mathematical problems. He viewed a curve as being made up of infinitely small segments, whereby the slope of the tangent could be calculated for each segment. Similarly, a curved surface could be seen as the sum of an infinite number of tiny rectangles. Leibniz thus recognized the relationship between differential and integral calculus. Newton, on the other hand, was more interested in solving a physics problem: how to determine the instantaneous speed of an accelerating object. He viewed a curve as a reflection of constant acceleration and imagined a point as an infinitely small segment of a line. The time interval between observations of an object's motion can be reduced to the point that the change in speed disappears. Thus, acceleration or deceleration can be calculated as the sum of the instantaneous speeds of the observed object.

Leibniz was later accused of stealing Newton's ideas from correspondence exchanged by the two, and the Royal Society of London, influenced by Newton, erroneously pronounced him guilty. However, Leibniz' system became the dominant form of calculus, thanks to its elegant notation and simplicity of calculation.

Isaac Newton (1643–1727) and Gottfried Wilhelm Leibniz (1646–1716).

milestones

see also: Applied mathematics, p. 419

integral calculus

Originally, integral calculus was developed to calculate the surface area underneath the curve of a function. In general, integral calculus is a method to calculate the sum of infinite elements of a function within a continuous space.

Using integral calculus, an infinitely large number of infinitely small areas are added to calculate the area under a graph. many rectangles of equal width. The area of each rectangle can be calculated simply by multiplying two measurements, its width Δx and its height $f(x1)$: $A1 = \Delta x \cdot f(x1)$. The area of the curve is found by adding together the areas of the rectangles under the curve and factoring out Δx: $A = \Delta x \cdot (f(x1) + f(x2) + \ldots)$. The narrower the rectangles are (the smaller Δx is), the closer the sum is to the actual area under the curve.

INTEGRAL CALCULUS AND VOLUME

in focus

A practical application of integral calculus is the determination of the volume of three-dimensional figures formed by rotating a function $f(x)$ around the x or y axis.

A cone, for instance, is formed through the rotation of a linear function, $f(x)=a\cdot x$ (a line that passes through the origin). To calculate its volume, we can divide the cone into extremely thin slices, which when added together approach the cone's volume. The thinner the slices (differentials, or dx), the more exact the approximation. As dx approaches zero, the sum of the slices will exactly equal the volume of the cone. The sum of this infinite number of slices is an integral, or a linear depiction which assigns a numeric value or function to a given domain of the integration.

The volume of this ice cream cone can be calculated using integral calculus.

Many narrow strips

A method of integration was first developed by Archimedes during ancient times, and later generalized by Leibniz and Newton. The area under the curve is divided into many rectangles of equal width.

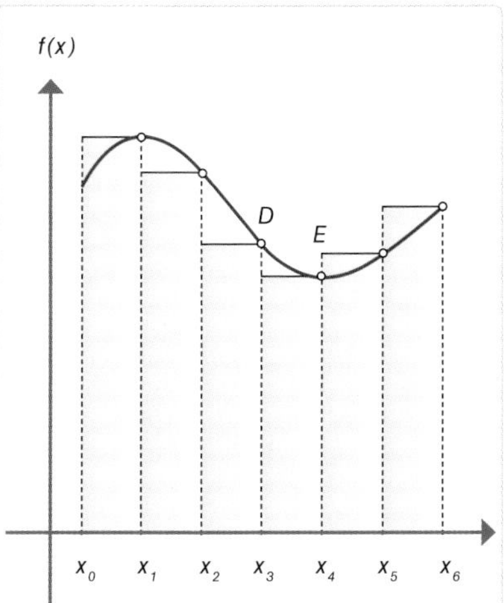

The approximate calculation of an area under a curve through summation of rectangles.

Primitive and definite integral

Leibnitz and Newton presented the idea that the derivative of the integral of a function F will lead back to the original function f: $F'(x) = dF(x)/dx = f(x)$. F is said to be the indefinite or primitive integral of the function f. This rule connects the two branches of infinitesimal calculus. The definite integral of f from a to b is: ${}_a\int^b f(x)dx = F(b) - F(a)$. This is the limit of the summation of the areas $f(x)\cdot dx$, from $x = a$ to $x = b$, as dx becomes infinitely small. The symbol of integration is a long drawn-out "S" from the Latin word *summa*.

Integral calculus can also be solved in various methods, such as writing it out.

Applications

In physics, when a force is constant, simply multiplying the force by the distance gives the mechanical work done on an object by a force acting over a distance. When the force varies over the distance, calculus must be used $W = {}_{s1}\int^{s2} F(s)\, ds$.

FOCUSING LIGHT WITH GLASS MICROSPHERES

practice

Transparent spheres with a diameter of just a few light wavelengths can act as lenses, focusing light. They also have a tendency to arrange themselves on a surface in a single compact layer. Thus, a layer of microspheres can focus the light of a single laser pulse to produce millions of similar structures, only a few dozens of nanometers in size. In addition to processing materials, microspheres play an important role in semiconductor technology and micro and nanomechanics. To take full advantage of their focusing properties, a special concept from calculus must be used—the Bessoid integral—which takes the wave nature of light into account.

Microspheres are so tiny that they must be analyzed using a microscope.

statistics

Once election polls have closed, pollsters are able to make relatively precise projections that forecast what the results of the final vote will be. They are able to do this using the tools of statistics.

Descriptive statistics are used to represent collected data. Very large quantities of information about individual events can be summarized according to certain variables in order to represent the data clearly, for example as a graphic display. Once data is collected, a rough list is created in order to identify the characteristics of units in a population. This is dependant on what type of information is sought. For example, determining the ages of people in a city. A representative sample can be used for a very large population. As each characteristic, in this case an individual's age, can come up repeatedly, a frequency table is created. The absolute frequency, or the actual counted number of times a characteristic occurs, is used in a frequency table, but the relative frequency is occasionally interesting and meaningful, too. For example, if 47 of 3,050 inhabitants in a city are 60 years old, the relative frequency of sixty-year-olds in the city is $f = 47 = 0.01541$, or 1.541 percent.

Evaluation of data

Statistics aims to summarize data with a minimal loss of information. The mean value of the data, or the average of the information collected, is an important parameter of statistics. Statistical spread is the observed difference of individual data from the mean value, which is used to calculate the standard deviation. When a specific promise cannot be made, such as a company's cellular battery lasting exactly 70 hours, the company can use testing and statistics to predict that the battery will last 70 hours, give or take five hours.

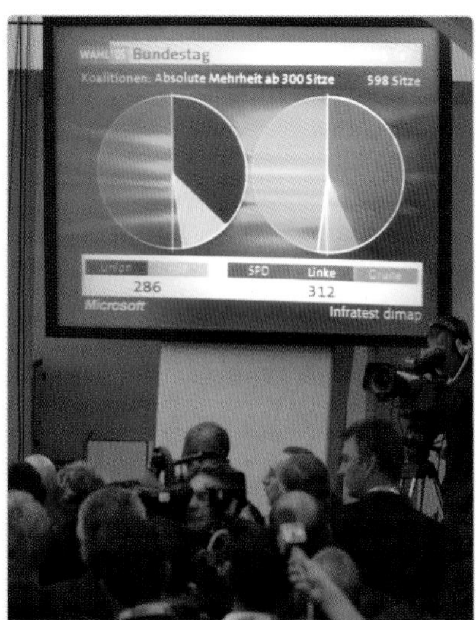

Pie charts showing the first results for an election after the closing of the polling-stations.

statistics | probability theory

PROBABILITY

As numbers themselves cannot lie, any harm or misunderstanding caused by numbers is the fault of the presenter or interpreter of the numbers. Probability theory and statistics help to describe events with numbers, allowing predictions and forecasts regarding the probability that certain events will occur to be made with a degree of certainty. Mathematical statistics are used in all the natural sciences, as well as in the scientific study of human society and social relationships.

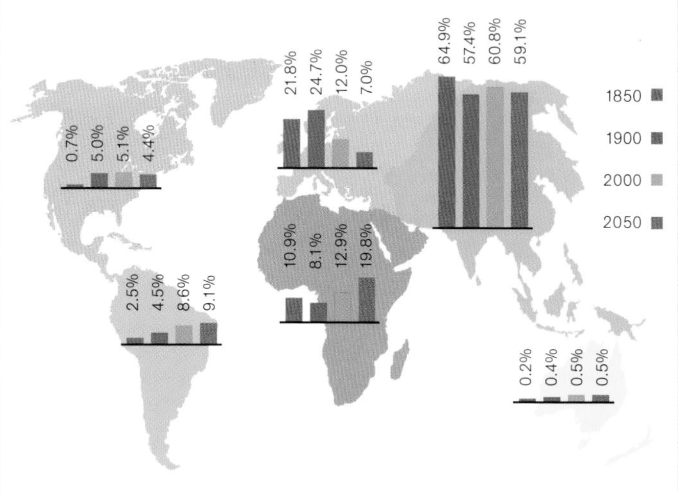

The percentage of people in individual continents in the total population of the world at different points of time can be represented with side-by-side bar diagrams.

in focus

DATA PRESENTATION

Presenting data as graphic displays is an important component of the field of statistics. However, a certain amount of interpretation may be involved. In preparing the figures, a particular weight must be given to each aspect thus making a truly neutral presentation nearly impossible. For instance, depending on how the scales are chosen for the x and y axes, a graph of changing values such as unemployment or GDP may appear to be a gradual or steep curve. False impressions can also be created when a portion of an axis is omitted.

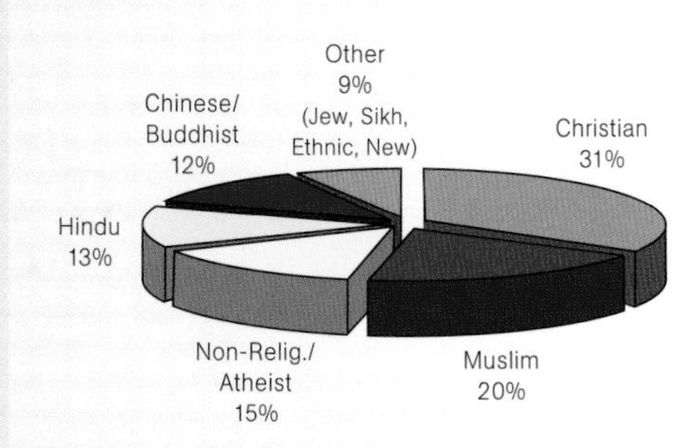

The most commonly used graphic displays for presenting data are bar graphs and pie charts.

probability theory

Probability theory, the mathematical study of the estimated consequences of future events, is useful for predicting costs and planning complex processes when not all variables are available.

If the consequences of a future event have to be estimated, probability theory is used to help make an informed decision. For example, a company insuring for accidents must cover the risk to the insurer, without high costs. If the risk of an accident occurring over one year is estimated at five percent, the yearly premium will therefore be at least five percent of the average cost of an accident.

Card games, relying on chance and players' skill, gained increasing popularity during the 19th century.

Probability theory was developed from questions about chance in gambling. The probability of an event is calculated by dividing the number of specific outcomes in that event by the number of all possible outcomes. For example, the probability of drawing a black card from a deck of cards is 0.50, or 50 percent (26 black cards out of 52 cards total).

If the outcomes are part of subevents, a decision must also be made as to whether they are dependent on each other. For instance, if you begin drawing cards from a single deck, the chance that you will draw a queen of hearts increases as the size of the deck decreases (1/52, 1/51, 1/50, and so on). To find the probability of several independent events occurring, you must multiply the individual probabilities. For example, the chance that you would roll two dice and both would come up "six" is $P(X=6) =$ 1/36 or 2.8 percent.

Probability theory is particularly useful for making responsible decisions in economic or social situations.

Limits of probability theory

There are some probabilities that cannot be calculated accurately. For example, when a person is faced with a choice between plain chocolate and chocolate with peanuts, the probability that the plain chocolate is chosen could be calculated at 0.5. However, other factors, such as an allergy to peanuts, may influence the decision.

practice

STOCHASTIC THEORY

Stochastic theory includes the fields of probability and statistics. It concerns itself with situations which are not fully predictable, such as economic developments, quantum mechanics, changes in animal populations, the effects of medications, and failure rates for mechanical equipment, in addition to natural weather predictions. All of these models must include parameters that can be adjusted. Stochastic theory is used to help researchers choose the most appropriate parameters and test their models.

The NYMEX exchange in New York: stochastic theory helps predict market trends.

milestones

THE GALTON BOX

A device developed in the 19th century by Sir Francis Galton (1822–1911) provides a model for experiments involving two options in a series of repetitions. It consists of a vertical board with a grid of pins. Balls dropped into the grid at its center strike the pins and have an equal probability of falling to the left or right. With eight pins, there are $2^8 = 256$ possible paths, most of which lead to the central tracks. The binomial distribution can be used to calculate the individual probabilities.

The Galton box is used to demonstrate the probability distribution of random trials.

The foundations of probabiltiy theory lie in games of chance, where high value bets may be placed on the probability of certain outcomes.

pure mathematics

The purpose of pure and abstract mathematics is to find solutions to mathematical problems for the sake of fundamental research into mathematics itself without any primary interest in practical applications.

The graphic images resulting from the solutions of equations can be surprisingly beautiful.

pure mathematics | applied mathematics

PURE AND APPLIED MATHEMATICS

During the 20th century, mathematics became a very abstract discipline that lost its connection to concrete mathematical problems somewhat. Despite a high level of abstract thinking, it often results in newly developed theories that are surprisingly useful for practical applications, for example, in the financial industries.

Pure mathematics may be difficult to understand but this may very well be the reason for its fascination. Ancient Greek mathematicians formed a distinction between pure and applied mathematics. For example they introduced the term logistics to describe basic arithmetic calculations to solve tangible problems, such as the organization of an army. Logistics was therefore considered its own subject area separate from mathematics and perhaps less interesting to the Greek mathematicians. They were much more interested in abstract principles, and the close connection between philosophy and mathematics was certainly a contributing factor. Modern numeric theory is based on this fundamental research which resulted in the development of various mathematical methods including the calculation of the greatest common factor and a proof of the existence of an infinite number of primes. After the fall of the Roman Empire, mathematics of the Middle Ages was mainly influenced by monasteries, for instance the development of a calendar, and by contact with Arabic cultures.

Hilberts program

Up to the 20th century the development of mathematics as a discipline was shaped by increased efforts into fundamental research and critical analysis. One of the central figures was the German mathematician David Hilbert (1862–1943) whose work and definitions of terms played an important role in determining the boundaries of mathematical knowledge. His legacy has been a great influence for mathematical sciences. In 1900 he proposed a list of 23 mathematical problems that were yet to be solved. One of the problems included assumed the existence of a fully consistent axiom system, the ultimate mathematical formula of the universe or the theory of everything, from which other scientific insights could be derived. Only a few years later a mathematician called Kurt Gödel (1906–1978) proved that such a system was impossible. Gödel's incompleteness theorem states that as soon as objects are related to each other, a system is created that is either incomplete or contradictory. This is why mathematics also studies the limits of knowledge. Another one of Hilberts approaches concentrated on the structural relationships in mathematics, for example, between geometry and algebra, as every bridge between these two disciplines that could combine the previous results of each, would count toward an enormous rise in knowledge.

in focus

FIELDS MEDAL

The Fields Medal is the highest possible award for a mathematician (apart from the Abel Prize first awarded in 2003), as the Nobel Prize does not include a category specifically for mathematics. Every four years the International Mathematical Union (IMU) awards two to four mathematicians during the International Congress of Mathematics. A recipients age cannot be more than 40 years at the time of their accomplishment.

Mathematicians awarded in 2006 were Terence Tao, Wendelin Werner, Andrei Okunkow, and Girgori Perelman, who was the first person to decline this award.

David Hilbert (1862–1943) was one of the most influential mathematicians of modern times.

see also: History of mathematics, pp. 406–407

applied mathematics

Although the fields of pure and applied mathematics overlap, the latter can be said to be most occupied with the solutions to problems and application of methods, in concert with other sciences such as physics or economics.

Due to its analytical properties mathematics can be used as a tool in various fields. Several sub areas and disciplines have also emerged on the basis of mathematical theories or methods.

Mathematics in the insurance industry

Insurances are based on the principle of taking on a communal risk where several insurance holders pay a monetary sum to ensure coverage of a random or relatively unlikely cash requirement. To make sure that the total sum of money paid by all members will be sufficient, the contribution required by each policy holder can be estimated by statistically analyzing the extent of damages. Here statistics and probability theory play an important role. Lets assume that a corporation decides to pay out the following amounts for a new three dice game: three dollars for one six, six dollars for two sixes, and ten dollars for three sixes. The money at stake is two dollars. In this case, the random value is the average net gain of a player and the question of whether the stake value covers payout. The average frequency of paid combinations appearing and the expected net gain (E(G)) can be calculated using probability theory (p. 408). The net gain depends on the number of sixes and is negative two, one, four, or eight dollars and the formula of the expected value coupling the probability with the corresponding net gain is: $E(G) = -2 \cdot 125/216 + 1 \cdot 75/216 + 4 \cdot 15/216 + 8 \cdot 1/216 = -0.51$. As a result, the player loses on average \$0.51. This is how an insurance company can, for example, calculate the premium for a life insurance plan. However, calculating the probability of death or fire is much more complicated due to the many factors that it depends on.

Cryptology

Information protection is becoming more and more important during the modern times of global networking. Personal data, company data, and military and strategic information must have limited access only.

The encryption machine Enigma was used by the German military during World War II for encrypting military information.

Cryptology deals with technical processes required to limit data access. This subject area developed when secret messages needed to be protected from the enemy. Some cryptographic procedures were already used by the Egyptians. Today's cryptology has many applications, for example, data encryption, user identification for computer systems and digital banking. These applications are based on mathematical theories such as algebra, numeric theory, and probability theory.

Mathematics
e.g.,
- Analysis
- Linear algebra
- Stochastics
- Numerics
- Optimization

Economics
- Business Administration
- Macroeconomics

Financial and insurance mathematics

Computing

Information systems

Computer science

The distribution of various disciplines in insurance and financial mathematics.

milestones

THE NOBEL PRIZE FOR A MATHEMATICIAN

The American John Forbes Nash (born in 1928) is one of the few mathematicians who has been awarded the Nobel Prize. After he completed his studies and several outstanding works in game theory, differential calculus, and partial differential equations, he was nominated for the Fields Medal. At the end of the 1950s, he was diagnosed with paranoid schizophrenia. It took him 30 years to manage his disease and, finally, he was able to return to mathematics. In 1994 he and two other mathematicians studying game theory were awarded the Nobel Prize for economics. In 2001 his story was released as a movie called *A Beautiful Mind.*

John F. Nash, born in West Virginia, worked at Princeton University.

see also: Practical application, p. 409

proofs

A mathematician is not satisfied until a statement is valid for all conceivable cases, not just hundreds, thousands, or millions of cases. A logical derivation that is true in all conceived circumstances is known as a proof.

proofs | modern mathematics

NEW MATHEMATICS

The foundation of mathematics is a logical proof based on pure thought, but at the same time, proven statements are generally accepted truths. A conflict exists in mathematics over whether it is a tool for science or an end in itself. While there have been strong movements to reduce mathematical objects to their fundamental characteristics—to achieve the highest possible level of abstraction—modern life has made increasing demands on mathematics.

According to astrophysicists, there are only about 10^{78} elementary particles in the universe. Why then do humans continue to try to prove a greater number of objects exist, or try to prove statements for infinitely more mathematical objects? The very nature of mathematics as a science is that it does not limit itself to finite entities. For practical purposes many scientists and mathematicians are content to make approximations to obtain results that may be limited but work well for everyday life. However, human beings also have a natural compulsion to know everything, whether or not the knowledge gained has a practical use.

Using known, proven, or assumed statements, one can advance step-by-step to a logical derivation known as a proof. Natural numbers have an inductive character that allows statements about them to be proved. Because every natural number n has the successor $n + 1$, a proof can be constructed based on the principle of falling dominoes. All the dominoes fall if two conditions are satisfied: one domino must fall, and every falling domino must hit the next one.

When describing an infinitely large number of mathematical objects, a particular characteristic is specified, as in the theorem by the sixth century B.C. Greek philosopher Thales, which states that a triangle is always a right triangle if its base is a diameter of a circle and its vertex is a point on the circle.

DIRECT PROOFS

Direct proofs are created using previously proven statements in combination with logical reasoning. For example, in this way the properties of angles can be used to prove that the sum of the angles of a triangle equals 180°. As a demonstration a line parallel to one side of a triangle is drawn through the point formed by the other two sides. The three angles next to each other together form a straight angle of 180°.

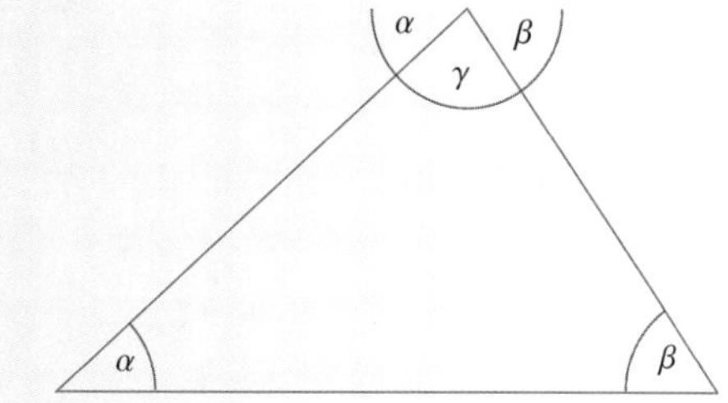

Drawing of a triangle with an added demonstration line.

in focus

Proof by contradiction

Changing the logical structure is another way of proving a statement—instead of proving the statement true, one proves the opposite statement is false. For example the theorem of the Greek mathematician Euclid, (born in Alexandria, Egypt, in 300 B.C.) that states there are infinitely many prime numbers can be proved by first hypothesizing the "greatest" prime number, and then proving that hypothesis false by finding another larger prime.

Pierre de Fermat's (1608–1665) last theorem, $x^2 + y^2 = z^2$, was not proved until 1993 by an English mathematician called Andrew Wiles.

THE WORLD OF PRIME NUMBERS

A prime number is a natural number that can be divided by only two particular numbers: itself and one. Although the fact that there are infinitely many primes was demonstrated by Euclid over 2,000 years ago, unsolved problems remain and are the focus of many modern-day mathematicians.

1	2	3	4	5	6
7	8	9	10	11	12
13	14	15	16	17	18
19	20	21	22	23	24
25	26	27	28	29	30
31	32	33	34	35	36
37	38	39	40	41	42
43	44	45	46	47	48
49	50	51	52	53	54

A list of prime numbers: those that can only be divided by themselves or one.

milestones

see also: New issues in physics, pp. 336–337

mathematics in the 21st century

Since the beginning of the new century mathematics has become an increasingly complex subject area. This is due to more and more conjunctive and innovative ways of carrying out mathematics.

Mathematics still provides us with models to describe natural phenomena. If there is such a thing as a universal formula to model all fundamental interactions in nature, however, this would certainly not be one single formula but rather some kind of complex mathematical structure. Not only would it have to confirm empirical data, it would also need to be consistent in a mathematical sense.

THE GOOGLE ALGORITHM

In recent years Google has established itself as the market leader of Internet search engines. One of the most important reasons, aside from its high performance and user friendliness, is the quality of its search results in comparison to those of other search engines. This is based on a procedure called PageRank. It is a specialized procedure, which evaluates the link popularity of a page. Link popularity is measured by the number of incoming links for each document or page. In contrast to the previous concept of link popularity, PageRank links the absolute number of incoming links with the corresponding page.

Google founders Sergey Brin and Larry Page succesfully utilized the PageRank algorithm.

milestones

Future outlook

In the 21st century more people will be traveling and more goods will be transported than ever before. Communication and data exchange has been changing considerably. These processes can be described mathematically in order to regulate them as efficiently as possible. Applied mathematics is also used by large companies in strategic decision making that depends on many factors. Project management can utilize computer-based planning and implementation.

The interaction of the most recent technological developments, for example, computer technology and mathematics are paving the way to solving many problems that previously were difficult for humans to solve. One example is the four-color map problem, which became the first mathematical mystery to be solved by computers. In the middle of the 19th century Francis Guthrie had posed the question of how many colors would be required to color adjacent areas so that they can still be distinguished from each other on any possible map. Humans were not able to establish a proof for this without the use of a computer.

Mathematical models are also applied in microbiology where they describe systems with vast amounts of numbers and relations. For example, models designed for data processing and statistics gave insights into the HIV virus and associated therapy. Another threat to human health is the resistance to drugs. In this area too, mathematical models can be used to help create more effective drugs.

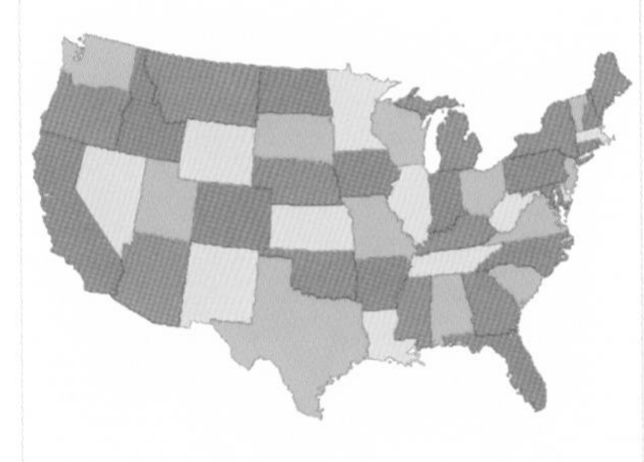

Four different colors are enough to differentiate regions of a map from one another.

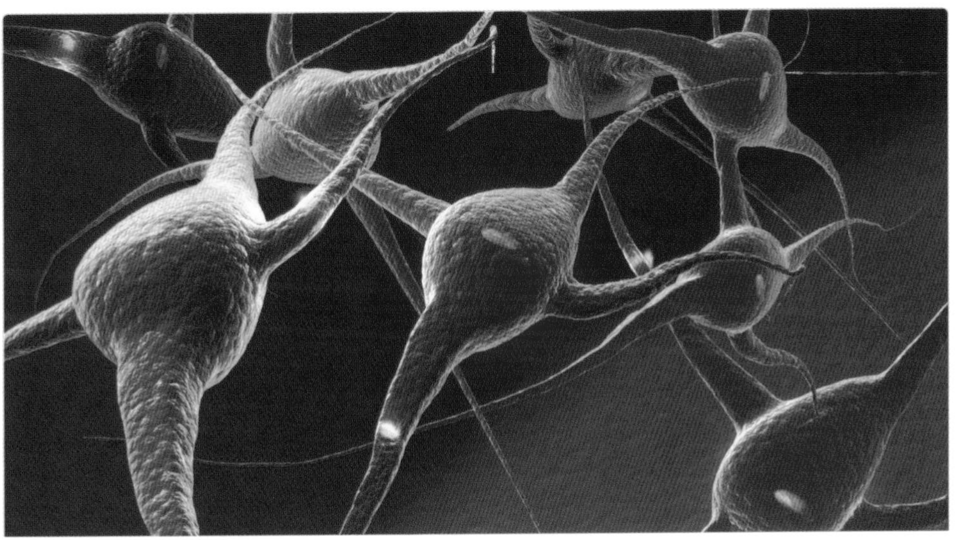

The nervous system: By using mathematic computer models, the formation of neural architecture during the development of the brain can be examined and explained.

see also: The internet, p. 387

APPENDIX

M145
R16
R14
R15
R5
R7
R8
R9

INDEX

Agassiz, Louis

May 28, 1807, in Motier, Switzerland–December 14, 1873, in Cambridge, Massachusetts

The Swiss-American geologist, zoologist, and naturalist Louis Agassiz is considered one of the most important developers of the modern American scientific tradition. Educated in both Switzerland and Germany, he emigrated to the U.S. in 1846 and became a professor of zoology and geology at Harvard University two years later. Agassiz's most widely recognized work includes groundbreaking research on glaciers and the great ice age, as well as work on both living and fossil fish. His works influenced many paleontologists and zoologists and made landmark contributions to the way the natural sciences were taught in the U.S. Agassiz was a lifelong detractor of Charles Darwin's (p. 439) theory of evolution.

Archimedes

ca. 287 B.C. in Syracuse, Sicily–ca. 212 B.C. in Syracuse, Sicily

Archimedes is considered one of the most important mathematicians and physicists of Greek antiquity. He was also a renowned astronomer, inventor, and engineer. While most of the finer details of Archimedes' life remain an enigma, it is generally agreed that he was born in Syracuse, on the eastern coast of Sicily, and educated in Alexandria, Egypt. He is credited with the discovery of the law of hydrostatics, which states that a body submerged in liquid loses weight equal to the weight of the amount of displaced liquid. In mathematics, he used the method of exhaustion to estimate the values of pi, which marked a major advance in ancient Greek mathematics. Archimedes is also credited with defining the principles of the lever as well as with inventing the compound pulley, the hydraulic screw, and various siege and countersiege weapons and machinery meant to defend Syracuse in wartime. Archimedes was allegedly killed by a Roman soldier while trying to protect his mathematical diagrams; however, this may be one of several embellishments to his legend.

Oil painting of Archimedes by the Sicilian artist Giuseppe Patania (1780–1852).

The many antique marble busts of Aristotle that have been found differ greatly in regard to their depiction of him.

Aristotle

384 B.C. in Greek Macedonia–322 B.C. in Chalcis, Greece

Greek philosopher and scientist Aristotle had a great impact on Western philosophy and theology, and was one of the founding fathers of science. In 367 B.C. he joined the Academy of Plato in Athens, staying for 20 years and writing on philosophy and early logic. He traveled to Asia Minor where he researched natural history, made especially detailed observations in zoology, and developed his own scientific ideas and natural philosophy. He tutored King Philip II of Macedonia's son (later known as Alexander the Great), then returned to Athens around 335 B.C. to found his own school. The Lyceum gave open, public lectures and divided the many topics of knowledge into disciplines such as engineering, architecture, politics, ethics, and theoretical sciences. Aristotle fled Greece in 323 B.C. when the political climate became dangerous and died a year later on the island of Euboea.

Avery, Oswald Theodore

October 21, 1877, in Halifax, Canada–February 20, 1955, in Nashville, Tennessee

The Canadian-born American Oswald Avery was a prominent bacteriologist, physician, and a pioneer in the field of immunochemistry. He studied humanities at Colgate University and—despite having no scientific background—was accepted into the College of Physicians and Surgeons in New York where he received his medical degree in 1904. Avery took a post at the Rockefeller Institute in 1913 and continued to work there for the next 35 years. His breakthrough came in 1944 when he made the monumental discovery that deoxyribonucleic acid (DNA) is the carrier of genes in cells and serves as genetic material, not protein as was previously assumed. His research was a milestone in DNA research and molecular biology.

Barnard, Christiaan Neethling

November 8, 1922, in Beaufort West, South Africa–September 2, 2001, in Paphos, Cyprus

Christiaan Neethling Barnard became world famous after performing the first successful human-to-human heart transplant in history. Born to a poor preacher and his wife in Beaufort West, South Africa, in 1922, Barnard studied medicine at the University of Cape Town. His discovery that intestinal atresia, a fatal defect of the intestine in newborn babies, was caused by an insufficient supply of blood to the fetus led to the development of a surgical operation to repair the abnormality. Barnard completed his doctorate on a scholarship to the University of Minnesota and returned to South Africa as a cardiothoracic surgeon in 1958. He established the first heart unit at Groote Schuur Hospital in Cape Town and for several years experimented with animal heart transplants. On December 3, 1967, Barnard's surgical team performed the world's first successful human heart transplant on Louis Washkansky, a 55-year-old patient suffering from diabetes and heart disease. The donor heart came from Denise Duvall, a young woman who had been killed in a road accident. Following the operation, Washkansky lived for 18 days, but later died of pneumonia induced by immunosuppressant drugs taken to prevent the body's rejection of the donor organ. Barnard's accomplishments in the field of heart surgery were a major breakthrough in life-extending surgery. His other innovative efforts include designing an artificial heart valve and performing the world's first "twin-heart" operation. In 2001 he died of asthma complications during a holiday in Cyprus.

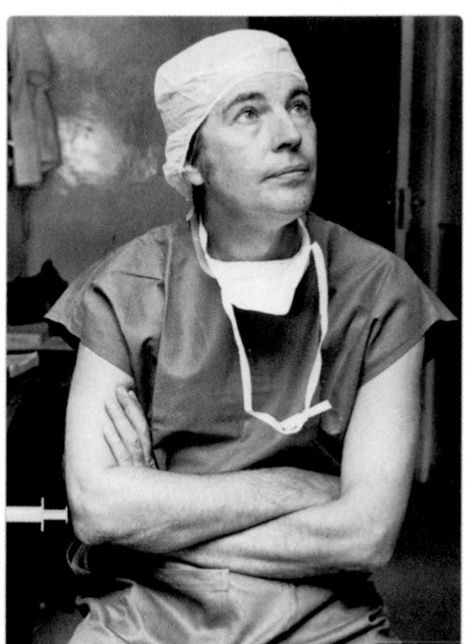

■ Christiaan Barnard was known as the "film-star surgeon" due to his fame and glamorous lifestyle.

Barton, Frederick Otis

June 5, 1899, in New York City, New York–April 15, 1992, place of death unknown

Inventor, engineer, and deep-sea diver Barton was attending Columbia University for engineering studies when he became interested in constructing a deep-sea device. His design for the bathysphere—a hollow, spherical deep-sea vessel that could be lowered into the ocean on a cable—was finally completed in 1930. In 1934, Barton and his partner, zoologist William Beebe (p. 434), made a record dive of 3,028 feet (923 m). This record stood until Barton broke it himself in a solo effort in 1949 when he descended to 4,500 feet (1,372 m), the deepest any human being had ever gone under the ocean. For this dive, Barton used the benthoscope, a new and improved bathysphere, which was heavier (weighing about 7,000 pounds or 3,000 kg) and sturdier in order to resist the immense pressure of Barton's ever deeper dives. Barton later worked on a fictional dramatic film about the bathysphere and his dives called *Titans of the Deep*, and engrossed the public with colorful tales of his adventures in the ocean depths. Even today, after immense leaps in technology, submersible research devices still use many of the principles originally developed by Barton. In later life, Barton became interested in the exploration of rain forests by airship, using a large gondola to access the jungle canopy for scientific study.

Beaufort, Sir Francis

May 7, 1774, in Navan, Ireland–December 17, 1857, in Hove, England

Sir Francis Beaufort was an Irish-British hydrographer and officer in the British Royal Navy. His father, Daniel Augustus Beaufort, a Protestant cleric and member of the Royal Irish Academy, published one of the earliest topographic maps of Ireland in 1792. Joining the East India Company in 1789 and remaining in active military service until 1812, Francis Beaufort dedicated his naval career to nautical charting. In 1805 he conducted his first hydrographic survey of the Rio de la Plata estuary of South America. During this expedition, he devised an objective scale for observing and classifying wind force at sea ranging from calm (0) up to storm (13). The Beaufort wind force scale was adopted by the British Admiralty and became mandatory for log entries in all ships in the Royal Navy by 1838. Beaufort commissioned further major expeditions, such as those of the *Beagle* with Charles Darwin (p. 439) and the *Erebus* with Joseph Hooker. In 1857 he died in Hove, Sussex, England, at the age of 83. During the 20th century, the Beaufort scale was adapted to steam power and later extended to meteorological observations through the addition of land-based descriptors.

■ Along with the Nobel Prize, Antoine-Henri Becquerel was awarded many accolades for his work.

Becquerel, Antoine-Henri

December 15, 1852, in Paris, France–August 25, 1908, in Le Croisic, France

Becquerel was a French physicist born into a distinguished family of scientists. In 1892 Becquerel became professor of applied physics in the Department of Natural History at the Paris Museum. From his father he had inherited a quantity of uranium salts, which phosphoresce when in contact with light, and in the late 1890s he decided to investigate whether there was any connection between x-rays—discovered recently by Wilhelm Röntgen (p. 459)—and phosphorescence. In 1896, while studying phosphorescence in the uranium salt, Becquerel found that uranium emits radiation without an external source of energy. He thus accidentally discovered the phenomenon of natural radioactivity, the spontaneous emission of radiation by a material. Becquerel was awarded the Nobel Prize in physics in 1903 for this discovery, along with Marie and Pierre Curie (p. 438). The SI unit for radioactivity, becquerel (Bq), is named in his honor.

Beebe, William

July 29, 1877, in New York City, New York–June 4, 1962, in Trinidad and Tobago

Together with Frederick Barton (p. 433), William Beebe—naturalist, explorer, and author of natural history books—was one of the first to scientifically explore the depths of the ocean. He was co-inventor of the bathysphere, the submersible vessel the pair used to make record dives. Beebe was the curator of ornithology for the New York Zoological Society from 1899 to 1952 and wrote many popular books about his adventures in the deep seas. He was, furthermore, a regular contributor to *National Geographic* magazine. In 1934 Beebe and Barton made a world-record dive of 3,028 feet (923 m) and Beebe would go on to complete a total of 35 dives in the bathysphere. Due to his lifelong interest in zoology, in 1950 Beebe bought land and set up a research station in the Trinidadian jungle where he died at the age of 85.

Benz, Karl

November 25, 1844, in Karlsruhe, Germany–April 4, 1929, in Ladenburg, Germany

Karl Benz was a German mechanical engineer who designed and built the first automobile to run on an internal-combustion engine. He was born to one of Germany's first railroad engine drivers, who died when Benz was a young boy, and consequently grew up in relative poverty. In 1883 Benz founded a factory in order to produce stationary two-stroke gas engines, which he designed. After failing to adapt this engine for use in vehicles, he managed to devise a water-cooled, two-stroke engine with electrical ignition with the gears designed to fit a three-wheeled vehicle. In 1886 Benz received the first patent for his gas-fueled three-wheeled vehicle, the Motorwagen, and began to commercially produce and sell his automobile, making it the first commercially available automobile in history. It was not until 1893 that a fourth wheel was added, the same year that the Benz Velo became the world's first inexpensive, mass-produced car. In 1896 Benz was granted a patent for his design of a flat engine, the first horizontally opposed engine—a design that continues to be used today in some high-performance racing vehicles.

Berners-Lee, Sir Timothy John

June 8, 1955, in London, England

English computer scientist Berners-Lee grew up as the son of two mathematicians and graduated from Oxford University in 1976 with a degree in physics. As an independent contractor at the world's largest particle physics laboratory CERN (European Organization for Nuclear Research) in Geneva, Berners-Lee first proposed a hypertext project to facilitate information sharing among researchers in 1980. With the help of Robert Cailliau and the staff at CERN, he created an Internet-based hypermedia initiative, to be known as the World Wide Web, in 1989. Berners-Lee wrote the first Web browser and the first standardized communication protocol—called Hypertext Transfer Protocol (HTTP). Furthermore, he developed the initial specifications of URLs and HTML. The release of the first text-based Web browser in 1991 marked the beginning of general public Internet use. In 1994 he founded the World Wide Web Consortium at MIT's Laboratory for Computer Science. He accepted a chair in computer science at the University of Southampton in 2004 to work on his new project, the Semantic Web: an evolution of the Web that promises many applications.

Berzelius, Jöns Jakob Freiherr von

August 20, 1779, in Linköping, Sweden–August 7, 1848, in Stockholm, Sweden

The Swedish chemist Berzelius is considered one of the fathers of modern chemistry. He enrolled at the Uppsala University where he studied medicine from 1796 to 1801. In 1807 Berzelius was appointed as a professor of chemistry and pharmacy at the Karolinska Institute in Stockholm. He made an incredible number of discoveries. Berzelius was the first person to distinguish between organic and inorganic compounds. He discovered the law of constant proportions, which states that an inorganic substance always contains exactly the same proportion of elements by weight. In 1828 he compiled a table of relative atomic weights including all known elements of his time and developed a system of chemical

■ Karl Benz (in front) at the wheel of his patent Motorwagen Model III: Although slow with wooden wheels and unable to climb steep hills without a push, it is historic as the very first commercially available automobile.

notation in which the elements were given simple written labels, for example, O for oxygen, with proportions noted by numbers. Furthermore, Berzelius is credited with identifying chemical elements such as silicon and selenium, as well as originating terms such as "catalysis" or "isomer."

Bjerknes, Vilhelm

March 14, 1862, in Christiania (Oslo), Norway–April 9, 1951, in Oslo, Norway

Vilhelm Bjerkes was a Norwegian physicist and meteorologist who paved the way for modern meteorology and the practice of weather forecasting. Assisting his father in mathematical research, Bjerkes was introduced to fluid dynamics at an early age. After studying mathematics and physics, he moved to Bonn, Germany in 1890. There he became assistant to the German physicist Heinrich Hertz and made substantial contributions to Hertz's work on electromagnetic resonance. He was appointed professor of applied mechanics and mathematical physics at the University of Stockholm in 1895, where he discovered the fundamental interaction between fluid dynamics and thermodynamics, which was applicable to large-scale motions in the atmosphere and oceans. In 1904 Bjerknes proposed the procedure now known as numerical weather prediction: with sufficient information about the current state of the atmosphere, scientists can use math formulas to predict future weather patterns. In the following years, he held chairs at the Royal Frederick University in Oslo as well as the University of Leipzig, and founded the Bergen Geophysical Institute and the Bergen School of Meteorology. From 1926 to 1932, he taught at the University of Oslo, where he died of heart problems in 1951. His son Jacob Bjerknes (1897–1975) was also an accomplished meteorologist.

Bohr, Niels

October 7, 1885, in Copenhagen, Denmark–November 18, 1962, in Copenhagen, Denmark

Niels Bohr completed his doctorate at the University of Copenhagen in 1911. He moved to Cambridge University for a short time, and then to Victoria University in Manchester to work with physicist Ernest Rutherford on his proposed Rutherford atomic model. Bohr applied relatively new quantum theory ideas of the time to Rutherford's research, with the result that he was able to construct a groundbreaking new model for the atomic structure. In his new model, electrons in stable orbits around the nuclei of atoms determined the characteristics of the elements, and these electrons only absorbed or emitted packets of radiation by changing orbits. This also accounted for the lines that had been observed in the spectrum of light of hydrogen atoms. Bohr returned to Copenhagen and founded the internationally renowned Institute of Theoretical Physics in 1921. He was awarded the Nobel Prize in physics in 1922 and later worked on the Manhattan Project in Los Alamos, where he was a strong advocate of peaceful and open international management of atomic weapons. These strongly held beliefs sometimes brought him into disagreement with the strict secrecy surrounding atomic research at that time. The developments he made in nuclear physics were extremely influential to the understanding of nuclear fission, and he was also instrumental in the creation of the European Organization for Nuclear Research, also known as CERN.

Painting of Niels Bohr aged 65, scientist and advocate for the peaceful use of atomic research.

Boltzmann, Ludwig

February 20, 1844, in Vienna, Austria–September 5, 1906, in Duino, Italy

Ludwig Boltzmann was an Austrian physicist famous for his founding contributions in the fields of statistical mechanics and thermodynamics. After receiving his doctorate from the University of Vienna, he worked as an assistant to Joseph Stefan who introduced him to James C. Maxwell's (p. 454) electromagnetic theory. In 1869 he was appointed professor of mathematical physics at the University of Graz. After working in Heidelberg, Berlin, and Vienna, Boltzmann returned to Graz, where he developed his statistical concept of nature. In 1885 he joined the Imperial Austrian Academy of Sciences and in 1887 he became the president of the University of Graz. Besides holding chairs in theoretical physics in Munich and Vienna, he also lectured on natural philosophy. Boltzmann supported the atomic theory when it was still highly controversial and applied the laws of mechanics and the theory of probability to the motions of atoms. Maxwell-Boltzmann statistics and the Boltzmann distribution over energies are considered the foundation of classical statistical mechanics. After 1900 Boltzmann suffered from severe mood shifts and eventually killed himself during an attack of depression in 1906. Shortly after his death, his studies were confirmed by new discoveries in atomic physics.

Ludwig Boltzmann—who strongly advocated the modern atomic model—pictured at age 31.

Boole, George

November 2, 1815, in Lincoln, England–December 8, 1864, in Ballintemple, Ireland

Born the son of a tradesman, the mathematician and philosopher George Boole was mostly self-taught and his skill in mathematics was not realized until his establishment of a school at Lincoln. Despite lacking an academic degree, he was appointed in 1849 as the first professor of mathematics at Queen's College in Cork, Ireland. In his work, Boole drew an analogy between logic and mathematics. He explained his general symbolic method of logical inference in Laws of Thought (1854). His invention of a two-valued algebra of logic, now known as Boolean algebra, remained relatively obscure until Claude Shannon demonstrated its electrical application in 1937. Today-Boole's discoveries are considered the basis of all modern computer arithmetic and used in telephone switching and electronic digital computing.

Brahe, Tycho

December 14, 1546, in Knudstrup, Denmark–October 24, 1601, in Prague, Bohemia (now the Czech Republic)

Before the invention of telescopes, Danish astronomer Tycho Brahe made the most accurate stellar and planetary descriptions of his time, correctly describing the positions of over 700 stars with the naked eye. He realized the proposed celestial systems of his day did not accurately predict events in the heavens so he began correcting all known data. Fredrick II of Denmark granted him the island of Hven in 1576, financing the construction of a large observatory named Uraniborg. The work completed here correctly updated most of the astronomical data available at the time, placing Uraniborg at the center of European astronomy until it lost its financial support with the death of the king in 1588. Brahe moved to Prague where he died in 1601, most likely from mercury poisoning although other legends are popular. His assistant Johannes Kepler (p. 449) inherited his astronomical data, which he used to discover the laws of planetary motion.

■ Tycho Brahe was a meticulous astronomer, and probably the first person to record a supernova.

Brown, Robert

December 21, 1773, in Montrose, Scotland–June 10, 1858, in London, England

After studying medicine in Scotland and entering the British Army as an ensign and surgeon, Robert Brown obtained the post of naturalist on a marine expedition bound to chart the Australian coast in 1800. Over three and a half years, he conducted intensive botanic research and collected about 3,900 species. In 1810 Brown published part of his findings in *Prodromus Florae Novae Hollandiae et Insulae Van Diemen*, thus laying the foundations of Australian botany and refining existing plant classification systems. In 1827 Brown transferred Sir Joseph Banks's botanical collection to the British Museum and later became appointed keeper of the museum's newly formed botanical department. During the same year, he discovered the phenomenon of movement in particles of pollen that came to be known as Brownian motion. In a paper read before the Linnaean Society in 1831, he named the recently discovered cell "nucleus." Brown was also the first to categorize seed-bearing plants into gymnosperms ("naked seed" plants) and angiosperms (flowering plants).

Buchner, Eduard

May 20, 1860, in Munich, Germany–August 13, 1917, in Munich, Germany

Buchner was a German chemist who made groundbreaking discoveries regarding the process of fermentation. He received his doctorate in chemistry from the University of Munich in 1888 and held several professorships in various universities across Germany, including the University of Berlin. Buchner's experiments with yeast cells led to his discovery that fermentation is the result of chemical action caused by a substance released by a yeast cell, not of action within a yeast organism itself. Moreover, he discovered the substance released by the yeast in 1897 and named it zymase. Thus, it was determined that the presence of yeast cells is not necessary for the process of fermentation as was previously assumed. In 1907 he was awarded the Nobel Prize in chemistry for his biochemical research and his discovery of noncellular fermentation. During World War I, Buchner served as a major in a field hospital in Romania where he was wounded on August 3, 1917. He died ten days later from the wounds he sustained after being brought back to Munich.

Calvin, Melvin

April 8, 1911, in Saint Paul, Minnesota–January 8, 1997, in Berkeley, California

The American biochemist Melvin Calvin was born into a Russian-Jewish immigrant family. After obtaining his Ph.D. in chemistry from the University of Minnesota in 1935 and spending two academic years at the University of Manchester in England, he became a faculty member at the University of California in Berkeley in 1937, where he was promoted to professor of chemistry in 1947. Along with Andrew Benson and James Bassham, Calvin was awarded the 1961 Nobel Prize in chemistry for the delineation of the chemical pathways of carbon during photosynthesis, now known as the Calvin cycle. In 1963 he was given the additional title of professor of molecular biology. His scientific work also included radiation chemistry, chemical evolution, and the study of oil-producing plants as renewable sources of energy. Until his retirement, he simultaneously directed the Berkeley Radiation Laboratory and the Laboratory of Chemical Biodynamics, which was renamed Melvin Calvin Laboratory in 1980.

Chargaff, Erwin

August 11, 1905, in Czernowitz, Ukraine (then part of Austria)–June 20, 2002, in New York City, New York

Through his research of DNA, this Austrian-American biochemist detected two rules, known as Chargaff's rules, which would become the basis for the discovery of the double helix structure of DNA by James Watson and Francis Crick. Chargaff received his Ph.D. from the University of Vienna in 1928 and then spent two years at Yale University. He returned to Europe, working in Berlin and Paris, before emigrating to the U.S. in 1935. Chargaff's first and best known rule reveals the occurrence of equal pairing of nucleotide bases in DNA cells. He observed that this equal pairing was consistently true for the DNA cells of all species, including human beings. Chargaff's second, less known rule showed that the composition of DNA varies from one species to another. When, in 1962, Watson (p. 462) and Crick (pp. 437–438) were awarded the Nobel Prize for their discovery of the double helix structure of DNA, Chargaff resented his exclusion. During his retirement, he wrote extensively and produced a total of 450 articles and 15 books before his death.

Copernicus, Nicolaus

February 19, 1473, in Torun, Poland–May 24, 1543, in Frombok, Poland

Nicolaus Copernicus dramatically transformed mankind's understanding of its place in the universe with his descriptions of Earth's relationship to other celestial objects, by removing its central role. Copernicus had left university in Poland for Italy in order to complete his studies in medicine, astrology, astronomy, and law and returned home in 1503 with a doctorate in canon law. He took up mainly administrative and medical duties for his uncle, Bishop Watzenrode, and practiced astronomy mainly in his spare time. His focus on the discrepancies that were apparent in existing astronomical works led to his realization that all the planets—including Earth—orbited the sun. With this insight, he was able to describe the year as one orbit around the sun, and the day as one complete revolution of Earth on its axis. His theories were recorded in 1514, but the final proofs were not published until he was near death in 1543. This collection of his theories, *De revolutionibus*, formed the basis for revolutionary work by Johannes Kepler (p. 449) and Galileo Galilei (p. 442) and were explained by Isaac Newton (p. 456).

■ Copernicus created the first mathematical model showing celestial objects and the Earth orbiting a stationary sun.

Coriolis, Gustave Gaspard

May 21, 1792, in Paris, France–September 19, 1843, in Paris, France.

The French mathematician and engineer Gustave Coriolis studied at the university École Polytechnique and later taught there. His 1829 book *On the Calculation of Mechanical Action* introduced the scientific terms "kinetic energy" and "work" still used today. In his 1835 paper *On the Equations of Relative Motion of Systems of Bodies* he described the inertial force that appears to affect the path of a moving body on a rotating surface, deflecting the path to one side or the other, depending on whether the rotation is clockwise or counterclockwise. This Coriolis effect is important in predicting atmospheric movements on spinning objects such as the Earth, other planets, and the sun. It aids in understanding the movements of wind and ocean currents, magnetic fields and sunspots, and is factored into ballistic missile and space vehicle launches. Coriolis became director of studies at École Polytechnique in 1838, but had suffered poor health throughout most of his life and died in 1843.

Cousteau, Jacques-Yves

June 11, 1910, in Saint-André-de-Cubzac, France–June 25, 1997, in Paris, France

Cousteau was a French ecologist, ocean explorer, filmmaker, and conservationist who studied the sea and its life forms and educated the general public on his pioneering findings through the mediums of film, books, and a television series. In 1930 Cousteau entered the French Naval Academy and later worked for the French Resistance during World War II. In 1943, together with the engineer Émile Gagnan, Cousteau developed the first self-contained underwater breathing apparatus, otherwise known as scuba or Aqua-Lung, which opened the sea to human beings as never before. During this period, he also shot the first underwater films ever made. In 1950 Cousteau converted a former minesweeper into an oceanographic research vessel, the *Calypso*, a kind of mobile laboratory with which he explored the world. From 1968 to 1976, he hosted the wildly popular TV series entitled *The Undersea World of Jacques Cousteau*, which introduced viewers to whales, sharks, and other oceanic forms of life and set a standard for future TV documentaries. In 1974 he founded the conservation organization called the Cousteau Society, which is still growing in global membership today. In his later years he became increasingly outspoken about human destruction of the seas and the need to conserve. He was recognized with countless awards and in 1989 elected into the French Academy, one of the most distinguished French societies.

Crick, Francis

June 8, 1916, in Northampton, England–July 28, 2004, in San Diego, California

Together with James Watson (p. 462), the molecular biologist Francis Crick discovered the double helix structure of the DNA molecule in 1953, which was the key to understanding how genetic information is transmitted. Crick studied physics at University College, London until 1939, then worked as a scientist for the British Admiralty during the war. He returned to his studies in 1947 to learn biology and organic chemistry at the University of Cambridge, working on x-ray diffraction of proteins. Watson joined him, and together with Maurice Wilkins (pp. 463–464) and help from the x-ray diffraction pictures of Rosalind Franklin (p. 441), they discovered the three-dimensional model of DNA. Crick left Cavendish Laboratories in 1976 to take up the position of distinguished professor at Salk Institute for Biological Studies, in San Diego, California, where he remained until his death in 2004.

Curie, Marie

November 7, 1867, in Warsaw, Poland–July 4, 1934, in Sancellemoz, France

The Polish-French physicist and chemist Marie Curie was the first female Nobel laureate and the first person honored with two Nobel Prizes. Born Maria Sklodowska in Warsaw, she went to Paris in 1891 to study mathematics, physics, and chemistry. At the University of Paris she met and married Pierre Curie, an instructor in the School of Physics and Chemistry. Inspired by Antoine-Henri Becquerel's (pp. 433–434) discovery of radiation, the Curies studied the radioactive properties of pitchblende and discovered the chemical elements polonium (named after Marie's native country) and radium in 1898. In 1903 Marie Curie completed her doctorate, becoming the first woman in France to receive such a degree. During the same year, she was awarded the Nobel Prize in physics together with her husband and Becquerel for their achievements in the field of radioactivity. After Pierre Curie's tragic death in a street accident in 1906, she succeeded him as professor of general physics at the University of Paris. She continued their work and was able to isolate pure radium, for which she was awarded the 1911 Nobel Prize in chemistry. In her later years, she remained dedicated to her research and founded Curie Institutes in Warsaw and Paris. Her death of aplastic anemia in 1934 was presumably related to her unprotected exposure to radioactive substances. Her elder daughter Irène Joliot-Curie (pp. 448–449) also became a Nobel laureate in chemistry in 1935.

■ Marie Curie's Nobel Prize portrait.

Curie, Pierre

May 15, 1859, in Paris, France–April 19, 1906, in Paris, France

The French physical chemist Pierre Curie was educated by his father, a general medical practitioner, before entering the Sorbonne as laboratory assistant in 1878. In his early studies conducted with his brother Jacques, Curie discovered piezoelectricity in crystals. His doctoral studies on magnetism formulated the relation between magnetic substances and temperature—now known as Curie's law. Together with his fellow scientist, Marie Curie—whom he married in 1895—he continued to examine radioactive substances. Their achievements in the field of radioactivity, in particular the discovery of radium and polonium, gained them the Nobel Prize in physics in 1903, which they shared with Antoine-Henri Becquerel (pp. 433–434). Furthermore, Curie examined the radiation emissions of radioactive substances. He was appointed professor of general physics at the Sorbonne from 1904 until 1906, his brief tenure ending after his death in a street accident. The curie, a unit of radioactivity, was named in honor of him and his wife.

■ Pierre Curie's Nobel Prize portrait.

Cuvier, Baron Georges

August 23, 1769, in Montbéliard, France–May 13, 1832, in Paris, France

Cuvier was a renowned naturalist and zoologist in the early 19th-century Parisian scientific circles. He was appointed professor of animal anatomy at Paris's National Museum of Natural History in 1795. Cuvier was the first to establish extinction as a scientific fact at a time when many scientists believed animals that were known only from fossils still survived in some remote parts of the Earth. Cuvier, however, produced a great number of studies illustrating the past existence of animals no longer found on Earth, including the Irish Elk—undeniable proof of the extinction theory. With his highly detailed fossil studies he effectively unleashed modern vertebrate paleontology and comparative anatomy. Cuvier was also an opponent of early theories of evolution, believing rather that organisms were functional wholes and any modification of an organism's anatomy would have resulted in its inability to survive.

Daimler, Gottlieb

March 17, 1834, in Schorndorf, Germany–March 6, 1900, in Stuttgart, Germany

Gottlieb Daimler was instrumental in developing the modern car. In 1852 he studied mechanical engineering at Stuttgart polytechnic institute and later worked for Nikolaus August Otto in a company designing internal combustion engines. In 1872 he became a director, during which time the alleged first motorcycle was developed by attaching one of these engines to a two-wheeled bicycle. In 1882 he left with designer Wilhelm Maybach to try to improve the engine by making it faster and lighter. They patented their first engine in 1885 and developed an early gasoline-injected carburetor (mixing air with the gasoline to use as fuel). This was attached to a four-wheeled carriage in 1886, making it one of the world's first four-wheeled automobiles. In 1889 they designed their first commercial automobile, and the following year founded Daimler-Motoren-Gesellschaft (DMG)—later to become Mercedes-Benz. They sold their first automobile in 1892.

Darwin, Charles Robert

February 12, 1809, in Shropshire, England–April 19, 1882, in Kent, England

Charles Darwin published many scientific works but is famous for proposing and providing evidence for the theory of evolution by natural selection, which explains the processes driving the evolution of species. He studied medicine at Edinburgh University but left in 1827 after only two years to study theology at Christ College in Cambridge. His interest in botany and geology led to his placement as a naturalist on H.M.S. *Beagle*. Darwin's writings on the five-year journey were published on his return, and in 1839 he was admitted to the Royal Society. He continued to formulate his ideas, and it was not until 1858 that he published his ideas about evolution from common descent, after receiving a letter from Alfred Russel Wallace with similar observations. Both theories were published together, and Darwin's book *On the Origin of Species* was completed in 1859. He received the Royal Society's Copley Medal in 1864, and published *The Descent of Man*, on human evolution, in 1871.

Einstein, Albert

March 14, 1879, in Ulm, Germany–April 18, 1955, in Princeton, New Jersey

Albert Einstein, who as a boy was told by a teacher that he would never amount to anything, appeared to have dim prospects throughout his schooling in Munich and later in the Swiss Federal Polytechnic School, Zurich. He took a job at the Swiss Patent Office after graduating in 1901, where he was completely unknown to the academic world and spent his spare time considering the behavior and speed of light. He received his doctorate in 1905 and at that time published several remarkable papers. Among them was a paper on the photoelectric effect (for which he won the Nobel Prize in physics in 1921) and special relativity, which revolutionized physics. In 1915 he completed the general theory of relativity. He received many awards, honorary doctorate degrees, and fellowships throughout Europe. He had to leave his professorship at the University of Berlin for Princeton, New Jersey, in 1933 due to the threat of the Nazi regime. He was at first in support of the atomic bomb to defend against this threat, but later warned against its use. In America, he continued work on unifying the theories of physics. His predictions in cosmology, such as black holes, continue to be verified long after his death.

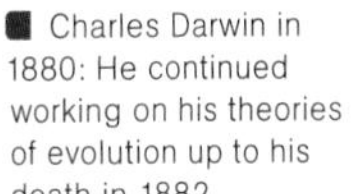

Charles Darwin in 1880: He continued working on his theories of evolution up to his death in 1882.

Albert Einstein has become a cult figure, with his name signifying genius and intelligence.

Euclid

ca. 300 B.C. in Alexandria, Egypt

Apart from his writing, hardly any biographical details are known about the life of one of the most important Greek mathematicians of antiquity, regarded as a founding father of mathematics. He may have been educated at Plato's Academy in Athens. During the Hellenistic period under the reign of Ptolemy, Euclid was active at the great Library of Alexandria, presumably founding a school at which he taught pupils. Until today, he is primarily known for *Elements*, a 13-volume treatise on geometry, which is considered the most influential textbook in the history of mathematics. It has not only set the standard for scientific methods of proof but also presents a model for philosophical treatises. Euclid integrates his mathematical propositions into a logical system of thinking. Today the geometrical system described by Euclid (Euclidean geometry) is distinguished from non-Euclidean geometries discovered during the 19th century. Among other works, Euclid's surviving writings include *Optics*, a theory on perspective, and *Phaenomena*, a treatise on spherical astronomy.

Fermat, Pierre de

August 17, 1601, in Beaumont-de-Lomagne, France–January 12, 1665, in Castres, France

Pierre de Fermat was significant in the history of mathematics. Born to a leather merchant of Basque origin, Fermat studied civil law at the University of Orléans. He worked as a lawyer in Bordeaux before he was appointed counselor at the Parlement in Toulouse in 1631. He held this office for the rest of his life. Besides his juridical knowledge, he was interested in languages, literature, and mathematics in particular. Independently of his contemporary René Descartes, Fermat discovered the fundamental principles of analytic geometry. His research proved helpful to Newton (p. 456) and Leibniz (pp. 451–452) in their development of the fundamental theorem of calculus. In collaboration with Blaise Pascal, he also laid the groundwork for the theory of probability. In number theory, he formulated one of the most famous mathematical problems, known as Fermat's last theorem. Despite his achievements, Fermat never considered mathematics his primary vocation and received very little scientific recognition during his lifetime.

Fibonacci, Leonardo

ca. 1170 in Pisa, Italy–1250 in Pisa, Italy

Fibonacci, also known as Leonardo of Pisa, was an Italian mathematician famous for introducing the Arabic numeral system to Europe. As a young boy, Fibonacci accompanied his father, who directed a trading post near Algiers. On his travels through North Africa, he became familiar with the Hindu-Arabic numeral system and studied arithmetic under the leading Arab mathematicians. Convinced of the advantage of Hindu-Arabic numerals over Roman numerals used at the time, he published the influential *Liber Abaci (Book of Abacus)* in 1202. The book explained the decimal number system (using digits 0 to 9) and the use of place value (the digit's position denotes a singular unit or tens, hundreds, and so on) and made clear how these could be applied for practical and commercial use. It also introduced a sequence of numbers, previously found in Indian mathematics, where each number after the first two is the sum of the previous two numbers. The Fibonacci numbers have many interesting properties and can be applied to the financial market, music, art, and architecture, and also appear in nature, such as in spiral shells, flower petals, and pinecones.

Fischer, Emil Hermann

October 9, 1852, in Euskirchen, Germany–July 15, 1919, in Berlin, Germany

The chemist and Nobel Prize winner Emil Hermann Fischer received his doctorate from the University of Strasbourg in 1874. He moved to the University of Munich in 1875, as an assistant to Adolf von Bayer, and in that year discovered the chemical compound phenylhydrazine. He became a professor of chemistry at Erlangen University, Bavaria, in 1882 and then Würzburg in 1885. Fischer's main contributions to organic chemistry include his significant works on the chemical structure of sugars and showing that compounds (which he named purines) found in vegetable sources, caffeine, and animal excrement were related through their nitrogenous structure. In 1892 he was appointed professor of chemistry at the University of Berlin, where he continued to study proteins. He published new research on enzymes and methods of identifying different amino acids. He received honorary doctorates from various notable universities and in 1902 became the first organic chemist to receive the Nobel Prize in physics. Fischer is considered the founding father of classical organic chemistry.

Fleming, Sir Alexander

August 6, 1881, in Lochfield, Scotland–March 11, 1955, in London, England

It is difficult to estimate how many millions of lives Fleming's discovery of penicillin has saved. He worked at a shipping office for several years, then studied at St. Mary's Hospital Medical School in London, where he excelled and gained his doctorate in 1908. While serving in the Army Medical Corps during World War I he witnessed soldiers dying of infection and became particularly focused on antibacterial agents. He returned to St. Mary's Hospital and discovered a mild natural antibiotic called lysozyme, but his great discovery was not made until 1928 when, famously, he noticed a fungus in a bacterial culture that he had not disposed of properly was producing an antibacterial substance.

Alexander Fleming was the discoverer of the lifesaving properties of penicillin.

A way to mass produce the substance, which he called penicillin, was finally discovered in the early 1940s, with the help of Howard Florey and Ernst Boris Chain. For this work, they shared the 1945 Nobel Prize in physiology or medicine with Fleming. He was elected to the Royal Society in 1943 and knighted in 1944.

Flohn, Hermann

February 19, 1912, in Frankfurt am Main, Germany–June 23, 1997, in Bonn, Germany

Hermann Flohn was an internationally renowned German climatologist and meteorologist who spearheaded international climate change research. As early as 1941, he began discussing anthropogenic (or human induced) global change of climate, the first scientist ever to do so. After being held as a prisoner of war in World War II, Flohn began working for the German Weather Services. Later, in 1961, he became a professor at the University of Bonn as well as the director of its Meteorological Institute. He led many scientific projects, including one on the environmental impacts of large-scale climate change for the North Rhine-Westphalian Academy of Sciences in Düsseldorf. He retired in 1977 and was awarded the International Meteorological Organization (IMO) Prize in 1986.

Foucault, Léon

September 18, 1819, in Paris, France–February 11, 1868, in Paris, France

Léon (also known as Jean-Bernard-Léon) Foucault, was a home-educated French physicist who made developments in microscopy and solar photography, improved telescopes, and invented a method to test their mirrors for surface defects. He also discovered eddies in electric current (called Foucault currents), which occur in conductive metal and electromagnetic fields moving in relation to each other. In 1850 he designed an experiment showing that light is slowed by water (and thus must act as a wave rather than a particle) and went on to make the most accurate measurement of the speed of light up until that time (only one half of one percent off today's calculations). In 1851, to demonstrate for the first time that the Earth is dynamic and not static, he suspended a pendulum from a long wire in the Panthéon in France. Over a period of 24 hours, the swinging pendulum appears to change its position relative to the ground due to the Earth's rotation on its axis. Foucault also improved mirrors, lenses, lamps, and telescopes. He was appointed to the position of physicist at the Paris Imperial Observatory, was elected to the Bureau des Longitudes in 1862, and the Royal Society of London in 1864.

■ Foucault first studied medicine, a subject he quickly abandoned once he discovered his interest in the field of physics.

Franklin, Rosalind

July 25, 1920, in London, England–April 16, 1958, in London, England

British physical chemist Rosalind Franklin played a groundbreaking role in the discovery of the exact structure of DNA, though she did not live to see any rewards for her work. While her father disapproved of university education for women, Franklin was intent on studying science and in 1945 received her doctorate in physical chemistry from Cambridge University. After her Ph.D., Franklin learned the techniques of x-ray diffraction, which she would later apply to the study of DNA. She photographed the DNA molecule, clearly revealing DNA's helical structure in a level of detail that was previously unattainable. Without her knowledge, her lab partner Maurice Wilkins (pp. 463–464) shared this important information with James Watson (p. 462) and Francis Crick (pp. 437–438) who used it to piece together a model showing the structure of DNA. Franklin died of ovarian cancer in 1958. Four years after her death, Watson, Crick, and Wilkins were awarded the 1962 Nobel Prize for their discovery of the double helix structure of DNA.

■ Joseph Fraunhofer died young, probably due to the poisonous substances used in glassmaking.

Fraunhofer, Joseph von

March 6, 1787, in Straubing, Germany–June 7, 1826, in Munich, Germany

Joseph Fraunhofer became an apprentice to a glassmaker shortly after he was orphaned at around 11 years old. This experience led him to take a job as an optician at an institution near Munich, where he became the manager in 1818. His design of scientific instruments and techniques for making glass, lenses, and mirrors gained him a worldwide reputation. His innovations included telescopes with clock mechanisms that enabled them to automatically track the sky and lens correction techniques that are still used today. He invented the spectroscope for light spectrum analysis and used it to plot over 500 of the constant, dark lines that had been observed in the spectrum of sunlight (now called Fraunhofer lines). It was later discovered that these lines are caused by the absorption of certain wavelengths of light and are characteristic of particular gaseous elements. Fraunhofer also invented a more precise method of measuring the wavelength of light (now called Fraunhofer diffraction). In 1822 he was awarded an honorary doctorate from the University of Erlangen, honorary citizenship to Munich, and was knighted in 1824.

Galilei, Galileo

February 15, 1564, in Pisa, Italy–January 8, 1642, in Arcetri, Italy

Galileo, referred to as the father of modern science by Albert Einstein (p. 439), studied medicine at the University of Pisa. Discovering his mathematical talent, he discontinued his medical studies to become a tutor of mathematics. During the following years, he taught astronomy, mechanics, and geometry as professor at Pisa and Padua. He was instrumental in developing the experimental scientific method describing the parabolic path of projectiles and the law of falling bodies in his research on motion and strength. He improved the telescope, famously turning it on the heavens in 1609 to discover new features such as four moons of Jupiter (now Galilean moons), sunspots, and the mountainous nature of the moon's surface. In 1623 he published *The Assayer* on the scientific method. His *Dialogue Concerning the Two Chief World Systems*, published in 1632, was banned because of its defense of the Copernican theory that the Earth orbited the sun against the favored geocentric worldview. He was forced to recant before the Holy Roman Inquisition and placed under house arrest for the remainder of his life. However, he managed to continue his work and smuggle texts to other countries for publishing. The heliocentric worldview was not accepted by the Catholic Church until 1758.

Galle's method of measuring the scale of the solar system was proved correct only after his death.

Galle, Johann Gottfried

June 9, 1812, in Gräfenhainichen, Germany–July 10, 1910, in Potsdam, Germany

Johann Galle, who identified the planet Neptune, was an assistant director under Johann F. Encke at the Berlin Observatory from 1835 on. He exchanged correspondence with the French astronomer and mathematician Urbain Le Verrier regarding the irregular orbit of Uranus. Le Verrier predicted that this may be due to the gravitational influence of an as yet undiscovered planet (also independently predicted by John Adams around the same time) and gave Galle instructions on what part of the sky to examine. With his student, Heinrich Louis d'Arrest, Galle discovered Neptune in 1846 after only an hour of searching. Although Galle and d'Arrest are often credited with the discovery of Neptune, Le Verrier in many ways was the discoverer due to his meticulous mathematical calculations. Galle and d'Arrest are correctly recognized as the first to identify Neptune through their scientific expertise and access to vital maps and telescopes. Galle became director of the Breslau observatory in 1851. Among other achievements, he discovered three new comets, published a list of 400 others with the assistance of his son, and suggested a successful method of using the parallax of asteroids to measure the scale of the solar system.

Galileo suffered much controversy throughout his life, enduring censure by the Catholic Church.

Galton, Francis

February 16, 1822, in Birmingham, England–January 17, 1911, in Haslemere, England

Galton was an influential scientist in several fields including meteorology, geography, psychology, anthropology, and statistics. He was also a tropical explorer, and the inventor of fingerprinting and the modern weather map. From 1840 to 1844, Galton studied mathematics at Trinity College, Cambridge but did not receive an Honors degree because he suffered a nervous breakdown before his exam. He began his scientific career with a voyage to tropical Africa and was subsequently elected to the Royal Geographical Society. The publication of Charles Darwin's *On the Origin of Species* (p. 439) in 1859 gave Galton the idea of studying variations in inherited human personality traits and mental abilities. He founded the field of eugenics—the study of improving the human population by means of encouraging reproduction by people presumed to have desirable inherited traits—which would go on to be disastrously used by the Nazis. Galton was general secretary of the British Association for the Advancement of Science from 1863 to 1867 and active on the council of the Royal Geographical Society for more than 40 years. Honorary degrees were bestowed on him by Oxford and Cambridge, and in 1909 he received a knighthood.

Gauss, Karl Friedrich
April 30, 1777, in Braunschweig, Germany–February 23, 1855, in Göttingen, Germany
Sometimes referred to as the "Prince of Mathematics," Gauss is responsible for the advancement of several mathematical and scientific disciplines including statistics, differential geometry, number theory, analysis, and optics. He is frequently assumed to be one of the most influential mathematicians of all time. In 1792 Gauss entered Brunswick Collegium Carolinum where he discovered the binomial theorem, the law of quadratic reciprocity, and the prime number theorem. In 1795 he began to study at Göttingen University but left three years later without a diploma. He had, however, made one of his most significant discoveries there: a regular heptadecagon constructed by a ruler and compasses. This is considered the most important advance in mathematics since Greek antiquity. It was published in Gauss's famous text *Disquisitiones Arithmeticae* in 1801, which he completed three years earlier at the age of 21. In 1807 Gauss became director of the Göttingen observatory. Two years later he published a significant two-volume treatise on the motion of celestial bodies. Throughout the 1820s, Gauss was involved with geodesy, and in 1828 he published his most renowned work on differential geometry. Gauss also claimed to have known of the existence of non-Euclidian geometry long before it was published in 1832, although he did not publish on the matter for fear of controversy. Gauss died in his sleep in 1855.

■ Gauss was renowned for his intellect, dedication and for his sometimes turbulent family life.

Geiger, Rudolf Oskar
August 24, 1894, in Erlangen, Germany–January 22, 1981, in Munich, Germany
The German climatologist and meteorologist Geiger is considered one of the founders of microclimatology, a subsection of climatology, which studies the nature of climatic conditions within only a few yards (meters) of the Earth's surface. In 1920 he received his doctorate from the University of Erlangen and worked as a research assistant at the Darmstadt University of Technology where he developed a passion for meteorology. In 1927 he wrote an important paper on the state of air at ground level. From 1937 to 1945, he held the post of director at the meteorological Institute of the University of Applied Sciences in Eberswalde. Rather than studying large-scale climate issues, he concentrated on compact vegetated areas for precise and meticulously detailed results. In 1950 he published *The Climate Near the Ground*, which is now considered a milestone in microclimatology. He also helped Wladimir Köppen (pp. 449–450) develop and extend his system of climate classification, which is still in use today.

Glauber, Roy Jay
September 1, 1925, in New York City, New York
Glauber, significant in the development of the new field called quantum optics, graduated from Bronx High School of Science in 1941 and began studying at Harvard University. At the age of 18, he was selected to join the Manhattan Project in Los Alamos, and left his schooling to work on calculations for the atomic bomb. In 1949 he completed his Ph.D. in physics at Harvard, where he currently holds the position of Mallinckrodt Professor of Physics. He was awarded one half of the Nobel Prize in physics with John L. Hall and Theodor W. Hänsch in 2005, for his application of quantum theory in explaining the interactions of light and matter, and the characteristic differences between thermal light and coherent light sources such as lasers. Glauber is the recipient of a number of awards and honors and also works on high-energy (particle) collision theory.

Gödel, Kurt
April 28, 1906, in Brno, Austria-Hungary (now the Czech Republic)–January 14, 1978, in Princeton, New Jersey
Gödel was an Austrian-American mathematician and logician whose works transformed the field of mathematical logic. His most important works, known as Gödel's incompleteness theorems, were published when he was only 25 years old, a year after completing his doctoral studies at the University of Vienna. His landmark theorems proved that an arithmetic system cannot be both consistent and complete;

in other words, it will always be possible to write a statement that is true, but not provable, in that system. Gödel emigrated to the U.S. with the help of his friend Albert Einstein (p. 439) in 1940 after the Nazi regime's annexation of Austria. There he began to work at the Institute for Advanced Study in Princeton, New Jersey. During the latter part of his life, he was plagued with mental instability and paranoia. Believing that someone was trying to poison him, he would not eat any food other than his wife's cooking. When his wife then became ill and was hospitalized—leaving Gödel with no one to cook for him—he refused to eat and died of malnutrition.

Goethe, Johann Wolfgang von

August 28, 1749, in Frankfurt am Main, Germany–March 22, 1832, in Weimar, Germany

Goethe is best known as one of Europe's literary geniuses for his work as a poet, playwright, and novelist, but he was also a natural philosopher and scientist. Largely home-tutored and self-taught, he knew several languages at a young age. His interest in science took hold later in life, and among the subjects he studied were alchemy, meteorology, and anatomy. He independently discovered the human intermaxiliary bone in 1782, and several years later he went to Italy to study geology and botany. There he climbed the volcano Vesuvius and wrote *Metamorphosis of Plants*, which considered the similar forms and structures of all plants and was published in 1787. He wrote many papers on the scientific method, favoring a more subjective approach rather than quantitative science. His 1810 paper *Theory of Colors*, which he is quoted as having described as his "single most important work," was written in response to Newton's (p. 456) theories and made significant insights into the way humans perceive light and color.

Goodall, Jane

April 3, 1934, in London, England

The British ethologist and primatologist Jane Goodall is most famous for her long-term and remarkably detailed study of chimpanzees in the Gombe Stream National Park in Tanzania as well as her strong involvement with environmental and conservation causes. The world's foremost authority on chimpanzees, Goodall arrived at Gombe in 1960 on an assignment from the renowned archaeologist Louis Leakey. Here she began a study that would last a quarter of a century and provide the scientific community with a wealth of new insight and data on the closest living relatives to humans. In 1965 Goodall earned her Ph.D. in ethology from Cambridge University, and soon afterward she resumed her studies in Tanzania, establishing the Gombe Stream Research Center. She was instrumental in revealing wild chimpanzee social learning, cognition, and culture. Goodall discovered that chimpanzee communities periodically engage in a primitive form of brutal "warfare," and that chimps hunt and eat bushpigs, and are capable of making tools, not only using them as was previously assumed. In 1977 Goodall founded the Jane Goodall Institute for Wildlife Research, Education, and Conservation to provide support for field research on wild chimpanzees. She is the recipient of countless humanitarian and environmental awards and was selected as an English UN Messenger of Peace.

Grzimek, Bernhard

April 24, 1909, in Neisse, Germany–March 13, 1987, in Frankfurt am Main, Germany

Bernhard Grzimek was a renowned German zoologist and animal conservationist. Born in Upper Silesia, he studied veterinary medicine at Leipzig and Berlin, receiving his doctorate in 1933. After World War II, he became director of the then ruined Frankfurt Zoological Garden, which he rebuilt into one of the largest zoos in Europe. He also served as president of the Frankfurt zoological society for over 40 years. Grzimek is most famous for the work he undertook for

Although primarily known as one of the greatest poets of his time, Goethe was also an accomplished scientist working with optics and natural sciences.

the conservation of the Serengeti ecosystem in Tanzania, East Africa. During the 1950s, Grzimek and his son Michael studied the wildlife of the Serengeti. They produced the nature documentary

Bernhard Grzimek was a popular television personality as well as a well-regarded zoologist.

Serengeti Shall Not Die, which gained enormous public appeal and won an Academy Award in 1960. During the filming, his son died in an air crash caused by a collision with a griffon vulture. On the topic of animal life, Grzimek published numerous books and an extensive encyclopedia. He also edited a popular magazine in collaboration with Nobel Prize winner Konrad Lorenz (p. 453), and a very popular television series. He died in Frankfurt am Main in 1987.

Gutenberg, Beno

June 4, 1889, in Darmstadt, Germany–January 25, 1960, in Los Angeles, California

The German Beno Gutenberg was one of the most important seismologists of the 20th century, making several major contributions to the fields of geophysics and seismology. Gutenberg earned his Ph.D. from the University of Göttingen in 1911 with a thesis on microseisms and then promptly turned to studying the Earth's interior. In 1913, he made the first precise determination of the location and depth of the Earth's core. Other important contributions of Gutenberg include the famed universal magnitude-frequency relation for earthquake distribution, which he developed with Charles Francis Richter, and which some people believe should have been named the Gutenberg-Richter scale. In 1930 Gutenberg became a professor of geophysics at the California Institute of Technology (Caltech) and later the director of Caltech's newly acquired seismological laboratory. Under his leadership, it became one of the leading centers in the world for deep Earth and earthquake studies. Gutenberg died of pneumonia in 1960.

Gutenberg, Johannes

ca. 1398 in Mainz, Germany–Febuary 3, 1468, in Mainz, Germany

Born around 1400 in Mainz as son of a patrician merchant, Johannes Gutenberg was trained as a goldsmith and gem cutter. Working in Strasbourg, he began to experiment on the development of a printing press around 1439. Woodblock and copper-type printing already existed in China and Korea; however, Gutenberg was the first to invent a practical, less expensive printing system that made the mass production of print media possible by combining the use of metal movable type, oil-based printing ink, and a mechanical printing press. His major printing project was the Bible, also known as Gutenberg Bible, published in 1455 with a print run of about 180 copies. Gutenberg's invention quickly spread across Europe, thus revolutionizing the distribution of knowledge. It remained the basis of modern typesetting until the late 20th century.

Johannes Gutenberg created two of the most revolutionary inventions in history: mechanical printing and movable type.

Hall, John Lewis

August 24, 1934, in Denver, Colorado

John Lewis Hall studied at the Carnegie Institute of Technology, receiving his Ph.D. in 1961. He worked at the Joint Institute for Laboratory Astrophysics and later taught at the University of Colorado in Boulder. In 2000 he and Theodor W. Hänsch completed the development of a high-precision laser-based technique, called the optical frequency comb technique, to measure the frequencies of visible light. Pulsing laser light was already in use but their technique brought it to a new level of precision. This method can be applied to the design of atomic clocks, improving global navigation systems, quantum computing and for calculating whether the constants of nature change imperceptibly over time. For their work, they shared the Nobel Prize in physics with Roy J. Glauber (p. 443) in 2005.

Halley, Edmond

November 8, 1656, in Shoreditch, England–January 14, 1742, in Greenwich, England

Halley was an English mathematician, physicist, and astronomer who is most famous for calculating the orbit of a comet later named after him. He was also a pioneer in the disciplines of social statistics as well as geophysics, especially due to his work on tides, trade winds, and the Earth's magnetism. Halley is, furthermore, known for his role in persuading Newton (p. 456) to publish his famous 1687 text *Philosophiae Naturalis Principia Mathematica.* Using Newton's theories, he was able to calculate the orbit of a major comet—since then known as Halley's Comet—which appeared in 1682, and also predicted its return in 1758. Halley's other astronomical contributions include making one of the first studies of compass

This famous portrait of Edmond Halley hangs in the National Portrait Gallery in London.

variations in the North Atlantic, determining the proper motion of the stars, and devising a method—involving the transit of Venus across the sun—that determined the precise distance of the Earth from the sun. In 1720 Halley was made astronomer royal at Greenwich and held the position until his death.

Hänsch, Theodor Wolfgang

October 30, 1941, in Heidelberg, Germany

Theodor Hänsch completed his doctorate at the University of Heidelberg in 1969 and then began teaching at Stanford University in the U.S. where he wrote many papers and began developing techniques to improve laser spectroscopy. In 1986 he returned to Germany and became director of the Max Planck Institute for Quantum Optics and professor at the Ludwig-Maximillians-University in Munich. With John L. Hall, he continued to develop his ideas, and in 2000 they completed the optical frequency comb generator. This laser-based technique is now used in laboratories worldwide for the extremely precise measurement of the frequencies of visible light. For their contribution to the field of laser spectroscopy, Hänsch, Hall, and Roy J. Glauber (p. 443) were awarded the Nobel Prize in physics in 2005.

Hawking, Stephen William

January 8, 1942, in Oxford, England

Theoretical physicist and cosmologist Stephen Hawking has conducted research on the origins of the universe, space and time relationships, and black holes. He enrolled at University College, Oxford in 1959 and received a first class honors degree after only three years, before moving to Cambridge University. Shortly after this he was diagnosed with an incurable, degenerative motor neuron disease that was expected to kill him within a few years. However, he went on to receive his Ph.D. and was elected the youngest Fellow of the Royal Society in 1974. During the same year, Hawking made the theoretical prediction that black holes emit thermal radiation due to quantum effects. However, the actual existence of the so-called Hawking radiation remains disputed in scientific circles because the phenomenon has not yet been observed in space. Hawking has been the Lucasian Professor of Mathematics at Cambridge since 1979. Despite being almost completely paralyzed and having lost his voice box, he continues to lecture and write using a voice synthesizer and portable computer. His books *A Brief History of Time* and *The Universe in a Nutshell* have become bestsellers. Hawking's media appearances, lectures to the general public, and his popular approach to science have made him an academic celebrity.

Helbing, Dirk

January 19, 1965, in Aalen, Germany

Dirk Helbing is a professor of sociology at the Swiss Federal Institute of Technology in Zurich, Switzerland. He studied physics and mathematics at the University of Göttingen in Germany; however, his interests were not only confined to established and well-researched sciences. After writing a master's thesis on the observed self-organization phenomenon in pedestrian crowds, he then went on to prizewinning Ph.D. work in the modeling of social interaction processes. A prolific author who is widely covered in the media, he has published more than 200 papers in journals such as *Science* and *Nature* and has diverse interests in fields such as traffic science and disaster management. Helbing's team has also developed traffic assistance systems and patented the principle of a self-organized traffic light control, which implements a massively parallel and decentralized control concept for the optimization of the flow of traffic.

Stephen Hawking is the Lucasian Professor of Mathematics at Cambridge University, an eminent and distingushed post once held by Isaac Newton (p. 456).

Herschel, Friedrich Wilhelm

November 15, 1738, in Hanover, Germany–August 25, 1822, in Slough, England

Friedrich Wilhelm Herschel was a German-born British astronomer who is best known for discovering the planet Uranus. He was born to a musical family, and at the age of 19 he left Germany for England, where he anglicized his name to Frederick William and worked as a music teacher, organist, composer, and conductor. Being a self-taught scientist fascinated by mathematics and astronomy, he constructed his own telescopic equipment and began cataloging his observations. His discovery of Uranus on the night of March 13, 1781, turned out to be the first identification of a planet in modern times. In 1782 Herschel was

Friedrich Wilhelm Herschel was a talented musician, mathematician, and astronomer.

appointed the king's astronomer. As a successful telescope maker, he constructed more than 400 telescopes and made significant improvements to the reflecting telescope. Herschel developed the theory of star evolution, discovered infrared radiation, and proved that the solar system moves through the galaxy. His meticulous catalog of nebulae helped him form his theory that nebulae are composed of stars. He discovered a total of 2,500 celestial objects during his lifetime and coined the terms "asteroid" and "binary star." Herschel was knighted in 1816 and in 1821, a year before his death, he became president of the Royal Astronomical Society.

Hillary, Sir Edmund

July 20, 1919, in Auckland, New Zealand–January 11, 2008, in Auckland, New Zealand

The adventurer and philanthropist Sir Edmund Hillary is best known for his record-breaking Everest expedition. As a boy, he was shy and preferred adventure books or hiking in the hills to school work, and only later discovered his remarkable aptitude for climbing and mountaineering. He served in the Second World War as a navigator in the Royal New Zealand Air Force, eschewing active military service for moral reasons. After the war, he joined the British expeditions that were focused on conquering the world's highest mountain, Mount Everest, on the China-Nepal border. In 1953 Hillary and Tenzing Norgay, a Sherpa mountaineer, became the first to reach the summit of Everest and descend safely, assisted by an expedition of over 400 people. Afterward he dedicated much of his life to helping the Nepalese Sherpa people, polar exploration, and occasional political campaigning.

Hubble, Edwin

November 20, 1889, in Marshfield, Missouri–September 28, 1953, in San Marino, California

The American astronomer Edwin Hubble is famous for proving that there are galaxies beyond our own and for providing the first evidence for the expansion of the universe, effectively setting off the big bang theory and founding the field of observational cosmology. As a youth, Hubble was a gifted athlete, but he turned his attention to mathematics and astronomy as an undergraduate. After attending Oxford University as a Rhodes scholar, Hubble returned to

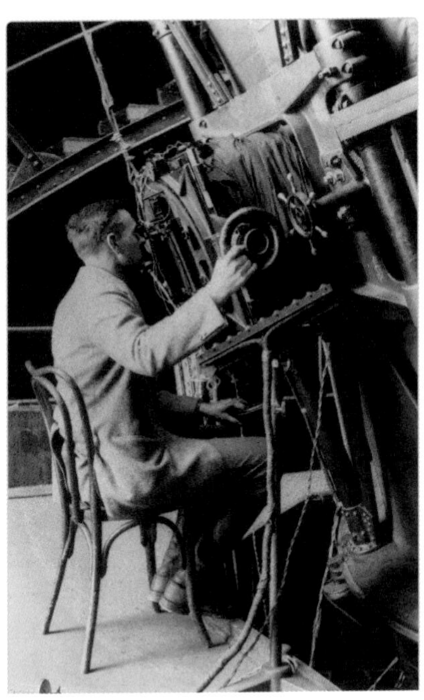

Edwin Hubble using the 100-inch telescope at the Mount Wilson Observatory in California.

the U.S. and resumed his passion for astronomy while undertaking graduate studies at Yerkes Observatory; he received his doctorate in 1917 from the University of Chicago. Hubble then began to work with the Mount Wilson Observatory in California. Between 1922 and 1924, he discovered Cepheid variable stars in certain nebulae. These were in fact other galaxies, which he determined were outside the Milky Way. Hubble also developed a system for classifying these newly found galaxies. In 1927 he made the extraordinary discovery that these other galaxies were receding from the Milky Way, implying that the universe is expanding. During Hubble's lifetime, the Nobel Prize committee did not consider astronomy a valid area of physics, for which there is an annual award. Thus, to his great disappointment, and despite having fundamentally changed our conception of the universe, Hubble was never awarded the prestigious prize.

Humboldt, Alexander von

September 14, 1769, in Berlin, Germany–May 6, 1859, in Berlin, Germany

Naturalist Alexander von Humboldt first distinguished himself as a mining official, improving mines and their conditions. He ended that career in 1797, when he decided to follow his passion as a naturalist explorer and began studying various fields of science. In 1799 he left on a five-year expedition to Latin America with botanist Aimé Bonpland. He made economic and political studies of Spanish colonies, confirmed the role of volcanoes in the formation of rocks and the Earth's crust, and researched the atmosphere, river systems, ocean currents, and in particular the relationship of the flora and fauna to their environment. His work was admired by Darwin and other intellectuals, and developed the fields of climatology, geomagnetism, and biogeography. Afterward he spent 21 years in Paris consolidating his work into a many-volumed encyclopedia. He became a popular hero, honored by academies, royalty, and the public alike. Returning to Berlin in 1827, he concentrated on researching the Earth's

The intense and detailed portrait of Alexander von Humboldt in 1843, painted by the German artist Joseph K. Stieler.

geomagnetic field. Humboldt was among the first to establish international cooperative ventures in pursuit of scientific knowledge and was instrumental in establishing meteorological observatories in British colonies around the globe. His work on the universe, *Kosmos*, was published in 1845.

Hutton, James

June 3, 1726, in Edinburgh, Scotland–March 26, 1797, in Edinburgh, Scotland

After a brief period as a lawyer's apprentice, James Hutton studied medicine in Paris and Holland, where he received his doctorate in 1749. He started a company producing sal ammoniac and took up farming. During this period, his interest in geology and the natural sciences grew, and he began to formulate theories about rock formation based on his observations and travels. Eventually he gave up farming but continued with his interests. His work on uniformitarianism was presented to the Royal Society of Edinburgh in 1785. It revolutionized geology as it proposed that rock formation is an ongoing process that has been occurring over millions of years through heat and pressure. This theory opposed the then accepted biblical age of the Earth, which is merely 6,000 years.

Huygens, Christiaan

April 14, 1629, in The Hague, Netherlands–July 8, 1695, in The Hague, Netherlands

The Dutch astronomer, mathematician, and physicist Christiaan Huygens is a significant figure of the scientific revolution who contributed to several branches of science, particularly astronomy. Huygens studied law and mathematics and received his doctorate in 1655. His many achievements include the discovery of Saturn's largest moon, Titan, and several interstellar nebulae and double stars, devising new and improved methods to grind and polish telescope lenses, as well as the invention of the first pendulum clock. He proposed a wave theory of light which was opposed by Newton's (p. 456) argument that light consists of small particles. Both theories were eventually proved correct by Thomas Young's concept of wave-particle duality (p. 464). Huygens also made an important contribution to mechanics when he proved that in a collision between two bodies, neither gains nor loses momentum. Although he designed the basic form of an internal combustion engine, fueled by gunpowder, he never successfully built one. Huygens was elected as a member of the Royal Society in 1663 and three years later moved to Paris where he held a position at the French Academy of Sciences.

Christiaan Huygens was known for his great intellect and many patents, inventions, and discoveries.

Joliot-Curie, Frédéric

March 19, 1900, in Paris, France–August 14, 1958, in Paris, France

After graduating from the School of Chemistry and Physics of the City of Paris, Frédéric Joliot became an assistant to Marie Curie (p. 438) at the Radium Institute. He married her daughter Irène in 1926 and both assumed the name Joliot-Curie. After obtaining a doctorate in science, he became a lecturer at the Paris Faculty of Science. In collaboration with his wife, he conducted research in the projection of nuclei. For their discovery of artificial radioactivity, the Joliot-Curies were awarded the 1935 Nobel Prize in chemistry. He received a chair in physics at the Collège de France in 1937 and continued to work on the requirements for the construction of an atomic reactor. During the German occupation of France, Joliot-Curie was part of the Resistance as president of the Front National. After World War II he served as high commissioner of atomic energy but was dismissed in 1950 for his communist affiliations. In 1956 he succeeded his wife as director of the Institute of Radium and died in 1958. Both children of the Joliot-Curies, Hélène and Pierre, have become acclaimed natural scientists.

Joliot-Curie, Irène

September 12, 1897, in Paris, France–March 17, 1956, in Paris, France

Following in the footsteps of her parents Marie and Pierre Curie (p. 438), the French scientist Irène Joliot-Curie continued their research on radioactivity and also became a Nobel laureate. Raised in an academically challenging environment, she decided to study mathematics and physics at the Sorbonne in Paris. During World War I, she worked as a nurse radiographer in mobile field hospitals established by her mother. Continuing her studies at the Radium Institute in Paris, she received a doctorate in physics in 1925. In 1926 she married the chemical engineer Frédéric Joliot with whom she conducted significant research in nuclear physics. They were awarded the Nobel Prize in chemistry in 1935 for their discovery of artificial radioactivity. Irène continued her academic career as professor at the Sorbonne's Faculty of Science and director of the Radium Institute. In addition, she was politically active in the promotion of women's education, as commissioner of atomic energy and member of the World Peace Council. In 1956 she died of leukemia, presumably caused by her previous accidental exposure to polonium.

James Joule was a mostly self-taught physicist, driven by a desire for precision and accuracy.

Joule, James Prescott

December 24, 1818, in Salford, England–October 11, 1889, in Sale, England

James Joule was a British physicist who is best known for his research in thermodynamics. He established the fact that various forms of energy—mechanical, electrical, and heat—are basically the same and can be changed into one another. It was through this discovery that he formed the basics of the law of conservation of energy (the first law of thermodynamics). Joule came from a wealthy family and was privately tutored until the age of 17, when he went to the University of Manchester and studied under the chemist John Dalton. In 1837 Joule went to work in the family brewery and began making scientific experiments in his basement, specifically with the electric motor. Through the 1840s, he verified the law of conservation of energy through extensive experimentation while studying the transfer of mechanical energy directly into heat energy, without any electrical steps. The value of the mechanical equivalent of heat is represented by the symbol J and the SI derived unit of energy—the joule—is named in his honor. Joule also studied magnetostriction, worked with the English physicist Lord Kelvin to develop the absolute scale of temperature, and discovered that the relationship between the flow of current through a resistance and the heat dissipated—now called Joule's law.

Kepler, Johannes

December 27, 1571, in Weil der Stadt, Germany–November 15, 1630, in Regensburg, Germany

Johannes Kepler, the mathematician and astronomer, not only inherited Tycho Brahe's (p. 436) astronomical data after his death, but also succeeded him as imperial mathematician and astrological adviser to Rudolph II (1552–1612), the Holy Roman Emperor. Kepler's work resulted in three new laws of planetary motion. Using the Copernican system (with the sun as the center), he described mathematical relationships between the motions of the planets and their distance to the sun, and discovered their elliptical orbits. Kepler also wrote on optics, making accurate explanations of the way light interacts with eyes, eyeglasses, and the newly invented telescopes. He was a supporter of Galileo Galilei's (p. 442) studies with this invention, and designed an improvement (the Keplerian telescope) in 1611, which is the basis for the modern refracting telescope.

Koch, Robert

December 11, 1843, in Hanover, Germany–May 27, 1910, in Baden-Baden, Germany

Along with Louis Pasteur, the German physician and Nobel laureate Robert Koch is considered to be one of the founders of bacteriology. Born to a mining official, Koch grew up with 12 siblings in relative poverty. In 1866 he received his medical degree from the University of Göttingen and four years later volunteered for service in the Franco-Prussian War. From 1872 to 1880 he was District Medical Officer for Wollstein, in Prussian Poland. It was there that, with extremely limited resources, no access to a library, and insufficient scientific equipment, he began his groundbreaking scientific work with bacteria. Koch became the first person to isolate the anthrax bacillus, observe its life cycle, and develop a preventative inoculation for it. He also isolated the tuberculosis organism and established its role in the disease, which at the time killed one in nine people. Koch was the first to prove a causal relationship between a bacillus and a disease and developed and improved methods of fixing, staining, and photographing bacteria.
In 1885 he was appointed professor of hygiene at the University of Berlin and director of its newly established Institute of Hygiene. In 1905 he was awarded the Nobel Prize for his foundational work in the field of bacteriology.

Robert Koch and his students discovered a multitude of disease-causing organisms.

Köppen, Wladimir Peter

September 25, 1846, in Saint Petersburg, Russia–June 22, 1940, in Graz, Austria

Köppen was a Russian-born German climatologist, meteorologist, and geographer best known for his systematized classification and mapping of the Earth's climate regions and advancement of the fields of meteorology and climatology. While he received his university education at Saint Petersburg, Köppen spent the majority of his life in Germany and Austria. In 1875 he became director of the meteorological research department of the German Naval Observatory in Hamburg. In the late 1890s, Köppen worked on oceanic wind charts, which he simplified and improved upon. He was also interested in the field of paleoclimatology and published his research in a text coauthored with Alfred Wegener (p. 463), the famed meteorologist and Köppen's son-in-law. This paper was significant in providing support for Milutin Milanković's theory on ice ages. First formulated in 1900 with his student Rudolf Geiger (p. 443), Köppen's world-renowned climatic system of classification, which bears his name, is still in use today. Köppen also coedited an important five-volume work with Geiger entitled *Handbuch der Klimatologie (Handbook of Climatology)*, which he began in 1927 and worked on until his death in 1940. Geiger continued to modify the climate classification system after Köppen's death.

Kossel, Albrecht

September 16, 1853, in Rostock, Germany–July 5, 1927, in Heidelberg, Germany

This German biochemist and medical doctor made many contributions to our understanding of the chemistry of nucleic acids and proteins. Although originally interested in the field of botany, Kossel was encouraged by his father to study medicine instead. In 1872 he began his studies at the newly founded University of Strasbourg, and in 1878 he received his medical degree from the University of Rostock. In 1883 Kossel did work at the Institute of Physiology in Berlin before moving on to a post as director for the Institute of Physiology at Marburg in 1895. Kossel was interested in investigating the constitution of the cell nucleus, in particular the substance nuclein discovered earlier by Johann Miescher (pp. 454–455). From 1885–1901, Kossel used hydrolysis as well as other chemical techniques to analyze nucleic acids and subsequently discovered the nucleic acids that are the bases of the DNA molecule: adenine, cytosine, guanine, thymine, and uracil. He also discovered the amino acid histidine, agmatine, and thymic acid. For these accomplishments Kossel was awarded the Nobel Prize in physiology or medicine in 1910.

The Nobel Prize-winning biochemist Albrecht Kossel, researcher into the properties of the cell nucleus.

Kühne, Wilhelm

March 28, 1837, in Hamburg, Germany–June 10, 1900, in Heidelberg, Germany

Kühne was a German physiologist whose work falls into two main categories: the physiology of muscle and nerve and the chemistry of digestion. He was also interested in the mechanics of vision and the functions of the eye. Kühne studied at the University of Göttingen under the chemist Friedrich Wöhler (p. 464) and received his Ph.D. in 1856. In 1863 he took charge of the chemical department of the pathology laboratory at the University of Berlin. He later held chairs of physiology at Amsterdam beginning in 1868 and at Heidelberg from 1871 until his retirement in 1899. Kühne worked on the problems of digestion and managed to isolate trypsin from pancreatic juice. In 1859, while working with sartorius muscle, he showed that nerve fibers are able to conduct impulses both ways. Kühne was also known for his research on vision and the chemical changes occurring in the retina under the influence of light, as well as for coining the word "enzyme."

Kuiper, Gerard Peter

December 7, 1905, in Harenkarspel, Netherlands–December 23, 1973, in Mexico City, Mexico

Kuiper was a Dutch-born American astronomer who is widely considered the father of modern planetary science, due to his extensive studies of the solar system and his numerous discoveries. Kuiper earned his Ph.D. from Leiden University in 1933 and in 1935 joined the Harvard College Observatory. He moved on in 1937 to the Yerkes Observatory of the University of Chicago where he remained until 1960. Among his many discoveries were Uranus's satellite Miranda, the carbon dioxide atmosphere on Mars, and Neptune's satellite Nereid. In 1951 he made his best known discovery: the existence of a region of innumerable comets and planetesimals outside Neptune's orbit. This region is now known as the Kuiper belt. In the 1960s, Kuiper also played an integral role in pioneering infrared airborne astronomy by using a Convair 990 aircraft in order to discover the physical and chemical properties of celestial objects, which cannot be done from the ground. NASA's Kuiper Airborne Observatory was named in his honor.

Landsteiner, Karl

June 14, 1868, in Vienna, Austria–June 26, 1943, in New York City, New York

Karl Landsteiner was an Austrian-American immunologist and pathologist who discovered the major blood groups and developed the modern ABO blood typing system, which enabled blood transfusions to become a routine medical practice. Landsteiner obtained a degree in medicine from the University of Vienna in 1891 before studying organic chemistry under Emil Fischer (p. 440). He returned to the University of Vienna in 1897 to pursue the emerging field of immunology and in 1901 published his discovery of the ABO blood group system from his identification of agglutinins in the blood. He researched why transfusions of blood between animals and humans would lead to cell destruction, causing dangerous shock and other ailments such as hemoglobinuria. In humans, destruction of the blood cells did not happen in every transfusion, and Landsteiner identified the existence of blood groups. In 1922 he began research at the Rockefeller Institute (now the Rockefeller University) and remained there for the rest of his life. Landsteiner also identified other blood factors, including the M, N, and P factors (1927) and the Rhesus (Rh) system in 1940. He moreover laid the foundation for the development of the polio vaccine and helped to identify the microorganisms responsible for syphilis. He received the Nobel Prize in physiology or medicine in 1930 for his work in identifying the differences in human blood. In 1943, while working in his laboratory, Landsteiner had a heart attack and died two days later in the hospital of the institute in which he had carried out his renowned research.

■ Karl Landsteiner developed the system of blood typing that made blood transfusion into a routine medical procedure.

Laughlin, Robert Betts

November 1, 1950, in Visalia, California

American physicist Laughlin, together with Daniel C. Tsui and Horst Störmer, received the 1998 Nobel Prize in physics. In 1972 Laughlin graduated from UC Berkeley and received his Ph.D. in physics from MIT in 1979. He then undertook research projects at Bell Laboratories in New Jersey until 1981 and at the Lawrence Livermore National Laboratory in California the following year. In 1983 he formulated an explanation to a puzzling discovery made by Tsui and Störmer, now known as the fractional quantum Hall effect, in which electrons in a strong magnetic field and at extremely low temperatures form a quantum fluid and have identifiable fractional charges. His model offers new insights into the nature and structure of matter. In 1985 he became an associate professor of physics at Stanford University and was appointed full professor there in 1989.

Leeuwenhoek, Antoni van

October 24, 1632, in Delft, Netherlands–August 26, 1723, in Delft, Netherlands

Antoni van Leeuwenhoek, considered the "father of microbiology," became an apprentice draper in Amsterdam at the age of 16, where he was first introduced to the method of using glass lenses for low-powered magnification. After returning to Delft and setting up his own successful textile business, he perfected techniques for crafting tiny lenses and began to use his improved microscope to examine nature. Leeuwenhoek kept the methods he utilized in making these tiny lenses a secret, leading people to believe that he was creating lenses by grinding, the predominant method of the day. Some of the methods he used, such as grinding lenses down from already tiny globules of glass, are known today. However, many of his techniques still remain a mystery. Leeuwenhoek took a degree in geography (completed in 1669) and had his first paper recognized by the Royal Society in 1673. The next year he published accounts of the "animalcules" (mircoorganisms) he had discovered, followed by the first descriptions of protozoa, bacteria, and spermatozoa. However, people were skeptical of his claims, and he was not made a fellow of the society until 1680. Leeuwenhoek gained worldwide fame and published hundreds of works during his lifetime. His research led to the discovery that muscle fibers are banded and he made the first accurate descriptions of red blood cells.

Leibniz, Gottfried Wilhelm

July 1, 1646, in Leipzig, Germany–November 14, 1716, in Hanover, Germany

A great philosopher, rationalist, and mathematician, Gottfried Leibniz taught himself subjects such as Latin and mathematics as a child. He studied law and philosophy at the University of Leipzig, but received his doctorate from the University of Altdorf before beginning a career in law. His aim

■ Gottfried W. Leibniz was interested in the simplification of thought and reasoning through the application of mathematics.

of reconciling variations of knowledge into shared basic elements led him to become a pupil of Huygens in Paris in 1672, where he designed an early calculating machine. He was elected to the Royal Society two years later, and began to develop calculus (independently of Newton (p. 456)). Leibniz's version of notation is used in today's differential and integral calculus. In 1679 he invented the binary system (the basis for modern computing) and, motivated by a broad interest in all sciences, traveled and corresponded with major intellectuals of his time. He was among the first in the field of geology and paleontology, and his ideas in dynamics, forces, biology, psychology, and other sciences are considered visionary for his time. His later years were marred by disputes with Newton over the invention of calculus, with allegations of theft being placed at Leibniz's feet.

Leonardo Da Vinci

April 15, 1452, in Republic of Florence (now in Italy)–May 2, 1519, in Cloux, France

The unmistakable face of Leonardo da Vinci, an inventor and scientist many years ahead of his time.

Leonardo da Vinci is considered not only visionary as a painter and sculptor, but also as a scientist. His work showed remarkable insight and inventiveness in many fields, including engineering, mathematics, anatomy, optics, botany, and the mechanics of flight. He spent the early part of his life in Florence where he began his distinguished career as an artist, and during this time also began applying his observational talents to technical and mechanical drawings. Motivated by a thirst for knowledge, he dissected both animals and humans in his goal to accurately portray the skeleton, muscles, and organs. He also studied the geometry of both animal and human forms (as depicted in his famous drawing *Vitruvian Man*). Da Vinci was convinced science and art were intertwined, and favored empirical evidence over speculation. His careful pictorial reproductions of the world paved the way for scientific illustration. He worked on many diverse projects throughout Italy, such as fortifications, hydraulics, mechanics, and architecture. During his life, he developed an interest in the invisible forces of nature, such as in currents and eddies of air and water, and believed these forces shaped the visible appearance of all material objects, from the growth of living things to the formation of rocks and clouds. He died as an honored, long-term guest of the King of France in 1519.

Linnaeus, Carolus (Carl von Linné)

May 23, 1707, in Råshult, Sweden–January 10, 1778, in Uppsala, Sweden

Linnaeus was a Swedish botanist who first developed binomial nomenclature to classify and organize plants and animals, and is considered the father of modern taxonomy. Linnaeus studied botany at the Uppsala University and began lecturing there in 1730, also undertaking an expedition to the then unknown northern Swedish area of Lapland. In 1735 Linnaeus left Sweden for the Netherlands to pursue a medical degree and later returned to Uppsala University, eventually acquiring the chair of botany in 1742, where he remained for the rest of his life. While living abroad, between 1735 and 1738, he published the first edition of *Systema Naturae*, his most important work. He became the first to develop principles for defining genus and species of organisms and a uniform system for naming them. This system, known as binomial nomenclature, was based mainly on flower parts, since they remain mostly unchanged during evolution. Linnaeus not only classified the plants and animal kingdoms, but also the mineral kingdom. This made it possible to rapidly place an organism in any named category. Linnaeus also made further expeditions to yet unexplored parts of northern Sweden and brought back significant scientific findings. When not traveling, he busied himself with expanding his classification system more thoroughly. Linnaeus's final years were plagued by poor health and after suffering two strokes he died in 1778.

Carolus Linnaeus in the traditional costume of the Lapland area of Sweden, where he made one of his earliest scientific expeditions.

Lorenz, Edward Norton

May 23, 1917, in West Hartford, Connecticut–April 16, 2008, in Cambridge, Massachusetts

Lorenz's research into the dynamics of weather patterns caused a revolution in science and shaped modern chaos theory. After earning his master's degree in mathematics from Harvard University, he became a weather forecaster for the U.S. Army Air Corps during World War II, a role that spurred his interest in meteorology. In 1948 he received his doctorate from MIT, where he was to continue a distinguished career. In 1963 he published a paper showing that accurate predictions of atmospheric activity over several weeks or more is impossible because tiny variables in initial calculations can result in varying and complex weather patterns. Chaos theory began to be applied in all sciences, particularly after his "butterfly effect" paper—showing a diagram of two diverging lobe patterns from one starting point—was published in 1972. He became head of meteorology at MIT in 1977, emeritus professor in 1987, and won many awards, including the 1991 Kyoto Prize for Basic Sciences with the prize committee stating that Lorenz "brought about one of the most dramatic changes in mankind's view of nature since Sir Isaac Newton."

Lorenz, Konrad

November 7, 1903, in Vienna, Austria–February 27, 1989, in Altenburg, Germany

This Austrian zoologist and animal psychologist is considered to have founded modern ethology, the study of animal behavior, along with Nikolaas Tinbergen. Lorenz took an interest in animals, their behavior, and the theory of evolution from an early age. While he later wanted to study paleontology, Lorenz complied with his father's wishes to pursue medicine, a decision he would not regret since it opened his mind to the comparative method he would later apply to behavior. Lorenz acceeded to the chair of the psychology department at the University of Königsberg in 1939. Two years later, he served as a doctor for the German army in World War II. He spent the years 1942 through 1948 as a prisoner of war. Lorenz's rediscovery and extensive study of the phenomenon of imprinting—the occurrence of a young animal learning the characteristics of its parents and following them around—was to prove foundational to the field of ethology. Lorenz was particularly interested in nidifugous birds (those that leave their nests shortly after they are born) such as geese. For his contributions to the study of animal behavior, Lorenz was awarded the Nobel Prize in physiology or medicine in 1973.

■ Konrad Lorenz is regarded as the founder of modern ethology: the study of animal behavior.

■ As well as working in the field of physics, Ernst Mach was also a noted philosopher, often combining the two subjects.

Mach, Ernst

February 18, 1838, in Brno, Austria (now Czech Republic)–Febuary 19, 1916, in Haar, Germany

The Austrian Ernst Mach, whose name is equated with the speed of sound, made important contributions to the fields of physics, physiological psychology, and philosophy. He was the first in physics to study supersonic motion in a systematic manner and was also involved in clarifying the Doppler effect. Furthermore, his criticism of Newton's conceptions of absolute space and time had a significant impact on Albert Einstein, who later acknowledged Mach's ideas as predecessors to the theory of relativity. Mach was educated at home until 14 years of age and only briefly attended a high school, before he entered the University of Vienna at 17 where he studied mathematics, physics, and philosophy. He received his doctorate in physics in 1860 and taught both mechanics and physics in Vienna until 1864, at which point he became a mathematics professor at the University of Graz, where he was increasingly interested in the psychology and physiology of sensation. Mach moved on to becoming a professor of experimental physics at the Charles University in Prague in 1867, where he would remain for the next 28 years. He developed optical and photographic techniques for the measurement of sound waves between 1873 and 1893. In 1887 he established the principles of supersonics and the Mach number. Mach returned to the University of Vienna in 1895 but after suffering two strokes retired in 1901 when he was appointed to the Austrian Parliament. Mach continued to write until his death in 1916.

Mandelbrot, Benoit

November 20, 1924, in Warsaw, Poland

This Polish-born French mathematician is regarded as the cofounder of the chaos theory and fractal geometry. Mandelbrot was born in Warsaw, Poland, into a Jewish Lithuanian family, who fled to Paris in 1936 due to the rising influence of Hitler's Germany. He was taught mathematics by his uncle, a famed mathematician by the name of Szolem Mandelbrojt. In 1945 he entered the École Polytechnique, one of the oldest and most prestigious engineering schools in the world. After leaving, he went on to study in the U.S. and then France, but left Europe in 1958 for a post in the research arm of IBM in New York. Using the early graphical computer systems of the time, he developed new mathematical ideas and computer programs that could represent visually the famous "Julia" fractal set. He is a strong proponent of the effect of fractals in many natural and artificial phenomena such as nature, coastlines, and stock markets. Mandelbrot received many honorary degrees and awards for his work, including the Wolf Foundation Prize for Physics in 1993 for his innovative work on fractals. He was the Sterling Professor of Mathematical Sciences, Emeritus at Yale University until 2005 when he retired.

Benoit Mandelbrot, known as the father of fractals, cofounded the chaos theory and fractal geometry.

Maxwell, James Clerk

June 13, 1831, in Edinburgh, Scotland–November 5, 1879, in Cambridge, England

Physicist Maxwell is best known for his pioneering work on the electromagnetic theory, which synthesized previously unrelated observations of electricity, magnetism, and optics into a unified theory and showed them to be manifestations of the same phenomenon: the electromagnetic field. This synthesis is considered to be one of the greatest advances in physics. Maxwell graduated with a degree in mathematics from Trinity College, Cambridge in 1854 and held professorships at Marischal College in Aberdeen (1856) and King's College in London (1860). He became the first Cavendish Professor of Physics at Cambridge in 1871. Between the years 1864 and 1873, Maxwell conducted his most significant research, developing four equations related to electric and magnetic fields and their interrelated nature. These partial differential equations appeared in his text *Electricity and Magnetism*, which he published in 1873. They have since become known as Maxwell's equations and are considered to be one of the greatest achievements of 19th-century physics. Maxwell also made important contributions to science through his study of Saturn's rings and the development of the kinetic theory of gases. He died of cancer at the relatively young age of 48.

Mendel, Johann Gregor

July 22, 1822, in Heinzendorf, Austria–January 6, 1884, in Brno, Austria (now Czech Republic)

Mendel, acknowledged as the founder of modern genetics, was not a famous scientist in his time, but rather was an Austrian Augustine monk who taught the natural sciences to high school students. Mendel entered the Augustine Abbey of Saint Thomas in Brno in 1843. He then studied mathematics and biology at the University of Vienna in 1951 and returned to the abbey two years later to begin teaching. In 1954 Mendel began to pursue the study of the variation in plants in the monastery's garden. Using edible peas for his research, it was these early experiments that led to his formulation of the basic principle of heredity. Between 1856 and 1863, he tested tens of thousands of pea plants and narrowed his conclusions down into two generalizations, which came to be known as Mendel's laws of inheritance. In 1865 he presented a paper called *Experiments on Plant Hybridization*, which was poorly received but is now considered a classic scientific text. His work remained largely unappreciated and discredited in his time, and after he died in 1884 his scientific papers were burned. However, his research was revived in 1900 by three botanists who had independently come to the same conclusions, resulting in the beginnings of the field of modern evolutionary biology.

Miescher, Johann Friedrich

August 13, 1844, in Basel, Switzerland–August 26, 1895, in Davos, Switzerland

The work of Swiss biologist Johann Miescher was a forerunner to DNA research and paved the way for the identification of DNA as the carrier of inheritance. He came from an academic family and studied medicine at the University of Basel before turning to physiological chemistry, feeling that his partial deafness would hinder him as a doctor. He spent the years 1868 through 1879 studying organic chemistry and physiology at Tübingen and Leipzig respectively, and in 1871 he was appointed professor of physiology of Basel. In 1869, while working on pus cells at Tübingen, Miescher discovered a substance containing both phosphorus and nitrogen in the nuclei of white blood cells, which he named "nuclein" because it seemed to originate in cell nuclei. Miescher succeeded in separating the substance into a protein and acid component; however, the role that the molecule, later named nucleic acid, played in the cell biology was not known until the structure of nucleic acid was revealed by James Watson (p. 462) and Francis Crick (pp. 437–438) in 1953. Miescher was also one of the earliest researchers to suggest that concentrations of carbon dioxide, rather than oxygen, in the blood regulate breathing. He died of tuberculosis at the age of 51.

Miescher isolated the phosphate-rich nucleic acids in blood cells, enabling the later discovery of DNA.

Milanković, Milutin

May 28, 1879, in Dalj, Croatia–December 12, 1958, in Belgrade, Serbia

A renowned Serbian geophysicist, Milutin Milanković developed some of the most important theories regarding the ice ages, long-term climate change, and variations of the Earth's movements. In 1904 he received his doctorate from the Vienna Institute of Technology in civil engineering, the first Serbian national to do so. During World War I, Milanković was confined to a camp in Hungary where he studied solar climates. Milanković would spend his entire professional life developing a climate theory based on the variations of radiation picked up by the Earth, known as insolation. His theory, referred to as the Milankovic´ Theory, was found to apply to all periods of geological time and his theory of insolation curves explained the variations in climate during the ice age, though this was not proved until after Milanković's death. Revealed in 1982 at an international symposium, Milanković's cycles proved to be a breakthrough in the field of geophysics.

Miller, Stanley Lloyd

March 7, 1930, in Oakland, California–May 20, 2007, in National City, California

Exobiologist Stanley Miller earned his bachelor's degree at UC Berkeley, then began his doctorate work at the University of Chicago. In 1953, with supervisor Harold Urey, he designed an experiment on the origins of life (now known as the Miller-Urey experiment). By setting up a system with the hypothetical conditions of early Earth (a body of warm water, an atmosphere of water vapor, methane, ammonia, and molecular hydrogen) and providing it with discharges of electricity to act as lightning, they found that carbon formed into organic molecules within a week. Some of these were amino acids, the essential building blocks of life. Although the atmosphere of early Earth is now thought to be different from these conditions, the experiment succesfully showed that biological molecules can arise from inorganic compounds. Miller became a member of the National Academy of Sciences in 1973 and was awarded the Oparin Medal in 1983. He spent the remainder of his career as a professor at UC Berkeley.

Mohorovičić, Andrija

January 23, 1857, in Volosko, Croatia–December 18, 1936, in Zagreb, Yugoslavia

Insightful meteorologist and geophysicist Andrija Mohorovičić spoke several languages from a young age. He was a physics student of Ernst Mach (p. 453) at the University of Prague and began a teaching career first at the Royal Nautical School in Bakar—where he set up a meteorological station—and then at the Main Technical School in Zagreb. He became director of the Zagreb meteorological observatory in 1897. By using data recorded there after a nearby earthquake in 1909 and comparing it with other stations, he saw that the velocity of seismic waves changed at a certain level below the Earth, implying a lower density layer (the Earth's crust) and a higher density structure (the mantle). This was later confirmed, and the discontinuity boundary between these layers was named after him. This layer called the Mohorovičić discontinuity is often shortened to just "Moho" due to the difficulty speakers of non-Slavic languages had in pronouncing his name. Mohorovičić also discovered the dual nature of seismic waves and developed early techniques for the construction of earthquake-protected buildings and for locating the epicenter of earthquakes as well as some early work in utilizing power generated by the wind.

Nash, John Forbes

June 13, 1928, in Bluefield, West Virginia

Mathematician John Nash graduated from the Carnegie Institute of Technology in 1948, receiving his master's and bachelor's degrees simultaneously. He received his Ph.D. from Princeton University at the age of 21, for his dissertation on noncooperative game theory (or Nash equilibrium theory).

John Nash won the Nobel Prize in economics in 1994 for his work on game theory.

By the age of 29 he was tenured at MIT, and in 1956 was offered an Alfred P. Sloan grant to spend a year working on solutions to classical problems in mathematics, which he completed at the Institute for Advanced Study at Princeton. However, in 1959 he was diagnosed with paranoid schizophrenia and resigned from his position at MIT. He rejoined Princeton on an informal basis as he struggled with his illness periodically until he finally recovered two decades later, and now continues to work on mathematical problems. In 1994 he shared the Nobel Memorial Prize in economic sciences for early work in game theory, and was awarded the Leroy P. Steele Prize in 1999.

Newton, Sir Isaac

January 4, 1642, in Woolsthorpe, England – March 31, 1727 in London, England

Sir Isaac Newton, father of classical mechanics, studied at Trinity College, Cambridge from 1661 where he developed theories that, along with the independent theories of G.W. Leibniz (pp. 451–452), were to form modern calculus. He recieved his degree in 1665, then—due to the Great Plague—researched in isolation for two years, developing his ideas on motion, optics, and the composition of light. In 1669 he became a professor of mathematics. His construction of the first reflecting telescope brought him to the attention of the Royal Society. At Edmond Halley's (pp. 445-446) insistance, he finally published *Philosophiae Naturalis Principia Mathematica* in 1687. It included his three laws of motion and law of universal gravitation and is considered one of the most important works in the history of modern science. After the publication of the Principia, Newton became very popular and gained many friends, including the Swiss mathematician Nicolas Fatio de Duillier. After a nervous breakdown, Newton retreated to study alchemy and the forces of attraction and repulsion in solitude.

Ockham, William of

ca. 1285–1349 in England

Philosopher and theologian William of Ockham was a member of the Franciscan order. He had a keen interest in logic and studied theology at Oxford. However, his opinions on applying logic and reason to theology offended some of the faculty, and he was refused his master's degree. He continued to teach rationality and reason as a way of understanding theology, even at the risk of being condemned, and is famous for his philosophy that plurality should not be assumed without necessity. This principle became known as Ockham's razor and is generally understood to mean that unecessary hypotheses, that make no difference to the explanation, or that add complexity, should be discarded. Although it was developed to debate theology, it can be applied in some cases to scientific explanations, such as discarding Ptolemy's cosmology for the less complex system of Copernicus. However, Ockham's razor is sometimes criticized in fields such as biology, where the models that offer the best explanations for processes may not be simple, and overapplication of Ockham's theory could lead to correct theories being disregarded.

Oort, Jan Hendrik

April 28, 1900, in Franeker, Netherlands– November 5, 1992, in Leiden, Netherlands

Jan Oort, a key figure in understanding the structure and dynamics of the Milky Way galaxy, studied at the University of Gronigen in the Netherlands and Yale University in the United States, then returned to his home country in 1924 to begin a long career at the Leiden University observatory. In 1927 his research on the velocity of stars confirmed Bertil

■ A painting of Newton by the English poet and artist William Blake. Here he is depicted as the "Divine Geometer," a satire on the omission of God in Newton's theories.

Lindblad's theory that the Milky Way is rotating. He also made early calculations of the mass of the galaxy, the distance and direction of the galactic center, and discovered that the sun is near the outer edge of the disk. He was among the first to realize the use of World War II radio technology for astronomy. After the war, he became full professor and director of the university observatory and later proposed that comets originate from a cloud of distant, orbiting bodies, now known as the Oort cloud. He continued carrying out influential research on many areas of astronomy for the rest of his life and received many awards and honors.

Pasteur, Louis

December 27, 1822, in Dole, France–September 28, 1895, in Jura, France

Louis Pasteur, one of the founders of microbiology, excelled in both his schooling and teaching career. When only 26 years old, he made his first discovery—on chemical compounds used in the fermentation process. Throughout the rest of his life, he transformed the knowledge of his day. He showed that fermentation does not require oxygen, invented a heating technique to kill microorganisms (pasteurization), and discovered that they were airborne, rather than appearing spontaneously. He developed the germ theory of disease, a major advancement in medicine and in 1881 designed methods to artificially cultivate weakened strains of viruses (which he knew could immunize against disease). In doing so, he created inoculations against anthrax and cholera. In 1885, while experimenting on cultivating weak strains of rabies from the dried nerve tissues of animals, he attempted to save a nine-year-old boy who was infected. This unexpected trial was a success, and three years later the Pasteur Institute for the prevention and treatment of infectious diseases was opened in Paris.

As well as developing the process that bears his name, Louis Pasteur created vaccines and cures for many diseases.

Pauli, Wolfgang

April 25, 1900, in Vienna, Austria–December 15, 1958, in Zurich, Switzerland

This Austrian physicist won the 1945 Nobel Prize in physics for his discovery of the Pauli exclusion principle, which states that in an atom, no two electrons can occupy the same quantum state simultaneously. Pauli received his early education in Vienna and—barely out of his teens—published an exceptional account of the theory of relativity. He then went on to attend the University of Munich where he obtained his degree in medicine in 1921, followed by one year at the University of Göttingen and another at Copenhagen under the famous physicist Niels Bohr (p. 435). Pauli spent several years as a lecturer at Hamburg before being appointed professor of theoretical physics at the Federal Institute of Technology in Zurich. Between 1935 and 1936, he was a visiting professor at Princeton and was elected to the Chair of Theoretical Physics there in 1940. He chose to return to Zurich after World War II after having become a naturalized citizen of the U.S. Pauli's exclusion principle shed light on the structure of the atom and helped to characterize the state of the electron. He established the spin theory and laid down the foundations for the quantum theory of fields. Pauli died in 1958 of pancreatic cancer.

Wolfgang Pauli worked mostly in the field of quantum physics, and was a noted perfectionist.

Pauling, Linus Carl

February 28, 1901, in Portland, Oregon–August 19, 1994, in Big Sur, California

Linus Pauling, the only person to win two unshared Nobel Prizes, revolutionized the field of chemistry, helped to establish the study of molecular biology and quantum chemistry, and is considered one of the most influential chemists of the 20th century. Pauling graduated from Oregon Agricultural College in 1922 with a chemical engineering degree and became interested in the molecular nature and structure of chemical bonds. He then enrolled at the California Institute of Technology (Caltech) where he earned his Ph.D. in chemistry in 1925. Pauling then spent 15 months in Europe on a Guggenheim Fellowship studying the basics of atomic structure. In 1931 he became full professor and head of the chemistry division at Caltech. Pauling conducted research with x-ray crystallography and modeling in crystal and protein structures, an approach used some years later to discover the double helix structure of the DNA molecule. Pauling's wife introduced him to issues of peace activism and social justice, and he was well known as a committed humanitarian. In 1963 he won the Nobel Peace Prize, nine years after winning the Nobel Prize in chemistry. Pauling then spent the next decade working for various universities and institutions before co-founding a research institute in California in 1973 studying and advocating vitamin C consumption. He died at the age of 93 from cancer.

Piccard, Jacques

July 28, 1922, in Brussels, Belgium

This Swiss explorer and oceanic engineer is renowned for making the deepest oceanic dive ever. He achieved this in the submersible vessel *Trieste*, which he designed with his father, Auguste Piccard. The *Trieste* was a bathyscaphe—an improved bathysphere—which measured 50 feet (15 m) long, was equipped with a steel cabin that could resist immense pressure, and could descend and ascend on its own unlike the former bathysphere. In 1953 father and son dove to a record 10,168 feet (3,099 m) off the coast of Italy. In 1960 Jacques and U.S. Navy Lt. Don Walsh (p. 462) shattered that record, descending to an incredible 35,810 feet (10,915 m) in the Pacific Ocean's Challenger Deep, the deepest known point in the oceans. The descent toward the ocean floor lasted five hours and set a new submarine depth record. Piccard went on to work for the U.S. Navy as a consultant of deep-diving navigation, and the *Trieste* was bought by the Navy and used to locate sunken submarines. Piccard also designed the first mesoscaphe, an observational submarine able to carry up to 40 passengers, as well as the submarine *Ben Franklin* (or *PX-15*), which was used to study oceanic currents. Since Piccard and Walsh's dive in 1960, no one has come within 10,000 feet (3,048 m) of their record.

Planck, Max

April 23, 1858, in Kiel, Germany–October 4, 1947, in Göttingen, Germany

The German physicist and founder of quantum theory, Max Planck, revolutionized the way we conceive of atomic and subatomic processes. Born into an intellectual family and musically gifted from an early age, the young Planck chose rather to pursue his interest in physics. He studied at the universities of Munich and Berlin and received his Ph.D. at the former in 1879. He then focused on thermodynamics and was especially interested in the nature of radiation from hot materials. He became a private lecturer at Munich from 1880 to 1885 and associate professor of theoretical physics at Kiel until 1889, when he undertook the post of full professor at Berlin University. He remained there until his retirement in 1926. In 1901 Planck devised a radical theory stating that energy does not flow in a steady continuum but rather is delivered in packets he later called quanta. His work marked a turning point in the history of physics and he was awarded the Nobel Prize in physics in 1918. During the Nazi regime's rule, Planck was openly critical of the government's policies, especially its persecution of Jews. He suffered tragic personal losses throughout his life beginning with the death of his first wife in 1909, his oldest son during the First World War, his twin daughters two years apart from childbirth complications, and his second son who was hanged in 1945 for an attempt on Hitler's life. Planck's house and much of his work was destroyed by bombing in World War II; he died two years after the war's end.

■ Like many early academics, there are no accurate paintings of Ptolemy. This depiction was painted in the 16th century.

Ptolemaeus, Claudius (Ptolemy)

ca. A.D. 100–A.D. 170 in Egypt and Greece

Ptolemy, the historically influential astronomer, mathematician, and geographer, wrote *The Algamest*. This text described the movements of the planets and the sun as if they were projected onto spheres against the backdrop of the stars, with the Earth at the center. He used complicated mathematic descriptions of epicycles and eccentricities to explain the apparently irregular motions of the planets, and included charts with almost 50 star constellations, the names of which are still in use today. Celestial movements could be accurately predicted with The Algamest, which remained in use for over a thousand years and was influential to both European and Islamic science. The geocentric view was replaced in 1543 when Copernicus (p. 437) proposed a heliocentric (sun-centered) model instead. Among other great works, Ptolemy also published Geographia, which set universal standard for mapmaking. His cartography technique placed north at the top and east to the right, and included symbols, scales, a legend, and coordinate system with a latitude and longitude grid. His resulting atlas of the Earth was incorrect, however, due to a miscalculation of the Earth's circumference.

Pythagoras

ca. 580–500 B.C. in Samos, Greece

The ideas ascribed to the Greek mathematician Pythagoras—originally steeped in mysticism—laid the foundations for later Western philosophy and mathematics. As Pythagorean principles were mainly expanded by his successors and no writings of his exist, there is controversy surrounding what should be credited to him personally. His ideas influenced Plato and Aristotle and Pythagorean geometry was later expanded by Euclid (p. 439) in the highly influential work *The Elements*. Pythagoras placed great significance on his discovery that musical notes could be represented mathematically and founded a brotherhood that sought spiritual insight through numbers, ratios, and the harmonics of nature. The teachings of the secretive academy were mainly oral, and members were required to take vows of silence, as well as to subscribe to a particular way of life that instructed their morals, politics, diet, and behavior. Pythagoras is acknowledged to have made the first proof of what is now known as the Pythagorean theorem, on the geometry of the right-angled triangle. He was among the first people to conjecture that the Earth is a spherically shaped object that rotates on its axis, with the planets all sharing this characteristic, and that they are all traveling around one central point.

Pythagoras was the first person to consider himself a philosopher and had an enduring influence.

Ramanujan, Srinivasa

December 22, 1887, in Erode, India–April 26, 1920, in Kumbakonam, India

Born into a poor family, Srinivasa Ramanujan developed his mathematical genius with hardly any formal education. By the age of 17, he was developing his own theorems on complex mathematical concepts. Failing his nonmathematical subjects, Ramanujan could not enter the University of Madras. Nevertheless, he published his research in the journal of the Indian Mathematical Society while working as an accounting clerk. In 1912, he began to send samples of his results to the University of Cambridge. Impressed by his skills, the British mathematician G. H. Hardy invited Ramanujan to study at Cambridge. After receiving a Bachelor of Science by Research in 1916, he was elected to the London Mathematical Society in 1917 and entered the Royal Society in 1918 as the second Indian and one of the youngest fellows. In his short lifetime, Ramanujan compiled nearly 3,900 highly unconventional results—the majority of them new breakthroughs—that have contributed to areas of mathematical analysis, number theory, infinite series, and continued fractions. After being diagnosed with tuberculosis, he returned to India in 1919 where he died at the age of 33.

Röntgen, Wilhelm Conrad

March 27, 1845, in Remscheid, Germany–February 10, 1923, in Munich, Germany

Wilhelm Röntgen received his Ph.D. in 1869 from the Federal Polytechnic University of Zurich and taught in prominent positions at various universities throughout Europe, finally settling at the University of Munich where he accepted the Chair of Physics. In 1895, during experiments with electrical discharges being passed through gases in low pressure tubes, he noticed that a plate coated in barium platinocyanide was becoming fluorescent, even when he placed different materials between it and the rays. A week later, he used these invisible rays to take the first x-ray picture of a human: his wife's hand against a photographic plate, through which her bones could be seen. He published a paper on these x-rays (or Röntgen rays) the following month and was the recipient of a great many awards, prizes, and honors, including the 1901 Nobel Prize in physics.

Röntgen did not patent his work, with the field of medicine owing a great debt to his modest altruism.

Smith, William

March 23, 1769, in Churchill, England–August 28, 1839, in Northampton, England

William Smith was a British engineer and geologist who is credited as the creator of the first nationwide geological map of England and Wales and as a founder of the field of stratigraphy. Of humble origins, Smith was a largely self-taught geologist who painstakingly recorded geological data on his walking tours of England as a surveyor. While surveying canal routes, Smith began studying fossils embedded in rock layers and found that the fossil-bearing strata followed a consistent and predictable sequence that can be found in other regional rocks and introduced the technique of using fossils for dating layers. In 1815 he produced his most memorable work to date: a geological map in which different geological strata were precisely colored. Though Smith was neither the first person to use fossils to identify rock layers nor to create a geological map, his feat of marking such a wide variety of rock formations over such a large area was unprecedented and set the style for modern geological maps. His work was subsequently plagiarized, instigating the financial troubles that would land Smith in debtor's prison. He was finally recognized for his work in 1931 when the Geological Society of London awarded him the Wollaston Medal, one of the highest honors for a geologist. He designed Scarborough's Rotunda Museum, which still stands today and has been renamed Rotunda—The William Smith Museum of Geology.

Steno, Nicolaus

January 10, 1638, in Copenhagen, Denmark–November 26, 1686, in Schwerin, Germany

Steno was a Danish anatomist and geologist, now widely considered the father of modern geology and a pioneer in the field of paleontology. In 1660 Steno began studying human anatomy in Amsterdam. Six years later, he was dissecting a large shark's head when he noticed a resemblance between the shark's teeth and stony objects found in certain rocks, called glossopetrae. Steno argued that glossopetrae, which many scientists still believed were merely stones, were in fact fossils of once-living sharks. Following this experience, Steno embarked on an intensive geological study of the region of Tuscany. In 1668 he published his most famous work, generally referred to as *Prodromus*, now considered a masterpiece of classic geology. In the text, he expounded the basic laws of stratigraphy (the study of rock layering), including the idea that the stacking of layers took place according to age and that fossils provide records of once living animals. After only three years, in which he contributed to the field enormously, Steno abandoned geology and devoted himself to religious affairs. He was canonized by Pope John Paul II in 1988.

Sullivan, Louis Henri

September 3, 1856, in Boston, Massachusetts–April 14, 1924, in Chicago, Illinois

Louis Henri Sullivan was an innovative, modern architect who influenced the structural design of some of the world's earliest skyscrapers. He was among the first to address issues regarding the composition of materials that should be used in tall buildings, and his designs took variations in stress to both the vertical and horizontal framework into account, paving the way for modern skyscraper design. He attended MIT in 1872 in the U.S.'s first architectural school, then traveled and studied in Europe before returning home to Chicago. In 1879 he formed a partnership with Dankmar Adler. Their firm (Adler and Sullivan) transformed modern architecture in Chicago by creating over a hundred unconventional and functional buildings with natural ornamentation and style. Sullivan continued to work alone after their partnership dissolved, designing smaller buildings that are still considered "Chicago's architectural jewels."

Sumner, James Batcheller

November 19, 1887, in Canton, Massachusetts–August 12, 1955, in Buffalo, New York

Sumner was an American chemist and Nobel laureate noted for having isolated the first enzyme in pure form. When he was 17 years old, Sumner's arm had to be amputated just below the elbow following a hunting accident. In 1906 he entered Harvard College and graduated in 1910 with a degree in chemistry. In 1912 he entered Harvard Medical College to study biochemistry and received his Ph.D. two years later. His supervisor at the time suggested that Sumner study law, since he believed that a one-armed man would have a hard time succeeding in chemistry. During World War I, Sumner held the position of assistant professor of biochemistry at Cornell Medical School. There he began his research into isolating and crystallizing enzymes, which had not been achieved before. He worked with the enzyme urease, extracted from the jack bean. Despite strong discouragement from his colleagues, he persisted with his research and in 1926 he was finally successful. He was made full professor in 1929 and in 1946 he won the Nobel Prize in chemistry for his discovery that enzymes can be crystallized. Two years later Sumner was elected to the U.S. National Academy of Sciences. He succumbed to cancer in August of 1955.

■ Thales as depicted in the *Nuremberg Chronicle.*

Thales of Miletus

ca. 634 B.C.–546 B.C., in Miletus, Greece (now Aydin, Turkey)

This pre-Socratic Greek philosopher is considered the founder of Greek science, mathematics, and philosophy. Though relatively little is known of Thales, there is reason to believe he visited Egypt and possibly Babylon, bringing back knowledge of astronomy and geometry. Before Thales, the Greeks attributed the origin and nature of the world to the actions of gods; Thales, however, attempted to find naturalistic explanations of the world, marking the beginning of scientific methodology. He invented deductive mathematics and several foundational mathematical theorems, including that a circle is bisected by a diameter. He also measured the height of pyramids using their shadows. Furthermore, Thales achieved great fame when he predicted the solar eclipse of May 28, 585 B.C. He is also credited with the discovery of the electrical properties of amber (Greek: electron), after he rubbed a piece of amber and noticed that it attracted small particles. This observation led to the present science of electricity. Thales' best known belief was his cosmological principle that the world originated from water. It is said that Thales died of dehydration while watching a gymnastics contest.

Thomson, Joseph John (J. J.)

December 18, 1856, in Cheetham Hill, England–August 30, 1940, in Cambridge, England

Discoverer of the electron, Joseph John Thomson received his B.A. in mathematics from Trinity College in Cambridge in 1880. He moved to Cavendish Laboratory to continue experimental research, where his work on electromagnetism earned him a fellowship to the Royal Society in 1884 and the chair of physics at Cavendish University. From experiments on the nature of cathode rays in 1897, he was able to conclude that the rays must be made up of tiny negatively-charged particles or "corpuscules," much smaller than the supposedly indivisible atom, and proposed that they also make up all other matter in the universe. Later experiments confirmed his theory, revolutionizing scientific understanding of the atom. He was awarded the Nobel Prize in physics in 1906, the same year that he discovered hydrogen has only one electron, and was knighted in 1908. Thomson also developed mass spectrometry—a way of characterizing different isotopes of an element—in 1913.

J. J. Thomson was awarded the Nobel Prize for his discovery of the electron.

Tombaugh, Clyde

February 4, 1906, in Streator, Illinois–January 17, 1997, in Las Cruces, New Mexico

Clyde Tombaugh, discoverer of the planet Pluto, was enthusiastic about astronomy as a teenager and used instructions from astronomy magazines and spare parts found around the family farm to construct several telescopes. In 1928, using one of his homemade telescopes, he made drawings of the surfaces of Mars and Jupiter and sent them to the Lowell Observatory in Arizona. The director invited him to work on a photographic survey of the sky in search of the proposed "planet X." Tombaugh began a painstaking and systematic search, taking two photographic plates of each section of the sky and spending thousands of hours analyzing the plates for signs that a star had appeared to shift its position. He is estimated to have compared well over a million stars before finding Pluto in 1930. He also discovered new clouds of galaxies, star clusters, and over 700 asteroids. Tombaugh earned his master's degree in astronomy from the University of Kansas in 1939 and was a distinguished astronomer and professor emeritus of New Mexico State University.

Turing, Alan

June 23, 1912, in London, England–June 7, 1954, in Wilmslow, England

Turing was an English mathematician, logician, code-breaker, and pioneer of computer science. He studied mathematics at King's College, Cambridge in 1931 and later also taught there. In 1936 Turing produced his most significant paper, "On Computable Numbers," in which he showed that automatic computation cannot solve all mathematical problems. He called this concept the Turing machine, a hypothetical computing device. He supposed that computers would eventually be capable of thought indistinguishable from human thought and his Turing test would assess this capability. His theories are considered foundational to the fields of artificial intelligence and modern computing theory. In the late 1930s, he began to work clandestinely for the British cryptanalytic department, the Government Code, and the Cypher School as a code-breaker. During World War II, he played a significant role in deciphering German codes. In 1952 Turin was arrested and tried for homosexuality, a criminal offense at the time, and forced to accept injections of estrogen in order to avoid prison. He committed suicide on June 7, 1954, by eating an apple laced with cyanide.

Van Allen, James Alfred

September 7, 1914, in Mount Pleasant, Iowa–August 9, 2006, in Iowa City, Iowa

James Alfred Van Allen was a pioneering American astrophysicist after whom the Van Allen radiation belts—two zones of radiation surrounding the Earth—are named. He received his Ph.D. in physics from the University of Iowa in 1939 and worked with the Johns Hopkins University Applied Physics Laboratory until 1942 when he was assigned to the U.S. Navy as an officer in World War II. In the mid-1950s, Van Allen was part of a team that developed the first plans for a comprehensive and systematic study of the Earth from space. On January 31, 1958, the first successful American satellite, Explorer 1, was launched. It provided scientists with a groundbreaking discovery: a region of encircling radiation trapped by the Earth's magnetic field. Van Allen was involved in the first four Explorer investigations as well as many other pioneering projects in outer space.

Walsh, Don

November 2, 1931, in Berkeley, California

Don Walsh is one of the only two people in the world to have reached the deepest part of the ocean in the Mariana Trench. After previously serving in the Navy as a submariner during the Vietnam and Korean Wars, Don Walsh became associated with ocean science, engineering, and marine policy for over 30 years. On January 3, 1960, Walsh and Swiss oceanic engineer Jacques Piccard (p. 458) descended to a depth of 35,810 feet (10,915 m) in the Pacific Ocean's Challenger Deep, the deepest known part of the Earth. They were aboard the bathysphere *Trieste* when they made their record-breaking dive. Walsh graduated from the U.S. Naval Academy in 1954 and later earned a Ph.D. in physical oceanography from Texas A&M University. In 1975 Walsh became dean of marine programs and professor of ocean engineering at the University of Southern California. Walsh also worked with NASA to determine how spacecraft could be used to study oceans; he became one of the first oceanographers to work with remote sensing from aircraft and spacecraft. Walsh currently works with the deep-sea tourism firm Deep Ocean Expeditions.

Wankel, Felix

August 13, 1902, in Lahr, Germany–October 9, 1988, in Heidelberg, Germany

The German engineer and inventor Felix Heinrich Wankel received neither a higher education nor an apprenticeship in engineering. However, since childhood he possessed an extraordinary three-dimensional imagination, taught himself technical subjects, and developed an interest in combustion engines. Engaged in engineering labors, he first conceived of the idea of a rotary-piston engine in 1924. Between 1936 and 1945, Wankel worked as a sealing specialist for the Reich Air Ministry (Reichsluftfahrministerium), developing rotary valves for aircraft and navy torpedoes in his own research facility. After World War II, he was briefly imprisoned by the Allies and prohibited from working. He completed the design of the Wankel engine in 1954 and the first unit was tested in 1957. Since 1967 the Japanese automobile company Mazda has successfully used the engine for several series of sports cars. In his later years, Wankel established the Felix Wankel Foundation dedicated to the protection of animals.

Warburg, Otto Heinrich

October 8, 1883, in Freiburg, Germany–August 1, 1970, in Berlin, Germany

Biochemist and Nobel Prize winner Otto Heinrich Warburg spent a large part of his life studying cancer and carcinogenic compounds. He was the first scientist to prove that smoking had a causal effect on the development of cancer cells, and that food additives had a carcinogenic effect. He developed the theory that cancerous cells could be destroyed by radiation, which was the starting point of the medical treatment of radiology. Furthermore, he recommended cancer treatments consisting of enzymes and vitamins. Warburg was both a gregarious and eccentric character, who was sometimes controversially dedicated to his work, even to the length of working for the Nazi regime, which he detested and criticized. However, he believed in the importance of his work and was unwilling to let the prevailing political enviroment affect his research. He was awarded the 1931 Nobel Prize in physiology or medicine for his work in cellular respiration, important and lifesaving research that discovered how cancer cells metabolize differently from normal cells, leading to new developments in treatments and detection.

Otto Warburg demonstrated the carcinogenic nature of food additives and cigarette smoking.

James Watson is strongly opposed to the highly controversial act of patenting gene sequences.

Watson, James

April 6, 1928, in Chicago, Illinois

James Watson is an American geneticist and molecular biologist who, together with Francis Crick (pp. 437–438), discovered the double helix structure of the DNA molecule in 1953. Watson was only 15 years old when he enrolled at the University of Chicago. After graduating, he went to Cavendish Laboratories at the University of Cambridge to learn x-ray diffraction and met Francis Crick in 1951. Together with Maurice Wilkins they discovered the structure of DNA, and shared the 1962 Nobel Prize for physiology or medicine. Watson became a professor of biology at Harvard University, publishing his famous textbook *Molecular Biology of the Gene* in 1965, and left in 1976 to pursue cancer research in his role as director of Cold Spring Harbor Laboratory. Watson later helped establish the Human Genome Project, with the goal of mapping the genes of the entire human body. His own genome was sequenced and published on the Internet in 2007, in order to encourage the development of "personalized medicine."

Wegener, Alfred

November 1, 1880, in Berlin, Germany–November 1930 in Greenland

Wegener was a German meteorologist and geophysicist who achieved fame for devising the first complete statement of the continental drift theory. In 1905 he earned his Ph.D. from the University of Berlin in astronomy but soon became interested in paleoclimatology. The following year, he took part in an expedition to Greenland to study polar air circulation. In 1912 Wegener presented his groundbreaking continental drift hypothesis to a stunned scientific community. Three years later he presented his evidence in the book *The Origin of Continents and Oceans*, claiming that 200 million years ago the continents were originally joined together, forming a large supercontinent, which he called Pangaea (Greek for "all Earth"). Wegener was also known for his work on lunar craters, which he correctly believed were the results of impacts rather than volcanism. While his theories on continental drift were greeted with disbelief at the time, they were resurrected during the sixties as part of the theory of plate tectonics. Wegener and his companion Rasmus Villumsen disappeared during his fourth expedition to Greenland. Wegener's body was found later, with the cause of death being suspected heart failure from overexertion.

The last known pictures of Alfred Wegener and Rasmus Villumsen before their ill-fated fourth expedition.

Weinberg, Steven

May 3, 1933, in New York City, New York

Steven Weinberg received his doctorate in physics from Princeton University in 1957. He worked at Columbia University, UC Berkeley, MIT, and at Harvard University. While a visiting professor at MIT in 1967, he began work on unifying two of the four fundamental forces in particle physics: electromagnetism and weak interaction. His electroweak theory, which described massive bosons and massless photons as two aspects of the same force, was developed with Sheldon Lee Glashow and Abdus Salam and won them the Nobel Prize in physics in 1979. This highly cited theory opened up new developments in the field of physics and was later confirmed when technology allowed experimental particle accelerators. Weinberg moved to the University of Texas in 1982 as professor of physics and astronomy. He has written two extremely influential physics textbooks as well as popular science books for the public and is an outspoken defender of the scientific method. Weinberg is also one of the most cited scientists on the H index: an index designed to quantitatively measure a person's relative scientific productivity and impact.

Wiechert, Emil

December 26, 1861, in Tilsit, Prussia (now the Russian enclave of Kaliningrad Oblast)–March 19, 1928, in Göttingen, Germany

Emil Wiechert grew up in Tilsit, which was then a province of the Kingdom of Prussia. After his father died, however, he moved with his mother to Königsberg in Germany, and then attended the University of Königsberg, receiving his Ph.D. in 1889. During his life, he made many significant and important discoveries, including the existence of the electron, the properties of streams of electrons, and the behavior of moving charges. His groundbreaking and most notable work was in seismology, where he wrote a well-respected paper on seismic waves, and developed improved models of seismographs. Wiechert was also the first scientist to detail the structure of the Earth's interior. Although he was relatively well known for his seismological studies, the scientific establishment did not particularly take much note of his work in the field of electrodynamics and electromagnetism.

Wilkins, Maurice

November 16, 1916, in Wairarapa, New Zealand–October 5, 2004, in London, England

Maurice Wilkins was born in New Zealand and when he was six years old his family moved to Birmingham in England. He studied physics at Saint John's College, Cambridge and then received a Ph.D. in 1940 from Birmingham University. During World War II, he worked on the development of advanced radar screens before transferring to the top-secret Manhattan Project in the U.S. After the war, and having experienced the success and eventual dropping of two nuclear bombs on Hiroshima and Nagasaki, Wilkins became strongly opposed to nuclear weapons. He decided to move into a field of study with more positive benefits, and so began various biophysical studies with physicist John Randal (renowned for his work on the development of radar). However, it is his later work on DNA that defined Wilkins's career. While at King's College in London he began x-ray diffraction experiments in order to discover the structure of DNA and sperm. With these experiments, combined with equations that were created by Francis Crick (pp. 437–438), he developed Crick's proposal that resolved the molecular structure of DNA. The amount of researchers working on the x-ray diffraction of DNA, however, was so extensive, that the discovery of DNA could not be credited to just Crick and Wilkins, but also Alexander Stokes, Herbert Wilson, Rosalind Franklin (p. 441), and Raymond Gosling. In order to ensure that credit was correctly apportioned, all the combined research was published separately in a historic edition of the journal *Nature*.

Wöhler, Friedrich

July 31, 1800, in Eschersheim, Germany–September 23, 1882, in Göttingen, Germany

Friedrich Wöhler was a German chemist best known as a pioneer in organic chemistry and as the first to synthesize urea. While Wöhler intended to study as a physician and took his degree in medicine and surgery in 1823 in Heidelberg he was persuaded to devote himself to chemistry instead. Wöhler resumed further studies in chemistry in Heidelberg; however, he was judged to be too advanced for the course. He was then sent to study with the Swedish chemist Jöns Jakob Berzelius (pp. 434–435) forming a strong friendship between the two scientists. After teaching at various technical schools, including in Berlin and Kassel, Wöhler was appointed to the chair of chemistry in the medical faculty at the University of Göttingen in 1836. He maintained this post until his death. In 1828 Wöhler broke down the barrier between organic and inorganic chemistry when he made the breakthrough of artificially synthesizing urea; this disproved the belief that a "life force" was needed to produce organic compounds. Wöhler is also known as the co-discoverer of beryllium and silicon.

■ Friedrich Wöhler managed to artificially create urea, a tremendous development for the time.

Young, Thomas

June 13, 1773, in Milverton, England–May 10, 1829, in London, England

Thomas Young was an English doctor and scientist whose encyclopedic knowledge and contributions to the fields of science, languages, medicine, music, and archaeology led him to be characterized as a genius and polymath. He grew up in rural England to a Quaker family and was comprehensively educated, leading to further study in medicine and physics. He began work shortly after as a medical doctor, funded by an inheritance from his granduncle Richard Brocklesby, who was well known for supporting many of the most outstanding scientists and literary men of the age. Throughout his life, and despite his many other accomplishments, Young regarded medicine as his primary profession, and when he published scientific and academic works he often did so anonymously so as not to effect his standing in the medical community. One of his most important theories was the concept of wave-particle duality, an important feature of the field of quantum mechanics. Young was truly prolific, with many concepts and theories bearing his name, such as Young's modulus and the Young-Laplace equation. As also befitting a medical doctor, he developed Young's rule, a method for defining the correct dosage of a drug for children, important in preventing over- and underdosage. Toward the end of his life, he spent many years in the difficult field of the translation of hieroglyphs, and although not always successful he laid the groundwork for breakthroughs by later scientists such as the so-called father of Egyptology, Jean-François Champollion.

■ Thomas Young was a renowned polymath, working in many scientific fields, sometimes simultaneously.

Zuse, Conrad

June 22 , 1910, in Berlin, Germany–December 18, 1995, in Hünfeld, Germany

Konrad Zuse was a German scientist who is credited with creating the first programmable computers (his Z series) as well as the first high-level computer language (Plankalkül). Although having strongly influenced the design and construction of modern computers and programming languages, his early work went unnoticed as he was working in Germany during the Second World War. As an added setback, his company and computers were destroyed in 1945 during the Battle of Berlin. However, despite all these problems, in 1950 Conrad Zuse unveiled the first commercially available computer, the Z4, built by his company Zuse KG. From then on, he built many more groundbreaking computers, completing a total of 251 and rebuilding a replica of his first computer, the Z1 between 1987 and 1989. In later life Zuse took up painting and postulated a radical theory that the universe is itself "computable," and that reality is either the output from a highly advanced computer or can be simulated by a sufficiently powerful computer. This theory gained a lot of interest from the scientific community as no evidence exists to disprove him.